For my family, and for everyone educating the next generation of palaeontologists

Contents

JOURNEY THROUGH THE CENOZOIC

TETRAPOD FAUNAS 66 MILLION YEARS IN THE MAKING

WRITTEN AND ILLUSTRATED BY WILLIAM J. TOOSEY

FOREWORD BY PROF. MICHAEL J. BENTON

Journey through the Cenozoic

ISBN 978-1-0369-5388-1

Printed and distributed by print-on-demand.

Foreword

Humans are mammals, and there have been many books about human evolution, but far fewer on the history of mammals more broadly. In the 1950s and 1960s, Finnish palaeontologist Bjorn Kurtén (1924–1988) wrote popular science books and even fiction, about his beloved Pleistocene mammals from Northern Europe; books such as *The Age of Mammals*, *Pleistocene Mammals of Europe*, and *The Cave Bear Story* enthralled readers worldwide. Then, in 1986, Bob Savage published his *Mammal Evolution: An Illustrated Guide*, a rapid romp through mammalian evolution illustrated with amazing artwork by Michael Long. Since the 1980s, there have been numerous books about the Cenozoic time scale, Pleistocene geology, human evolution, and collections of papers about particular groups of Cenozoic vertebrates, such as turtles, lizards, birds, horses, rhinos, rodents, kangaroos, and the rest.

This book then fills a major gap. Since 1986, an enormous amount of new work has been done on the Cenozoic and its faunas, and the information is widely available to academic specialists. The new work has especially expanded our knowledge outside the well-studied continents of Europe and North America, with rich new fossil finds, some of them quite unexpected and astonishing, from China, other countries in Southeast Asia, Australia, New Zealand, South America, and Antarctica. For example, in trying to understand the conundrum of why we find marsupials today in Australasia and South America, and how they communicated, fossils have now been found in Antarctica showing how the ancestors of the modern forms trekked across that continent before it became ice-covered.

The book is comprehensive—everything is here, from the human-sized penguins of New Zealand to the Eocene rhinoceroses of Germany, from the armoured meiolaniid turtles of Australia to the horse-like litopterns of Brazil. The book steps through all the tetrapod groups in the first half, and then a sequence on examples of key assemblages through the Paleogene and Neogene, with even-handed inclusion of examples from all continents, including classics from Europe (Messel, Quercy, Creswell Crags) and North America (Bridger, Agate, Rancho La Brea), as well as newer finds from the other continents, such as Qasr el-Sagha and Nawata from Africa, Cerrejón and Chapadmalal from South America, Riversleigh and Naracoorte from Australia, and La Meseta from Antarctica.

William Toosey has written and illustrated the book. The text is comprehensive and covers background themes such as dating the rocks and reconstructing phylogeny, as well as a very useful chapter on how to reconstruct fossil vertebrates, with sections on reconstructing skeletons and musculature, and the external appearance such as skin, fur, horns, and colouration. The accounts of all the families of amphibians, reptiles, birds, and mammals start with highly illustrated cladograms showing the relationships, and key species are illustrated. In his survey of key faunas, all the current literature is brought to bear, and this enables him to explore the Pleistocene in some detail and look forward to the present day and the future.

A key feature of the book is the illustrations. In fact, the author is a natural history artist and scientific illustrator as well as a writer. For the book, he has prepared hundreds of paintings, mainly showing the ancient beasts in restored life appearance. Two highpoints of the book are the cladograms, which consist of branching evolutionary trees ornamented by his own paintings of the relevant animals, and the scenes of the 43 faunas through the Cenozoic. Each of these ecosystem images shows a wide range of animals in a reconstructed scene of vegetation of the time, and with detailed accounts of all the evidence.

Fossils have never been more popular than they are today, and it's not just with kids. Admittedly, the huge popularity of dinosaur books for children under ten does not persist at such a high level to adult audiences. However, in the past ten years, many publishers have risked publishing books about dinosaurs and other fossils, and the history of life in general, and they have found that there is an appetite among adults for knowledge about deep time and origins. The professional field is booming, with new fossils being excavated and described constantly, as well as, importantly, the application of ever more sophisticated laboratory and computational methods to understanding the evolution, palaeoecology, and functional biology of extinct plants and animals.

In the popularisation of science, it is important that the writer and illustrator shows all the details, the strengths and weaknesses of the new knowledge, and William Toosey has done this here. For readers interested in Cenozoic life, and especially the evolution of mammals in the past 66 million years, it's all here.

Michael J. Benton
University of Bristol
March 2025

Acknowledgements

I am very much indebted to my friends and colleagues, much too numerous to list exhaustively, but who have been of help through many conversations during the lengthy process of this book's preparation. I give my most sincere gratitude to every one of them. In many ways, this book is a dream of mine come true, and it could not have become a reality without their help. In addition, they have made palaeontology among the most rewarding communities to be involved with. Regarding the sources of inspiration for this book, these include Bob Savage's *Mammal Evolution: An Illustrated Guide* (1986), and Donald Prothero's *Age of Mammals* (2006), both reviews of Cenozoic mammal evolution which remain a good read to this day.

I express my sincere thanks to two anonymous reviewers, who generously reviewed early drafts of this book for general proofreading, along with providing helpful advice and support. I am also thankful to the following people, for you having been generous with your time to give some additional palaeontological expertise on my drafts of text, or provided some other form of helpful input such as sending me up-to-date papers or commenting on the accuracy of some of my artistic reconstructions, assisting me in avoiding egregious errors in various topics in which your expertise surpasses mine: Neil Adams, Mike Archer, Steve Brusatte, Darin Croft, Christine Janis, Spencer Lucas, Jin Meng, Jack Wilkin and Thomas Williamson. Any errors that may remain in the text are my responsibility, but I have tried my very best to keep them to as much a minimum as possible.

I am grateful to the online palaeontological community and those who follow it; your support, interest and enthusiasm helps make books like this one possible. I hope you all enjoy it. Furthermore, I am indebted to Steve Brusatte, Sarah Shelley and Tom Williamson for taking me on as a master's student and introducing to me the interesting world of Paleocene mammals. I thank them all for their support, guidance and mentoring, and they have served as valuable mentors and role models for me as my career has progressed.

Additionally, I wish to acknowledge the people from whose property the fossils of the extinct tetrapods featured in this book have been recovered, and give thanks to the palaeontologists themselves. If it were not for the innumerable hours of study carried out by the palaeontologists worldwide who have decoded the rock record such that it can be interpreted at all, neither this book nor any of my artistic reconstructions of extinct animals within it would be possible. Their professionalism and commitment has enabled myself to produce up-to-date and plausible illustrations of long-extinct animals.

In this book, the literature in the further reading section alone refers to the research of many hundreds of individual palaeontologists. Among them, the following researchers are cited most frequently: Mike Archer, Christine Argot, Robert Asher, Robin Beck, Michael Benton, Jonathan Bloch, Darin Croft, Demberlyn Dashzeveg, John Flynn, Gene Gaffney, Emmanuel Gheerbrant, Philip Gingerich, Dave Grossnickle, Christine Janis, Asier Larramendi, Michel Laurin, Maureen O'Leary, Adrian Lister, Gerald Mayr, Malcolm McKenna, Christian de Muizon, Mike Novacek, Marcelo Reguero, Kenneth Rose, Guillermo Rougier, Ross Secord, Mary Silcox, Floréal Solé, Richard Tedford, Hans Thewissen, Kenny Travouillon, Sergio Vizcaíno, Xiaoming Wang, Lars Werdelin, John Wible and Graham Wilson. To everyone who has ever excavated, dissolved, screened, or CT-scanned prehistoric marvels, whether featured in this book or not, I express my gratitude.

I also thank Michael J. Benton for having been kind enough to write a foreword for this book. His work on topics such as the large-scale evolution of major vertebrate groups, the quality of the fossil record, and the impacts of environmental change and mass extinctions on the evolution of tetrapods—including reptiles, mammals, and birds—has contributed greatly to our understanding of the history of life.

Last but not least, I would like to thank my family for their patience, support, and understanding as they endured my many long months at the computer writing, illustrating, and compiling this book.

Preface

The Mesozoic Era, or "Age of Reptiles", is hugely popular in the media, with many books and television documentaries such as *Prehistoric Planet* reconstructing the interesting lives and times of dinosaurs and their contemporaries. In contrast, the ensuing Cenozoic Era, or "Age of Mammals", has received far less attention than the Mesozoic. Moreover, there are still relatively few contributions for a window into the Cenozoic for the non-specialist, and I have never come across a book that gives a good overview on the evolution of tetrapod faunas around the world throughout that era, filled with numerous high-quality colour reconstructions. This book intends to fill that gap in the literature, and is intended to reach general readers and palaeontology enthusiasts who may have an interest in post-Mesozoic life or science more broadly, but not necessarily a university-level background in palaeontology or Earth sciences. At the same time, it provides sufficient depth for higher-education students in palaeontology or life sciences to benefit from its content. Balancing accessibility with scientific rigour is challenging, but I have aimed to present complex concepts as clearly as possible.

I have written this book at an intermediate level, with concepts fully modernised and aligned with the latest advances in cladistics. Where possible, I have used familiar anatomical terminology to make the text approachable, while introducing more technical terms in context. My goal is to make this book accessible not only to amateur palaeontology enthusiasts but also to students enrolled in vertebrate evolution undergraduate courses, where prior training in systematics or anatomy is not required. This is a challenging undertaking, since the majority of subjects outlined in this book require a deeper understanding of anatomy or systematics. *Journey through the Cenozoic* is essentially a tour of the Earth charting the changes it underwent during its geologically recent past, and how tetrapods have, or have not, adapted. We will visit a series of fossil sites throughout the Cenozoic to immerse ourselves in the ancient environments, view the flora and fauna, and learn what we can about today's world from these vanished **ecosystems**. Through exploring fossil sites and their faunas with the perspective of a time-travelling safari-goer, it is my intention to bridge the gap between the present and the past. When an environment is vividly reconstructed, it is easier to obtain a perception of the frequently familiar ways in which animals live, feed, reproduce and compete.

My artistic reconstructions in this book will act as an educational tool and a means of making accessible to readers the anatomy, ecology and diversity that the fossil record of Cenozoic tetrapods has disclosed to palaeontologists following many decades of study. Fossils are excavated solely via arduous fieldwork, ensued by painstaking cleaning and preparation of the fossils and countless hours of analyses, CT scan image-refining, or whichever method is used to obtain data from fossil material. Should the information so strenuously gathered be hidden away in scholarly literature, inaccessible to inquisitive lay readers? I disagree. An individual fossil can teach us a lot about function and shape, about morphological variability, and the things an animal can do with basic adjustments to a generic developmental toolset. However, no fossil, be it a microorganism, fungus, plant or animal, lived isolated. Every fossil formed part of an ecosystem, an interplay between countless species and the environment, an intricate mix of biota, chemistry and weather also reliant on the minerals in the water or the soil, the arrangement of the landmasses, the Earth's rotation, and the constraints brought about on a region by its former residents. Reconstructing the worlds where fossils settled, and where their makers existed, is a task that scientists have been endeavouring to resolve since the eighteenth century, gaining ground in both detail and speed during the past several decades.

Recent advances in physical and computational technologies have provided insights into aspects of ancient life that were previously believed impossible. By probing deep into the structural details of fossils, we are now able to reconstruct the colours of fur, feathers and snake scales, senses including smell and vision, and the life histories of fossil animals such as placental mammals from the Paleocene of New Mexico. In drawing comparisons between them and modern animals, we can determine their interactivity in food webs, their skull strength or bite force, and their reproductive behaviours and social structure. The landscapes of the fossil record are not solely accumulations of taxonomic lists of names and bunches of impressions on rocks anymore. The most recent studies have uncovered flourishing and lively **communities**, the remnants of living, genuine animals residing in worlds which followed the same biological laws and processes as those of today.

It is my hope that the reader will find this book intelligible, accurate and up-to-date, including all the most recent thinking and findings of the intriguing extinct tetrapods featured within it. As a palaeontologist, author, natural history artist and scientific illustrator, my role is to encourage, inform and educate all of us to feel the same appreciation for extinct animals as I do myself. By recognising the uniqueness of those species and the intricate connections of those ancient ecosystems, we can, in the same way, treasure the fragile creatures and environments around us today, understanding how things came to be as they are and how they may also cease to be.

1. Introduction

The science of palaeontology (the study of all extinct life on Earth prior to the emergence of human history, through the examination of fossils) is currently going through a golden age. New discoveries continue to emerge, ranging from fossils that reveal previously unknown branches of the tree of life, to ancient DNA recovered from a 12,000-year-old South American mammal, and newly uncovered footprints that provide insights into the lifestyles of extinct animals. In the modern world, where access to information is at our fingertips, palaeontology is more openly available than ever before. This broad scientific discipline encompasses the study of all forms of past life, from fossilised plants (palaeobotany) to **terrestrial** arthropods (palaeoentomology), and from prehistoric fish (palaeoichthyology) to **primitive** humans (palaeoanthropology), to mention just several. For many, this initial exposure develops into a deeper fascination, revealing the immense scale and complexity of the history of life.

The scope of this book covers the fossil tetrapods but only those from the Cenozoic Era, the most recent geological era in the Earth's history spanning from about 66 million years ago to the present day. The Cenozoic is divided into three periods—the Paleogene, Neogene and Quaternary, respectively (Figs. 1.1, 1.2). In turn, each period is subdivided into narrower units called epochs. The Paleogene (formerly the Early Tertiary) comprises, from oldest to youngest, the Paleocene, Eocene and Oligocene epochs. The Neogene (formerly the Late Tertiary) comprises the Miocene and Pliocene, and the Quaternary comprises the Pleistocene and Holocene. Given that epochs can span tens of millions of years, they are frequently subdivided by applying the terms 'Early', 'Middle' and 'Late'. Additionally, epochs can be subdivided into narrower units called ages or stages (Fig. 1.2). (Other than epochs, palaeontologists who study Cenozoic fossil mammals apply another series of terms relating to geological time called Land Mammal Ages, based on changes in fossil mammal assemblages between different rock layers, and with systems varying in name, age and duration for North America (NALMA), South America (SALMA), Europe (ELMA) and Asia (ALMA), but those are not addressed here.)

The term "tetrapod" translates as "four-footed" in Ancient Greek, but its scientific meaning is broader than the name suggests. In evolutionary biology, it refers to all vertebrates other than fish, encompassing the earliest amphibians and every lineage descended from them. This includes an extraordinary range of forms, from early reptile-like animals and dinosaurs to birds, mammals, and ultimately humans. Although the term may seem misleading—given that whales have flippers instead of limbs, birds possess wings rather than four legs, and snakes lack limbs entirely—each of these groups evolved from four-limbed ancestors and therefore belongs within the tetrapod lineage. The Cenozoic, in particular, has a rich and fascinating tetrapod fossil record, especially among the mammals, which are frequently preserved much more completely and commonly in comparison to dinosaurs. Therefore, mammal fossils can be used to scrutinise complicated prehistoric ecological **communities** or interpret their evolutionary trends across time. Sadly, the majority of fossil mammals are almost completely unknown to most people. This is due, in part, to the fact that many diverse and well-represented lineages have neither close modern relatives nor surviving descendants. It is therefore challenging to draw conclusive remarks concerning the habits of these creatures or infer their life appearances, but nevertheless, hypotheses for those details have been provided by expert analysis of the fossils.

This book communicates the latest scientific understanding of these and other Cenozoic tetrapods by employing informative text, along with detailed colour illustrations that I have produced digitally, of the diverse tetrapod lineages discussed, in order to document a rich evolutionary history. Subsequent to this book's release, our understanding of all the fossil **taxa** (a species or a specific grouping of species) featured within it will almost certainly advance, as innovative methodologies, further studies and the discovery of additional fossils are made. For the bulk of this book, guided by the fossil record, we will visit various fossil sites and their faunas on a global journey through time. The course of this journey will pass from the Cretaceous-Paleogene extinction event of 66 million years ago forwards in time, with life and geography becoming increasingly familiar along the way, right up to the present day. The great majority of fossil sites highlighted in this book are geological formations, the exceptions being several cave sites, pitfall traps and tar pits. Formations are specific bodies of geological strata that have distinct and consistent physical characteristics, time and location. Over hundreds of thousands to millions of years, the Earth's surface—along with all the animal and plant life it supported—gradually became buried beneath new layers of sediment and the remains of newly evolved animals and plants. As time progresses, each new layer upon the last serves as a geological snapshot of each time interval, with the earliest at the bottom up to the most recent at the top.

How do fossils form? After organisms die, their durable components (such as wood, shells, and bones) must be buried rapidly to avoid destruction. When conditions are favourable, these remains are carried underground and covered by layers of loose sediment. Over time, sandy deposits are compacted and bound together by minerals carried in groundwater, forming sandstone, while fine clay and mud particles are compressed and cemented into hard, brittle rocks known as **mudstone** or shale. These sedimentary layers may later be buried even deeper within the Earth's crust. During prolonged burial, plant or animal remains gradually mineralise and become fossils as groundwater moves through them and deposits minerals that replace the original organic material. After millions of years, powerful tectonic processes can raise these ancient rocks, folding them upward as continental plates collide to create mountain ranges. In other cases, the layers may be tilted, allowing erosion to expose the buried sediments. Once exposed, erosion can bring fossils to the surface, where wind, frost, heat, and rain gradually break down both the fossils and the surrounding rock. This cycle has operated for millions of years, and countless fossils that were once buried for vast spans of time have sadly been uncovered by erosion and destroyed long before any human could discover and study them.

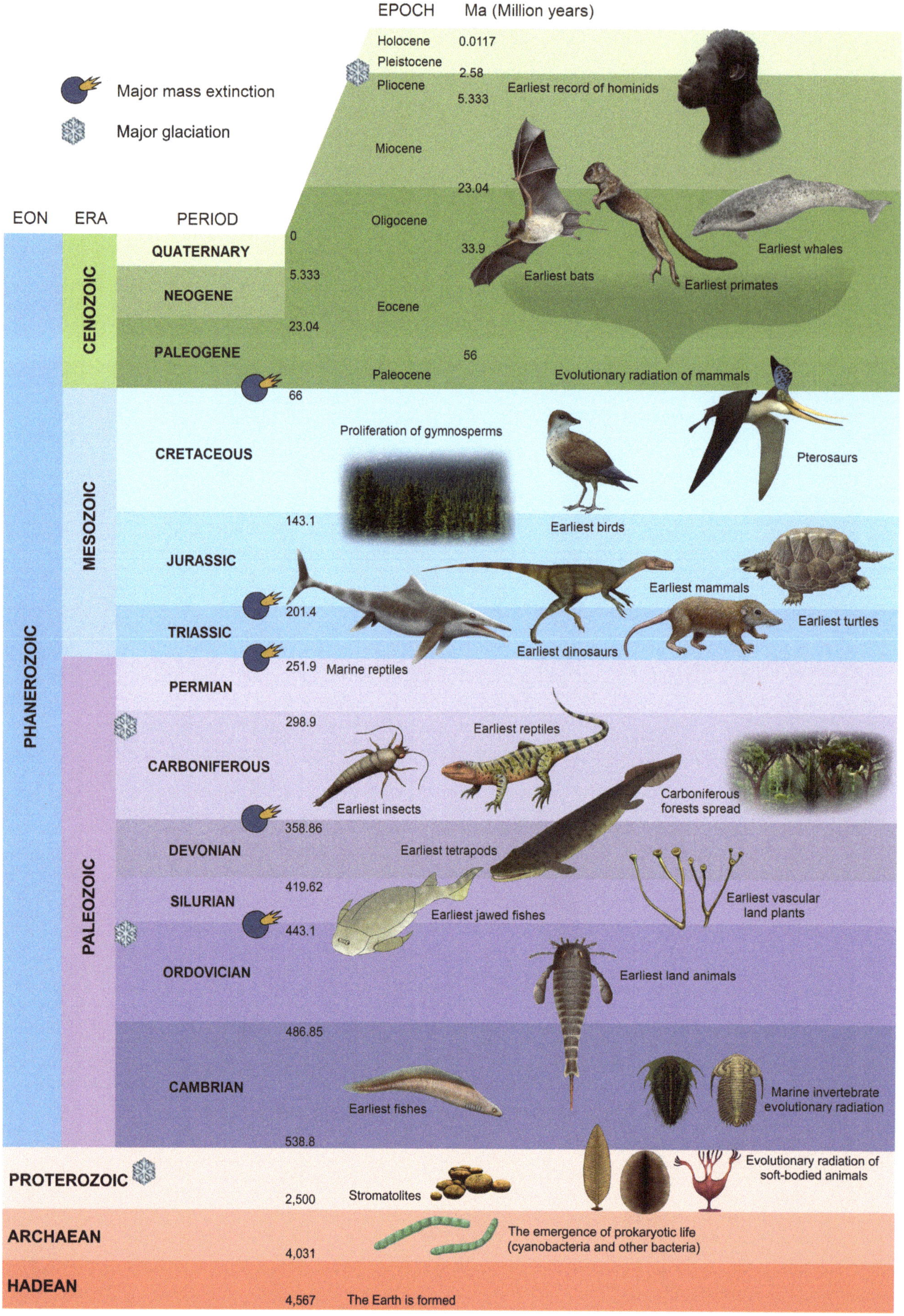

Fig. 1.1. Geological time scale, highlighting some of the most significant events in the history of life on Earth, with age ranges following those reported by the International Commission on Stratigraphy (ICS) as of 2024.

Once a fossil is lucky enough to be recovered, and its locality details documented, the following stage is identifying what it was. As with today's plants and animals, scientific names are given to fossils too. Although some people may find scientific names to be somewhat lengthy and difficult to pronounce, they are necessary for communication within the scientific community, since most fossil organisms lack a common name. Every organism has a two-word (**binomial**) scientific name, most often derived from Latin or ancient Greek. The first word is the genus name (plural is "genera") and is capitalised at all times, and either italicised (in print) or underlined (hand-written). A genus usually comprises a number of species. Two different species or genera may be distinguished by minute anatomical features. While the species name is always italicised or underlined, it is never capitalised. Additionally, the genus name can be abbreviated. For example, the scientific binomial of lions is *Panthera leo*, but there are other species of *Panthera*, such as *P. tigris* (tigers), *P. pardus* (leopards), *P. onca* (jaguars) and *P. uncia* (snow leopards). Moreover, the genus name can be written in full and italicised while the species name is abbreviated as either "sp." ("species"), used when referring to a single, unspecified/unknown species within a genus, or "spp." ("species pluralis"), used when referring to multiple species within a genus (and to the genus as a whole). Both "sp." and "spp." are not italicised. In taxonomy, "cf." (the Latin word "confer", meaning "compare to") means a specimen looks like a certain species, but scientists are not completely sure and suggest comparing it to that species.

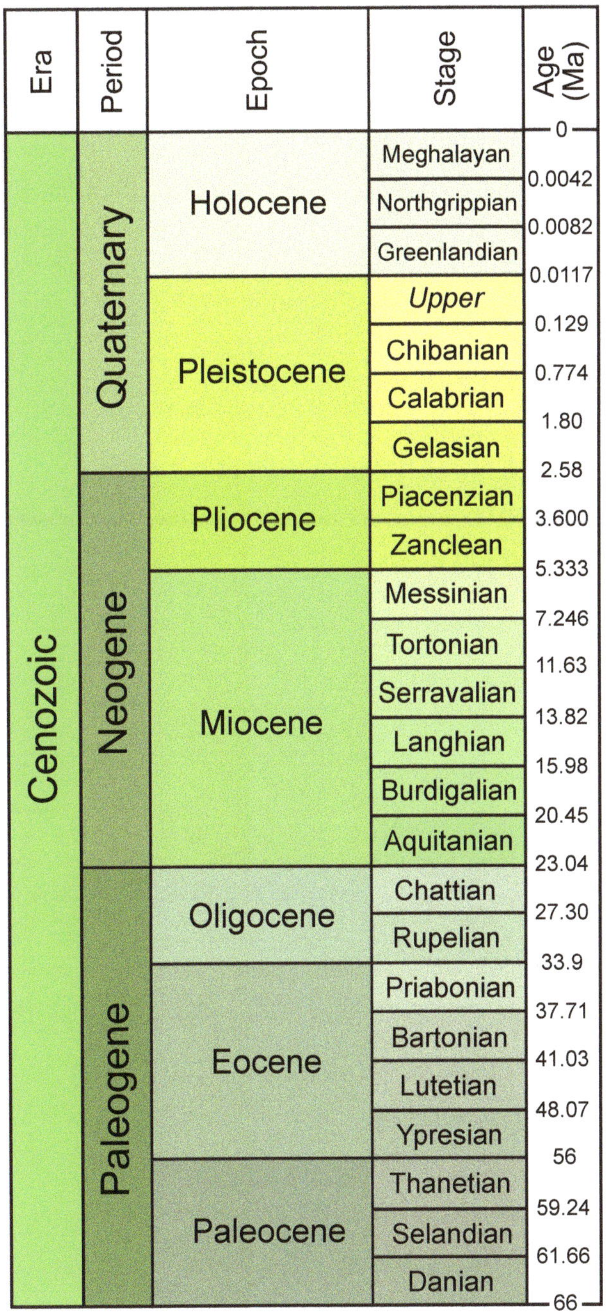

Fig. 1.2. Geological timescale discussing the individual periods, epochs and stages within the Cenozoic Era. Age ranges follow those reported by the International Commission on Stratigraphy (ICS) as of 2024.

Many readers may be aware of the Linnaean System of biological classification where organisms are traditionally grouped into seven major **ranks**: kingdom, phylum, class, order, family, genus and species. The species is the basic unit of biodiversity, as it is through natural selection on populations within species that new species evolve. The genus is somewhat more arbitrary, depending on the scientists' discernments regarding which species are clumped together. Genera are clumped into larger groups called families. For instance, the genus *Panthera* belongs to the family Felidae, along with other genera such as *Neofelis*, *Acinonyx* and *Felis*. For animals, all family names end in the suffix -idae, which is a useful hint when someone comes across an unfamiliar name. (For plants, families end in -aceae, so Poaceae is the plant family that consists of grasses.) Families are clumped into larger groups known as orders. For instance, dogs (Canidae), cats (Felidae) and bears (Ursidae) are several of the families that form the order Carnivora.

Orders are clumped into classes. Among vertebrates, mammal families are clustered into class Mammalia, whereas the birds (class Aves), the Reptilia, the Amphibia, and so forth are classes. Classes are clumped into a larger group known as a phylum (plural is "phyla"). Vertebrates form part of the phylum Chordata. The next highest rank is the kingdom. We belong to the kingdom Animalia, but kingdoms also exist for the plants, the fungi, and so forth. Additionally, numerous other ranks lie between these seven major ranks, many formed by adding a prefix to indicate whether it is marginally more inclusive (higher in the hierarchy) or less inclusive (lower in the hierarchy). For instance, a subfamily (ending in the prefix -inae) is below a family and a superfamily (ending in the prefix -oidea) is above. An example of what the hierarchy of the traditional major taxonomic ranks looks like is outlined in Fig. 1.3. One issue with classification when identifying splits in the evolutionary tree is that there are currently more branching points that require names than there are names available. For instance, ranks such as "superorder" and "infraclass" fall between order and class, yet there are relatively few consistently applied ranks between superorder and order. Consequently, many traditional taxonomic ranks above genus and species are being de-emphasised or abandoned by some biologists, since ranked categories often impose

KINGDOM	Animalia	Animalia	Animalia
PHYLUM	Chordata	Chordata	Chordata
CLASS	Mammalia	Reptilia	Aves
ORDER	Carnivora	Crocodylia	Passeriformes
FAMILY	Felidae	Crocodylidae	Passeridae
GENUS	*Panthera*	*Crocodylus*	*Passer*
SPECIES	*leo*	*niloticus*	*domesticus*

Fig. 1.3. Some examples of how the hierarchy of groups within groups appears, including, for each column from left to right, the lion, the Nile crocodile, and the house sparrow.

artificial structure on the inherently complex and branching process of evolution, and numerous methods exist for portraying evolutionary branching patterns without designating formal ranks for every evolutionary split. Almost all fossils are preserved in a single category of rock known as sedimentary rocks. These are rocks composed of loose grains of gravel, sand or mud, or other particles that weather away from the hard bedrock and settle in floodplains, rivers or the seabed. The pattern of rocks and weathering in a formation can be used to reconstruct the climate and environment that occurred in that time and location—whether it was a desert, a floodplain, or a river, for example. When animal and plant fossils are discovered within a particular formation, we can accurately presume they lived at roughly the same time and location in this ancient environment, or palaeoenvironment. Thus, a formation typically represents a kind of fossilised biome—a snapshot of life and environmental characteristics at a specific time and location in prehistory. Similarly, cave sites, pitfall traps, and tar pits often provide good glimpses of past local ecosystems. I hope that the journey through the Cenozoic provided by these snapshots of fossil tetrapods and their **communities** will help foster a broader familiarity with the amazing extinct tetrapods that inhabited this Earth from the aftermath of the extinction of the non-avian dinosaurs to just before the present day.

About this Book

The purpose of *Journey through the Cenozoic* is to give the reader a compelling but intelligible overview of the astonishing diversity and evolutionary history of tetrapods around the world throughout the Cenozoic Era. Of course, this is a rather tall order, and it would be impossible for me to cover every fossil site or mention and illustrate every known animal in the fossil record of Cenozoic tetrapods, especially in a single volume. Instead, I have chosen to restrict myself by examining essentially the most important fossil sites and the better known **taxa**, although some more obscure taxa are also mentioned to provide a more comprehensive picture. I commence by introducing the science of phylogeny and how to interpret evolutionary trees (which this book is dominated by) in the last section within this chapter. Then in Chapter 2 (by far the largest chapter), I provide an up-to-date account of the origins of the seven major **clades** of tetrapods and their evolutionary history during the Cenozoic, in a comprehensive manner, albeit not exhaustive. This is mainly so that the reader can refer to the explanations of the various groups of mammals, birds, reptiles and amphibians when reading about the fossil sites and their faunas in Chapters 5 to 7.

In Chapter 2, the tetrapod lineages are illustrated through stylised evolutionary trees, to make the relationships between various taxa easy to understand. For the trees in Chapter 2, and for any tree in Chapters 5 to 7 that includes both living and extinct species, if a dagger symbol (†) is placed next to the name of a taxon, then that means it is extinct. All images of extant taxa in the phylogenetic trees of Chapter 2, and in Chapter 3, are photographs I took at zoos, wildlife parks, aquariums or in the wild, except a few images were illustrated by myself: the manatee (*Trichechus*, p. 71); the giant otter shrew (p. 63); the saola (p. 119); the baleen whales (p. 123); the beluga and Yangtze river dolphin (p. 125); the aye-aye (p. 151); and the cebid, atelid, aotid, and colobine primates (p. 153). In Chapter 2, I also explain how certain animal groups were affected by climate change and shifting landmasses, and why some did not survive to the present day. In Chapter 3, I outline the detailed process behind my reconstructions of the extinct animals in this book, covering the general methods employed to assemble skeletons, and reconstruct ancient tissues and life appearances. After that is the main section of the book, which starts with Chapter 4 reviewing the mass extinction that marks the start of the Cenozoic, other major changes to the general ecosystem that were in the works for tens of millions of years prior to that extinction, what the earliest Cenozoic faunas were like, and what favoured the rise of early placental mammals. The final paragraphs provide a summary of the Cenozoic Era itself, concerning the geology, climate and general development of life. Chapters 5 to 7 outline in detail how the world's physical geography and faunal compositions have changed over time, broken down into one or more fossil sites each for all or most of the continents for each geological epoch (except the Holocene) within each geological period. The only exceptions are Australia for the Paleocene, Asia for the Pliocene, and Antarctica for all epochs other than the Eocene, due to highly sparse or inadequately represented tetrapod fossil records.

I highlight a range of fossil tetrapod species present at each of 42 especially well-known sites from the geological past in the style of a field guide. These Cenozoic sites were selected since they have provided particularly revealing glimpses of how life existed and interacted in the past, greatly advancing palaeontological science, and they have yielded important fossils of extinct tetrapods. These sites are arranged chronologically within three different geological periods, all three each being broken down into separate chapters from the Paleogene (Chapter 5), when mammals began to dominate the world, to the Neogene (Chapter 6), when grasslands spread at the expense of forests, and culminating in the Quaternary (Chapter 7), which covers the great Ice Ages and the tiny sliver of geological time that represents the present day. The great majority of life reconstructions for the fossil sites in this book are of mammals, but also include various birds and reptiles, and several amphibians. For sites with a much richer fossil record and/or in order to fit neatly with the text, I have drawn some stylised phylogenetic trees in order to condense illustrations of lots of notable species into several pages for that site, showcasing how the species are classified and interrelated, but not drawn to scale. At the end of each branch is my illustration of a species accompanied by information such as the estimated body size (in length for **quadrupedal** and aquatic animals, in height for **bipedal** animals, and in wingspan for flying animals) (Fig. 1.5). For phylogenetic trees anywhere in the book, a dotted branch means the proposed relationships of a species are highly contentious and remain in flux. The abbreviation "cont." means that a phylogenetic tree, or a section on a particular fossil site, is continued from the preceding two pages.

At the end of the main section (at the end of Chapter 7), is a summary of the Holocene, the current geological epoch that we are in, highlighting the rise of human civilisation, and the recent human impact on the world's wildlife, environment and climate. The "Further Reading" section features a list of relevant publications and scientific literature for each section of the book that should be referred to for extra information regarding the sites, tetrapod clades and other topics highlighted in this book. While this bibliography is extensive, it is selective instead of exhaustive; many of the references offer great starting points for anyone who intends to pursue more information on certain topics. The Glossary can be referred to for definitions of terms appearing in bold within the text (and also within figures when possible). The extensive illustrations of fossil species I have produced for Chapter 2 and Chapters 5 to 7, both enhance the text and bring the geologically recent development of the world's fauna to life. In many instances, my illustrations were largely based on closely related species that are represented by more complete fossil material. For every species, I inferred external features including colouration and fur length on the basis of its putative ecological **adaptations** along with its modern relatives or **analogues** (when feasible). Most importantly, my illustrations throughout this book are all scientifically accurate, derived from the latest fossil evidence and careful scientific research as opposed to simple assumptions. However, my artistic reconstructions of animals based on limited and/or fragmentary fossils should be seen mainly as educated interpretations instead of absolute fact. In some cases, not all fossil taxa could be illustrated for every fossil site covered because the material for some taxa was too limited or ambiguous, as well as because of limited scope in this book. For some fossil sites (and elsewhere in the book), certain species have been portrayed in a suitable environment with regard to climate and overall vegetative structure.

For any palaeoenvironmental reconstructions in which species are numbered and labelled below, the abbreviation "indet." stands for "indeterminate", meaning a fossil cannot be confidently assigned to a more specific taxonomic category (e.g. genus or species). Lastly, any important fossil species that are not present in any of the fossil sites highlighted are illustrated in the evolution chapter (Chapter 2).

Time			
Ma	Mega annum (translates as one million years)	**BCE**	Before the Common Era. Equivalent to BC (Before Christ). Years before 1 CE are numbered backwards (e.g. 1 BCE immediately precedes 1 CE)
ka	Kilo annum (translates as one thousand years)	**CE**	Common Era. Equivalent to AD (Anno Domini). Referring to the year 1 (1 CE) and subsequent years of the Gregorian calendar
ca.	circa/about		
Metric		**Imperial**	
m	metres	**ft**	feet
cm	centimetres	**°F**	degrees Fahrenheit
°C	degrees Celsius		

Fig. 1.4. Abbreviations used in this book.

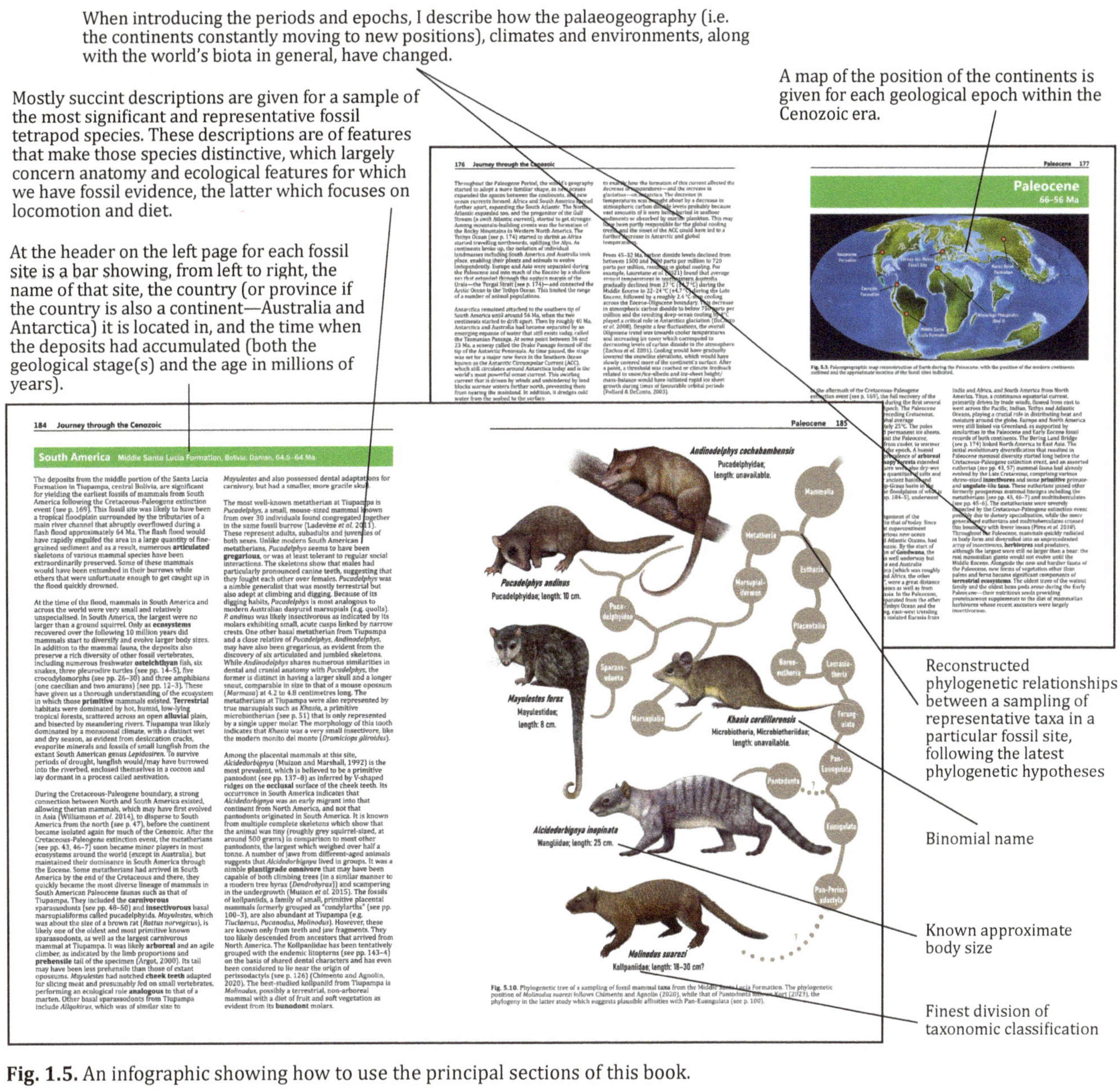

Fig. 1.5. An infographic showing how to use the principal sections of this book.

Phylogenetic Hypotheses

Phylogeny is the study of evolutionary relationships among species and their position within the "tree of life". It is based on the phylogenetic hypothesis that all living things descend from a common ancestor. The relationships of **taxa** are depicted via phylogenetic trees, or cladograms, which are diagrams, often computer-generated, that visually represent the predicted evolutionary pathways of organisms. The structure of phylogenetic trees is influenced by shared traits among various organisms. The tree-like branching represents diverging lineages from a common ancestor. When interpreting a phylogenetic tree, it is vital to understand terms including nodes, branches, **clades** and **grades** as well as taxa. Nodes are points on a phylogenetic tree where branching occurs (Fig. 1.6), representing the most recent common ancestor of the descendant lineages. A node does not depict a single species giving rise to another, but rather the divergence of one ancestral lineage into two or more descendant lineages. Branches are the lines that represent these lineages over time, connecting ancestors to their descendants. The tips of branches terminate in terminal taxa, which may be extant or extinct.

The study of phylogenetic relationships often challenges traditional classifications, which makes it important to understand the limitations of **rank**-based taxonomy. In reality, most "higher" ranks (such as classes and orders) originate from "lower ranking grades" (species), making these ranks somewhat arbitrary. For example, in the Linnaean system, Class Reptilia includes the Order Dinosauria and suborder Theropoda, which is further divided into infraorders and families—one of which gives rise to another class, Aves—followed by its own series of orders and suborders. Thus, scientists today often prefer to use clades: unranked groupings of taxa united by shared morphological and genetic features. Nonetheless, ranks such as order, family and subfamily remain useful for communication and will occasionally be used in this book for convenience. Sister taxa, or sister groups, are pairs of terminal taxa and/or clades that split from the same node and are typically regarded as each other's closest relatives. Grades are groups of taxa united by shared features or adaptations but do not necessarily reflect true evolutionary relationships. Since they often exclude some descendants of a common ancestor, grades are typically **paraphyletic** ("artificial" groupings), and in some cases, they may be **polyphyletic**, combining taxa from different evolutionary lineages. Quotation marks around taxon names (e.g. "Condylarthra") indicate informal groupings that are not natural in a strict phylogenetic sense.

Within phylogenetic trees, **stem groups** are taxa which have shared ancestry with, but are not members of, a particular **crown group** (Fig. 1.7). Since crown groups consist of all extant species, stem groups are by definition extinct. Stem groups are always paraphyletic, comprising lineages outside the crown group. Furthermore, stem groups can exist within larger crown clades; for example, stem-birds are positioned within the crown group Archosauria. Crown groups, on the other hand, are clades defined by extant organisms. They comprise all extant species of that group, their last common ancestor, and all of that ancestor's descendants, even if they are extinct. Understanding the difference between stem and crown groups is essential for interpreting how traits evolved within a lineage and for accurately classifying organisms within a cladistic framework. Cladistics is a method of classification based on evolutionary relationships, grouping organisms according to shared **derived** characters, or **synapomorphies**, as determined by morphological and molecular similarities. The main assumptions of cladistics are

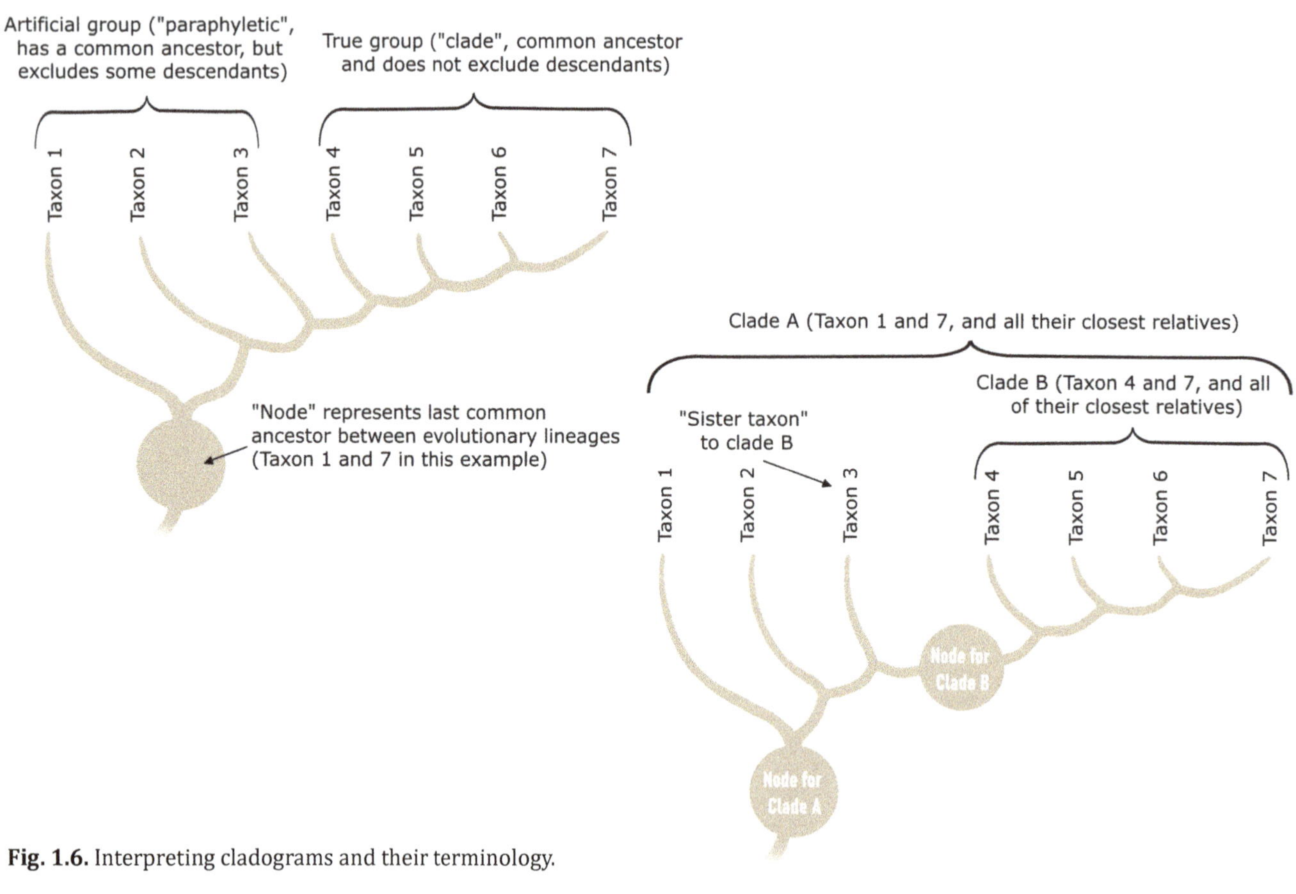

Fig. 1.6. Interpreting cladograms and their terminology.

that all organisms share a common ancestor, that new organisms arise when existing populations are separated into two or more groups, and that over time, lineages undergo modifications in characteristics. As an example of synapomorphies, primates are characterised by nails rather than claws, hands adapted for grasping, and forward-facing **orbits** with stereoscopic vision. However, lineages within primates are distinguished by their own synapomorphies, so members of the family Cercopithecidae, the Old World monkeys, share anatomical traits including a downward-facing nose, nostrils situated close together, two premolars with sharp cusps, a non-grasping tail and numerous other skeletal characteristics. The presence of nails instead of claws would not be used to define the cercopithecids, since that feature is a shared **ancestral** character, or symplesiomorphy, among primates. The presence of hair is also a symplesiomorphy, but for mammals as a whole, it is a synapomorphy (Fig. 1.8).

Historically, certain animal lineages were grouped together on the basis of a combination of synapomorphies and symplesiomorphies, and scientists have been striving to avoid clusterings of species that are "artificial" wastebaskets based on very generic traits. This has been an issue with mammals for a long time. For instance, numerous mammal clades were previously enormous **wastebasket taxa** including animals grouped together solely by shared **primitive** characters. Such wastebasket taxa encompass the now-abandoned order "Insectivora" (see p. 75) and the "Condylarthra" (see pp. 100–3), the latter which was a term applied to any extinct hooved mammal that did not belong to any modern lineage. Following the abandonment of "Condylarthra" as a meaningful term, substantial progress was, and is still being, made to resolve the phylogenetic relationships between all extinct mammals with hooves. Beginning in the 1960s, significant breakthroughs occurred in the understanding of classification and its methodology. Numerous taxonomists gradually came to concur with the emphasis on shared evolutionary novelties and bypassing wastebasket taxa defined solely by shared primitive characters. Any branch on a phylogenetic tree consisting of all the species that descended from a most recent common ancestor is called a **monophyletic** group, also called a clade or a natural group. This has proved challenging for those used to the conventional groups of species. A key principle is that if classification reflects only evolutionary branching, then every monophyletic group should consist of all descendants within it, or else it is an arbitrary, "artificial" paraphyletic assemblage. For instance, as previously mentioned, the birds, or Aves, constitute a lineage within Dinosauria, rather than a separate class distinct from the class Reptilia as two parallel taxa of the same rank. If birds are excluded from Reptilia, this makes "reptiles" a wastebasket term for any tetrapod that is not avian, thus highlighting why traditional classifications such as "Reptilia" are considered paraphyletic (Fig. 1.9).

Similarly, "Amphibia" is often used today to refer specifically to frogs and salamanders (and caecilians), although reptiles actually evolved from within early tetrapods that were traditionally grouped under "Amphibia". This makes "Amphibia" another "artificial" wastebasket taxon. These old "wastebasket" taxa, though well-known, are "artificial" and are gradually being abandoned in favour of new names based solely on shared evolutionary novelties. Therefore, the Tetrapoda encompasses all amphibians, reptiles and their descendants. The Amniota (tetrapods that undergo embryonic development within an amniotic membrane (see p. 11)) consists of all reptiles, birds and mammals. Outdated notions of the traditional terms "Reptilia" and "Amphibia", characterised as clusterings of animals based on shared primitive traits and excluding some of their descendants, are slowly disappearing from taxonomy.

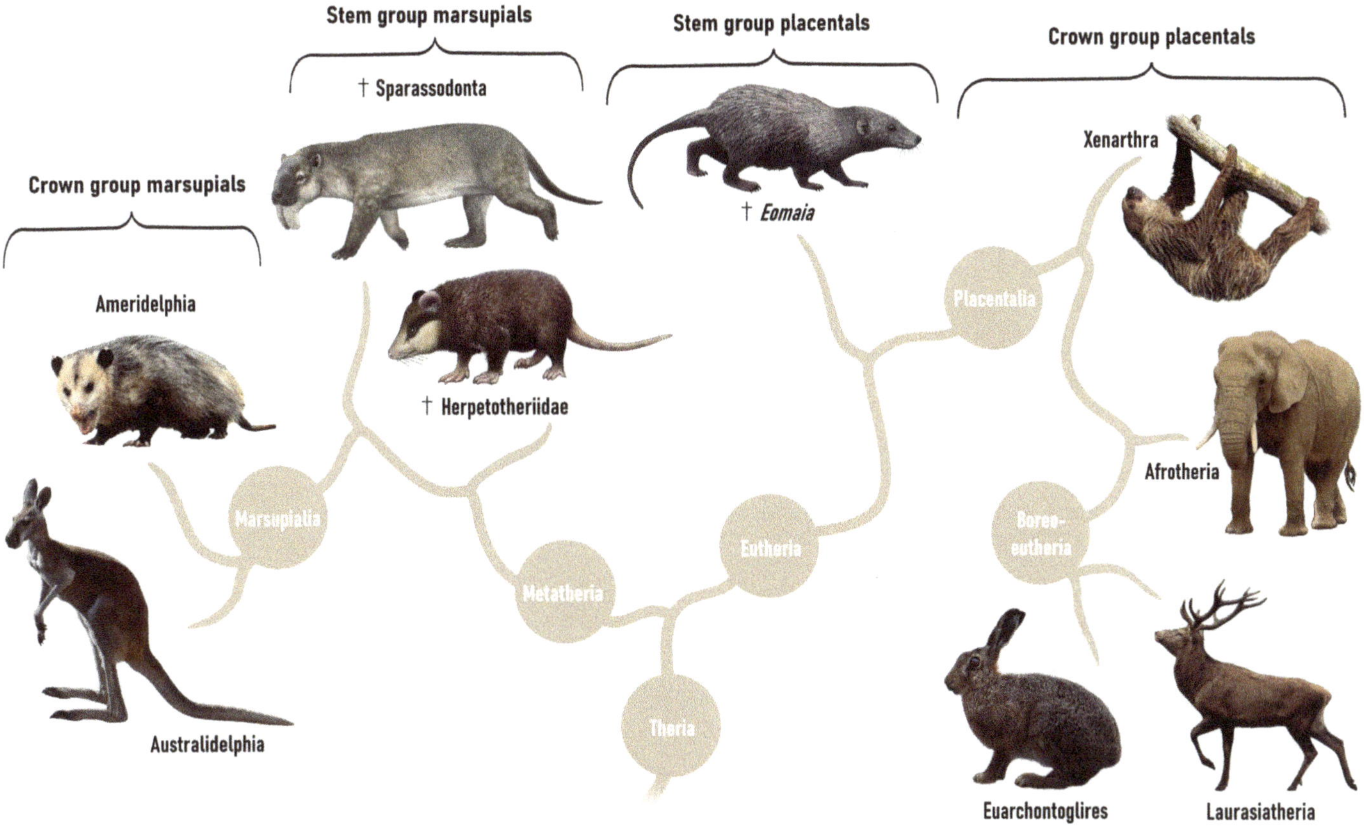

Fig. 1.7. An example of stem and crown groups.

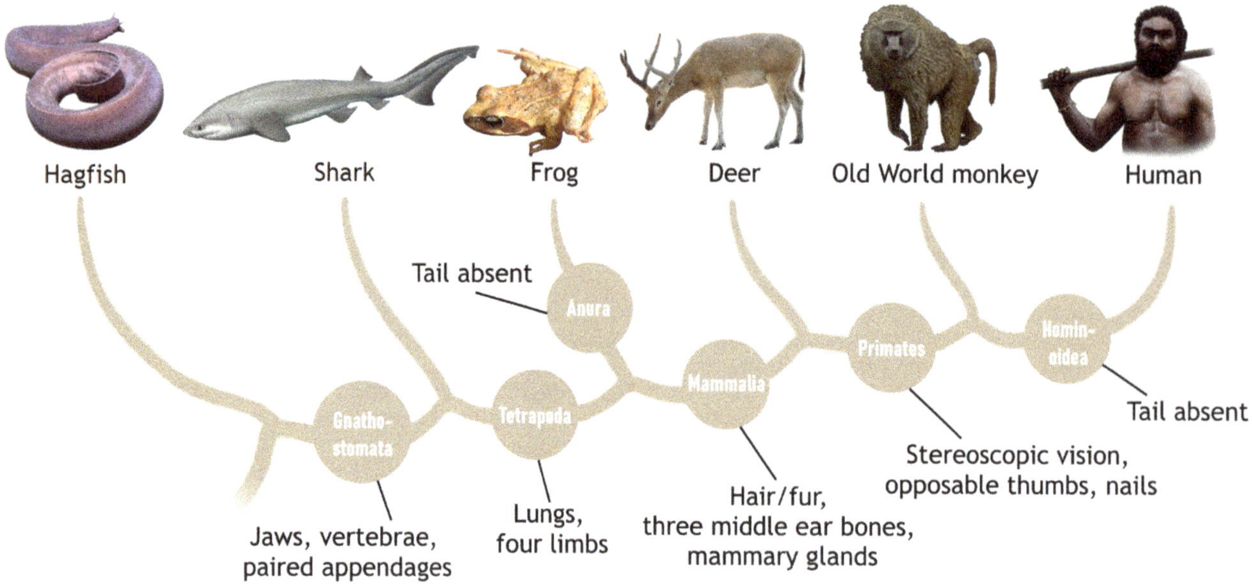

Fig. 1.8. A phylogenetic tree of the relationships for a select few vertebrates illustrating that particular traits are derived, or specialised, for certain lineages (e.g. four limbs in tetrapods) albeit primitive in others. This example demonstrates that mammals are characterised by hair/fur, three middle ear bones, and mammary glands, rather than by more primitive traits such as four limbs.

How are all the major tetrapod clades related to each other? At first, the only source of evidence was from their morphology, particularly in the teeth and skeleton, but the whole anatomy of all body systems (such as the nervous and muscular systems) can also be applied. This data was initially analysed thoroughly by scientists during the nineteenth century, who identified most of the morphological traits which remain in use for grouping tetrapods today. However, numerous issues remained, including problems with clusterings of mammals that exhibited scant evidence regarding what they were most closely related to. The most significant issue is **convergent evolution**. For example, with hooves, no reason exists that their evolution had to occur on just one occasion, in the last common ancestor of the entirety of modern hooved mammals. Rather, they may have evolved independently multiple times, when various distantly related mammals each needed to survive by sprinting more quickly over open terrain.

During the 1980s and 1990s, a new data set arose from material that biologists were unaware of before the 1960s. This was molecular biology-derived data, particularly from complex organic compounds (nucleotides in DNA and amino acids in proteins) forming each gene present in all living creatures. When sequencing the genomes (the complete set of DNA in an organism) of a group of species, aligning them and comparing them, a phylogenetic tree can be produced by clustering species together with the most comparable DNA. This involves grouping species according to shared DNA mutations, which they have but are absent in other species. Palaeontologists were able to determine if a shared morphological trait between two mammals was possibly a result of their common ancestry (and therefore can be applied to construct phylogenetic trees) or just a deceptive consequence of convergence. The issues of morphology were completely overcome when using DNA to accurately build the phylogenetic trees. Among the most notable examples are the earliest DNA-based phylogenies of placental mammals that were published during the late 1990s and early 2000s by evolutionary biologist Mark Springer, which disproved a lot of previous phylogenetic hypotheses among placentals, shown to be illusions caused by morphological convergence. In 1945, George Gaylord Simpson published a study called "*The principles of classification and a classification of mammals*". Morphological similarities formed the basis of this work, which was mostly reflected in the milestone phylogeny published by Novacek (1992). The morphology-based family tree depicts a basal divergence between an edentate cluster comprising xenarthrans (anteaters, armadillos and sloths) and pangolins, and a larger cluster comprising the remaining orders of placental mammals. The latter cluster consists of Insectivora, Anagalida (elephant shrews, rodents and rabbits), Archonta (primates, bats, colugos and treeshrews), Ungulata (all hooved mammals) and Carnivora (see pp. 81, 83–4). Simpson and Novacek were aware of the problem of convergent evolution, but lacked the technology for disentangling shared ancestry from convergence. By the late 1990s, aspects of this topology were questioned by recent fossil discoveries as well as by innovative technology for sequencing genes.

For instance, genes revealed that pangolins actually cluster with cats and dogs rather than sloths and anteaters. Bats are not close relatives of primates or colugos, but belong to a much larger group with pangolins, cats and dogs, along with the perissodactyls, or odd-toed **ungulates** (see p. 126), and the artiodactyls, or even-toed ungulates (see p. 104). Both the latter lineages possess hooves, although other mammals with hooves are scattered around the topology as well, including the hyraxes, closely related to elephants. Certain **insectivores** including elephant shrews, tenrecs and golden moles really cluster with elephants and hyraxes—a most peculiar relationship that no one had ever anticipated from morphology. Therefore, insectivory and the specialised molars associated with it were redeveloped on multiple occasions by many distinct mammal groups. This topology has replaced that of Simpson's as the norm and within it, the placentals are categorised into four major subdivisions, or superorders. Close to the tree's base, two groups diverge from the others. One early splitting group is the aforementioned largely South American Xenarthra, and the other is that surprising assemblage of aardvarks, tenrecs and golden moles, plus hyraxes, "sea cows" and elephants, together forming the Afrotheria, the latter group which was

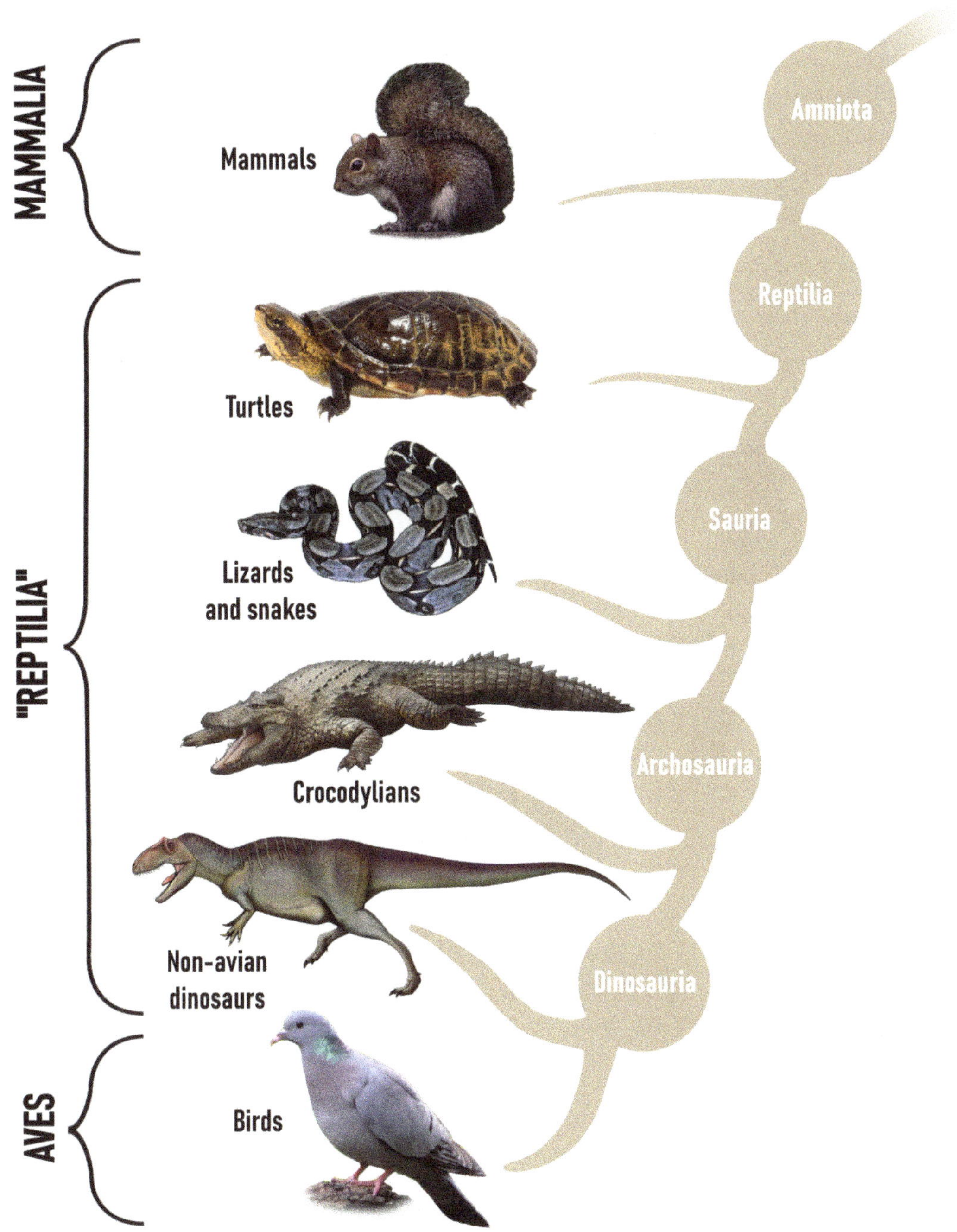

Fig. 1.9. Various methods of classifying the same lineage of tetrapods. Traditional classifications (left) tend to emphasise the enormous evolutionary diversification of mammals and birds and group them in classes of their own, equivalent in rank to all other amniotes, collectively placed in the paraphyletic "Reptilia". A cladistic classification (right) prioritises phylogeny, recognising only monophyletic groups and strictly follows evolutionary branching, grouping organisms according to shared ancestry and derived traits rather than other factors such as **divergent evolution**. This phylogenetic tree shows that birds form a subgroup of dinosaurs and in turn, archosaurs, saurians and reptiles.

originally confined to Africa. The crown of the tree is formed by two diverse lineages ranging broadly across North America and Eurasia, but also includes species living in the Southern Hemisphere. Those lineages are the Laurasiatheria (the assemblage of bats, pangolins, cats, dogs, perissodactyls and artiodactyls) and the Euarchontoglires, the latter which includes us and our primate relatives, along with rodents and rabbits. Both crown lineages cluster together in the Boreoeutheria. The relationships between Xenarthra, Afrotheria and Boreoeutheria has been a subject of controversy, because all of the competing hypotheses had some evidence to support them, and three hypotheses have been proposed regarding which lineage is basal or initially split from other placentals (Fig. 1.10):

1. The Epitheria hypothesis—Xenarthra as the sister group to Afrotheria plus Boreoeutheria (Fig. 1.10A) (Waddell *et al.* 2001; Kriegs *et al.* 2006; Churakov *et al.* 2009; O'Leary *et al.* 2013).

2. The Exafroplacentalia hypothesis—Afrotheria as the sister group to Xenarthra plus Boreoeutheria (Fig. 1.10B) (Murphy *et al.* 2001; Waddell and Shelley, 2003; Nishihara *et al.* 2007; Meredith *et al.* 2011; McCormack *et al.* 2012).

3. The Atlantogenata hypothesis—a monophyletic (natural group) clade comprising Xenarthra and Afrotheria as the sister group to Boreoeutheria (Fig. 1.10C) (Murphy *et al.* 2007; Hallström *et al.* 2007; Hallström and Janke, 2008; Meredith *et al.* 2011; Song *et al.* 2012).

Of the three hypotheses, the one I follow in this book (for phylogenetic trees including both xenarthrans and afrotherians) is the Atlantogenata hypothesis, because molecular analyses and several supertree studies (e.g. Liu *et al.* 2001, Beck *et al.* 2006, Bininda-Emonds *et al.* 2007) consistently recover xenarthrans and afrotherians as sister groups. I also favour this hypothesis because it reconciles molecular, morphological, and biogeographic evidence, explaining otherwise puzzling early placental distributions. While the precise origin of this ancestor remains uncertain, some interpretations link the split to the separation of Africa and South America during the Early Cretaceous, more than 100 Ma.

In summary, scientists who conduct research on phylogeny are always aware of the fact that phylogenetic trees are hypotheses; they are explanations for data that can be tested and modified in response to new evidence, or as reinterpretations take place or errors are noticed. Through building phylogenetic trees that incorporate as much data as is available, we can hope to reconstruct evolutionary trends with accuracy. Only with these trends before us can we hope to comprehend the true changes that tetrapods underwent throughout their evolutionary history.

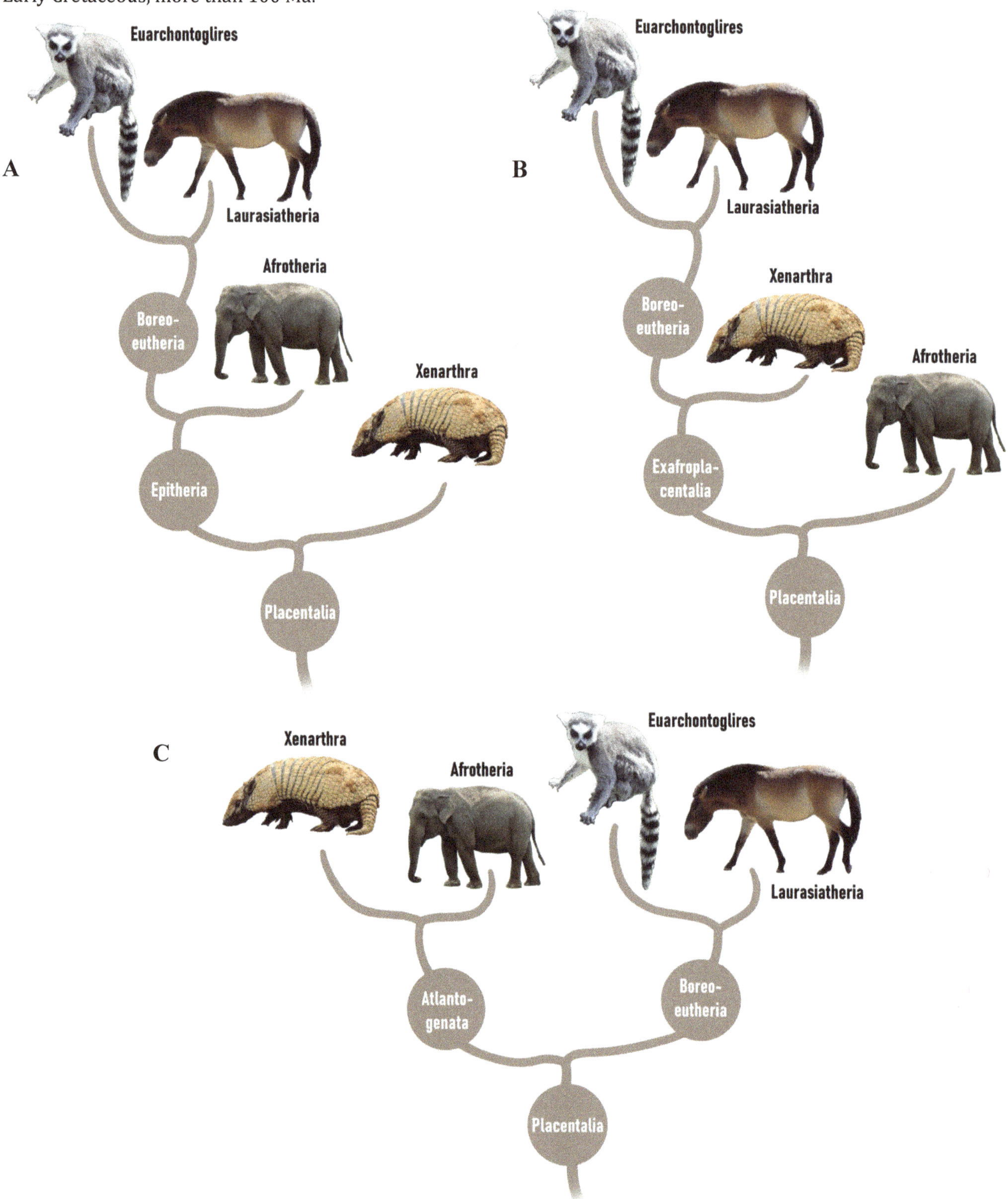

Fig. 1.10. The three principal hypotheses for the supraordinal relationships among placental mammals. (A) the Epitheria hypothesis; (B) the Exafroplacentalia hypothesis; (C) the Atlantogenata hypothesis.

2. The Evolution of Cenozoic Tetrapods

Evolution relates to heritable change that occurs in organisms over many generations. As organisms change over time to adapt to new environments, they give rise to new species. Evolutionary change is not aimed towards an end goal, neither does it rely exclusively on natural selection to influence it. The sole thing that is ever-improving is how well populations are adapted to their given surroundings. We can never say for certain about which fossil species was directly ancestral to another, although transitional fossils are frequently used as models for such ancestors.

The close ancestors of tetrapods were completely aquatic and were members of a bony fish group called lobe-finned fishes (the Sarcopterygii). The sarcopterygians include the coelacanths, lungfish and tetrapodomorphs. Marine sarcopterygians that never evolved specialisations for living in shallow water represent the coelacanth group. The sarcopterygians diverged into two main subgroups: the Actinistia (including coelacanths) and the Rhipidistia (consisting of extinct branches of lobe-finned fishes that led to lungfish and tetrapodomorphs). Tetrapodomorpha is the **clade** that includes the common ancestor of all tetrapods. Tetrapods originated from a tetrapodomorph lobe-finned fish during the Middle Devonian, around 390 Ma (see p. 2). The earliest tetrapods seem to have been entirely aquatic. The lissamphibians (modern amphibians), which arose from older lineages, are typically **semi-aquatic**. The majority of Cenozoic tetrapods are amniotes, largely consisting of **terrestrial taxa** whose lineage diverged from more **primitive** tetrapods (amphibious reptiliomorphs) during the early Late Carboniferous, around 312 Ma (see p. 2). The main innovation distinguishing amniotes from amphibians is the amniotic membrane, which allows the eggs to retain their aqueous components terrestrially instead of having to remain in water. Throughout the evolutionary history of amniotes, shelled eggs have been lost multiple times: many reptiles give birth to live offspring, while nearly all mammals lack shelled eggs. Certain amphibians, including caecilians, and certain reptiles such as snakes, lost all or part of their limbs via further evolution after they became useless; certain species possess vestigial skeletal elements as a remnant of the limbs of their distant progenitors. Other tetrapods either went back to being amphibious or having partly or entirely aquatic lifestyles, such as whales, seals and several lissamphibians. In addition, among tetrapods, only certain amniotes evolved powered flight, including birds among the dinosaurs and bats among the mammals. The amniotes branched into two major clades. The diapsids gave rise to the reptiles, which includes testudinatans, crocodylomorphs, dinosaurs (including birds), lepidosaurs and extinct cousins such as choristoderes. Within Diapsida, the group Archosauria, which originated during the Early Triassic (ca. 248 Ma), encompasses the last common ancestor of extant crocodylomorphs and birds, and all of its descendants, whereas the synapsids gave rise to the mammals (Fig. 2.1). The diapsids and synapsids can be differentiated by their skull morphology and specifically the number of **temporal fenestrae** behind each **orbit**. There is a single **fenestra** in synapsids, and most diapsids have two, the exceptions being lizards, which have lost one fenestra, and snakes and turtles, having lost both fenestrae, but still classified as diapsids based on their ancestry. It is the evolutionary history of these amniote clades, in addition to lissamphibians, that will be discussed on the following pages.

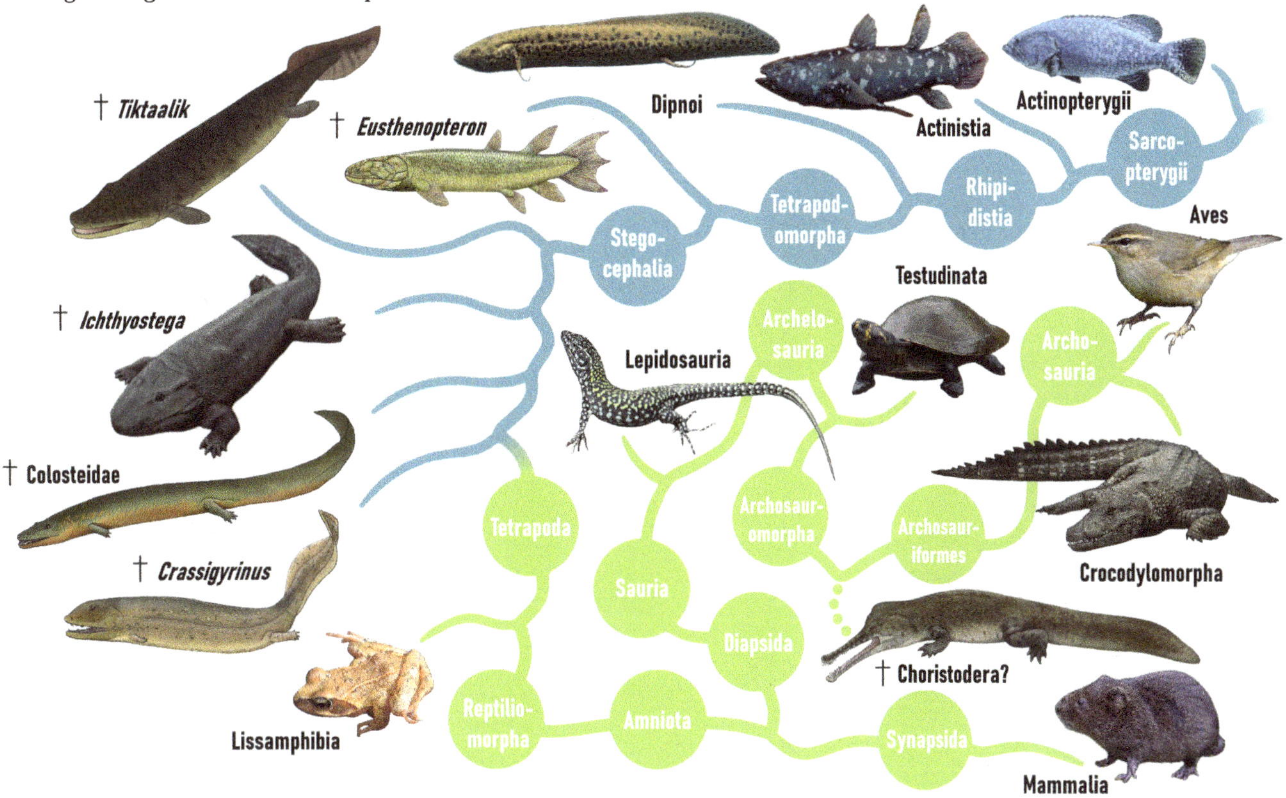

Fig. 2.1. Phylogenetic tree of tetrapod relationships after Coates *et al.* (2008), Swartz *et al.* (2012), Sengupta *et al.* (2017), Simões *et al.* (2022), and Wolniewicz *et al.* (2023). The phylogenetic position of Choristodera follows Jiang *et al.* (2023).

Lissamphibia

Where do modern amphibians, or the lissamphibians, come from? The origins and phylogenetic affinities among the three major groups of lissamphibians—the frogs (Anura), newts and salamanders (Urodela), and caecilians (Apoda)—remain somewhat controversial, although many molecular analyses suggest a Late Carboniferous to Early Permian range for the basalmost split among lissamphibians (Mauro, 2010; Marjanović and Laurin, 2007) timing for the basalmost split among lissamphibians. The frogs, salamanders and caecilians form a **monophyletic** group according to many morphological and molecular studies on modern amphibians. The Lissamphibia most probably originally evolved from somewhere within an order of **primitive** amphibians called temnospondyls, which have the primitive amphibian body plan, and characteristic teeth with their crowns resting on pedestals, a trait which all modern toothed amphibians possess.

The Salientia **crown group** consists of anurans along with their closest fossil relatives, the Early Triassic "**stem**-frogs" or "proto-frogs". The Caudata is the **clade** including urodeles and all fossil amphibians more closely related to them than to anurans. The oldest batrachians—the group including Salientia and Caudata—are known from the Early Triassic, around 250 Ma. However, molecular analyses estimate the divergence between Anura and Caudata at around 292 Ma during the Early Permian, with Apoda diverging by roughly 239 Ma (Pyron, 2011).

The earliest anurans are known from the Early Jurassic (ca. 200 Ma), but by the Cretaceous, frogs appeared almost entirely modern anatomically and had already radiated into most of the modern groups. Genetic analyses suggest that the clade Natatanura, superfamily Hyloidea and family Microhylidae (making up nearly 90% of extant frogs) quickly radiated simultaneously shortly after the Cretaceous-Paleogene extinction event of 66 Ma (see p. 169). The beginnings of **arboreal** habits, as seen in Natatanura and Hyloidea, followed this extinction and the subsequent forest recovery (Feng *et al.* 2017). Cenozoic frog fossils are known from every continent, including those of a Late Eocene (ca. 40 Ma) helmeted frog on Seymour Island near the Antarctic Peninsula (see pp. 200–3), suggesting that this area was inhabited by frogs related to those existing in Chilean forests today (Mörs *et al.* 2020). The palaeobatrachids were an ancient lineage of entirely aquatic frogs, resembling modern clawed frogs (*Xenopus*) and dating back to the Late Cretaceous (ca. 70 Ma) (Fig. 2.2). The Cretaceous-Paleogene extinction event had a relatively small impact on the palaeobatrachids, and they continued to thrive in Europe throughout the Cenozoic up until the Middle Pleistocene (ca. 500 ka), becoming extinct because of their specialised habitat preferences (swampy environments) and inability to adapt to the colder, drier conditions of the Ice Ages.

Fig. 2.2. Reconstructed life appearance of the Early Pliocene to Middle Pleistocene frog *Palaeobatrachus langhae* from Europe, possibly the last species of the genus. Reconstructed length: 10 cm.

The first urodeles are known from the Late Jurassic (ca. 160 Ma), and by the Cretaceous, fossils representing some of the modern families of salamanders and newts are known from many regions around the globe. Extant salamanders are represented by two lineages: the Cryptobranchoidea (comprising giant and Asian salamanders) and the Salamandroidea (comprising all other extant salamanders) (Fig. 2.3), which had originated in Asia during the Middle and Late Jurassic, respectively. Among the features distinguishing salamandroids from cryptobranchoids are the use of internal fertilisation and the inability to perform asymmetrical suction-feeding (where just one side of the lower jaw is depressed at a time). Many or all of the extant salamander families had possibly evolved by the Late Cretaceous.

Today, lissamphibians are incredibly diverse, with more than 8,000 species so far documented, over 7,600 of which are frogs and toads, while there are only over 800 species of newts and salamanders. Additionally, there are over 220 species of extant caecilians, which are limbless amphibians that live a **fossorial** lifestyle in the tropical soils of Asia, Africa and Latin America. They possess very small eyes capable of distinguishing between dark and light, and certain species have their eyes positioned on the ends of sensory tentacles, although many are blind. Caecilians virtually resemble giant earthworms to the non-specialist. The evolutionary history of caecilians is poorly understood, especially during the Cenozoic, since their fossil record is very scanty. Only isolated fossil vertebrae, which possess distinctive features also present in extant species, have been discovered in Paleocene deposits, and nothing is so far known about their fossil ontogeny.

A fourth major group of Cenozoic amphibians known as albanerpetontids first appeared during the Middle Jurassic (ca. 166 Ma). They are typically regarded as being slightly more closely related to urodeles and anurans than to caecilians, although they also may not be true lissamphibians and are sometimes placed immediately outside the group instead as evolutionary "cousins". It is somewhat complicated as we are still not entirely certain which specific lineage of primitive amphibians the lissamphibians diverged from.

Albanerpetontids are known from Africa and across the Northern Hemisphere, and resembled small salamanders. Albanerpetontids were typically about five centimetres long. Similar to certain modern small lizards such as ground skinks, albanerpetontids are believed to have spent much of their lives in leaf litter hunting for small invertebrates. Their defining feature is a mosaic-like covering of small scales on their bodies, but contrary to reptile scales, these were bony structures beneath a layer of skin, being structurally more similar to fish scales, and they may not have been conspicuous in life. The albanerpetontids became primarily confined to Europe after making it through the Cretaceous-Paleogene extinction event, where they appear to have persisted for almost the whole Cenozoic. *Albanerpeton* was among the last known genera of this group, surviving until the Early Pleistocene (ca. 2.5 Ma). After surviving numerous mass extinctions throughout the course of their 160 million year history, the cooling Ice Age climate appears to have ultimately driven the albanerpetontids to extinction.

The amphibians alive today vary widely in size, from the minute New Guinean frog (*Paedophryne amanuensis*), measuring just 7.7 millimetres (0.3 inches) in length, to the enormous Chinese giant salamander (*Andrias davidianus*) more than one metre (39 inches) long. Newts and salamanders retain the simple cylindrical body shape with simple front and rear limbs, and a lengthy tail, a condition also present in the most basal amphibians (from the Early Carboniferous). Of all the extant amphibians, however, the frogs are the most remarkably divergent from this primitive body plan. While toads and frogs lack tails as adults, their larvae, or tadpoles, are born with tails that are later resorbed into their bodies as they grow older. Frogs have short, broad heads exhibiting a blunt snout which enables them to open their jaws wide while catching food, frequently using an elongated, adhesive tongue. Their highly elongated, muscular hindlimbs enable them to swim with considerable power and leap enormous distances, which they use to both evade predators and capture prey. The frog skeleton also has a more compressed torso, with highly reduced to absent ribs, and possesses long, narrow pelvic bones to support the hindlimb musculature. Rather than using ribs, frogs breathe via an inflatable throat pouch that can force air in and out, which is used to produce a range of sounds too. Frogs vary greatly in size, from the aforementioned New Guinean frog to the mighty Goliath frog (*Conraua goliath*) of West Africa, which reaches up to 32 centimetres (12.6 inches) long, and is three kilograms (seven pounds) in weight. It is so large that it preys on birds, small mammals and even other amphibians, in addition to invertebrates.

Unfortunately, many extant amphibians are under threat from a number of environmental factors, including habitat loss (particularly in tropical forests), climate change, various diseases that are dispersing through and decimating frog populations, and rising acidity in freshwaters caused by acid rain, which amphibians are unable to withstand due to their permeable skins.

Fig. 2.3. Phylogenetic tree of Lissamphibia after Schoch (2019), and Jones *et al.* (2022).

Testudinata

The Testudinata consists of modern turtles (**crown group** turtles, or testudines) along with their extinct shelled cousins (**stem**-turtles). Due to their anapsid skull condition (i.e. having no **temporal fenestrae**), testudinatans were long regarded as some of the most **primitive** extant reptiles. However, molecular and genome-scale analyses have revealed that turtles are nested within Diapsida, most likely as the sister group to Archosauromorpha, rather than being primitive reptiles. This placement helps explain shared developmental and anatomical features—including limb morphology, the shoulder girdle, and cranial ossification—that are more consistent with diapsids.

The characteristic shell of testudinatans is composed of dermal and skeletal bone—the dorsal surface forms a domed **carapace**, whereas the ventral surface forms the flattened **plastron**, or belly shield. The carapace is primarily formed of expansions of the back ribs. It is possible that, during the Late Triassic, the entire shell originally evolved by integrating dermal ossifications into the rib cage. Modern turtles are split into two main **clades** depending on how they retract their head into the shell: pleurodires and cryptodires. This split has existed for a long time, with the oldest fossils of both dating from the Late Jurassic (although it is estimated to have existed as far back as the Late Triassic).

The pleurodires (meaning "side joint"), or side-necked turtles, bend their necks in a horizontal plane like closing a Swiss Army knife, protecting their heads beneath an overhanging lip on the front margin of the carapace. As well as being rare and threatened, the pleurodires are currently restricted solely to the **Gondwanan** remnant landmasses including Australia, Africa and South America. The Pleurodira consists of three extant families which are all known from the Cretaceous onwards: the Chelidae, the Pelomedusidae and the Podocnemididae. Fossil evidence shows pleurodires had a considerably more diverse history during the Cenozoic, with a global distribution.

Closely related to the extant podocnemidids, the bothremydids were the most widely distributed of the pleurodire families, with fossils across Africa, the Middle East, Europe, India, North America and South America. Fossils date from the Cretaceous, passing the Cretaceous-Paleogene extinction event (see p. 169) and through to the Miocene. In some fauna locations (such as Late Cretaceous Europe) they were the most common turtles. Unlike modern pleurodires, which are exclusively freshwater turtles, bothremydids inhabited freshwater, marine and brackish environments. This family went extinct during the Miocene, possibly due to global cooling. In comparison to the Bothremydidae, there is a rather conservative series of South American successive outgroups in the evolution of the Podocnemididae, with a highly persistent skull and shell anatomy, starting in the Late Cretaceous, continuing with the Paleocene *Lapparentemys* and leading up to several modern **taxa** including *Podocnemis* (Gaffney *et al.* 2011). The podocnemidids include the largest land turtles ever, including *Carbonemys*, from the Paleocene of Colombia (see pp. 186–7), and the monstrous *Stupendemys*, from the Miocene of Venezuela (see pp. 268, 270).

The cryptodires are the better known and most diverse turtle lineage on Earth, consisting of over 260 extant species. The name cryptodire translates as "hidden joint" and derives from the fact that when they retract the head back into the shell, the neck is vertically flexed into an S-shape within the front of the carapace. Cryptodires diverged from the lineage leading to pleurodires and began evolving terrestrial forms by the Late Jurassic, eventually dominating freshwater environments (Pereira *et al.* 2017). The cryptodires started evolving into **terrestrial** species by the latest Jurassic, while almost entirely replacing the pleurodires in freshwater environments.

Within the Cryptodira, four extant superfamilies are represented: the Trionychia, the Testudinoidea, the Chelydroidea and the Chelonioidea. Within Chelydroidea, the Chelydridae forms a sister group to the Kinosternoidea. The Chelydroidea and Chelonioidea are placed in the clade Americhelydia which, along with Testudinoidea, is grouped within Durocryptodira (Fig. 2.4). The Americhelydia and Durocryptodira likely diverged during the Early Cretaceous.

The Trionychia includes two families: the Trionychidae (including multiple species of softshell turtle known from Africa, North America and Asia) and the Carettochelyidae (which today is solely represented by the pig-nosed turtle (*Carettochelys insculpta*), found in New Guinea and northern Australia). These families likely split in the Late Jurassic (Pereira *et al.* 2017). In softshell turtles, the evolutionary loss of the hard shell was likely to enable increased manoeuvrability and swimming speeds, as well as to sustain aerobic respiration for extended durations of time (Caspermeyer, 2017).

The Testudinoidea encompasses four families: the Geoemydidae, the Testudinidae (e.g. tortoises, which are present on every continent except Australia and Antarctica), the Platysternidae (a **monotypic** family including only the extant big-headed turtle (*Platysternon megacephalum*) of Southeast Asia) and the Emydidae (a mostly North American family including terrestrial, **semi-aquatic** and freshwater forms such as terrapins). These families quickly diversified simultaneously during the Early Paleogene. The earliest split among crown testudinoids took place during the Late Cretaceous (ca. 95 Ma). The dating of this diversification provides evidence that testudinoids underwent long-distance dispersal during the Paleocene to Early Eocene and also supports the presence of various land bridges connecting the northern continents at the time (Lourenço *et al.* 2011). While the majority of fossil turtles are somewhat small, having an approximately equal size range to the modern ones, giant tortoises exist on remote islands, including the Aldabras and Seychelles in the Indian Ocean, and the Galápagos west of Ecuador. These tortoises independently evolved enormous body dimensions from smaller common ancestors due to the absence of island predators and considerably less competition for resources.

The chelydrids have a lengthy evolutionary history, dating back to the Late Cretaceous, with fossils of extinct taxa known from North America and across Eurasia, well beyond their modern distribution, in which they are presently found only in the Americas. Among well-represented chelydrids in the fossil record are *Protochelydra zangerli*, known from the Late Paleocene of North Dakota (Danilov and Parham, 2008). Compared to the modern snapping turtle *Chelydra*, *P. zangerli* has a higher-domed shell,

Fig. 2.4. Phylogenetic tree of Cenozoic testudinatans after Thomson *et al.* (2021).

a feature hypothesised to be for protection against predation from large, turtle-eating crocodylomorphs (see pp. 26–30). Today, South American chelydrids are solely present in the northwestern region of the landmass, a testament to their recent immigration from Central America during the Great American Biotic Interchange, or GABI (see pp. 252, 295). Witihin the superfamily Kinosternoidea, kinosternids are characterised by a highly domed carapace with a prominent keel along its centre, whereas the dermatemydids are distinguished by a buttressed shell with no keel on the carapace. The close phylogenetic affinities between both families is based on several skeletal features such as the cervical vertebrae morphology. The Kinosternidae and Dermatemydidae diverged during the Late Cretaceous (Iverson *et al.* 2013). Fossil dermatemydids were distributed in Europe and East Asia as well as North America. Among members of Chelonioidea, extant sea turtles comprise one radiation that split off from all other testudines during the Early Cretaceous (at least 110 Ma)(Kear, 2006), rather than being derived from more than one of the lineages of extinct sea turtles. The brains and limbs of sea turtles have evolved to suit their dietary needs. Their limbs evolved primarily for locomotion, but more recently evolved to assist with foraging.

Sea turtles use their limbs to corral, swipe, and hold their food, aiding in more efficient feeding (Fujii *et al.* 2018). Sea turtle skulls have a cheek region enclosed in bone (Jones *et al.* 2012; Chatterji, 2020), likely a recently evolved condition. Sea turtles are the largest extant turtles, with their enormous sizes maintained by the buoyancy of the water in which they inhabit. The largest of these is the leatherback sea turtle, which is also the third largest of all extant reptiles. Large specimens of leatherbacks can grow to over 2.2 metres (7 feet) in length and reach weights of 700 kilograms (1,540 pounds). The leatherback is named after the significant reduction of its bony shell, and the skeleton of its back is enclosed by a layer of tough, rubbery skin. This bony armour reduction prevents the animal from becoming too dense and sinking too quickly, as its skin is sufficiently thick to ward off many predators, and mature leatherbacks have extremely few predators. Even though extinct giant terrestrial turtles could not quite grow this large without the support of buoyancy in water, they nevertheless exceeded all extant giant tortoises in size. Giant tortoises (those with a carapace more than 70 centimetres long) have been abundant in continental **ecosystems** since at least the Oligocene. From the Oligocene of France is *Taraschelon*, with a carapace about 80 centimetres long. Among the largest tortoises ever was *Megalochelys atlas* (with a carapace measuring over 2.7 metres (9 feet) in both length and width, and weighing around 1,000 kilograms (2,200 pounds) or so), which ranged from Europe to South Asia to Indonesia between the Late Miocene and Middle Pleistocene. It essentially resembled an immense version of the Galápagos tortoise. It also coexisted with a standard mix of continental mammalian megafauna (Fig. 2.5) and Eastern Europe was also home to similarly large tortoises during the Early Pliocene, which are possibly extra *Megalochelys* specimens (Boev, 2008). Other behemoths included *Cheirogaster*, which lived in Greece from the Miocene to Pleistocene Epochs, with a carapace length that could exceed 1.5 metres. *Hesperotestudo* lived in North America during the Plio-Pleistocene (see pp. 306–7). While this short listing is by no means exhaustive, huge, completely terrestrial tortoises inhabited numerous continental habitats throughout the Cenozoic.

If ever there is proof of nothing new in nature, then one need only look to the **convergently evolved** lifestyles of meiolaniid testudinatans with the similarly extinct ankylosaurid dinosaurs or glyptodontine armadillos (see pp. 59–60). Although each group approached the niche from different angles, all three are armoured from head to tail, complete with a mace-like tail for defence. The end result is that one is well-protected from the slashing sharp blades of teeth, whilst returning fire with a hefty club of the tail. Although meiolaniids resembled modern turtles, remarkably they fall into an extinct clade called the Perichelydia, which date back to the Jurassic and are the sister group to all modern testudines. Thus, the meiolaniids are most likely stem-turtles. Meiolaniid fossils have been found in Australasia and South America, between the Eocene

Fig. 2.5. Some representatives of the Early Pleistocene fauna of Sulawesi, Indonesia, including the extinct giant tortoise *Megalochelys atlas*, along with a pair of extinct four-tusked dwarf elephants of the species *Stegoloxodon (Elephas) celebensis*, and a small group of the modern lowland anoa (*Bubalus depressicornis*).

through to the Early Holocene, where the last few species hung on at several islands in the Southwestern Pacific. It is quite possible that modern humans interacted with living meiolaniids and also, sadly, rendered the entire clade extinct. Some genera including the Pleistocene *Ninjemys* (Fig. 2.6) were colossal, reaching around 200 kilograms in weight (similar to modern giant tortoises).

The examples of aforementioned extinct testudinatan species provide just a tiny glimpse of the massive evolutionary radiation of turtles, tortoises and stem-turtles.

Fig. 2.6. Reconstructed life appearance of the meiolaniid testudinatan *Ninjemys oweni*, from the Late Pleistocene of northern Australia. Reconstructed length: 1 m.

Choristodera

Modern crocodylians dominate the predatory niche in freshwater/estuarine environments. They did not always have it to themselves, with other groups regularly filling a similar role. One such group, the choristoderes (Fig. 2.7), included long-snouted **piscivorous** forms and had something else in common with the crocodylians: both survived the Cretaceous-Paleogene extinction event (see p. 169). Ranging across the Northern Hemisphere from the Middle Jurassic up until the Late Miocene (a duration of over 150 million years!), choristoderes were one of the longest surviving lineages of tetrapods. All choristoderes exhibit some morphological **adaptations** to living in freshwater environments. Among these adaptations are laterally-compressed tails, which were likely their primary means of propulsion. In more derived choristoderes, the nasal openings are merged and placed near the tip of the snout, which enabled those animals to use their nostrils as a snorkel while keeping the remainder of their bodies submerged (Acorn, 2007).

While they were certainly diapsids, the exact phylogenetic affinities of choristoderes with other **taxa** are unclear, despite there being hundreds of well-preserved specimens. This is because they lack definitive osteological characters associating them with any major reptile lineage. Choristoderes have been placed as basal lepidosauromorphs (i.e. closer to lizards), basal archosauromorphs (the lineage of reptiles leading to archosaurs, including crocodylomorphs (see pp. 26–30) and birds) or even basal diapsids. Although the skulls of choristoderes are vaguely lepidosauromorph-like, they do not possess the same **quadrate** configuration. The results of recent analysis on the ossification stages of their embryos suggest a possible placement of choristoderes as basal archosauromorphs (Jiang *et al.*, 2023), although this remains debated. If true, then there is a missing record of nearly 60 million years, between the last known Early Triassic relatives of choristoderes and the earliest choristodere fossils in the Middle Jurassic.

The crocodylian-like neochoristoderes, or champsosaurs, are derived choristoderes and are the most well-known members, having survived into the Cenozoic. Champsosaurs had the thin, elongated snout typical of piscivores, making them superficially resemble the narrow-snouted gharials, or gavialids (see p. 29). Unlike crocodylians, their limbs were gracile and paddle-like, which suggests they were primarily aquatic **ambush predators** rather than fast **terrestrial** hunters. While it was once thought that champsosaurs were absent from habitats inhabited by crocodylians, this is untrue. There was no overlap with long-snouted forms, although champsosaurs were regularly found alongside blunt-nosed crocodylomorphs such as *Borealosuchus* (see pp. 26, 28). Champsosaurs were virtually fully aquatic species, with only females coming on to land to lay eggs and were more tolerant of marine/polar environments (which would have been temperate at the time). Genera ranged from the Late Cretaceous, across the Cretaceous-Paleogene boundary (*Champsosaurus*) and through into the Eocene (*Simoedosaurus*). *Champsosaurus* exhibits dietary

Fig. 2.7. Phylogenetic tree of Cenozoic choristoderes after Dong *et al.* (2020).

modifications following the Cretaceous-Paleogene extinction event (Matsumoto, 2015), after which the genus underwent an **adaptive radiation** as the world recovered, resulting in several species such as *C. gigas*, the largest known species of the genus at approximately 3.5 metres (11 feet) in length. *Simoedosaurus* had wider and more powerful jaws than other champsosaurs like *Champsosaurus* (Matsumoto and Evans, 2016). *Champsosaurus* is known from North America and Europe, whereas *Simoedosaurus* is known from Europe and West Asia. Champsosaurs persisted until the earliest Eocene (ca. 56 Ma), and the reasons for their extinction remain unclear. Their extinction is sometimes blamed on competition with crocodylians, although both lineages lived alongside each other throughout the Late Cretaceous, Paleocene and Eocene of the Northern Hemisphere. Furthermore, in freshwater environments, neochoristoderes appear to have dominated over the **longirostrine** crocodylians (Matsumoto and Evans, 2010; Matsumoto *et al.* 2014). Factors including changes in freshwater habitats and resource availability may have contributed to the decline of champsosaurs.

While there were a few species of large champsosaurs during the Early Cenozoic, the choristoderes as a whole were long believed to have gone extinct by the Early Eocene. Later, sampling from the Paleocene, Oligocene and Miocene strata of Europe yielded fossils of the basal, small-bodied (about 30 centimetres long), superficially lizard-like *Lazarussuchus inexpectatus*, the only known surviving choristodere after the Eocene. Although it was among the last, *Lazarussuchus* possessed a relatively **primitive** skull morphology, with a snout that is considerably shorter and more tapered than that of the champsosaurs, measuring approximately 27 centimetres (12 inches) in length. The nearest fossil relatives of *Lazarussuchus* are known from around the Late Jurassic or Early Cretaceous, and are absent from the fossil record from approximately 120–25 Ma, a gap in the fossil record of nearly 100 million years. Thus, it was named for the figure of Lazarus in the Bible, who Jesus raised from the dead. The species name *inexpectatus* reflects the unexpected presence of an Early Cretaceous relic in Paleocene, Oligocene and Miocene strata without any record in between.

The choristoderes are especially well-known for the lengthy temporal gaps in their fossil record, including the gap between their earliest known Middle Jurassic fossils and the hypothetical Early Triassic relatives, and the gap between the Early Cretaceous relatives of *Lazarussuchus* and the earliest appearance of fossils from that genus in the Middle Paleocene. The reason for these gaps is poorly understood and highly disputed, although being mainly limited to lake deposits and present only in specific geographic areas possibly played a role in their rarity in the fossil record. With the disappearance of the European *Lazarussuchus*, the choristoderes as a whole finally went extinct by the Late Miocene (ca. 11.6 Ma). While climate change and competition with more advanced reptiles likely contributed to their extinction, the exact reasons remain unclear due to the incompleteness of the fossil record.

Lepidosauria

The Lepidosauria (lizards, snakes and relatives) are by far the most abundant and diverse lineage of extant reptiles. Over 11,500 species of lizards and snakes are alive today, constituting the largest **clade** of tetrapods. The diversity of the other extant reptile lineages is nowhere near as close. Extant lepidosaurs comprise the myriad lizards and snakes (order Squamata) along with the tuatara, a "living fossil" of a Mesozoic reptile lineage (order Rhynchocephalia) which superficially resemble lizards and consists of one species, *Sphenodon punctatus*, distributed across several islands off the New Zealand coast (Fig. 2.8A).

The lepidosaurs are distinguished by several unique traits. The term Lepidosauria translates as "scaly reptile" in ancient Greek, and Squamata means "scale" in Latin. Thus, scales are a key trait in all lepidosaurs. Their bodies are coated in rather thin scales underlying **keratinous** layers. The majority of lepidosaurs shed their outermost layer of scales in flakes when they become worn out, while snakes shed their whole skin in a single large, continuous sheath. In contrast, crocodylians and turtles typically have considerably thicker scales and are covered by horny plates known as scutes (see p. 163). Contrary to scutes, lepidosaur scales frequently overlap in a shingled manner similar to roof tiles. Among other lepidosaur characteristics are a cloacal slit that is orientated transversely along the body axis, in both males and females, whereas the remaining reptiles have one cloacal opening situated parallel to the body axis. The majority of lepidosaurs (particularly lizards but several snakes too) are capable of discarding their tails in order to divert a predator's attention to the tail while the remainder of the reptile escapes. The tail has specialised vertebrae with fracture planes that enable it to break off. Lastly, characteristics such as a forked tongue are used to pick up scent molecules from their surroundings and transport them to the vomeronasal organ immediately above the hard palate in which they are "tasted", assisting in behaviours including hunting and mate selection. Lepidosaurs are the sole group of reptiles that have forked tongues.

A **pleurodont** dentition is present in the majority of **primitive** lepidosaurs, such as the oldest relatives of the tuatara, and the majority of lizards and snakes. In primitive rhynchocephalians, the bone beneath the lower **temporal fenestra** of their diapsid skull is not complete, a condition also present in squamates. This bone is fused back together in later rhynchocephalians. A knee joint in which a lateral recess on the femur articulates with the fibula (a distinctive trait among tetrapods), and the presence of **gastralia**, are two characteristics that set rhynchocephalians, including the tuatara, apart from squamates. Furthermore, the lepidosaurs exhibit various distinctive cranial characteristics, among the most notable which is a shell-shaped capsule within the cranium called a "conch" that supports the eardrum and aids in hearing. Lastly, many molecular studies have consistently demonstrated that Lepidosauria is a **monophyletic** group consisting of both the squamates and the tuatara, so the validity of this lineage is firmly established.

However, for various reasons, lepidosaurs are frequently overlooked in the fossil record and in the general public's perception of ancient life. The most apparent reason is that most of them were small-bodied, and therefore had delicate bones which were easily fragmented and left behind scarcely any fossils. By comparison, the majority of dinosaurs left behind massive bones that could fossilise with ease, the skeletons of crocodylomorphs (see pp. 26–30) are often relatively large and sturdy, and turtle shells are prevalent fossils in a number of deposits. Jawbones and portions of the limb girdles can be used to distinguish certain fossil lizards, although many are

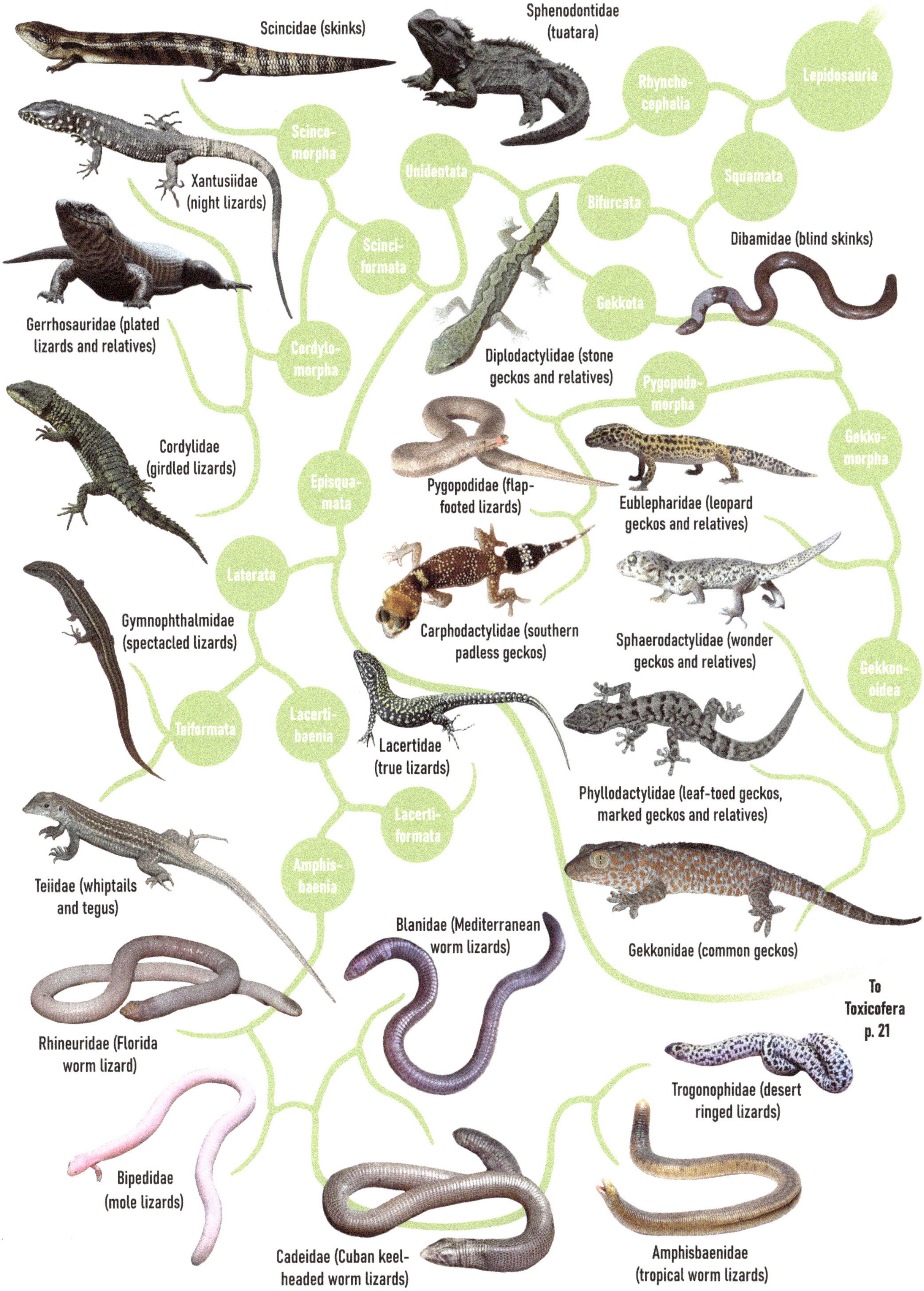

Fig. 2.8A. Phylogenetic tree of Lepidosauria, particularly **taxa** outside Toxicofera, after Wiens *et al.* (2012), and Zheng *et al.* (2016).

represented by small, partial skeletal elements that make it impossible to identify them by genus or species. Snake fossils are even less likely to be well-preserved as their skeletons consisting of brittle vertebrae and ribs, along with their delicate skulls consisting of narrow struts of bone, can disintegrate and separate very easily. Of course, there are a small number of fossil snakes and lizards represented by exceptionally preserved complete skeletons, although an overwhelming majority are represented solely by fragments. It is unsurprising that they have a limited fossil record given their low chances of fossilisation. Lepidosaurs are most closely related to archosaurs (see p. 11). The earliest lepidosaurs date back to the Early Triassic, although they have a sparse fossil record over much of the Mesozoic, while the archosaurs conquered the world and dominated it even subsequent to the demise of the non-avian dinosaurs 66 Ma (see p. 169).

Rhynchocephalia

Exactly when the Rhynchocephalia ("beak heads" in ancient Greek) diverged from Squamata is contentious, but the earliest fossils are known from the Middle Triassic (ca. 240 Ma). The oldest rhynchocephalians were small-bodied, but the lineage had attained a broad variety of body sizes by the Late Triassic (Herrera-Flores *et al.* 2022). In fact, the majority of lepidosaurs during the Late Triassic and the Jurassic were rhynchocephalians. The rhynchocephalians peaked in morphological diversity during the Jurassic, consisting of specialised aquatic and **herbivorous** species (Herrera-Flores *et al.* 2017). Why the rhynchocephalians declined throughout the Cretaceous remains uncertain, but has frequently been linked to competition with advanced lizards and mammals.

The fossil record of rhynchocephalians becomes highly sparse by the Cenozoic Era, culminating in the "living fossil" tuataras (*Sphenodon punctatus*) of New Zealand. The most recent known rhynchocephalian fossils beyond New Zealand belong to *Kawasphenodon peligrensis* from the Early Paleocene of Argentina, immediately after the Cretaceous-Paleogene extinction event (see p. 169). Some partial lower jaws of an indeterminate rhyncocephalian preserving the dentition are known from the Early Miocene (19–16 Ma) of New Zealand, being very similar to those of the modern tuatara. The progenitors of the tuatara are unlikely to have reached New Zealand through oceanic dispersal. The prevailing scientific view is that they already existed in New Zealand before it split from Antarctica 82–60 Ma (Jones *et al.* 2009). Tuataras, the sole surviving species of rhynchocephalian, are greenish-brown reptiles with an unusual appearance, looking like a thickset lizard. Tuataras reach up to 80 centimetres (31 inches) long, possessing a stubby snout with an overhanging "beak" on the upper jaw, a conspicuous "third eye" (pineal gland) for sensing light and seasonal change, and a series of spines down the centre of the back and tail. They predominantly feed on small invertebrates, particularly slugs, snails, earthworms, insects and grubs, and occasionally bird's eggs. Tuataras exhibit a sequence of wedge-shaped teeth lining the outer margin of the jaw, which combined with extra teeth on the palate, assist in grinding these organisms. The jaw apparatus facilitates a slight fore-aft movement of their teeth and jaws, which assists in processing foodstuffs. Tuataras were once widely distributed across New Zealand, but were largely extirpated when rats were introduced by Polynesian settlers, and are currently restricted to 32 nearshore islands, where they are regarded as threatened with extinction.

Squamata

As previously mentioned, all the 11,500+ extant lepidosaur species, excluding the tuatara, are squamates—lizards, snakes, and a typically limbless lineage known as amphisbaenians, or "worm lizards". Besides the unique characteristics of their scales, and their shedding process, also aforementioned, the most notable characteristic of squamates is the loss of the bone at the base of the lower **temporal fenestra**, resulting in the **quadrate**, which articulates with the lower jaw, being able to swing freely in a fore-aft movement, a condition called streptostyly. This extends the lower jaw and increases gape, enabling squamates to swallow particularly large prey.

This condition is especially pronounced in snakes. These have the ability to fully swallow prey animals larger than their skull diameter. In order to enable this, the majority of their cranial bones, apart from the braincase, have been reduced to thin, elongated splints **articulated** to one another and linked by flexible ligaments and tendons. While they begin to enclose their jaws around a larger prey item, they extend their whole mouth over it, with their **posteriorly** directed teeth in the upper jaw moving independently in a series of ratchet movements over the prey gradually. They ultimately extend their whole mouth around the prey, then gradually process it down their oesophagus and stomach, which forms a prominent bulge in the snake's body during digestion, which can last for several weeks. Snakes frequently undergo a state of torpor during the challenging process of digesting an entire unchewed carcass. As the prey is digested, the bulge of the prey is visible travelling down their bodies.

Fig. 2.9. Reconstructed life appearance of the iguanian lizard *Barbaturex morrisi*. Reconstructed length: 6 ft (1.8 m).

Fig. 2.8B. Phylogenetic tree of Lepidosauria, particularly taxa within Toxicofera, after Wiens *et al.* (2012), and Zheng *et al.* (2016).

What is the origin of squamates? The fossil record, along with genetic and morphological studies, all prove that they diverged from a common ancestor with Rhynchocephalia, dating to the Middle Triassic. According to fossil evidence, **crown** squamates are thought to have possibly evolved during the Early Jurassic (Jones *et al.* 2013). Fossils of extant lizard taxa, such as skinks and geckos, have been documented from the Middle Jurassic onwards, and varanoids (monitor lizards) and iguanians first appeared during the Cretaceous.

Nowadays, almost all scientists acknowledge several distinct **monophyletic** groups among the squamates. The amphisbaenians encompass over 200 extant species of limbless worm-shaped squamates that have evolved to live an entirely subterranean existence as **fossorial** animals. They were formerly classified alongside snakes and lizards as a distinct but equal group, but are now understood to be merely a lineage within the squamates. In addition to tropical geckos, which can grip onto walls and other surfaces thanks to their adhesive toe pads, the Gekkota also consists of pygopodids, or flap-footed lizards, a different family of squamates that became functionally limbless and specialised in burrowing. In addition to the **herbivorous** iguanas, the Iguania also comprises the chamaeleons, agamids (such as bearded dragons, thorny devils and spiny-tailed lizards), collared lizards, basilisks, horned lizards (or "horny toads"), anoles (New World lizards that can change colour), and numerous other poorly-known groups (Fig. 2.8B).

Within Iguania, two major groups are accepted, based largely on distribution and where the teeth were attached to the jawline—the Old World Acrodonta (chamaeleons, agamas) and the mostly New World Pleurodonta (iguanas, etc). Although a lack of commonality of morphological features makes grouping the two **clades** problematic, molecular analyses have shown Iguania to be monophyletic. Acrodontans (both fossil and living) are almost exclusively found in the Old World (Africa, Madagascar, Europe and Asia), apart from one basal form (*Gueragama*) from Late Cretaceous Brazil. In addition to generalised **insectivores**, they included large herbivores such as *Barbaturex* from the Eocene deposits of Myanmar (Fig. 2.9).

Whilst most pleurodontans are found throughout the Americas, there are several outliers currently living in Madagascar (oplurids) and Pacific Islands (iguanids). Oddly, they were most related to the **arboreal** and tropical forest-dwelling anoles (*Anolis*) and woodlizards (hoplocercids). The remaining families are linked within the Iguania and include iguanids, basilisks, tropidurid ground lizards (*Leiocephalus*), horned lizards and oplurids. Although not as diverse in body forms as the acrodontans, pleurodontans include the only sea-going lizard, the marine iguana (*Amblyrhynchus cristatus*).

The varanoids, or monitor lizards, are an incredibly successful group of lizards. The earliest genera (*Ovoo*, *Aiolosaurus* and *Telmasaurus*) are known from Late Cretaceous Mongolia and range through to the modern day distribution across Africa, Eurasia and Australasia. Early genera such as the long-tailed *Saniwa* (Fig. 2.10) even successfully spread to North America in the Eocene. In general, they have a conservative body plan, with all species roughly similar despite the differences in size. Based on all modern species being active pursuit hunters (apart from a few **frugivorous** species) with a high aerobic capacity, we can assume that this was possibly true for extinct relatives too. Almost all varanoids from the Miocene to modern species fall within the genus *Varanus*. The tiny 23 centimetre Dampier Peninsula monitor (*V. sparnus*) is the smallest species ranging up to the gigantic Pleistocene *Megalania* (*V. priscus*), the largest known lizard to walk on land (see pp. 332, 335).

The Laterata, or Lacertoidea, comprises the true lizards (Lacertidae, the primary lizard family in Africa and Eurasia), the spectacled lizards, and the whiptails or tegus. The Anguimorpha consists of Gila monsters, knob-scaled lizards, anguids (slow worms, glass lizards and alligator lizards), anniellids (such as California legless lizards), earless monitor lizards, and true monitor lizards (such as goannas, Komodo Dragons). In addition to skinks, the Scinciformata includes night lizards, plated lizards and girdled lizards too. Lastly, there are also more than two dozen snake families. I will not delve into further detail about all these lineages since that is beyond the scope of a book like this, but is something all herpetologists gain an understanding of early in their career.

Due to the limited fossil record of these rather fragmentary specimens, many scientists had to concentrate on the wealth of information found in the soft tissues of the 11,500+ extant squamate species in order to determine their phylogenetic relationships. However, despite our many analyses, this has not yet provided us with a definitive answer. The standard idea (dating back to studies by Richard Estes in 1988 and before), stemming from fossils and anatomy, regarded iguanians and their relatives as the earliest branch of extant squamates, whereas the remaining squamates were placed in Scleroglossa (translated as "hard tongue" in ancient Greek), consisting of skinks,

Fig. 2.10. Reconstructed life appearance of the varanid lizard *Saniwa ensidens*. Reconstructed length: 1.3 m.

Fig. 2.8C. Phylogenetic tree of Lepidosauria, particularly taxa within Serpentes, after Wiens *et al.* (2012), and Zheng *et al.* (2016). The phylogenetic position of Madtsoiidae follows Scanlon (2006), while that of Palaeophiidae follows Folie *et al.* (2021).

anguid lizards and all other lizards (along with amphisbaenids and snakes). Iguanians employ their tongues to seize prey items (like basal lepidosaurs including the tuatara do), while the scleroglossans use teeth-and-jaw prehension, and the tongue is less significant (because thin, forked tongues are typically used to sense the air rather than for eating). This classification seemed to be supported by a molecular analysis by Lee *et al.* (2004). However, since 2005 (Vidal and Hedges, 2005), DNA sequence studies appear to have bene providing a different result from the molecular findings by Lee *et al.* (2004). According to this evidence, geckos are placed as the oldest lineage, followed by skinks, lacertoids, amphisbaenians, snakes, anguimorphs, and lastly the iguanians as the most advanced lineage. This topology also appears to be supported by more recent studies (e.g. Reeder *et al.* (2015); Zheng and Wiens (2016)). Although the relationships remain unresolved, this appears to be the consensus from molecular analyses conducted in the previous decade.

Serpentes

Snakes are undoubtedly among the most prosperous and diverse **terrestrial** animal lineages. In spite of their very specialised predatory behaviour (feeding exclusively on live animals), there are more than 3,900 extant species, grouped in around 30 distinct families and several hundred genera. They exist from Scandinavia in the north to Australia in the south, and inhabit all continents excluding Antarctica. They live at elevations of up to 4,900 metres (16,000 feet) in the Himalayas, and the sea snakes are entirely aquatic animals. Snakes are absent in numerous islands, including Iceland, Ireland, Hawaii, New Zealand and most of the South Pacific. These islands were most probably isolated from mainland regions where snakes were present, so they were unable to arrive at these locations, even when sea levels fell during the Last Glacial Period (see p. 313) and many terrestrial mammals could stroll or swim to remote islands. During the Pleistocene, certain islands, including Ireland, were nearly totally blanketed by ice sheets, whereas some, such as Hawaii and numerous other distant islands, were too far away for snakes to travel from any continent.

While some snakes have excellent vision, most merely have blurred vision and are typically adept at sensing movement; several species are blind. Many snakes "taste" odours in the air with their forked tongues rather than their eyes, before transporting the chemicals to the vomeronasal organ immediately above the hard palate. Furthermore, most snakes can sense the presence of **endothermic** animals (including their own prey and potential predators) thanks to heat-detecting pit organs on their snouts. External ears were lost in snakes, relying on their lower jaws instead for sensing airborne sounds.

There are between 100 and 400 vertebrae **posterior** to the skull, with the number increasing as they mature. By comparison, humans have just 33 vertebrae with none ever added, and many long-tailed animals possess around 50. An intertwined network of muscles shroudng the ribs enables the snake to push itself through a series of sinuous curves and regulate its movement. Much of the body is formed of a highly elongated ribcage and a shortened tail. There is only one right lung within this elongated ribcage, and the left lung is significantly smaller to accommodate their narrow bodies and allow more space for large prey. To fit within their elongated, narrow bodies, all the remaining paired organs, including the gametes and kidneys, are staggered with one portion ahead of the other. Small remnants of the pelvic bones and femurs of the most ancient snakes (particularly the boas and their cousins) are still present. These bones are no longer used as limbs, although they remain useful for courtship and **intraspecific competition**. These reflect their ancestry amongst four-limbed tetrapods.

Snakes exhibit a remarkable diversity of sizes considering their restrictions in body structure. The Barbados threadsnake (*Tetracheilostoma carlae*), which is the tiniest snake, is a leptotyphlopid (Fig. 2.8C) only around 10 centimetres (4 inches) long and can neatly coil up on a 10p coin. Many snakes are around 1 metre (3.3 feet) in length, which is sufficiently large to overcome their typical prey of small mammals, birds and sometimes other snakes. At the opposite end of the spectrum are the anacondas (*Eunectes*) and reticulated python (*Malayopython reticulatus*), two immense boa constrictors which belong to the family Boidae, which in turn is grouped with most other modern snakes in the **clade** Alethinophidia (Fig. 2.8C). Lacking venom, boids kill prey by constriction. The body is quickly coiled around prey and pressure is applied, causing lack of blood flow to vital organs. Prey is swallowed whole and depending on size may take between days and weeks to digest.

Anacondas are boids adapted for swimming, dragging prey underwater while squeezing the life out of it. Anacondas can grow up to 5.2 metres (17 feet) long, and weigh up to 80 kilograms (176 pounds). The reticulated python is more lightweight, although it can grow slightly longer, attaining 7.3 metres (24 feet). Due to their enormous size, these two snakes can ingest a variety of massive prey, including capybaras, sheep, goats, small cattle and other larger tetrapods. However, these are dwarfed by ancient giants including *Titanoboa* (see pp. 186–7). Compared to modern boids, most of the early genera were far more modest in size but are known from limited isolated vertebrae. Fortunately, there are several Eocene boids known from amazingly well-preserved complete skeletons such as the tiny *Boavus* from the Green River deposits in Wyoming (see p. 204) and *Eoconstrictor* from the Messel Pit in Germany (see pp. 211, 216).

Reptiles including snakes have reentered the marine environments multiple times throughout their evolutionary history. Among snakes, hydrophiine sea snakes and sea kraits **convergently evolved** for a marine lifestyle from different elapid lineages. Of all extant marine reptiles, only the true sea snakes have no need to come on to land, being capable of giving birth to live young at sea. Several other snake families will frequent saltwater environments including the mostly aquatic file snakes (Acrochordidae). These snakes can reach a length of over two metres long, rarely travel on land and are **piscivorous ambush predators** from the bottom of rivers and estuaries. Interestingly some fossil relatives were even more saline-tolerant and reached larger sizes. One early radiation of snakes, the extinct palaeophiids, are known from only a handful of species that ranged from the Late Cretaceous through to the Early Eocene. They all show some **adaptations** to an aquatic lifestyle, although this varies greatly between species. A dense vascular network is present, a feature typically seen in large snakes, indicating a high growth and metabolic rate (Houssaye *et al.* 2013). The more speciose *Palaeophis* have species which fall into two **grades**: '**primitive**' forms have only slightly modified vertebrae for aquatic life, while the 'advanced' snakes have more specialised compressed vertebrae. In general, species of *Palaeophis* were found in near shore, estuarine and probably freshwater environments, as indicated by the weakly compressed vertebrae and lack of **pachyostosis**. The smallest known species, *Pa. casei* (Fig. 2.11A), was 1 to 1.3 metres long. Several species were massive, with the largest known, *Pa. colossaeus* (Fig. 2.11B), estimated to have reached 8.1 to 12.3 metres long. The highly specialised, four metre long *Pterosphenus* (Fig. 2.11C) was even more laterally compressed and possibly had a flattened tail paddle, allowing it to range into more open water environments. Complete skeletons are not known for the large species, but the amazing fossils of small *Archaeophis* (Fig. 2.11D) show skin impressions.

Among other extinct snake families, one particular group of relatively obscure primitive snakes has an unusual record. With species found between the Late Cretaceous and the Late Pleistocene, the largely **Gondwanan** and Trans-Tethys (see p. 174) madtsoiids have the longest known temporal period for any family of snakes. Their position within Serpentes has been difficult to pin down with some placing them outside the clade. However, most consider madtsoiids as a very basal family within Serpentes. Much of the problem stems from incomplete skeletal material including only four skulls. Two very basal snakes from the middle Late Cretaceous of South America have regularly been linked with madtsoiids. *Dinilysia* (probably a burrower) and *Najash* (retained well-developed hindlimbs) show some features in common, but due to the fragmentary nature of fossil material these features may also be in common with other clades of snakes. Instead, all confirmed madtsoiids are known from various locations from the latest Cretaceous onwards. There is little argument that madtsoiids dispatched prey by constriction in a similar manner to boas and pythons. However, their choice of prey was limited by a lack of adaptation for swallowing large items. Madtsoiids include one of the largest known snakes, *Gigantophis* (see p. 221). Following the end of the Eocene, madtsoiids disappeared from all regions apart from Australia. Two of the best known species are the Oligo-Miocene *Yurlunggur* (see pp. 292, 294) and the Plio-Pleistocene *Wonambi* (see pp. 334–5).

Ever since the beginning of the Cenozoic, snakes have exhibited remarkable success and adaptability. How did snakes first evolve? How did a four-limbed lizard transform into a snake? Are there any transitional fossils showing this process in action? In fact, evolving limblessness is the simplest aspect of the evolutionary process. Losing limbs is relatively easy for a straightforward reason. A set of Tbx and Hox genes regulates the growth of limb buds and ultimately the limbs; hence, limbs can be eliminated by deactivating the genes that give the order to generate limbs. However, the likelihood of discovering a fossilised snake in the process of losing its limbs is very low. Given that they are formed of hundreds of tiny, fragile and typically **disarticulated** ribs and vertebrae, snake skeletons rarely fossilise. Only a small number of fossil snakes have been identified from partial or complete **articulated** skeletons. Since most fossil snakes are solely represented by a handful of vertebrae, the diagnostic features of those animals must be based on minute anatomical details of the vertebral column.

In spite of all these challenges, the fossil record has yielded a remarkable collection of material demonstrating the transition from four-limbed reptiles to limbless snakes. A broad range of fossil snakes retain small, functionless rear limbs while lacking forelimbs, providing clear evidence of gradual limb reduction. One notable example is *Najash rionegrina*, a terrestrial basal snake from the Late Cretaceous of Argentina, dated to around 90 Ma. *Najash* preserves a well-developed pelvic girdle, sacral vertebrae, and robust hindlimbs, illustrating that early snakes retained remnants of their ancestral limbs even after losing forelimbs. Numerous fossils of marine snakes from Cretaceous strata in the Eastern Mediterranean, specifically in Slovenia, Lebanon, and Israel, demonstrate that some lineages later adapted to fully aquatic environments. Fossil evidence such as *Najash* strongly supports the view that snakes originated on land, with early limb loss potentially linked to **fossorial** habits, though some researchers have argued for an alternative aquatic ancestry, and the precise ecological context of snake origins remains debated. In this light, traits such as fused transparent eyelids and the absence of external ears are often interpreted as fossorial adaptations, protecting the eyes and ears during burrowing, although these features also proved compatible with aquatic lifestyles in some lineages. Collectively, the fossil record reveals a rich mosaic of transitional morphologies, highlighting both terrestrial and secondary aquatic adaptations in snake evolution.

Fig. 2.11. A sampling of the diversity of palaeophiid snakes (drawn to scale). (A) *Palaeophis casei* (Early Eocene of the Southern United States); (B) *Palaeophis colossaeus* (Early Eocene of Mali); (C) *Pterosphenus schucherti* (Late Eocene of the Southeastern United States); (D) *Archaeophis proavus* (Early Eocene of Italy). Scale bar: 1 m.

Fig. 2.12. Phylogenetic tree of Crocodylomorpha after Lee and Yates (2018), Massone *et al.* (2019), Rio *et al.* (2021), and Hekkala *et al.* (2021).

Crocodylomorpha

The archosaur branch Crocodylomorpha (which includes crocodylians and all their extinct relatives), first evolved in the Late Triassic, and has a long and distinguished history. During the Early Cenozoic (as well as throughout the preceding Mesozoic), the crocodylomorphs were a far more diverse and adaptive group of reptiles than today, having evolved into a huge variety of sizes and forms. Today, crocodylomorphs are **semi-aquatic ambush predators** largely restricted to freshwater or estuarine environments. Crocodylomorphs reached their highest diversity during the Mesozoic, dominating freshwater environments, but were also well-represented on land and in marine environments.

However, following the Cretaceous-Paleogene extinction event (see p. 169), there were only three crocodylomorph groups: eusuchians, sebecians and dyrosaurids (Fig. 2.12). The latter two groups of crocodylomorphs, which were not closely related, survived largely in the **Gondwanan** remnant continents. Of the three, only the semi-aquatic eusuchians (crocodylians) are still alive today, with origins dating back to the Early Cretaceous, while the sebecians, which are deep-snouted **terrestrial** hunters, died out in the Miocene (see p. 30). The third group, dyrosaurids, would have probably looked very similar to extant crocodylians and are a distant sister group to the eusuchians. The eusuchians and dyrosaurids form a group known as the Neosuchia. The Eusuchia includes numerous extinct genera that are closely related to the extant Crocodylia, the order

that consists of crocodiles, alligators, caimans and gharials, along with their closest extinct relatives. Recent Crocodylia consists of three families: the Crocodylidae (true crocodiles), which has three extant genera and many extinct genera, the Alligatoridae, comprising two species of alligator and six species of caiman alive today, and the Gavialidae, which contains two extant species and several extinct relatives.

Eusuchia

Modern eusuchians fall into three broad categories: the crocodyloids, alligatoroids and gavialoids. The first two are undeniably closely related while there is some confusion as to exactly where the long-snouted gharials fit in. The crocodile-gharial lineage (Longirostres) is believed to have diverged from the superfamily Alligatoroidea in the Late Cretaceous, around 87 Ma. A secondary bony palate allows modern crocodylians to safely breathe in through their nostrils while the rest of the head (including the mouth) remains submerged. This structure reaches its greatest elaboration among eusuchians, in which the internal **nares** are completely surrounded by the pterygoid bones (flat, paired bones forming part of the palate in many vertebrates).

Crocodyloidea

The crocodylids (true crocodiles), which first arose in the Eocene around 46 Ma, is cladistically defined as a **crown group** formed of the last common ancestor of the Nile crocodile (*Crocodylus niloticus*), the Dwarf crocodile (*Osteolaemus tetraspis*), and all of its descendants.

True crocodiles form one of the three surviving branches, amongst various extinct side twigs, of the order Crocodylia. Within the Crocodylidae, there are a number of subgroups of crocodile that show some distinct differences to the far more familiar modern crocodile genus *Crocodylus*. Two lineages have their origins in Africa and appear to be each other's closest relatives, in spite of very different appearances. The mecistopins are longer snouted fish hunters and include the gigantic 10 metre long *Euthecodon* (see p. 284). Osteolaemins have a more generalised snout and their similarity to *Crocodylus* has caused some confusion about identity in the past. Several genera, such as *Voay* from Pleistocene Madagascar, are notable for potentially interacting with modern humans. Both mecistopins and osteolaemins have at least one extant genus each (slender-snouted crocodiles (*Mecistops*) and African dwarf crocodiles (*Osteolaemus*) respectively, although modern analysis suggests both are in fact multiple species). Additionally, the fossil record reveals a surprising amount of crocodile diversity in Africa, such as *C. anthropophagus* (Fig. 2.13).

It is another lineage of crocodylians distinct from the *Crocodylus* group which have a much closer tie to Australia. The diverse mekosuchines have been found in Australia since the Eocene, and eventually spread out to islands in the Pacific. They are considered to be a separate subfamily of crocodylids or possibly even a distinct family (depending on taxonomy) due to differences mostly in skull bones. Recent research places them as an independent group within or closely related to Longirostres. To the casual person, mekosuchines would have appeared similar enough to other crocodiles, although they do vary in appearance. Some such as *Harpacochampsa* had a long snout reminiscent of gharials, and has sometimes been considered not even a true mekosuchine (Fig. 2.12). Some genera such as *Kambara* (see p. 199), *Baru* (see p. 294) and *Paludirex* (see pp. 296–7) were all large genera that were typical **semi-aquatic ambush predators**, but they also had a straighter humerus compared to *Crocodylus*, conducive to improved **terrestrial** locomotion. The Plio-Pleistocene genus *Quinkana* (see pp. 334–5) was the most terrestrial of all. Mekosuchines underwent a drastic decline in Australia beyond the Miocene, with all genera, except for *Quinkana* and *Paludirex* becoming extinct in Australia by the end of the Pliocene. After the demise of *Paludirex* and *Quinkana* by during the Quaternary extinction event (see p. 312)), the final mekosuchines survived on Southwestern Pacific islands including Vanuatu and New Caledonia. Humans evidently interacted with them, with the last species of *Mekosuchus* (*M. inexpectatus*) only dying out in the last 3,000 years.

Alligatoroidea

Alligatoroids (as opposed to the crocodyloids), based on the known fossil record, probably had its origins in North America, and were around from the Paleocene onwards. Alligatoroid fossils are known from across Eurasia since land bridges between both the Bering Strait and the North Atlantic have linked Eurasia to North America during the Paleogene and Neogene Periods. Alligators and caimans, the extant species within the alligatoroid line, diverged in North America during the Paleogene (ca. 65–53 Ma), with caimans arriving in South America during that time, prior to the closing of the Panama Isthmus during the Late Neogene (see pp. 252, 295).

There are also a bunch of fossil species which do not fit cleanly into either the alligator or caiman subfamilies. Fossils of these early alligatoroids started to appear during the Late Cretaceous and persisted over the Cretaceous-Paleogene extinction event (see p. 169). *Leidysuchus* was a medium-sized genus and fairly typical crocodylian living in Cretaceous waterways in North America. It once consisted of a number of different species, but a reclassification found that most were not actually alligatoroids but basal crocodylians renamed *Borealosuchus* (Fig. 2.12), which survived up to the Early Eocene, about 47.8 Ma. Various fossil basal alligatoroids, alligators and basal crocodylians overlapped in both time, location and habitat between the Late Cretaceous and the Eocene, adding to the taxonomic confusion. The last of the basal alligatoroids were Miocene species of the small caiman-like *Diplocynodon* (see pp. 210–1), and these were amongst the last crocodylians in mainland Europe (there is some evidence of Nile crocodiles (*Crocodylus niloticus*) being present in Mediterranean islands during the Early Holocene). In modern times only one species of alligatoroid, the American alligator (*Alligator mississippiensis*), is found on the continent, while the vast majority of the rest of the group are found to the south, plus the Chinese alligator (*Alligator sinensis*). This largely South American group derives from a different lineage to the alligators, and are known as caimans. However,

Fig. 2.13. Life reconstruction of *Crocodylus anthropophagus*, known from the Pliocene to Pleistocene of Tanzania. It reached up to 7.5 metres in length. This large horned crocodile may have been one of the largest and most dangerous predators encountered by the early hominin *Homo habilis* (see pp. 158–9).

they were not the only crocodylomorphs in South America post-Mesozoic. Alligatorines consist of only two extant species (in the Southern United States and East China), but extinct relatives once ranged across Eurasia and North America.

The position of early alligatorids has undergone numerous revisions with species bouncing between being basal alligatoroids (outside the Alligatoridae), basal alligatorids or even early caimanines. They generally had blunt snouts with a **heterodont** dentition including large crushing teeth in the rear of the jaws. One suggestion is that some of the early species were a side branch to the true alligators/caimans and include both American (*Stangerochampsa*, *Brachychampsa* and *Ceratosuchus*) and Asian orientalosuchine (*Krabisuchus*) species. Fossils of *Brachychampsa* (Fig. 2.12) are known from either side of the Cretaceous-Paleogene extinction event, along the old coastlines of the Western Interior Seaway. Rather than pointed tips, their teeth were distinctively rounded with pommel-like caps to crush molluscs. Although physically like the other early alligatorids, *Ceratosuchus* had a pair of flattened triangular plates at the back of the skull, a feature also seen in the unrelated extinct Madagascan crocodyline *Voay* (Fig. 2.12) and the extant Cuban *Crocodylus* crocodiles. Very distinct from the extant Chinese alligator (a much later immigrant from North America, which diverged from the American alligator around 33 Ma), the early Asian *Krabisuchus* showed some features in common with more **terrestrial** crocodylians.

True alligatorines show up in the fossil record from the Eocene and were generally small, snub-nosed species from Europe and North America (*Hassiacosuchus* and *Allognathosuschus*). The more familiar flat-snouted alligatorines of the genus *Alligator* descended from a line of blunt-snouted species, and was the only surviving genus from the Oligocene onwards. It was during the Paleocene that caimans arrived from the north, although several species such as the Eocene *Tsoabichi* later recolonised North and Central America. It has been suggested that early caimans had an advantage being descended from more cold-tolerant stock, enabling them to prosper after the Late Eocene global cooldown (see p. 197). It was only during the Miocene when caimans really diversified into a broad range of niches, including the terrifying monster caiman *Purussaurus* (see pp. 268–9). Among basal members of the caimanines such as *Globidentosuchus* (see p. 268), it is possible that a specialised shellfish diet and dentition was **ancestral** to the group. In contrast, all modern species have a dentition and hunting style for a generalised aquatic **ambush predator**.

Gavialoidea

Crocodylian snouts are a clear indicator of their dietary choice. Species with a pointed snout, such as most *Crocodylus*, have a generalised diet—a balance of length and breadth for strength. Short, broad snouts give a more powerful bite and, as seen with some caimans, useful for breaking into hard-shelled molluscs or turtles. However, for fast-moving prey, one needs narrow, elongated jaws (**longirostrine**) and interlocking teeth. This feature has evolved multiple times among crocodylians (five species are alive today), although none is as highly adapted to fish-hunting as the gavialids. Today, only one species represents this family, the Indian gharial (*Gavialis gangeticus*). Highly endangered and very limited in distribution, this contrasts with the near global fossil record of gavialids. Much confusion seems to surround gavialids and other longirostrine crocodylians. Tomistomines and the Late Cretaceous-Early Paleocene *Thoracosaurus* have been closely linked, and at least with the latter species the features have proven to be **convergent**. The earliest true gavialids appear during the Eocene in Africa, and was possibly ancestral to the distinct South American lineage, the gryposuchines. Early examples from this group, such as the Caribbean and South American *Aktiogavialis* (see pp. 268, 271), were small, but it was during the Miocene that they diversified and grew to some impressive sizes. Most species were estuarine or fully marine, such as the large narrow-snouted *Piscogavialis* which, amazingly, lived alongside penguins, aquatic ground sloths (see pp. 61–2), giant sharks and *Livyatan* whales (see pp. 124–5). The more robust *Gryposuchus* (see p. 268) was one of the largest known crocodylians. Another South American coastal species, *Ikanogavialis* (see pp. 268–9) may have spawned a species that managed to cross the Pacific Ocean. Fossils of a small gavialid have been found on the distant Murua or Woodlark Island, off the coast of Papua New Guinea. Surprisingly, this species only recently became extinct, and although the exact dating is unknown, it is possible that *Ikanogavialis* lived well into the Holocene.

Following on from the predominantly **piscivorous** gavialids is the other family of gavialoids: the false gharials, or tomistomines. Sometimes, a popular name can tell a lot about a species. The modern extant false gharial (*Tomistoma schlegelii*) is just that—they are not the "real" gharial. However, what exactly *Tomistoma* and their fossil relatives are can be a bit harder to answer. Morphologically, they are only convergent with the "true" gharial and are actually closer in physical features to other crocodylids (true crocodiles). This leaves gavialids as the sister **taxon** to all other crocodylians, which split during the Late Cretaceous, approximately 70.6 Ma. Conversely, the molecular evidence points to a sister taxa relationship between tomistomines and gavialids which diverged during the Eocene. While this closely links the two piscivorous crocodylian lineages, it is complicated by the aforementioned morphological similarities with crocodylids. Unfortunately, this is not easily resolved by comparison of ancestral crocodylians, and currently the taxonomic situation sits in limbo.

Interestingly, modern gharials are found exclusively in freshwater environments, while both fossil tomistomines and gavialoids are regularly found in marine environments. The earliest tomistomine fossils, such as *Kentisuchus* and *Megadontosuchus*, have been found in coastal environments around the Tethys Ocean (see p. 174) (Europe and North Africa) during the Eocene. *Megadontosuchus* departs from the slender jaws of other tomistomines with its more generalised, robust **rostrum** and large teeth. Tomistomines crossed into the isolated Indian subcontinent, East Asia and North America during the Eocene and Oligocene, showing a remarkable tolerance to saltwater. Ranging along the shallow coastline of Eastern North America was the large *Thecachampsa* which lasted through the Miocene. Another group of tomistomines reinvaded Asia during the Miocene, which includes the giant *Rhamphosuchus* and ancestors of the current extant species. Once thought to be one of the largest known crocodylians, the size of *Rhamphosuchus* has undergone re-evaluation and dropped from an estimated maximum of 18 metres to a still respectable 11 metres. Its dentition suggests that it had a more generalised diet than other tomistomines. Tomistomines as a group are fairly uniform, differing mostly in size. Although they have very slim snouts suggestive of a mainly piscivorous diet, extant false gharials tend to be more generalist with their diet. Several verified attacks on humans and even a cow should suggest there be no reason to assume that fossil species were not capable of tackling large prey.

Sebecia

While the eusuchians evolved in **Laurasia**, another sister **clade**, the Ziphosuchia, are mostly known from **Gondwana**. Ziphosuchians (which include notosuchians and sebecians) evolved into a wide range of largely **terrestrial** forms and included several groups of specialised **cursorial** predators. Unique **adaptations** such as large blade-like teeth along with more elongated limbs, made ziphosuchians formidable predators. One family of sebecians, the sebecids, continued through to the Middle Miocene after having survived the Cretaceous-Paleogene extinction event (see p. 169). All known sebecids were native to South America and Southern Europe. *Iberosuchus* and *Bergisuchus* (see p. 211) are both known from Eocene deposits throughout Western Europe, and are often considered the closest relatives to the largely South American clade. These two outlying species did not descend from a lineage of Cretaceous European sebecosuchians, but appear to be part of a separate invasion from South America possibly via North Africa. Interestingly, a third European species has recently been found. Strangely, *Ogresuchus* was a definite member of the Sebecidae. This one metre long species is also the smallest and earliest known sebecid, dating from the Late Cretaceous.

Early Paleocene sebecids such as *Lorosuchus* differed from their Cretaceous relatives by less specialised dentition, indicating that they adopted a mixture of **semi-aquatic** and terrestrial lifestyles alongside alligatoroids in the same environments. After the Eocene, sebecids developed the taller, more compressed skull previously seen in other Late Cretaceous sebecosuchians. Blade-like teeth were edged with denticles capable of slicing flesh, a feature also seen in many theropod dinosaurs. This would suggest a similarity in the manner of consuming prey. **Articulated** postcranial material of the Eocene *Sebecus*, possibly the most well-known species, has provided further evidence of terrestrial behaviour, with proportionally more elongated limbs than extant crocodylians. While modern crocodylians have **orbits** raised above the top of the skull, sebecid orbits were on the side of the skull. The largest known genus and possibly the largest terrestrial predator of the whole Cenozoic was *Barinasuchus* (Fig. 2.14). To put its size in context, a portion of its skull resembles large, Late Cretaceous tyrannosaurid dinosaurs. Ironically, it was changes to the large lake system across South America during the Middle Miocene which saw the end of the terrestrial crocodylomorphs. Uplift of the Andes caused major ecological changes to the continent, which had a massive effect on the susceptible megafauna, in particular, the unique South American native **ungulates** (see p. 139), which were the main prey of sebecids.

Dyrosauridae

The dyrosaurids first evolved during the Late Cretaceous (ca. 83.5 Ma). Freshwater and marine forms of dyrosaurids would have looked similar to modern crocodylians, while the sebecids were deep-snouted land predators. Dyrosaurids had the crocodylomorph body plan, but they differed in important details. The nostrils were not on the tip of the snout, but the back end of the snout, enabling their body to be mostly submerged while they hunted or swam. Proportionally, dyrosaurids had more musculature around their necks and forelimbs than crocodylians, and may have primarily used the hindlimbs for steering. Additionally, their tails were relatively longer and deeper. Some genera showed subtle **adaptations**. Dyrosaurids were **semi-aquatic** hunters (studies reveal they had **osteoporotic** bone structure which is consistent with fast swimming) and that the majority of genera were from coastal marine and estuarine environments. However, some genera (*Anthracosuchus* and *Cerrejonisuchus* (see pp. 186–7)), plus unnamed remains from Sudan and Pakistan) have been found in inland riverine locations, while the juveniles of marine species are also known from freshwater environments. This latter observation suggests that young dyrosaurids migrated downriver towards the coastal environments as they matured. Generally, most dyrosaurids are recognisable by a proportionately elongated snout, reaching about 70% of the total length of the skull. However, the basal forms had shorter snouts (e.g. *Chenanisuchus* (see pp. 188–9), *Anthracosuchus*, *Cerrejonisuchus*), while the longer-jawed species (e.g. *Arambourgisuchus* (see pp. 188–9), *Dyrosaurus* (Fig. 2.12)) probably evolved multiple times. Most of the earliest fossils are known from Late Cretaceous North and West Africa, with a few from Europe. They had also spread to North America before the end-Cretaceous. However, it was during the Paleocene and Eocene when dyrosaurids were most diverse, with species distributed across North/West Africa, Middle East, South Asia and South America. Dyrosaurids did not survive beyond the end of the Eocene, possibly due to a global cooling.

Fig. 2.14. Reconstructed life appearance of the sebecian crocodylomorph *Barinasuchus arveloi*, known from the Middle Eocene to Middle Miocene of Argentina, Peru and Venezuela. Reconstructed length: 6 m.

Aves (Birds)

The birds are diapsid tetrapods distinguished by feathers, beaked jaws lacking teeth, have a robust albeit lightweight skeleton, a four-chambered heart, production of hard-shelled eggs and are **endothermic**. The birds are, in fact, feathered theropod dinosaurs which comprise the sole surviving dinosaurs, and their nearest extant relatives are crocodylomorphs (see pp. 26–30). Birds diverged from the **primitive** avialans (Fig. 2.15), the latter which originally evolved during the Middle Jurassic (ca. 166 Ma) and consist of **taxa** including *Archaeopteryx*. Widely recognised as among the first transitional fossils to be discovered, the Late Jurassic *Archaeopteryx* provided strong support for the theory of evolution from 1861 onwards. *Archaeopteryx* is the first discovered fossil to exhibit both distinctly classic reptilian features (clawed **digits**, toothed jaws and an elongated, bony tail) along with wings possessing flight feathers comparable to those of living birds. *Archaeopteryx* was not directly ancestral to birds, but may be closely related to the actual common ancestor (Mayr *et al.* 2007).

Avialans subsequently underwent a tremendous diversification throughout the Cretaceous. Numerous lineages still possessed rudimentary features such as teeth and clawed wings. However, certain avian lineages such as modern birds lost the former independently. Among more derived avialans, the innovation of the **pygostyle** bone in the lineage Pygostylia led to shortened tails. During the Cretaceous, the avialans gradually developed many of the characteristics which are familiarly avian: complex **pennaceous feathers**, increased brain size, elongation of forelimbs, a move away from **hypercarnivory**, fused tail vertebrae, and an open pelvis. It was only with the later ornithothorace avialans that flight capabilities were refined by a loss of skeletal features (including the grasping hand), an enlarged sternum and a semi-rigid rib cage. Ornithothoraces are divided between the euornithes (which includes all extant avians plus close fossil relatives) and the so-called "opposite birds", the enantiornithes, which mysteriously did not survive into the Cenozoic. They are called “opposite birds” because of an odd shoulder joint (a reversal of the **articulation** between the scapula and coracoid bones, totally opposite from the condition in all Aves).

The ornithuromorphs were a major bird lineage during the Mesozoic, consisting of the closest Cretaceous relatives of the modern bird diversification and their modern descendants. A more derived subgroup of ornithuromorphs called the ornithurines arose during the Late Cretaceous, comprising the tern-like ichthyornithes and the loon-like hesperornithes. It is hard to determine how suddenly these primitive birds became extinct at the end of the Cretaceous since the fossil record of Mesozoic birds is so sparse. Nonetheless, only the Aves persisted into the Cenozoic.

There are numerous morphological characters unique to Aves and not present in their Mesozoic forebears, such as the total absence of teeth and the total fusion of the metatarsal structures with the **distal** tarsus to form the tarsometatarsus. From a handful of ornithurine-related survivors, a renewed proliferation of modern bird families took place throughout the Paleocene, leading to tremendous bird diversity by the Eocene, largely consisting of modern lineages. As such, the diversification event of Paleocene birds is similar to the massive **adaptive radiation** of mammals during the Paleocene, following the demise of the non-avian dinosaurs which opened up new niches for large **terrestrial** tetrapods. Put another way, the avian

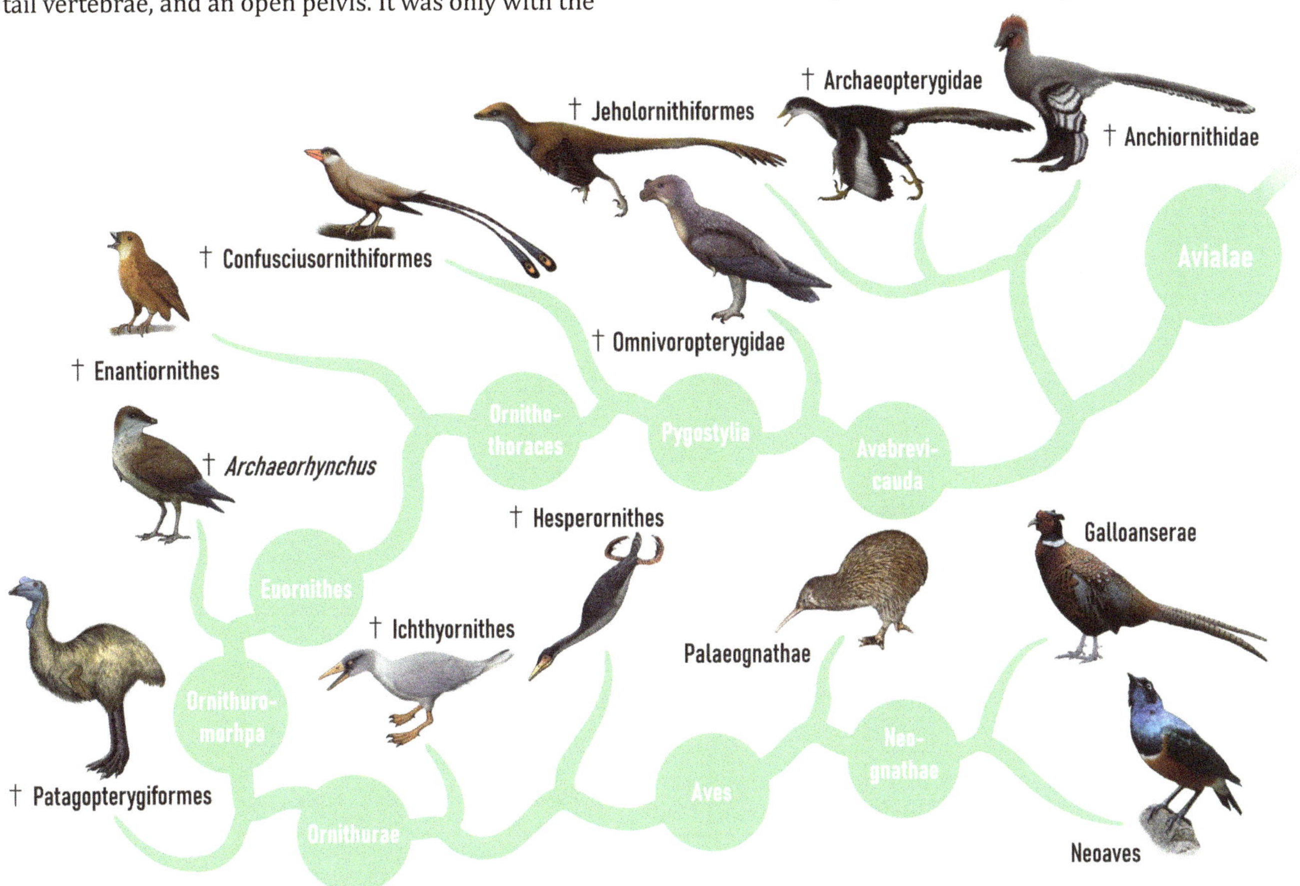

Fig. 2.15. Phylogenetic tree of Avialae after Wang *et al.* (2016), McLachlan *et al.* (2017), and Pei *et al.* (2020).

dinosaurs experienced a massive adaptive radiation following the extinction of the non-avian dinosaurs.

The majority of fossil birds are very difficult to examine, since their somewhat brittle, hollow, thin-walled skeletal elements easily disintegrate, so complete, **articulated** fossil bird skeletons are exclusively preserved in **Lagerstätten**, such as the Messel Formation of Germany (see pp. 210–7). Only a small amount of fragmented bones are most commonly preserved and thus, reconstructing the remainder of the skeleton is necessary. In other deposits, the bones are dissociated and scattered all over, such as within the Rancho La Brea tar pits (see pp. 322–7). Since the majority of fossil birds are solely represented by a handful of bones (particularly the tarsometatarsus, or lower hindlimb bone, the most robust element of a bird's skeleton), scientists must base their comparisons on the available material they have.

Palaeognathae

The earliest evolutionary divergence amongst the extant bird lineages takes place between the Palaeognathae, comprising the modern flightless birds called ratites, and the Neognathae, which includes the rest of the modern birds. The ratites consist of the rheas from South America, the ostriches from Africa, the kiwis from New Zealand, and the cassowaries and emus from Australasia. As well as the modern palaeognaths, some extinct behemoths included the moas (or the dinornithiforms) from New Zealand, reaching up to 3.6 metres (12 feet) tall, and were hunted to extinction by the Māori Polynesians by the fifteenth century CE, only around 100 years after the latter's arrival on those islands. However, the largest birds ever were the aepyornithiforms, which filled a similar ecological role to the moa and encompass the well-known "elephant birds" such as *Aepyornis*, the giant form *Vorombe titan*—although recent genetic evidence (Grealy *et al.* 2023) suggests it may represent a giant female form of *A. maximus* rather than a separate genus—and the smaller, more gracile *Mullerornis*, all of which lived in Madagascar during the Pleistocene and Holocene. At an estimated 700 kilograms, *Aepyornis* is amongst the heaviest known birds while the eggs are the largest known for any amniote (up to 34 centimetres long). Elephant birds were also hunted to extinction by humans after the latter's initial colonisation over 2 ka.

According to a 2021 molecular analysis of every bird order, the palaeognaths diverged from the neognaths during the Late Cretaceous or earlier, before 70 Ma (Kuhl *et al.* 2021). However, the emergence of all extant palaeognath orders can only be traced back to the Late Paleocene and thereafter, with ostriches splitting during the Late Paleocene (ca. 57.7 Ma), rheas during the Early Eocene (ca. 50.7 Ma), kiwis (and likely elephant birds) immediately afterwards (ca. 49.9 Ma), and lastly the casuariiforms (emus and cassowaries) and tinamous (and likely moas) splitting from each other during the Middle Eocene (ca. 45.1 Ma).

Ratites belong to the order Palaeognathae as they possess several distinctive characters in addition to flightlessness (tinamous are still somewhat capable of flying). The palaeognathous palate is more **primitive** and differs markedly relative to the palate observed in many extant birds (neognathous), although it possesses several specialised characters indicating that palaeognaths are a **monophyletic** group, rather than merely a **wastebasket taxon** of large, primitive flightless birds. Among other unique traits that distinguish ratites are the arrangement of the pelvis. The ischium and pubis are longer than the ilium, so those elements protrude under the tailbones, and unlike neognaths, a large **fenestra** is absent between the ischium and ilium.

Traditionally, it was believed that the flightless ratites first evolved on the southern supercontinent of **Gondwana**, based on their striking distribution pattern across the Gondwanan remnant continents today. When the continents split up and drifted apart, they carried the ancestors of the current species (e.g. ostrich, emu, cassowary, kiwi, rhea) with them. However, the idea was too simplified, with the relationships between the various families being more complicated. Extinct groups from the northern continents, once thought to be linked to other birds, were closely allied to the ratites. Plus, the **clade** includes flighted forms, with the poorly-flighted tinamous most closely related to the wingless moa. This was surprising when those groups lived alongside other ratites in South America (rhea) and New Zealand (kiwi), respectively. The ancestors of ostriches once lived throughout Eurasia before expanding into Africa. Ratites appear to have had a worldwide distribution throughout the Early Cenozoic, so that only a few modern relics are restricted to the Gondwanan remnant continents.

The earliest known palaeognaths were the lithornithids from the Paleocene and Eocene, and found only from the Northern Hemisphere. Although quite diverse in forms, lithornithids generally resembled tinamous in shape, had slender bills, exhibited rhynchokinesis (ability to flex the beak) which suggests that they foraged by probing leaf litter, and curved toe bones for perching. Endocasts from specimens of the speciose *Lithornis* have revealed enhanced **olfactory** lobes consistent with **nocturnal** forest-dwellers. Several species including *Lithornis* (Fig. 2.16) had powerful flight muscles for soaring flight. It is possible that the palaeognaths originated in **Laurasia** as evident from the occurrence of the anatomically most primitive fossil **taxa** (including the lithornithids) in the Northern Hemisphere. The successive colonisations of the southern continents by flying species that eventually became flightless, and often underwent gigantism (ostriches, rheas and emus as well as elephant birds and moas), would have ultimately resulted in the current distribution, near exclusively Gondwanan (Yonezawa *et al.* 2019).

The struthioniforms, which today only consists of the struthionids (ostriches), originated during the Early Eocene, ranged all over Laurasia and diversified into several clades as they lost the ability to fly early on, such as the palaeotids of Europe and the eogruids of Asia, the latter which is sister to the struthionids. The struthionids possibly first evolved in Asia (Mayr and Zelenkov, 2021), before immigrating to Africa by the Early Miocene (ca. 21 Ma), then Eurasia by the Late Miocene (ca. 12 Ma) (Widrig and Field, 2022). From the Late Pliocene to Middle Pleistocene of Eurasia, *Pachystruthio* was among the largest birds ever, with some species (*P. dmanisensis*) reaching weights of 450 kilograms (990 pounds).

Sometimes, comparable features between unrelated groups can lead to confusing results. The eogruids were a small family of **cursorial** birds which had features in common with two distantly related modern bird groups—cranes and ostriches. Generally, it was thought that they were closer to cranes within Gruiformes (hence the name), while the slender limbs and a reduction of toes (only two were present in

Fig. 2.16. Phylogenetic tree of Palaeognathae after Sangster *et al.* (2022). The phylogenetic position of Palaeotidae follows Mayr and Zelenkov (2021).

later species) were an **adaptation** for running and considered **convergent** with ratites. Several researchers (dating back to 1951) had considered that eogruids were actually closely related to ostriches, but the consensus stood. The recent discovery of a partial skull revealed that eogruids had the "primitive" palaeognathous palate and that some limb material was very closely **analogous** to ostriches. So eogruids actually had convergent features with cranes! Eogruids are known from Eurasian grasslands between the Middle Eocene and Early Pliocene. Early genera were small enough to potentially be capable of flight, while later genera such as *Ergilornis* (Fig. 2.16) were much larger, more cursorial and were functionally flightless.

Rheas (the rheiforms) have a poor fossil record, but putative members occur in the Paleocene and Eocene of Antarctica and South America. They do not seem to have expanded beyond these landmasses. The small, **terrestrial** tinamiforms (tinamous) originated in the fossil record during the Miocene Epoch (ca. 10 Ma), and their position within the ratite radiation, according to genetic studies (Phillips *et al.* 2010), suggests basal ratites were capable of flight. The persistence of flight among the tinamiforms could be attributed to rheas occupying South America prior to the arrival of tinamou ancestors. Tinamous were forced to continue living as small, **omnivorous**, and flight-capable animals since the ecological niche for large, flightless **herbivores** had already been filled (Mitchell *et al.* 2014).

Although their exact affinities are unknown, flight-capable lithornithids from the Early Paleogene appear to have been the most morphologically similar precursors to the tinamiforms and were possibly ancestral to both them and all other palaeognaths (Phillips *et al.* 2010). The oldest definite tinamiform fossils are known from the Miocene (Bertelli and Chiappe, 2005), yet flightless birds from the Paleocene resembling ratites could also be included in this lineage (Agnolin *et al.* 2017). The casuariiforms have a sparse fossil record. Casuariiforms are best known from Australia but the oldest known member, *Diogenornis* (see p. 228), comes from South America. It remains unclear whether *Diogenornis* and emus/cassowaries lost the ability to fly independently or if they crossed Antarctica into Australia (or vice versa).

Recent DNA studies suggest that kiwis are the nearest modern relatives of elephant birds. Given how specialised elephant birds were, they could have been on Madagascar since the Paleocene, so the still volant kiwi ancestors probably crossed the Indian Ocean with India as a stepping island, then Australia before arriving in New Zealand in the Miocene. Or they could have both come from either Asia and migrated southwards, or Antarctica and inversely migrated northwards. The absence of Australian fossils is of particular interest. The earliest kiwis in the fossil record are the Miocene *Proapteryx* from New Zealand, which was smaller than the modern species and could possibly fly, lending support to the theory that kiwi progenitors and moas arrived at New Zealand independently, the latter being already larger and flightless when the former colonised the islands (Worthy *et al.* 2013).

Neognathae

All the other modern birds exhibit the advanced neognathous palate. Neognaths consist of a vast radiation of over 10,900 extant species, almost all of which arose and radiated within the Cenozoic. Although the phylogenetic affinities and evolution of these modern bird **clades** remain highly disputed, it appears that a consensus has been reached concerning the avian molecular phylogenetic tree. Over the years, various phylogenetic trees of the modern birds have been produced based on morphology, although with no consensus. Then in 1990, DNA hybridisation was a pioneering and early molecular approach used in studying bird phylogeny, which yielded a totally different framework of bird relationships. However, these results have been questioned and are now out of date, having been superseded by direct genome sequencing. There is still some disagreement among various DNA phylogenies today, given that numerous bird orders seem to have split so quickly during the Early Cenozoic that there are only minor molecular differences between them.

Despite some conflicting findings, the DNA has consistently proven that there is a distinct divergence between a **primitive** lineage, the Galloanserae, comprising the Galliformes (the assorted landfowl, consisting of turkeys, chickens, grouse, pheasants, guineafowl and relatives) and the Anseriformes (the variety of waterfowl, encompassing screamers, ducks, swans, geese and relatives) versus the remainder of the birds (Fig. 2.18A). The earliest known galloanseran fossils are a collection of fragmented material unable to be placed into a particular lineage within Galliformes or Anseriformes. The most complete and well-examined of these fossils are of the bird *Vegavis*, from the latest Cretaceous of Vega Island, Antarctica. Although *Vegavis* seems to have resembled anseriforms in having elongated hindlimbs with webbed feet, it is actually a basal relative of both galloanseran lineages. Additional similar-looking bird genera from the Late Cretaceous, all vegavids, comprise *Polarornis* of Seymour Island in Antarctica, and *Neogaeornis* of Chile. Across the Southern

Fig. 2.17. Reconstructed life appearance of the presbyornithid *Presbyornis pervetus*. Reconstructed height: 1 m.

Fig. 2.18A. Phylogenetic tree of Neognathae, particularly all taxa outside Elementaves, after Stiller *et al.* (2024).

Hemisphere, these birds were certainly diverse prior to the Cretaceous-Paleogene extinction event (see p. 169), after which they rebounded and radiated into the Galloanserae throughout the Cenozoic.

Fossils that clearly belong within the orders Galliformes and Anseriformes are known from the Middle Eocene onwards. The duck lineage consists of numerous fossils that can be assigned to modern groups (including screamers, swans, ducks and geese). Some possible early fossil relatives include the presbyornithids. A very common fossil, *Presbyornis* is very recognizable as the "flamingo-duck" from the Early Paleocene through to the Eocene (Fig. 2.17). A unique combination of a duck-like bill and the long slender features of flamingoes confused scientists for some time into thinking there was an evolutionary connection between the two. Genetic studies later showed that flamingoes were very distantly related and that waterfowl were an old lineage linked to landfowl. This left the presbyornithids as either a specialised subset of the anseriforms or a basal lineage. The discovery of *Conflicto* from the Early Paleocene of Antarctica helped to clear up some of these queries. Much like *Presbyornis*, it resembled a stretched fowl with long legs and a narrow duck-like bill. *Conflicto* also had small bony bumps on the skull (seen in other waterfowl) and some skeletal features in common with galliforms. In common with the similarly aged *Anatalavis*, they represent the most basal members of the anseriforms, while presbyornithids were also basal forms. Thousands of *Presbyornis* bones have been found, particularly at the Green River deposits in the United States. This suggests that they were very social birds, flocking together like modern waterfowl. Fossils are known only from freshwater deposits where they fed by filtering food.

Fig. 2.19. Reconstructed life appearance of the pelagornithid seabird *Pelagornis sandersi*. Reconstructed wingspan: 6 m.

Among the earliest **stem** relatives of anseriforms were gigantic birds called gastornithids, which lived during the Early to Middle Eocene. Only three genera have been named: the European *Gastornis* (see pp. 211–2), the Chinese *Zhongyuanus*, and *Diatryma* from North America. Differences between the three are minimal and now all are subsumed under the first named genus *Gastornis*. These birds reached heights exceeding 2.5 metres (6.6 feet) and possessed a massive skull with a deep, sharp beak more than 0.5 metres (1.5 feet) long. Gastornithids possibly weighed up to 175 kilograms (385 pounds). Like many flightless birds, the gastornithids possessed vestigial wings, and also had thickset legs with sturdy claws. The gastornithids were previously allied to the order Gruiformes, but recent research has shown gastornithids to be more closely related to the Australian dromornithids. Why the gastornithids went extinct is currently unclear, but competition with large mammals and changes in food sources may have played a significant role.

Dromornithidae, also known as mihirungs (an aboriginal name for a giant emu), were a radiation of large, flightless birds that have been nicknamed the "thunder birds". Many dromornithids, such as *Dromornis stirtoni* (see p. 292), reached massive sizes for a bird. Like ratites, dromornithids were mostly **herbivorous** with a straight-edged beak. Additional evidence for herbivory include hooved **digits** instead of sharp claws, and a large stomach for food fermentation. Dromornithids originated in Australia during the Oligocene (with virtually no Cenozoic record before this), and survived up to the Late Pleistocene, dying out along with all the other Australian megafauna (see p. 335). One other fascinating lineage are the Pelagornithidae, or "pseudo-toothed birds", with **keratinous** tooth-like points along the beak margin, which are not true teeth but outgrowths from the jaw bones, which like true teeth, would function to grasp slippery, soft-bodied prey with their beaks. Although their appearance is similar to other soaring seabirds such as albatrosses and pelicans, pelagornithids are surprisingly grouped within the ancient Galloanserae lineage. The pelagornithids dominated the world's oceans in virtually the whole Cenozoic, living from the Late Paleocene to the Early Pleistocene, only to be superseded by marine soarers such as albatrosses. The pelagornithids were among the largest flying birds, with *Pelagornis sandersi*, from the Oligocene beds in South Carolina, being the current record holder, with a six to seven metre wingspan when fully feathered (Fig. 2.19). That is the largest wingspan of any known bird. *P. sandersi* had the form of an enormous albatross, with a wingspan almost twice the width of a wandering albatross (*Diomedea exulans*).

Neoaves

All the other modern non-galloanseran neognathous birds belong to the lineage Neoaves. As indicated by the consensus phylogeny of Stiller *et al.* (2024), based on an outstandingly broad taxonomic coverage and elaborate genomic sequence data of all extant bird species, the subsequent branch point within the Neoaves was an assemblage of four main strongly-supported supraordinal **clades** (Fig. 2.18): a grouping of grebes and flamingos called the Mirandornithes, among the earliest lineages to evolve (this first clade was long disputed, although it is backed up by molecular evidence); a clade comprising the pigeons

Fig. 2.18B. Phylogenetic tree of Neognathae, particularly taxa within Elementaves, after Stiller *et al.* (2024).

and doves (Columbaves); a newly proposed clade called Elementaves, encompassing diverse ecological niches, including shorebirds, hummingbirds and tropicbirds; and a clade of land birds called Telluraves. The study proposes that birds began radiating long before the extinction of the non-avian dinosaurs, with a common ancestor for Neoaves estimated to have lived around 130 Ma. The early Neoaves experienced rapid diversification into at least ten major clades, including the "magnificent seven" and "three orphans", encompassing 95% of extant species. The "magnificent seven" include the landbirds, waterbirds, tropicbirds, cuckoos, nightjars, doves and flamingos, while the "three orphans" are shorebirds, cranes and hoatzins. The term "orphans" is used because the evolutionary relationships of those clades to the other Neoaves remain poorly understood, and their placement within the broader avian phylogeny remains uncertain. Nevertheless, this new family tree resolves some long-standing debates over the relationships among bird species and lays a crucial foundation for studying bird evolution and trait development.

While the study by Stiller *et al.* (2024) has built upon and refined previous avian phylogenies (such as that of Hackett *et al.* (2008) and Prum *et al.* (2015)) by incorporating new data and analytical techniques, it has faced criticism regarding certain aspects of its topology, particularly concerning the placement of major clades, with alternative topologies emerging from other molecular analyses (e.g. Wu *et al.* (2024)). It would be impossible in a book of this scope to review this thoroughly, especially since there is ongoing debate regarding the positioning of the bird clades. Neither is there room to cover the evolutionary history of all 37 orders and almost 11,000 species of Neoaves. Rather, let us focus on a selection of spectacular and peculiar neoavian **taxa** in the fossil record.

One remarkable group of birds are the penguins (Sphenisciformes), the only birds that not only are flightless but also have turned their wings into flippers for swimming. Penguins are a very ancient group, with more than two dozen fossil species currently known. The first known species lived only a mere five million years following the Cretaceous-Paleogene extinction event (see p. 169), and basal forms, such as *Waimanu* (Fig. 2.20A), were already specialised for aquatic life. The wings showed an **adaptation** for propelled swimming but were not quite the short, flat flippers of modern penguins. In body shape, they probably resembled a cormorant with a spear-like beak, common to early penguins, indicating a more **piscivorous** diet.

From the Late Paleocene onwards, penguins were a dominant predator in the southern oceans, and had spread to most of the southern continents by the Eocene, all much closer together during that time. Some modern penguins get really large, such as the king penguin (*Aptenodytes patagonicus*), and the emperor penguin (*Aptenodytes forsteri*), which reaches 1.2 metres (4 feet) in height, and weighs up to 45 kilograms (100 pounds). Both of these are dwarfed by the gigantic penguins *Kumimanu* from the Paleocene and several genera from the Late Eocene, such as *Palaeeudyptes* and *Anthropornis* (see pp. 201–2), which all vie for the largest genera known. It is difficult to compare each species as isolated bones are known and proportions differ. These were not a **monophyletic** group separate from other penguins, rather a succession of different species. While penguins are usually linked to cold climates, this was not the case during the reign of the giants. Ice caps would have been absent or at least limited during the Eocene/Oligocene and some genera such as *Icadyptes* (Fig. 2.20B) flourished in balmy climates on the ancient Peruvian coastline.

Long spear-like beaks are an **ancestral** feature of early penguins and it is assumed that they employed a different hunting strategy to that used by modern species with broader beaks (exaggerated by the outer rhamphotheca layer). The giant penguin beaks can be incredibly long, up to two-thirds of the skull's length. Neck musculature was strongly developed to resist impacts and it is thought that they used the bill like a

Fig. 2.20. A selection of fossil penguins. (A) *Waimanu manneringi*, from the Paleocene of New Zealand; (B) *Icadyptes salasi*; (C) *Inkayacu paracensis*. The latter two species are known from the Late Eocene of Peru. Scale bar: 1 m.

spear on larger prey. A larger body size and robust flippers would have also allowed the giant penguins to dive deeper and remain underwater for longer periods for hunting. Incredibly, fossilised feathers belonging to *Inkayacu* (Fig. 2.20C) revealed melanosomes of reds and greys, a contrast from the predominately white and black colouration that we typically see in penguins. Giant penguins declined during the Oligocene, coinciding with the arrival of seals and early toothed whales. However, later penguins were able to survive in the niche of much smaller hunters. For various reasons, penguins have never managed to make much progress north of the Equator. However, as a classic example of **convergent evolution**, several other groups have taken on the "penguin" role in the north. During the Oligocene (but also extending back to the Late Eocene and through to the Early Miocene) the plotopterids closely resembled penguins in both form and behaviour. All species were flightless with dense bones and were sea-going, but like all sea-going birds plotopterids would have needed to occasionally make landfall to mate, lay eggs and moult feathers.

They were related to cormorants, gannets and darters within the order Suliformes, but unlike these other groups which relied on foot propulsion when swimming, plotopterids used their shortened wings to propel when diving. Plotopterids were some of the largest seabirds known, with some genera such as *Copepteryx* (Fig. 2.21) reaching a length of two metres. Fossils have been restricted to sites in the North Pacific, specifically in Japan, Canada and the United States. In a final comparison with penguins (specifically the giant forms), their downfall seemed to coincide with the rise of seals and dolphins during the Miocene. Despite the similarities, the alcids (auks) are not closely related to penguins. In fact, being highly specialised for "flying" underwater has made the evolutionary relationships of penguins with other bird groups unclear. In addition to auks, the grebes, pelecaniforms and petrels have all been suggested as possible relatives. Recent genetic analysis has confirmed a link with the latter group. Auks have a long fossil history, with early specimens dating to the Eocene. However, it was not until the Miocene when auks started to be prominent in the fossil record. Auks underwent an even greater expansion of diversity in the Pliocene. For example, there were up to six different species of *Alca* in the Early Pliocene Atlantic (only one species, the razorbill (*A. torda*) survives in modern times). Alcids declined during the Middle Pliocene, facing increasing competition from a diversity of marine mammals and a gradual warming of the seas.

Fig. 2.21. Reconstructed life appearance of the plotopterid *Copepteryx hexeris*, from the Late Oligocene of Japan. Reconstructed height: 1.3 m.

Among the most instantly recognisable birds are the parrots (Psittaciformes), split into three uneven superfamilies—the New Zealand parrots (strigopoids), cockatoos (cacatuaoids) and the true parrots (psittacoids). Among the strigopoids, they are restricted to New Zealand and consist of only a handful of extant and extinct species. Destructive keas (*Nestor notabilis*) and critically endangered kakapos (*Strigops habroptilus*) are the best known of the extant species, but a recently described fossil species grabbed the attention in 2019. *Heracles inexpectatus*, known from only two leg bones from the Miocene, was described as the largest parrot ever known (Fig. 2.22). Almost certainly flightless, the fossils were originally attributed to a raptor species. Strigopoids are considered the most basal group of parrots, which also suggests that all parrots evolved in Australasia.

Fig. 2.22. Reconstructed life appearance of the giant parrot *Heracles inexpectatus*. Reconstructed height: 90 cm.

Predators of the Skies

I have mentioned the aerial birds with the largest wingspan, the albatross-esque *Pelagornis sandersi* (see p. 36), although this was closely followed by the enormous condor-like bird *Argentavis magnificens*, the "magnificent Argentine bird" (see pp. 272–3). *A. magnificens* was not a condor, vulture or eagle, but part of an extinct family of birds known as teratorns, which possessed characteristics of both condors (broad, long wings for soaring on thermals (a rising body of heated air)) and eagles (large skulls with structures indicating that they typically consumed their prey whole without tearing it apart). All teratorns may have had fully feathered heads like large eagles. Since they are extinct, we do not know exactly what teratorns fed on, but with their size they would have been opportunistic **scavengers** like modern vultures and condors, driving off the predators from their kill when necessary. *Teratornis merriami* was among the last surviving teratorn species (see p. 327). As well as the teratorns,

Fig. 2.23. The largest known accipitrid, the Haast's eagle (*Hieraaetus moorei*), preying on a little bush moa (*Anomalopteryx didiformis*). Early second millennium CE of New Zealand.

there are large early relatives of the California condor (*Gymnogyps californianus*), and over 20 other species of scavenging and predatory birds, such as cousins of the turkey vulture (*Cathartes aura*), Egyptian vulture (*Neophron percnopterus*), black vulture (*Coragyps atratus*), rough-legged buzzards (*Buteo lagopus*) and caracaras (*Caracara plancus*), 10 species of owl, 15 other species of kites, falcons and hawks, and four species of eagles, out of approximately 140 species of birds.

When trying to understand how predatory birds hunt, it is best to start looking at the feet. Talons are specialised for grabbing prey, although they differ greatly between the main groups. Owls do not land killing blows with their feet, but use their short strong talons to crush prey prior to swallowing it whole. Falcons disable prey with a swift strike, either in midair or landing from height (i.e. kestrels). As a result, their talons are usually shorter, with prey being killed by a break of the neck with their beak. The osprey (*Pandion haliaetus*) plucks fish from the water's surface using its long, curved talons to hold on to slippery prey. The secretary bird (*Sagittarius serpentarius*) spends more time walking than any other predatory birds and, as a result, has short talons. It usually deals with prey by rapid stomping before dispatching with the beak (a similar method was probably employed by other long-legged predators such as the phorusrhacids (see pp. 41–2)). Accipitrids (traditional birds-of-prey including hawks, buzzards and eagles) have large talons on two toes which gives the birds a fierce grip on struggling prey. Generally, the prey dies from organ failure or massive blood loss while being dismembered by the beak. Vultures appear to be the exception to the rule—as they do not usually hunt prey, the claws are used to grip onto carcasses while feeding.

Up until DNA studies in 2008, accipitrids were considered part of a more inclusive Falconiformes. Falcons and their closest relatives are now considered closely allied to parrots and passerines. Acciptrids are the sister family to secretary birds and ospreys, and more distantly related to the cathartids (New World vultures). The first accipitrids are known from during the Eocene, although most early fossils tend to be fragmentary in nature. By the Oligocene, accipitrids were diversifying into a variety of niches. The Pliocene and Pleistocene saw a diversification of the more open environment-living hawks and booted eagles, including some of the largest known accipitrids. Best known of the predators is the extinct Haast's eagle (*Hieraaetus moorei*), a species of aquiline or booted eagle that is the largest known accipitrid. The smaller males weighed about nine kilograms (equal to the heaviest known current species of eagle, the harpy eagle (*Harpia harpyja*)), while females probably reached 12 kilograms. As opposed to the long soaring wings of *Aquila* eagles, Haast's eagles had short, broad wings (wingspan of 2.6 metres) for flight amongst woodland, indicating that it was a largely forest-dwelling species, like harpy eagles. All birds from medium-sized waterfowl up to the largest giant moa (around 250 kilograms) were potential prey (Fig. 2.23).

Among other birds of prey, modern owls are split across two distinct families, the strigids and tytonids, with an additional two extinct families and a selection of unresolved basal forms. Some frogmouths and potoos have a passing resemblance to owls and, up until recently. were considered the sister group, but that is no longer considered correct as it was based on similar features due to **nocturnal** hunting. As to which group the owls are closely related is a bit of a mystery; they fall into an unresolved group which includes the day-time raptors. Strigid owls are by far the more speciose owl family, with 230 of the 250 extant owl species. Being very diverse means that specific characteristics of strigids can be difficult to find. A rounded head, shorter beak and larger coloured eyes are a few features which differentiate strigids from tytonids. While tytonids often hunt while on the wing, strigids usually hunt from a perch (although there may be some exceptions to this rule). Several very early strigids are known from the Early Miocene in Europe. Little is known about them as most remains consist of leg bones. However, they were diverse in size, ranging from the very small *Intutula* up to medium-sized *Mioglaux*. Most later fossil owls are closely linked to modern genera. The largest known strigiform was *Ornimegalonyx oteroi* from the Late Pleistocene of Cuba (Fig. 2.24). They were equally flight-impaired or, at best, capable of short glides, with bones so large that they were originally mistaken for phorusrhacids. Known from several skeletons, *O. oteroi* is closely related to modern wood owls (*Strix*) despite having proportions similar to modern burrowing owls (*Athene cunicularia*). The legs and feet of *O. oteroi* were large and powerfully built, and it was capable of hunting large rodents and juvenile ground sloths.

Tytonid owls, on the other hand, are known from earlier deposits than the strigids, with species known from the Eocene. The most obvious feature that unites tytonids is the presence of a heart-shaped facial disc which helps to amplify sounds towards the ears. There was a diversity of species from the Eocene to the Miocene, but unfortunately most fossils are known from a few bones and thus we know little about them. It remains poorly understood as to why the strigids began to displace the tytonids in the Miocene, and in modern times we are left with only two genera of small and medium-sized species. *Tyto* (barn owls) first appeared in the Miocene, and by the Late Miocene, several different-sized species were found on the Italian Gargano Peninsula.

Fig. 2.24. Reconstructed life appearance of the strigiform owl *Ornimegalonyx oteroi*. Reconstructed height: 1.1 m.

Phorusrhacidae ("Terror Birds")

In certain regions, the birds did not totally give up the niche of large **terrestrial** predators to **carnivorous** mammals. Throughout the Cenozoic, in places where large predatory mammals were absent, birds frequently filled that niche. Following the demise of the great non-avian theropod dinosaurs, it took until the Oligocene for a family of giant cariamiforms (see p. 35), namely phorusrhacids, or "terror birds", to retake the mantle of top carnivores in an **ecosystem**. Within the isolated (at that time) continental island of South America, the phorusrhacids grew to a size large enough to compete against terrestrial crocodylians such as sebecians (see p. 30), and marsupial-like sparassodonts (see pp. 48–50).

Phorusrhacids had powerful neck muscles to support a long skull armed with a wickedly hooked beak tip. The birds used their strong feet to hold or disable relatively large prey, and then employed a deadly downward peck to make a kill. The phorusrhacids possessed small vestigial wings that were useless for flight, but their long hindlimbs allowed them to run quickly enough that they did not need to fly. Similar to ratites (see pp. 32–4), phorusrhacids gave up flight in order to grow larger and concentrate on running. Their strong, sturdy feet with sharp claws were certainly employed to strike and slash at prey, pinning it to the ground.

During much of their reign between the Oligocene and Pliocene, South America was home to a wide range of rodents and small notoungulates (see pp. 140–3), which were ideal prey for the phorusrhacids. The earliest known phorusrhacids date back to the Eocene, not in South America, but from Europe and possibly North Africa. We do not know when the family made its way to South America, but other groups including platyrrhine primates (see p. 152), caviomorph rodents (see pp. 146–8) and hoatzins were able to cross the proto-Atlantic during the Paleocene. Phorusrhacids were a lineage of the largely terrestrial cariamiforms, a group which first evolved in the Eocene. Several putative basal phorusrhacids were discovered from France and North Africa, although the fragmentary nature of the remains is not conclusive. In South America, early species were able to quickly evolve into large forms (e.g. *Paraphysornis*, from the Late Oligocene to Early Miocene of Brazil) and they dominated apex predator roles throughout the Miocene.

During the whole Cenozoic Era in South America, the phorusrhacids exhibited remarkable diversity, with a minimum of 17 species across five distinct subfamilies (Fig. 2.25). They initially appeared with the small (one metre tall) *Palaeopsilopterus* in the Early Eocene (53–50 Ma) (see p. 228), then radiated in diversity during the Eocene and Oligocene. All five subfamilies had evolved by the Middle Miocene, and existed in a variety of forms and sizes. Some were tall and gracile (subfamily Phorusrhacinae), at up to 3.3 metres (10 feet) tall. The Patagornithinae included intermediate-sized (1.6 metres (5.4 feet)) and had a gracile body with elongated hindlimbs, indicating that they were agile, specialist runners. The Mesembriornithinae included comparatively small phorusrhacids, standing just 1.3 metres (4.4 feet) tall. Lastly, the "**archaic**" Psilopterinae were the phorusrhacid subfamily from which the others possibly originated, and never grew taller than 1 metre (3.3 feet), despite existing from the Paleocene to the Early Pliocene (ca. 60–3 Ma), the whole temporal range of the phorusrhacids.

From the Middle Miocene of Argentina, *Kelenken* is the largest known phorusrhacid. It possessed the largest skull of any known bird, measuring 71 centimetres (28 inches) long, 45.7 centimetres which includes the beak. *Kelenken* possessed a rather long, slender beak with a hooked tip similar to that of an eagle's, instead of an enormous deep skull present in many phorusrhacids. *Kelenken* was about 3 metres (10 feet) tall, and with its elongated hindlimbs, is estimated to have attained top speeds of 30 miles per hour (48 kilometres per hour). Phorusrhacids declined during the Late Miocene to a small number of species prior to the Great American Biotic Interchange, or GABI, of the Late Pliocene (see pp. 252, 295), which saw the southern continent colonised by a host of potential mammalian competitors such as felids (see pp. 86–9) and canids (see pp. 92–3). However, one genus, *Titanis*, was actually a rare northward migrant early on during the GABI and lived successfully alongside sabre-toothed cats and large canids in the Southern United States during the Early Pleistocene, persisting until about 1.8 Ma. *Titanis* differed from most phorusrhacids that remained in South America, possessing a larger skull, a shorter, stockier neck, and a more robust physique overall than the speedier phorusrhacids that stayed behind in South America. The largest *Titanis* individuals were formidable birds, standing 2.5 metres (8.2 feet) tall and weighing 150 kilograms (330 pounds). It was the last of the giant phorusrhacids, although unidentified remains of a smaller psilopterine appear to have existed into the Late Pleistocene. Additionally, the cariamids (seriemas), terrestrial birds native to South America, are now the sole surviving cariamiforms, which was once a far more diverse and widespread group (including many other extinct families), especially during the Paleogene. By the Middle Miocene, the cariamids were restricted to South America. The extinct bathornithids are considered the closest North American **analogues** to the (largely) South American phorusrhacids. In addition to the long legs and short wings, they too evolved large skulls with a relatively deep, hook-tipped bill.

This is only a tiny sampling of the astonishing avian fossil record, particularly among the nearly 11,000 species of neoavian birds that have radiated throughout the Cenozoic. Although they may be the last surviving remnants of the massive diversification of dinosaurs, in numerous aspects, the birds continue to dominate the globe. More bird species exist than amphibians, reptiles or mammals, and their diversity is nearly equal to that of the vast radiation of **osteichthyan** fish that existed during the Cretaceous and Cenozoic. While the Cenozoic is commonly referred to as the "Age of Mammals", birds remain the most species-rich group on both land and in the air. In this sense, one could argue that we are still living in an "Age of Dinosaurs".

Fig. 2.25. Phylogenetic tree of Phorusrhacidae after Degrange *et al.* (2015).

Mammalia

The mammals are ubiquitous, with over 6,400 extant species, occupying every major habitat on Earth, exploiting a diverse range of dietary and ecological niches and ranging in size from the Etruscan shrew (*Suncus etruscus*), weighing less than two grams, to the blue whale (*Balaenoptera musculus*), weighing over 100 tonnes (Wilson and Reeder, 2005; Luo, 2007). Extant mammals are divided into three major groups: the monotremes (the egg laying species), the marsupials (species which give birth to live but poorly-developed offspring which then develop further in a pouch on the abdomen) and the placentals (species which give birth to live, well-developed offspring). Extant mammals are defined by several features, including the presence of milk-producing mammary glands in females for feeding young, fur or hair, a four-chambered heart, neocortex in the brain, and three bones that form the middle ear. Mammals and their closest extinct relatives have a long evolutionary history, having evolved from a synapsid ancestor during the Late Triassic (Kielan-Jaworowska *et al.* 2004; Luo, 2007). The fossil record shows successive periods of the diversification of mammalian lineages throughout their evolutionary history.

Major divisions of mammalian groups occurred during the Mesozoic. The Theriimorpha, which includes Theria (all extinct and extant marsupials and placentals plus their closest relatives (Rougier *et al.* 1998)) and all their closest relatives, diverged from Australosphenida (all extinct and extant monotremes and their closest relatives) during the Middle Jurassic, around 160 Ma according to genetic analysis (Luo, 2007; Graves and Renfree, 2013). Theriiformes, the **clade** including the most recent common ancestor of Allotheria (multituberculates and gondwanatherians) (see pp. 45–6)) and Cladotheria, originated slightly later. Cladotheria is characterised by features such as a specialised jaw structure for rotational chewing, and includes a series of nested mammal clades (such as the meridiolestidans (see p. 46)) culminating in Tribosphenida (Fig. 2.26). The Tribosphenida, which includes Theria and their closest relatives, and have tribosphenic molars (molars formed into a "reversed triangle" pattern of three cusps), first appeared in the Northern Hemisphere during the Late Jurassic. Among the earliest tribosphenidans are the aegialodontians, known from the Early Cretaceous of England, whose teeth provide key evidence of the transition to fully tribosphenic molars and the early evolution of therians.

Within Theria, Metatheria (all extinct and extant species more closely related to marsupials than to any placental (Rougier *et al.* 1998)) diverged from Eutheria, (all placentals and extinct species more closely related to them than to any metatherian (Rougier *et al.* 1998)), during the Early Cretaceous (Luo, 2007). It is widely acknowledged that mammals evolved moderate diversity during the Mesozoic. Compared to extant forms, some Mesozoic mammals, such as the allotherians, were generally small and relatively generalised (Luo, 2007). It was not until after the Cretaceous-Paleogene extinction event of 66 Ma (see p. 169) that mammals started to attain the taxonomic and morphological diversity observed today. The Cretaceous-Paleogene extinction event was a critical event in mammalian evolutionary history. Although mammals were severely affected by the extinction, their small size and relatively generalised morphologies likely permitted their survival whilst comparatively more specialised animals, including the non-avian dinosaurs, became extinct.

Fig. 2.26. Phylogenetic tree of Mammaliaformes after Luo *et al.* (2015).

Monotremata (Monotremes)

The fact that monotremes lay eggs is a **primitive adaptation** probably shared with many basal mammaliaforms (i.e. close extinct relatives of true mammals). Monotremes have caused controversy since their initial discovery by European scientists. The first platypus (*Ornithorhynchus anatinus*) specimen was met with much scepticism, with scientists even attempting to pry the bill away from the dried pelt. Hardly surprising as the duck-like bill is decidedly unmammal-like. Echidnas (e.g. *Zaglossus, Tachyglossus*) are also far from mainstream, with a dense coat of spines and a tubular snout. Both also share odd features—a primitive cloaca (combined opening for reproductive organs and expelling waste), electrosensitive snout and spurs on the hind feet (venomous in the case of the male platypus). As mammals go, the egg-layers are truly weird.

All known monotremes, including extinct forms and the five extant species, are restricted to Australasia, apart from one South American exception. Unfortunately, the fossil record is sparse, with species being restricted to the Early and Late Cretaceous, Paleocene, Miocene and Pleistocene. Most are known from few remains (largely dental). Furthermore, no living monotreme possesses teeth as an adult (although young platypus do possess molars) and toothlessness is believed to be linked to the development of electrolocation. Given that modern platypus and echidnas fill highly specialised niches, any realistic interpretation of fossil species must be treated with caution. Monotremes may be part of a larger **clade** (Australosphenida) which includes several other obscure **Gondwanan** mammalian groups (from Madagascar and South America) on the basis of the distinct tribosphenic (see p. 43) molars. Since some monotreme fossils are known from the latest Cretaceous and Paleocene of southern South America, one theory is that monotremes evolved in Australia during the Late Jurassic or Early Cretaceous, and that some migrated to South America via Antarctica, all continents which were merged at the time (Chimento *et al.* 2023).

All Cenozoic fossils show features in common with the extant species. Little is known about the Paleocene *Monotrematum* (Fig. 2.27A), as only several teeth have been found. They do show a close link to the later platypus from the Miocene despite it being the only monotreme from South America. From the Miocene onwards, relatives of modern platypus and echidnas are present, with slight differences. Earlier echidnas, such as *Megalibwilia* (Fig. 2.27B) were more robust than the extant *Tachyglossus*. This reached the pinnacle in the Pleistocene with the giant *Murrayglossus* (Fig. 2.27C), the largest known monotreme, at about the size of a sheep. It had longer limbs than the New Guinean *Zaglossus* (which it was formerly named under) and a shifted centre of gravity to the rear probably allowed the giant echidna to assume a vertical stance and even permit semi-**arboreal** behaviour for access to tree-nesting termites. In comparison to the Cenozoic forms, the Mesozoic monotremes were quite diverse, but are known from limited teeth, jaw bones and postcranial material. They ranged in size from the tiny 10 centimetre long *Teinolophus* up to the one metre long *Kollikodon* (one of the largest known Mesozoic mammals). The latter is a disputed member of the monotremes and occasionally placed with the haramyids (Late Triassic to Early Cretaceous basal mammaliaforms whose phylogenetic affinities remain contentious). Evidently, there was some specialisation amongst the early species: the broad arm bones of *Kryoryctes* are indicative of a **fossorial** lifestyle, whereas *Teinolophus* would have behaved and looked much like a modern shrew, hunting in the leaf litter.

Fig. 2.27. A selection of fossil monotremes from the Cenozoic. (A) *Monotrematum sudamericanum*, Early Paleocene of southern Argentina; (B) *Megalibwilia owenii*, Late Pleistocene of South Australia and Tasmania; (C) *Murrayglossus hacketti*, Late Pleistocene of South Australia. Scale bar: 20 cm.

Allotheria (Multituberculates and Gondwanatherians)

One major branch of mammals that is now extinct, placed within Theriiformes and outside of Theria, is a group called Allotheria, which originated during the Middle Jurassic. One lineage of allotherians that are well-represented in the fossil record are the Late Jurassic to Late Eocene multituberculates. Contrary to many Mesozoic mammal groups (apart from the extant monotremes), which were poorly diverse during a portion of the Mesozoic and then disappeared, the multituberculates persisted throughout the whole Early Cenozoic, disappearing near the close of the Eocene (ca. 35 Ma). Furthermore, they survived through the Cretaceous-Paleogene extinction event (see p. 169), and during the Paleocene and Early Eocene, went on to become even more diverse, common and successful in the rainforests of Europe and North America. Indeed, they are so common that they rank among the most plentiful fossils yielded from those **strata**, and their diversity accounts for at least 200 species.

In size and form, most of the multituberculates were superficially similar to squirrels, while others were more **terrestrial** and possibly had beaver- or marmot-like lifestyles. The majority of multituberculates had a rodent-like skull with chisel-like incisors isolated from the crushing **cheek teeth** by a broad **diastema**. However, the molars of multituberculates were unlike those of any other mammal group, because they were rather narrow and elongated, exhibiting multiple longitudinal rows of tubercles or cusps (hence the name). What is even more striking is that in most multituberculates, the last lower premolar developed into a characteristic massive shearing blade resembling a chisel. This is true of the Cimolodonta, but less so of the "Plagiaulacida". Numerous species of Paleocene multituberculates can be identified solely by this one blade-like tooth. Multituberculates thrived across the Northern Hemisphere throughout the Jurassic and Cretaceous Periods. Additionally, a fourth premolar and two lower jaw fragments pertaining to the genus *Corriebataar* from Australia's Early Cretaceous strata confirms the presence of multituberculates on that landmass too. Following the Cretaceous-Paleogene extinction event, the multituberculates ruled a variety of niches, including the squirrel-like niche of small nut-eating and **granivorous arboreal** forms in the Paleocene rainforests of the northern continents (e.g. the European kogaionids and the North American ptilodontids). They are so common that they are present in virtually any strata of this age, with their characteristic blade-like lower premolars being present in large quantities. The largely Paleocene superfamily Taeniolabidoidea (there are controversial records from the Late Cretaceous), characterised by molars as large or larger than the premolars (Rose, 2006), included the taeniolabidids, which were considerably larger than the other multituberculate families. The taeniolabidids are characterised by features including a reduced upper fourth premolar with several cusps (Rose, 2006). Taeniolabidids were among the earliest mammals to reach large sizes following the Cretaceous-Paleogene extinction event. The taeniolabibids also remained restricted to North America during the Early Paleocene and did not migrate into Asia, contrary to the other family of taeniolabidoids, the lambdopsalids.

The lambdopsalids were exclusively Asian, and were largely **fossorial**. They also had more complex dental **adaptations** to cope with grass than the taeniolabidids. The emergence of lambdopsalids in the Late Paleocene of Asia indicates that during the Early to Middle Paleocene, their ancestors dispersed to that continent from North America. The lambdopsalids were fossorial according to many anatomical characters, which include a robust humerus, fused cervical vertebrae and incisors with a tough outer layer of enamel. A triangular skull that housed an uncoiled cochlea, along with the structure of the middle ear bones, suggests that the lambdopsalids were suited for low frequency sound reception, another indicator of fossorial behaviour (Miao, 1988). The kogaionids are multituberculates more basal to the taeniolabidoids (Smith *et al.* 2021) (Fig. 2.28).

Fig. 2.28. Phylogenetic tree of Cenozoic multituberculates after Chimento *et al.* (2015) and Smith *et al.* (2021).

The kogaionids originally evolved as island **endemics**, on Haţeg island, during the Late Cretaceous, in what is now Romania. Similar to certain extant shrews, some Late Cretaceous kogaionids possessed markedly red iron-pigmented teeth, indicating they were specialised **insectivores**. Strangely for island **taxa**, being typically ecologically sensitive and vulnerable, the kogaionids made it through the Cretaceous-Paleogene extinction event thanks to their insectivorous lifestyle. Throughout the Paleocene, as conditions shifted and Haţeg became connected to the rest of Europe, the kogaionids quickly expanded their range and became abundant all over the continent (Smith *et al.* 2015). The kogaionids were potentially outcompeted by the first rodents to reach Europe during the earliest Eocene, and went extinct soon afterwards. After the Paleocene, all multituberculates began to decrease in diversity and abundance, becoming scarce by the Middle Eocene, and disappeared entirely by the Late Eocene. The causes of this remain uncertain. There were climate shifts which resulted in new types of vegetation, from lush rainforests during the Middle Eocene to sparse woodlands and shrubland during the Oligocene, so their preferred habitat could have been vanishing. This correlates with a decrease in the diversity of strepsirrhine primates (see pp. 150–2) during that time, which also required dense jungles to survive. Other hypotheses include the evolution of new predators and that the multituberculates were driven to extinction by the quickly radiating rodents through competitive exclusion, which were possibly better suited as the small nut- and seed-consuming **herbivores**. Following a lengthy and diverse existence spanning over 130 million years, the multituberculates finally disappeared.

In addition, there was also a lineage of obscure, marmot-like allotherian mammals called gondwanatherians, which are known almost exclusively from the Southern Hemisphere (hence the **Gondwana** part of their name) and are characterised by deep, robust snouts. We know gondwanatherians were allotherians thanks to recent discoveries of well-preserved skulls and postcrania from the Late Cretaceous (70–66 Ma) of Madagascar. These include *Vintana* (Fig. 2.26), which is represented by a single skull. It was a specialised herbivore with a rodent-like dentition and massive, powerful jaw muscles attached to downward-projecting cheekbones—a feature **convergently** present in several other mammals including glyptodontines (see pp. 59–60). *Adalatherium*, a long-bodied, fossorial animal, is known from a complete skeleton. One unique feature is a hole at the top of its snout positioned between the nasal bones and surrounded by many foramina, which possibly supported a soft-tissue sensory structure. Gondwanatherians survived past the Cretaceous-Paleogene extinction event, only to become extinct during the Miocene, potentially due to competition with caviomorph rodents (see pp. 146–8) (Rougier *et al.* 2021).

Meridiolestida

Originating in the Late Cretaceous, the Meridiolestida was a lineage of basal cladotherian mammals **endemic** to South America, distingushed by features including a long **rostrum** and an absence of tribosphenic molars (see p. 43). Meridiolestidans possibly evolved from a common ancestor with the dryolestids, a lineage of more primitive cladotherians found primarily in the Northern Hemisphere during the Middle Jurassic to Early Cretaceous. This suggests that some descendants of that common ancestor may have dispersed to South America from the north, giving rise to Meridiolestida. Despite being known mostly from teeth and jaw fragments, the meridiolestidans appear to have been a highly diverse lineage of mammals during the Late Cretaceous, radiating into a range of sizes and ecological niches varying from small, squirrel-like **insectivores** to larger, **herbivorous** forms. The majority of these went extinct during the Cretaceous-Paleogene extinction event (see p. 169), the known exceptions being the small dog-sized, herbivorous *Peligrotherium* from Argentina, persisting until the Late Paleocene (59 Ma), and one genus that survived considerably later in the Cenozoic. After a gap of more than 40 million years in the fossil record, the meridiolestidans reappear in the Early Miocene (ca. 21–17.5 Ma) of Argentina, represented solely by the late-surviving *Necrolestes* (see pp. 266–7). Subsequent to *Necrolestes*, further evidence of meridiolestidans existing any closer to modern times is absent. Similar to the late-surviving gondwanatherians they coexisted with, these final meridiolestidans possibly became so specialised that they were unable to adapt to abrupt environmental changes, and the Middle Miocene Climate Transition (see p. 253) could have been the final blow to this once successful lineage.

Metatheria

Almost everyone is aware of the kangaroo, koala, opossum and the Tasmanian devil. These animals are all marsupials, grouped within the **clade** Metatheria (Fig. 2.29). Nowadays, marsupials are the predominating mammals solely on the isolated landmass of Australia, where native placental mammals are scarce. Whenever marsupials have had to compete with placentals in prehistory, they have frequently lost that struggle for survival, which has made some people suggest that marsupials are biologically inferior to and less advanced than placentals. In fact, marsupials are not inferior; rather, their body structure and evolutionary tactics are highly distinct. The primary distinction lies in their reproductive strategy, which functions very differently to that of placentals. Subsequent to fertilisation, the embryo develops for just several weeks, before the foetus is born at an early stage of development, with solely the mouth and forelimbs fully functioning. Those appendages are crucial since the foetus needs to pull itself along the fur of its mother's abdomen after birth in order to locate the pouch. In the great majority of marsupials, the newborn joeys are born as small as a grape, or even tinier. When it enters the pouch, it wriggles inside and attaches itself onto a nipple, where much of the joey's development takes place.

In placentals, the embryo is enveloped by an organ called the placenta. The placenta is formed by the allantoic and chorionic membranes outside the amnion (innermost membrane), shielding the foetus throughout development. Additionally, via hormone regulation, the placenta stops the mother's immune system from rejecting the foetus as it develops. The marsupials lack this kind of protection, which results in premature birth of the foetus before it can experience immune rejection. Marsupial reproduction enables numerous generations of offspring to be reared simultaneously. The generational changeover can occur relatively quickly in marsupials, as they can raise a third generation that is still living in close proximity, while carrying a foetus in the uterus and an infant in the pouch. A marsupial mother can abandon the young in her pouch with little danger to herself if she is at risk of starvation or predation, and survive to reproduce. A placental mother is unable to abort her foetus without significant danger to herself, so she has to carry it to term. Put another way, marsupial mothers bear less risk since they invest less energy and resources in raising each of their offspring. The statement about marsupial reproduction being a drawback due to a smaller neocortex at birth and prolonged dependency is generally inaccurate because the short gestation period and **altricial** young are **adaptations** that allow for flexibility in resource availability and environmental conditions. On the other hand, some placentals including rodents and rabbits have the ability to decrease the time it takes for their offspring to be weaned, which allows them to reproduce more quickly than many marsupials. Some placentals including giraffes, wildebeest or zebras give birth to young that are relatively mature and nearly functional since, within hours of their birth, they have to be capable of galloping alongside their mother to avoid becoming food for predators.

Identifying an animal's reproductive system solely from fossilised teeth and bones is not possible, so we have to search for other lines of evidence. A key osteological feature differentiating the metatherians from the eutherians (see pp. 43, 57) is in their dentition: metatherians have four pairs of molars in each jaw, while eutherians always have no more than three pairs (Rose, 2006). Based on this feature, *Sinodelphys* was formerly regarded as the oldest known metatherian, found in the Early Cretaceous beds of China (ca. 125 Ma). Thus, it can be considered a contemporary of certain **primitive** eutherian **taxa** discovered in the same region (Hu *et al.* 2010). However, Bi *et al.* (2018) revised *Sinodelphys* as a primitive genus of eutherian. The earliest undisputed metatherians are currently dental fossils from Western North America dating back 110 million years. Throughout the Late Cretaceous, metatherians were widely distributed across North America and Asia, including the marsupialiforms (the clade including true marsupials and all their closest evolutionary "cousins"). During the Cretaceous-Paleogene extinction event (see p. 169), metatherians experienced a more severe reduction in diversity than the contemporary multituberculates (see pp. 45–6) and eutherians, and they recovered from this loss more slowly. During the earliest Paleocene, marsupials dispersed from North America to South America, perhaps by way of the Aves Ridge (a plateau in the Eastern Caribbean now submerged underwater), after which they underwent a major diversification. In addition to the progenitors of modern marsupials, the Metatheria also consisted of the extinct polydolopimorphians (see p. 48) and sparassodonts (see pp. 48–50), the former which most probably consumed a variety of foods (Eldridge *et al.* 2019). The earliest known Australian marsupials are from the Early Eocene and are thought to have emigrated from South America via Antarctica.

Herpetotheriidae and Peradectidae

Following the Cretaceous-Paleogene extinction event, only opossum-like basal marsupialiforms persisted through most of the Cenozoic in the northern continents. During the Cenozoic, the taxonomic and morphological diversity of metatherians in the Northern Hemisphere remained lower than that of eutherians. The small, opossum-like herpetotheriids and peradectids were the two main families of Cenozoic metatherians in the northern continents, having originated in the Late Cretaceous (ca. 72–66 Ma). Herpetotheriids were distributed across Eurasia, North America and Africa. Although the herpetotheriids were more **terrestrial** than the opossums they resembled and were suited for running on the ground instead of climbing, they too were likely opportunistic **omnivores** that consumed a broad range of foodstuffs. They appear to have been rather abundant across the Northern Hemisphere during much of the Cenozoic. All the native metatherians of North America vanished during the Middle Miocene (ca. 16 Ma), although the landmass would subsequently be repopulated by opossums from South America during the Great American Biotic Interchange, or GABI (see pp. 252, 295). The African herpetotheriids went extinct at approximately the same time, although the European and Asian metatherians persisted slightly

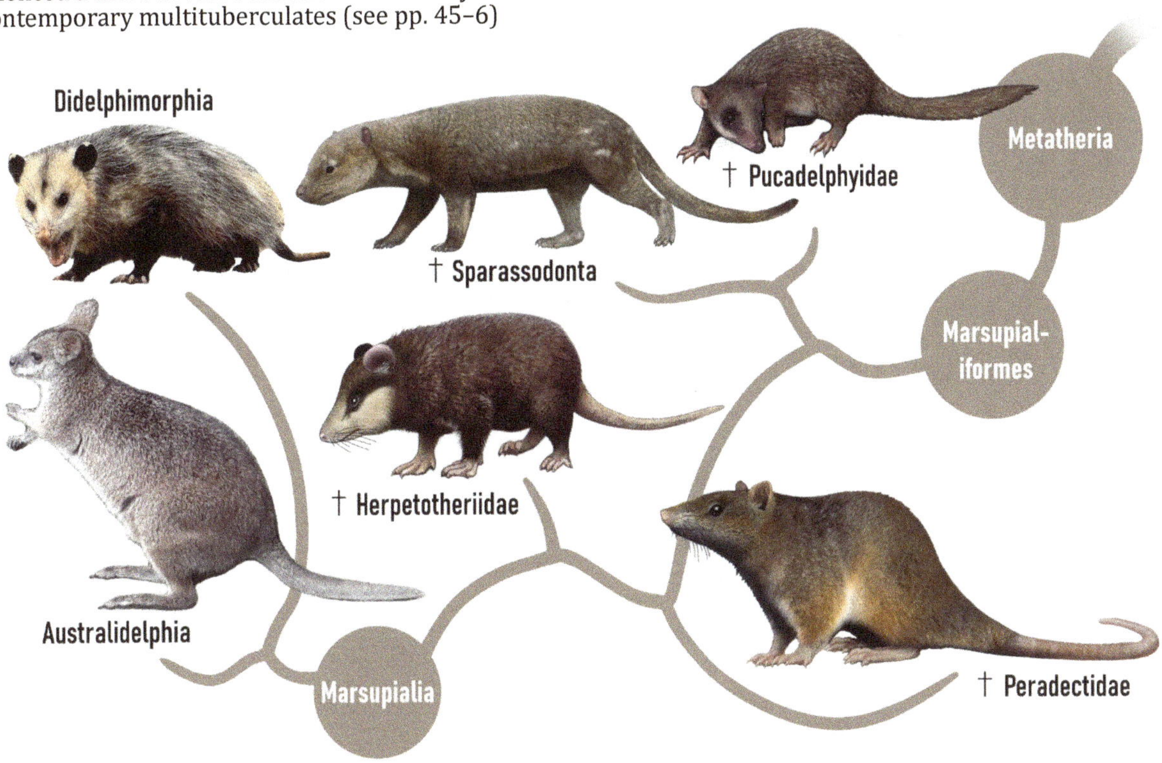

Fig. 2.29. Phylogenetic tree of Metatheria after Beck (2023).

longer into the Middle Miocene. The namesake of the family, *Herpetotherium* (see pp. 235, 239), lived across North America from the Early Eocene to the Early Miocene, and survived in Europe up until the Middle Miocene. It is among the last known herpetotheriids.

The opossum-like peradectids, such as "*Peradectes*" (see pp. 212, 216), have been regarded as a **primitive** phase in the evolution of opossums, even though they are no longer considered to be true marsupials. (Marshall and Muizon, 1988). In North America, during the Cretaceous, a peradectid or peradectid-like metatherian was possibly ancestral to didelphids (New World opossums). While placental mammals have historically been considered responsible for the extinction of northern metatherians, the fact that they coexisted peacefully with placentals for more than 45 million years indicates that something else drove them over the edge. In fact, the demise of northern metatherians could be more associated with the continuing cooling and drying trends throughout the second half of the Cenozoic. The disappearance of Herpetotheriidae appears to have occurred shortly after the Middle Miocene Climatic Optimum (see p. 253), an abrupt shift towards cooler temperatures that possibly changed their local habitats too fast for them to adapt. The peradectids survived until at least 11 Ma, with a number of the last species known from Southeast Asia, although they could have also been wiped out by further pulses of worldwide cooling that occurred during the latest Miocene and Early Pliocene.

Anatoliadelphyidae

There was also another extinct family of metatherians called the Anatoliadelphyidae, which includes *Anatoliadelphys*, a specialised **carnivore** from the Eocene of Turkey, with powerful jaws and bone-crunching teeth **analogous** to those of the Tasmanian devil (*Sarcophilus harrisii*). *Anatoliadelphys* was about the size of a domestic cat (three to four kilograms), and possessed a **prehensile** tail and opossum-like, gripping **digits** that suggest it was an adept tree climber. The phylogenetic position of Anatoliadelphyidae within Metatheria is uncertain. It was originally believed to be closely related to peradectids, although more recent studies (e.g. Métais *et al.* 2018) have suggested that it is possibly more closely related to the Polydolopimorphia, an extinct order known primarily from South America. It is unknown exactly how its ancestors arrived in Turkey, although they may have initially immigrated to Africa when the continents remained considerably closer than they are now, and later travelled north. The major issue with this hypothesis is that there are so far no African polydolopimorphians in the fossil record, although little is known about the Eocene fauna of Africa, so there are possibly a number of metatherian fossils out there that could close this gap.

Polydolopimorphia

During much of the Cenozoic, while the opossum-like herpetotheriids and peradectids (see pp. 47–8) persisted across the Northern Hemisphere, a greater range of metatherians was found in the southern continents. For more than 60 million years, from the Paleocene to Pliocene Epochs, the polydolopimorphians coexisted in South America alongside the sparassodonts and true marsupials. Despite being known largely from isolated teeth and fragmentary jaws, the polydolopimorphians appear to have been a highly diverse order which filled numerous ecological roles such as **herbivores** and **insectivores**, and specialised **frugivores** and **granivores**. Their precise phylogenetic relationships within Metatheria remain relatively uncertain and debated, with various studies yielding conflicting findings. Polydolopimorphians were possibly marsupialiforms, placed in a more basal position than marsupials and the herpetotheriids, although some scientists regard them as **crown** marsupials, being close relatives of either microbiotherians or shrew opossums. Others adopt both positions, suggesting that they were not a **monophyletic** group, but **polyphyletic**, meaning that certain species were true marsupials whereas others were marsupialiforms.

The oldest confirmed polydolopimorphian fossils date to the earliest Paleocene of South America (ca. 66 Ma). Although a few isolated fossils from North America have been tentatively linked to polydolopimorphians, their identification remains uncertain, and the group's Late Cretaceous presence is not definitively established. By the Late Eocene (ca. 40–33 Ma), polydolopimorphians had also reached Antarctica, likely via dispersal from South America before Antarctica became glaciated and isolated. However, there is no evidence to date that they dispersed into Australia, unlike some other marsupial lineages. Polydolopimorphians reached their peak diversity during the Paleogene, gradually declining in abundance before going extinct by the Late Neogene. Their extinction was probably driven by climatic cooling and increased aridity, which could have altered their habitats and reduced the availability of key food resources.

Sparassodonta

When South America separated from Antarctica during the Paleogene, the continent started a journey north unconnected to any other landmass for over 50 million years. The metatherians, in particular, diversified into the **carnivorous** roles left in the absence of the placental carnivorans (see pp. 81, 83–4) in South America during its "splendid isolation". These included the current families of marsupial opossums and caenolestids (see p. 51), but more impressive was the diversification of an extinct order known as Sparassodonta (Fig. 2.30). Despite having to share the large carnivore niches with both the phorusrhacids (see pp. 41–2) and the sebecian crocodylomorphs (see p. 30), the sparassodonts still became the dominant predatory mammals of the continent. As a case of **convergent evolution**, the sparassodonts had similar sharpened shearing teeth to carnivorans, and species resembled modern civets, foxes, otters, bears and cats in form and behaviour. Sparassodonts are not considered marsupials as such, but more of a sister group within the larger Metatheria. Were they pouched like marsupials? No one is so sure. Fossil specimens lacked the epipubic bones seen in marsupials, but in some species such as thylacines, the epipubic bones were cartilaginous. So maybe they did raise offspring in pouches.

The earliest undisputed sparassodont fossils are known from the Early Paleocene (ca. 64 Ma), although they may have first evolved sometime during the Late Cretaceous prior to the Cretaceous-Paleogene extinction event (see p. 169). Like the marsupials, it is possible that sparassodonts originated in North America and subsequently immigrated to South America before it became isolated. The early history of the Sparassodonta is poorly understood, since most Paleocene and Eocene representatives of this lineage are known solely from isolated teeth and jaw fragments. The most basal sparassodonts, such as

Fig. 2.30. Phylogenetic tree of Sparassodonta after Engelman *et al.* (2020).

Mayulestes (see pp. 184–5), were very small and relatively unspecialised forms which resembled dasyurids (see p. 52) and opossums, and have dental **adaptations** for **omnivory**. Among the most **primitive** and generalised families of sparassodonts are the hathliacynids (Early Eocene-Early Pliocene), which were relatively slender-bodied, possibly **scansorial**, ranged from weasel-sized (e.g. *Pseudonotictis* from Early Miocene Argentina, which had possible **arboreal** capabilities) to thylacine-sized (e.g. *Acyon* (see pp. 263–4)), and had elongated, fox-like snouts and teeth well-adapted for carnivory. However, the molars had highly reduced grinding surfaces, indicating some flexibility in diet (Prevosti and Forasiepi, 2018).

The basal family Hondadelphidae (Middle to Late Miocene) may have been more omnivorous and are currently represented by two genera. The small-bodied (3.7 kilograms) *Hondadelphys*, from Middle Miocene Colombia, was likely arboreal and exhibited strongly developed shearing modifications along with grinding specialisations (Goin, 2003). *Stylocynus*, from Late Miocene Argentina, was a large (15–30 kilograms), **terrestrial** hondadelphid probably with a **mesocarnivorous** diet (Prevosti and Forasiepi, 2018). From the Eocene, the borhyaenoids became the dominant group of sparassodonts, becoming larger-bodied and stronger-jawed. Basal borhyaenoids (Early Eocene to Late Miocene), such as *Lycopsis* and *Prothylacinus* (see pp. 263–4), were borhyaenoids that cannot be easily assigned to the families Borhyaenidae, Proborhyaenidae or Thylacosmilidae, and varied in size and shape. The last surviving basal borhyaenoids date to the early Late Miocene (*Lycopsis viverensis* and *Pseudolycopsis cabrerai*), and were subsequently at least partly replaced by large basal sparassodonts including *Stylocynus*. It has been suggested that this change in dominance was due to the more omnivorous lifestyle of basal sparassodonts, which were possibly better able to exploit the more seasonal climates of South America during the Late Miocene.

The borhyaenids (Early to Late Miocene) were more dog-like, characterised by robust, powerful jaws and had features of a **cursorial** terrestrial predator, although they were less cursorially specialised than modern carnivorans or thylacines. Including some of the largest known carnivorous metatherians were the proborhyaenids (Early Eocene to Late Oligocene), which were powerfully built, terrestrial, wolverine-like forms. Proborhyaenids are distinguished by **hypselodont**, grooved upper and lower canine teeth, and massive bone-crushing jaws. The majority of proborhyaenids had a strongly-built, hyaena-like skull, while that of *Callistoe vincei* (from Early Eocene Argentina) was long, narrow and more thylacine-like. *Arminiheringia* (also from Early Eocene Argentina) had unusually long, protruding lower canine teeth, the purpose which is unknown.

The thylacosmilids, also a terrestrially specialised family, are both familiar and the least typical members of the borhyaenoids. Their specialised sabre teeth may have been inherited from their common ancestor with other large sparassodonts such as the proborhyaenids. Originating in the Early Miocene (ca. 20–15 Ma), the thylacosmilids developed massive, elongated upper canine teeth that resembled those of placental sabre-toothed cats. Among the most primitive thylacosmilids were the small-bodied (18 to 23 kilograms) *Anachlysictis* from the Middle Miocene (ca. 13.8–11.8 Ma) of Colombia. It was anatomically less specialised than later relatives such as *Thylacosmilus* (see pp. 272–3) in having shorter mandibular flanges (since the sabre teeth are less proportionately elongated) and a rim around the **orbit** was absent, giving the skull a flattened profile. A more advanced thylacosmilid from around the same time (ca. 15.5–11.8 Ma) of Patagonia, *Patagosmilus* was intermediate between *Anachylsictis* and *Thylacosmilus* in skull morphology and upper canine length, albeit still less specialised in various details than the latter. However, *Patagosmilus* did have hypertrophied upper canine teeth and a bowed molar series as in *Thylacosmilus*, and the molar morphology indicates a close phylogenetic relationship with that genus. *Thylacosmilus* was the last, largest (jaguar-sized) and most highly specialised of the thylacosmilids, existing from the Late Miocene to Late Pliocene (ca. 9–3 Ma).

Although it is commonly assumed that similar placental carnivores that arrived in South America during the Great American Biotic Interchange, or GABI (see pp. 252, 295), outcompeted the last sparassodonts such as the thylacosmilids, this does not seem to have been the case. The overall decrease in sparassodont diversity from the Late Miocene to the latest Pliocene, when they became extinct, may be associated with the cooling climates subsequent to the Middle Miocene Climatic Optimum (see p. 253) and during the beginning of the Quaternary Ice Ages (see pp. 311–3), which limited the availability of their favoured prey. The extinction of sparassodonts left numerous vacant ecological niches for various immigrant carnivorans to fill during the GABI.

Marsupialia

Fossils of the **clade** Marsupialia (true marsupials) do not show up in the fossil record until the Early Paleocene. Marsupials entered Australia via land connections through South America and Antarctica sometime in the Early Eocene (ca. 55 Ma), as the continents began to separate. This likely represents a single dispersal event by an australidelphian ancestor, with a subsequent rapid evolutionary radiation in Australia. Among the features differentiating true marsupials from all the other metatherian groups are the presence of multiple cusps on the stylar area (outer area of the upper molar crown) and the lack of a **posterior** cingulid (a shelf-like ridge around the outside of a lower molar) (Beck *et al.* 2022).

"Ameridelphia"

The Marsupialia was once divided into two main groups: Ameridelphia (which included the didelphimorphians and paucituberculatans), and Australidelphia (consisting of the huge Australian radiation of marsupials along with South American microbiotherians). However, recent phylogenetic analyses (e.g. Beck, 2023) show that "Ameridelphia" is not a true **clade** but a **paraphyletic grade**.

Didelphimorphia

Whereas the **carnivorous** sparassodonts (see pp. 48–50) are extinct, the New World opossums remain highly diverse today. They comprise many species such as the mouse opossums (*Marmosa*), bushy-tailed opossums (*Glironia venusta*), four-eyed opossums (*Philander*) and water opossums (*Chironectes minimus*). The majority typically resemble the Virginia opossum (*Didelphis virginiana*) in size and form, while some species such as the mouse opossum are more diminutive. Opossums are distributed across across much of South America, being most diverse in the Amazon jungle and the Andean forests. The earliest known **crown**-didelphid fossils date to the Oligocene (ca. 26 Ma, Bolivia), and the most recent common ancestor of all extant opossums probably lived in the Late Oligocene or Early Miocene, not solely in the Miocene. Molecular and fossil evidence suggest that opossums diversified in northern South America, where they began their first radiation (Castro *et al.* 2021). They were already important components of mammal **communities** by the Early Neogene, although sparassodonts occupied many carnivorous niches until their decline. The idea that large-bodied opossums such as *Didelphis* became progressively larger as sparassodont diversity decreased is possible but speculative, since sparassodonts survived until the Late Pliocene and overlapped with large opossums. A number of didelphid lineages, such as the sparassocynins (see p. 298), and some extinct paucituberculatans, evolved carnivorous specialisations during the Late Miocene and Pliocene, before the entry of placental carnivores in South America. Many of those lineages are extinct, apart from the lutrine opossum (*Lutreolina*). One study has proposed that the size and morphology of the lower jaw in ancestral didelphids would have been most **analogous** to that of mouse opossums.

Paucituberculata

Another lineage, the shrew opossums or caenolestids, evolved in South America during the Paleogene and remain a relic of its once vast metatherian diversity. The caenolestids are mouse-sized and restricted to the high altitudes of the Andes mountain range, consisting of just three genera and seven species. They belong to the order Paucituberculata, which dates back to the Late Paleocene (ca. 58.7 Ma). At least seven genera of caenolestids existed as late as the Early Miocene (ca. 20 Ma). Some researchers once hypothesised that their decline was due to competition with placental shrews that entered South America in the Miocene, but more recent work suggests this is unlikely to be the sole cause. Instead, their reduced diversity may reflect a combination of ecological and climatic changes. As their name suggests, many shrew opossums resemble shrews or mice in size and appearance, possessing an elongated, tapering snout, gracile limbs and a hairy, shrew-like tail. They are predominantly **nocturnal omnivores**, although contrary to most nocturnal mammals, they have tiny eyes and rather poor vision, instead using whiskers to sense prey.

Australidelphia

Australidelphia is primarily native to Australia, as the name implies. These marsupials are unified by a suite of molecular and morphological similarities, and also by their geography. In addition to the Australian **clades**, Australidelphia also includes the South American Microbiotheria.

Microbiotheria

The microbiotherians, a surviving branch of marsupials in South America, are crucial for understanding the evolutionary links between South America, Antarctica and Australia. While microbiotherians are rat-sized, they somewhat resemble a cross between a monkey and an opossum. As in certain **nocturnal** primates such as tarsiers, they possess large **orbits** for nocturnal activity, and grasping **digits** for capturing invertebrates and jumping between branches. As in certain opossums and New World monkeys, they have a long, somewhat **prehensile** tail including a naked underside that assists with clinging onto branches. A small radiation of microbiotherians persisted into the Neogene, and today a single species, *Dromiciops gliroides*, is extant, although recent studies have proposed at least three distinct lineages (*D. gliroides, D. bozinovici, D. mondaca*) based on genetic and morphological evidence. These peculiar and elusive South American marsupials are called the monito del monte. They persist in the Andean forests of Argentina and Chile, where they are confined to high altitudes. The fossil record of microbiotherians in South America extends back to the Early Paleocene, and they have also been documented from the Early Eocene beds of Seymour Island in the Antarctic Peninsula (see pp. 200–3). Microbiotherians are evident of the connection between South America, Antarctica and Australia over 50 Ma.

Australian Marsupials

When it comes to mammals, marsupials have always dominated Australia. There is currently no fossil evidence for the establishment of placentals in Australia, apart from one isolated tooth of a probable Eocene placental called *Tingamarra* (see p. 199). In the absence of placental competitors, Australian marsupials radiated to occupy most of the ecological niches that placentals filled elsewhere, resulting in some striking parallels. Marsupial **analogues** exist for placental anteaters (numbats), mice (dasycercids), rabbits (bilbies), ground-dwelling **herbivores** (wombats), moles (marsupial moles, or notoryctids), flying squirrels (sugar gliders), kinkajous (cuscuses), cats (quolls), dogs (the Tasmanian tiger or thylacine), hyenas (Tasmanian devils), and large-bodied herbivores such as wallabies and kangaroos. A few specialised forms exist as well, including koalas, which lack any close placental equivalent apart from perhaps tree sloths. The primary herbivorous marsupials are kangaroos, but they move around via **saltatorial** locomotion, something that large hooved placentals never mastered, although numerous lineages of small placentals including springhares and kangaroo rats did. These represent only the modern Australian marsupials. Throughout the Neogene and Quaternary, the marsupials radiated into some striking forms, including those from the Miocene limestone deposits of Riversleigh, Queensland (see pp. 288–94).

Dasyuromorphia

Similar to how sparassodonts (see pp. 48–50) constituted most of South America's predatory mammals, dasyuromorphs constitute a large portion of Australia's predatory mammals. Although many dasyuromorphs are small-bodied, several extinct species (e.g. *Thylacinus potens*) were wolf-sized. The dasyuromorphs are divided into four families: Dasyuridae (the dasyurids, consisting of a broad range of predators from five-gram planigales to dunnarts, quolls, and Tasmanian devils), Myrmecobiidae (numbats), Malleodectidae (a **monotypic** family consisting of the specialised, **durophagous** marsupial *Malleodectes* (see pp. 288–9)) and Thylacinidae (the dog-like thylacinids). There is only one recent species each within Myrmecobiidae and Thylacinidae, while Dasyuridae contains numerous (at least 74) species. The one species of thylacinid, the Tasmanian tiger or thylacine (*Thylacinus cynocephalus*), is extinct, although occasional unconfirmed sightings have been reported. Dasyurids and thylacinids have closer phylogenetic affinities to each other than to numbats. Hailing from the Middle Miocene of Queensland, Australia, the relationships of the Malleodectidae to the extant families of the Dasyuromorpha remains uncertain. The earliest dasyuromorphs are known from southern Queensland, dated to at least 55 Ma, although their evolution between the Early Eocene (55 Ma) and the Late Oligocene (24 Ma) remains poorly understood. Their fossil record is rich from the Early Miocene (23–16 Ma), during the warm and humid greenhouse interval, which led to high overall diversity in rainforest **ecosystems**. From about 15 Ma, when Australia began cooling and drying, dasyuromorph diversity increased steadily, but family-level diversity subsequently declined due to increasing aridity (Armati *et al.*, 2006; Wilson and Reeder, 2005).

Among the dasyurids, devils were once broadly distributed across Australia—a giant Pleistocene species called *Sarcophilus laniarius* was roughly 15% larger and 50% heavier than the modern species (*Sarcophilus harrisii*). The extinction of mainland devils was likely caused by climatic changes induced by the Last Glacial Period (see p. 313). In Tasmania, they have undergone population declines due to devil facial tumour disease, and are currently estimated at about 18,000 to 20,000 individuals. Compared to Tasmanian devils, all other dasyurids are considerably smaller-bodied, ranging from cat-sized quolls to over 60 species of rat- and shrew-like marsupials. The majority of small-bodied dasyurids are **insectivores**, while larger forms also consume small vertebrates and fruit. Among modern Australian marsupials, dasyurids are often considered relatively **primitive**, though some species show ecological specialisations not present in other lineages such as macropodids (see p. 54). A full pouch is absent in most dasyurids, and females possess simple skin folds on the sides of the abdomen.

The numbats (Myrmecobiidae) slightly resemble miniature versions of placental anteaters in possessing an elongated snout and a long, bushy tail. As in other insectivorous mammals, they possess long, slender tongues and multiple ridges on the palate for scraping insects off the tongue before swallowing. They lack massive front claws for tearing open termite nests. The fossil record of numbats is limited to a few Pleistocene specimens. Genetic studies suggest that numbats diverged from other marsupials during the Late Eocene (32–42 Ma).

While very little is known about the fossil record of dasyurids and numbats, the family Thylacinidae, whose most famous member is the modern thylacine, is surprisingly diverse in species, ranging back to the Late Oligocene. Earlier genera such as *Badjcinus* (see pp. 244–5) were generally smaller and resembled quolls or civets in form, size and possibly behaviour, although indications of an increasingly **carnivorous** diet can be observed as far back as the Early to Middle Miocene, in genera such as *Nimbacinus* and *Muribacinus* (see pp. 288–9). Later species in the genus *Thylacinus* had the more familiar dog-like body shape. However, *Thylacinus* hunted more as an **ambush predator** and was a long distance runner like modern canids (see pp. 92–3). Species of the genus *Thylacinus* show a marked increase in both body size and in the expression of dental **adaptations** for carnivory, with the largest species, the Miocene age *Thylacinus potens*, being wolf-sized (Fig. 2.31). The modern thylacine was widely distributed (albeit never populous) all over Australia and New Guinea during the Late Pleistocene and Early Holocene, which had declined and survived only in Tasmania by around 3.5 ka, possibly due to climate change. The last known surviving individual of the modern species died in 1936 in a Tasmanian zoo, thus likely rendering the species, and family, extinct due to a combination of factors including disease, competition with dogs, and bounty hunting since they were perceived as a threat to livestock.

Peramelemorphia

The Peramelemorphia comprises seven extant genera and at least five recognised fossil genera from the Late Oligocene to Pleistocene Epochs. The bandicoots are primarily small, rat-sized to rabbit-sized mammals with large ears, an arched vertebral column, relatively short limbs, and other **convergences** on rabbits. Contrary to rabbits, they have a pointed snout and a thin, mouse-like tail. The bilbies have softer fur, and more elongated ears (standing straight as in jackrabbits) and tails relative to the bandicoots. Among the earliest known fossil peramelemorphians are the Late Oligocene genus *Bulungu palara*. The earliest known peramelid is the iconic fossil bandicoot *Crash bandicoot* (see pp. 288–9), named after the eponymous video game character as a lighthearted homage by the authors. The rapid radiation of modern bandicoots likely

Fig. 2.31. Reconstructed life appearance of the thylacinid *Thylacinus potens*. Reconstructed length: 1.5 m.

Fig. 2.32A. Phylogenetic tree of Marsupialia, particularly **taxa** outside Phalangerida, after Black *et al.* (2012), May-Collado *et al.* (2015), Upham *et al.* (2019), and Álvarez-Carretero *et al.* (2022).

occurred following the Middle Miocene, when Australia became increasingly arid, a habitat favoured by extant species (Travouillon *et al.* 2014). Bilbies are believed to have split from their nearest relative, an initially **carnivorous** bandicoot, during the Early Miocene (ca. 20 Ma). The oldest known bilby fossil is dated to around 15 Ma, consisting of a lower jaw with teeth more **brachydont** than in modern bilbies, possibly adapted for **frugivory**. Extant bilbies have evolved **hypsodont** teeth adapted for excavating pits in the desert for preying on invertebrates as Australia became increasingly arid. The phylogenetic affinities of Peramelemorphia to the remainder of the Australian marsupials have long been contentious. They retain a relatively **primitive** morphology, with minor differences from the dasyurids. On the other hand, they have two **digits** joined together (**syndactyly**), a specialised characteristic also present in Diprotodontia. The syndactylous foot consists of reduced second and third digits firmly conjoined to the base of the claws, while other digits remain separate. Another modification is the loss of the first digit (hallux) and the expansion of the fourth and fifth digits. Current consensus suggests syndactylous feet in Peramelemorphia and Diprotodontia likely represent convergent evolution, arising independently via different developmental pathways (Weisbecker & Nilsson, 2008; Beck *et al.*, 2023).

Diprotodontia

All of the major groups of **herbivorous** marsupials are linked under the order Diprotodontia by specialisations including short jaws and a prominent pair of lower incisors working against three pairs of upper incisors. These **adaptations** are associated with a primarily herbivorous diet, facilitating the processing of tough vegetation. Within Diprotodontia are three main suborders representing the macropodiforms (wallabies and kangaroos), the vombatiforms (wombats and koalas), and the phalangeriforms (possums, gliders and cuscuses). Vombatiforms are generally **terrestrial**, although one family, the koalas, took to climbing back into the trees. As well as the broad scope of extant species, the Diprotodontia has a rich fossil record stretching back to the Late Oligocene (ca. 28 Ma). The most dominant and populous Australian marsupials have always been diprotodontians, resulting in dozens of species having been discovered from the well-known Oligo-Miocene Riversleigh deposits (see pp. 244–5, 288–94) and in many other fossil localities ranging from the Miocene to Pleistocene.

Macropodiformes

The macropodiforms are synonymous with Australia, but are also found in New Guinea. There are four separate families: the extinct fanged balbarids, the hypsiprymnodontids (mostly fossil with only one extant species, the musky rat-kangaroo (*Hypsiprymnodon moschatus*)), the potoroids (potoroos and bettongs) and the macropodids (true kangaroos and wallabies) (Fig. 2.32B). Whilst all extant species can hop, this is not necessarily the two-legged bounce that we are most familiar with. In the rainforests of far northeastern Australia lives the most **primitive** macropodiform, the musky rat-kangaroo (the aforementioned extant representative of the Hypsiprymnodontidae). At slow speeds, it employs a pentapedal gait using all four limbs and the tail, while at faster speeds it bounds using all four limbs, somewhat resembling the bounding and leaping movements of **arboreal** possums in terms of limb coordination. It is believed that the earliest macropodiforms also used a similar method to move about. This combination of pentapedal walking and **quadrupedal** bounding likely provided both stability and efficiency in dense forest undergrowth. Macropodiforms split into their respective families during the Late Oligocene. Early macropodids included *Cookeroo* (see pp. 288, 293). They prospered alongside the **browsing** balbarids, which did not make it past the Miocene as environments began to dry out. From then on, the fully **bipedal** hopping macropodids expanded into most **terrestrial herbivorous** niches largely taken by **ungulates** on other continents. One place that forest-living macropodids continued to flourish was in New Guinea. The primitive *Watutia* had features in common with early sthenurine kangaroos, an extinct browsing lineage which included giant forms. While large bodied species of *Protemnodon* on the mainland resembled the typical *Macropus* kangaroos, those in New Guinea were morphologically more like the short-footed tree-kangaroos (*Dendrolagus*). Like small forest ungulates and herbivorous rodents, compact body shapes were important for negotiating the forest floor.

Wallaby is a common name given to a diverse range of smaller macropodids. Hare-wallabies are the smallest extant macropodids and, like much of the smaller mammalian fauna in Australia, have suffered due to changes in land management, predation (introduced foxes and cats) and competition (introduced rabbits and livestock). Taxonomy within the macropodids splits the family unequally into three subfamilies (plus a few basal forms). The vast majority of all known extinct and extant macropodid genera are members of the macropodines. All typical kangaroos, wallabies, tree-kangaroos, pademelons, rock-wallabies and hare-wallabies (with one exception, the Musky rat-kangaroo) are included in this group. They range in size from 800 gram hare-wallabies (*Lagorchestes*) up to 90 kilogram (in males) red kangaroos (*Osphranter rufus*). Although this species of *Osphranter* is the largest extant marsupial, it is dwarfed by several close fossil relatives. During the Late Pleistocene, there were a number of giant macropodids, including a recently discovered species (yet to be officially named) that reached heights of up to three metres and had an estimated body mass of nearly 300 kilograms, likely representing a giant sthenurine rather than a typical macropodine.

Tree-kangaroos are an interesting twist in macropodiform evolution. All macropodiforms show some of the **ancestral adaptations** of living in trees (**syndactyl** toes on the hind foot being the best example). When they descended to the ground, they evolved longer feet. However, when some wallabies moved back into the trees, the feet reverted back to being shorter and broader for perching on branches. Some of these features are seen in rock-wallabies, which are probably close to the ancestor of tree-kangaroos. Like other macropodines, the tree-kangaroos also had some giant relatives. Most surprisingly, *Bohra* was native to southern and eastern Australia, far away from the tropical rainforests of modern tree-kangaroos. Lagostrophines consist of only one extant species, the small Banded hare-wallaby (*Lagostrophus fasciatus*). Confusingly, it is not closely related to the other hare-wallabies. The best known of the fossil macropodids are possibly the short-faced sthenurines. Only a handful of genera are known but they were as diverse in size as the macropodines. Up until the discovery of the aforementioned unamed giant macropodine, the sthenurine *Procoptodon goliah* (see pp. 332–3) was

Fig. 2.32B. Phylogenetic tree of Marsupialia, particularly taxa within Phalangerida, after Black *et al.* (2012), May-Collado *et al.* (2015), Upham *et al.* (2019), and Álvarez-Carretero *et al.* (2022).

the largest known macropodid—nearly twice the size of any extant species. The last sthenurine died out in the Late Pleistocene, after human arrival to Australia.

Vombatiformes

Three species of wombat (family Vombatidae) and the koala (family Phascolarctidae) are the sole surviving members of a formerly huge radiation of vombatiform marsupials. This suborder also consisted of the largest Australian marsupial **herbivores**, the diprotodontids. Like Australian marsupial families in general, our understanding of the divergences between the various diprotodontian families has been obscured by the temporal gap in the known fossil record between the Early Eocene (ca. 55 Ma) and the Late Oligocene (ca. 25 Ma). All vombatiform families had already diverged by the Late Oligocene, according to recent molecular phylogenies (Beck 2023). These included various obscure extinct wombat-like families such as the herbivorous wynyardiids and ilariids. The former is characterised by semi-**lophodont** molars, whereas the latter is distinguished by an intricate enamel folding pattern on the molars. Neither family survived past the Middle Miocene likely because of climate change-induced losses of their favoured habitat (dense tropical forests).

Diprotodontidae

While remaining conservative in shape, the extinct diprotodontids were the most speciose vombatiforms and were Australasia's **analogues** of **ungulates** between

Fig. 2.33. Reconstructed life appearance of some representative small diprotodontids. (A) *Kolopsis torus*, from the Late Miocene of northern Australia; (B) *Hulitherium tomasetti*, from the Late Pleistocene of New Guinea. Scale bars: 0.5 m.

sheep and rhinoceros in size. The family were **plantigrade** and first appeared during the Late Oligocene, with members that were mostly sheep-sized, such as *Neohelos* (see pp. 290–1) and *Kolopsis* (Fig. 2.33A), and substantially diversified from the Late Miocene, eventually reaching an apex of diversity during the Pliocene with seven genera, likely due to the increase of open forested landscapes. Later genera had modifications to the skull, such as the flaring cheekbones of *Euryzygoma* (see pp. 296–7). Among the last and largest members of the diprotodontids were *Zygomaturus* and *Diprotodon* (see pp. 333–5). Diprotodontids were not just restricted to Australia, as several genera were found on the large neighbouring island of New Guinea. Found in the montane bamboo forests and of similar size, *Hulitherium* (Fig. 2.33B) is regularly dubbed the marsupial panda, having become extinct only around 40 ka, subsequent to human arrival in Australasia.

Palorchestidae

The extinct palorchestids include the marsupial tapirs (one of two families constituting the **clade** Diprotodontoidea, the other family being the aforementioned diprotodontids). Vombatids are the nearest modern relatives of palorchestids (Beck *et al.* 2020). The oldest palorchestids such as the Miocene *Propalorchestes* had rather generalised forelimbs, unlike the clawed forelimbs of the final representative of the family, *Palorchestes* (see pp. 333–4). All palorchestids are regarded as **browsers**. The body size of palorchestids became progressively larger over time, with *Propalorchestes* weighing approximately 150 kilograms, and the last surviving species, *Palorchestes azael*, possibly weighed over one tonne (Richards *et al.* 2023).

Vombatidae

Little is known about the evolutionary history of vombatids. Vombatids are thought to have split from other diprotodontians approximately 25 Ma, although some estimates place this split as far back as 40 Ma. In vombatids, scratch-digging specialisations arose before the evolution of **hypselodont cheek teeth** according to phylogenetic analysis (Beck *et al.* 2020). The size of certain fossil vombatid genera was much greater than that of extant vombatids. *Phascolonus gigas* is the largest known vombatid, reaching 360 kilograms (790 pounds) and surviving until about 46–40 ka (Hocknull *et al.* 2020; Louys *et al.* 2022).

Phascolarctidae

Koalas and wombats are each other's closest relatives, which both shared a common ancestor during at least the Late Oligocene. The earliest **arboreal** koalas may have originated from a **terrestrial** progenitor resembling a wombat, possibly to exploit a food source not being fed on by others. Koalas used to be much more prevalent across Australia during the Miocene, represented by fossil species that lived in the southern, central and western regions of the continent when rainforests were more extensive. There are currently eight genera and 17 species of koalas in the fossil record, which all come from either Queensland or South Australia. The Late Oligocene fossil koalas *Madakoala* and *Perikoala* are placed in a more basal position on the koala family tree relative to *Nimiokoala* (see pp. 290–1). The genus *Nimiokoala* is currently considered the **primitive** sister **taxon** to a **clade** consisting of *Litokoala* (see pp. 290–1) and the modern genus *Phascolarctos*. *Phascolarctos* originated during the Late Miocene or Early Pliocene, with three fossil species so far documented: *P. yorkensis* from the Late Miocene to Pleistocene of New South Wales and South Australia, and *P. maris* and *P. stirtoni* from the Early Pliocene to Pleistocene of South Australia. *P. stirtoni* was roughly twice as large as modern koalas. The sole extant species, *P. cinereus,* originated during the Middle Pleistocene, at least 350 ka. What the modern koala evolved from and the reason one species survived while the giant species disappeared remains unknown. One possible cause of those species' extinction is climatic changes.

Thylacoleonidae

For all the potentially dangerous species that inhabit Australia today (think sharks, crocodiles, blue-ringed octopus, box jellyfish, stonefish, funnelweb spiders and venomous snakes), one particular aspect of the fauna is missing: large predatory mammals. Of course, the southern continent was once home to the dog-like thylacines, but these were far from apex predators. Early indigenous settlers arriving on the continent around 60 ka had more to worry about from the giant madtsoiid snakes (see pp. 23, 25), mekosuchine crocodylians (see p. 28) and the giant monitor lizard *Varanus priscus* (see pp. 332, 335). Even so, one large marsupial hunter would have sent shivers down the backs of the original human inhabitants. Dasyuromorphs are the dominant group of **carnivorous** marsupials, with most other marsupials being the largely **herbivorous** diprotodontians, including macropodiforms (see pp. 54–6), vombatiforms (see p. 56) and possums. However, one extinct group of vombatiforms bucked this vegetarian trend and became the largest known marsupial carnivores, the thylacoleonids.

The thylacoleonids were a predominantly **arboreal** group of predators with specialised **hypercarnivorous** dentitions. Large, flat, shearing blades in the rear of the mouth were actually specialised premolars and **convergent** with the **carnassial** teeth found in carnivoran placentals (see pp. 81, 83–4), while the incisors at the front of the mouth were more **caniniform**. Together, the dentition and muscle attachments gave thylacoleonids a stronger bite force than comparably-sized carnivores. Heavily muscled forelimbs and retractable claws allowed them to climb trees, and it has been suggested that thylacoleonids could jump from a height onto the back of potential prey (whether this was actually possible is debatable). Smaller genera showed similar **adaptations** to possums and koalas for life among the trees in the structure of the arm and shoulder. Thylacoleonids are first known from the latest Oligocene and earliest Miocene ranging in size from diminutive (*Microleo*, *Lekaneleo* (see pp. 290–2)) tree-dwellers up to small dog-sized genera (*Wakaleo* (see pp. 291–2)). From the Pliocene until late into the Pleistocene, the only surviving genus was the larger *Thylacoleo* (see pp. 296–7, 333, 335).

Phalangeriformes

The phalangeriforms are a **paraphyletic** assemblage of small- to medium-sized **arboreal** diprotodontians which originated during the Oligocene. The phalangeriforms include cuscuses, gliders and Australian possums, and are generally **nocturnal**. Certain species of phalangeriforms are called possums due to their similarities to the New World opossums. The name "opossum" is applied to the New World species ("possum" is a frequent abbreviation), whereas similar Australian marsupials are called "possums" proper. The phalangeriforms range in size from the tiny Tasmanian pygmy possum (*Cercartetus lepidus*), which measures up to 7.5 centimetres (3 inches) long, to the bear cuscus (*Ailurops*), which is 61 centimetres (24 inches) long and weighs up to 10 kilograms (22 pounds). It is a generally accepted hypothesis that the anatomical divergence of the bear cuscus from other cuscuses was caused by its isolation on the Indonesian island of Sulawesi during the Miocene. Little is known about the evolutionary history of phalangeriforms, with the most significant fossil evidence coming from a distinctive extinct family called ektopodontids, which have a fossil record extending from the Late Oligocene to Early Pleistocene. The ektopodontids, which may have resembled cuscuses, were likely a specialised lineage of possums in the **clade** Phanlangeroidea. The progenitors of ektopodontids have yet to be discovered, but possibly resembled brush-tailed possums. Ektopodontids are characterised by subtriangular to rectangular, **brachydont cheek teeth**. One trend in ektopodontid evolution was increasingly elaborate crests and cusps on the teeth. Among other possums, *Burramys* (which today includes the mountain pygmy possum) was relatively common in the lowland rainforests across Australia during the Early to Middle Miocene, but their modern relatives survive only in the high altitudes of southeastern Australia (Archer *et al.* 1989).

Eutheria and the Placental Radiation

Within Eutheria (see p. 43), Placentalia is defined as the **crown group** which includes all extant placental mammals, their most recent common ancestor, and all extinct species that also descended from that ancestor (Rougier *et al.* 1998; Archibald, 2012). Placentalia constitutes the most taxonomically and morphologically diverse subdivision of Mammalia, comprising around 95% of all known living and fossil mammal species. The key division between most extant mammals is the marsupials and the placentals. Among some defining features of placentals, and eutherians, is the absence of the epipubic bone in the pelvic joint, and the presence of four premolars as well as three molars. In comparison to marsupials, their upper molar crowns have a much smaller outer region adjacent to the lips (stylar region), with a larger emphasis on the inner part of the crown. These and various other characteristics enable numerous fossils to be assuredly recognised as placental or marsupial. The earliest known undisputed eutherian is *Juramaia* from the Late Jurassic of China (ca. 160 Ma), exhibiting the distinctive placental teeth and other morphological characters. The second oldest known fossil eutherian is *Eomaia* (Fig. 1.7), from the Early Cretaceous of China (ca. 125 Ma), which was approximately shrew-sized and highly advanced for its time.

Eutherian evolution accelerated during the Late Cretaceous, with many complete skeletons of "**archaic**" lineages called kennalestids and asioryctids known from China and Mongolia. The earliest fossils of some of the major lineages of placentals are known from the latest Cretaceous and earliest Paleocene, such as primates (*Purgatorius* (see pp. 170–1)). Following the extinction of the non-avian dinosaurs at the end of the Cretaceous (see p. 169), placental mammals saw an unprecedented **adaptive radiation**, resulting in the evolution of almost every modern order by the Early Eocene (including many extinct ones). Among these are rodents, miscellaneous **insectivores**, true carnivorans (see pp. 81, 83–4), and numerous **ungulate** orders, along with mammals as disparate as cetaceans and bats. This has long been considered one of the largest adaptive radiations ever. Molecular phylogenies published over the past few decades have proposed major subdivisions of placental mammals that were not previously recognised from morphological evidence alone. As more studies accumulated, the molecular case for these clades became increasingly strong, and a huge majority of mammal palaeontologists now accept them as part of the standard classification. In most cases, these groupings are supported primarily by molecular data, and direct morphological support is limited or ambiguous. Some morphological characters do not align with the molecular relationships, and these discrepancies remain unresolved. Nevertheless, the molecular evidence—particularly from large-scale genomic analyses—has become robust enough that the majority of researchers accept and use these subdivisions: Xenarthra, Afrotheria, and Boreoeutheria (see pp. 9–10). Molecular clock estimates suggest that the divergences among these three lineages occurred during the Late Cretaceous.

Xenarthra: Anteaters, armadillos, and sloths form a distinctive and long-recognised **monophyletic** group. Fossil and molecular evidence indicate that xenarthrans diverged early from other placental mammals (McKenna, 1975). Molecular clock analyses (*e.g.* Romiguier *et al.* 2013; Nishihara *et al.* 2009) place their origin between roughly 100 and 110 Ma. Some phylogenomic studies suggest they may be the sister group to Afrotheria, though the exact branching order remains debated.

Afrotheria: This clade includes a variety of mammals currently living in Africa or of African evolutionary origin. The best-known members are the tethytherians—proboscideans (elephants) and sirenians (sea cows)—along with the extinct embrithopods, with hyracoids typically grouped alongside them. The term Tethytheria was originally introduced by McKenna (1975) for this subset. Molecular data place Afrotheria's split from other placentals at around 105 Ma (Springer *et al.* 2003), broadly coinciding with the Cretaceous isolation of Africa.

Boreoeutheria: This superorder is estimated to have diverged from Xenrathra and Afrotheria between roughly 90 and 105 Ma (e.g. Tarver *et al.* 2016). Within Boreoeutheria, the clades Laurasiatheria and Euarchontoglires are estimated to have split between about 87 and 91 Ma (Springer *et al.* 2003; Hallström *et al.* 2007).

Xenarthra

The oldest fossils of the superorder Xenarthra (Fig. 2.34), formerly called "edentates", date back to the Late Paleocene (ca. 60 Ma). While the word "edentate" suggests that they lack teeth, that condition only occurs in anteaters, as armadillos and sloths possess simple peg-like teeth. Xenarthra means "strange joints" in ancient Greek, referring to the vertebrae of anteaters, armadillos and sloths all exhibiting multiple spines with unusual additional joints and articulations. Xenarthrans are highly **primitive** placentals as evident from their anatomical characters, devoid of most of the specialisations present in the remainder of placentals. For instance, a uterus simplex is present in female xenarthrans, being partitioned by a septum and lacking a cervix. The metabolic rate of xenarthrans is much lower and more poorly regulated than in all other placentals. Xenarthrans also have the least advanced brains and neural development of all placentals.

Uniquely among placentals, in xenarthrans, the stapes (one of the three middle ear bones that transmits sound from the eardrum to the inner ear) have a primitive rod-like shape typical of early amniotes, lacking a hole at the base for the stapedial artery. As well as additional vertebral articulations, xenarthrans exhibit strange fusions of the pelvic region with the sacrum and tail vertebrae as well. Xenarthra, along with Afrotheria (see p. 62), is positioned basally on the placental family tree, or occasionally nearer to Afrotheria than the rest of the placentals. This **clade** comprising Xenarthra along with Afrotheria has been proposed as "Atlantogenata" (see pp. 9–10) since the afrotherians first evolved and radiated on the African side of the Atlantic while xenarthrans were confined to South America throughout much of their evolution. Their Cretaceous progenitors were some of the first mammals to evolve on South America while it was largely cut off from all the other landmasses, and as a result, the continent supported a broad range of xenarthrans throughout the Cenozoic.

Cingulata

Armadillos are among the best known of the extant groups of armoured mammals. The armour is composed of multiple bony scutes called **osteoderms** (which grow from the dermis layer of the skin) and are covered by leathery **keratin**, with several solid shields usually separated by flexible bands. Rigid sections called bucklers cover behind the head and shoulders with a second over the hips, although one notable lineage, the glyptodontines, had a single solid **carapace**. Further protection covers the tail. All species had an unprotected underbelly, but only the extant three-banded armadillo (*Tolypeutes*) species are flexible enough to roll up into a completely protected ball. Extant armadillos range in size from the pink fairy armadillo (*Chlamyphorus truncatus*), only 13 centimetres (5 inches) long, to the giant armadillo (*Priodontes maximus*), up to 150 centimetres (59 inches) long and weighing up to 54 kilograms (119 pounds). Armadillos form the order Cingulata (meaning "girdle") which is composed of around five to six families (only two of which contain extant species, the dasypodids and chlamyphorids). The most diverse armadillo family, the chlamyphorids, include both the smallest and largest known species. In the oldest armadillos, both front and back bucklers may have been absent, and the carapace was most probably formed of loosely-**articulated** osteoderms which were very mobile in relation to one another. Over time, those osteoderms became arranged into mobile bands and ultimately into the solid bucklers that define modern armadillos.

Peltephilids (*Peltephilus* (see pp. 263, 267)) are the most distinct group and are considered among the most basal of armadillos. Strangely, peltephilids were **convergent** with another extinct horned **fossorial** mammal, the rodent *Ceratogaulus* (see pp. 260–1). Peltephilids include some of the earliest cingulates from the Eocene, but are best known from the latest species during the Miocene. The most familiar of all extant armadillos are the dasypodids, or long-nosed armadillos, and this is largely due to the widespread nine-banded armadillo (*Dasypus novemcinctus*). Traditionally, all extant armadillos were incorporated within the family. However, comparisons with extinct

Fig. 2.34. Phylogenetic tree of Xenarthra. The phylogeny within Cingulata and Vermilingua follows that of Slater *et al.* (2016) and Delsuc *et al.* (2016), whereas the phylogeny within Folivora follows that of Presslee *et al.* (2019).

forms showed that based on morphological features, dasypodids were the sister group to other cingulates, apart from peltephilids. They show a full spectrum of sizes and diets from the smaller specialist **myremecophage** *Stegotherium* (see pp. 263, 267) up to the large **herbivorous** *Eutatus*. They include some of the earliest armadillos, with *Pucatherium* present during the Eocene. The pachyarmatheriids are a poorly-known family which include only two genera (*Neoglyptatelus* and *Pachyarmatherium*) ranging from the Miocene to the Pleistocene. The notable features include very thick osteoderms, a lack of interconnecting flexible bands between the shields, and a long narrow skull without teeth. In common with extant *Dasypus*, pachyarmatheriids were **insectivorous**.

Additionally, one group of giant extinct relatives of armadillos are the pampatheriids. These creatures had a shell made of bony plates, or osteoderms, with distinctive round or hexagonal shapes, and three broad bands of hinged armour plates across their back that gave them a bit of flexibility. After originating in South America, they got larger through their evolution. They can be traced from the relatively small *Vassallia* (Miocene to Pliocene of South America (see pp. 272–3)) to the medium-sized *Kraglievichia* (Pliocene to Early Pleistocene of South America and Florida) to the larger *Pampatherium* (Middle to Late Pleistocene of South America). Including some of the earliest fossils discovered in South America, pampatheriids (*Holmesina* (see pp. 318–9)) were large-bodied herbivorous armadillos. Some species had a deep lower jaw, powerful jaw musculature and **hypselodont** teeth consistent with grinding coarse vegetation in arid environments, while others fed on softer plants in more humid locations. Pampatheriids can be distinguished from other groups by the three flexible bands of osteoderms across the body. The earliest pampatheriids date from the Miocene and continue right up to the end of the Pleistocene.

One remarkable extinct subgroup of chlamyphorids are the glyptodontines, which had a large, tortoise-like dome of armour over their backs, but their osteoderms were fused into a single shield that was not flexible. They also had rings of bone covering their tails and some, such as *Doedicurus clavicaudatus* (see pp. 328–9), even had mace-like spikes on the tip for striking opponents. Some glyptodontines were the size of a SmartCar, with the largest species, *D. clavicaudatus*, measuring over two metres long and weighing two tonnes (including 400 kilograms of bony armour). The glyptodontines first evolved in the Early Miocene of South America (*Parapropalaeohoplophorus* from Chile), and many of the dozens of genera are known only from the south of Panama. A few glyptodontines (as in the pampatheriids and armadillos) headed north over the Panama Isthmus (Fig. 6.2) during the Early Pliocene, but got only as far north as Texas and Florida in the United States. They apparently preferred moist, subtropical habitats and may have been good in the water. Chemical analysis of their teeth indicates that they were **grazers** of typical **savannah** grasses.

Pilosa

The Pilosa is separated into the two lineages of non-armoured xenarthrans: the anteaters, or Vermilingua (Latin for "worm tongue"), comprising four extant species in three genera, and the sloths, or Folivora (Latin for "leaf eaters"), comprising seven extant species in two genera.

Vermilingua

Anteaters have a highly sparse fossil record. This is due, in part, to the fact that these mammals typically have low population densities and are solitary, thus restricting the amount of their remains that can ultimately be fossilised. Additionally, anteaters are limited to tropical and subtropical environments with year-round access to termites and ants. A large number of South America's palaeontological sites are situated in temperate regions which were possibly never ideal for anteaters. Lastly, termites and ants are distributed unevenly, and consuming these necessitates a multitude of anatomical **adaptations**. Therefore, any given area can usually only support one or two **myrmecophagous** animals. All anteaters have developed strong digging forelimbs, highly elongated, toothless snouts and long, sticky tongues for foraging on termite nests.

From the Early to Middle Miocene of Patagonia, *Protamandua* (see pp. 266–7) is the earliest known fossil anteater. It was possibly ancestral to both modern tamanduas (*Tamandua*) and giant anteaters (*Myrmecophaga tridactyla*) (Hirschfeld, 1978). *Neotamandua* (meaning "New Tamandua") is known from the Miocene to Pliocene of Colombia and Argentina (Fig. 2.35). The name is rather misleading as the genus is considerably older than the earliest tamandua fossils documented, which are of Pleistocene age. What's more, *Neotamandua* likely has closer phylogenetic affinities with the giant anteater than with modern tamanduas. Therefore, it should have really been called *Palaeomyrmecophaga*. *Neotamandua* was intermediate in size between modern tamanduas and giant anteaters. While its hands and feet somewhat resembled those of modern tamanduas, it lacked a **prehensile** tail. *Neotamandua* possibly preyed on insects, **terrestrially** and in trees.

Fig. 2.35. Reconstructed life appearance of the anteater *Neotamandua borealis*. Reconstructed length: 2.1 m.

Additionally, a fossil relative of the silky anteater (*Cyclopes didactylus*) is *Palaeomyrmidon* from the Middle to Late Miocene of Argentina which, unlike its extant relative, was terrestrial rather than **arboreal**. *Palaeomyrmidon* is represented by a single skull that is more elongated than that of silky anteaters and superficially resembles that of the tamanduas. *Palaeomyrmidon* is possibly a sister genus to the modern silky anteater.

Folivora

The sloths are the third great branch of the Xenarthra. Today, there are three species of the three-toed sloth *Bradypus*, and two species of the two-toed sloth *Choloepus*. These small animals are legendary for hanging upside down from trees their whole lives, moving very slowly, and eating and digesting leaves slowly between long naps. While modern sloths are relatively small **arboreal** mammals, the fossil record shows that they were previously a far more diverse group of animals, dating back to the Early Eocene. Sloths are split into eight families, consisting of 86 genera and hundreds of species of extinct ground sloths. Only two sloth families, the Bradypodidae and Choloepodidae, survive today. Interestingly, the two modern families are not even each other's closest relatives, nor on the same side of the major split within Folivora (sloth suborder). The two extant families diverged approximately 40 Ma and are not that closely related, which indicates that many of their similarities are due to **convergent evolution**. One of these families, the Choloepodidae, are part of a larger group called the Mylodontoidea and allied with two other families of ground sloth—mylodontids and scelidotheriids.

The Mylodontidae includes large to very large ground sloths from North and South America, that lived between the Early Miocene through and up until the Early Holocene (coexisting with early humans in South America). There are numerous examples of subfossil remains, dung, fur and burrows dug by mylodontids. **Osteoderms** have been found on the preserved skin on a number of species. Mylodontid sloths have noticeably broad snouts indicating a broader diet of grasses and browse.

The Scelidotheriidae includes large ground sloths from South America that lived from the Late Oligocene to the Late Pleistocene. Previously thought to be the sister group to the former family, actually they are less closely related than the extant two-toed sloths are to the mylodontids. They have a much narrower slender snout, which suggests that they were **selective feeders**. Prior to a recent revision of the various families of sloths, the megalonychids were a mixed bag from small to very large ground-dwellers across Central and South America. They include the extant two-toed sloths and also some species that had migrated further afield, populating North America and various islands in the Caribbean. However, a molecular analysis of all major groups found the megalonychids to be **paraphyletic**, with species being split off and moved to different branches of the phylogenetic tree. Despite some morphological similarities, the Caribbean sloths were found to be basal to all other sloths and placed in their own family, the Megalocnidae. The arboreal two-toed sloths were placed within a larger Mylodontoidea with other ground sloths including the mylodontids. Finally, the slimmed down megalonychids were placed in the other major **clade**, the megatheroids, notable for some giant species including the megatheriids.

Megalonychids have been found mostly in Central and South America from the Oligocene onwards. In addition a possible specimen has been found from the much earlier Eocene in Antarctica, which was then connected to South America. They also migrated into North America via swimming short distances or land bridges with numerous genera being found in Central America (e.g. *Pliometanastes*, *Xibalbaonyx*, *Zacotzontli*). The heavily built *Megalonyx* (see pp. 323, 327) is the best known and largest genus of megalonychid, surviving until the end of the Late Pleistocene as far north as Alaska and Canada. It was probably the most northerly known sloth and must have had some **adaptations** for the colder climate. Surprisingly, one megalonychid took to swimming in rivers (*Ahytherium*), a feature that it shared with several other unrelated sloths (such as *Thalassocnus* (Fig. 2.36)). *Ahytherium*, unlike the other megalonychids, was found primarily in freshwater environments. *Ahytherium* had several specialised features such as a dorsoventrally flattened tail, robust body and tall, shortened skull. Its remains, found in an underwater cave in Brazil, are dated to the latest Pleistocene.

The megatheriids are divided into three subfamilies: small schismotheriines, giant megatheriines and the aquatic thalassocnines. In truth, not all ground sloths actually stayed on the ground. Multiple small fossil species across all sloth families were arboreally adapted, although they probably were not quite as full-time tree-dwellers as modern sloths. One well-known and common genus was *Hapalops* (see pp. 263, 267) which had an extensive range across western and southern South America during the Miocene. Split between 26 species with some being named after fossil fragments, it is not surprising that the validity of all of these species is in doubt and a taxonomic review is long overdue. Unlike its larger relatives, *Hapalops* had a bone structure that closely resembled the current tree-dwelling sloths. This correlates with the idea that *Hapalops* was at least partially arboreal.

Fig. 2.36. Reconstructed life appearance of the nothrotheriid ground sloth *Thalassocnus natans*. Reconstructed length: 2.1 m.

The most impressive sloths are perhaps the megatheriines due to their massive bulk, the largest xenarthran which is *Megatherium* (see pp. 328–9). Up to six species range from the Pliocene up to the end of the last Ice Age (ca. 11.7 ka), with *M. americanum* being the largest, latest and best known species. There is plenty of evidence that *Megatherium* was hunted and butchered by humans. It lived in temperate semi-arid woodland in the southern half of South America, while the related *Eremotherium* (see pp. 318–9) was found in northern South America ranging as far north as the United States. Very large species such as proboscideans and rhinoceroses are usually sparsely haired due to the susceptibility of overheating. So it is unlikely that large megatheriid sloths had the thick furred coat inferred from hair remains in other species (I have reconstructed *Megatherium* and *Eremotherium* with a very fine coat rather than hairless).

Nothrotheriids were a small group that have previously been considered part of the megatheriids. Following the review of relationships within Folivora (all sloths), they are now considered a separate family although still closely allied to the giant species. In general they were similar to other ground sloths, although they mainly consisted of small to medium-sized species. Only the poorly-known *Lakukullus* was comparable to large species in other families. Coprolites of the small Brazilian *Nothrotherium* show that it fed on xerophytic (dry-adapted) leaves and fruit suggesting that it was at least partially arboreal. *Nothrotheriops* is the most familiar genus, better known as the Shasta ground sloth (see pp. 323, 327). The most amazing of the nothrotheriids was the marine *Thalassocnus*, with five known species. All five of those species gradually showed increasing aquatic adaptations with time, including dense heavy bones, elongated tibia bones, negative buoyancy, a wider snout, grinding teeth and increased tail vertebrae. Thalassocnines were previously grouped with the related nothrotheriids, and the similarities are probably due to some basal features of the swimming sloths. As **herbivores**, the swimming sloths were restricted to nearshore habitat, mostly walking along the seafloor to reach seagrasses. A loss of seagrass meadows due to global cooling at the end of the Pliocene no doubt caused the extinction of the thalassocnine sloths.

The last surviving family of ground sloths was the Megalocnidae. Although it may come as a surprise that sloths may be capable of rafting to the islands (as is the suspected mode of travel), sloths are very capable swimmers and living in a raft of branches for a period would be entirely plausible. All Caribbean sloths belong to the Megalocnidae, but they were previously allied to the two-toed sloths and grouped within the Megalonychidae. However, recent molecular studies have shown them to be the most basal group of sloths. A small family with only a handful of genera, the number of species recognised is currently in flux due to probable over-splitting of species. Nonetheless, multiple species were found throughout the Greater Antilles (Cuba, Hispaniola and Puerto Rico). Strangely, no species have been recorded from Jamaica. Sloths were present during the Miocene, and probably crossed the sea during the Early Oligocene when sea levels were low. Apart from a femur found in Puerto Rico from the Oligocene, currently only *Imagocnus* is known from Miocene deposits in Cuba. All other fossils date from the Pleistocene and Holocene. At 90 kilograms, *Megalocnus* is the largest of the Caribbean sloths. Based on subfossil remains, it is possible that this genus survived in the highlands in Cuba up until as recently as the sixteenth century, although actual fossilised remains are dated to 5 ka. *Neocnus* was far smaller, and had features which suggested more arboreal activity. At less than 10 kilograms, *Neocnus toupiti* was smaller than modern tree-dwelling sloths. Long after the giant sloths of the continental Americas had died out, the ground sloths of the Caribbean followed them to extinction with the arrival of the first human settlers around 5 ka.

Xenarthrans were among the few South American natives to successfully march north across the Panama Isthmus in the Plio-Pleistocene against the tide of North American mammals heading south. Ground sloths, armadillos and glyptodontines (see pp. 59–60) were all common in the Pleistocene of Southern North America.

Afrotheria

The phylogenetic affinities between the main modern lineages of African mammals (particularly elephants, "sea cows", hyraxes and aardvarks) and several minor lineages (including elephant shrews, golden moles and tenrecs) were long contentious. Nevertheless, various palaeontologists had observed similarities in morphology between Sirenia ("sea cows"), which predominantly inhabit the subtropical and tropical seas across the globe, and the Proboscidea (elephants and fossil kin). The Tethytheria, a subgroup of Afrotheria, was named since the proboscideans, sirenians and embrithopods appeared to have evolved within and near the ancient Tethys Ocean (see p. 174). A wealth of evidence from fossils and morphology support the **clade** called Tethytheria (Fig. 2.37). Several peculiar specialisations are shared between sirenians and proboscideans, particularly in the configuration of the skull bones surrounding the **orbits**, and in the middle ear bones, and both groups exhibit **bilophodont cheek teeth**. Following the formal naming of Afrotheria in 1998 (Stanhope *et al.* 1998), based exclusively on molecular evidence and DNA sequence data, the Afrotheria hypothesis has been consistently supported by subsequent molecular phylogenies and is now a typical aspect of mammal phylogeny and taxonomy.

However, some biologists remain hesitant to accept Afrotheria. In spite of thorough investigation, we have not yet discovered any clear anatomical traits shared by all afrotherians (apart from the convincing evidence of the affinities amongst tethytherians). A number of traits have been proposed, including the increased count of thoracic and lumbar vertebrae, and unusual tooth eruption patterns, although a lot of these are present in other mammal lineages too; moreover, every one of those traits are found in the most basal afrotherians, so relying on them for evidence of Afrotheria's **monophyly** is problematic. Still, multiple fossils that appear to be highly **primitive** tethytherians or afrotherians have lately been discovered from Early Paleogene strata in Africa. Among these fossils are: *Hadrogeneios*, *Ocepeia* and *Abdounodus* from the Middle to Late Paleocene of Morocco (see pp. 190–1), basal afrotherians unable to be grouped within any modern order; the basal proboscideans *Eritherium* and *Phosphatherium* (see pp. 190–1); a possible "**archaic**" elephant shrew called *Chambius*, from the Early Eocene of Tunisia; a petrosal bone from the Eocene of Tunisia that represents the oldest known African sirenian fossil, but remains unnamed (Benoit *et al.* 2013); *Stylolophus*, the oldest known embrithopod (see p. 68); and *Seggeurius*, the oldest known hyrax (see p. 70).

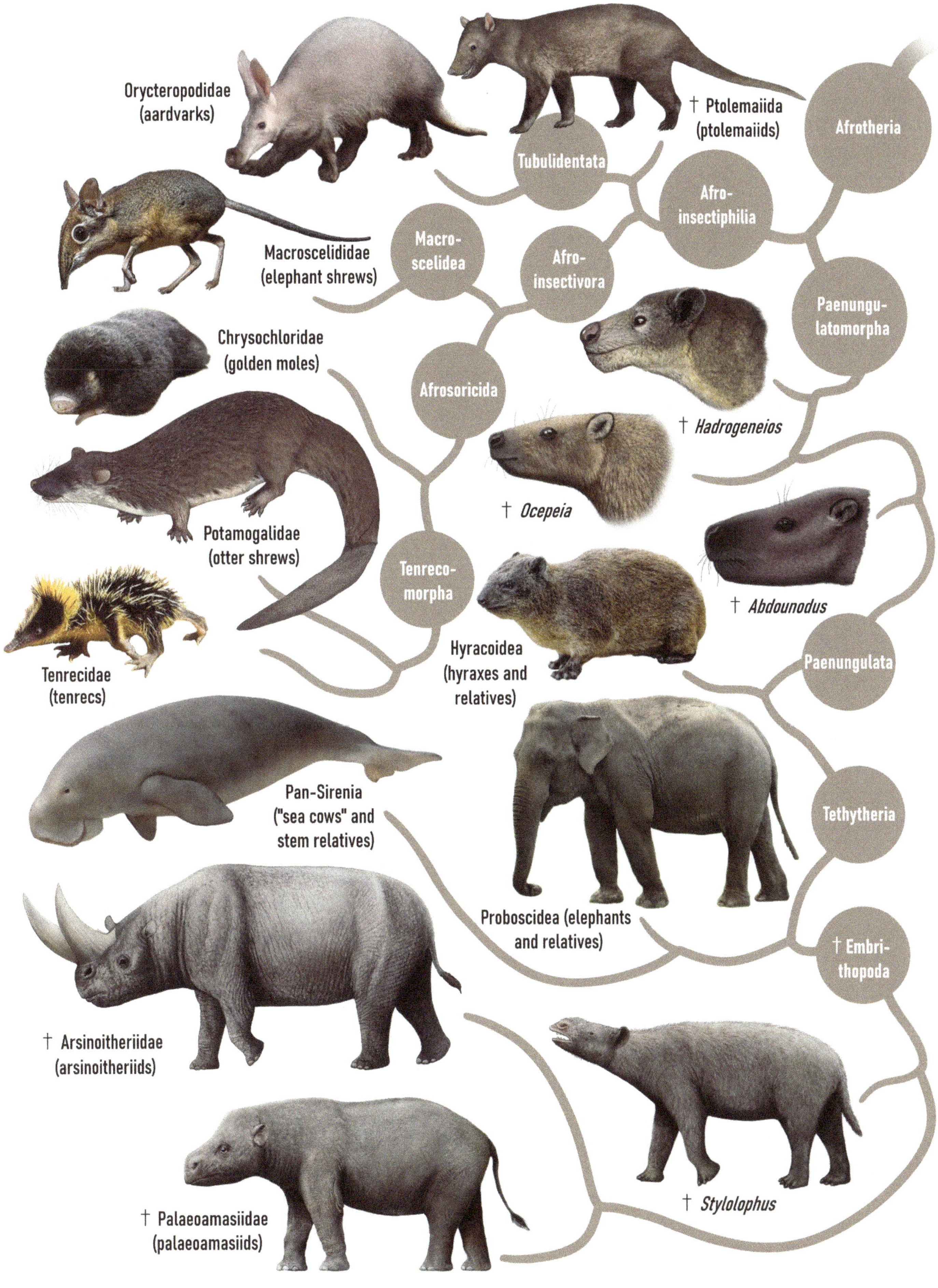

Fig. 2.37. Phylogenetic tree of Afrotheria after Tabuce *et al.* (2008). The phylogenetic position of *Hadrogeneios*, *Ocepeia* and *Abdounodus* follows Gheerbrant (2022), while that of Embrithopoda follows Gheerbrant *et al.* (2018).

The **endemic** evolution of the entire Afrotheria while Africa was an isolated continent during the Cretaceous is supported by this fossil data. Molecular clock analyses indicate that the divergence of Afrotheria into its various lineages began in the Late Cretaceous, at around 80 Ma (Springer, 2022). Throughout the Paleogene, Africa was home to a mammal fauna composed predominantly of afrotherians after which some immigrant **taxa** arrived from Eurasia. Many of the afrotherians were confined to Africa prior to the Oligo-Miocene, when proboscideans, hyraxes and several other lineages dispersed to Eurasia. The majority of afrotherians, including golden moles, elephant shrews and tenrecs, were always confined to Africa. Shortly after their emergence in the Early Eocene, the aquatic sirenians spread all the way from Africa across the Tethys Ocean and dispersed to primarily tropical seas across the globe.

Proboscidea

At present, elephants are the largest extant **terrestrial** animals on Earth, comprising one species in South and Southeast Asia, and two in sub-Saharan Africa. The **clade** they constitute, the Proboscidea, has a vast fossil record, stretching as long ago as the Late Paleocene of Africa (Fig. 2.38). They were confined to Africa for much of their evolutionary history until the Early Miocene (ca. 19 Ma), when the continent docked with Asia and resulted in the closure of the Eastern Mediterranean Sea, enabling them to disperse to Eurasia and later to the Americas. The earliest members of the Proboscidea were very small and showed less of the **derived** features that are familiar in later species. One notable feature which changed was the decrease in the number of incisors with later groups. Early forms had up to three pairs in the upper jaw and two in the lower jaw, while later species (deinotheriids and elephantiforms) only had a maximum of a single pair of tusks in either jaw. There is some argument as to whether the tusks of later groups are derived from incisors, but this seems to be the most logical conclusion. Other **adaptations** to the skull saw a shortening of the face and a retraction of the nasal opening. As proboscideans grew larger, their stance also changed to a **graviportal** one supported by columnar limbs. Early proboscideans had relatively shorter limbs and smaller feet that were more **plantigrade**.

The earliest, smallest and most basal known member of the Proboscidea was the rabbit-sized *Eritherium* (see pp. 190–1) from the Late Paleocene (60 Ma) of Morocco, described in 2009 (Gheerbrant and Tassy, 2009). It stood about 20 centimetres at the shoulder (Larramendi, 2016) and its molars were **bunodont**, with low, rounded cusps. Its teeth exhibit the four rounded cusps generally found in **primitive** tethytherians, with a subtle indication of the cross-crest pattern observed in more derived proboscideans. The **orbits** were placed forward in the cranium, a characteristic that shifted increasingly towards the rear in later proboscideans. The second oldest known proboscidean is the fox-sized *Phosphatherium*, from the latest Paleocene (ca. 56 Ma) of Morocco (see pp. 190–1). It had many proboscidean features, such as molars with better-defined cross-crests, or **lophs**, compared to *Eritherium*.

The following stage is *Daouitherium* from the Early Eocene (ca. 53 Ma) of Morocco, then the slightly later (ca. 51 Ma) *Numidotherium* from Algeria. While the known skull of *Numidotherium* is fragmentary, it already exhibits indicators of the high, vertical forehead typical of elephants. Unlike the low-slung body shape of other basal proboscideans, *Numidotherium* was more terrestrial with relatively long limbs and would have most resembled a tapir. An elongated nasal opening shows that *Numidotherium* had a short **proboscis**, while the enlarged incisors protruded forwards. The earliest large proboscideans appeared during the Late Eocene and Early Oligocene, many of which were evolutionary offshoots rather than direct ancestors. Among these were the hippopotamus-sized beast *Barytherium* (see p. 220), which turned out to be a slightly more advanced proboscidean with fully-developed cross-crests on the molars, a tall forehead and a short trunk. Unlike later proboscideans, it retained the plantigrade feet (Gheerbrant and Tassy, 2009). The most spectacular of the early large proboscideans is the Late Eocene *Moeritherium* (see p. 220), which had a barrel-shaped body and was the size and shape of a pygmy hippopotamus (*Choeropsis liberiensis*) or a tapir, and similar to those animals in height, at about 0.7 metres (2.3 feet) tall, but it was nearly 3 metres (10 feet) in length.

The proboscideans had diverged into several branches by the Early Oligocene. Among these was a family known as Deinotheriidae. Distinctive downward pointing tusks have made deinotheriids a familiar, if not unusual, small branch within Proboscidea. They were not closely related to modern elephants, but to barytheriids, splitting off early on. Only three genera are known and all are very similar in appearance, apart from size, sharing similar features distinctive of the family. Deinotheriids had elongated limbs suggestive of an efficient walking gait, and proportionally small heads. The tusks were probably used in pulling branches and bark from trees. The oldest known deinotheriid is *Chilgatherium* from Ethiopia, which is only known from several molars, but is estimated to have been fairly small, about as tall as a human (Larramendi, 2016). By the Early Miocene, deinotheriids had grown to the size of a small elephant, all belonging to the genus *Prodeinotherium*, and had spread to Eurasia (Larramendi, 2016). During the Middle Miocene, *Prodeinotherium* (see pp. 275–6) was gradually replaced by the much larger *Deinotherium* (see pp. 275–6, 301) in both Eurasia and Africa, which persisted into the earliest Pleistocene (Göhlich, 1999). The family went extinct by this time possibly due to climate change, after which they were replaced by their distant cousins, the mammoths and straight-tusked elephants, as the largest terrestrial mammals.

Two early elephantiforms, *Phiomia* and *Palaeomastodon* (see pp. 240–1), were slightly younger proboscideans than *Moeritherium*, living during the Early Oligocene in what is now Egypt. It was this lineage that would go on to produce "gomphotheriids", mammutids (mastodonts) and elephantids (elephants). The subsequent stage in the evolution of proboscideans was the divergence of Mammutidae from the remainder of the Proboscidea. Elephant teeth are pretty interesting. Aside from the pair of tusks, true elephants only have one molar per side of each jaw at any time. These molars are like solid bricks, flattened on top with narrow ridges. When the jaws slide forwards and backwards, the ridges efficiently grind abrasive vegetation such as grass. Luckily for the elephant, they have six pairs of the molars to last them through a lifetime. A very different type of tooth was present in the distantly related mammutids. The teeth were zygolophodont (the **occlusal** surface having deep cone-shaped cusps aranged in pairs and connected by ridges) an adaptation for crushing tough branches and leaves. Such is the distinct shape of the cusps that the name mastodon means "breast tooth". Mammutid skeletons

Fig. 2.38. Phylogenetic tree of Proboscidea after Benoit *et al.* (2022).

greatly differ from modern elephants, having shorter limbs, longer bodies with a heavy torso and a long, low skull. They form the most basal lineage within Elephantimorpha, which also includes the "gomphotheriids", amebelodontids, stegodontids and true elephants.

The earliest mammutids, such as *Losodokodon* and *Eozygodon*, were modest in size and found in East Africa during the Oligocene and Early Miocene. Notably, they had a longer lower jaw with a small pair of tusks. Later genera had relatively shorter skulls and some lost the lower tusks, with genera increasing in size. By the Middle Miocene, the mammutids had evolved into *Zygolophodon*, which had spread beyond Africa by about 19 Ma, and by 18 Ma, had spread across Eurasia and had even crossed the Bering Land Bridge (see p. 174) to North America. Never diverse, only several genera of mammutids are known, with most species being split between *Zygolophodon* and *Mammut*. The best known mammutid is the American mastodon (*Mammut americanum*), the last member of the family which lived across North America until the end of the last Ice Age (ca. 11.7 ka). Although usually depicted as a large, brown, shaggy-haired beast, much like the unrelated woolly mammoths, it would seem that this image is baseless. While some guard hairs are known, there is no good evidence of preserved soft tissue.

Some of the confusion is probably due to the similarity between the names *Mammut* and "mammoth". In a strange twist, some of the fossil "hair" discovered with *Mammut* bones has proven to be a filamentous algae! The long upper tusks were strongly curved, while the lower jaw had vestigial tusks in some individuals. The large tusks were evidently used in aggressive sparring by males as fossilised damage shows annual irregularities to the **dentin** growth patterns. Like other members of the family, species of *Mammut* were proboscideans of closed forest environments, avoiding the open plains. Fossilised gut contents reveal they were likely **selective feeders**, eating low herbaceous vegetation, bark and twigs. During the Pliocene, several very large mammutids have been found with remarkably large tusks. One of the largest known proboscideans was the Eurasian Borson's mastodon (see p. 308). Mammutids failed to disperse into South America (while other groups were able to) and it is believed that crossing grassland was the main barrier. As woodland species, an absence of **browsing** species would have left the mammutids without a suitable food supply.

The main diversification of Proboscidea can be traced from *Phiomia* to the "gomphotheriids" of the Late Oligocene to Early Holocene. "Gomphotheriids" were a very widespread and long-lived **paraphyletic** group of proboscideans. Among later elephantiforms, the "gomphotheriids" would have resembled modest-sized elephants with a barrel-shaped body and shorter limbs. However, it is the shape of the skull which is the most interesting. Early genera from the Miocene include *Gomphotherium* (see pp. 275–6). The unusual *Gnathabelodon* retained a long lower jaw but had lost the lower tusks. Later genera were **convergent** with true elephants in having a shorter skull, and had either absent or vestigial lower tusks. "Gomphotheriids" have traditionally included a much wider range of species, but three aberrant groups have since been separated off. The shovel-tuskers, or amebelodontids, are now considered the sister group to "gomphotheriids". More distantly related, the choerolophodontids are now considered a basal group within Elephantimorpha. Although very similar, the basal elephantoids *Anancus* (see p. 308) and *Tetralophodon* (see pp. 278, 280) had teeth more like those of true elephants. "Gomphotheriids" had three cusps on their teeth as opposed to four seen in elephantids, *Anancus* and *Tetralophodon*. There is a popular misconception that Miocene "gomphotheriids" were found only in swampy environments, using their lower tusks to scoop up aquatic vegetation (an argument also levelled at amebelodontids). However, fossils have been found largely in wooded environments and, based on their high-ridged teeth, "gomphotheriids" were **mixed feeders**. Later "gomphotheriids" from the Pleistocene such as *Cuvieronius* (see pp. 329–31) and *Notiomastodon* adapted to cooler environments and changed to a more specialist diet including **grazing**. The "gomphotheriids" of South America were the last remnants of this group, and this was probably due to lack of competition from mammoths and mammutids which failed to migrate past the Panama Isthmus. There is evidence in Mexico dating back more than 13 ka of early Clovis culture humans regularly hunting "gomphotheriids", while a recently discovered *Notiomastodon* skull in Brazil was complete with a human artefact, raising the possibility that later humans also interacted with "gomphotheriids" in the Americas.

The most distinctive proboscidean group is perhaps the amebelodontids, the so-called shovel-tuskers. The amebelodontids possessed a pair of flat tusks in the lower jaw in addition to the upper pair. The purpose of the lower tusks has caused much confusion. Channelling the old view of sauropod dinosaurs, shovel-tuskers were historically believed to be swamp dwellers, wading through shallow water to scoop up aquatic vegetation. This has since proven to be untrue. Wear patterns including the bevelled edge on the teeth reveal microscopic damage caused by rough vegetation including tree bark. A short, broad fleshy trunk is also an erroneous idea. Amebelodontids used a long, flexible trunk, much like other proboscideans, in combination with both sets of tusks to efficiently strip trees and uproot vegetation. Different genera varied in the shape of the lower jaw and tusks. Some had narrow, elongated jaws with scoop-shaped tusks (*Amebelodon*), while others had short, broad jaws with square tusks (*Platybelodon*). The giant *Konobelodon* had especially long, diverging lower tusks in addition to the long upper pair. As always, one genus was an exception to the rule, with *Aphanobelodon* being tuskless in the upper jaw. How this affected the browsing habits of these proboscideans is not known, although female elephants regularly cope without tusks. Amongst male elephantiforms, the feature was unique, and convergent with the distantly related deinotheriids. Certain genera (e.g. *Platybelodon*, *Aphanobelodon*) are known from multiple specimens in different age groups, giving us a good picture of growth patterns.

The last major branch of Proboscidea to emerge was the Elephantoidea, consisting of the Elephantidae, which in turn comprises the extinct mammoths and the extant elephants. In order to develop the massive pair of upper tusks and have the lower tusks lost completely, elephantids underwent a new evolutionary trajectory, which involved a raising of the **sagittal crest**, and shortening of the facial region and lower jaw. Simultaneously, their molars evolved a series of densely-folded dentin plates and enamel which constitute a massive grinding surface suited for processing coarse plant material including grasses. Their short facial regions and jaws led to a unique debtal arrangement where just one or two molars were present on both sides of the upper jaw and lower jaw at any given time. Unlike the majority of mammals, which have baby teeth that are replaced by adult teeth emerging from below, the teeth of proboscideans emerge from the rear of the jaws. As the molars are worn, they are thrusted forward in the jaws until the earliest worn molars are forced out the **anterior** portion of the jaw. This is called horizontal tooth replacement, and is present in sirenians (see pp. 69–71) as well. Basal elephantoids including *Tetralophodon* and *Anancus* are characterised by flattened, elongated skulls, and upper tusks that were straight, elongated and approached 3 metres (10 feet) long. Stegodontids were a successful family (sister to the Elephantidae) found across Africa and Asia from the Miocene until the latest Pleistocene. While some species such as the Chinese *Stegodon zdanskyi* were some of the largest known proboscideans, others were much smaller. Across islands in Japan, Taiwan, the Philippines and Indonesia (see pp. 336–7), stegodontids have been found showing a propensity for smaller individuals due to **insular dwarfism**. Least familiar of the Southeast Asian dwarf proboscideans was *Sinomastodon*. Despite the name, it was an Asian member of the brevirostrine (short-jawed) "gomphotheriids" and would have resembled a shorter-limbed elephant. During the Early Pleistocene, the Sunda Islands was flooded and then isolated. One species of *Sinomastodon* living on Java was less than half the size of the mainland species.

Stegotetrabelodon (see pp. 285–7) represented the next evolutionary step, which possibly evolved from a *Tetralophodon*-like ancestor. The modern family Elephantidae is the latest evolutionary stage. They can be traced back to *Stegotetrabelodon*, the earliest and most primitive known elephantid, and then to *Primelephas*, a Late Miocene genus known from Chad and East Africa which, as in "gomphotheriids", retained short lower and upper tusks. This lineage evolved into *Mammuthus*, the mammoths, which arose in the Late Miocene of Africa before dispersing northwards and spreading across the Northern Hemisphere; *Loxodonta* (which includes the African elephant) during the Late Miocene; *Elephas* (which includes the Asian elephant), first appearing in the latest Miocene or Early Pliocene of Africa; and *Palaeoloxodon*, the straight-tusked elephants, which appeared in the Middle Pliocene of Africa before diversifying into an array of Eurasian species (all now extinct), including various island dwarf species (Lister, 2004). In Africa, the extant genus *Loxodonta* had replaced *Palaeoloxodon* by around 500 ka. All four genera developed comparable feeding morphologies through **parallel evolution** to enable specialised grazing. The molars of basal elephantoids evolved an array of small cusps that became merged into a sequence of parallel, **hypsodont** ridges in those of elephantids (Agustí and Antón, 2002); the gaps between the ridges were then infilled with dental cement like other grazers.

Mammoth molars gradually increased in size and complex, featuring an increase in the number of enamel and dentin **folds**. The earliest mammoth species, *Mammuthus subplanifrons*, gave way to numerous subsequent species such as steppe mammoths (*Mammuthus trogontherii* (see pp. 338, 341)) that dispersed north, covering much of Eurasia before spreading to the Americas during the Middle Pleistocene. The massive Columbian mammoth (*Mammuthus columbi*) (see pp. 323, 326–7) represents the culmination of mammoth evolution in North America. The margins of the Pleistocene ice sheets were populated by the slightly smaller woolly mammoths (see pp. 342–3), which originated in East Asia by around 400 ka (Lister *et al.* 2005), possibly from steppe mammoths. Woolly mammoths are represented by many mummified specimens frozen in the Arctic permafrost of Alaska and Siberia, which reveal not only how they appeared in life, but also details of their diet, which included a variety of forbs (herbaceous **angiosperms**) in addition to grasses.

Lastly, at the end of the last Ice Age, the entire continental population of mammoths had died out possibly due to a warming world brought about by climatic fluctuations as the vast steppe they called home disappeared. Yet, an isolated population of dwarfed woolly mammoths persisted on Wrangel Island off the coast of Siberia, largely separated from the mainland around 12 ka and surviving up until over 4 ka. The Wrangel Island population was not the only isolated "dwarfed" population, with another off the coast of Alaska on St Paul Island, dying out around 5.7 ka. Causes of extinction differed, with the St Paul's mammoths having a lack of fresh water, while the Wrangel mammoths succumbed to a combination of extreme weather events, fertility issues and possibly the arrival of humans. Island-dwelling woolly mammoths were not the only miniature mammoths, with a (sheep-sized) true dwarf present on the Northern Channel Islands off California during the Late Pleistocene, called *Mammuthus exilis* (Fig. 2.39). Those island dwarfs were descended from the massive four metre tall Columbian mammoth which made the six kilometre sea crossing. Rising sea levels and reduced living space encouraged a rapid decrease in the overall size of the mammoths. There is evidence for hunting by humans, which along with wildfires and habitat stress, were the likely causes of its extinction.

Fig. 2.39. Reconstructed life appearance of the Channel Islands dwarf mammoth (*Mammuthus exilis*). Reconstructed shoulder height: 1.72 m.

A range of proboscideans have also inhabited Japan in the past, including woolly mammoths during glacial periods of the Late Pleistocene. Alternating during warmer interglacials was a relatively small species (two metre shoulder height) of the straight-tusked elephant *Palaeoloxodon* which inhabited the temperate forests (*P. naumanni*). That species was descended from the massive *P. namadicus* (Fig. 2.40A), probably the largest terrestrial mammal that ever lived, with an estimated shoulder height of 4.35 metres and a weight of 22 metric tonnes. Humans coexisted with those elephants from about 40 ka. Despite being regularly hunted, it is thought that a return to colder conditions sealed their fate. During glacial periods, sea levels drop as available water is locked up in vast ice sheets, while during warmer interglacials sea levels rise. Thus, during these warmer periods, islands were created when land bridges were submerged. The Mediterranean Sea is dotted with islands which have fluctuated between being connected and split from the mainland on and off throughout the Pleistocene. Fossils of several small proboscideans have been located on a number of these islands. Only two elephant genera are known from Pleistocene mainland Europe, and all insular species derive from either the European mainland straight-tusked elephant (*Palaeoloxodon antiquus*) or a mammoth. The ensuing stage was probably *Mammuthus meridionalis* (see p. 338).

Fig. 2.40. Life reconstructions of extinct giant and dwarf elephants. (A) *Palaeoloxodon namadicus*; (B) *Palaeoloxodon falconeri* (encountering the giant swan *Cygnus falconeri*); (C) *Palaeoloxodon cypriotes*; (D) *Mammuthus creticus*.

Most dwarf elephants are generally believed to have originated from the giant *Palaeoloxodon*. There are at least eight named species and possibly an additional seven more forms that were found on the Mediterranean islands. Due to the cyclic nature of the islands forming and reconnecting with the mainland, some of the larger islands were colonised by different forms. Sicily has evidence of three separate colonisations of elephants. Some islands exhibit the most extreme insular dwarfing seen in proboscideans. The best known example is *P. falconeri* (Fig. 2.40B) from Sicily and Malta, possibly the smallest known elephant. While males reached just short of one metre at the shoulder, the tuskless females were even more diminutive standing at a paltry 80 centimetres. They were roughly comparable to modern baby elephants in size, but not in shape. Sometimes referred to as dachshund-like, the tiny elephant had a short, rounded skull, long torso and short limbs. Even with the larger species, size reduction varied on parts of the body. Weight reduction in *P. cypriotes* (Fig. 2.40C) was down to just two per cent of the 10 tonne mainland species, but they retained molars 40% of the size of its ancestor.

There is no obvious reason for the size of any dwarf elephant species, with the smallest and largest species being found on the same islands at different times, suggesting that surface area of islands is not a simple link. One possible suggestion was that sea levels were higher at different times, creating smaller palaeo-islands and encouraging smaller elephants to adapt. A species on the island of Tilos has been dated to around 4,000 BCE (Holocene) and was the last elephant to survive in Europe. Two mini-mammoths are also known from Mediterranean islands. However, they were originally thought to be yet more island forms of *Palaeoloxodon*. A reevaluation of the fossil material, particularly the teeth, shows that they were members of *Mammuthus*. The earlier Cretan species was less than half the size of the mammoth found on Sardinia, and comparable to the smallest *Palaeoloxodon*. There is some uncertainty of the age of *M. creticus* (Fig. 2.40D), and it is possible that it descended from an earlier *Mammuthus* species.

Embrithopoda

Whereas some embrithopods such as *Arsinoitherium* (see pp. 241–2) were superficially similar to rhinoceroses, not every embrithopod, particularly the **primitive** forms, were horned. Despite their appearance, embrithopods are considered close relatives of proboscideans (see pp. 64–8) rather than perissodactyls (see p. 126). Some defining features of the embrithopods include a **compete dentition**, a closed tooth row with no **diastema**, and highly **plantigrade** feet (more so than in proboscideans). The oldest known embrithopod fossils are from the earliest Eocene of Morocco, supporting an African origin for that mammal order and suggesting that it diverged earlier than previously thought (Gheerbrant *et al.* 2018). These belong to *Stylolophus*, a sheep-sized animal larger than the other contemporary mammals. Special skull characters suggest that it had a well-developed nasal sinus system, and the snout was likely hornless. The dentition of *Stylolophus* shows that embrithopods were by the Early Eocene already highly specialised and quite distinctive with respect to other afrotherians. Its **cheek teeth** were also much less **hypsodont** than later embrithopods including the significantly larger *Arsinoitherium*.

Some Early to Middle Eocene fossils from Turkey named *Palaeoamasia* and *Hypsamasia*, and of the dentition of a peculiar Late Eocene fossil from Romania called *Crivadiatherium*, induced a study by McKenna and Manning (1977) proposing that all of these fossils represented embrithopods and that they are morphologically similar to basal proboscideans. All three genera belong to the family Palaeoamasiidae, the skull material which suggests that they too did not develop horns. Court (1990) thoroughly examined the skull, inner ear structure, and brain cavity of those genera and discovered additional proof that embrithopods belong to Tethytheria, being distant relatives of sirenians and proboscideans. It is now widely acknowledged that the embrithopods were an extra lineage of tethytherians which, throughout the Eocene, inhabited primarily Africa and the Middle East. Embrithopods increased in size over time, with *Stylolophus* possibly weighing 20 to 128 kilograms, the more derived *Palaeoamasia* weighing around 275 kilograms, and the youngest member, *Arsinoitherium*, reaching 1.5 to 2.4 tonnes. With the extinction of *Arsinoitherium* from the rainforests of Africa and the Arabian Peninsula by the Early Oligocene, the whole of Embrithopoda became extinct too, possibly because of drastic climatic changes which led to the loss of their preferred forest habitat via cooling and drying, and in turn, their food sources.

Desmostylia

Ungulates that lived mostly at sea? With limbs that do not end in fins? Forward-pointing incisors/canine teeth and column-like molars? What sort of weird "made-up" mammals are these? Hippopotamuses, elephants, "sea cows" and walruses all show some similarities in either behaviour or physiology. Fossils of desmostylians have been found only in coastal sites around the North Pacific and are restricted to the Oligocene and Miocene (Fig. 2.41). With nostrils and **orbits** placed high on the skull, there is no doubt that desmostylians spent a lot of time in the water. They were **herbivorous**, using the front teeth for digging up and grabbing plant material, but uniquely among marine herbivores, they suctioned up food with strong throat muscles (walruses similarly suction-feed on shellfish). Unlike other marine mammals such as cetaceans (see pp. 121–6), pinnipeds (see pp. 96–8) and sirenians, desmostylians retained limbs with proportions that suggest regular **terrestrial** locomotion. However, some studies indicate that they spent the majority of their lives in water. *Desmostylus* was an active swimmer which preferred shallow waters, while other genera were slow bottom 'walkers' which would venture further out into deeper waters. Swimming locomotion was by forelimb propulsion in a similar manner to polar bears (*Ursus maritimus*).

So what exactly were desmostylians? The easy answer is that nobody is too sure. They have been linked to Afrotheria due to skeletal similarities with "sea cows" and elephants. However, all desmostylian fossils have been found as far away from Africa as it is possible, forming doubts that this is a valid grouping. Desmostylians also have similarities to the extinct anthracobunids (see p. 102) and have been linked to a wider perissodactyl grouping (see p. 126). Whether they are closely related or not, the extinction of desmostylians in the North Pacific left the door open for sirenians to become the dominant herbivores of the seas. Although this was not to last either.

Pan-Sirenia

Sirenian evolution is distinguished by the emergence of numerous features present in all living and fossil species, such as gently downturned snouts, high, retracted nasal openings and dense, heavy ribs providing ballast, preventing the animal from floating too high in the water. Pan-Sirenia is a broad term that encompasses both the ancient **stem**-sirenians and **crown** Sirenia, the latter specifically referring to the lineage leading to the extant species of dugongs and manatees (Fig. 2.42). Recent biogeographic and phylogenetic analysis suggests that Pan-Sirenia arose in the latest Paleocene of the Tethys Ocean region (see p. 174), and that the earliest evolution of stem-sirenians was concentrated there until the latest Eocene, with only one (*Eosiren imenti* (see p. 240)) persisting into the Oligocene. The earliest stem-sirenians lived in estuarine and nearshore marine environments (Domning, 2001).

Fig. 2.41. Reconstructed life appearance of some representative desmostylians. (A) *Palaeoparadoxia tabatai*, from the Late Oligocene to Late Miocene of Japan and the Western United States; (B) *Desmostylus hesperus*, from the Middle to Late Miocene of Japan, eastern Russia, Mexico and the Western United States. Scale bar: 1 m.

During the Early Eocene (ca. 50 Ma), a trans-Atlantic immigration from the Tethys Ocean to the Caribbean region gave rise to the prorastomids of Jamaica (such as the sheep-sized, **quadrupedal** genera *Prorastomus* and *Pezosiren*), but it seems that prorastomids originated in the Tethys before this dispersal. Prorastomids lacked a large, muscular tail for propulsion in the water, and had a skeletal anatomy indicative of spinal undulations as a primary method of propulsion in the water, akin to otters. Thus, prorastomids likely had a hippopotamus-like **semi-aquatic** lifestyle, transitioning to aquatic life with a different propulsion method than later sirenians. A second immigration from the Tethys probably gave rise to the fossil record of "sea cows" in eastern North America during the Middle to Late Eocene. No fossil evidence exists that any of those stem-sirenian lineages persisted in the Western Hemisphere after the Eocene. The protosirenids, such as *Protosiren*, are believed to have been the first sirenians to have fed on seagrasses, and the fairly weak sacroiliac joints indicate that those animals were unable to properly lift their bodies on land. Thus, *Protosiren* is regarded as mainly or fully aquatic (Zalmout and Gingerich, 2012). It may have propelled itself in the water by **caudal** undulation, paddling with its hindlimbs to move along the seabed of nearshore marine environments when foraging. *Protosiren* appears to have had a nearly worldwide distribution as evident from specimens found in the United States, Europe and India.

Subsequent to the likely extinction of prorastomids and protosirenids at the end of the Eocene, the region including northern European latitudes is thought to have been the birthplace of a key trans-Atlantic dispersal of sirenians, towards the Caribbean and the northern-surrounding West Atlantic during the latest Eocene, giving rise to crown Sirenia. The stem-sirenians of the Eastern Hemisphere became extinct before the latest Oligocene due to the drastic climatic changes of the Eocene-Oligocene boundary. The origin of crown Sirenia in the Western Hemisphere during that boundary led to the divergence of Dugongidae and Trichechidae. The oldest known trichechids come from the Early Miocene of Peru, implying that trichechids persisted as a "ghost lineage" throughout the Oligocene. Trichechids had dispersed to Western Amazonia, specifically within the Pebas Lake and associated wetlands, by the Early Miocene. This dispersal is linked to the formation of the Amazon Basin and its connection to the Caribbean Sea, which was facilitated by higher sea levels. Manatee evolution occurred primarily there up until the Pliocene. The forming of the Amazon drainage into the South Atlantic during the Pliocene probably enabled crown trichechids to leave South America. By then, the West Indian manatee (*Trichechus manatus*) had arrived in the Caribbean, while the dugongids there went extinct.

Shortly after the divergence of crown Sirenia, dugongids underwent a swift expansion of diversity, reaching its zenith during the Early Oligocene. The trans-Atlantic dispersal of dugongids to the Eastern Hemisphere likely occurred in the Late Oligocene and potentially the earliest Miocene, the first of which entered the Tethys Ocean before spreading to the Indian Ocean—giving rise to the South Asian genus *Rytiodus*. *Rytiodus* was well-adapted to the seagrass meadows, using elongated tusks to rip out the plants. Two early Miocene dugongid lineages, one leading to the extinct Hydrodamaliinae and the other to the modern Dugonginae, both originated from the same West Atlantic region and immigrated to the Pacific Ocean. The diversity of hydrodamaliines peaked near the close of the Early Miocene, with *Dusisiren* being a prominent member, and this lineage would persist in the Pacific with the Steller's sea cow, surviving until the eighteenth century. The Steller's sea cow (*Hydrodamalis gigas*) grew to a length of 8 to 9 metres (26 to 30 feet) long and weighed about 8 to 10 tonnes. The Steller's sea cow appears to have been an unfortunate bystander to a wider environmental issue. Among the sea life around the Commander Islands in the North Pacific are sea otters (*Enhydra lutris*), which were highly prized for their fur. The otters are a keystone species which prey on the sea urchins which also eat the kelp. Removal of the otters caused a catastrophic ripple effect. An explosion in urchin numbers decimated the kelp beds, thus eliminating the main food item of the Steller's sea cows. A warming of the seas during the Mediaeval Warm Period (see p. 345) likely put stress on the kelp beds. During its discovery in 1742, the entire global population of Steller's sea cows of less than 1,500 individuals lived along the shallow shorelines and in the mouths of freshwater rivers of the Commander Islands. It was completely docile and unafraid of humans, since no Europeans had ever hunted it, and the limited population was easily slaughtered for meat by Russian fur trappers. By 1768, just 27 years after its discovery, the largest of all known sirenians sadly became extinct. Ultimately, among dugongids, only the modern dugong survived the fluctuating sea temperatures, finding suitable seagrass habitat throughout the tropical regions of the Indo-Pacific. The tusks of modern-day dugongs may have originally evolved for digging, but they are now primarily used for social interaction.

Hyracoidea

The hyraxes are small, furry animals from Africa and the Middle East that somewhat resemble a woodchuck, until you look more closely at their anatomy or their behaviour. Despite their woodchuck-like appearance, hyraxes do not behave like rodents. Many live in rocky outcrops in the African grasslands and in the Middle East, where they can hide in crevices and caves from predators; one genus of hyrax, *Dendrohyrax*, lives in trees. Their feet have soft pads so that the animals can move very quickly in the rocks and even cling to steep surfaces as they climb. They feed mostly on grass, with some leaves and shoots, but they do not have the chisel-like incisors for cropping as in most **grazers**, so they must nip off vegetation with the side of the mouth.

There are only five extant species of hyrax, which all look pretty similar. However, their fossil record reveals a far richer history dating back to the Eocene. The order Hyracoidea first appears in the fossil record at a site in Algeria in the form of *Seggeurius*, from the Early Eocene (ca. 50–56 Ma). During the Paleogene, hyracoids were the most diverse **terrestrial** '**ungulates**' in Africa, filling many of the niches in the absence of the artiodactyls (see p. 104) and ranging in size from small rabbit-sized species through to giants. Extant hyraxes share several features in common with the other paenungulates, elephants and "sea cows": males have internal genitalia, females have teats near their armpits (although hyraxes have additional ones near the groin), flattened nails and tusk-like incisors. Ancestors of all three groups probably resembled each other closely. Hyracoids are generally split into four valid families, with all modern species being members of the Procaviidae. The vast majority of fossil species are from four other extinct lineages (the **polyphyletic** "Saghatheriidae", the Geniohyidae, the Pliohyracidae, and the Titanohyracidae) (Fig. 2.43).

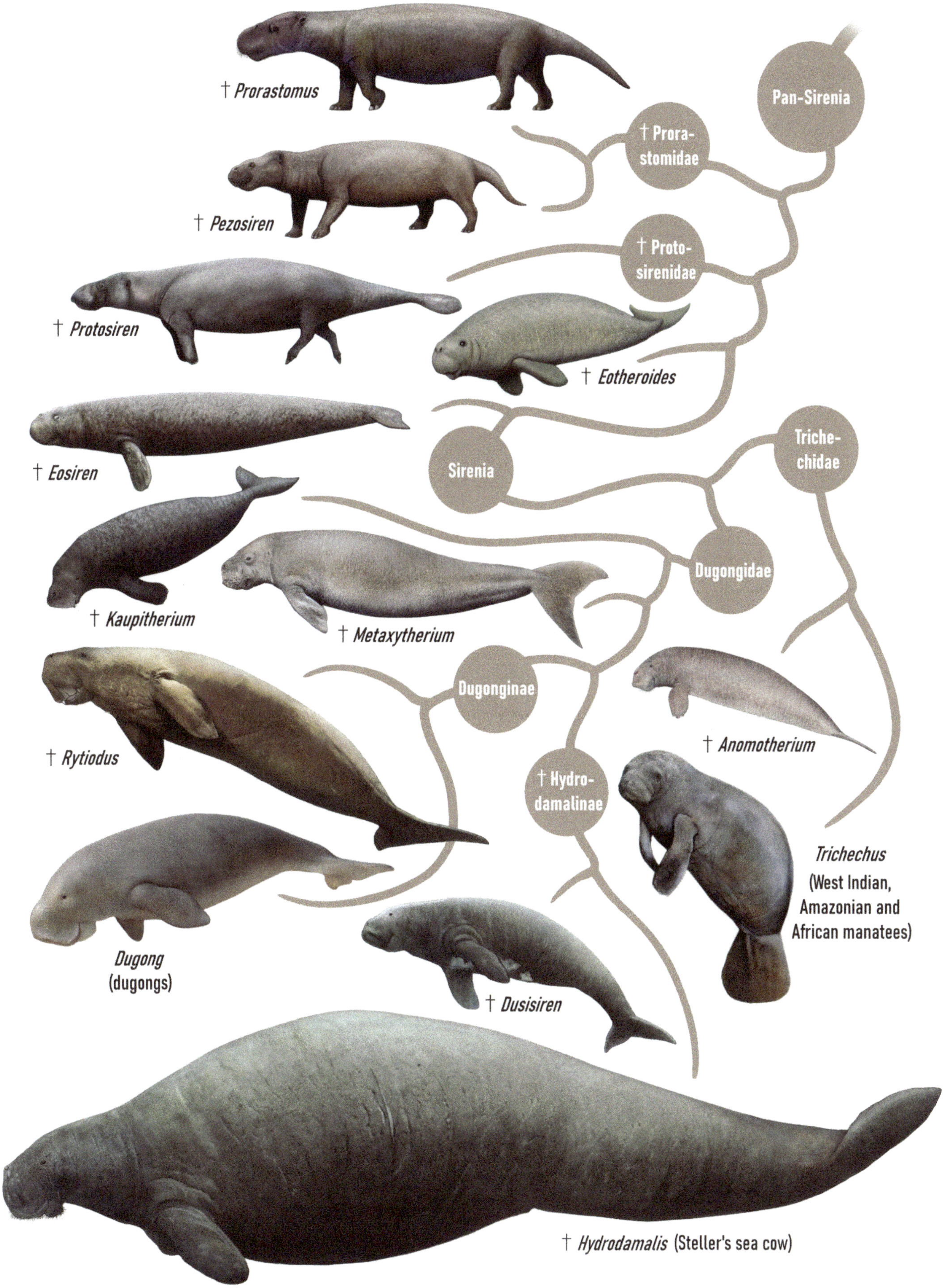

Fig. 2.42. Phylogenetic tree of Pan-Sirenia after Kerber *et al.* (2021), and Heritage and Seiffert (2022).

The earliest hyracoids were diminutive species, such as the three kilogram *Microhyrax* from the Early to Middle Eocene of Algeria, but rapidly diversified during the Late Eocene, with a variety of heavily built, **semi-aquatic**, cursorial and agile tree-climbing forms. Many of the differences between the earliest forms of geniohyids, "saghatheriids", pliohyracids and titanohyracids are down to slight differences in tooth shape. Much like miniature pigs, *Geniohyus* (see pp. 241–2) had a generalist dentition for crushing hard fruit and tubers. The "saghatheriids" were the most diverse of the early hyracoids, ranging in size from the smallest species of *Thyrohyrax* (*T. tantulus*, at about 23 kilograms) up to the tapir-sized genus *Megalohyrax* (see pp. 241–2). *Saghatherium* (see pp. 241–2) is one of the best known fossil hyracoids, with complete skeletons known from Libya. Toes and limbs were adapted to **cursorial** behaviour, a **primitive** trait for all early hyracoids.

Taking the cursorial **adaptation** a step further was the distinctive titanohyracid *Antilohyrax* (see pp. 240–2). Teeth shape varied as much as one might see among the hooved mammals in modern Africa. Hyracoid species varied between **omnivorous**, **bunodont** teeth, gracile, ruminant-like **selenodont** teeth, and **hypsodont**, **lophodont** grazers. However, hyracoids never developed the radial enamel found in large hypsodont perissodactyls (see p. 126) and artiodactyls. As a result, during the Miocene, the rise of efficient grazers and **browsers** (specifically the bovids (see pp. 116, 118–9)) displaced hyracoids into specialised niches. The pliohyracids had teeth and foot bones in common with procaviids. Some pliohyracids, such as *Pliohyrax*, spread into Southern Europe during this epoch. The environment in Southern Europe, through to East and Southern Africa consisted largely of dry woodland and riverine forests. Diversity within Pliohyracidae remained high

Fig. 2.43. Phylogenetic tree of Hyracoidea after Tabuce *et al.* (2017).

up until the end of the Pliocene. One unusual genus from Eastern Europe was the robust *Kvabebihyrax*. Thought to be semi-aquatic due to the high-placed small eyes, *Kvabebihyrax* also had short nasals suggestive of a mobile, short trunk. Despite being largely displaced by bovids, the hyracoids as a whole remained widespread and diverse as late as the end of the Pliocene, with representatives throughout most of Africa and Eurasia. Climbing behaviour is a later specialisation seen only in procaviids. The procaviids were mostly small- to medium-sized species. Despite the name, the Pliocene *Gigantohyrax* was only modestly larger than the extant *Procavia* species, being about three times larger than the modern rock hyrax (*P. capensis*). Why only the procaviids survive today is likely due to a combination of factors, including their ecological adaptability and ability to thrive in diverse environments.

Ptolemaiida

One enigmatic extinct lineage of possible afrotherians is the Ptolemaiida, known from the latest Eocene to Miocene of Africa. The origins and classification of ptolemaiidans have long been debated. The type genus, *Ptolemaia* (see pp. 241–2), was first interpreted as a primate. Later discoveries of elongate skulls with prominent canine teeth led to their reinterpretation as hyaenodonts (see pp. 82–3), or alternatively as large **carnivorous** relatives of the extinct pantolestans (see pp. 77–8). More recent evidence, however, suggests that Ptolemaiida belonged within Afrotheria, being restricted to Africa and exhibiting cranial features shared with aardvarks. It remains unclear whether ptolemaiidans represent the sister group to Tubulidentata or a **paraphyletic grade** leading to them. Nonetheless, their close phylogenetic affinities indicate that ptolemaiidans are best placed within Afroinsectiphilia.

Tubulidentata

The aardvark (*Orycteropus afer*) is a pig-sized **fossorial** mammal presently native to Sub-Saharan Africa. Aardvark translates as "Earth Pig" in Afrikaans. It is largely **nocturnal** and has powerful forelimbs with shovel-like claws used for tearing open termite mounds and excavating burrows for shelter. *O. afer* is the only surviving species of the order Tubulidentata, meaning "tubule-toothed" in Latin, after the rounded, peg-like molars consisting exclusively of small tubules of **dentin** compressed together, lacking enamel. Aardvarks have a rather scant fossil record. The earliest undisputed fossil aardvark is *Myorycteropus*, known from the Miocene of Kenya, so there is no fossil evidence of how the lineage originally split from the rest of Afrotheria during the Early Paleogene. The oldest known species of the modern genus *Orycteropus* is *O. mauritanicus*, from the Middle Miocene beds of Algeria. During the Middle to Late Miocene, aardvarks were distributed across most of Europe and the Middle East, even in Samos Island, Greece, where an exceptionally preserved skeleton of *Amphiorycteropus* was discovered. Fossils of the modern aardvark are sometimes yielded from the Plio-Pleistocene deposits of East Africa, alongside remains of **primitive** humans.

Macroscelidea

Elephant shrews are tiny, **insectivorous** mammals that today live exclusively in Africa, where they are called "sengis". They closely resemble shrews, as implied by their English name, with the exception of an elongated, mobile snout used for sniffing out invertebrates in leaf litter. Today, Africa is home to six genera and 19 species. Over a dozen fossil genera are known (all African), the oldest being *Chambius* from the Early Eocene of Algeria and Tunisia. A large radiation of extinct macroscelidean species took place during the Paleogene. A number of those fossil elephant shrews, such as *Myohyrax* (Fig. 2.44), resembled hyraxes in dentition and diet, whereas some, including *Mylomygale*, had a more rodent-like appearance. The macroscelideans are believed to have split from all other afrosoricidans during the Late Paleocene (ca. 57.5 Ma), and the radiation of extant elephant shrews seems to have started when the modern *Rhynchocyon* branch diverged during the Early Oligocene (ca. 33 Ma).

Fig. 2.44. Reconstructed life appearance of the macroscelidean *Myohyrax oswaldi*, from the Early to Middle Miocene of Southern and East Africa. Reconstructed length: 15 cm.

Chrysochloridae

The golden moles constitute another African **endemic** family, consisting of 21 extant species in 10 genera. Their **fossorial adaptations** for "sand swimming" are similar to those of Australia's marsupial moles (see p. 53), with strong digging forearms, snouts suited for shoveling through dirt, small eyes fully concealed by skin, and dense hair to ward off soil and water. The oldest known golden mole fossils are of Early Miocene age, discovered in Kenya (*Prochrysochloris*), and seem to have changed little over time.

Tenrecomorpha

Tenrecs form the highly diverse family Tenrecidae (34 species in 10 genera), **endemic** to Madagascar, where they are the main **insectivorous** mammals (hedgehogs, moles and shrews never arrived on the island). The tenrecs are estimated to have diverged from their nearest relatives, the potamogalids, during the Early Eocene (ca. 53–47 Ma). Today, the potamogalids are represented by the **semi-aquatic** giant otter shrew (*Potamogale velox*), which inhabits the rainforests of West Africa. The earliest known fossil tenrecs are three genera from the Early Miocene of Southern Africa. Therefore, tenrecs had a lengthy evolutionary history in Africa, including Madagascar. It is thought that Madagascan tenrecs diverged from just one species that arrived on the island between 42 and 25 Ma. The main hypothesis of how tenrecs got to Madagascar implies that a handful of individuals may have travelled across the Mozambique Channel (the sea separating Madagascar from mainland Africa) via rafts of floating vegetation (Olson, 2013). Upon establishment, tenrecs radiated to fill many ecological niches on Madagascar. A diversity of species have evolved to **converge** on opossums, hedgehogs, mice and shrews.

Laurasiatheria I: Carnivorans, Bats, Insectivores and their Relatives

Molecular phylogenetic studies of extant mammals reveal an unexpected cluster of **taxa**, collectively known as Laurasiatheria since they first evolved in the northern landmasses (Laurentia, or North America, along with Asia), and their evolutionary history is largely confined to the Northern Hemisphere. Nowadays, the Laurasiatheria encompasses the true **insectivores** (eulipotyphlans), the bats, the scaly insectivorous pangolins, plus all the **carnivorous** mammals and the hooved mammals (the artiodactyls (see p. 104) and perissodactyls (see p. 126)). Furthermore, there are multiple extinct lineages related to these extant mammals, such as multiple lineages of "**archaic**" hooved mammals, the carnivorous "creodonts" (see pp. 81–3) and a huge diversity of extinct insectivorous mammals, some potentially related to eulipotyphlans. Laurasiatheria comprises the majority of modern mammalian diversity, from the smallest (shrews) to the largest (whales), and lifestyles ranging from **fossorial** (moles) to marine (whales) to aerial (bats). While there is still a lack of robust morphological evidence for the phylogenetic affinities between these taxa, they are consistently corroborated by molecular data, so that over time this classification has gained support among mammalogists and mammal palaeontologists (Fig. 2.45). As a result, this implies that the Lauriasiatheria comprises the majority of mammal diversity, in addition to rodents. Given the immense diversity, I am unable to review all the families and genera in depth; instead I will concentrate on the more important, fascinating and peculiar examples.

Eulipotyphla and Other Insectivorous Mammals

The majority of Mesozoic mammals had dentitions ideal for processing invertebrates with tough cuticles, which have always provided a plentiful food supply for small mammals. Certain mammals developed sharp, blade-like teeth for shearing flesh or **bunodont** cusps for feeding on soft plant material only at later evolutionary stages. Therefore, nearly all mammals originated from "**insectivorous** mammals", an extensive category for any mammal engaging in insectivory.

Eulipotyphla

As previously mentioned, a broad range of different kinds of mammals with teeth suitable for eating insects have evolved, but most are not closely related

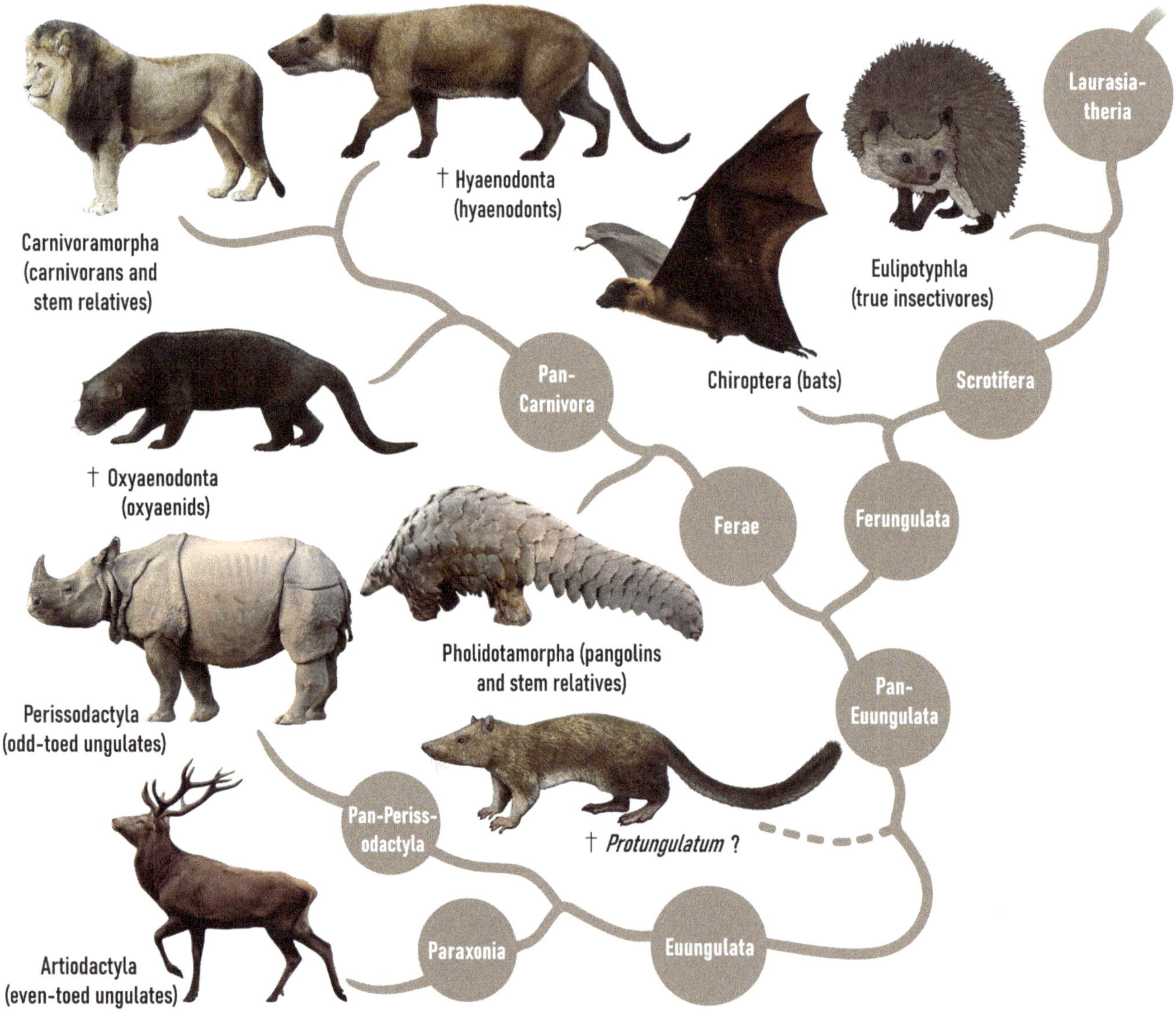

Fig. 2.45. Phylogenetic tree of Laurasiatheria after Welker *et al.* (2015), and Xue *et al.* (2021). The phylogeny of Pan-Carnivora follows Flynn *et al.* (2020), while the phylogenetic position of *Protungulatum* follows the general consensus (e.g. Upham *et al.* 2021, Velazco *et al.* 2022, Carlisle *et al.* 2023).

in the phylogenetic sense. "Insectivora" was long used as a **wastebasket taxon** for miscellaneous **insectivorous** mammals. In general, insectivorans were small, conservative in body shape and (as the name suggests) largely insectivorous. With such vague connections between the modern species, it is not surprising that Insectivora became a dumping ground for any small **primitive**-looking fossil species. Today, the "Insectivora" is no longer considered a natural (**monophyletic**) group. There is a group of insectivorous mammals that are indeed closely related, and they are called the Eulipotyphla to avoid confusion with the old wastebasket use of the name "Insectivora" (Fig. 2.46). As to where the eulipotyphlans sit in the broader Laurasiatherian **clade** is uncertain, but gone is the thought they are among the most basal mammal groups.

Modern eulipotyphans form four well-defined families: talpids (moles and desmans), soricids (shrews), erinaceids (hedgehogs and moonrats) and the oddball solenodons. Most of these mammals are tiny in body size, because their prey is small. The smallest extant eulipotyphlan, and mammal, is the Etruscan shrew (*Suncus etruscus*), reaching only 3.5 centimetres in length (barely over one inch). Even smaller was the extinct *Batodonoides vanhouteni* from the Early Eocene (53 Ma) of Wyoming. Based on tiny molars, it was the smallest mammal that ever lived, at 1.3 grams (0.05 ounces), smaller than the living Etruscan shrew (average weight 2.5 grams (0.09 ounces)) and bumblebee bat (weight two grams (0.07 grams)). *Ba. vanhouteni* was so small that it was barely larger than the rubber eraser on a pencil. Once a palaeontologist focuses on microvertebrate fossils by washing lots of fossiliferous sediments through a screen, and trap their tiny bones, teeth, and jaws, then shrews, moles, and hedgehogs have an excellent fossil record.

Many of the earliest eulipotyphans were small, nimble, insectivorous mammals that would have looked much like shrews. A number of distinct families are distinguishable from each other, but due to a paucity of material, the relationships between them and true soricids are vague. Probably the most basal eulipotyphans were the micropternodontids, which lived between the Paleocene and the Miocene. Although mostly known from skull material, some species had a very broad humerus similar to moles. *Prosarcodon* possibly burrowed amongst the loose leaf litter. Most fossil digging moles looked very similar to modern species, suggesting that the **adaptations** have changed little since the ancestral moles of the Eocene. Those adaptations to subterranean life include short arms and broad spade-like hands, small feet and reduced eyes. Not all moles are actually true "moles". Talpids are the only exclusively extant burrowing group within Eulipotypha, but there are other species from Australia (marsupial mole) and Africa (golden moles) descended from completely unrelated lineages, as previously mentioned. However, there were some other fossil burrowing eulipotyphans which were not talpids either. *Mesoscalops* is one of the best preserved proscalopids, a group which had a unique digging motion using the widened head and flattened muzzle to help displace soil. In this regard, they more closely resembled the unrelated Afrotherian chrysochlorids (see p. 73) than talpids. However, morphological features place them as true eulipotyphlans.

Fig. 2.47. Reconstructed life appearance of the desman-like eulipotyphlan *Dimylus paradoxus*. Reconstructed length: 15 cm.

Dimylus (Fig. 2.47) was a close relative of the modern desmans, an aquatic member of the mole family. Some Miocene environments in Central Europe were obviously very favourable for dimylids, with multiple

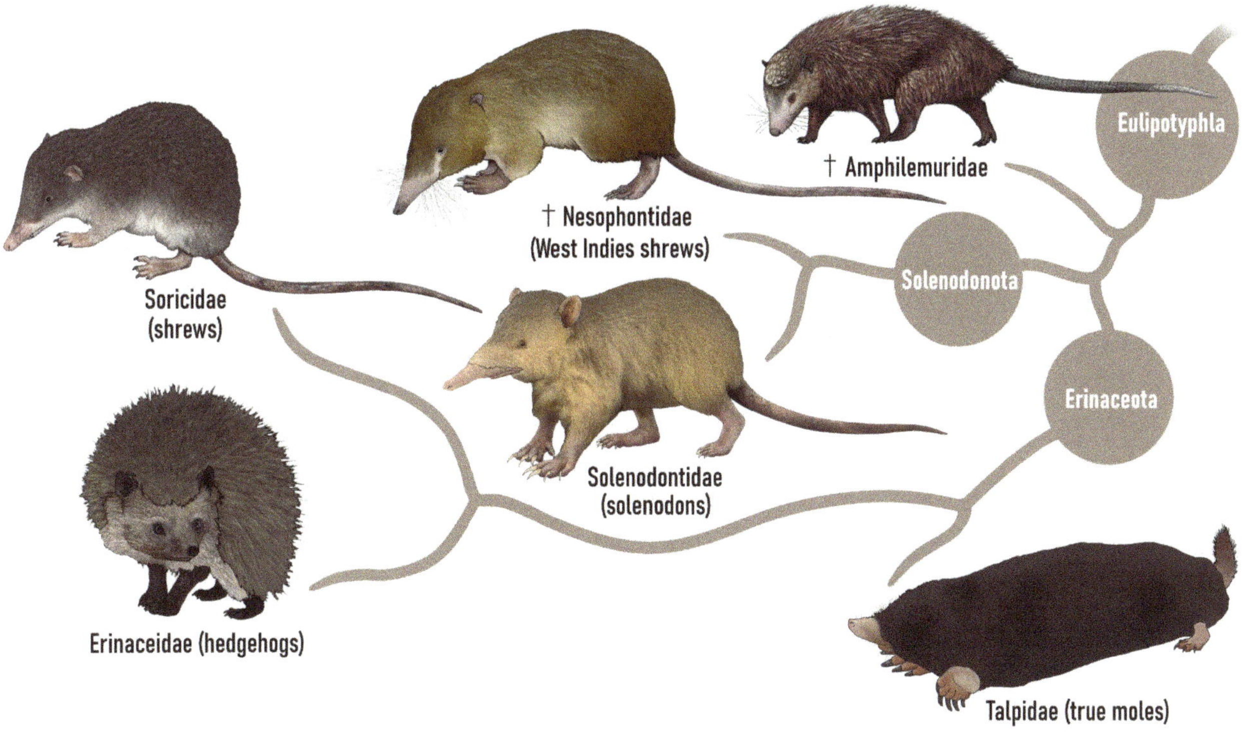

Fig. 2.46. Phylogenetic tree of Eulipotyphla after Roca *et al.* (2004), and Brace *et al.* (2016).

species being found in the same fossil strata. They are noted for an unusual dentition, rounded for crushing crustaceans. Shrews are by far the most diverse and successful eulipotyphlans, accounting for three-quarters of all extant insectivorous species (over 300) and are nearly cosmopolitan. Many of the features typical of shrews, such as enamel pigmented by the deposition of iron, are already present in the earliest unequivocal shrews during the Middle Eocene. Although a number of shrew genera were present in Europe and North America by the Oligocene, it was not until the Miocene that the family became truly diverse and widespread. During the Pleistocene, several closely related giant shrews called *Nesiotites* were found on islands throughout the Mediterranean. It is believed that ancestral shrews colonised the Mediterranean during the Messinian Salinity Crisis in the Late Miocene (see p. 253), and were subsequently stranded following the reflooding of the sea. One large species, *N. hildago*, survived on the Balearic Islands well into the Holocene.

Although modern hedgehogs and gymnures are restricted to Eurasia and Africa, the earliest erinaceids (true hedgehogs) are from North America and Europe. Those two groups had an early split during the Eocene. An early representative of the spiny hedgehogs was the tiny, thumb-sized *Silvacola*. Fossil hedgehogs have been found throughout Eurasia and generally appear to be similar to modern forms, suggesting the cape of spines to be an effective evolutionary trait. *Exallerix* was from a group of large hedgehogs noted for short robust jaws, possibly enabling it to tackle larger prey. Galericines (known as moonrats, gymnures or hairy hedgehogs) are poorly known and largely differ from hedgehogs with a lack of spines and a longer tail. Modern species are restricted to Southeast Asia, but gymnures have a much deeper fossil diversity, with species found throughout Eurasia, Africa and North America. Numerous divergent lineages occurred throughout the Oligo-Miocene. Some such as the small *Schizogalerix* were amongst the most abundant small eulipotyphans within their **ecosystem**. Their impressive lower jaws allowed for a more **herbivorous** diet compared to other species. Although most remained tiny, some eulipotyphlans were unusually large. Through the Miocene, the Gargano Peninsula in Italy was home to a number of unique species. Several species of moonrats evolved in this environment. In a classic case of **insular gigantism**, *Deinogalerix koenigswaldi* (Fig. 2.48) was not only the largest species, but also one of largest **terrestrial** predators of the Gargano fauna, and probably the largest eulipotyphlan (at about 10 kilograms). The extant solenodons have often been considered the most basal of all eulipotyphlans and it must be said that they "shout" primitive. Features such as venomous saliva, a unique mobile ball-and-socket joint snout, and with a distinct lack of close extant relatives has led many people to this view. Two species currently barely hang on to existence on Cuba and Hispaniola, while another two species (strangely, one each smaller and larger than the extant species) died out on the islands during early human colonisation.

Venom is found in very few mammals, but one notable example other than the solenodons are the extant short-tailed shrews of the genus *Blarina*. The recently extinct shrew genus *Beremendia* also possessed similarly grooved incisors similar to *Blarina*, which probably enabled them to deal with relatively large prey. *Be. fissidens* was evidently successful, spreading from Western Europe in the Early Pliocene across to Siberia by the Late Pliocene. It managed to hang on in Europe also into the Pleistocene.

In addition to the extant families, there were also various small families that consisted of extinct members only. The Geolabididae, which includes the aforementioned *Ba. vanhouteni*, was a primitive family dating back to the latest Cretaceous. It has an uncertain relationship with other eulipotyphans, although they show some similarities with the equally obscure nyctitheriids. Postcranial remains including foot bones show that they were capable climbers. A group of slightly larger shrew-like species were the plesiosoricids, found across Eurasia and North America from the Eocene to Miocene Epochs. A particularly large, robust species from Southeast Asia, *Siamosorex* showed features typical of specialised **carnivores** such as extant weasels and extinct "miacids" (see p. 84). Roughly the size of a small

Fig. 2.48. Reconstructed life appearance of the erinaceid *Deinogalerix koenigswaldi*. Reconstructed length: 60 cm.

weasel, it was able to hunt relatively large prey such as small vertebrates and arthropods. Intriguingly, *Siamosorex* was possibly venomous, a feature seen in some extant shrews such as water shrews (*Neomys*), due to the presence of large grooved canine teeth in the lower jaw. One early family of eulipotyphlans were the amphilemurids such as *Pholidocercus* (see pp. 212–3) and *Macrocranion* (see pp. 212, 217). Most species have been found in Eocene deposits in Europe and North America. Some were spiny like hedgehogs, whereas others were nearly devoid of spines or had bristly fur on the back, possibly as a defence mechanism. Most species are known only from teeth.

Apternodontids were also found in the Eocene, but their relationship with other eulipotyphans is not well-understood. They have been previously linked with other early "insectivores" and also solenodons, but this is possibly due to shared primitive features (plesiomorphies). Some features such as pigmented teeth suggest a close relationship to shrews. Peculiar rounded canine teeth in some *Apternodus* species are thought to have been used in crushing the hard shells of arthropods or crustaceans.

Until recently, several species of nesophontes or West Indies "shrews" (not true shrews) inhabited islands throughout the Caribbean. Considerably smaller than their closest relatives, the extant solenodons, it is believed that they were no match for the invading rats aboard Spanish vessels arriving in the sixteenth century. Some of the remains were courtesy of very well-preserved owl pellets!

Miscellaneous Extinct Insectivorous Groups

As previously mentioned, there are numerous extinct families of **insectivorous** mammals, although how they are related to Eulipotyphla remains uncertain. Many fossil genera are represented solely by jaws and isolated teeth, so few details can be inferred about their body shapes or mode of life, nor can they be confidently restored to look more dissimilar than an extant shrew.

Leptictida

Humans are pretty unique among mammals—we are one of the few **bipedal** species which is able to walk upright. For most bipedal vertebrates, the back is horizontal and the **anterior** part of the body is balanced by a long tail. Among mammals, small forms often resort to the more efficient bipedal hopping, or **saltatorial** movement, with kangaroos being an obvious exception to the rule! One very **primitive** group which shows evidence of saltatorial hopping are the leptictidans. They were among the earliest groups of Cretaceous **insectivorous** mammals to survive into the Cenozoic.

Leptictidans typically resembled modern insectivorous mammals and were originally members of the old "Insectivora" (now considered **paraphyletic**) (see p. 75), and then later connected to rodents, to primates and then to afrotherians (see p. 62) for resembling elephant shrews. Current thinking is that they are placentals, possibly **stem**-afrotherians. However, it is also possible that the group is actually a paraphyletic assemblage of species. There are three recognised families: pseudorhynchocyonids, leptictids and gypsonictopids. Perhaps the best known form is *Leptictidium*, a pseudorhynchocyonid (see pp. 212–3). There are eight recognised species (ranging in size from 60 centimetres to 90 centimetres) found in the subtropical forests of Eocene Europe. Leptictidans are known from the Late Cretaceous (*Gypsonictops*) through to the Early Oligocene (*Leptictis* (see pp. 238–9)). Late Cretaceous leptictidans such as *Gypsonictops* suggest that they were among the oldest groups to branch off from the earliest radiation of placentals. This is apparent in their highly primitive skulls and skeletons, with the typical insectivorous dentition and cranial bones. In other ways, they were specialised, as is demonstrated by the presence of a peculiar pair of ridges along the top of the skull. Many of the species were generally **quadrupedal**, having well-developed forelimbs for digging. Features in the skull also suggest an elongated, mobile snout for probing invertebrates in the leaf litter. A large genus, *Megaleptictis*, was the size of a large domestic cat (nearly 90 centimetres (35 inches) long), and had a broader snout and tall teeth which is indicative of consuming different food, possibly small vertebrates or plant material. The North American leptictid *Leptictis* is known from well-preserved skeletons and, despite similarities with *Leptictidium*, there were obvious physical differences in the skeleton which prevented *Leptictis* from moving in the same manner. *Leptictis* is the last known leptictidan, disappearing in the Early Oligocene around 30 Ma. Why the Leptictida became extinct (without having left any descendants) was possibly because they were adapted to live only in forest **ecosystems** and were unable to adapt to the opening plains of the Oligocene.

"Cimolesta"

"Cimolestans" were among the major groups of mammals to make it through the Cretaceous-Paleogene extinction event 66 Ma (see p. 169). "Cimolestans" are considered a **polyphyletic wastebasket taxon** of multiple **"archaic"** placental lineages with uncertain evolutionary relationships. The "cimolestans" blossomed in diversity during the Paleogene and quickly evolved into a broad variety of specialised forms. Some were shrew-like and largely **carnivorous** (cimolestids), whereas members of the subgroup Pantolesta included **terrestrial** forms, some which possessed digging **adaptations** and were **semi-aquatic** (pantolestids), whereas other forms were very likely **arboreal** (paroxyclaenids). The cimolestids are distinguished by **cheek teeth** with transversely broad upper molars and tall, pointed cusps, and originated in North America during the Late Cretaceous. *Cimolestes* is the most **primitive** known cimolestid, which survived the Cretaceous-Paleogene extinction event presumably for having a generalised diet, with teeth adapted for crushing, puncturing and shearing. A handful of other cimolestid genera are known from the Paleocene to Eocene of Europe and North America, Early Eocene of Asia, and Late Paleocene of Africa.

The family Pantolestidae originated in North America during the Late Cretaceous, where they consisted of a variety of genera exhibiting only slight dental differences from other **insectivorous** mammals. Soon afterwards, they evolved into a range of **piscivorous** predators that became increasingly **convergent** with the unrelated extant otters and otter shrews in their body plan, from the Middle Paleocene *Bessoecetor*, to the Middle Eocene *Palaeosinopa* from the Green River Basin in Wyoming, to the Middle Eocene *Buxolestes* from Europe (see pp. 212, 214, 217). Their relatively robust canine teeth and powerful jaw musculature are possible adaptations to a **durophagous** diet consisting of hard-shelled molluscs. The paroxyclaenids converged with raccoons and are known solely from the Eocene of Europe. Many paroxyclaenids are known exclusively from jaws and teeth, while the raccoon-sized *Kopidodon* (see pp. 213–4) is represented by a few complete skeletons from the Messel Formation of Germany. While pantolestans were never particularly common or diverse, they were rather widespread, extending their distribution from their origins in North America across to Europe and eventually into Asia. A cooling and drying climate by the Eocene-Oligocene boundary appears to have caused the extinction of most pantolestans—although several Asian genera (such as the pantolestid *Gobiopithecus*) survived slightly longer as the very last of the "cimolestans", with the most recent fossils dated to the Early Oligocene (ca. 28 Ma).

Apatemyidae

The small, **insectivorous** apateymids have been proposed to be basal euarchontoglires (see p. 144), (Silcox *et al.* 2010), whereas other studies (Meredith *et al.* 2011) suggest that they were part of Laurasiatheria (see p. 74). The apatemyids were dormouse- to rat-sized mammals whose fossil record is best represented in the Paleocene and Eocene beds of North America, (although fossils from Florida reveal that they persisted into the Oligocene). In the Eocene, they made it to Europe too, including forms such as *Heterohyus* (see p. 212) and *Jepsenella*. Among the last known apatemyids is *Sinclairella*, documented from the Late Eocene to Early Oligocene strata of the North American badlands. Its description was based on one specimen comprising the jaws and cranium (now lost) that implied it resembled a sabre-toothed shrew, having a skull length of roughly 7.5 centimetres (three inches). Both its upper and lower incisors were elongated and projected forwards, the former pair which exhibit indications of abrasion to form chisels at their ends. Many apatemyids seem to have **converged** on modern lemurs or squirrels in body plan and lifestyle. The extinction of apatemyids was possibly caused by a cooling, drying climate.

Chiroptera

Many people have unfavourable opinions about bats, but these are actually a rather successful group of mammals. Bats are the only mammals capable of powered flight, having evolved wings by developing a skin membrane between all five **digits** of the forelimb. The birds and insects evolved wings in entirely different ways. Bats diversity is also very high, with over 1,400 extant species in 186 genera and 18 families. Therefore, when it comes to mammal diversity, bats are ranked second only to rodents. Today, bats constitute approximately 20% of all the world's mammal species. Since bats have such small, light and delicate skeletons, they are seldom fossilised. All except several fossil bat genera are represented only by jaws and teeth, and palaeontologists estimate that up to 90% of their fossil record is currently missing. Bats are significantly more prevalent in tropical regions, but those places are infamously bad at fossilising tiny, fragile bones, given the humid climate and corrosive groundwater which rapidly disintegrates bone within the sediment. The bat fossil record is largely limited to Pleistocene cave deposits, in which numerous cave-dwelling bats are preserved, although these comprise a microscopic portion of the diversity that once existed. Bats are not limited to small **insectivores** that evolved to echolocate (the likely **paraphyletic**

Fig. 2.49. Phylogenetic tree of Chiroptera after Tsagkogeorga *et al.* (2013). The phylogenetic positions of the fossil bats follow those reported by Jones *et al.* (2024).

microbats). There are also the fruit bats (family Pteropodidae in the order Yinpterochiroptera (Fig. 2.49)), which are among the largest known bats, with some species attaining wingspans of up to 1.7 metres (5.6 feet) and weights of 1.45 kilograms (3.2 pounds). They are colloquially called "flying foxes" since their heads resemble those of modern foxes in possessing pointed ears and a long, narrow snout. Contrary to microbats, fruit bats are mostly **diurnal** mammals with good vision. They are tree-roosting instead of cave-roosting, and do not echolocate (except the Egyptian fruit bat (*Rousettus aegypticus*)), since they are mainly **frugivorous** (also consuming flower nectar). There are over 197 extant species of fruit bat, the majority which live in the tropics of Africa, Asia, Australia, and on islands in the South Pacific and Indian Ocean. Most islands in the South Pacific are home to **endemic** species of fruit bat, since it took a single founding colony to arrive on each island, where they later radiated from their mainland progenitors.

Given their extremely sparse fossil record, we only get hints as to how bats diverged from non-flying mammals. The earliest known bats in the fossil record are from the Late Paleocene and Early Eocene, such as *Wyonycteris*, but are limited exclusively to teeth, as no deposits exist that could keep such fragile skeletons intact. The most well-preserved fossils of **primitive** bats are known from lake strata with exceptional preservation, such as the Early Eocene *Onychonycteris* and *Icaronycteris* from the Green River Basin in Wyoming, and *Palaeochiropteryx* from the Messel Formation in Germany (see pp. 210–7). Therefore, only a limited early fossil record exists, and when bats do emerge, they already possess wings like modern bats. However, upon closer inspection, you will notice that these fossils are in fact highly primitive, lacking bat-like characteristics other than their wings. These had rather unspecialised skulls, primitive tooth crowns and lack the inner ear structures (e.g. large cochlea) necessary for echolocation, suggesting flight may have evolved before echolocation. Their first digit was not fused to the wing membrane as in extant bats. They possessed a long tail (unlike in extant bats) and lacked a skin membrane connecting the ankles with the tail (uropatagium) present in the majority of modern bats. The vertebrae and pelvic bones were not fused as in extant bats, and an expanded keel on the sternum for powerful flight muscles was absent. One other bat from the Green River deposits, *Onychonycteris*, is the most primitive transitional fossil so far known, exhibiting less advanced characteristics than in any extant or fossil bat. All five of its digits were clawed (modern bats possess claws on two digits), it had short, broad wings ideal for a combination of gliding and fluttering instead of prolonged flight, and short forelimbs and elongated hindlimbs, indicating that it developed flight from a lifestyle of habitually climbing trees.

Pholidota

Pangolins can be confusing animals, not helped by the fact that most people are unlikely to ever see a live individual due to their secretive natures. Possibly the best description is of an animated pinecone which curls into a ball (pangolin is Malay for "one who rolls up"). They have some attributes seen amongst other groups: a long, tube-like snout and a sticky tongue to facilitate a primarily ant/termite diet, powerful forelimbs and long front claws to break into mounds, and some species climb trees with the aid of **prehensile** tails. They are mostly **nocturnal**, and they have poor eyesight, relying on sound and smell for getting around instead. The most obvious feature is the unique covering of hard, **keratinised** scales overlapping to provide a near impervious armour, making pangolins vaguely resemble scaly anteaters. All species have long tails which can provide a counter-balance in larger species, allowing them to stand and even walk **bipedally** over short distances. Due to common **adaptations** of a diet of ants and termites, pangolins were originally linked in a **clade** with aardvarks and anteaters. It was a lack of skeletal similarities that determined that it was an "artificial" grouping, specifically a **polyphyletic** one. It took until an examination at the molecular level to look past the specialised nature of pangolins to find that they are actually the sister group to all modern carnivorans (see pp. 81, 83–4).

Pangolins are scarce animals with a limited fossil record, and their evolutionary history was very poorly understood before the 1970s. While fossils of *Necromanis* are known from the Oligocene and Miocene of France, they are highly fragmentary. In 1970, a fossil was discovered from the Late Eocene deposits of Wyoming, described by Robert Emry as a pangolin and published as *Patriomanis* (Emry, 1970). This demonstrated that pangolins were formerly present in the New World. Then in 1978 (Storch, 1978), an exquisitely preserved skeleton of a whole pangolin (even consisting of scales) from the Messel Formation of Germany was named *Eomanis* (see p. 212), proving that the pangolin body form had already evolved by around 47 Ma. Therefore, high quality fossils of pangolins show that they were dispersing all over the Northern Hemisphere by the Middle Eocene. Other early pangolins have also been found in North America and China, while later species are restricted to Eurasia and Africa. The giant *Manis palaeojavanica* was one and a half times larger than the largest modern species, the giant pangolin (*Smutsia gigantea*) of Central and West Africa, and survived alongside humans in Indonesia during the Pleistocene. The remaining fossil record of pangolins is sparse, but there are eight species alive today, four in Africa and four in Asia.

Palaeanodonta

Another group, the extinct **insectivorous** palaeanodonts, were previously grouped within the Xenarthra (see p. 58), or within the Pholidota, and their phylogenetic affinities have long been disputed and unclear. Despite their ecological similarities, the palaeanodonts are extremely unlikely to be xenarthrans since they lack the specialised vertebrae with extra joints. The most recent studies indicate that they are a sister **taxon** to pangolins (both grouped within the **clade** Pholidotamorpha) (Fig. 2.50). The palaeanodonts lived in North America and Eurasia from the Middle Paleocene to Early Oligocene. Among the earliest palaeanodonts were *Ernanodon*, from the Middle to Late Paleocene of Mongolia and China (see pp. 190–1). The family Metcheiromyidae consists of armadillo-sized genera incuding *Propalaeanodon* from the Paleocene, *Palaeanodon* from the Early Eocene, and *Metacheiromys* from the Middle Eocene (see pp. 205, 209). The features palaeanodonts shared with anteaters and armadillos include a narrow, tubular snout, and powerful front claws for digging up invertebrates.

The Epoicotheriidae was a palaeanodont family characterised by shorter, broader skulls and retained a greater number of teeth, and their skeletal anatomy

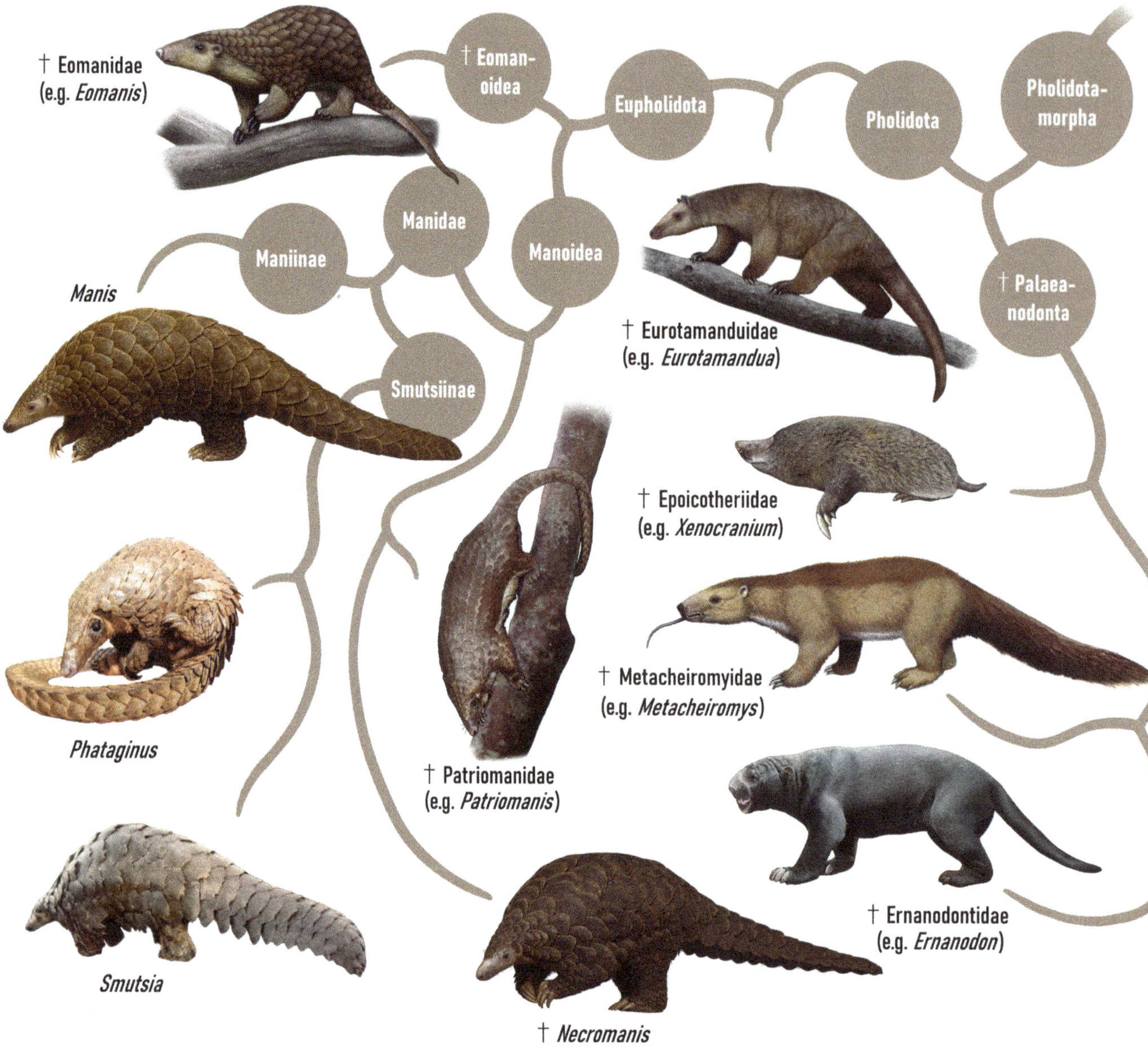

Fig. 2.50. Phylogenetic tree of Pholidotamorpha after Gaudin *et al.* (2009), and Kondrashov *et al.* (2012).

virtually resembles that of moles in exhibiting extraordinary **fossorial adaptations**. In addition to very powerful forelimbs, accommodating well-developed digging musculature, the upturned snout of the strangest representative, *Xenocranium*, had a wide scoop on the tip, indicating it scooped away dirt with its snout while burrowing. Following the extinction of palaeanodonts from Eurasia and North America, it would be some time before they were replaced by other burrowers. The anteater niche woud never be occupied in North America again. Very few animals filled that niche in Eurasia before the Late Miocene. In North America, the fossorial niche would ultimately be occupied by rodents including gophers, and in Eurasia by other rodent lineages, although only moles have occupied the insectivorous and fossorial niches simultaneously.

Carnivorous mammals (Oxyaenodonts, Hyaenodonts and Carnivorans)

A vital component of the food web are predators, since they control the populations of prey animals. A diverse array of animals have taken on the predatory niche since the dinosaurs left that niche vacant by the end of the Mesozoic. The majority of predatory mammals throughout the Paleocene and Early Eocene were small, while huge snakes and crocodylomorphs (see pp. 26–30) filled the large predator niche. The largest predatory mammals in South America during the Miocene and Pliocene were roughly wolf-sized, whereas birds of the family Phorusrhacidae evolved independently into large apex predators (see pp. 41–2). Numerous lineages within different mammal orders have evolved to occupy that niche. In Australia, where the pouched marsupials were the sole native mammals, various lineages of cat-like, lion-like and dog-like predators diverged from possum-like progenitors. This was also the case in South America, where predatory marsupials radiated to strikingly **converge** on sabre-toothed cats (the thylacosmilids), and hyaenas and wolves (the borhyaenids) (see pp. 48–50). The sabretooth niche was actually filled not just by marsupials, but by oxyaenodonts (see pp. 81–2) and two distinct carnivoran lineages (see pp. 82–3) as well, such as several felids (see pp. 86–9). Across the northern continents, there were basal **ungulates** of the order Mesonychia which became the largest mammalian predators during the Paleogene.

During much of the Cenozoic, three major mammal lineages have filled the niches of predators. These encompass the extinct "**archaic**" predators of the orders Oxyaenodonta (see p. 83) and Hyaenodonta (see pp. 82–3), formerly grouped as "Creodonta", and the modern order of predators, the Carnivora (dogs, cats, hyaenas, bears, weasels, raccoons, seals, and numerous others). In a phylogenetic sense, the total lineage that includes Carnivora and all its extinct close relatives is termed Pan-Carnivora, while the modern group itself—comprising the most recent common ancestor of all living carnivores and its descendants—is referred to as **crown** Carnivora. When talking about all these mammals, we need to use terminology carefully. The term "**carnivore**" applies to any flesh eater (even encompassing carnivorous plants such as sundews), while a "carnivoran" is part of the order Carnivora. Furthermore, not every carnivoran is carnivorous. The red panda (*Ailurus fulgens*) and giant panda (*Ailuropoda melanoleuca*) are **herbivores** that live virtually on bamboo, while fruit forms 90% of the diet of the kinkajou (*Potos flavus*). As well as exhibiting prominent canine teeth for gripping prey, nearly all mammalian predators have evolved **cheek teeth** specialisations for cutting flesh. Therefore, the teeth have developed from the simple **bunodont** teeth of their progenitors into teeth forming a series of cutting blades which work against the opposing blades to produce a scissor-like motion. Furthermore, a shared feature between all hyaenodonts, oxyaenodonts and carnivorans (different from what is observed in other carnivorous mammals) is the development of a specialised pair of massive cheek teeth called **carnassials**. These paired upper and lower teeth function as the main slicing and crushing components of the jaw. These are particularly well-suited for shearing meat and tough sinews, and in numerous carnivores, for cracking and crushing bone. These vital teeth are immediately noticeable when looking at the mouth of any dog or cat.

While carnassials are present in oxyaenodonts, hyaenodonts and carnivorans, their main distinction is their location. Apart from seals and sea lions, which have developed all their teeth into simple cone-shaped pegs for holding slippery prey, the carnassials of all carnivorans are formed from the fourth upper premolar and first lower molar. In the jaws of "creodonts", on the other hand, the carnassial slicing pair is located more **posteriorly**: formed either from the first upper molar and second lower molar in oxyaenodonts, or the second upper molar and third lower molar in hyaenodonts. The arrangement of these vital teeth is essential for determining the order to which a fossil belongs. It could also shed light on why carnivorans superseded the "creodonts" and became so prosperous. The forward location of their carnassials allowed for increased evolutionary flexibility in skull shape and feeding habits, enabling the facial region to be shortened and other components of the dental row to be reduced without compromising the vital carnassials. However, the carnassials and post-carnassial molars of certain carnivorans, such as bears, have evolved to specialise in functions such as grinding. In comparison, the carnassials of oxyaenodonts and hyaenodonts were positioned in the rear of the jaws, which restricted the functions their teeth could perform, and thus retained the same stereotyped dentition, and did not evolve into bear-like or felid-like forms (felids possess shortened **rostrums**, and virtually lack molars posterior to the carnassials), or other extremely specialised **taxa**.

"Creodonta"

Some fossil species have the misfortune of looking vaguely familiar, but in other ways like nothing else. Within the mammals, this is most apparent in those early species which have **primitive** features. While resembling one **taxa**, a species might actually belong to a totally different group. If they do not fit obvious groupings, then difficult species usually end up in **"wastebasket" taxa**. Two notable wastebasket taxa are the **herbivorous** "condylarths" and the **carnivorous** "creodonts". Membership to each group has swelled and shrunk over time with better knowledge. Now reassigned to other groups are the "miacids" (**stem**-carnivorans) (see p. 84), "arctocyonids" (now split between various groups) (see p. 101) and mesonychians (often considered stem-**ungulates**) (see pp. 102–3). Left remaining under the "creodont" banner were the oxyaenodonts and the hyaenodonts. These were the first major groups of Cenozoic mammals to adopt the predatory lifestyle. Originally mistaken for carnivorans, or considered to be ancestors of carnivorans, more recent analyses show that oxyaenodonts and hyaenodonts were an early experiment in predator evolution that was eventually replaced by carnivorans. They were relatively primitive in their body form, without the wide range of different shapes (like a bear, a dog, a weasel or a cat) seen in carnivorans. Both oxyaenodonts and hyaenodonts do not really show any commonality of features beyond having **carnassial** teeth, which they also share with carnivorans. Not a ringing endorsement for relatedness. So what is the modern consensus?

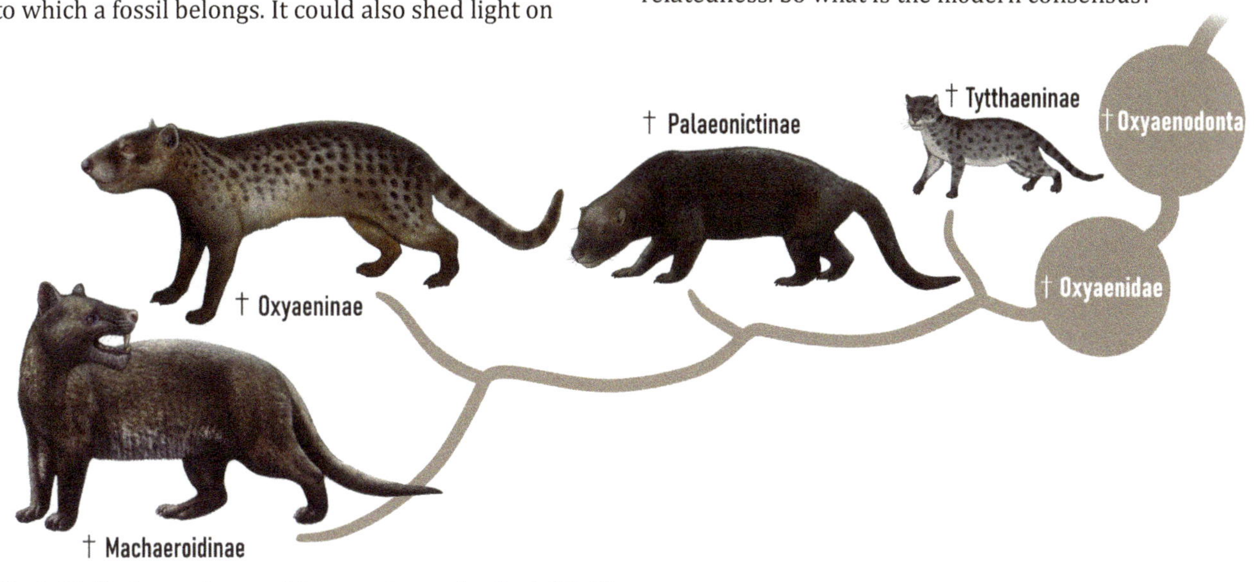

Fig. 2.51. Phylogenetic tree of Oxyaenodonta after Zack (2019).

Well, there is no consensus really. Many of the old "creodonts" have been loosely linked up again under the Ferae banner (see p. 74) which also includes carnivorans and the Pholidota (see pp. 79–80).

Oxyaenodonta

Within the order Oxyaenodonta, Oxyaenidae was not a diverse family and can be described as being more cat-like, compared to the dog-like hyaenodonts, with a fair amount of mustelid about them too (otter-like is often mentioned) (Fig. 2.51). They included some of the earliest large mammalian **carnivores** across North America and into Eurasia, with fossils found between the Late Paleocene and the Middle Eocene. The skulls of oxyaenids were short, broad and robustly built, strongly muscled to provide a tremendous bite force. The oxyaenids also had unspecialised skeletons with small brains, long tails, and generally short, robust limbs, so they were **ambush predators** rather than fast runners. The oxyaenid *Oxyaena* (see pp. 205, 208), from the Late Paleocene and Early Eocene of North America, was built like a wolverine (*Gulo gulo*), with powerful robust limbs and a flexible body. *Patriofelis* from the Middle Eocene of North America (see pp. 205, 208), was roughly lion-sized and had a huge, robust skull. The teeth of oxyaenids were designed for crushing rather than cutting (**convergent** with hyaenas). Although not fast hunters, the feet of oxyaenids were broad, which would have allowed them to creep silently, whereas some species were capable of climbing into trees. While most oxyaenids were generalist carnivores, the teeth of *Palaeonictis* suggests that its diet was more **omnivorous**. Although most oxyaenids were stereotyped in the heavy-bodied lion-like predator shape, one specialised subfamily called Machaeroidinae includes *Machaeroides* (see pp. 205, 208) and *Apataelurus*, which both developed sabre-like canine teeth and converged on the other two lineages of sabre-toothed carnivorous placentals. With those elongated sabre teeth, they were possibly capable of hunting prey much larger than themselves. The last surviving oxyaenid, and oxyaenodont, was the genus *Sarkastodon* (see pp. 222–3).

Hyaenodonta

Due to few alternatives, carnivorans are the first choice when thinking of a **carnivorous** mammal (the name alone confirms it) as they occupy the majority of roles within various **ecosystems**. Disregarding the mammalian lineages which are airborne or entirely aquatic, the only other diverse group of carnivores are relatively small marsupials. However, that has not always been the case. Throughout the early part of the Cenozoic, it was not carnivorans that ruled **terrestrial** ecosystems, but the now extinct hyaenodonts (Fig. 2.52). Their name is misleading, since they were unrelated to hyaenas, which are members of the order Carnivora (see pp. 83–4). Hyaenodonts had more delicate wolf-like skeletons and slender limbs compared to hyaenas, or especially compared to oxyaenodonts.

So successful were the hyaenodonts, that it was not until the end of the Miocene that the carnivorans had finally assumed dominance. Long considered a member of the "creodonts" with the rather different oxyaenodonts, their true position within mammalian phylogeny remains uncertain. Hyaenodonts appear to be members of an expanded Ferae, which includes extant carnivorans, bats and pangolins, along with a host of other extinct groups including the oxyaenodonts. Alongside carnivorans, hyaenodonts were the longest-lived radiation of carnivorous mammals, ranging from the Early Paleocene to the Late Miocene. A remarkably diverse order, with species ranging from a couple of kilograms up to possibly half a tonne in weight, hyaenodonts spread throughout Eurasia, Africa and North America. They were previously thought to have had origins in Africa (leading to a fanciful idea that hyaenodonts were allied to the Afrotheria (see p. 62)). However, a revision of the cladistics within Hyaenodonta has shown that the most basal forms were European in origin. A major group of mostly Eurasian hyaenodontoids (hyaenodontids and proviverrids) split early from the rest of the hyaenodonts. A second major **clade** were the largely Afro-Arabian hyainailouroids, which featured some of the latest and largest species. Several un-**ranked** clades of smaller species (from Eurasia and North America) fit between these two major groups.

As the name suggests, the proviverrids, such as *Lesmesodon* (see pp. 213–4), were **convergent** with modern day civets. Similarities between *Hyaenodon* (the type genus) and hyaenas is limited to both being carnivorous mammals. While some modern hyaenas have powerful **carnassials** capable of crushing bones, *Hyaenodon* had teeth only capable of shearing meat. *Hyaenodon* was very diverse, with species ranging from five to over 350 kilograms. *Hyaenodon* was a highly successful predator from the Middle Eocene to the Late Miocene, a span of about 26 million years (longer than just about any other fossil mammal). Hyaenodontids were among the most common predators in the Early Oligocene of North America, but also Eurasia. By the Late Oligocene, *Hyaenodon* had vanished from both North America and Europe, pushed out by the more advanced carnivorans. However, hyaenodonts persisted as major predators in Asia and Africa well into the Late Miocene.

The remaining hyaenodonts do not fit major clades, but various side branches. They all had **adaptations** for forest-living, having proportionally short limbs that allowed them to hunt in undergrowth. A typical member was *Sinopa* (see p. 208). It and many other hyaenodonts disappeared at the Eocene-Oligocene boundary, coinciding with a reduction in forest cover as the planet cooled. There was a large separate radiation known as hyainailouroids, centred in Afro-Arabia. This was a significant group that included massive species, some of which were amongst the largest known carnivorous land mammals. The taxonomy is very muddled and hyainailouroids can either be split into several families or subfamilies, depending on which approach one uses. They include: the hyainailourids, which includes the **semi-aquatic** apterodontines and the large to giant hyainailourines; two small obscure species known as prionogalids; and a long-lived lineage known as teratodontids. Fossils of hyaenodonts are a bit hit-and-miss in Africa during the Paleogene. During the Late Eocene and Early Oligocene, the Fayûm Depression in what is now central Egypt was home to a complex and diverse range of hyaenodonts including small, gracile **mesocarnivores** (*Masrasector* (see pp. 242–3)) and large **hypercarnivorous** species, with at least seven genera known.

Several large species were part of Early Eocene migrations into Europe (*Kerberos*) and then into North America (*Hemipsalodon*). However, it was not until the latest Oligocene that landbridges between the Arabian Peninsula and Southwestern Europe allowed migration and diversification of hyainailouroids into Eurasia. At the same time, the

Fig. 2.52. A possible phylogeny of Hyaenodonta after Solé *et al.* (2020).

northern carnivorans were finally able to disperse throughout Africa. Carnivorans replaced hyaenodonts largely in the small mesocarnivorous roles, but hyaenodonts were able to specialise in hypercarnivorous niches. The two clades of predators lived alongside each other throughout Africa and Eurasia. However, very few genera made it through into the Miocene: hyainailourines such as *Simbakubwa*, *Hyainailouros* and *Megistotherium*, the teratodontid *Dissopsalis* and the hyaenodontid *Hyaenodon*. The Asian *Dissopsalis* was the only known genus to almost reach the end of the Miocene. Hyainailourines commonly had long faces, even when compared to other hyaenodonts, and it is suspected that prey preference had a factor in the adaptation. Large **herbivores** such as deinotheriids, mammutids (see pp. 64–6), suids (see pp. 108–10), rhinocerotids (see pp. 130–2) and "anthracotheriids" (see pp. 120–1) coexisted with the largest hyaenodonts. The largest genus known genus is *Megistotherium*, from the Early Miocene of Africa, with a 66 centimetre long skull and a clearance of 30 centimetres between the tips of the canine teeth. *Megistotherium* was potentially able to fully encircle the limb of a proboscidean (see pp. 64–8) in its mouth. Alongside the largest genus were smaller hyainailourines which suggests that they were adapted for different prey and hunting styles. Although oxyaenodonts were extinct before 37 Ma, the hyaenodonts persisted long after them, vanishing about 11 Ma in Africa and Asia. The cause of their extinction remains unknown. All hyaenodonts lacked true post-canine crushing molar teeth, thus limiting their ability to process flesh. It was probably their inability to adapt to other food items (e.g. bone material), or **omnivory**, which put the group at a disadvantage with the more advanced carnivorans in many parts of the world, while the oxyaenodonts and hyaenodonts were stuck without much evolutionary flexibility in their jaws, teeth, or limbs.

Carnivora

Today, most **carnivorous** mammals belong to the order Carnivora. They comprise at least 291 extant species in 12 families, and approximately 10 times as many fossil species. Of all the mammal orders, the carnivorans have the greatest size range, from the least weasel (*Mustela nivalis*), up to 10 inches long and weighing 25 grams, to the massive polar bears (*Ursus maritimus*) which can weigh over 1,700 pounds, to the enormous elephant seals (*Mirounga*), which can attain lengths of 5 metres (16 feet) and weigh up to 4,000 kilograms (8,800 pounds). As previously mentioned, carnivorans exhibit various specialisations for predatory habits, from the pointed canine teeth for puncturing and grasping prey, to the **cheek teeth** suited for shearing flesh. Additionally, the majority of carnivorans possess long, sharp claws for functions such as combat and slashing, keen senses (particularly hearing, vision and smell) for finding prey and pursuing it, and digestive adjustments for a meat-based diet. Carnivorans are split into two major groupings according to morpological and molecular evidence: the Feliformia (cats, mongooses, hyaenas and kin) and the Caniformia (dogs, weasels, raccoons, bears and kin, along with the seals and sea lions). Both lineages diverged in the Early Eocene, or possibly even the Paleocene. They also evolved from the **primitive stem**-carnivorans, or carnivoramorphs, of the Paleocene and Eocene, grouped into the poorly-known **monophyletic** family Viverravidae (the fossils closest to the Feliformia, largely from the Paleocene) and the **wastebasket taxon** "Miacidae" (the fossils closer to the Caniformia, largely from the Early to Middle Eocene). Whilst closely resembling current extant species, the carnivoramorphs retained the primitive condition of a higher tooth count. Most "miacids" (*Paroodectes*, *Messelogale* (see pp. 214, 216), *Vulpavus*, *Miacis* (see pp. 205, 208), *Dormaalocyon*) and viverravids (*Protictis* (see pp. 179, 183), *Viverravus* (see p. 208)) were small creatures about the size and proportions of weasels or mongooses (Fig. 2.53). Both groups were generally small hunters with long bodies and tails, which occupied niches both in the trees and on the ground. All fossils between the two groups are known from Eurasia and North America. Given their small size, prey was limited to mostly insects and small vertebrates. Throughout the Paleocene and much of the Eocene, these small carnivoramorphs were overshadowed by the much larger oxyaenodonts (see pp. 81–2) and hyaenodonts (see pp. 82–3), and their radiation into the modern carnivorans was delayed until the decline of many of the oxyaenodonts and hyaenodonts in North America and Eurasia during the Late Eocene and Oligocene.

Feliformia

The Feliformia diverged into numerous families probably approximately 50.6–35 Ma. While they share certain characteristics (such as retractable claws in most species, an often shortened **rostrum** and reduced **cheek teeth posterior** to the **carnassials**), the Feliformia is distinguished by uniquely specialised traits in the parts of the skull including the braincase. The validity of this grouping has been confirmed since the 1990s, when the close phylogenetic affinities between all feliforms were verified by molecular studies (Fig. 2.54). Among basal feliforms, the "stenoplesictids" are an extinct **polyphyletic** group of civet- and genet-like animals from the Old World that are generally considered basal feliforms. They appear to be close to the base of civets and felids, as evidenced by the presence of a double-chambered **auditory bulla**. Among the most **primitive** "stenoplesictid" genera were *Asiavorator* (see pp. 224–5) from the Late Eocene of Mongolia. Their sudden appearance in Europe immediately after the Eocene-Oligocene extinction event (see pp. 229–30) indicates that they migrated, possibly from Asia. They went extinct by the Middle Miocene. One other group are the palaeogalids, an extinct **monotypic** family of small, marten-like predators that only includes *Palaeogale* (see pp. 248–9). The ancestry of palaeogalids remains enigmatic. The

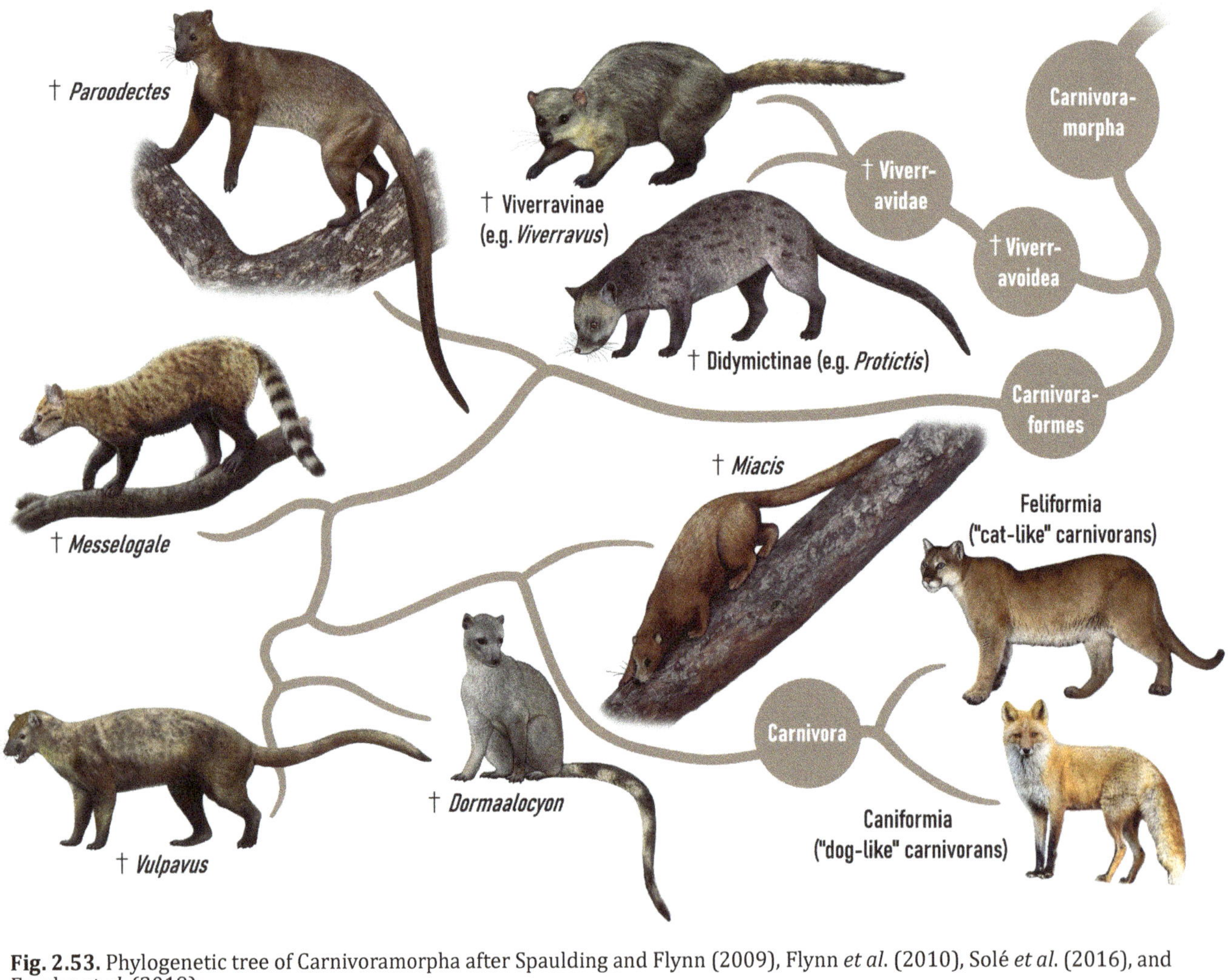

Fig. 2.53. Phylogenetic tree of Carnivoramorpha after Spaulding and Flynn (2009), Flynn *et al.* (2010), Solé *et al.* (2016), and Faurby *et al.* (2019).

palaeogalids appeared in Europe by 32 Ma, after the Grande Coupure (see p. 230), but 35 to 36 million year old specimens from Pipestone Springs, Montana, are the oldest known. Palaeogalids survived until the early Early Miocene of East Asia and the late Early Miocene of Europe. Habitat changes and competition with other carnivores were possible causes of the extinction of both "stenoplesictids" and palaeogalids.

Nimravidae

The nimravids (or "false sabretooths") were a series of cat-like animals that appeared in the Late Eocene (ca. 40 Ma) and were an early off-shoot which filled the vacant role of large **ambush predators** during the Eocene and Oligocene. There is some variation within the group. Early nimravids, such as *Maofelis* (Fig. 2.54), *Eofelis*, *Nimravus* (see pp. 237–8), *Dinaelurictis*, and *Quercylurus* (see pp. 231–2), had shorter canine teeth and longer limbs, while one genus, *Dinaelurus*, is best defined as a conical-toothed cat, exhibiting some skull similarities with the modern cheetah (*Acinonyx jubatus*). The better known later genera (such as *Hoplophoneus* (see pp. 237–8), *Pogonodon* and *Eusmilus* (see p. 230)) were imitated later on by the famous sabre-toothed machairodontine felids (see pp. 86–7). The sabre teeth were rounded in cross-section and protected by bony sheaths on the lower jaw when the mouth was closed. Short, powerful limbs would wrestle down and hold prey while the teeth were employed. Nimravids were evidently **terrestrial** hunters, having short feet, and they walked either in a **plantigrade** or semi-plantigrade posture. The ear structure of nimravids differed from that of the more advanced felids in lacking the septum that divides the two chambers in the **auditory bullae**. Juvenile nimravids were largely dependent on adults to provide food, in much the same manner as machairodontine felids. Sabre teeth erupt late and the **deciduous** teeth were not suitable for killing prey. This has potential implications on the sociability of nimravids. The Nimravidae developed in Asia, with *Maofelis* as one of the most **primitive** forms. The nimravids reached North America by the Late Eocene and Europe by the Early Oligocene. As they evolved, the nimravids continued to perform the cat-like role of ambush predators until felids evolved. Nimravids enjoyed a warm Late Eocene climate that was trending towards a cooler, drier environment into the Oligocene. They flourished as the dominant cat-like predator in the Late Eocene and Oligocene. The nimravids peaked in diversity during the early Late Oligocene (ca. 28 Ma) in what is now the Great Plains of North America, and the Big Badlands of South Dakota. Some were the size of domestic cats, while most were as large as leopards or even lions. As woodland was replaced by more open grassland, slower predators including nimravids were unable to keep pace with the fleeter-footed prey and they died out in North America around 26 Ma, going extinct altogether by the Late Miocene (ca. 7.2 Ma).

Barbourofelidae

Barbourofelids were robust, cat-like feliforms that are distinguished by features such as short, high skulls (particularly in advanced species), **posteriorly** serrated upper canine teeth with vertical grooves, and short limbs and metapodials. Most barbourofelids were

Fig. 2.54. Phylogenetic tree of Feliformia after Werdelin *et al.* (2010), and Morales *et al.* (2019).

specialised sabretooths. The fossil record of Barbourofelidae dates back to the Early Miocene of Africa (ca. 20.5 Ma). The most **primitive** forms such as *Ginsburgsmilus* are represented only by dental and mandibular material. The land bridge that linked Africa to Eurasia by the end of the Early Miocene enabled barbourofelids to disperse from Africa to Europe on at least three occasions. Some of the first European forms included species of the genus *Prosansanosmilus*. *Afrosmilus*, known from sub-Saharan Africa, is somewhat more derived than earlier barbourofelids.

Sansanosmilus (see pp. 274–5) originated in Europe, while other barbourofelids dispersed eastwards across Eurasia and entered North America during the Late Miocene. The European genera such as *Sansanosmilus* generally evolved into larger and more sabre-toothed forms, although their specialisations were less extreme than those of the North American *Barbourofelis* (Werdelin and Yamaguchi, 2010) (Fig. 2.54). The latter somewhat resembled a cross between a bear and a sabre-toothed cat, and is distinguished by hypertrophied upper canine teeth, a **postorbital bar** and the shortening of the cranium behind the **orbits**. It also had massive **carnassials**, ideal for efficiently processing a carcass. *Barbourofelis* had semi-**plantigrade** limbs showing very large muscular insertions, and its back was shortened, with limited lateral movements. The Barbourofelidae were a short-lived family, having vanished by the early Late Miocene (ca. 9.5 Ma) possibly because of competition from sabre-toothed felids.

Felidae

Felids are familiar to us, wild and domestic, fossil and living, and occur worldwide except in Antarctica (and also not native to Australia). In general, the felids have radiated into fast, flexible and powerful predators, adapted to actively pursue prey on the ground and climb trees. There are 41 species of felids in eight genera alive today, ranging in size from the rusty-spotted cat (*Prionailurus rubiginosus*), measuring up to 50 centimetres long and preying on various kinds of small agile vertebrates, to large **ambush predators** such as tigers (*Panthera tigris*), which weigh several hundred kilograms and can dispatch prey larger than themselves. The Felidae is regarded as a sister **taxon** to the extant civet-like Asiatic linsangs (Prionodontidae), both groups which may have diverged about 35.2-31.9 Ma (Late Eocene to Early Oligocene).

The earliest known true felid in the fossil record is *Proailurus* (see pp. 248–9), which originated in Eurasia during the Late Oligocene (ca. 25 Ma). It was possibly ancestral to *Pseudaelurus* (see pp. 274–5), a more advanced felid that lived in Eurasia during the Miocene, from about 20–10 Ma. *Pseudaelurus* is a **paraphyletic** genus, with different species groups being ancestral to different felid lineages. When *Pseudaelurus* immigrated to North America via the Bering Land Bridge (see p. 174) around 18.5 Ma, it ended the "cat gap" and started the massive evolutionary diversification of felids on that continent (concurrent with the diversification in Eurasia). *Pseudaelurus* was somewhat larger than a domestic cat, although its limbs were relatively shorter and it had an elongated torso like a civet, indicating that it was **scansorial**. From *Pseudaelurus*-**grade** species, the branching of all the felid lineages occurred in the Old World. The modern lineages, the subfamilies Felinae and Pantherinae, possibly split roughly 16.4 Ma, during the Middle Miocene. The Felinae comprises many of the modern cats familiar to us today such as cougars, bobcats, lynx, cheetahs and various small cats, whereas the Pantherinae includes lions, tigers, leopards, jaguars and snow leopards. One extinct lineage, the subfamily Machairodontinae, constitutes the diverse range of sabre-toothed cats, which first evolved in Europe during the Middle Miocene, before migrating northwards during the Late Miocene, then eventually down to South America once the Panama Isthmus formed around 3.5 Ma. *Miomachairodus*, from the Miocene of Turkey and China, is the oldest known machairodontine. With their large, serrated upper canine teeth, machairodontines were adapted to prey on large-bodied **megaherbivores**. Not all sabre-tooths were the same, with some variation in the size and form of the upper canine teeth. This also had an impact on hunting style. Generally, machairodontines fall into four tribes.

The Machairodontini (e.g. *Miomachairodus*, *Hemimachairodus*, *Machairodus* (see pp. 278–80)) originated during the Middle Miocene and are distinguished by long upper canine teeth (though shorter than other more famous sabre-toothed cats such as *Smilodon* (see pp. 318–9, 322–4, 329–30)), elongated limb bones, and medium-length tails that did not reach the ground. The functional morphology of the killing bite in *Machairodus*, and characters related to this behaviour, were considerably more **primitive** than in later machairodontins from the Eurasian Late Miocene. The Metailurini is characterised by upper canine teeth longer than in felines, but smaller than in true sabre-toothed cats (smilodontins). The Metailurini was clearly an Old World group, with the evolution of *Metailurus* (see pp. 282–3) centred in Eurasia and that of *Dinofelis* (see pp. 284, 302, 316–7) in Africa. Both of these genera originated in the Miocene and survived into the Plio-Pleistocene; *Metailurus* was mainly a Miocene genus, while *Dinofelis* had its main radiation in the Pliocene.

Smilodontini, the "dirk-tooths", had elongated, narrow upper canine teeth, a wide gape, a stocky body for wrestling prey to the ground, and shortened tarsal bones. The group includes *Smilodon*, a classic dirk-tooth with a body shape not dissimilar to that of a bear, along with the broadly distributed genus *Megantereon* (see p. 302). *Megantereon* originated in Africa during the Late Miocene (ca. 5.7–7 Ma), having later dispersed to Eurasia, then the Americas, where it probably evolved into *Smilodon* by the start of the Pleistocene. *Smilodon* inhabited the Americas from approximately 2.5 Ma. The Homotherini, or "scimitar-tooths" (e.g. *Homotherium* (see pp. 323–4, 343–4), *Xenosmilus* (see pp. 320–1), *Dinobastis*), is estimated to have split from the Smilodontini around 18 Ma. The homotherins had shorter, broader upper canine teeth, and a body and limbs adapted for running. They were more diverse and less specialised than the dirk-tooths. They were generally leaner cats, yet include some of the largest felids known in genera such as *Amphimachairodus* (see pp. 282–3) and *Homotherium*. The homotherins spread across Africa, Eurasia and North America. The machairodontinae became extinct shortly after the end of the last Ice Age (ca. 11.7 ka), with the extinction of *Homotherium* and *Smilodon*. If sabre-toothed predators were specialists in hunting the largest prey, which other **carnivores** could not kill, then their disappearance less than 10 ka may be associated with the extinction of all the megaherbivores that lived during the last Ice Age (see pp. 312, 327).

Within Pantherinae, the genus *Panthera* is sister to *Neofelis* (clouded leopards) (Fig. 2.55A). The

Fig. 2.55A. Phylogenetic tree of Felidae, particularly **taxa** outside Felinae, after Piras *et al.* (2013), Li *et al.* (2016), and Hemmer (2023).

geographic origin of pantherines is most probably Central or Northern Asia. *Panthera blytheae*, known from a single skull discovered in the Tibetan Plateau and dated to the Late Miocene (ca. 5.95–4.1 Ma), was originally considered the oldest known *Panthera* species. However, one recent study (Hemmer, 2023) suggests that this species plus *"Miopanthera" pamiri* are more closely related to *Neofelis* (clouded leopards) than extant pantherines, and both species have been placed in a new genus called *Palaeopanthera*. The study also suggests that modern clouded leopards are surviving and specialised "palaeopantherines"—closely related to *Palaeopanthera*. *Panthera blytheae* was likely not the common ancestor of all pantherines.

Fossil evidence suggests that Africa played a central role in the early diversification of the genus *Panthera*, even if the genus itself likely originated in Eurasia. One of the earliest known African representatives is the lion-sized *Panthera principialis* from the Late Pliocene of Laetoli (see p. 302), which may have descended from a Eurasian "palaeopantherine" ancestor that dispersed into Africa. In this view, the earliest large-bodied *Panthera* species began spreading across Africa's increasingly open landscapes well before the evolution of modern lions and leopards, highlighting Africa as a key centre of *Panthera* evolution and adaptation during the Late Pliocene. The genus had likely dispersed into Africa by around 3.3 Ma or earlier. Within this framework, the tiger (*Panthera tigris*) and snow leopard (*Panthera uncia*)—estimated to have diverged between 3.7 and 2.7 Ma—are considered sister taxa. The extinct species *Panthera gombaszoegensis*, or European "jaguar" (see p. 341), is the oldest known pantherine cat to have inhabited Europe (ca. 1.95–1.77 Ma). It was probably closely related to the modern tiger, which makes sense as its range was much closer to the tiger than the jaguar. *Panthera shawi*, known from southern Africa, possibly predates the *P. leo-P. pardus* divergence (ca. 2–3 Ma), and is often regarded as a primitive lion or a close **stem**-relative. During the Early Plesitocene, jaguars (*P. onca*) evolved in the Old World before dispersing into North America by about 1.2 to 1 Ma, and then South America by the Middle Pleistocene (ca. 0.5 Ma). Lions and leopards are considered the final *Panthera* species to disperse out of Africa, likely by the late Early Pleistocene.

The earliest undisputed leopard fossils are from East Africa, at about two million years old. Leopards originally entered Eurasia sometime between 1.2 and 0.6 Ma (late Early to early Middle Pleistocene). The most recent European species, the European Ice Age leopard (*Panthera pardus spelaea*), originated during the early Late Pleistocene and persisted until the Early Holocene in the Iberian Peninsula. Lion-like pantherines are thought to have evolved in East Africa, with the first fossils dating from around 1.4 Ma (Early Pleistocene). The earliest lions outside of Africa were found in Western Siberia around 610 ka. Known as *Panthera fossilis*, or Mosbach lions, these lions were very robust in proportions and differed in cranial morphology from modern-day lions. Mosbach lions were the common ancestral stock (and very similar morphologically) for other steppe lions in the Late Pleistocene, the Eurasian cave lion *Panthera spelaea* (see pp. 343–4) and the American lion *Panthera atrox* (see pp. 323–4). Genetic data from fossils indicates that *Panthera spelaea* and *Panthera atrox* are sister taxa that diverged about 340 ka. *Panthera atrox* and *Panthera spelaea* became extinct at the end of the Pleistocene without leaving any descendants on other continents. Modern lions may have been widespread across Africa during the Middle Pleistocene, with the Late Pleistocene seeing the beginning of their divergence in sub-Saharan Africa.

Regarding the subfamily Felinae (Fig. 2.55B), all extant species are descended from a common ancestor that arose in Asia during the Late Miocene, according to DNA analysis. All genera of the Felinae split between 10.67 and 4.23 Ma (Li *et al.* 2016). They dispersed to Europe, Africa and the Americas throughout at least 10 waves of migration within the span of 11 million years (Johnson *et al.* 2006). These dispersals were facilitated by climatic fluctuations and sea level falls. Recent genetic analysis of the Felinae suggests that the common ancestor of the modern genera *Leopardus*, *Lynx*, *Puma*, *Prionailurus*, and *Felis* had entered the Americas via the Bering Land Bridge 8–8.5 Ma. These genera subsequently split off in that order. Then around 2–4 Ma, North American felines arrived in South America during the Great American Biotic Interchange, or GABI (see pp. 252, 295), after the Panama Isthmus formed.

The American "cheetah" *Miracinonyx* (see pp. 324, 326) was more closely related to the cougar (genus *Puma*) than the cheetah. The similarities between *Miracinonyx* and cheetahs are a result of **parallel evolution**. *Miracinonyx* and the cheetah are thought to have diverged from a cougar-like ancestor during the Late Pliocene (ca. 3 Ma). Some studies propose that the cheetah lineage split off from the cougar lineage in the Americas before the former dispersed back to Asia and then Africa, whereas other studies propose the cheetah split off in the Old World instead. The earliest cheetah fossils are known from Southern and East Africa, and are 3 to 3.5 million years old. The giant cheetah (*Acinonyx pardinensis*), considerably larger and slower relative to the modern cheetah (*Acinonyx jubatus*), lived in Eurasia, and Southern and East Africa, during the Plio-Pleistocene. The modern cheetah originated in Africa during the Early Pleistocene (ca. 1.9 Ma), with a fossil record limited to that continent. The genus *Felis* originated approximately 12 Ma and later radiated into numerous modern small cats such as the African wildcat (*F. lybica*), the wild ancestor of domestic cats. Among the earliest modern *Felis* species was *F. lunensis*, which originated during the earliest Pleistocene (ca. 2.5 Ma), and was possibly ancestral to the European wildcat (*F. silvestris*).

Viverridae

The Viverridae includes 33 species of civets, genets and binturongs, found largely in the tropical regions of Africa and Asia. Although their size and body proportions are similar to those of mongooses or weasels, viverrids are largely adapted for an **arboreal** existence instead of the **terrestrial** lifestyle of herpestids, since they possess semi-retractile claws and long, frequently **prehensile** tails. It is generally accepted that viverrids, particularly the palm civets (*Nandinia*), possess morphological traits that are considered **primitive** and similar to those of the common ancestor of Feliformia. The African palm civet is really a basal side branch within the Feliformia, and not closely related to all other feliform families, as evident from molecular data. The fossil record of viverrids is rather sparse, dating back to the Eocene of Eurasia, and is represented in Africa.

Hyaenidae

After felids, the most familiar feliforms are the hyaenids and their close relatives, the percrocutids. A common cultural stereotype of hyaenas is that they are greedy, stupid, cowardly and ugly. This is not just

Fig. 2.55B. Phylogenetic tree of Felidae, particularly taxa within Felinae, after Piras *et al.* (2013), and Li *et al.* (2016).

a Western viewpoint; some stories stem from native African folklore, but it differs greatly between ethnic groups. **Scavenging** and robbing graves certainly does not tend to bestow a good name! But the reputation is misplaced as hyaenids include active apex predators, complicated clan dynamics, perform cooperative problem-solving, include secretive **insectivores**, and most famously, species which can crack open bones to access the marrow. Today, only four species constitute the entire hyaenid family (one of the smallest mammalian families) (Fig. 2.56). However, there is an extensive fossil record, with widespread diversity across over 100 named species.

Bone-cracking is a relatively recent development in hyaenids. The earliest forms from the Early Miocene originated in Europe and were insect/small vertebrate hunters. Genera such as *Protictitherium* (see pp. 279–81) and *Plioviverrops* were more like civets or mongooses, even capable of climbing trees. During the Miocene, hyaenids split into two main niches: the agile dog-like forms and the heavier bone-crackers. The dog-like hyaenids, such as *Ictitherium* (see pp. 284–5), were very successful during the Miocene, but showed a massive decline prior to the Pliocene, as canids (see pp. 92–3) migrated west across the Bering Land Bridge (see p. 174). It was not completely one-way traffic, as hyaenids also migrated in the other direction. One species of the so-called “running hyaena” *Chasmaporthetes* (see pp. 305–7) lived alongside canids in North America. It only died out following the arrival of large wolf-like canids of the modern genus *Canis* from Eurasia. Today, we are left with only one of the dog-like hyaenids, the specialised, termite-eating aardwolf (*Proteles cristatus*).

Not directly competing with canids allowed the bone-cracking hyaenas to maintain a dominance that continues today (some borophagine canids were also specialised bone-crackers but restricted to North America). The best known genus outside extant forms is *Pachycrocuta*. It was as heavy as a lion (*Panthera leo*), despite standing only as tall as a modern spotted hyaena (*Crocuta crocuta*). Various human species interacted with *Pachycrocuta*, such as at the famous Dragon Bone Hill site in China where half of the *Homo erectus* (see pp. 158–60) bones show signs of hyaena gnawing. It was a less **cursorial** genus compared to spotted hyaenas, and probably spent relatively more time scavenging carcasses than hunting. Across its range, *Pachycrocuta* was replaced by the more versatile modern genus *Crocuta* during the Middle Pleistocene.

Fig. 2.56. Phylogenetic tree of Hyaenidae after Turner *et al.* (2008).

Percrocutidae

One family which has been a puzzle for some time are the large, hyaena-like percrocutids (*Dinocrocuta* (see pp. 282–3) and *Percrocuta*), which lived from the Middle Miocene to Late Pliocene. Regularly placed as a sister family to hyaenids, or linked to the basal "stenoplesictids" (see p. 84), percrocutids were long considered a **convergent durophagous** lineage to hyaenas. A recent study of morphological features in *Dinocrocuta* appears to show that they are likely a side branch to the hyaenids, occupying the niche of large **carrion** eaters before true hyaenas.

Herpestidae

In addition to felids and hyaenids, other feliforms include a family called the Herpestidae (mongooses, meerkats, and kin), consisting of abundant small-bodied predators from Africa and Asia divided into 34 different species in 15 genera. Although not particularly abundant as fossils, the oldest known herpestids date back to the Early Miocene of Europe (ca. 21.8 Ma), and they also have a rich fossil record in Africa and Asia. Forming a sister **taxon** to Herpestidae is the family Eupleridae from Madagascar, comprising the cat-like fossa (*Cryptoprocta ferox*), the Malagasy civet (*Fossa fossana*), and various other mongoose-like mammals **endemic** to the island. Although these were formerly regarded as herpestids, recent molecular studies have revealed that they actually belong to a distinct family which independently evolved on Madagascar, whose ancestors diverged from those of mongooses during either the Late Oligocene or Early Miocene.

Caniformia

All carnivorans that do not belong to the cat line (Feliformia) are placed in the dog line, the Caniformia. In addition to dogs, these comprise the bears, weasels, raccoons and their relatives, and pinnipeds (see pp. 96–8), along with extinct families such as the amphicyonids ("bear-dogs") (Fig. 2.57). While they may not seem to have much in common, these mammals exhibit certain specialised characteristics of the braincase and inner ear region (e.g. single-chambered **auditory bulla**) that are unique to the Caniformia. In the 1990s and 2000s, the validity of the Caniformia was supported when molecular analyses revealed that all the **taxa** in that suborder are closely related to each other and are distant relatives of feliforms.

Amphicyonidae

Between the Eocene and Pliocene, the amphicyonids, or "bear-dogs", were a major group in Eurasia, North America and to a limited extent in Africa. Despite appearances, amphicyonids were neither bears nor dogs, but were their own distinct family. They had features in common with other caniforms. They were also probably basal to the other groups within Caniformia. The earliest members, such as *Gustafsonia* and *Cynodictis* (see pp. 230–1), were small and civet-like, while later lineages saw dog-sized genera such as *Daphoenus* (see pp. 237–8) from the Late Eocene and Oligocene, and the Early Miocene *Daphoenodon* (see pp. 254–6). Huge, bear-sized genera subsequently evolved in both North America and Eurasia, the largest being *Ischyrocyon* and especially *Amphicyon* (see pp. 274–5), the former estimated to have weighed up to 550 kilograms (1,212 pounds) and the latter up to 773 kilograms (1,704 pounds). These were the largest predatory mammals seen up to that time in North America. Smaller amphicyonid genera were **digitigrade** while large genera, such as *Amphicyon*,

Fig. 2.57. Phylogenetic tree of Caniformia after Flynn *et al.* (2005), Law *et al.* (2018), and Paterson *et al.* (2020).

reverted to a more weight-bearing **plantigrade** posture. As they got larger, amphicyonids switched from the more slender dog-like build, walking on the tips of their toes, to a fully robust, bear-like build, walking on the palms of their feet and hands. It is unknown whether they utilised pack behaviour during hunting, but amphicyonids substantially declined after their peak in the Miocene, probably because of increased competition with canids.

Canidae

The earliest canids are known from the Late Eocene (ca. 40 Ma) of North America, represented by a small, weasel-shaped form named *Prohesperocyon*, whose fossils have been discovered in southwestern Texas. The diagnostic canid features within *Prohesperocyon* include the absence of the upper third molar (contributing to a shift towards an increasingly shearing bite), and an expanded **auditory bulla** in the middle ear region. The limbs of *Prohesperocyon* were possibly slightly more elongated than its precursors, and had **digits** that were parallel and closely spaced, unlike the splayed toes of ursids. Shortly afterwards, the Canidae split into three subfamilies, all estimated to have branched during the latest Eocene: the Hesperocyoninae, Borophaginae and Caninae (Fig. 2.58). The Caninae is the sole subfamily that persists today, comprising jackals, foxes, coyotes, wolves and domestic dogs. Within each subfamily, there is a general trend of increasing body size over time. All three canid subfamilies are known in North America's fossil record from the Early Oligocene. Hesperocyoninae was the oldest and most basal of these lineages, consisting of the coyote-sized *Mesocyon* (see pp. 237–8) from the Early Oligocene to Early Miocene (ca. 30–20 Ma). These **primitive** canids were possibly adapted as speedy pursuit predators in grassland biomes and looked similar to extant civets. Hesperocyonines ultimately vanished by the Middle Miocene due to competition with borophagines and felids (see pp. 86–9). The primitive hesperocyonine genus, *Hesperocyon* (see pp. 237–8), was possibly ancestral to *Archaeocyon* (see pp. 237–8) and *Leptocyon* (see pp. 255, 258). These latter two lineages gave rise to the borophagine and canine diversifications, respectively.

The borophagines, or "bone-crushing dogs", evolved robust jaws and teeth (as in modern hyaenas). These culminated in bear-sized genera such as *Epicyon* (see pp. 260–1) and *Borophagus* (see pp. 305–6), and as well as being the major North American predators throughout the Miocene, occupied the niche of hyaenas too. Borophagines went extinct during the earliest Pleistocene, perhaps because of competition with felids and members of the extant subfamily Caninae, having been replaced by the latter. During the Late Miocene (ca. 8 Ma), species of the genus *Eucyon* migrated from North America to Asia once the Bering Land Bridge (see p. 174) emerged, and went on to populate Europe. The genera *Vulpes*, *Urocyon* and *Canis* evolved from canids originating in North America, where the Caninae started diversifying. The evolution of lower **carnassials** suited for both chewing and shearing was essential to the success of these genera. During the Early Pliocene (ca. 5 Ma), certain species of *Eucyon* from the Old World gave rise to the earliest species of *Canis*. *Canis lepophagus* (see pp. 305–6) originated in North America during the Early Pliocene (ca. 4–5 Ma) and is hypothesised to have been ancestral to the modern coyote (*Canis latrans*).

Once the Panama Isthmus formed around 3 Ma, North American canids immigrated to South America and later radiated there, although the youngest common ancestor of South American canids existed in North America around 4 Ma, and because more than one lineage exists in South America today, more than one invasion possibly occurred across the Panama Isthmus. The extinct dire "wolf" (*Aenocyon dirus*) (see pp. 323–4) and the extant grey fox (*Urocyon cinereoargentus*) are two North American lines present in South America. The genus *Aenocyon* forms a highly divergent lineage, separating from the last common ancestor of all living canids around 5.7 Ma. Thus, *Aenocyon* is not closely related to modern grey wolves and evolved in isolation in North America. There are also South American **endemic** species, including the South American foxes (*Lycalopex*), the bush dog (*Speothos venaticus*), the crab-eating fox (*Cerdocyon thous*), the short-eared dog (*Atelocynus microtis*), and the maned wolf (*Chrysocyon brachyurus*). These species form a **monophyletic** group as supported by molecular studies.

The lineage of North American wolves originated during the Early Pleistocene, with the emergence of *Canis edwardii* (see p. 320), easily recognisable as a wolf,. The Armbruster's wolf (see p. 320) originated in North America approximately 800 ka. This wolf roamed across North and Central America before being replaced by *A. dirus*, which later immigrated to South America by the Late Pleistocene. Various subspecies of grey wolf (*Canis lupus*) had evolved by around 300 ka and dispersed across Eurasia. Throughout the Late Pleistocene, the grey wolf dispersed across the Bering Land Bridge into North America at least three times, with each arrival comprising one or several distinct lineages of Eurasian grey wolf. According to DNA analyses, there are at least four modern lines of *Canis lupus*. The grey wolf coexisted with the dire "wolf", the latter which went extinct by around 9.5 ka. One likely factor was the loss of the **megaherbivores** on whose carcasses it depended upon, as it was mostly a **scavenger**.

A 2015 genetic analysis of Eurasian and African canids proposed that during the Pliocene and Pleistocene, modern wolf-like canids dispersed from Eurasia to Africa on at least five occasions, which corresponds to fossil data indicating that the majority of canids in Africa descended from the arrival of Eurasian progenitors (Koepfli *et al.* 2015). This likely corresponds to climatic shifts between humid and arid intervals during the Plio-Pleistocene. The study also suggested that the Eurasian golden jackal (*Canis aureus*) split off from the coyotes and grey wolves around 1.9 Ma, and then the African wolf diverged about 1.3 Ma. The Ethiopian wolf (*Canis simensis*) separated shortly before.

Ursidae

Bears appeal to us. Why is not exactly certain, but humans have coexisted with bears throughout Eurasia and North America for thousands of years, and are a familiar part of our culture. These animals are also the largest **terrestrial** carnivorans on every continent on which they live. Some of the earliest genera that were thought to be ursids, such as *Parictis* (an amphicynodontid), are now accepted as **stem**-pinnipedimorphs (i. e. seals; see pp. 94, 96). It is unsurprising that there was confusion about correct placement as the ancestral ursoids, pinnipeds and musteloids are closely related (Fig. 2.57). Instead, the earliest forms were Miocene ursavines, such as *Ballusia* and *Ursavus*, and the more divergent hemicyonines. *Ursavus* proliferated into multiple species in Asia, ranging from cat- to wolf-sized, and

Fig. 2.58. Phylogenetic tree of Canidae after Tedford *et al.* (2009), and Koepfli *et al.* (2015).

was part of the ancestral stock from which modern bears evolved. The hemicyonines first date from the Oligocene, with the Eurasian dog-like *Cephalogale* (see pp. 230–1). Later hemicyonine genera such as *Hemicyon* (ca. 16–13 Ma) and *Plithocyon* (ca. 15 Ma) (see pp. 274–5) migrated from Eurasia back to North America. Often known as "dog-bears" (which is confusingly similar to the unrelated amphicyonid "bear-dogs" (see pp. 91–2)), hemicyonines were the size of a small bear, but with the proportions of a large felid (see pp. 86–9). The hemicyonines chased after prey on **digitigrade** feet. True ursids are divided across three main subfamilies: ailuropodines, tremarctines and ursines (Fig. 2.59). A further aberrant form was *Agriotherium*, a large but lightly-framed ursid from the Late Miocene to Late Pliocene of Eurasia, Africa and North America. It was thought to be more predatory than the typical *Ursus* and statistical analysis of the short-snouted skull shows that *Agriotherium* had an exceptionally strong bite that may have been used for cracking bones. Giant pandas (ailuropodines) have not always been considered true bears, and were traditionally linked to the much smaller red panda (*Ailurus fulgens*), the latter which is now more closely allied to the mustelids (see p. 99)). They are amongst the most aberrant bears, having numerous **adaptations** for their mostly **herbivorous** diet. While close fossil relatives such as *Kretzioarctos* (known from the Middle to Late Miocene of Spain), show some similar features, most ailuropodines, such as the Late Miocene genus *Indarctos* (see pp. 279–80), were generalists.

Tremarctine short-faced bears include some of the largest known ursids and terrestrial carnivorans. The tremarctines diverged from the ursines after a dispersal event into North America during the Middle Miocene (ca. 13 Ma). They arrived in South America by 1.2–2.5 Ma, after the Panama Isthmus formed. The oldest known tremarctine from North America is *Plionarctos*, living about 10–2 Ma. *Plionarctos* was possibly ancestral to 900 kilogram giants including the North American and South American short-faced bears (*Arctodus* (see pp. 320–3) and *Arctotherium* (see p. 330), respectively), and to the spectacled bears (*Tremarctos*). Spectacled bears consist of fossil species from the Pleistocene of North America (*T. floridanus*), and the modern species (*T. ornatus*) from the mountainous regions of South America, which is the sole surviving tremarctine. Most extant ursids belong to the ursines, and range between the small tropical sun bears (*Helarctos melayanus*) all the way up to massive polar bears (*Ursus maritimus*). The ursines underwent a significant evolutionary radiation in the Late Miocene and Pliocene (ca. 5.3–4.5 Ma), correlating with notable climate shifts. This interval also marks the emergence of the earliest species within the genus *Ursus*. The sloth bear (*Melursus ursinus*) is a surviving relic of one of the first branches to split during this diversification; it evolved its strange anatomy, associated with a diet of ants and termites, by the Early Pleistocene at the latest. The species *U. minimus* originated in Europe around 3–4 Ma; except for body size, it is similar to the extant Asiatic black bear (*U. thibetanus*) in skeletal structure.

U. minimus was most probably ancestral to all modern ursines, except for perhaps the sloth bear. From *U. minimus*, two branches diverged: the brown bears (including the polar bear) and the black bears (comprising the American black bear (*U. americanus*), the Asiatic black bear and the sun bear). The fossil bear *U. etruscus* (see pp. 309–10) seems to have descended from *U. minimus*, the former which gave rise to both extant brown bears (*U. arctos*) and an extinct, specialised lineage of robust Pleistocene cave bears (there were several species). Largely restricted to boreal forest across Eurasia, cave bears lived up to their name, with populations being closely linked to the same locations every winter. Compared to other bears, it appears that cave bears were communal hibernators. Being largely herbivorous, their diet was restricted during winter, leading to both sexes hibernating. Brown bears coexisted with cave bears and avoided competition at this time through being more **carnivorous** than current populations. The opposite occurred in North America when living alongside *Arctodus*. With the extinction of the other species, brown bears shifted to a more generalised **omnivorous** diet. In the Early Pliocene, as far back as 4 Ma, certain ursine species underwent recurrent migrations from Eurasia to North America. The latest bear species to evolve is the polar bear, which diverged from a population of brown bears that ended up being isolated in northern latitudes as a result of glacial episodes, about 343–479 ka.

Pinnipedimorpha

Although well-adapted to a life at sea, it does not take a huge imagination to see that pinnipeds (seals, sea lions and walruses) are closely related to other carnivorans. But that is where it gets confusing. True seals look and behave very differently from sea lions and fur seals, and to a lesser degree with walruses. For a start, each group swims using a different set of limbs, so a popular hypothesis advocated that all pinnipeds were not closely related—the sea lions and walruses originating from a bear-like ancestor, while the true seals came from the mustelid line (see p. 99). Morphologically, this was a poor argument and was not backed up by molecular evidence. Pinnipeds are actually **monophyletic**, with physical features such as the inner ear morphology being very similar across the families. The second problem was that the earliest known pinnipedimorphs (such as *Enaliarctos*, from the Late Oligocene to Early Miocene of North America (Fig. 2.57)) were already recognisably "seal-shaped". Without a doubt, they are members of the Caniformia, although their origins remain a mystery. This was until the recognition of some fossil species, some of which were originally passed off as extinct musteloids (see p. 98), made things clearer.

Soft-tissue in the sinuses of fossil pinnipedimorphs *Enaliarctos* and *Pinnarctidion* show remarkable similarity to extant ursids and the extinct amphicynodontids, revealing that **stem-pinnipedimorphs** were closely linked to the ursoids. Many of the amphicynodontids, such as *Amphicynodon* (see pp. 230–1, 233) and *Amphicticeps* (see pp. 248–9), and the basal ursoid *Parictis* (Fig. 2.57), have been previously considered early ursids, but in life, they probably resembled modern-day martens, being small, **terrestrial** or semi-**arboreal** predators. One notable amphicynodontid from the Miocene of North America was much larger (1.2 metres in length) and usually depicted as bear-like, the odd *Kolponomos* (Fig. 2.57). It was a powerful beast with strong neck muscles. The skull had some interesting features: stereoscopic vision, a narrow, downturned snout, a deep, robust lower jaw with a prominent chin, heavy molars, and an anchoring fulcrum (the hinge for the lower jaw) in the jaw providing exceptional power. Many of these attributes are **convergent** with sea otters (*Enhydra lutris*), while the jaw fulcrum is also seen in the similarly muscle-bound machairodontine felids (see pp. 86–7). The conclusion is that *Kolponomos* was a coastal **molluscivore** with a crushing bite. Several freshwater genera (known as semantorids) show

Fig. 2.59. Phylogenetic tree of Ursidae after Qiu *et al.* (2014), Mitchell *et al.* (2016), and Kumar *et al.* (2017).

strong similarities to otters in body form and it is not surprising that *Potamotherium* (see pp. 258–9), the oldest known genus, was long considered to be a musteloid. It was the more recent discovery of *Puijila*, from Canada, which was able to fill in the blanks. Unlike otters, *Puijila* was probably not particularly graceful in water, having to use all four limbs when swimming. The skull shows more pinniped-like features. *Puijila* is known from Early Miocene deposits, much later than the earliest known true seals, so it was not the direct ancestor of the true pinnipeds.

Pinnipedia

The Pinnipedia, or seals, sea lions, and walruses, is an important lineage within Caniformia. Their origins were long contentious, but now morphological and molecular data reveals that they are undoubtedly related to the most **primitive** ursids (see pp. 92, 94–5). There are 34 extant species in 22 genera, divided into three modern families. Firstly, the Phocidae comprises the true seals, or earless seals, which are characterised by a lack of external ears and cannot walk with their rear flippers. Secondly, the Otariidae comprises sea lions and eared seals, distinguished by external ears and the ability to rotate their rear flippers forward while walking. Thirdly, the Odobenidae consists of walruses, along with more than 50 fossil species, and lack external ears. Despite their ungainly appearance on land, pinnipeds are incredibly graceful in the water. Phocids and otariids have very different propulsion methods underwater: phocids use their broad hind feet, while otariids swim with long pectoral fins. The odobenids combine being able to walk on land like a sea lion, but use their hindlimbs for swimming.

The progenitors of the Phocoidea and the Otarioidea split during the Late Oligocene (ca. 25 Ma). Among the earliest lineages of large-bodied pinnipeds to evolve was the extinct family Desmatophocidae, appearing during the earliest Miocene (ca. 23 Ma). Desmatophocids are characterised by long skulls, the ability to swim with both pairs of flippers, and the lack of dental and cranial specialisations present in modern seals. Desmatophocids are more likely to be closely related to otariids and odobenids than to phocids (Fig. 2.61). The fossil record of Phocidae dates back to the latest Oligocene or earliest Miocene of Italy. Modern phocids of the subfamilies Phocinae and Monachinae are estimated to have diverged during the Middle Miocene (ca. 15 Ma), having first evolved in the North Atlantic before dispersing all over the world's oceans, and even into the Arctic Ocean, Mediterranean Sea, Caspian Sea and Black Sea. Among the most basal monachines was the "swan-necked seal" *Acrophoca* (Fig. 2.60). Both subfamilies possibly entered the Pacific by way of the Central American Seaway (see p. 174), prior to the formation of the Panama Isthmus. The monachines radiated southwards into the Southern Hemisphere, whereas the phocines largely remained in the Northern Hemisphere. Within Otaroidea, the families Otariidae and Odobenidae diverged around 20 Ma.

The oldest known otariids in the fossil record come from the Middle Miocene (ca. 11 Ma) of the North Pacific, represented by genera such as *Thalassoleon*. The earliest extant subfamily to diverge was the Arctocephalinae, followed by the Otariinae, and eventually the rest, which spread to the Southern Hemisphere oceans. The odobenids originated approximately 18 Ma. Walruses have a long and interesting fossil record, far beyond those tusks. Where odobenids fit in with other pinnipeds has been fraught with argument, but molecular studies are pretty clear cut that they are the sister family to the otariids. So perhaps unsurprisingly, the early odobenids looked very much like sea lions. The odobenids overlapped with otariids and the desmatophocids through the North Pacific during the Middle Miocene, although it was the walruses which were the smaller species. Only during the Late Miocene did the odobenids diversify and increase in body size.

There are three main groups of odobenids: an informal conglomeration of basal species ("imagotariines"), dusignathines and odobenines. Although not immediately recognisable as walruses, the basal "imagotariines" show skeletal features in common with other odobenids. One of the earliest "imagotarines" was the tiny *Prototaria*, from the Middle Miocene. *Prototaria* was only the size of a common seal (*Phoca vitulina*), with a distinctive flat skull and a short snout. The Late Miocene *Imagotaria* was a larger genus with larger canine teeth (as opposed to tusks). *Imagotaria* also displayed extreme **sexual dimorphism**, in which the males were considerably larger in body size and had proportionately larger teeth than the females. The teeth of *Imagotaria* were not worn down chewing on hard items, suggesting that they were dietary generalists. The most impressive odobenid was also one of the largest pinnipeds, approaching modern elephant seals (*Mirounga*) in size. With a massive 60 centimetre skull featuring a tall crest at the rear, the Late Miocene to Pliocene genus *Pontolis* had basal features but on a far more robust scale.

Dusignathines were a small group notable for a pair of forward-protruding canine teeth in both jaws. In the large Late Miocene genus *Gomphotaria*, these teeth were four thick but short, horizontal tusks. While the modern walrus uses suction on shellfish, *Gomphotaria* used the tusks to smash prey. Extreme wear on postcanine teeth shows that they were crushing up prey before ingesting. Far more familiar are the two-tusked odobenines, which include the only surviving species (*Odobenus rosmarus*). Walrus tusks, which can reach one metre in length, are used largely in male-male interactions. Territorial fighting during breeding season can be as violent as those used for hauling onto sea ice and stirring up sediment when foraging.

Fig. 2.60. Reconstructed life appearance of the lobodontin pinniped *Acrophoca longirostris*, known from the Late Miocene of Peru and Chile. Reconstructed length: 1.5 m.

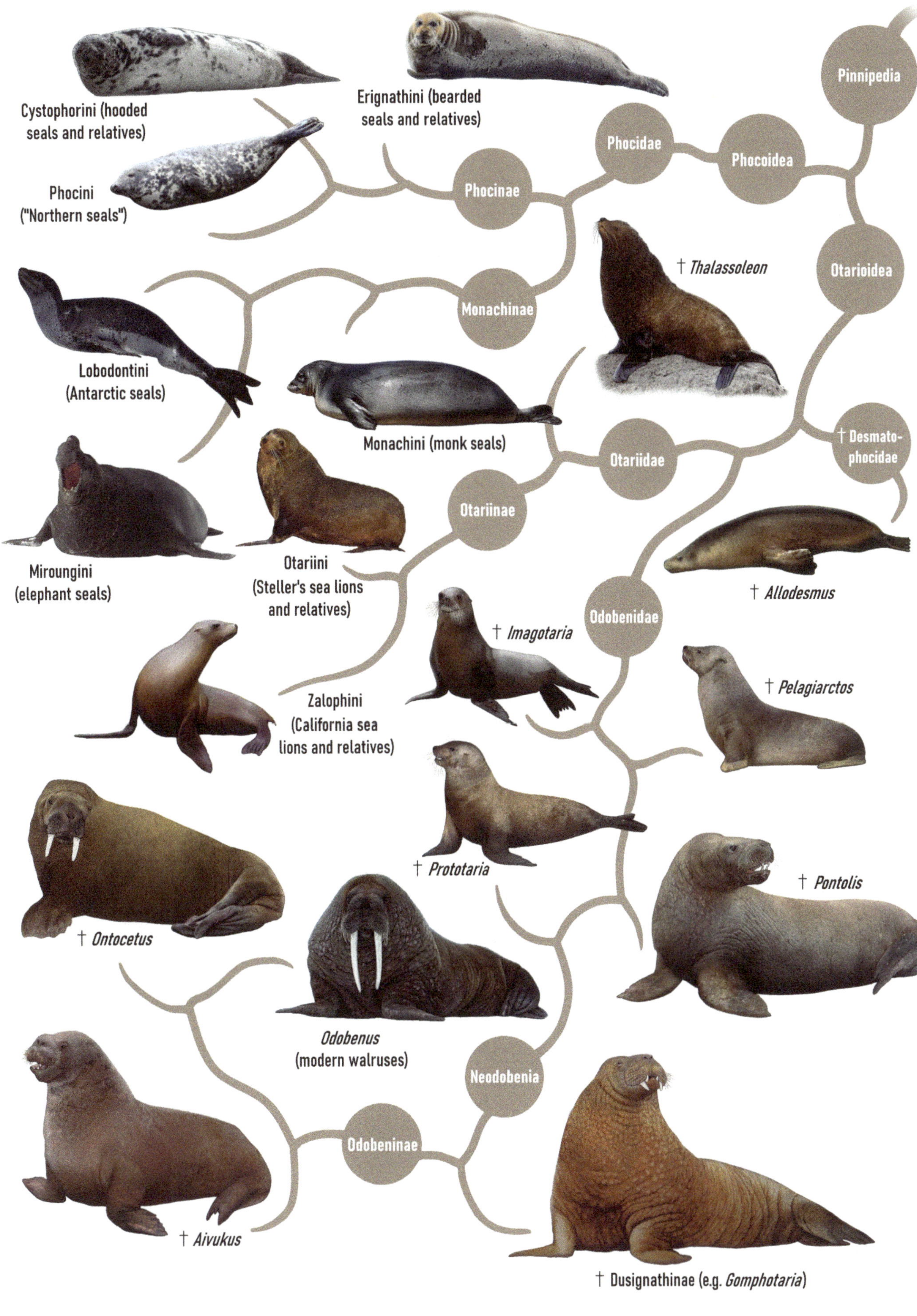

Fig. 2.61. Phylogenetic tree of Pinnipedia after Paterson *et al.* (2020) and Esteban Núñez *et al.* (2023).

Long tusks were only feasible with a move towards suction-feeding, and this has also occurred **convergently** in tusked cetaceans (narwhal, beaked whales and the extinct *Odobenocetops*). Amongst fossil odobenids, there is some variation in tusks: the Early Pliocene genus *Protodobenus* had canine teeth barely larger than sea lions, and the Plio-Pleistocene genus *Ontocetus* had forward-inclined tusks, while they were absent in the Miocene genus *Aivukus*. Odobenines were the only walruses which were found in the Pleistocene, and also colonised the Atlantic Ocean earlier during the Pliocene.

Musteloidea

Lastly, various smaller lineages of caniforms, namely the raccoons and their kin (family Procyonidae); the red panda (family Ailuridae), not a close relative of the famous giant panda (*Ailuropoda melanoleuca*); and the large family Mustelidae, are now placed in a lineage called Musteloidea as supported by recent molecular evidence, and thorough morphological analysis of the braincase and middle ear region. Skunks were formerly grouped within Mustelidae, but are now considered a distinct family called the Mephitidae.

Ailuridae

Fossil ailurids consisted of at least eight genera and 23 species. The fossil record of Ailuridae dates back to the Late Oligocene of Europe, with the oldest known representative, *Amphictis*, likely being a generalised **carnivore** according to its dental anatomy. Ailurids later migrated to Asia and North America by the Middle Miocene. *Simocyon* (see pp. 279, 281) lived across Eurasia, and in Africa and North America, from the Middle Miocene to the Early Pliocene. The Ailurinae, which originated in the Middle Miocene of Europe and encompasses the extant red panda (*Ailurus fulgens*) along with the fossil genera *Parailurus* and *Pristinailurus*, evolved specialised **cheek teeth** with blunt cusps, forming an efficient chewing surface for masticating vegetation (Salesa *et al.* 2022).

Procyonidae

Our familiarity with the procyonid family extends beyond the smart raccoons to include the monkey-like olingos and kinkajous, the long-snouted coatis of the Latin American rainforests, and the elusive ringtails and cacomistles of the North American deserts (a genus which diverged during the Late Miocene, around 10 Ma, and has remained largely unchanged) (Fig. 2.62). The procyonids are generally small, **arboreal** and **terrestrial** animals that are **omnivorous**, the structure of their **cheek teeth**, with their emphasis on crushing and grinding, being similar to that seen in bears. Kinkajous were the earliest lineage to diverge from the common ancestor of procyonids, and are considered the most peculiar. Kinkajous possess numerous arboreal **adaptations** such as long, fully **prehensile** tails, and exhibit dental modifications for a diet of primarily flowers and fruit, and sometimes eggs, insects and small vertebrates.

The fossil record of procyonids is limited, consisting mostly of teeth and jaws. Among the oldest known procyonids are the early ringtail *Probassariscus* from the Middle Miocene of the United States. Certain procyonids lived in the Old World up until the Pliocene as exemplified by *Stromeriella*, known from France and Germany. Procyonids have persisted as a New World family since the Middle Miocene,

Fig. 2.62. Phylogenetic tree of Procyonidae after Law *et al.* (2018). The phylogenetic position of *Cyonasua* and *Chapalmalania* follows Forasiepi *et al.* (2014).

diversifying into numerous genera during the Late Miocene, and into the extant genera of coatis (*Nasua* and *Nasuella*) and raccoons by the Pliocene. During the Late Miocene, around 7.3 Ma, procyonids are believed to have dispersed from Central America to South America via a series of islands before the Panama Isthmus formed during the Pliocene, giving rise to genera including the Late Miocene to Middle Pleistocene *Cyonasua* (see pp. 272–3) and a relative from the Late Miocene to Early Pleistocene roughly twice as large (the size of a spectacled bear (*Tremarctos ornatus*)), named *Chapalmalania* (see pp. 272–3). Much of recent procyonid evolution has occurred in Central America, where they have the highest diversity today.

Mustelidae

The largest family of carnivorans are mustelids, with more fossil species known than extant ones. There are at least 66 extant species in 22 genera, and typically have elongated bodies, short limbs, a short **rostrum**, and thick fur. Some mustelids have evolved to hunt a variety of prey by pursuit (wolverines), whereas others are better suited to prey on invertebrates and fish underwater (otters), or to hunt prey in burrows and on the ground (weasels), in trees (fishers and martens), or hunting small mammals by digging (badgers) (Fig. 2.63). Mustelids originated in the Early Oligocene of North America, as shown by the Oregon fossil *Corumictis*, before dispersing to Eurasia during the Late Oligocene (ca. 33 Ma), as indicated by fossils of *Plesictis*. Fossil mustelids underwent significant diversification in both North America and Eurasia throughout the Miocene, leading to a broad range of forms and sizes of otter-like, badger-like, wolverine-like and weasel-like species. Among the most familiar is *Sthenictis*, which lived in North America and Eurasia from the Middle to Late Miocene. Certain genera were rather striking, including the massive, bear-sized *Megalictis* from the Early Miocene of North America (see pp. 254–5), and the leopard-sized mustelid *Ekorus* from the Late Miocene of Africa (see pp. 284–5). Among the most spectacular is the massive "bear otter", *Enhydriodon*, which coexisted with **primitive** humans in the Pliocene of East Africa. It was up to 2.6 metres (8.5 feet) in length and is estimated to have weighed around 200 kilograms (440 pounds). Isotope studies of its teeth suggest that it was more **terrestrial** than modern otters. Throughout the Miocene, mustelids frequently travelled to and fro between Eurasia and North America, making them very good time markers since each migratory event represents a distinct time plane.

Mephitidae

Skunks are well-known for defensively spraying a pungent, foul-smelling odour from their anal scent glands. The mephitids consist of 12 extant species in four genera: two species of stink badgers from Indonesia and the Philippines; the four species of hog-nosed skunk, distributed in both the Americas; and the four species of spotted skunks and two species of striped skunks, only from North America. The mephitids appear in the fossil record in the Early Miocene of Eurasia, and they spread to North America by the Middle Miocene and became diverse there too.

Fig. 2.63. Phylogenetic tree of Mustelidae after Koepfli *et al.* (2008), and Law *et al.* (2018).

Laurasiatheria II: The Ungulates (Hooved Mammals)

Following rodents and bats, the hooved mammals, or **ungulates**, constitute the third largest lineage of placentals. Approximately one-third of all living and fossil mammal genera are hooved mammals, and ungulates account for almost all the large **herbivores**. As indicated by phylogenies based on morphological and fossil data, and molecular biology, they comprise the even-toed artiodactyls (such as camels, pigs, peccaries, giraffes, antelope, deer, cattle, sheep, goats, hippopotamuses, and whales), and the odd-toed perissodactyls (tapirs, horses, rhinoceroses, and various fossil relatives). The **crown group** of true ungulates, Euungulata, consists of the last common ancestor of living Perissodactyla and Artiodactyla, and all of its descendants. Pan-Euungulata refers to the total **clade** that includes Euungulata and all extinct mammals more closely related to Euungulata than to any other living **taxa** outside it. Not only have ungulates ruled the large herbivore niche for much of the Cenozoic, they also include the largest filter feeders and marine predators in the form of whales, and some (the mesonychians) even experimented with **carnivory**. While certain ungulates, such as horses and antelope, have elongated, gracile limbs for **cursorial** locomotion, some are massive, with robust limbs (including hippopotamuses, rhinoceroses, and numerous extinct lineages). Some are even capable of tree-climbing. As a result of the constraints of their diet and body size, ungulates have filled a broad range of ecological roles.

"Condylarths" ("Archaic" Ungulates)

The taxonomy of Paleocene and Eocene hooved mammals has always been problematic due to generalised dentitions and body forms, and historically many have been lumped together as the "condylarths". The term "Condylarthra" is a **polyphyletic** ancestral **"wastebasket" taxon** which often serves as an uninformative catch-all term for any "**primitive**" **ungulates**, meaning the group consists of species sharing primitive characteristics of ungulates as well as eutherians (see pp. 43, 57) in general (Archibald, 1998). The consensus is that many "condylarths" are part of the larger **taxon** Euungulata (which today includes artiodactyls and perissodactyls), with some turning out to be afrotherians too.

The focus on primitive characteristics in this classification masks key divergences in ancestral taxa, especially as the "condylarths" are not adaptively unified—the dentition of *Arctocyon* (see pp. 194–5) is indicative of **omnivory** towards **carnivory**, while *Phenacodus* (see pp. 206–7) was likely **herbivorous**—and their size varies significantly, from the rat-sized *Paschatherium* (see pp. 194–5) to the large (bear-sized) "arctocyonid" *Mentoclaenodon*. For these reasons, the term "**archaic**" ungulates is favoured when describing these animals, with "archaic" simply denoting the age of the taxa rather than the presence of less **derived** features (Archibald, 1998). With the acceptance that different groups of ungulates acquired hooves independently, the interrelationships between various members is uncertain. As a result, it is now much less common for taxonomists to use the grouping "Condylarthra" as a formal term. The more carnivorous mesonychians were previous members of the "condylarths" and are now generally considered basal to the artiodactyls. The herbivorous pantodonts (see pp. 137–8) were probably not closely related to any of the ungulate **clades** (although some studies suggest otherwise). The main placental divergences must have occurred rather early, as evidenced by the oldest ungulate fossils, which date to the early Late Cretaceous (ca. 85 Ma) of Uzbekistan. Despite these Mesozoic ungulates being represented largely by isolated teeth, jaw fragments and skeletal elements, they still exhibit diagnostic ungulate characteristics. Their molars are quadrate and more **brachydont**, exhibiting more rounded cusps, for processing plant material instead of insects, and the ankles already exhibit characteristics which are identifiably ungulate. The ungulates diverged into numerous distinct lineages during the Paleocene. The very early ungulate *Radinskya* (Fig. 2.64) was named too recently to have been considered a "condylarth". Known only from a partial skull from Paleocene **strata** in China, this tiny ungulate has previously been linked to the embrithopods within Afrotheria (see p. 68), but is probably a basal perissodactyl.

Fig. 2.64. Reconstructed life appearance of the basal ungulate *Radinskya yupingae*. Reconstructed length: 15 cm.

The phenacodontids of North America and Europe (e.g. *Tetraclaenodon* (see pp. 179, 181), *Phenacodus*, *Meniscotherium* (Fig. 2.67A)) consisted largely of small (domestic cat-sized) genera but ranged up to to the size of sheep. Whilst adapted for running, phenacodontids show some primitive features such as a long, heavy tail, **pentadactyl** feet (although they walked on the three largest) and large, pointed canine teeth. They do show some surprising specialisations for early herbivores in having molars with ridges of low-cusps. Fossils of the small *Meniscotherium* are often found in high concentrations, which suggests that they were **gregarious** and susceptible to mass burials. The phenacodontids may be close to the ancestry of perissodactyls, but that remains uncertain.

Possibly related to the phenacodontids (although this may be due to **convergence** on some features) were the small pleuraspidotheriids of Europe (e.g. *Hilalia* (Fig. 2.67B) *Orthaspidotherium* and *Pleuraspidotherium* (see pp. 194, 196)). They show a mosaic of features: the flat skull is similar to the ancestral artiodactyl, while the dentition are convergent with perissodactyls. The **cheek teeth** have advanced crescent ridges suggesting that pleuraspidotheriids were exclusively herbivorous. The ankle joint and **plantigrade** feet suggests that they had an ambling gait and were distinctly different from modern ungulates. While larger genera lived a full-time **terrestrial** existence, smaller genera such as *Orthaspidotherium* were possibly capable of some **arboreal** climbing.

Previously nicknamed "tube-sheep" due to their long cylindrical bodies, the hyopsodontids of North

America and Eurasia were typically small animals with primitive features. *Hyopsodus* (see pp. 204, 206–7) is the most familiar genus, with over 20 species named across North America, Asia and Europe. In fact it was very abundant and accounts for up to 30% of fossils in some locations. While *Hyopsodus* and *Tricuspiodon* (see pp. 194–5) were probably **fossorial**, some relatives such as *Paschatherium* were adapted for life in trees. Judging by their **complete dentition**, hyopsodontids were generalist omnivores, consuming vegetation and insects. Molecular analyses place them as an early branch of or basal to the perissodactyls.

One of more familiar families of Paleocene mammals were the "arctocyonids", which are commonly regarded as a **paraphyletic grade**. As their name suggests (beardog-like), "arctocyonids" were superficially similar to carnivorous mammals particularly with regard to their impressive canine teeth. While long canine teeth are a common feature in carnivorous mammals for holding prey, they are also used in **intraspecific competition** by non-predatory mammals (e.g. hippopotamuses, baboons, muntjac). Based on the unspecialised molars, it is possible that "arctocyonids" were omnivorous. They were generally robust in body form with a long tail, relatively short limbs and plantigrade feet which were not suited for running down prey, and that it is more likely that "arctocyonids" **ambushed** smaller creatures. Large forms (e.g. *Arctocyon* (see pp. 194–5), *Mentoclaenodon*) were mostly terrestrial and they filled a similar role to modern bears. While the lower canine teeth were long in all genera, the upper canine teeth were especially long in *Mentoclaenodon*, bordering on sabretooth-like in morphology. Smaller genera such as *Loxolophus* and *Chriacus* (see pp. 180–1) were possibly arboreal. Fossils of "arctocyonids" are known from the Early Paleocene through to the Early Eocene. Closely related were the herbivorous periptychids such as *Ectoconus* and *Periptychus* (see pp. 179–81), the family which is known solely from North America (Fig. 2.65).

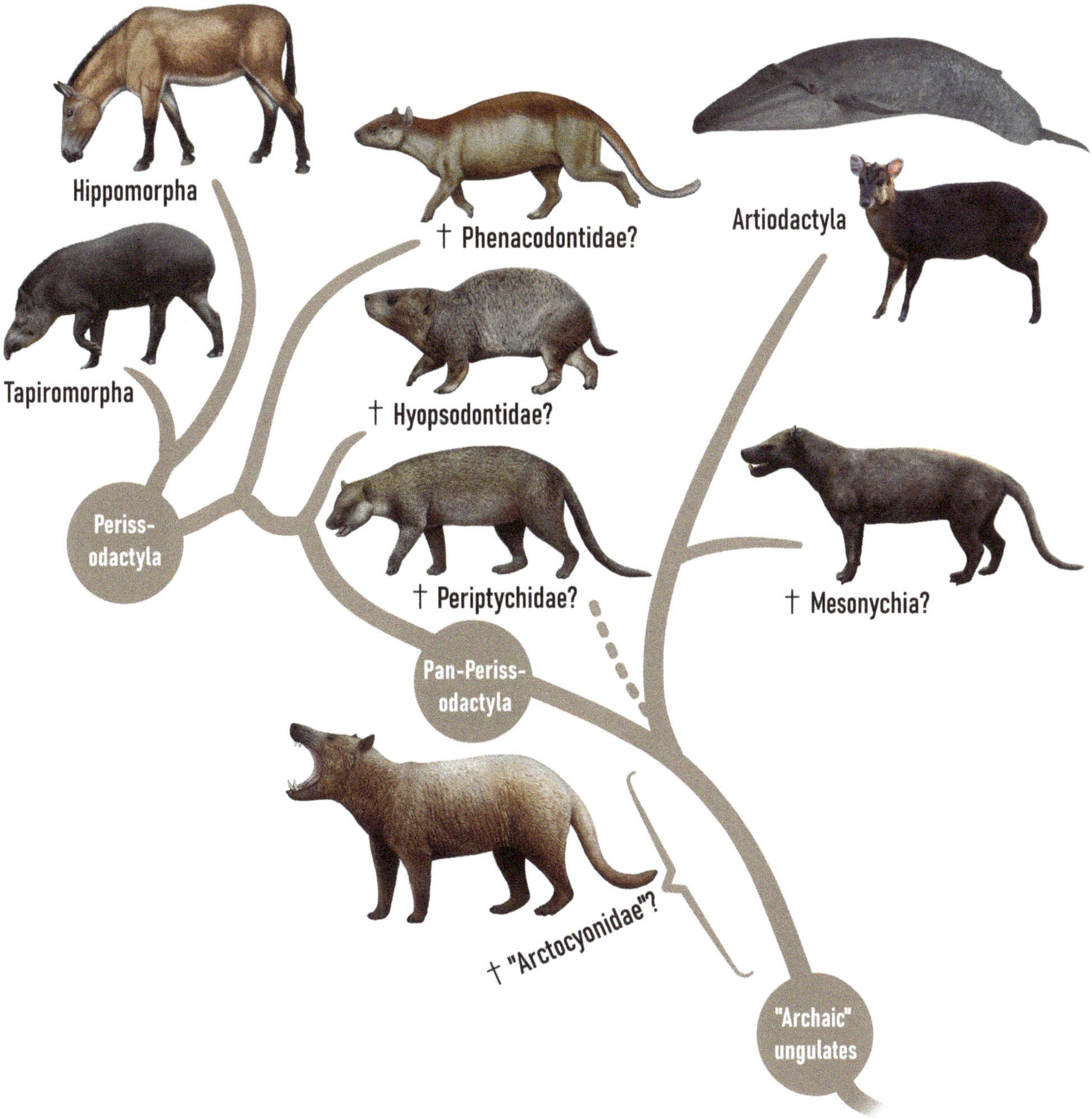

Fig. 2.65. The possible phylogenetic relationships among some representative groups of "archaic" ungulates. The phylogenetic position of Phenacodontidae follows Cooper *et al.* (2014) and Bai *et al.* (2021), while that of Periptychidae follows Puschel *et al.* (2019), that of Hyopsodontidae follows Orliac *et al.* (2012), and that of Mesonychia and "Arctocyonidae" follows Bast and Smith (2013).

Mostly known from teeth, the anthracobunids are believed to be very basal perissodactyls (or close to) with some features linking them to the bizarre marine desmostylians (see p. 69). Several genera have been found in Eocene deposits from around the Tethys coastline (India and Pakistan) (see p. 174)), including *Anthracobune* (Fig. 2.66), and isotope analysis of bones show that anthracobunids were possibly amphibious but feeding on terrestrial plants.

Fig. 2.66. Reconstructed life appearance of the anthracobunid *Anthracobune pinfoldi*. Reconstructed length: 2 m.

Also found in the same area were the closely related cambaytheriids. A lot is known thanks to hundreds of fossils of *Cambaytherium* (Fig. 2.67C) discovered in a coal pit mine in India. Features such as short toes and a **digitigrade** stance suggests that they were more **cursorial** in nature. Wear patterns on the teeth show that they were feeding on abrasive vegetation.

The most surprising of these former "condylarths" is an order of hooved predators known as mesonychians. They were the first group of mammals to become specialised meat-eaters, appearing in the Middle Paleocene before more specialised carnivorous mammals (the oxyaenodonts (see pp. 81–2) and hyaenodonts (see pp. 82–3), and eventually the carnivorans (see pp. 81, 83–4)) occupied that niche. Most mesonychians were the size and shape of large wolves or bears (although some were as small as foxes). The mesonychians had a heavy, robust skull armed with sharp canine teeth, and huge, round-cusped molar teeth suitable not only for eating meat, but also for bone crushing. Their body was also very wolf-like, with a long tail and limbs. Like many modern carnivorans, they were digitigrade instead of plantigrade. However, despite all these carnivorous **adaptations**, mesonychids were derived from hooved mammals. The proof is in their toes, which had hooves rather than claws. Compared to any other mammalian group, whales and dolphins show an exceptional specialisation to the aquatic environment. As a result, their early evolutionary history was uncertain until the use of molecular analysis. The "ideal" proto-whale was a **semi-aquatic** carnivore, with the mesonychians heavily promoted as close relatives of cetaceans. A link between both groups was the similar, unusual, triangular molar teeth (seen in "archaeocetes" (see p. 122)). Recent research has shed more light on the origins of cetaceans, and are now considered highly derived artiodactyl ungulates, with their closest extant relatives being hippopotamuses.

In general, mesonychians superficially resembled canids (see pp. 92–3), with most species being omnivores or carnivores. However, the mesonychians also share features in common with ungulates, such as blunt, hoof-like nails. Hence, they are often dubbed "wolves with hooves". The odd combination of attributes has traditionally left mesonychians lumped in with other early beasts in the wastebasket taxon "Condylarthra". Even now, mesonychians are sometimes placed as basal artiodactyls or within a wider Ferae (including carnivorans and possibly numerous fossil groups (see p. 74)).

Mesonychians were the first specialised mammalian carnivores and they dominated predatory niches throughout Eurasia and North America during the Early Paleocene to the Early Oligocene. Generally, mesonychians were larger than the hyaenodonts, oxyaenodonts and the "miacid" carnivoramorphs (see p. 84) which shared the same environments. There are three separate mesonychian lineages, the relatively robust, basal "triisodontids" (possibly a paraphyletic grade on the line to Mesonychia), small hapalodectids (see pp. 192–3) and the more familiar mesonychids (Fig. 2.68). Although the "triisodontids" reached sizes of only up to a large dog, they were among some of the largest (e.g. *Eoconodon*, *Triisodon* (see pp. 181–2)) during the Early Paleocene. "Triisodontids" are also among the oldest known

Fig. 2.67. Reconstructed life appearance of a sampling of herbivorous "archaic" ungulates. (A) *Meniscotherium chamense*, a phenacodontid from the Early Eocene of the United States; (B) *Hilalia sezerorum*, a pleuraspidotheriid from the Middle Eocene of Turkey; (C) *Cambaytherium thewissi*, a cambaytheriid from the Early Eocene of India. Scale bar: 20 cm.

Fig. 2.68. Phylognetic tree of Mesonychia after Shelley *et al.* (2015).

large carnivorous placentals, as indicated by distinct shearing edges on their **cheek teeth**. As evident from available postcrania, the "triisodontids" appear to have been nimble, but also powerful and capable of rapid terrestrial movement akin to a precursor of cursoriality. The "triisodontids" as a whole shared much of this locomotory adaptability to differing degrees, enabling the lineage to exploit a variety of niches also explored by early mesonychids. The potential for competition between these two groups may have played a factor in the apparent extinction of the "Triisodontidae" completely (by around 62 Ma), with mesonychids emerging as the dominant competitors. However, this is highly speculative, as no evidence for competition between these groups has been found.

Mesonychids were far more diverse and different in proportions. While the limb proportions may indicate cursorial behaviour, mesonychids had relatively stiff backs and thick limbs, which suggests that they they were not fast runners. In particular, the early genus *Dissacus* (see pp. 181, 183, 192–5) had the flexible elbow joints of a cursorial genus. *Dissacus* is considered to be a possibly paraphyletic genus, distributed across the whole Northern Hemisphere. Early mesonychids had plantigrade feet, although later genera developed a more digitigrade stance. Large heads with strong jaw muscles were a character also seen in the hyaenodonts. Blunt nails are not useful for holding prey, so mesonychids were very mouth-orientated. Their teeth were not as efficient for slicing meat, as the laterally compressed molars were often blunt. While *Dissacus* was an omnivore, the larger *Sinonyx* (Fig. 2.69A) and *Pachyaena* (Fig. 2.69B) were **hypercarnivores**. However, their strategies of hunting probably differed as *Sinonyx* had a long, narrow muzzle, a feature seen in hunters of small speedy prey, while *Pachyaena* had more robust jaws. Later genera such as *Harpagolestes* were capable of cracking bones with their heavily developed jaws and large teeth. Some genera such as *Pachyaena* show no differences between sexes, and were probably solitary in nature. By the Middle Eocene, the mesonychids rapidly declined in North America and Eurasia, where they had once dominated. The reason for this decline is not exactly clear, but may have been due to competition from hyaenodonts and the ancestors of modern carnivorans, and significant climatic shifts that likely impacted their habitats and food sources. Whatever the causes, the mesonychids were very rare by the late Middle Eocene, and they disappeared from North America by the end of the Middle Eocene, and mostly from Asia during the Late Eocene. Suriving into the Oligocene was the last and one of the largest known genera, *Mongolestes* (Fig. 2.69C).

Fig. 2.69. Reconstructed life appearance of a sampling of mesonychids. (A) *Sinonyx jiashanensis*, from the Late Paleocene of China; (B) *Pachyaena ossifraga*, from the Late Paleocene of the United States; (C) *Mongolestes hadrodens*, from the Early Oligocene of China. Scale bar: 1 m.

Artiodactyla

The artiodactyls, or even-toed **ungulates**, were among the earliest ungulate lineages to diverge. They are called "even-toed" because their hands and feet are **paraxonic**, so they typically have an even number of **digits**, numbering either two or four. In addition, the astragalus (ankle bone) of artiodactyls is unique in exhibiting a pulley-like facet on the dorsal and ventral surface. This greatly increases free rotation of the feet in a forward-backward motion for fast running, although it makes lateral rotation of the ankle more limited than that of more generalised mammals (Arifakis, 2022). Artiodactyls comprise the largest order of modern ungulates, encompassing at least 349 extant species, including the majority of domesticated ungulates (e.g. pigs, sheep, goats, cattle, camels) (Fig. 2.70).

The oldest known artiodactyls (including the extinct dichobunids such as *Diacodexis*) come from the earliest Eocene of Pakistan, and soon afterwards they dispersed across the remainder of Eurasia and North America. These small, **primitive**, fast-running forms were of slim build, looking like a long-tailed chevrotain, and some had such elongated hindlimbs relative to the forelimbs that they possibly had a bounding gait. The dichobunids had four or five toes on each foot, each ending in a small hoof, and **complete dentitions** suggesting a **browsing** diet. Their skulls and dentitions already had most of the distinctive specialisations present in artiodactyls. Where the artiodactyls came from remains uncertain. There is a hypothesis that, during the Paleocene, the artiodactyls evolved in isolation in India, when the subcontinent was an isolated landmass, before abruptly dispersing all over Eurasia and North America by the earliest Eocene once India had begun ploughing into Asia. However, no fossils have yet been found in India to back up this hypothesis, and it does not seem to be evident from the earliest Eocene fossils of South Asia. In 2007, a fossil called *Gunungulatum* was discovered from Late Paleocene deposits in Mongolia, exhibiting striking similarities to basal artiodactyls. This indicates that the origin of artiodactyls may predate the India-Asia collision, potentially in East Asia.

Subsequent to their dispersal all over Eurasia and North America, these early artiodactyls rapidly radiated into a broad range of families throughout the Eocene, including the **omnivorous**, heavily-built, Old World pigs; the pig-like New World peccaries; the **semi-aquatic** hippopotamuses; and various extinct pig-like **taxa** such as the enigmatic helohyids (see p. 204) and the bizarre entelodontids (see pp. 118, 120).

Miscellaneous basal artiodactyls

During the Middle and Late Eocene, when Europe remained an island archipelago, various **endemic** families of European artiodactyls evolved, eventually becoming extinct by the Early Oligocene because of competition with Asian immigrants including the more advanced ruminants (see p. 110), once both continents merged. These endemic European artiodactyls had radiated into a variety of body types.

The sheep-sized anoplotheriids, which had long limbs and tails, and camel-like heads, showed a tendency towards becoming increasingly **cursorial**, as evident from the side toes on the limbs being either highly reduced or lost. Thus, *Diplobune* and most species of *Anoplotherium* were **tridactyl**, whereas *A. commune* was **didactyl**. As evident from its skeletal anatomy, *Anoplotherium* was possibly adapted to **browse bipedally**, using its long, robust tail for support, which would have significantly lessened **interspecific competition** with contemporary **herbivorous** mammals. The gazelle-sized xiphodontids, such as *Xiphodon*, were small, highly cursorial artiodactyls which resembled the most **primitive** camels and had long, gracile limbs adept at fast running. As seen in other contemporary basal artiodactyls such as anoplotheriids, those limbs indicate a tendency towards the development of more open environments at the expense of dense Late Eocene forests in Europe.

The cainotheriids were a family of small rabbit-like artiodactyls characterised by a small skull, short snout, proportionately elongated hindlimbs, and four non-reduced toes terminating in long, claw-like hooves. One fossil trackway from Spain left by *Cainotherium* during the Early Miocene suggests that cainotheriids had a locomotion similar to that of small ruminants. Cainotheriids originated in Europe during the Late Eocene, where they were endemic, and survived until the Middle Miocene (ca. 14 Ma), disappearing because of a cooling, drying climate. Some Eocene artiodactyls from Europe, such as the cebochoerids and choeropotamids, looked somewhat similar to the oldest relatives of pigs. Other families including the more obscure amphimerycids and mixtotheriids, as well as the cainotheriids, still possessed the more deer-like limbs of the ancestral artiodactyls.

Oromerycidae

The **pentadactyl** oromerycids were a small family (just six genera) comprising early relatives of camels that lived during the Middle to Late Eocene. The oromerycids, such as *Protylopus* (less than one metre tall), were long regarded as camelids, but as evidenced by more complete fossil discoveries, the oromerycids are now recognised as a distinct North American family and close relatives of the oldest camelids. Oromerycids were similar to camelids in various aspects, such as having **hypsodont** dentition (albeit generally less hypsodont than in camelids), although they were highly **primitive** in skeletal anatomy, possessing short limbs, the retention of side **digits**, and the fusion of the hindlimb and toe bones, unlike the more **derived** features seen in early camelids. Oromerycids went extinct during the Eocene-Oligocene boundary, in addition to numerous other lineages unable to make it through the transition from the rainforests of the Late Eocene to the drier Early Oligocene **savannahs**.

Camelidae

Modern camelids fall into two closely related lineages. The true camels (Camelini) are found across a wide range that naturally extended from inland China through to the Middle East and Northern Africa, being restricted to areas of high aridity. Due to human intervention, camels have spread into Mediterranean Europe, West and East Africa, and much later into the interior of Australia. Llamas and their close relatives (Lamini) are found along the mountainous spine of the South American Andes and into the cold south of Patagonia. One continent that they are entirely absent from is North America, but in a strange twist of fate, this was actually their ancestral home. It was only by around 11 ka that the last of the North American camelids died out.

Fig. 2.70. Phylogenetic tree of Artiodactyla after Spaulding *et al.* (2009), Chen *et al.* (2019), Upham *et al.* (2019), and Weppe *et al.* (2020). The phylogenetic position of Protoceratidae follows that of Spaulding *et al.* (2009), while that of Dromomerycidae follows Janis (1987). The phylogeny within "Tragulina" follows Mennecart *et al.* (2021), except that the phylogenetic position of Lophiomerycidae and Tragulidae follows Janis and Theodor (2014).

Camelids are (surprisingly) the basalmost group of extant artiodactyls (Fig. 2.70). Most basal artiodactyl families outside Artiofabula are poorly known, emerging way back in the Eocene, and remained conservative in body shape and lifestyle. The camelids were the only one of those families to survive beyond the Miocene. The earliest confirmed camelids were the sheep-sized poebrotheriines from the Eocene and Early Oligocene, such as *Poebrotherium* (see pp. 235–6). Stenomylines, such as *Stenomylus* (see pp. 256–7), were smaller and comparable to fleet-footed gazelles of today. At the other end of the spectrum, *Capricamelus*, from the Pliocene of the Western United States, was an odd member of the miolabines, which had particularly short limbs. The limb proportions and joints in the feet are consistent with those of mountain goats (*Oreamnos americanus*), klipspringers (*Oreotragus oreotragus*) and takins (*Budorcas taxicolor*), so it is unsurprising that the genus was restricted to the dry mountainous terrain of the Southwestern United States. It also possessed a deep lower jaw and enlarged incisors, presumably an **adaptation** to deal with coarse vegetation. Camelids were also found in tropical climates, with the floridatragulines found across subtropical Southern United States and Central America during the Early to Middle Miocene. The most distinctive feature of *Floridatragulus* is its extremely long and shallow skull. Long extremities

Fig. 2.71. Phylogenetic tree of Camelidae after Scherer *et al.* (2013), and Geraads *et al.* (2021).

seem to be a regular recurrence in camelid anatomy. While modern camels have elongated necks and limbs, members of the Aepycamelinae had a particularly long neck and are known as the giraffe-camels. *Oxydactylus* (see pp. 256–7) was an early genus of that subfamily, and was able to avoid competition with other Miocene camelids and **browsers** by reaching higher into the trees. An early offshoot of the Camelinae were small species of the tribe Protolabidini. Most genera would have closely resembled small modern laminoids such as vicuña (*Lama vicugna*), although *Tanymkyter*, from the Early Miocene of the United States, had a longer, shallower skull.

Camelid genera are divided across five subfamilies (Fig. 2.71), but only the Camelinae survived past the Pliocene and migrated out of North America. Members of the Camelini (true camels) show a tendency towards gigantism. Throughout the Pliocene and Pleistocene, several genera of giant camels were found in temperate North America. They are characterised by **hypsodont** teeth, which contrasts strongly with those found in modern *Camelus*. This suggests a diet of tough, gritty vegetation from open environments such as grasses. The last genus of camel in North America was *Camelops* (see pp. 325–6), a sister genus to the similarly-sized *Camelus*. Modern *Camelus* are the smallest genus within the Camelini and are restricted to arid environments. They descend from the only lineage, *Paracamelus* (see p. 251), to migrate out of the Americas during the Late Miocene. In a striking example of **convergent evolution**, one group of camelids followed a lifestyle much like modern giraffes (*Giraffa*). Sometimes, the aepycamelines, or giraffe-camels (e.g. *Aepycamelus* (see pp. 260–1)), are regarded as true camelines and sometimes as the sister group to the tribe Lamini. Members of the Lamini have never been as diverse as the Camelini. Genera have been found across the Southern United States and also in South America, where extant genera now live. *Palaeolama* (see pp. 330–1), and/or possibly *Hemiauchenia* (see p. 305) (some confusion between the identity of these genera exists), had shorter limbs and a larger body than modern genera.

Merycoidodontoidea

It is quite remarkable that the most commonly represented group of mammals in the North American fossil record during the early part of the Cenozoic is so anonymous to most people. Amazingly, oreodonts are possibly the most abundantly preserved mammalian family in the entire fossil record. Upon hearing the

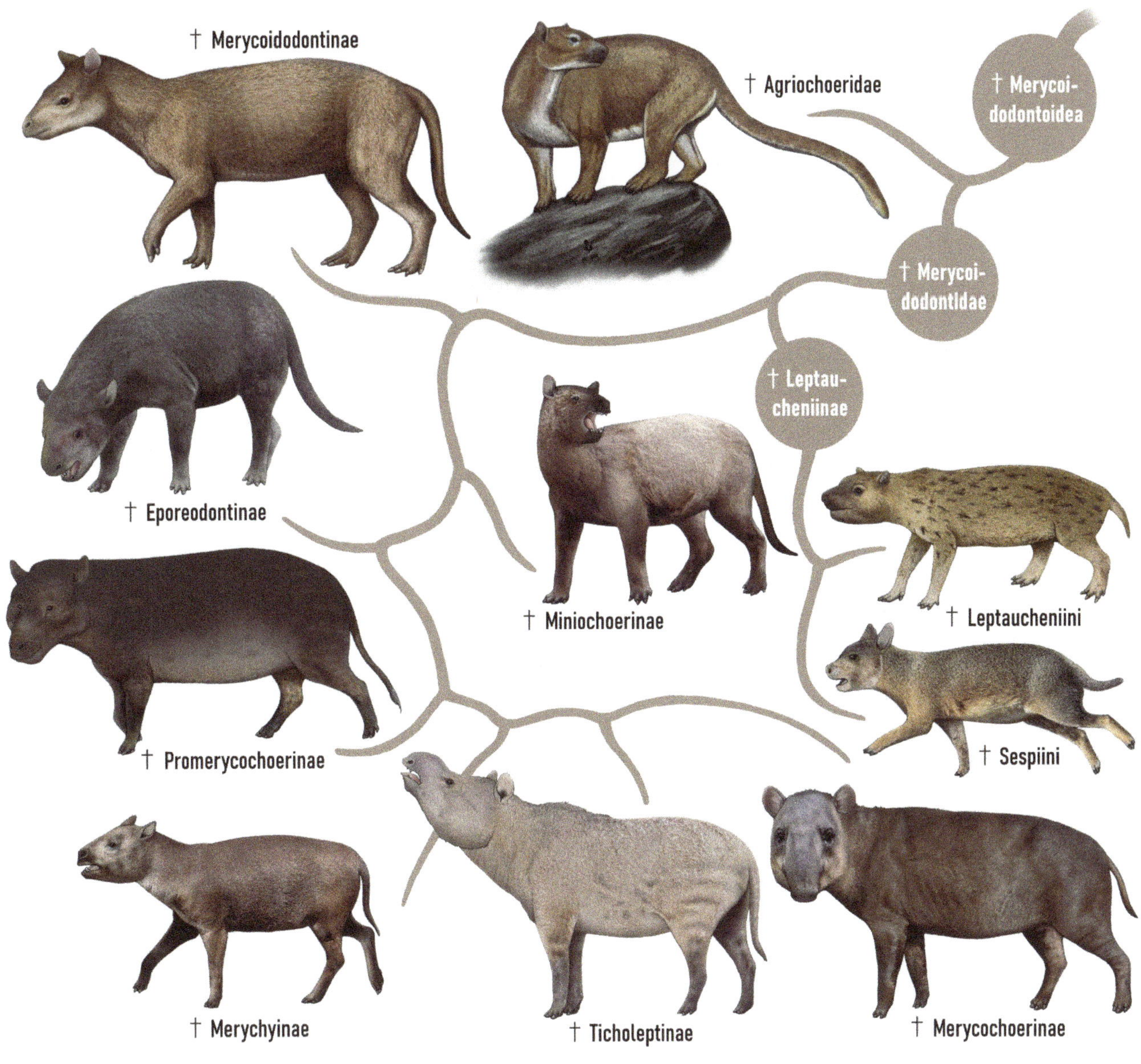

Fig. 2.72. Phylogenetic tree of Merycoidodontoidea after Mihlbachler and Solounias (2006).

name, people would be forgiven for thinking that the name has some reference to a type of biscuit! However, this informal name for the Merycoidodontoidea is based on the ridges on the molars (oreo-dont being Greek for "mountain tooth"). The Merycoidodontoidea was a diverse group of ungulates, ranging from the Late Eocene right up to the Early Pliocene (ca. 35–5 Ma). Physically, they were an incredibly varied group. Where the oreodonts sit taxonomically has caused much disagreement, with camels, Suina (pigs) or among the ruminants (see p. 110) being the usual suspects; currently, the evidence points to the former.

Despite a fair variation in morphology, oreodonts all had some **primitive** features in common such as relatively long tails, a **complete dentition** with only a small **diastema**, and enlarged canine teeth/premolars. Other features were more **derived** and present only in some species: retracted nasal bones indicating the presence of a trunk, facial depressions which may have accommodated scent-producing organs, and claw-like **digits** for climbing or digging. Early species had **brachydont** teeth for **browsing** leaves, while those from the later Miocene tended towards the harder-wearing **hypsodont** teeth of **grazers**. Unlike other **ungulates**, oreodonts did not adapt to **cursorial** life and probably were mostly forest dwellers. Only present in North America and as far south as Panama, oreodonts are known from the Middle Eocene through to late into the Miocene. With an abundant fossil record, over 50 genera of oreodonts have been named. The number of genera/species has been taxonomically oversplit and the true number is probably closer to 30. There are two closely related families—the basal agriochoerids (with two subfamilies) and the more derived merycoidodontids (around 12 subfamilies, although it depends on which taxonomy is used). Agriochoerids were either small- to medium-sized generic browsers (the Middle to Late Eocene *Protoreodon*) or possible **arboreal** genera (the Middle Eocene to Late Oligocene *Agriochoerus* (see p. 234)). They overlapped with the merycoidodontids, but died out during the Late Oligocene.

Ranging in size between a rabbit and pigs in size, the merycoidodontids were far more adaptive, a major diversification which occurred during the Late Oligocene. Leptaucheniines (the Oligocene to Early Miocene *Leptauchenia* (see p. 234) and the Late Oligocene *Sespia* (Fig. 2.72)) had high-placed **orbits** and extremely hypsodont teeth, with fossils commonly being found in dry environments. The most common and standard oreodonts were the generalist merycoidodontines (the Middle Eocene to Middle Miocene *Merycoidodon* (see pp. 234–5)) and the smaller miniochoerines (the Late Eocene to Early Oligocene *Miniochoerus* (see p. 234)). Predominantly browsers, it is thought they probably lived in small herds. Merycoidodontines show some features in common with the later lineages of oreodonts, and were possibly ancestral to these other groups. It is thought that the larger forms possibly descended from the merycoidodontines. Nasal shortening was notable in the merycochoerines (the Early Miocene *Merycochoerus* (see pp. 256–7)) suggesting that they had a mobile upper lip. The promerycochoerines (the Early Miocene *Promerycochoerus* (see pp. 256–7)) had low, barrel-shaped bodies with short limbs, regarded as an **adaptation** for an amphibious lifestyle. Both **clades** were closely related (in fact the two often get confused) and included some of the more robust oreodonts. More generalised in form were the eporeodontines (the Oligocene *Eporeodon* (see p. 234)), ticholeptines and merychyines (the Early Miocene *Merychyus* (see pp. 256–7)). While the merychyines included relatively small species, some species of *Eporeodon* were among the largest oreodonts. Amongst the most specialised oreodont genera was the Middle to Late Miocene merycocheoerine *Brachycrus* (Fig. 2.72). It had a bizarrely tall skull with a deep lower jaw and very reduced nasals, presumed to have sported a **proboscis**. Notable for their long skulls, moderately hypsodont molars, and fan-shaped **occipital** region, the closely-related ticholeptines (e.g. the Middle Miocene *Ustatochoerus* (Fig. 2.72)) were among the last known oreodonts. Why the oreodonts became extinct is a mystery. Maybe the brushy and swampy habitats favoured by later oreodonts had disappeared? Or perhaps they were outcompeted by other ungulates, or were easy prey for the many new predators that invaded North America from Asia during the Miocene?

Suidae

So familiar are suids/pigs, they do not really require much of an introduction. They are distinctive compared to other artiodactyls, with compact bodies, four **digits** on each foot, heavy heads, prominent, tusk-like canine teeth, a disc-shaped nose and small eyes. Pigs are sociable, intelligent and adaptable. Unlike other artiodactyls, pigs produce a litter of offspring and rear them in a specially prepared nest. Along with the extant tayassuids/peccaries and poorly known fossil sanitheriids/doliochoerids, they comprise the suborder Suina. This group was historically grouped with hippopotamids (see p. 121) until molecular analysis showed the latter were more closely linked to cetaceans (see pp. 121–6). Peccaries and pigs are thought to have diverged during the Late Eocene. However, it was not until the Oligocene that early fossils of suids have been found (in Pakistan). Identifiable genera are first known from the Early Miocene (*Hyotherium*, known from France and Germany). By the Middle Miocene, suids had split into four distinct subfamilies—listriodontines, cainochoerines, hyotheriines and tetraconodontines, with a fifth, the suines emerging in the Late Miocene (Fig. 2.73). Only the tetraconodontines and suines survived into the Pliocene.

Listriodontines (e.g. *Listriodon* (see pp. 276–7), from Europe, and *Kubanochoerus*, from Eurasia) were possibly the only totally **herbivorous** lineage of suids, based on the large, sharp-edged teeth used to clip leaves (the tusks were conversely small). All genera were large-bodied and generally long-limbed. One remarkable form was *Kubanochoerus*, notable for the long, bony protrusion rising from the forehead in males. It also possessed two small horns above the **orbits**, and the combined cranial display was probably used in territorial head-on jousting. Cainochoeriines (e.g. *Cainochoerus*, from the Late Miocene to Early Pliocene of Southern and East Africa, and *Albanohyus*, from the Middle to Late Miocene of Eurasia and possibly Africa) are a poorly-known group which have been occasionally described as miniature listriodontines or even peccaries based on the teeth. They were very small (pygmy hog-sized) and **cursorial**. They included some of the last of the early suids, surviving right up to the Miocene-Pliocene boundary. Hyotheriines (e.g. *Hyotherium*, and the Late Miocene *Chleuastochoerus* from China) were the least specialised and feature some of the earliest known suids. Although generalised **omnivores**, some species possessed **hypsodont** tubular teeth similar to tubulidentates (see p. 73), which led to the confusion that they were in fact aardvarks! *Chleuastochoerus* was possibly a

Fig. 2.73. Phylogenetic tree of Suidae after McKenna and Bell (1997), Liu (2003), and Harris and Liu (2007).

late surviving hyotheriine, with some species surviving into the Pliocene (although other studies suggest it may have belonged to another **clade**). The unusually strong jaw muscles indicate that hard-shelled food (such as nuts) were part of the diet. Tetraconodontines are notable for having robust lower jaws and a variety of dental adaptations (including thickened enamel on the teeth). They evolved in Eurasia but died out during the Miocene, although they survived in Africa until the Pliocene. The largest known suids included the Miocene to Pliocene genus *Nyanzachoerus* (see pp. 285–7).

The most diverse subfamily and the only surviving one today are the suines. Fossil and modern species were/are widespread across Eurasia and Africa (and currently further afield due to feral domesticated pigs). They are well-represented in the modern Indo-Malayan region, although little is known about their origins and migrations. While *Sus* are widespread across the islands, the unusual babirusa (*Babyrousa quadricornua*) is restricted to the island of Sulawesi. Previously a third genus, *Celebochoerus* was also present on Sulawesi and also Luzon in the Philippines, with fossils dating between the latest Pliocene and Early Pleistocene. Intriguingly, it was most closely related to the African potamochoerins (bushpigs) although its origins appear to be from Taiwan. *Celebochoerus* was a large pig and is notable for the enormous tusks in both jaws. Many reconstructions show the long upper tusks as standing vertical to the skull, but like other potamochoerins, they were probably positioned horizontally along the skull. The Plio-Pleistocene *Kolpochoerus* was a more standard-looking potamochoerin, with species across Africa (see pp. 315–6). Also restricted to Africa were the phacochoerins, best known for including warthogs (*Phacochoerus*). A large relative was the Plio-Pleistocene *Metridiochoerus* (see pp. 315–6), the so-called "giant warthog", and it probably resembled its smaller modern counterparts in appearance and lifestyle. It is mostly known from the earlier deposits in the Pleistocene, but there is some rare evidence that it survived towards the end of that epoch, overlapping with modern humans.

Tayassuidae

Before human intervention, true pigs/suids were only present across Eurasia and Africa. The Americas were pig-free up until domestic pigs set foot in the "New World" in 1539, courtesy of the first Spanish explorers. However, one family of pig-like relatives, the peccaries, were already well-established in both North and South America. Oddly enough, it was long suggested that the very early tayassuids were present in Europe from the Early Miocene. It is now accepted that all forms were restricted to North America until the Pliocene. Following the Great American Biotic Interchange, or GABI (see pp. 252, 295), peccaries were one group which colonised South America (such as *Brasiliochoerus*). While peccaries have similar features to pigs, such as the distinctive cartilaginous disc on the end of the snout, a major difference is the shape of the tusks. Pig tusks are curved and grow out, while the tusks of peccaries are vertically straight with both sets interlocking. By rubbing the tusks against each other, peccaries can make a chattering noise to deter predators. The earliest fossils are of the Eocene *Perchoerus*, although it is a controversial member of the family (they are sometimes considered basal suoids). The oldest species, *P. minor*, was the size of a domestic cat. Tayassuids became very diverse during the Miocene. *Skinnerhyus* was a distinctive genus from that epoch, with wing-shaped cheekbones and a long, low skull (Fig. 2.70). They retained a high diversity into the Pleistocene, with *Platygonus* and *Mylohyus* being familiar North American forms (see pp. 318, 320–1). Peccaries are now mostly found in the warm southern part of North America, with the group marginally more diverse in South America.

Ruminantia

During the Late Eocene and the Oligocene, one significant innovation took place when a branch of artiodactyls, the ruminants, evolved a complex, four-chambered stomach. After swallowing, ruminants allow their meal to ferment in the rumen, the first stomach chamber, where microorganisms digest cellulose and aid in the breakdown of plant material. During periods of rest, ruminants regurgitate partially digested plant material from the rumen to rechew the "cud", further breaking it down. Ruminants maximise nutrient absorption from every bite of plant material because once the cud is ingested again, there will be better absorption of nutrients in the intestines. In comparison, many other mammalian **herbivores** are hindgut fermenters, lacking a specialised fermentation compartment in the foregut, and thus rely on microbes in the hindgut (caecum and large intestine) to break down cellulose, extracting limited nutrients as food quickly passes through. Thus, hindgut fermenters require consuming larger amounts of vegetation than ruminants, making them less versatile or efficient in nutrient extraction. Thanks to this evolutionary breakthrough, the ruminants ultimately became the dominant **ungulates** of the Neogene and Quaternary, and possibly outcompeted numerous other lineages, including horses.

The oldest ruminants are represented by teeth and cranial material of rat-sized to hare-sized animals that lived in Asia during the early Middle Eocene (ca. 45 Ma). These comprise fossils (such as a well-preserved skull) of *Archaeomeryx* from China and Mongolia, along with numerous other Eocene **taxa** known from China (*Lophiomeryx*, *Miomeryx* (see pp. 224–5), *Zhailimeryx*, *Xinjiangmeryx*, *Notomeryx*), Thailand (*Krabimeryx*, *Archaeotragulus*) and Myanmar (*Indomeryx*). Some highly **primitive** hare-sized ruminants including *Hypisodus*, *Hypertragulus* (see p. 236) and *Leptomeryx* (see pp. 235–6) existed in North America by the late Middle Eocene (ca. 40 Ma). These three taxa are also prevalent in the Late Eocene and Early Oligocene strata of the North American badlands. In the Early Oligocene, a broad range of hornless, basal ruminants resembling the North American *Leptomeryx* had superseded the native Eocene artiodactyls of Europe.

Tragulidae

Tragulids are the sole surviving members of a large, **paraphyletic** group of diminutive, hornless ruminants that diversified during the Early Oligocene, called "Tragulina". Tragulids have narrow skulls, large **orbits** for navigating the dense, dimly-lit rainforests of Asia and Africa, and long, slender limbs for dashing relatively quickly in thick undergrowth. Nowadays, the tragulids consist of 10 species and three genera: *Hyemoschus aquaticus* from Central Africa, three species of *Moschiola* from South Asia, and six species of *Tragulus* from Southeast Asia. The fossil record of tragulids is relatively sparse, and the family was prevalent and successful throughout the Oligocene and Miocene, but has changed very little and persists today as an example of a **primitive** ruminant form.

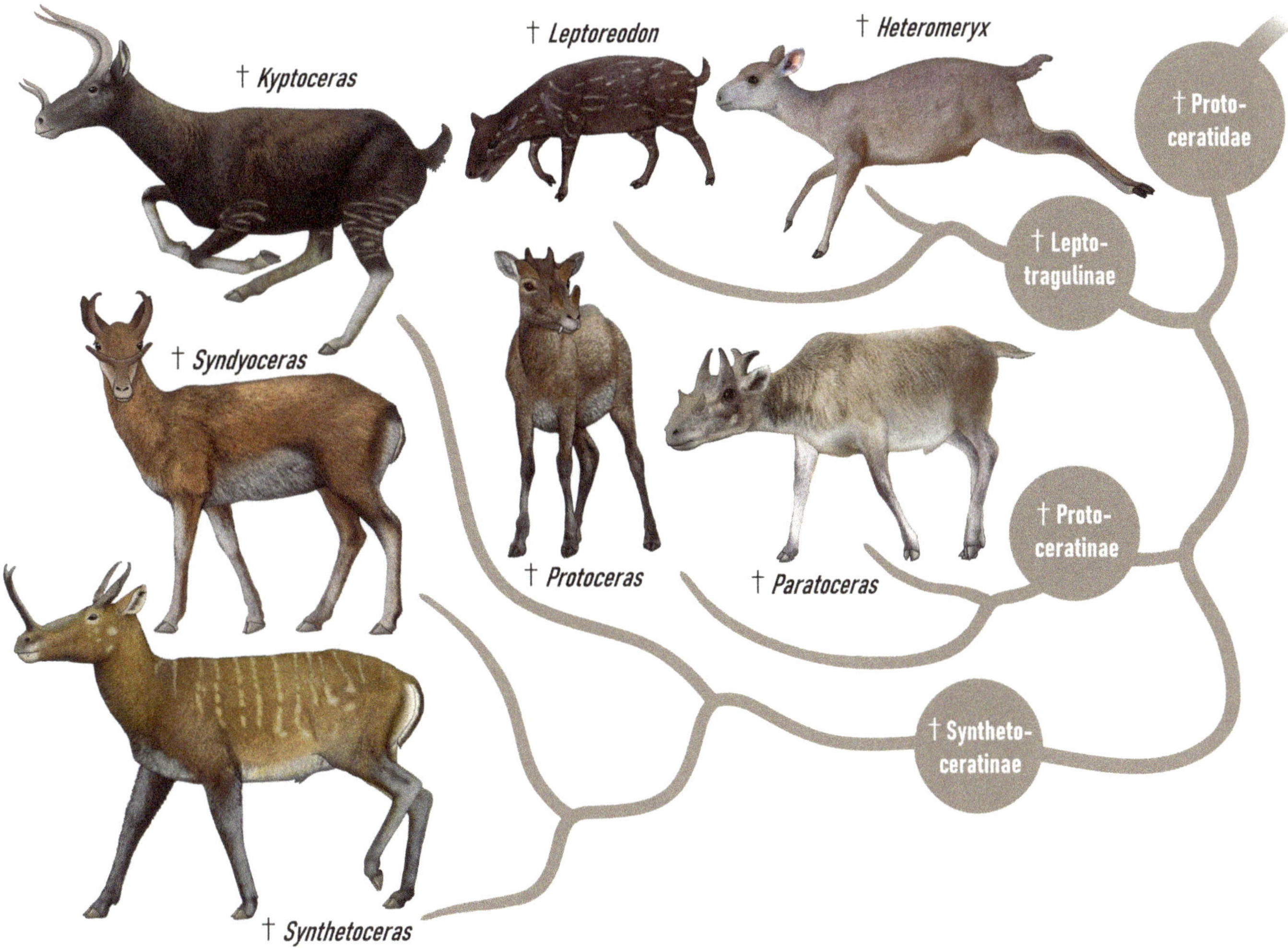

Fig. 2.74. Phylogenetic tree of Protoceratidae after Robson *et al.* (2022).

Protoceratidae

Convergent with deer and antelope, protoceratids are best known for sporting an exaggerated **rostral** bone on the snout in addition to the pair of horns above the **orbits**. However, these were not true "horns" sheathed in **keratin**, instead being bony **ossicones** covered in tough skin (as in the case of giraffes). Not all genera had horns, with some of the earlier genera, such as *Leptoreodon* and *Heteromeryx* (both of Late Eocene age), being hornless (Fig. 2.74). The protoceratids share features with various early unspecialised **ungulates**, so their closest relatives are often debated, but there are enough common features to suggest a surprising relationship with the tragulids (see p. 110).

Protoceratids were a successful yet rare family that was restricted to North America, living between the Eocene and the earliest Pliocene. Unlike deer and antelope, they were not **cursorial** species, having unfused metapodials and relatively short limbs. However, protoceratids were probably still capable of leaping, with some genera being capable of rearing up on their hind limbs to reach higher branches. Moving swiftly was probably not so important for life in forests or marshy environments. Their dentition was similar to deer and bovines, suggesting a diet of rough vegetation, but a major difference is the presence of canine teeth. They were long enough in some genera including *Syndyoceras* (see pp. 256–7), from the Miocene, to be used in rooting through the soil. Genera with more elaborate horns were **sexually dimorphic**, females being either hornless or, in the case of *Protoceras* (see pp. 256–7), from the Late Oligocene to Early Miocene, having only one pair of small horns. The largest genus, the Late Miocene *Synthetoceras*, had a long Y-shaped prong on the face which was used for sparring in a similar fashion to deer antlers. The angle of both sets of horns would safely interlock, allowing for a wrestling match based on strength. The last surviving genus was *Kyptoceras*, which lived in the humid forests of the Southeastern United States during the earliest Pliocene. As to why protoceratids went extinct, it can only be suspected that they were unable to adapt to a drying out environment alongside competition with other ungulates.

Antilocapridae

The modern pronghorn (*Antilocapra americana*) is the only surviving member of an **endemic** North American family with a rich fossil record which peaked in diversity during the Middle Miocene. It is during the Early Miocene that the ancestors of antilocaprids split from the giraffids.

Antilocaprids have a unique horned structure, being neither **ossicone**, horn or antler. It resembles the bony core seen in bovids (see pp. 116, 118–9), although the horny covering is **deciduous**, regrowing annually. Horn structure in all **ungulates** is heavily linked to lifestyle. Simple horns are used for display in smaller-bodied, pair-bonded genera found in wooded habitats. Larger-bodied genera with more complex structures tend to live in herds in open environments. Modern *Antilocapra* have a single, short, unbranched horn core,

Fig. 2.75. Phylogenetic tree of Antilocapridae after Janis and Manning (1998).

with a pronged, **keratinous** sheath. Extinct forms show a remarkable diversity of horn shape and size (Fig. 2.75). Basal forms, which are grouped as merycodines, had horns covered in skin, rather than a horny sheath. The earliest known genus, *Merriamoceros* from the Early Miocene, had short, palmate horns, while other basal genera, such as *Ramoceros* and *Merycodus* (both of Miocene age), resembled deer with tined horns. *Ramoceros* had some of the most unique horns of any artiodactyl. Both horns had three tines, although one horn was always longer than the other (it could be either left or right). It is not understood how the asymmetry affected their ability to wrestle during territorial fighting.

Later genera form a **monophyletic** group known as Antilocaprinae. They are generally larger-bodied and show **adaptations** for more open plains-living. Some lineages of antilocaprines lost the **anterior** prong of the horn, with the main shaft of the horn tightly twisting. The spiral horns of *Illingoceros*, from the Late Miocene, were **convergent** with tragelaphin antelopes (see p. 116, 119), and it was originally described as one. Strangely, the tips of the horns were forked. Other forms had very short horn shafts so that the tines appear to be multiple horns, with "four-horned" (*Tetrameryx*, from the Pleistocene) and "six-horned" (*Hexameryx*, from the Pliocene) genera known. Skin covered the base of the tines while the tips were separately sheathed. Not all of the later antilocaprids were large-bodied. During the Plio-Pleistocene, species of *Capromeryx* (see pp. 325–6) gradually shrunk, until those from the latest Pleistocene were 60 centimetre tall dwarfs. Likewise, the horns were smaller, reduced to just small spikes. *Capromeryx* was one of the few genera with dual horn cores (i.e. they were actually four-horned). Modern pronghorns are considered one of the fastest extant **terrestrial** animals. Second only to cheetahs, pronghorns can maintain top speeds for far longer periods. All antilocaprids were highly adapted to **cursorial** life. Aside from the cloven hooves, the lateral **digits** were completely lost. Antilocaprids also have very efficient respiratory and circulatory systems which allows for speedy long-distance travel. The Pleistocene species were likely hunted by the similarly fast felid *Miracinonyx* (see pp. 88, 324, 326).

Giraffoidea

Why giraffes have long necks has long puzzled scientists. Several suggestions have been put forward, with the most popular (favoured by evolutionary biologist Charles Darwin) being of competitive pressure to access higher levels of vegetation. While it is the case that no extant mammal can reach quite as

Fig. 2.76. Phylogenetic tree of Giraffoidea after Solounias (2007), Sánchez *et al.* (2015), and Ríos *et al.* (2017).

high as a giraffe can, the advantage is possibly questionable given that giraffes rarely stretch to their fullest height to feed, often feeding with a horizontal neck! The other popular theory is that the long necks are a secondary sexual characteristic for a specialised mode of fighting (known as necking). Little difference in neck length relative to body size between males and females, and the fact that non-fighting females also have long necks, muddy the argument. Interestingly, the lengthening of the neck started early on with the most basal giraffids, such as *Canthumeryx* (known from Saudi Arabia and East Africa), having a relatively long neck. This occurred during the Miocene, a period of drying out when forests gave way to grassland in many parts of the world. It would seem that the competitive edge in **browsing** was probably the driving factor, but at the same stage the thickening of the skull allowed the head and neck to be used as a weapon by males against each other. Another suggestion is that the long neck gave early giraffids a better view in a habitat with thinning trees. Those genera more like the extant and (relatively) short-necked okapi (*Okapia johnstoni*), which live in closed forests, have no use for long-range surveillance. It is possible that all three theories were partially responsible. Being tall has its own complications and giraffes have a range of remarkable physiological **adaptations** to deal with gravity and such a long neck: disproportionate lengthening of the cervical vertebrae, left laryngeal nerve which controls the voice box is five metres in length, large heart and muscular arteries to pump blood, valves in veins to prevent backflow of blood when the head is lowered, elastic ligaments that reduce the muscle effort to hold the head aloft, tight skin on the limbs act like compression socks to regulate blood pressure, and strong esophageal muscles for regurgitating food to the mouth for rumination.

Giraffids are most closely related to several extinct groups including Prolibytheriidae (which includes genera such as *Prolibytherium*, in which males uniquely had wide, flattened cranial appendages compared to the slender, horn-like form seen in females), Climacoceratidae (whose members, including *Climacoceras*, had long, thin, branched **ossicones** that resembled deer antlers) and Palaeomerycidae, and to the extant pronghorn. The giraffids are divided into various subfamilies which include the medium-sized palaeotragines (e.g. *Palaeotragus* (see pp. 282–3), from the Miocene to Early Pleistocene of Eurasia and Africa) and samotheriines (e.g. *Giraffokeryx*, from the Miocene of Eurasia, and *Samotherium*, from the Miocene to Pliocene of Eurasia and Africa), long-necked forms similar to extant giraffes (e.g. *Bohlinia* from the Middle Miocene of Greece), and the large to giant, robust sivatheriines (e.g. *Bramatherium*, which ranged between Turkey and India from the Late Miocene to the Pliocene, *Decennatherium* (see pp. 278, 281), from the Late Miocene of Europe, and *Sivatherium* (see pp. 314–5), from the Late Miocene to Early Pleistocene of Eurasia and Africa) (Fig. 2.76). The Middle Miocene samotheriines had two long ossicones that curved backwards from the skull. From Eurasia, *Giraffokeryx* had two sets of V-shaped ossicones over its **orbits** and ears. The earliest fossils that lead to the ancestors of the modern genus *Giraffa* are known from the Middle Miocene. Most of these fossil genera had short necks, although *Samotherium* had a neck that falls halfway between an okapi's neck and a modern giraffe's neck in length and structural development. This means the giraffid evolutionary record includes not only fossils representing a huge radiation of short-necked giraffes, but also the transitional fossils that show how and when giraffes developed their long necks.

During the Middle and Late Miocene, there was a great radiation of sivatheriines in both Africa and South Asia, including the huge, burly, short-necked giraffid *Sivatherium*. There are suggestions that one species of *Sivatherium* lasted until 7–8 ka based on rock paintings, although the fossil evidence does not back this up, with the latest known fossils being of Early Pleistocene in age. Additionally, one especially weird prolibytheriid was *Tsaidamotherium* (Fig. 2.77). It had an enlarged, saiga-like nasal cavity, but it is the asymmetrical "horns" for which it is (relatively) best known. The right one was always larger than its diminutive partner, and varied in shape between species. The reason for these odd structures is difficult to decipher, as they do not appear to have been useful for head clashes. Historically, *Tsaidamotherium* was considered an aberrant caprine (see p. 116) related to the extant musk ox (*Ovibos moschatus*) based on cranial morphology. However, the structure did not develop in the manner of horncores in bovids (see pp. 116, 118–9), and was far more like giraffid ossicones.

Fig. 2.77. Reconstructed life appearance of the prolibytheriid *Tsaidamotherium brevirostrum*, from the Late Miocene of northwestern China. Reconstructed length: 2 m.

Palaeomerycidae

Previously linked closely with the New World dromomerycids, the palaeomerycids, such as the European *Ampelomeryx* (see pp. 274–5), are now considered to be a different branch within the **ungulate** tree. Giraffids are now considered to their closest relatives, which probably is not surprising given the similarities towards modern okapis. The cranial horns of palaeomerycids consisted of two types: a pair of short horns over the **orbits** and a larger prong that juts out from the back of the skull (this is where the similarity to dromomerycids comes in). The horns were part of the skull, and like giraffes, they were probably covered in short hair. Palaeomerycids lived across Eurasia from the Early to Late Miocene. One reason why palaeomerycids went extinct may be competition with deer, cattle and giraffes for resources and territory.

Dromomerycidae

The dromomerycids lived only in Miocene North America (one species probably survived just into the Pliocene) and is probably the sister group to modern deer (Cervidae). All dromomerycids had a variety of bony horns, which do not appear to be either **deciduous** (like deer antlers) or covered in a

keratinised sheath (like bovids (see pp. 116, 118–9)). The best guess is that they were hair-covered in a similar fashion to giraffe **ossicones**. Some smaller/ earlier species (from Aletomerycini) had long upper canine teeth, while larger/later species (from Cranioceratini) had a third horn which projected out of the back of the skull. The third group (Dromomerycini) were more conservative, with a pair of horns. It is not known for certain if males and females were both horned. However, it is thought that individuals with longer horns were males. Only a few species are known to have had hornless females with certainty. The long tusks of the Aletomerycini were present in both sexes. The dromomerycids went extinct probably because of increasing aridity and cooler temperatures that reduced the availability of suitable environments for these browsing mammals.

Moschidae

Most ruminants have a horn-like feature on the top of the head (at least among the males of the species). They are primarily used in seasonal territorial fights but are also handy for deterring predators. The shape and form of the horns vary considerably, with composition falling into three main categories: true horns with a bony core/hard **keratin** outer sheath, skin-covered bony **ossicones** (giraffes), and antlers which consist of bare bone which is shed annually (deer). However, a few ruminants lack the decorative headpieces and rely on enlarged canine teeth as their weapons. They include several cervids, tragulids (see p. 110) and the odd moschids.

Extant members of the moschids are better known as musk deer. Although they have a vaguely deer-like appearance, moschids are not particularly closely related to cervids. In fact genetic analysis has shown them to be the sister group to bovids. These secretive animals are probably best known for the strong-smelling scent glands which are used in perfume production, and hunting is now a threat to the long term survival of the group. Moschids are split into three subfamilies—dremotheriines, blastomerycines and moschines, with only the latter containing extant species. Early moschids originated in Eurasia during the Late Oligocene to Early Miocene, with later migrations during the Miocene into North America. However, moschids declined rapidly at the end of the Miocene, leaving only the extant *Moschus* in Asia. Dremotheriines such as *Dremotherium* and *Pomelomeryx* were some of the earliest known genera. Notable for very enlarged fangs, *Dremotherium* was evidently a **high browser** as the cervical vertebrae were elongated. Blastomerycines were unique to North America, and included some of the largest known moschids (*Parablastomeryx*) and also the diminutive *Machaeromeryx*. An important part of the North American Miocene fauna, this group has been incorrectly associated with two other **Nearctic** groups in the past—pronghorns. The last surviving blastomerycine was *Longirostromeryx* (see pp. 260–1).

Cervidae

Deer are mostly well-known for their antlers—and for good reason. Although this lineage gave rise to some incredible beasts, they arose from small, antlerless and, surprisingly enough, tusked ancestors. Deer are superficially similar to antelope, but unlike them, they do not have true horns. Instead, they possess antlers, which are grown and shed every year, while horns are permanent structures. The numerous modern species vary greatly in size and live in a wide range of habitats around the world, from tropical rainforests to harsh tundra. The uplift of the Alps and the Himalayas, which led to substantial changes in geography, was the major cause of the massive radiation of deer-like ruminants and the origin of cervids from the Early Miocene to the Early Pliocene. The majority of Miocene deer lacked the fully developed antlers present in extant deer. *Dicrocerus* (see pp. 274, 277) and *Heteroprox* (see pp. 274, 277), which originated in Europe during the Middle Miocene, were possibly among the earliest antlered deer. The Miocene genus *Procervulus* may have had antlers that were not periodically shed, but unlike extant cervids, there was no clear divide between the permanent and non-permanent portions.

During the Late Miocene, Eurasian deer had radiated into dozens of species in various genera, some possessing small, simple antlers while others had massive, intricate antlers (Fig. 2.78). The Cervinae was the earliest lineage of modern cervids to evolve, originating in Central Asia during the Late Miocene (ca. 7–9 Ma). The tribe Muntiacini (muntjacs) had evolved by approximately 7–8 Ma, with *Muntiacus leilaoensis*. The size of **primitive** muntjacs ranged from that of rabbits to fallow deer. Those muntjacs possessed antlers for defence and tusks for **intraspecific competition**. Capreolinae (e.g. water deer and roe deer) appeared shortly afterwards, while Alceini (moose and extinct kin) emerged 6.4–8.4 Ma. The Tethys Ocean (see p. 174) had vanished around this time to be replaced by huge expanses of grassland. This gave deer access to a plentiful supply of proteinaceous plants, which contributed to the growth of ornamental antlers and enabled populations to thrive and colonise regions. While antlers became more prominent, the canine teeth either became much reduced or absent, possibly because diet was no longer primarily browse-based and antlers were more complex display structures. The canine teeth along with antlers are small in tufted deer and muntjac.

During the Pliocene, as a cooling global climate brought about extensive glaciation and a decrease in sea levels, grasslands became rich in nutritious fodder. This led to a fresh surge in deer populations. *Cervocerus* is the earliest known representative of Cervini, originating in Eurasia around the Miocene-Pliocene boundary (4.2–6 Ma). *Cervavitus* probably arose in wooded regions of Eastern Europe before spreading to East Asia and Western Europe during the Miocene, occupying the damp woodlands of Eurasia at the time. However, due to geographical changes such as the Himalayan uplift, which caused parts of Asia and Europe to become increasingly arid from the Pliocene onwards, these deer were driven to seek refuge in South China, where they either evolved into or were supplanted by the modern deer genera *Rusa* (sambar deer) and *Axis* (axis/chital deer) (Dong, 2011). Fossils representing the Cervini have been discovered from the Himalayas and China, dated from the Early Pliocene to Pleistocene. The genera *Dama* (fallow deer) and *Cervus* emerged in the Late Pliocene (less than 3 Ma), whereas *Axis* originated during the Middle Pleistocene. The tribes Rangiferini and Capreolini originated approximately 4–7 Ma.

In the Early Pliocene (ca. 5 Ma), the extinct genera *Eocoileus* and *Bretzia* (which had palmate, moose-like antlers) of the tribe Rangiferini were the first cervids to enter North America via the Bering Land Bridge (see p. 174). Among the deer that evolved in North America during the Plio-Pleistocene were *Odocoileus*, the genus comprising the modern white-tailed deer

(*O. virginianus*) and mule deer (*O. hemionus*). Upon formation of the Panama Isthmus, deer immigrated to South America in the Late Pliocene (ca. 2.5–3 Ma), during the Great American Biotic Interchange, or GABI (see pp. 252, 295). Thir success was partly due to the limited amount of native, competing ruminants in South America at the time. This resulted in a diversity of six **endemic** modern South American genera and dozens of species such as the marsh deer (*Blastocerus dichotomus*), brocket deer (*Mazama*), pampas deer (*Ozotoceros bezoarticus*), pudus (*Pudu*) and South Andean deer (*Hippocamelus bisulcus*). By the Early Pleistocene, giant deer possessing spectacular antlers had emerged, possibly because of plentiful food resources to fuel evolution. The Eurasian genus *Eucladoceros* was an Early Pleistocene deer similar in size to the modern moose (*Alces alces*). *Eucladoceros* had enormous, highly complex antlers spanning over 1.7 metres (5.6 feet) across, with each antler sporting over a dozen tines. *Megaloceros*, which lived from the late Early Pleistocene to Middle Holocene, is the most well-known of the Eurasian mega-deer (see pp. 343–4). *Megaloceros* was a close relative of the Central and East Asian genus *Sinomegaceros* (the latter which is characterised by antler brow tines with a palmate morphology) and probably other mega-deer genera. These giant deer became extinct primarily due to climate-induced habitat changes, with forests expanding and reducing the open grasslands and parklands they favoured. Meanwhile, reindeer (*Rangifer tarandus*) and moose spread from Siberia into North America (Breda and Marchett, 2005).

Bovidae

Bovids are one of the most successful families of large mammals. The key to their success has been their stomach. In common with other ruminants, bovids have a four-chambered stomach giving their owners a remarkably efficient digestive system. Bovids can be very social animals and also very territorial. Permanent, unbranched and sheathed horns are present in all species, although often smaller in females, and these are an important tool for maintaining territories, herd structure and defence. Modern bovids range in size from three kilogram royal antelope (*Neotragus pygmaeus*) to one tonne gaur (*Bos gaurus*). Modern bovids are split between 10 subfamilies: bovines (cattle and spiral-horned antelope), antilopines (gazelles and dwarf antelope), nesotragines (Suni and Bate's antelope), cephalophines (duikers), oreotragines (klipspringer), reduncines (waterbuck and allies), aepycerotines (impala), caprines (sheep and goats), hippotragines (**grazing** antelopes such as oryx and allies) and the alcelaphines (hartebeest and wildebeest) (Fig. 2.80). What is very obvious from this list is that the term "antelope" refers to a variety of small, medium and large-bodied horned **ungulates**, and not one taxonomic unit.

Bovids split from the cervids and giraffids (see pp. 112–4) during the Early Miocene. The earliest bovids, whose presence in Africa and Eurasia in the latter part of the Early Miocene (ca. 20 Ma) has been ascertained, were small animals, somewhat similar to modern gazelles, and possibly lived in woodland environments. Among the oldest known bovids was *Eotragus* (see pp. 274, 277), which possibly originated in Asia. Early in their evolutionary history, the bovids split into two main **clades**: Boodontia (of Eurasian origin, with a single subfamily, the Bovinae) and Aegodontia (of African origin, composed of all other bovid subfamilies). This early split between Boodontia and Aegodontia has been attributed to the continental divide between these landmasses. When these continents were later rejoined, this barrier was removed, and both groups expanded into the territory of the other. The tribes Bovini and Tragelaphini (both within Bovinae) split during the Early Miocene. Bovids are known to have reached the Americas during the Pleistocene by crossing the Bering Land Bridge (see p. 174). The present genera of Alcelaphinae appeared in the Pliocene. Alcelaphines are a large form of antelope, which include the familiar and abundant wildebeest (*Connochaetes*) that live on open **savannahs**. Their heavy forequarters, Roman nose and shaggy manes give wildebeest a very different appearance to became rich in nutritious fodder. Therefore, a fresh surge in other antelope. A very close relative from the Pleistocene, *Rusingoryx* is notable for its large nasal dome which was used for amplifying vocalisations (Fig. 2.80). Many specimens have been found alongside human butchering tools. The extinct alcelaphine genus *Paramularius*, which was the same in size as the hartebeest, is believed to have come into being in the Pliocene, but became extinct in the Middle Pleistocene. One particularly large alcelaphine was the late living *Megalotragus* (see p. 315). Several genera of Hippotraginae are known since the Pliocene and Pleistocene. This subfamily likely diverged from the Alcelaphinae in the late Early Miocene.

The Bovinae (cattle species) are believed to have diverged from the rest of the Bovidae during the Early Miocene. The bovine tribe (Bovini) split about 5–10 Ma into the buffalos (the water buffalo *Bubalus* and African buffalo *Syncerus*) and a group leading to bison and taurine cattle (including aurochs, the wild ancestor of modern domestic cattle (see pp. 342–4)). Bovines include the largest bovids in size. Even modern-day species are some of the largest **terrestrial** species on the planet. However, these would have been equalled or surpassed by several extinct species. The giant buffalo (*Syncerus antiquus*) was a close relative of the modern African buffalo (*S. caffer*), but has previously been considered a species of *Pelorovis* (see pp. 315–6). It was the huge sweeping horns that made the giant buffalo distinct (Fig. 2.80). As the **keratinised** sheaths do not fossilise, the estimated span of the horns tip to tip must have been nearly four metres! A tribe of smaller bovines with a good fossil record are the Boselaphini. Modern nilgai (*Boselaphus tragocamelus*) and four-horned antelope (*Tetracerus quadricornis*)) are the only survivors from a group which date back to the Miocene. *Tragoportax* was a large genus with the build of a fast runner and was found in open woodland habitat. The Boselaphini became extinct in Africa in the Early Pliocene; their latest fossils were excavated in Langebaanweg, South Africa, and in Lothagam, Kenya (see pp. 284–7). Caprines, ("goat-antelopes") are a familiar group that diversified into quite different forms, some of which do not particularly resemble domestic sheep and goats. Two groups of large-bodied species evolved separately and vaguely resemble cattle. Restricted to the bleak northern tundra are musk ox (*Ovibos moschatus*), the sole surviving species of a group (Ovibovini) which once ranged across Eurasia and North America in far more temperate environments, including genera such as *Bootherium* and *Euceratherium* (see pp. 325–6). The second group is the takins of the Himalayas. While similar in some features, takins are more closely related to either goats or sheep, rather than musk ox.

The explosion of small- and medium-sized horned ungulates during the Miocene have caused confusion and species are often lumped into various groups until they can be reexamined. A case in point is *Palaeoreas*, a medium-sized antelope from the Late

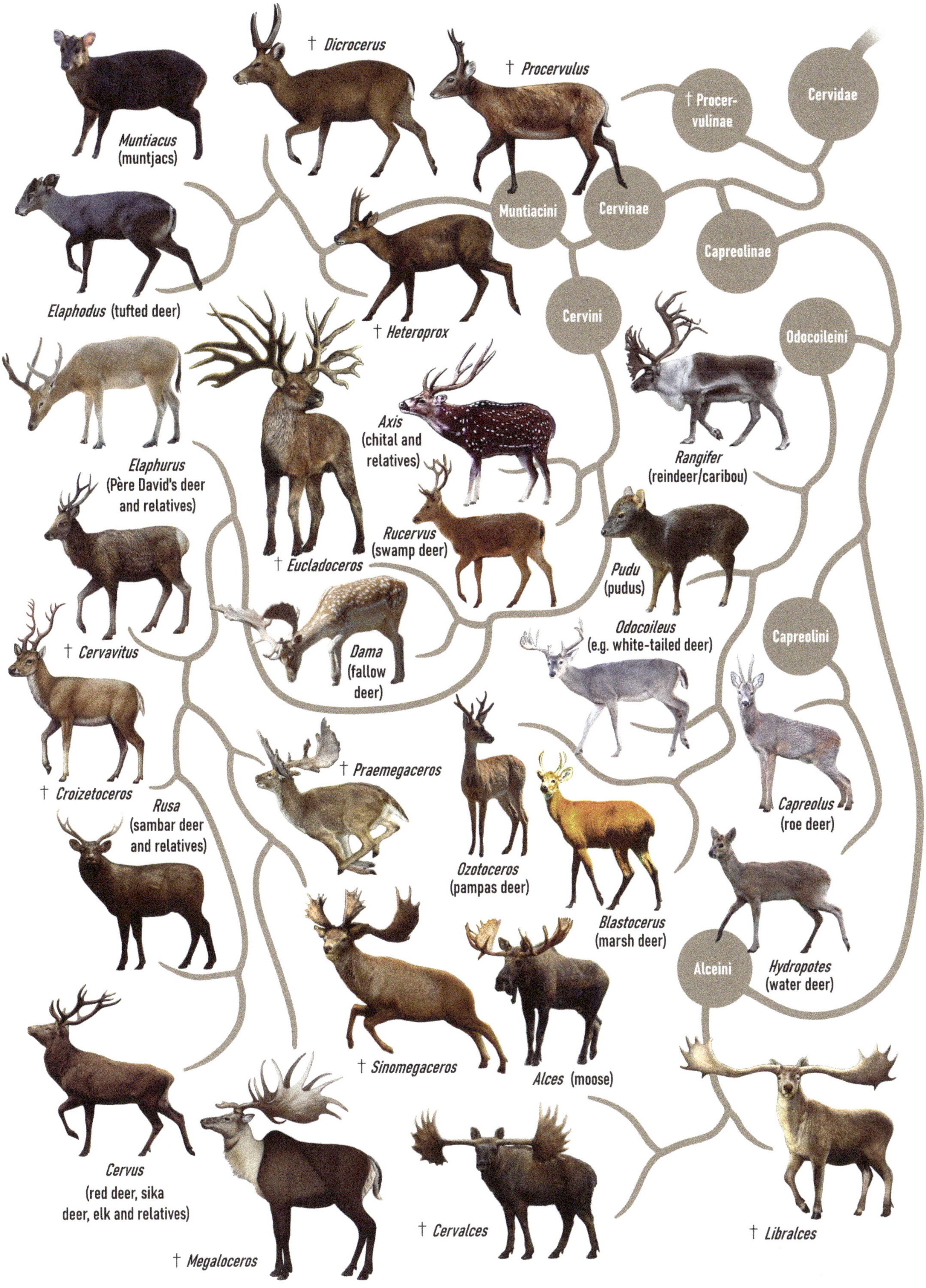

Fig. 2.78. Phylogenetic tree of Cervidae after Gilbert *et al.* (2006).

Miocene, strikingly similar to the modern Indian blackbuck (*Antilope cervicapra*). Males had robust, tightly-twisted horns, while females lacked horns. A depression in front of the **orbits** is indicative of **preorbital** glands that produced secretions used to mark territories. All of these features are typical for many antelope species, but are not necessarily seen in one group. Thus, at this point, *Palaeoreas* was a probable antilopine, but sufficiently distinct. The Middle Miocene marked the spread of the bovids into China and the Indian subcontinent. The Caprini emerged in the Middle Miocene, and seem to have been replaced by other bovids and cervids in Eurasia. The earliest fossils of the antilopines are from the Middle Miocene, though studies show the existence of the subfamily from the Early Miocene. **Speciation** occurred in the tribe Antilopini during the Middle or Late Miocene, mainly in Eurasia. By the Late Miocene (ca. 10 Ma), the bovids rapidly diversified, leading to the creation of 70 new genera. This Late Miocene radiation was partly because many bovids became adapted to more open, grassland habitats. The Aepycerotinae first appeared in the Late Miocene, and no significant difference in the sizes of the **primitive** and modern impala (*Aepyceros melampus*) has been noted. *Oioceros*, from Eurasia and Africa, was a contemporary genus which also shared some similarities with modern gazelles. However, *Oioceros* and its nearest relatives (such as *Samotragus*, *Hispanodorcas* and *Paraoioceros*) had a remarkable variety of horn shapes from short, stabbing spikes to broad, ramming horns. The later *Gazellospira* was a true antilopine and very similar to blackbuck apart from the widely diverging horns.

Also found in the Pleistocene was a medium-sized relative of the modern Indian nilgai called *Duboisia*, from Indonesia. Unlike its extant relatives, this genus was a **browser** of closed forests. Although forest-living is unusual for bovids, some modern species such as bongo (*Tragelaphus eurycerus*) and saola (*Pseudoryx nghetinhensis*) would be good **analogues**. Modern wild sheep are widespread but limited to mountainous areas of Eastern Europe, Asia and Western North America. Until the discovery of *Protovis* (Fig. 2.80), there was uncertainty as to the ancestral home of modern sheep. Living on the Tibetan Plateau during the Pliocene, *Protovis* combined some basal features distinguishable from other known genera. Gazelles are **cursorial** ungulates, well-adapted to life on open plains or deserts. Some species were more adventurous. The short nasals of *Gazella psolea* (Fig. 2.80), from the Late Pliocene, indicate a probable inflatable nose (like that seen in the unrelated saiga (*Saiga tatarica*)), an **adaptation** for migrations up into the cold, dry environment of the Atlas Mountains. At the opposite spectrum to speedy gazelles is possibly the most unusual bovid. Another example of island life (this time on the Balearic Islands), was a small goat-like genus known as *Myotragus*. It had a wealth of bizarre adaptations: an **insular dwarf** with forward-facing orbits, perennially-growing incisors, short, inflexible limbs, pronounced hump on the shoulders, and intermittent bone growth which changed due to food and water availability or temperature fluctuations. The first humans arrived on Majorca less than 5 ka and evidence shows they utilised *Myotragus* as a food supply. Amazingly, there is also evidence that some specimens were kept alive until adulthood, but the genus failed to be domesticated. Ultimately, hunting pressure pushed the genus into extinction. The bison and taurine cattle lineages are estimated to have split from the Bovini during the Early Pleistocene (ca. 2.5–1.65 Ma). During the Pliocene, the cooling global climate triggered an expansion of open grassland, driving the evolution of large grazers such as *Bos acutifrons* (Fig. 2.79), the oldest known species of *Bos* which is also a probable progenitor of the aurochs (*Bos primigenius*). A fossil skull of *B. acutifrons* is known from the Siwalik Hills of Northern India and dated to the Early Pleistocene (ca. 2 Ma). The earliest auroch fossils are known from early Middle Pleistocene beds (ca. 0.78 Ma) in Tunisia, suggesting that aurochs possibly originated in Africa before dispersing to Eurasia during the Middle Pleistocene.

Fig. 2.79. Life reconstruction of a herd of the bovid *Bos acutifrons*, in what is now the Siwalik Hills, Northern India, during the Early Pleistocene.

Bison are commonly thought to have descended from a lineage comprising the extinct genus *Leptobos* during the Late Pliocene to Early Pleistocene in Asia, and indeed the skull and horns of *L. vallisarni* very much resembled those of bison. *Leptobos* represents an intermediate stage between the antelope **morphotype** and the bovine morphotype, being medium-sized and more robust than antelopes and less robust than recent bovines. The oldest representatives of the bison lineage, known to have lived in China (*Bison palaeosinensis*) and India (*Bison sivalensis*) during the Late Pliocene to Early Pleistocene (ca. 3.4–2.6 Ma), belong to the subgenus *Eobison*. More advanced species within the genus belong to the subgenus *Bison (Bison)*, which originated by the late Early Pleistocene (ca. 1.2 Ma), with primitive representatives of the subgenus comprising the abundant *Bison schoetensacki* (see pp. 339–40). The steppe bison (*Bison priscus*) (see pp. 342–3) originated in East Asia during the mid-Middle Pleistocene, and later spread all over Eurasia. The steppe bison entered North America via the Bering Land Bridge during the late Middle Pleistocene (ca. 195–135 ka), giving rise to various extinct North American forms including the long-horned bison (*Bison latifrons*) and the ancient bison (*Bison antiquus*) (see pp. 325–6), followed by the modern American bison (*Bison bison*). The European bison (*Bison bonasus*) originated in Europe during the late Middle Pleistocene. Why the steppe bison went extinct remains highly contentious.

Entelodontidae

Nicknamed terminator/killer/hell pigs, it gives one an immediate idea that the entelodontids were beasts not to be treated lightly. Although pig-like in form, entelodontids were more closely related to the hippopotamus/cetacean line of artiodactyls, the so-called whippomorphs. Immediately obvious is the long, robust skull decorated with various bony bosses and flanges. These were absent in younger animals, only increasing in size with maturity. The teeth were not bladed for slicing flesh, but the interlocking front

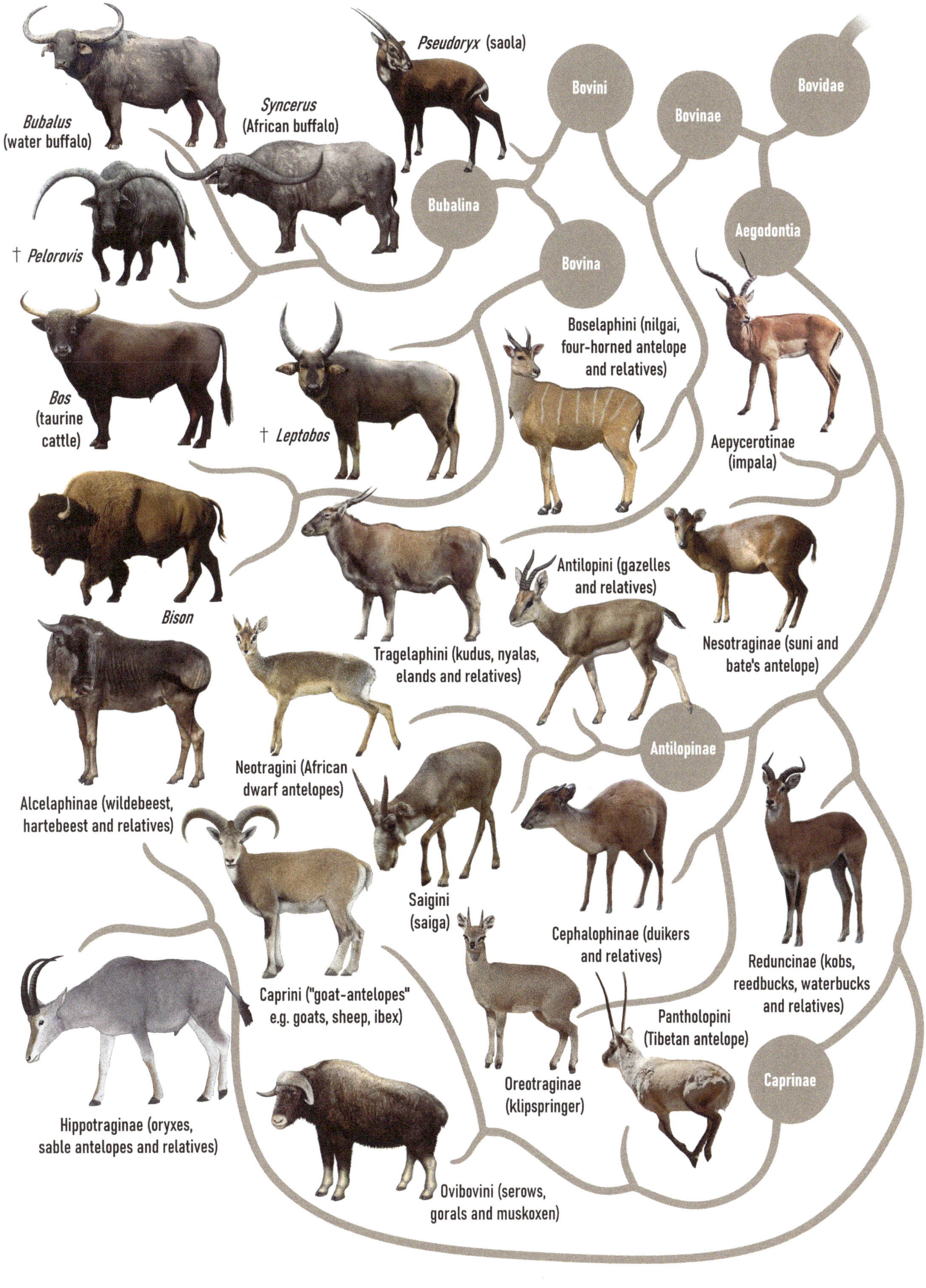

Fig. 2.80. Phylogenetic tree of Bovidae after Yang *et al.* (2013), and Calamari (2021).

teeth were definitely used for stripping leaves or bark off trees. With a high jaw strength for dealing with hard foods such as nuts or bones, entelodontiids represent a classic **omnivore**. Modern peccaries would be a good yet smaller **analogue**. In entelodontids, the body was large and the limbs were slender and **cursorial**. Weight was supported on two hooves, but the other foot bones were unfused which allowed the toes to spread out on soft ground. Entelodontids were never abundant and probably not herding animals, with males and females showing no obvious **sexual dimorphism**. The earliest fossils, such as *Brachyhyops* (see pp. 224–5) and *Proentelodon*, are known from the Middle Eocene of Central Asia and North America (ca. 40 Ma). According to fossil evidence, *Brachyhyops* most probably first evolved in South Asia, before dispersing northwards and later spreading eastwards into the Western United States (Tsubamoto *et al.* 2004). Later, in the Late Eocene and the Oligocene, entelodontids were spread across Eurasia and North America, and they grew larger, resulting in genera such as *Archaeotherium* (see pp. 235–6) and culminating in the massive *Daeodon* (see pp. 256–7). It is possible that entelodontids were unable to adapt to a cooling, drying environment during the Miocene.

Whippomorpha

As evident from genetic and molecular studies, hippopotamuses, whales and their close extinct relatives are clustered together in a lineage known as Whippomorpha. Whippomorphs are estimated to have separated from other artiodactyls in the Middle Paleocene (ca. 59 Ma) while the hippopotamus/whale split is estimated to be 55 Ma (Early Eocene). The aquatic features in hippopotamuses and whales most likely evolved independently rather than being of shared ancestry (Tsagkogeorga *et al.* 2015), as modern hippopotamuses showed up in the fossil record considerably later than whales.

"Anthracotheriidae"

The relationships between the various groups of whippomorphs is poorly substantiated, despite many fossil specimens. However, a lack of fossil hippopotamuses before the Miocene created a headache for taxonomists and palaeontologists. Frequently linked to hippopotamuses, the little known **paraphyletic** "anthracotheriids" have helped to fill this gap (Fig. 2.81). On first impressions, "anthracotheriids" look much like a skinny hippopotamus with a narrow skull. Many of the first fossils were found in coal deposits, which are reliable evidence of fossilised wet forests, and ultimately coal (known as anthracite) gave the beasts their name. Despite some similarities, "anthracotheriids" never developed the same physical **adaptations** to an amphibious life seen in hippopotamuses. The "anthracotheriids" had their early origins in

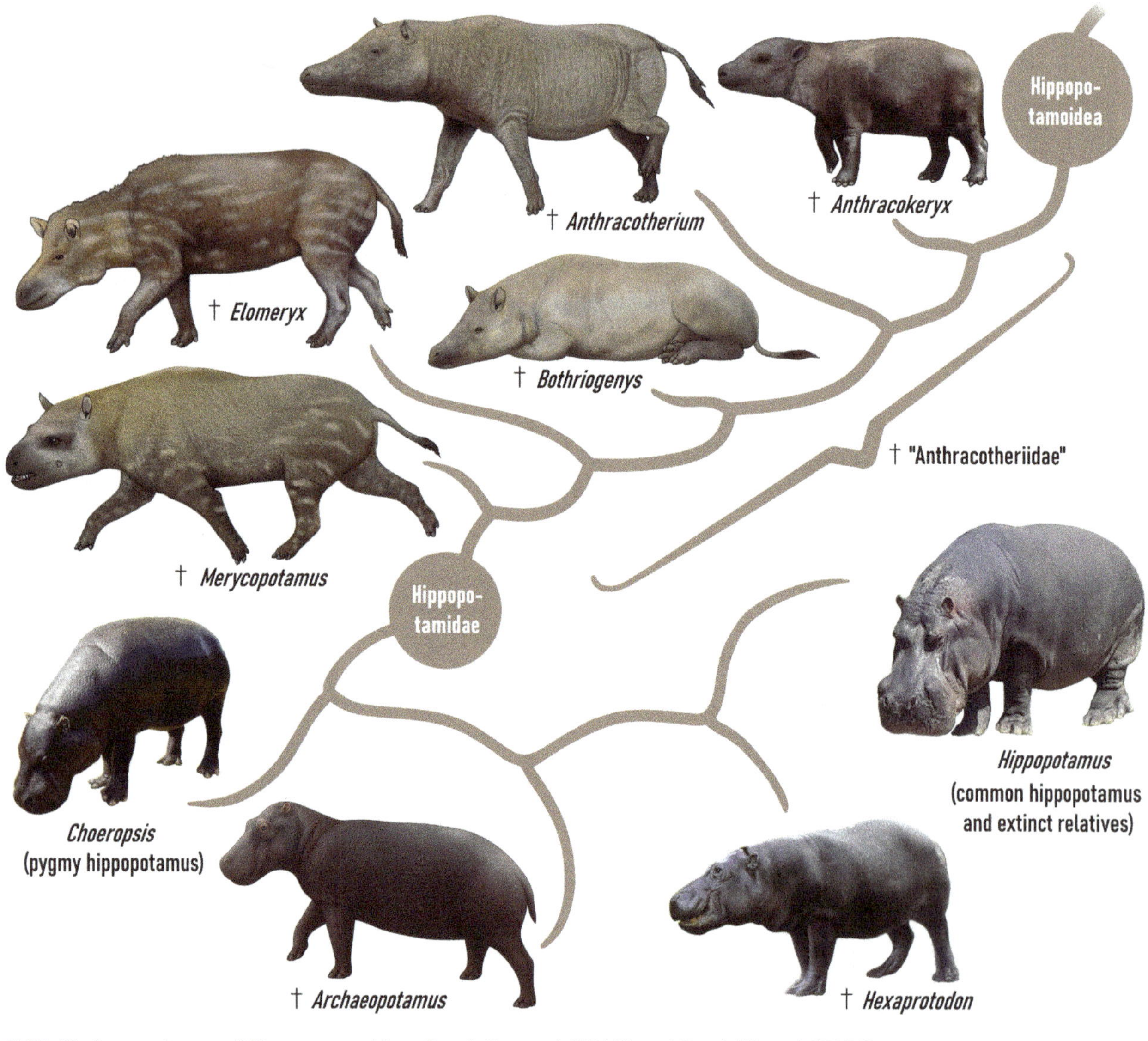

Fig. 2.81. Phylogenetic tree of Hippopotamoidea after Orliac *et al.* (2010), and Pandolfi *et al.* (2020).

Southeast Asia from the Middle Eocene (*Anthracokeryx*, the size of a domestic cat), with multiple migrations into Africa, Europe and North America during the Late Eocene and Miocene, ending with the last surviving genus (*Merycopotamus*) surviving in India during the Early Pliocene. An interesting observation is that the demise of the "anthracotheriids" during the Late Miocene coincides neatly with the earliest hippopotamuses. A reappraisal of the teeth of a small African "anthracotheriid" from the Oligocene, *Epirigenys*, shows strong similarities to those of true early hippopotamuses.

Hippopotamidae

Many hippopotamuses known in the fossil record were small, **primitive** forms, looking most similar to the modern pygmy hippopotamus. The hippopotamuses diverged from the "anthracotheriids" possibly during the Early Miocene, with the oldest known forms such as *Kulitherium* and *Morotochoerus* being known from this time in Africa. The kenyapotamines (e.g. *Kenyapotamus*, *Palaeopotamus*) arose during the Middle Miocene and diversified before eventually being replaced by the extant subfamily Hippopotaminae during the Late Miocene (ca. 8 Ma). Hippopotamuses originally spread to Eurasia around 6 Ma. *Archaeopotamus*, which lived in Africa and the Middle East from the Late Miocene to the Early Pleistocene (ca. 7.5–1.8 Ma), was among the earliest hippopotamines (see p. 286), but whether it descended from kenyapotamines remains unknown given the limited fossil material. *Archaeopotamus* possibly gave rise to the genera *Hippopotamus* and *Hexaprotodon*.

Hexaprotodon is an extinct **paraphyletic** genus (previously comprising numerous fossil hippopotamuses now considered unrelated), which lived from around 6 Ma (Late Miocene) until roughly 15 ka (Late Pleistocene). *Hexaprotodon* is known mainly from Asia, with the oldest records coming from the Siwalik Hills in Northern India, although several species are known from Africa too, suggesting that the genus dispersed to Africa from Asia. *Hexaprotodon* went extinct in its final stronghold of India possibly because of drought-induced habitat fragmentation. The extant pygmy hippopotamus has sometimes been classified in the genus *Hexaprotodon*, but actually differs a lot from these fossil hippopotamuses, for instance, in having only two front teeth per jaw. After a rich evolutionary history, *Choeropsis liberiensis* and *Hippopotamus amphibius* are the only two hippopotamids that persist today. These are the last surviving members of two main lineages, the pygmy hippopotamus and common hippopotamus, respectively. Both lineages are estimated to have split during the Late Miocene (ca. 8 Ma). *Choeropsis* is more basal than *Hexaprotodon* and *Hippopotamus*, and therefore looks much more similar to the ancestral hippopotamus. From the Early Pleistocene, the genus *Hippopotamus* dispersed broadly all over Europe (including Great Britain), represented by the extinct European hippopotamus (*H. antiquus*), before being replaced by the extant common hippopotamus (*H. amphibius*) during the late Middle Pleistocene. The genus eventually disappeared everywhere outside its African homeland by the end of the Pleistocene. Additionally, extinct dwarf hippopotamuses colonised numerous islands during the Pleistocene, such as Crete (*H. creutzburgi*), Malta (*H. melitensis*), Cyprus (*H. minor*) and Sicily (*H. pentlandi*), and included several species in Madagascar. Compared to mainland hippopotamuses, island hippopotamuses had to survive on a considerably more limited resource base.

Cetacea (Whales)

Until the 1980s, our understanding of cetacean evolution was merely guesswork. Fossils of early forms such as *Protocetus* have been known since the early twentieth century, but were already well-adapted to aquatic life. Several morphological features (including triangular molar teeth) seemed to link cetaceans to the mesonychians (see pp. 102–3), and they were generally considered to be the **terrestrial** ancestors of whales. However, phylogenetic analyses in the late 1990s showed that cetaceans actually formed a sister grouping with hippopotamuses and their relatives, the "anthracotheriids", and thus the fully aquatic **clade** was embedded within the hooved artiodactyls. A bone in the basal cetacean ankle, the talus, has a double-pulley shape only shared with artiodactyls. Common features with mesonychians are probably a result of the prevalent **ancestral** condition in basal **ungulates**. The cetaceans have evolved hairlessness likely for lowering friction along with enhancing locomotion in the water, and had developed increasingly specialised ear bones

Fig. 2.82. Life reconstruction of the pakicetid cetacean *Ichthyolestes pinfoldi*. Early Eocene of Northern India. Reconstructed length: 1.5 m.

for improved underwater hearing. The aquatic lifestyle of cetaceans started out on the Indian subcontinent around 50 Ma, having fully evolved over a period of less than 10 million years. The earliest known transitional whale is the fox-sized *Indohyus*, an extinct chevrotain-like ungulate from the Early Eocene (ca. 48 Ma). It belonged to an extinct family of artiodactyls called Raoellidae, which is believed to be the closest sister group of Cetacea (Fig. 2.84A). The ear region of *Indohyus* had many features present only in whales. *Indohyus* also had very dense limb bones, which provided ballast to help it wade or dive underwater without floating out of control. Chemical analysis of the bones show that *Indohyus* was aquatic, although the chemistry of its dentition shows that it ate land plants.

The earliest "archaeocetes" (a **paraphyletic** assemblage representing the most **primitive** stage in cetacean evolution) are all known from Eocene deposits in South Asia (along the ancient Tethys coastline (see p. 174)) and would have looked very different to modern cetaceans. Slightly younger than *Indohyus* were the pakicetids (e.g. *Pakicetus*, *Ichthyolestes* (Fig. 2.82)), which were typically wolf-like in form with long limbs, **digitigrade** feet, a flexible neck and a full complement of **heterodont** teeth. While largely tied to the land, pakicetids had a thickened **auditory bulla** for hearing underwater, like modern cetaceans. They were poor swimmers and thought to have mostly waded in shallow water when foraging. Within two million years, the "archaeocetes" had adapted to amphibious lifestyles. Some including the ambulocetids (e.g. *Ambulocetus*) were vaguely crocodylian-like, with long, robust jaws, high-placed **orbits**, a restricted movement of the limbs, and broad feet. Enough of the skeleton is known to show that early cetaceans flexed their spine up and down while swimming, using the webbed hind feet for propulsion. Isotopic analysis proves that ambulocetids moved between nearshore and estuarine environments.

In comparison, the remingtonocetids (e.g. *Remingtonocetus*, *Kutchicetus*) are known only from coastal waters and were no longer dependent on freshwater. They were strangely elongated, with long, narrow jaws, short limbs and small orbits. A long tail indicates that remingtonocetids swam primarily by **caudal** undulation. Thus, it was only during the latter stages of the Eocene that the basal cetaceans were actually able to spread across the globe. It is likely that they were not adapted for open water swimming and probably migrated along shallow coastal water. The most diverse of the early "archaeocetes" were the protocetids. Unlike their predecessors, protocetids had a range of amphibious features such as relatively large limbs and a strong tail (it is uncertain whether it possessed flukes). Some genera, such as protocetines (e.g. *Peregocetus*, *Rodhocetus*, *Maiacetus*), could support the body out of water, while the separation of the pelvis from the vertebral column in georgiacetines (e.g. *Georgiacetus*, *Aegicetus*) would have prevented terrestrial walking. Other features include the presence of hooves on feet, giving birth head first (as evidenced from a fossilised foetus in *Maiacetus*), nostrils placed further up the snout (able to breathe at the surface with a horizontal head), and non-masticating/chewing teeth. Tooth wear in at least some genera shows that when large prey was captured, it was ripped apart in a similar manner to crocodiles and sharks. Protocetines such as *Rodhocetus* were likely efficient swimmers and may have been cumbersome on land like today's seals (Thewissen *et al.* 2007). One unusual outlier was *Makaracetus*, which had an angled **rostrum** and expanded musculature around the nasal area, similar to both dugongs and tapirs. However, unlike both, *Makaracetus* was definitely **carnivorous** and one suggestion is that it behaved much like a walrus foraging on the seabed for **benthic** molluscs. The protocetids are not a natural (**monophyletic**) grouping, being paraphyletic due to the inclusion of the final group of "archaeocetes", the basilosaurids, plus the modern neocetes (true whales).

By the Middle Eocene (ca. 40 Ma), basilosaurids had finally abandoned returning to land for a fully marine existence. Although resembling modern whales, basilosaurids were much more primitive in lacking a blowhole on the top of the head, and the melon head characteristic of toothed whales (Odontoceti). Many toothed whales have a relatively homogenous dentition, although basilosaurids retained the well-defined range of teeth seen in other mammal groups used for chewing their prey. This consisted mostly of fish, squid and other marine mammals. Some basilosaurids (e.g. *Basilosaurus* (see pp. 218–20)) were long serpentine animals due to an elongation of the lumbar vertebrae. This facilitated a movement via tail propulsion similar to that seen in modern whales. However, some genera such as *Zygorhiza* (Fig. 2.83) and *Dorudon* (see pp. 218–20)), had a more compact, porpoise-like body shape. Like the early proto-whales, basilosaurids still possessed hindlimbs, but these were tiny and not used for locomotion. They are thought to have been used as claspers for mating.

By the Late Eocene, the roots of the modern whale families were evolving as the ancient "archaeocetes" declined. The latter may have become extinct from a combination of a cooling climate and competition with the more advanced cetaceans. One major cetacean group which uses echolocation is the toothed whales. By passing air from the blowhole

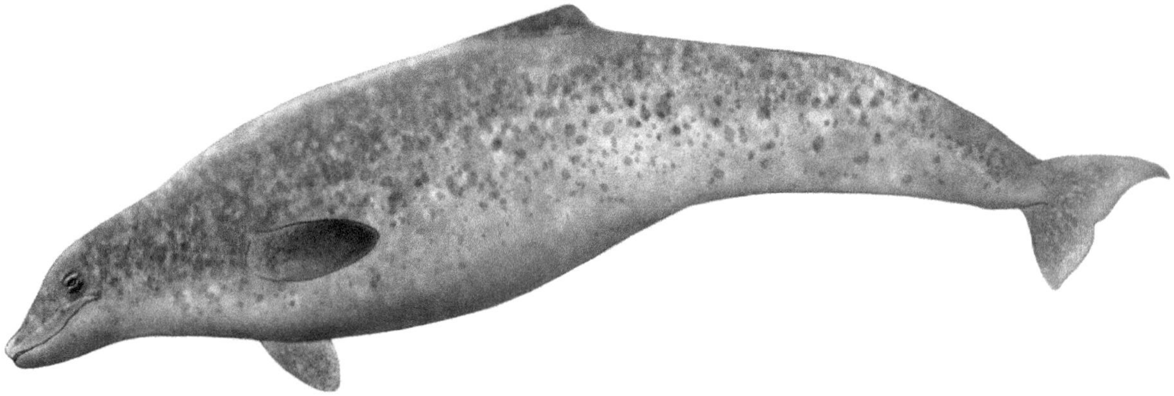

Fig. 2.83. Reconstructed life appearance of the basilosaurid cetacean *Zygorhiza kochii*. Reconstructed length: 5.2 m.

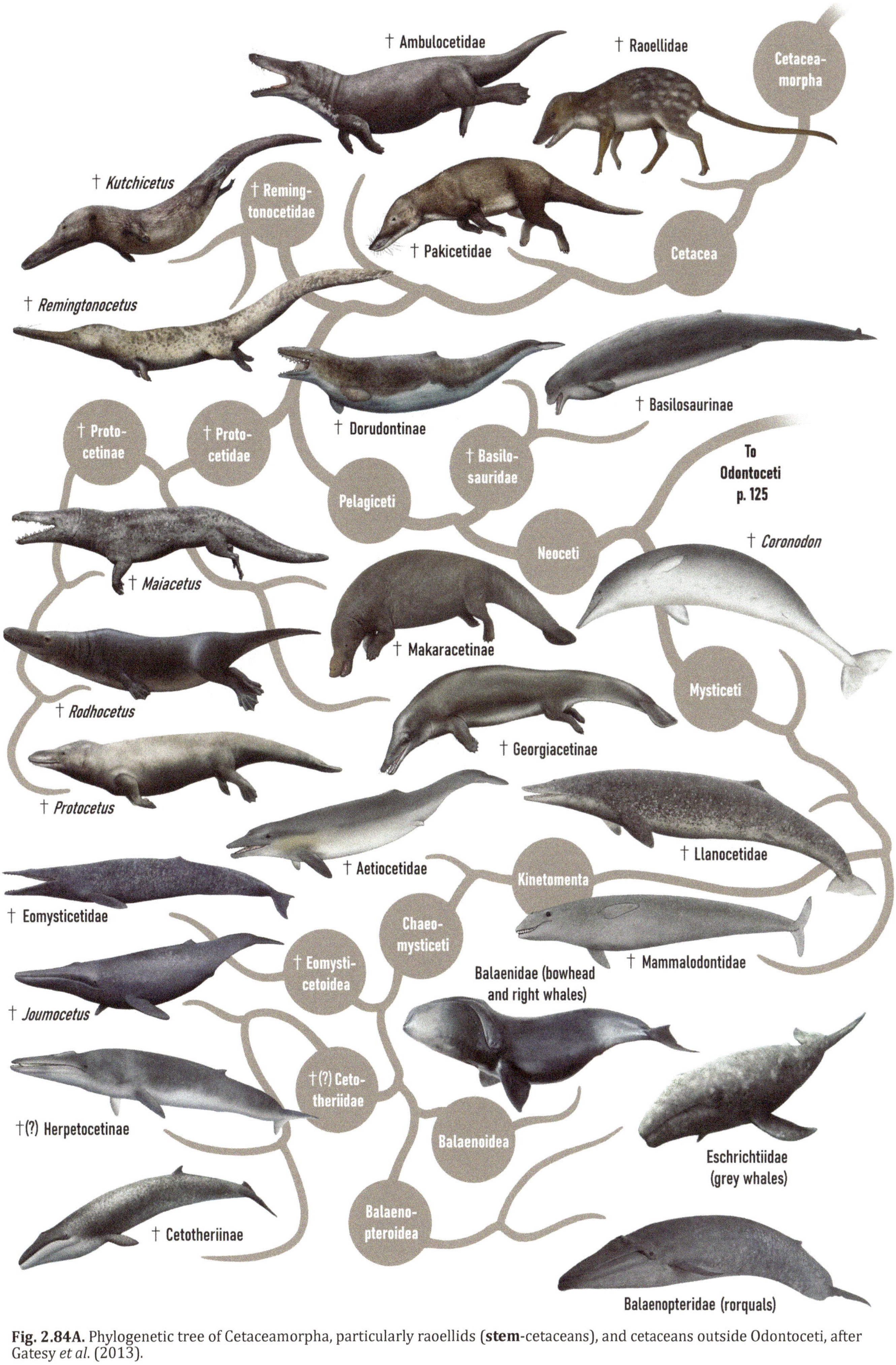

Fig. 2.84A. Phylogenetic tree of Cetaceamorpha, particularly raoellids (**stem**-cetaceans), and cetaceans outside Odontoceti, after Gatesy *et al.* (2013).

through a unique structure called the phonic lips (which act like human vocal cords), the air is vibrated creating a sound beam. Focusing the sound beam through the melon at the front of the head gives it shape and direction. Returning sound beams are picked up by the throat and passed to the inner ear. Lower frequencies are used for long distance, while higher frequencies give better definition to an object. So versatile is echolocation, that whales are able to determine distance, size, shape, speed and vector of movement, composition between two similar objects, and avoiding interference from other pod members. Echolocation is not known in the primitive "archaeocetes", or the mysticetes. As to why it developed in odontocetes is hard to pinpoint, but may be linked with cooler waters in the Oligocene (formation of the Southern Ocean), which created opportunities such as hunting in turbid rivers and below the photic zone in oceans.

Early odontocetes underwent a wide diversification and specialisation following the Late Eocene extinction of the basilosaurids (Fig. 2.84B). While they generally resembled modern dolphins and beaked whales, the basal forms had more primitive features, such as a bulky caudal peduncle (the base to the tail flukes; indicative of slower swimmers), smaller and more forward-placed melons, and often had long snouts with interlocking, double-rooted teeth. The oldest well-known fossil odontocete is *Squalodon*, the "shark-toothed whale" from the Middle Oligocene to Middle Miocene (ca. 33–14 Ma), although it was likely not ancestral to modern toothed whales. It had a much shorter neck than any Eocene cetacean. *Squalodon* shows evidence for the first hypothesised appearance of echolocation. Several commonalities with modern toothed whales include a well-compressed skull for housing the melon, and the rostrum telescoped outward into a beak.

Some primitive odontocetes, such as *Ankylorhiza*, had an array of large teeth, with those at the front pointing forwards. It has been suggested that this formidable predator behaved much like a modern orca/killer whale (*Orcinus orca*), ramming prey with the teeth causing fatal injuries. *Inticetus* had a heterodont dentition, with pointed teeth at the front and multi-lobed teeth towards the rear. A lack of wear on the teeth means that this cetacean did not chew food, possibly sieving smaller prey. In contrast, *Inermorostrum* and *Simocetus* are notable for shortened rostrums. *Inermorostrum* went a step further, lacking any teeth, and is thought to have preyed upon invertebrates through suction-feeding.

The first oceanic dolphins, such as kentriodontids, evolved in the Late Oligocene and diversified greatly during the Middle Miocene. By the Middle Miocene, many of the odontocete families were established, although with much more primitive members than their modern-day descendants. These include the sperm whales (Physeteroidea), which can be traced to fossils including *Ferecetotherium* from the Oligocene of Azerbaijan, and to a dozen other genera by the Middle Miocene, the most impressive being *Acrophyseter*. On a similar scale to the modern deep sea squid-hunting sperm whale was *Livyatan*, a very large macropredator from the Late Miocene to Early Pliocene. It had a short and broad rostrum (10 feet (3 metres) in width), which gave the whale the ability to inflict major damage on large struggling prey, such as other early whales.

The first fossil cetaceans near shallow seas (where porpoises inhabit) were found around the North Pacific; genera such as *Semirostrum* were found along California (in what were then estuaries). These animals spread to the European coasts and Southern Hemisphere only much later, during the Pliocene. One extinct group of toothed whales to evolve a long rapier-like rostrum, used in combination with high speed bursts of swimming, were the eurhinodelphinoids. At first glance, eurhinodelphinoids would have largely resembled typical delphinid dolphins apart from the hyperelongated snout (a feature which has evolved multiple times within odontocetes). Typically, most genera had complex hearing, an asymmetrical brain (also seen in modern delphinids (true dolphins)), and small, conical, peg-like teeth suited for hunting small fish and cephalopods. The long, sword-like snout had a toothless tip which is thought to have been used to stir sediment or stun fish in a tight-packed school.

It was the true eurhinodelphinids (e.g. *Eurhinodelphis, Xiphiacetus, Schizodelphis*) which had the iconic, long upper rostrum equating to almost a third of the full body length. However, not all of the eurhinodelphinoids had unequal-sized rostra. Two groups, the argyrocetids (e.g. *Macrodelphinus, Chilcacetus*) and *Eoplantanista*, had equally long jaws. *Macrodelphinus* was an especially large form, similar in size to a modern orca/killer whale. Relationships between the families have not been clearly resolved, and it may be that they were a loose collection of clades related to the platanistoids (which includes the Ganges and Indus river dolphin), delphinoids (dolphins and porpoises) and ziphioids (beaked whales). By the end of the Miocene, eurhinodelphinoids were replaced by delphinids. A cooling of the seas resulting in changes to prey choice may have been the cause.

The baleen whales (Mysticetes) possibly split from the toothed whales during the Late Eocene (ca. 34 Ma), according to molecular phylogeny. The development of baleen plates was one of the more intriguing **adaptations** that occurred during the evolution of cetaceans. With no teeth, mysticete whales use fringed, **keratinaceous** plates to sieve plankton and other small organisms. This led to further adaptations such as lunge-feeding in the rorquals, which resulted in stream-lined giants. Balaenids are slow, bulky swimmers which skim the water at a more leisurely pace, while the cetotheriids employed a variety of techniques such as skimming and suction-feeding. In comparison, early mysticetes had far more in common with the toothed whales. Evolutionary relationships between the three main families of basal toothed baleen whales and the first edentulous forms (eomysticetids) is still uncertain.

The earliest group were the llanocetids (e.g. *Mystacodon, Llanocetus* (see p. 201)) from the Eocene, giant suction-feeders known only from Southern Hemisphere deposits. **Anterior** teeth were pointed for grabbing prey, while those in the side of the mouth had rake-like structures for straining invertebrates, much like modern crabeater seals (*Lobodon carcinophaga*), also shared with the later *Coronodon*. Endocasts of the brain indicate that they were deep-divers. Also restricted to waters off Australasia during the Oligocene were the small mammalodontids (e.g. *Janjucetus*). The two known genera were **convergent** with modern leopard seals (*Hydrurga leptonyx*), possessing large eyes and interlocking teeth suggesting a raptorial or suction-feeding diet.

Sharing more features in common with the edentulous mysticetes were the aetiocetids (e.g. *Aetiocetus, Fucaia*) which possessed teeth alongside

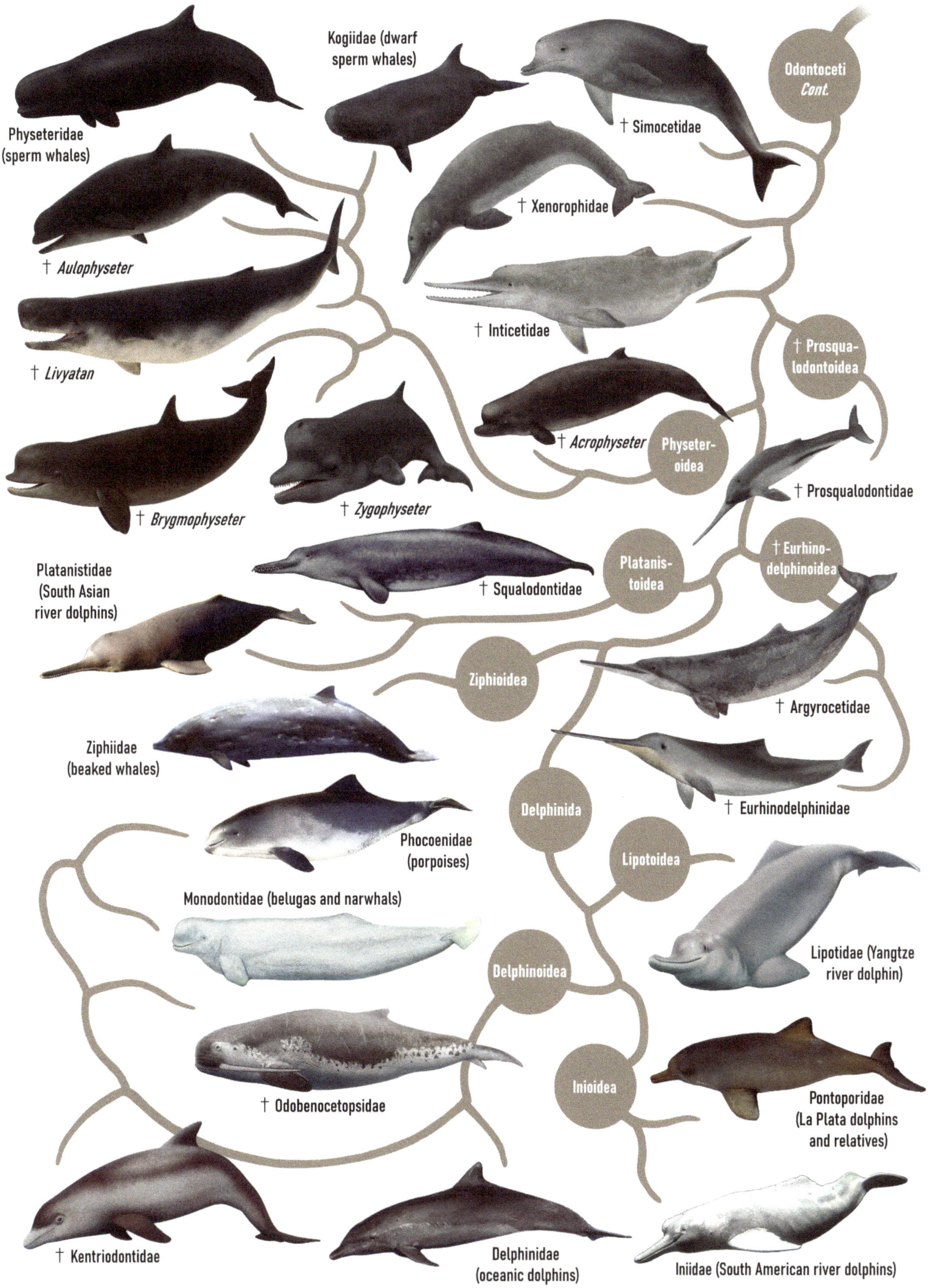

Fig. 2.84B. Phylogenetic tree of odontocete cetaceans after Gatesy *et al.* (2013).

baleen, plus a telescoping of the skull (nostrils placed further back). They are mostly known from the Late Oligocene and the Northern Hemisphere, although the smallest known member, the two metre long *Fucaia*, was present in Early Oligocene deposits. The recently discovered Early Oligocene *Maiabalaena*, which lacked any dentition (Peredo *et al.* 2018) may indicate that tooth loss was accompanied with the evolution of suction-feeding before baleen arose. Therefore, teeth and baleen may not have overlapped. The evolution of filter feeding could have resulted from physical changes in the oceans and global environmental alterations. Straining the smallest organisms from the water was a novel food item in the Miocene and mysticetes of the family Cetotheriidae were among the first whales to utilise this different food source. A major disadvantage was constantly being near the surface as this drew attention from the large predators such as sharks as well as predatory physeteroid whales. Fossil evidence has shown that the sharks and the physeteroids (e.g. *Livyatan*, *Brygmophyseter*, *Acrophyseter*) would attack from below, quickly stunning or killing the whales. The whales had wide ribs, to the point of overlapping in *Caperea*, and this may have provided some protection from a predator's attacking bite. Cetotheriids are the sister group to the rorqual whales, such as the modern blue whale (*Balaenoptera musculus*) and humpback whale (*Megaptera novaeangliae*).

Ranging from the Miocene onwards, the Cetotheriidae has, unfortunately, been used as a **wastebasket taxon** for any obscure small baleen whale fossils. Recent work has slimmed the group down to those with similar morphological features. All genera were relatively small (for a whale), at less than eight metres long, and the group are occasionally referred to as dwarf baleen whales. Cetotheriines from the Paratethys region (a shallow, inland sea stretching from Central Europe to Central Asia, existing from the Oligocene to Pliocene), such as *Cetotherium* and *Tranatocetus*, were generally larger genera and are found in Miocene deposits. The larger size is attributable to a more pelagic lifestyle and probably targeted fish. The herpetocetines of the Late Miocene and the Pliocene were smaller, such as *Piscobalaena* and *Herpetocetus*, and suction-fed on smaller benthic prey. Rorqual whales began to diversify at this time, and the cetotheriids were able to avoid competition with the pelagic whales. Most cetotheriids disappeared during the Pliocene, although one species of *Herpetocetus* made it into the Pleistocene. All four families of mysticetes were well-established by the Middle Miocene, which are generally hypothesised to have originated independently from different lineages within the broader group of cetotheriids. These are Balaenidae (right whales (*Eubalaena*) and bowhead whales (*Balaena mysticetus*)), Neobalaenidae (pygmy right whales (*Caperea marginata*), possibly relic cetotheriids), Eschrichtiidae (grey whales) and Balaenopteridae (rorquals such as humpback whales, fin whales (*Balaenoptera physalus*), minke whales (*Balaenoptera acutorostrata* and *Balaenoptera bonaerensis*), sei whales (*Balaenoptera borealis*), and the immense blue whale, the largest animal ever). The gigantism in mysticetes is directly related to their feeding mechanism—mysticete size is dependent on the amount of baleen used to filter prey.

Perissodactyla

The order Perissodactyla, or the **herbivorous** "odd-toed" **ungulates**, are the other main lineage of hooved mammals, encompassing the modern horses, asses, zebras, rhinoceroses, tapirs, and their fossil relatives. They are distinguished by various unique specialisations, although their main diagnostic characteristic is their forefeet and hind feet. In general, perissodactyls possess either one **digit** (horses) or three digits (rhinoceroses and tapirs) on each foot, and are **mesaxonic**. They are split into two major lineages: the Hippomorpha (horses and their fossil relatives) and the Tapiromorpha (rhinoceroses and tapirs, and their fossil relatives) (Fig. 2.86). As well as the familiar modern rhinoceroses and tapirs, the Tapiromorpha also includes a family of bizarre perissodactyls called chalicotheriids (see pp. 128–9). Fossil evidence suggests that perissodactyls possibly originated in or near India during the earliest Eocene (ca. 54.5 Ma), before India's collision with Asia. The cambaytheriids (see p. 102), known from India, appear to be close to the perissodactyls. All the main lineages of perissodactyls had diversified during the Early Eocene, and immigrated to Europe and North America from Asia. Prior to the Oligocene, the largest and most prevalent hooved mammals in North America and Eurasia were brontotheriids (see pp. 135–6) and **primitive** tapirs. Following the extinction of these lineages, horses and rhinoceroses were the most abundant perissodactyls, being highly diverse in species and body shape. Both lineages experienced reductions in diversity during another extinction event at around 5 Ma, and only several species of horses, asses and zebras, five species of rhinoceros and four species of tapir survive today. The perissodactyls were largely replaced by the ruminant

Fig. 2.85. Reconstructed life appearance of some representative fossil tapirids. (A) *Miotapirus harrisonensis*, from the Early to Middle Miocene of the United States; (B) *Tapirus (Megatapirus) augustus*, from the Middle to Late Pleistocene of China and Southeast Asia. Scale bar: 1 m.

Fig. 2.86. Phylogenetic tree of Perissodactyla after Holbrook and Lapergola (2011).

artiodactyls (see p. 110), such as deer, cattle, antelopes, and their kin, as the dominant large hooved herbivores. After originating in Asia, the hippomorphs dispersed all over the Northern Hemisphere. The oldest members of Tapiromorpha, and its subgroup Ceratomorpha (rhinoceroses and kin,) are known from Early Eocene strata and are nearly identical to the oldest horses. From this generalised ancestry, various primitive tapir-like animals branched.

Tapiroidea

Tapiroids (tapirs and their extinct cousins) originated and radiated during the Early Eocene, inhabiting North America (Isectolophidae, Helaletidae) and Asia (Deperetellidae, Lophialetidae), with some evidence suggesting movement between the continents via the Bering Land Bridge (see p. 174). During the Middle Eocene, the isolated islands of the European archipelago were also home to **endemic** tapiroids of the family Lophiodontidae. Following a significant Middle to Late Eocene radiation in North America and Asia, tapiroids experienced a notable decline in both diversity and abundance by the Oligocene. Among the last tapiroids to retain a **primitive** skull lacking a **proboscis** was the Late Eocene to Early Oligocene helaletid *Colodon*. *Hesperaletes*, known from late Middle Eocene beds close to San Diego, California, is the oldest known tapiroid to exhibit the characteristic retracted nasal bones and intricate facial structures suggestive of a well-developed proboscis. The nasal bones of the primitive Early Miocene tapirids *Protapirus* and *Miotapirus* (Fig. 2.85A) were further retracted. The fossil record of Tapiridae (true tapirs) extends back to the Early Oligocene in North America. Throughout the Miocene and Pliocene, the tapirids rarely show up in the fossil record of North America and Asia, possibly for being limited to forested environments that were disappearing during the cooling and drying Neogene climate. Among Pleistocene tapirs was the massive *Tapirus augustus* (Fig. 2.85B), which was nearly rhinoceros-sized. With their extinction at the end of the Pleistocene, the Malayan tapir (*Tapirus indicus*) was left as the only surviving Old World tapir.

Chalicotheriidae

Some prehistoric species have modern **analogues** with a host of common features, even if they are not remotely related. Others present a problem of being

Fig. 2.87. Phylogenetic tree of Chalicotheriidae after Coombs *et al.* (2001).

related to extant animals but not actually resembling any. Chalicotheriids are the classic conundrum, having various anatomical similarities in common with horses, giraffes, bears, anteaters and primates. They were actually true **ungulates**, despite possessing clawed **digits** instead of hooves. In fact, early chalicotheriids in Eocene Asia were difficult to tell apart from the forest-dwelling ancestral horses and brontotheriids (see pp. 135–6), their closest relatives. Some fossils have moved between groups due to incorrect placement.

All of the family were **browsers**, never developing **hypsodont** teeth for abrasive grasses. Consequently, they are restricted to forested habitats, and were never very diverse in genera. The earliest undisputed chalicotheriids are known from the Middle Eocene (ca. 46 Ma) of Asia, which is where they are believed to have originated. They dispersed to North America later in the Eocene. During the Late Oligocene, they had split into two major branches based around differences in locomotion: schizotheriines and chalicotheriines (Fig. 2.87). Those branches thrived for many millions of years, and peaked in diversity during the Miocene. More **primitive** chalicotherioids basal to the chalicotheriids are typically grouped within the family Eomoropidae (e.g. *Eomoropus*, *Grangeria*, *Litolophus*, *Palaeomoropus*), from the Early and Middle to Late Eocene beds of Asia and North America, respectively. It remains uncertain if the eomoropids were clawed or how the two families split. Subsequent to the extinction of chalicotheriids from North America by the end of the Middle Eocene, they remained primarily confined to Asia; genera including the schizotheriines *Metaschizotherium*, *Schizotherium* (see pp. 224–5), *Phyllotillon* and *Borissiakia* lived in small populations in Mongolia, China, Pakistan and Kazakhstan, respectively, and sometimes immigrated to Europe. Derived schizotheriines (represented by *Moropus* (see pp. 255–6)) travelled across the Bering Land Bridge (see p. 174) back to North America around the Oligocene-Miocene boundary. The chalicotheriine *Chalicotherium* became broadly distributed in Eurasia and Africa during the Early Miocene.

Schizotheriines had the more conventional body shape and walked on the soles of their feet. These included a dome-headed form called *Tylocephalonyx*, which lived in North America during the Early to Middle Miocene. Strong hindlimbs and an elongated pelvis allowed schizotheriines to rear upright for added height. A lack of upper incisors and canine teeth, **convergent** with bovids (see pp. 116, 118–9), and other **adaptations** to the skull, is highly suggestive of a protrusible tongue. The chalicotheriines were even more bizarre and had gorilla-like body proportions. Chalicotheriines were also more forelimb-dominated, pulling branches close towards their mouths. In the past, it was thought that the long claws were used to dig roots and tubers from the ground, although the lack of wear disproved this idea. It is now believed that the claws were similar to those of the extinct ground sloths and used to drag foliage to their mouths. Not all chalicotheriids ate the same foods, though. The African *Ancylotherium* (see pp. 302–3), and the Asian *Nestoritherum* and *Hesperotherium*, were among the only genera to survive into the Pleistocene. Chalicotheriids went extinct by the Middle Pleistocene possibly because they were too specialised for **high browsing**, making them vulnerable when their favoured habitats changed and competition from artiodactyls (see p. 104) increased.

Rhinocerotoidea

Within Perissodactyla is a subdivision comprising various families closely related to the extant rhinoceroses (family Rhinocerotidae). As well as the modern family, these include four extinct families: the Amynodontidae, Hyracodontidae, Paraceratheriidae and Eggysodontidae. These all possessed the distinctive upper molars, although the proportions and size of the teeth had numerous variations. These families are most easily differentiated by the configuration of the crests on the upper third molar. The Rhinocerotoidea diverged from a group of animals that included *Hyrachyus* (see pp. 216–7), an early Middle Eocene genus widespread across North America and Eurasia. All five rhinocerotoid families diverged during the early Middle Eocene, becoming established in North America and Asia.

Amynodontidae

The amynodontids are largely unnoticed, usually portrayed as amphibious, hippopotamus-like rhinocerotoids. This is not entirely true as only the metamynodontins such as *Metamynodon* (see pp. 235, 238) probably spent any time in water, with early forms (*Sharamynodon*, from the Middle to Late Eocene of Central and East Asia) and the snouted cardurcodontines being **terrestrial browsers**. An Oligocene fossil trackway from Arizona has been attributed to the Late Eocene to Early Oligocene North American genus *Metamynodon*, and although the prints lack details, they show that this rhinocerotoid was capable of walking across very soggy ground. In general, cardurcodontins such as *Cadurcodon* (see pp. 224–5), a Middle to Late Eocene genus from Mongolia and China, would have looked and behaved much like modern tapirs. Amynodontids are first known from the Middle Eocene and became very diverse in Asia and North America. They survived until the Late Oligocene in Europe and Pakistan (e.g. *Cadurcotherium*).

Hyracodontidae and Eggysodontidae

The hyracodontids are occasionally referred to as "running rhinos", since their limbs were considerably more elongated and gracile than in many other fossil rhinocerotoids. After having originated in Asia during the early Middle Eocene, the hyracodontids dispersed to North America later in the Middle Eocene, with the genus *Triplopus* (see pp. 222–3). The evolutionary history of hyracodontids in North America culminated with *Hyracodon* (see pp. 235, 238), which is a highly abundant fossil from the Big Badlands of South Dakota. *Hyracodon* originated during the Middle Eocene as a considerably smaller form, then became progressively larger-bodied, with premolar complexity increasing, until the Late Oligocene (ca. 26 Ma). It was the last known surviving hyracodontid. One distinct family of extinct rhinocerotoids that resembled hyracodontids are the eggysodontids, which arose during the latest Eocene and lived across Eurasia throughout the whole Oligocene. Eggysodontids were small-bodied **browsers** which are more closely related to true rhinoceroses than to hyracodontids or paraceratheriids.

Paraceratheriidae

In Asia, the paraceratheriids split from the hyracodontids in the Middle Eocene (ca. 47 Ma) and went on to become enormous. The beagle-sized *Forstercooperia* is known from the late Middle Eocene of Eurasia (see pp. 222–3). During the Late Eocene, paraceratheriids grew larger and evolved longer limbs and necks, exemplified by the horse-sized *Juxia* from the Late Eocene of China (Fig. 2.88). According to its **brachydont** molars and prominent, pointed incisors, *Juxia* was most likely a **browser**, stripping leaves off branches out of the reach of many other contemporary **herbivores**.

Fig. 2.88. Reconstructed life appearance of the paraceratheriid *Juxia sharamurenensis*. Reconstructed shoulder height: 2 m.

By the Early Oligocene, there were gigantic paraceratheriids in Asia, such as *Urtinotherium*, although the culmination of this family is the immense *Paraceratherium* (see pp. 249–50), the last and largest known paraceratheriid. It is thought that all paraceratheriids were mainly browsers. *Paraceratherium* survived in Asia throughout the whole Oligocene, but disappeared by the earliest Miocene (ca. 20 Ma). Its extinction remains mysterious, although theories include the emigration of large proboscideans (see pp. 64–8) from Africa, which could have significantly modified the habitats they entered by destroying trees and transforming woodland into grassland. Once their food source declined and in turn their numbers, populations of *Paraceratherium* would have been more susceptible to other threats. Giant predators such as the "bear-dog" *Amphicyon* (see pp. 274–5) and the hyaenodont *Hyainailouros* (see p. 83) also invaded Asia from Africa during the Early Miocene (ca. 23–16 Ma); these possibly preyed on *Paraceratherium* calves. Once *Paraceratherium* became extinct, the family Paraceratheriidae disappeared forever.

Rhinocerotidae

The rich fossil record of true rhinoceroses dates back to the Late Eocene (ca. 40 Ma), documented not only from Asia and Africa, but from Europe too and most notably from North America. The oldest fossil rhinocerotids lacked horns; indeed, according to fossil evidence, many rhinocerotids had no horns throughout their history. It is very apparent when looking at the skull of a rhinoceros that the horn is not part of the bony skeleton (in fact, by definition, it is not even a true horn). Composed of a dense mat of **keratin** fibers (the same material which makes up hair and fingernails), the horn actually develops from subcutaneous tissues. The horns are typically not preserved in fossil rhinocerotids, so we have to estimate their size and location from the rugosities on the dorsal surface of the skull where they were fixed.

There was a striking pattern of radiation, intercontinental dispersals, localised extinctions, and ultimately final confinement to only several regions around the tropics of Asia and Africa. True rhinoceroses first evolved in East Asia (the earliest fossils are known from Mongolia and China) from a shared ancestor with hyracodontids during the late Middle Eocene (ca. 40 Ma). The oldest rhinocerotids subsequently dispersed to North America during the Late Eocene, when they were similar in size to a Great Dane and almost indistinguishable from their **primitive** horse or tapir relatives. Several early rhinocerotids, such as *Uintaceras*, show similarities to the early hyracodontids (a basal rhinocerotoid group). Although it was a slender genus with long limbs, *Uintaceras* was not a fast sprinter. All other rhinocerotids are split into two subfamilies, the elasmotheriines and rhinocerotines, which are sister groups that diverged during the Eocene (Fig. 2.89). The most primitive rhinocerotids subsequently increased in body size, typified by the cow-sized, hornless early forms of elasmotheriines such as *Subhyracodon* (see pp. 235, 238) and *Trigonias* from the Late Eocene to Late Oligocene of North America, which were tapir-like **browsers**. Early elasmotheriines such as the aforementioned genera had elongated skulls supporting a cornified pad or horn over the nose. In later genera, this gradually shifted up the forehead as the skull shortened. It is unknown if rhinocerotids lived in Europe during the Eocene, although forms such as *Epiaceratherium* and *Ronzotherium* (see pp. 231–2) invaded the continent in the Early Oligocene during a faunal turnover called the Grande Coupure (see p. 230).

Following the extinction of brontotheriids (see pp. 135–6) at the end of the Eocene, rhinocerotoids emerged as the largest **terrestrial** mammals in both North America and Eurasia, a position they held until mastodonts emigrated from Africa around 20 Ma. Throughout the period of "rhino rule" that occurred approximately 34–20 Ma, North America was ruled by the diceratheriin rhinocerotids with two parallel horns on their snouts, including *Diceratherium* and *Menoceras* (see pp. 254–5). At the same time, various primitive rhinocerotid genera predominated across Eurasia, such as *Protaceratherium*, *Mesaceratherium*, *Aceratherium* and *Brachydiceratherium* of the **prehensile**-lipped lineage Aceratheriini. Additionally, numerous other rhinocerotid lineages, particularly the hippopotamus-like Teleoceratini along with the Rhinocerotini, originated in Eurasia too. The Aceratheriini and Teleoceratini dispersed to North America during the Early Miocene (ca. 19 Ma). The aceratheriins and teleoceratins were major components of many North American and Eurasian faunas during the Middle to Late Miocene, typically occurring in pairings of a **grazer** (usually a teleoceratin) or a browser (usually an aceratheriin)

A major difference between aceratheriins and the other major groups of rhinocerotines was that the former were hornless. Instead, they possessed elongated, tusk-like incisors in the lower jaw. These teeth were used primarily for defence and fighting, notably being much larger in males. Aceratheriins occasionally shared habitats with other rhinocerotids. In Miocene Florida, the long-limbed aceratheriin *Aphelops* overlapped with *Teleoceras* (see pp. 260–1), the latter being a well-known member of the short-limbed teleoceratins. Their different proportions were **adaptations** for feeding on different types of vegetation (*Aphelops* was largely a browser, while *Teleoceras* was more of a **mixed feeder**). Towards the end of the Miocene, rhinocerotines diverged into three separate tribes, each having extant species. The earliest representatives of the Sumatran rhinoceros group (dicerorhinins), and the Indian and Javan

Fig. 2.89. Phylogenetic tree of Rhinocerotidae after Tissier *et al.* (2020), and Lu *et al.* (2023).

rhinoceros group (rhinocerotins), evolved in Eurasia, while the tandem-horned rhinoceros group (the dicerotins) evolved in Africa, becoming the extant white rhinoceros and black rhinoceros. Formerly placed in *Rhinoceros* with the extant one-horned Indian rhinoceros (*R. unicornis*) and Javan rhinoceros (*R. sondaicus*), two recently extinct rhinoceroses from the Philippines and Taiwan have been placed in a new genus called *Nesorhinus*, with the small Philippine rhinoceros (*N. philippinensis*) showing some evidence of **insular dwarfism** (shorter limbs). Dicerorhinins also include a number of genera which were once distributed all over Eurasia such as *Stephanorhinus* (see p. 338), which first evolved in Western Europe in the Pliocene before migrating into eastern Eurasia during the Early Pleistocene. The most famous genus is the woolly rhinoceros *Coelodonta* (see pp. 342–3). Some of the relic "**archaic**" groups (Teleoceratini, Aceratheriini) disappeared around the Miocene-Pliocene boundary, leaving North America without any rhinocerotoids. This extinction was possibly triggered by climatic changes to cooler and drier conditions due to the drying of the Mediterranean. Meanwhile, the Dicerorhinini diversified during the Pliocene and Pleistocene, evolving into the woolly rhinoceros of the Ice Ages and represented by the extant Sumatran rhinoceros (*Dicerorhinus sumatrensis*). The Elasmotheriini also diversified across Eurasia, from Spain to China, culminating in the elephant-sized Eurasian form *Elasmotherium*, the best known genus. *Elasmotherium* includes the largest known rhinocerotid (*E. caucasicum*) and the last surviving member of the subfamily Elasmotheriinae (*E. sibiricum*). It was huge, standing over 2.5 metres high at the massive hump over the shoulders, which supported the heavy head.

Traditionally, *Elasmotherium* has been depicted with a tall horn projecting out of its forehead, which gave rise to the suggestion it was linked to unicorn mythology. The giant horn was assumed to protect the unusual hollow dome of bone on the forehead, which is now suspected to act as a resonating sound chamber. As rhinoceros horn is not attached to bone, there are no fossilised remnants to prove exactly how large the keratinous covering was. Recent studies have suggested that a large horn was impractical and that the thin-walled dome was probably protected by a smaller horn. *Elasmotherium* was a grazer, using a prehensile lip to rip up grasses and sedges which were processed by **hypsodont** molars. A cornified pad above the nostrils was useful for foraging on tubers. Bulky muscles on the neck indicate that *Elasmotherium* employed a wrestling and shoving style of fighting. *Elasmotherium* may have sported a longer coat in cooler locations, particularly during winter. Some specimens have been located in Siberia, dated to as recently as 35 ka, which meant there was potential overlap with humans. The earlier and related *Iranotherium* (see pp. 282–3) possessed a domed boss, like *Elasmotherium*, but like all other elasmotheriines, it was located on the **anterior** end of the snout. *Iranotherium* likely gave rise to *Sinotherium* during the Early Pliocene. Certain palaeontologists even prefer to place the genera *Iranotherium* and *Sinotherium* within *Elasmotherium*. *Stephanorhinus*, *Coelodonta* and *Elasmotherium* had all vanished before the end of the Pleistocene, likely due to climate change that disrupted their habitat, leaving only the Dicerotini in Africa and the Rhinocerotini in South and Southeast Asia.

Equidae

The evolution of true horses was centred in North America. Since the earliest Eocene, horses have seen many **adaptations**, migrations out of North America, and extinctions, leading to only one extant genus, *Equus*. The earliest equids, also known as dawn horses, were small, dog-sized **browsers** that existed in forested environments. Flexible bodies and relatively short limbs allowed the dawn horses to dart into undergrowth and run on uneven ground. On each foot, they had multiple **digits**, with four on the forefeet and three on the hind feet, with the longer toes spreading out on wet ground. The relationship between dawn horses and modern *Equus* is often portrayed as a straight line with a number of transitional genera. However, the truth is that horse evolution is full of various side branches which left no extant descendants (Fig. 2.90). Moreover, dawn horses were diverse in number, with multiple genera found throughout Eurasia and North America during the Eocene. Confusion reigns on the identity of some dawn horses, in particular, whether some of the North American genera are considered *Eohippus* or *Hyracotherium*. This may depend on whether two separate fossils are considered the same, or if some are even true equids (the related palaeotheriids (see pp. 134–5) are a possibility).

Protorohippus (previously called *Eohippus* or *Hyracotherium*) was a terrier-sized dawn horse that existed during the Early Eocene. Its descendants developed into numerous distinct lineages that co-existed for a period. One of these was *Orohippus* (see pp. 204, 207); in which the first premolar was reduced, the last premolar became **molariform**, and the crests on the **cheek teeth** were more marked. These factors enhanced the grinding efficiency of the teeth in *Orohippus*. *Orohippus* may have given rise to *Epihippus* by the Middle Eocene (ca. 47 Ma), the latter which had a shoulder height of about 60 centimetres and carried on the evolutionary trend of progressively effective grinding cheek teeth. *Epihippus* possessed five grinding, **brachydont** cheek teeth exhibiting well-developed crests. It was previously thought that the greyhound-sized, **tridactyl** horse genera *Mesohippus* and *Miohippus* (see pp. 235–6, 238), both from the Late Eocene to Early Oligocene, represented successive stages on a direct, unbranched line of equid evolution, but new evidence suggests they coexisted for several million years and that *Miohippus* evolved from *Mesohippus* through a branching, or cladogenetic, process. Horses then radiated from *Miohippus*-like progenitors to fill a wide range of ecological niches. As environments started to dry out in the Oligocene, equids began to show features for living on open plains. Teeth became higher-crowned, with complex crests for processing abrasive grasses, and as a result, skulls lengthened. Longer limbs were better adapted for **cursorial** behaviour and equids adapted to walking primarily on their central toe. Equids diversified into several distinct lineages during the Miocene.

Among the first of the **grazing** horses and members of the subfamily Equinae was the tridactyl, pony-sized *Merychippus*. As it shows many of the distinctive features of modern horses, *Merychippus* is considered a pivotable genus in equid evolution. In the Early Miocene, equines split into two major tribes: Equini, which includes all extant species, and the tridactyl Hipparionini. The *Merychippus* line probably gave rise to lineages including the dwarf horse *Calippus* (see p. 304), from the Middle Miocene to Early Pleistocene, *Astrohippus*, from the Middle Miocene to Early Pliocene, and *Protohippus* (see pp. 260–1), from the Late Miocene. Like the **primitive** basal forms, some later equids remained forest specialists. Anchitheriines were a distinct lineage of equids with

Fig. 2.90. Phylogenetic tree of Equidae after MacLaren (2021).

brachydont teeth designed for browsing, found in subtropical and temperate forests. Certain anchitheriines, such as *Megahippus*, were nearly as large as a modern Przewalski's horse (*E. przewalskii*). A remarkably successful group, able to live alongside grazing genera, anchitheriines were the first group of equids to migrate out of North America and into Europe, and survived through into the Pliocene. True equines are distinguishable by a complete reduction of the side toes, and these are divided into two tribes: the extinct hipparionins and the surviving equinins. They were fleet-footed runners of open plains, diversifying during the Miocene and eventually migrating out of North America. Hipparionins were a highly diverse and widely successful side branch of equids which evolved highly specialised teeth, and exhibited a characteristic concavity around the front of the facial region. Hipparionins lived throughout the Miocene and Pliocene, with genera in North and Central America (including *Cormohipparion* (see pp. 260–1), *Hipparion* (see pp. 282–3) and *Nannippus* (see p. 304)), and then migrating into Eurasia (including *Hippotherium*, *Cremohipparion* and *Proboscidapparion*) and Africa (*Eurygnathohippus* (see pp. 285–7)). Unfortunately, there is considerable confusion regarding names as *Hipparion* and *Hippotherium* (see pp. 278, 281) have been interchanged regularly (so that it is hard to know which name to use).

Numerous genera coexisted in different size ranges and niches throughout a broad spectrum of habitats. Both browsing and grazing genera could be found within the same genus. Those genera adapted to forested and woodland environments, such as *Hippotherium*, were classic **mixed feeders**. They had "deer-like" proportions, and were leapers rather than sustained sprinters. One large Asian genus, the long skulled *Proboscidipparion*, had extremely retracted nasals indicative of either a short **proboscis** or very mobile upper lip. Others were primarily grazers, including one species of the European *Cremohipparion* which was only 80 centimetres at the shoulder. *Eurygnathohippus* is known to have survived into the Early Pleistocene in Africa. Following the extinction of the hipparionins, the only remaining **clade** of horses were the equinins.

Within Equini, two separate lineages each represented by the tridactyl genera *Pliohippus* and *Dinohippus* independently evolved **monodactyly**, both genera which appeared during the Middle Miocene. Previously considered ancestral to modern horses was the slim-limbed *Pliohippus* (see pp. 260–1). During the Early Pliocene, many of these tridactyl and monodactyl horse lineages vanished, with only *Dinohippus* surviving and evolving into the modern horse genus, *Equus*. The tribe Equini was the main horse lineage to survive the latest Miocene to Early Pliocene extinctions. *Equus* was prevalent across the Northern Hemisphere. More recently split as a separate genus was *Haringtonhippus*, known as the stilt-legged horse. The slim lower limb bones strongly resembled those seen in wild asses, although the two lineages arose separately as *Haringtonhippus* is closer to the stouter-legged caballine horses (which includes the precursors of the domestic horse (*Equus ferus caballus*)).

Following the connection with South America, horses from North America were part of the Great American Biotic Interchange, or GABI (see pp. 252, 295), at the end of the Pliocene. Several lineages of horses radiated into a number of **endemic** forms. *Hippidion* had a large nose and short limbs, an adaptation to the cooler environment of Patagonia and the Andes (see pp. 329–30). With the first fossils dating to the Early Pliocene on either side of the Bering Strait, it seems that *Equus* left North America early on, quickly spreading throughout Eurasia and Africa. One **monophyletic** lineage led to zebras, which are now restricted to Africa. The largest known species of zebra was the Cape zebra, *E. capensis*. It had a build not unlike a small draft horse, with a shoulder height over 1.5 metres tall. It was one of the few African casualties during the Pleistocene-Holocene transition. At the end of the last Ice Age (ca. 11.7 ka), horses became extinct in the New World, possibly due to a combination of climate change and human hunting, but returned there in 1519 courtesy of the Spanish conquistadors landing in Mexico.

Palaeotheriidae

The Palaeotheriidae is an extinct family of vaguely horse- and tapir-like animals that lived in the dense tropical forests of the European archipelago throughout the Eocene. Palaeotheriids have frequently been confused with horses, and some palaeontologists still refer to them as horses, although thorough analysis reveals that the Palaeotheriidae is a sister **taxon** to the Equidae, the latter which was restricted to North America during the Eocene. In Europe, the palaeotheriids occupied the niche of true horses. Early palaeotheriids had short, stout limbs and are thought to have been **frugivorous**, whereas later, larger and longer-limbed forms are thought to have been **browsers**, feeding on leaves from low-hanging vegetation and plant shoots from ground level. *Hyracotherium* was one of the smallest, earliest and most **primitive** palaeotheriids (Fig. 2.91A). *Hyracotherium*, along with other early palaeotheriids including *Propalaeotherium* (see pp. 213, 215–6) were characterised by **brachydont** and **bunodont** teeth. Despite bearing a strong resemblance to early equids (see pp. 132–4), early palaeotheriids were characterised by a different development of the dentition. In the **cheek teeth**, the transverse ridges ran obliquely backwards, forming a vaguely W-like shape, whereas in equids, the structure of the ridges forms a pattern resembling the Greek letter π (pi). However, the palaeotheriids of the subfamily "Pachynolophinae" (e.g. *Plagiolophus* (Fig. 2.91B)) never developed the W-shaped **molariform** structure. "Pachynolophinae" has been considered a **paraphyletic** assemblage that gave rise to Palaeotheriinae (Danilo *et al.* 2013; Bai, 2017).

The palaeotheriines, such as the Early Eocene to Early Oligocene genus *Palaeotherium* (Fig. 2.91C), had a dentition that was **selenolophodont**, with higher-crowned cheek teeth. There was a trend towards progressively molariform premolars in *Palaeotherium*, culminating in the largest species, *P. magnum*, which was over 1.45 metres (4 feet 9 inches) in shoulder height, and is known from a near complete and beautifully preserved **articulated** skeleton discovered in Mormoiron, France. This shows that it possessed a horse-like skull, an elongated neck, relatively long forelimbs, and **tridactyl** forefeet and hind feet. The forelimbs were proportionally longer than the hindlimbs. *Palaeotherium* had somewhat heavyset limbs, suggesting it was a slow-moving browser. Palaeotheriids were greatly affected by the extinction event around the Eocene-Oligocene boundary (see pp. 229–30), triggered by a cooling climate and in turn, the loss of the dense forests that characterised the Eocene. *Palaeotherium* was the sole palaeotheriid to persist into the Oligocene, after which the family went extinct.

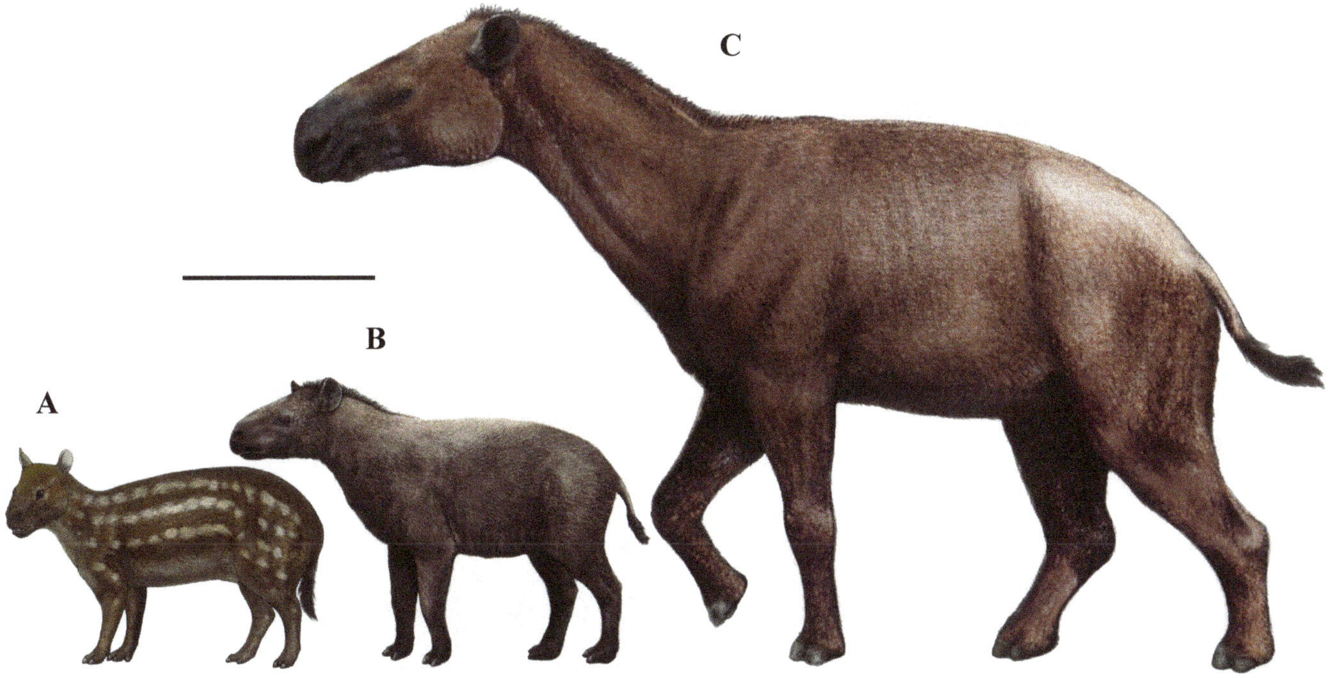

Fig. 2.91. Reconstructed life appearance of a sampling of palaeotheriids. (A) *Hyracotherium leporinum*, from the Early to Middle Eocene of England and France; (B) *Plagiolophus annectens*, from the Middle to Late Eocene of England, France, Spain and Switzerland; (C) *Palaeotherium magnum*, from the Middle to Late Eocene of England, France, Spain, Portugal and Greece. Scale bar: 0.5 m.

Brontotheriidae

Although vaguely resembling rhinoceroses, the brontotheriids, otherwise known as titanotheriids, were more closely related to horses (Fig. 2.86). Brontotheriids lived only during the Eocene Epoch and were restricted to North America and Asia, with the western-most genera barely reaching Eastern Europe. Their distinctive appearance and size easily grabs the attention of the public (Fig. 2.92). Early perissodactyls started small, with the first horses, tapirs, rhinocerotoids (see p. 129) and brontotheriids all being confusingly similar in appearance. They were short **herbivores**, skulking in thick undergrowth in a similar manner to modern duikers (Cephalophinae), pudu (*Pudu*) and agoutis (*Dasyprocta*). *Lambdotherium*, from the Early Eocene of North America, fits this classic early perissodactyl mould.

The evolutionary history of the Brontotheriidae is well-known due to an excellent fossil record in North America, where it most likely began around 53 Ma and lasted nearly 20 million years. The brontotheriids peaked in diversity during the Middle Eocene, with about two dozen genera known. The strong body size increase was striking. All early brontotheriids, such as the North American genera *Eotitanops* and *Palaeosyops* (see pp. 204, 207), were rather small, no more than one metre in height, and hornless. Some North American genera, such as *Sphenocoelus*, evolved highly elongated skulls, with the eyes located far forward. Over time, brontotheriids diversified in shape and generally got much larger, with horn-like skull appendages.

Clear horn formations were first observed in the late Middle Eocene, but the development took place in several stages. In some forms (*Telmatherium*), the **frontal** bone first began to grow over the nasal bone with small, triangular-shaped bony outgrowths. Later, small bony outgrowths formed, as demonstrated in the Asian *Rhinotitan* and possibly **semi-aquatic** *Aktautitan*, and the North American rhinoceros-sized forms *Diplacodon* and *Protitanotherium*. Only then did the classic horns form. Possibly the most impressive was *Megacerops*, from the Late Eocene of North America. *Megacerops* was previously known as *Brontops* and *Brontotherium*, and the origin of the Thunder horse myth. This was a giant brontotheriid at over three tonnes in weight and at 2.5 metres (8 feet) tall, as large as some modern elephants. On the end of the long skull was a Y-shaped bony prong, covered in skin like a giraffe's **ossicones**. Impressive for display (males had the largest horns), they were equally damaging as offensive weapons. Skeletons of large male *Megacerops* have been found with healed ribs, fractured during a fight with another male.

Overall, the brontotheriids were the most diverse group of large mammals in both North America and Asia during the Middle Eocene. What is remarkable is that there was no independent radiation on either continent; rather, the phylogenetic development is strongly interlinked. Some researchers believe there were up to a dozen independent intercontinental dispersal waves that ran in both directions. In the Late Eocene, the brontotheriids gradually declined in diversity. About 10 genera are known from this interval, but most can be placed in the early phase of it. All brontotheriids from that time were horn-bearing and mostly very large. In particular, two different lines emerged: those with two widely-spaced horns, which were closely related to *Megacerops*, and those with two closely-spaced or fused horns, which were closely related to an Asian lineage of late brontotheriids (from the Middle to Late Eocene) called the embolotheriins, which included the impressive *Embolotherium* (see pp. 224–5). It is suspected that the shift towards a cooler, drier environment at the end of the Eocene was the main cause of the extinction of brontotheriids. As plants altered to adapt to the new environments, the **brachydont** teeth of brontotheriids were not capable of dealing with the tougher vegetation.

Fig. 2.92. Phylogenetic tree of Brontotheriidae after Mihlbachler (2008), and Mader (2010).

Miscellaneous Mammals With Hooves

As well as the modern lineages whose phylogenetic affinities are now well-established, there are various extinct lineages of large, **herbivorous** hooved mammals that may or may not be relatives of true **ungulates**. Their affinities have long been controversial, although much of the evidence to date remains insufficiently strong to convince everyone. A consensus has yet to be made on what mammals they were related to, or if they were ungulates.

Fig. 2.93. Phylogenetic tree of Pantodonta after McKenna and Bell (1997), and Alroy (2002).

Pantodonta

Across Eurasia and North America, one group of early mammals, the pantodonts, were among the first groups of mammals to evolve large body sizes after the Cretaceous-Paleogene extinction event (see p. 169). The pantodonts were so successful that some genera were among the largest known animals during the Early Paleogene. The phylogeny of pantodonts remains highly contentious, with some studies (e.g. Kort, 2023) suggesting that they are members of Pan-Euungulata (see p. 100), whereas other studies (e.g. Halliday *et al.* 2017) recovered pantodonts at the base of Laurasiatheria (see p. 74). In general, all pantodonts were very conservative in body shape, with the early North American genus *Pantolambda* being a typical example (see pp. 178–9). Mostly heavy-bodied with a long tail, their feet had five well-developed toes and **digits** that ended in nails that were not claws or hooves. The heavy head contained an unspecialised **complete dentition**. Pantodonts are divided into four broad groups: the basal forms, the **primitive** bemalambdoids, the generalised pantolambdoids and the massive coryphodontoids (Fig. 2.93). Prior to the radiation of the pantolambdoids and the later coryphodontoids, several small groups of relatively unspecialised **herbivores** paved the way during the Paleocene for the pantodonts. While most pantodonts had roots in Eurasia or North America, interestingly, one of the earliest and most basal genera, *Alcidedorbignya* (see pp. 184–5), is the only known representative from South America. While later forms are best known from North America and Europe, most basal forms have been discovered in China. *Bemalambda*, from the Early Paleocene, and *Hypsilolambda*, from the Early to Middle Paleocene, were typical medium-sized genera with short, low skulls, broad snouts and robust bodies. The teeth of *Bemalambda* indicate an **omnivorous** diet. One small, rat- to squirrel-sized pantodont was *Harpyodus*, a generalised omnivore from the Middle Paleocene.

Although not a speciose group, pantolambdoids consisted of around seven different families. Genera from North America tended to be larger and more robust. Sheep-size pantolambdids from the Middle Paleocene gave way to the larger barylambdids in the Late Paleocene. Heavyset and with a vague resemblance towards ground sloths, *Barylambda* is the most familiar of all pantodonts. At around 650 kilograms, it was the largest mammal of the Late Paleocene and sheer size meant that it was probably immune from predation. Like ground sloths, the solid tail allowed *Barylambda* to rear up on its hind limbs in order to **browse** higher branches. Differing from other pantodonts due to the presence of strong claws was the bear-like *Titanoides*. Although the **cheek teeth** were similar to other pantodonts, the North American *Titanoides* possessed a pair of sabre-like upper canine teeth which were possibly used for territorial battles. Although fierce-looking, *Titanoides* is thought to have been an accomplished digger searching for roots. The familiar *T. primaevus* was medium-sized compared to other members of the genus, with species decreasing in size from the middle towards the end of the Paleocene. Fossils of pantolambdodontids, such as the Asian *Pantolambdodon*, are found up until the Middle Eocene, characterised by slender jaws with smaller teeth. *Pantolambdodon* was a medium-sized pantodont that had a high nasal opening indicating the presence of a tapir-like **proboscis**. The lower jaw was thin, elongated and gracile in appearance, unlike other smaller forms such as *Archaeolambda* (see pp. 192–3). The pantolambdodontids were among the last of the pantolambdoids, but not the last of the pantodonts.

Among the best known of the Early Cenozoic mammals, *Coryphodon* had a widespread distribution across Eurasia and North America. Stretching over a long period of the Paleocene/Eocene transition ranged various species of differing sizes. While many have been named, the validity of most are questionable. Like all of its closest relatives, *Coryphodon* was heavy-bodied with short limbs, and notably had a very small brain-to-body mass ratio (so not the most intelligent of mammals!). Although their preferred habitat was warm wetlands, some populations lived north of the Arctic Circle and thus spent long hours in extended darkness during the winter. Isotopic studies of the teeth of *Coryphodon* revealed diets which switched between the seasons. **Angiosperms** were eaten during the summer, but as these were not available in the winter, they switched to leaf litter, twigs, conifer needles and fungi. *Coryphodon* had notably enlarged tusks, although these were limited to large individuals, suggesting **sexual dimorphism**. **Polygynous** herding is a typical social structure of herbivores where males are larger than females, and this can also be inferred from the fossil material. Later coryphodontoid genera were restricted to East Asia during the Late Eocene, the knobbly-skulled *Eudinoceras* and the massive *Hypercoryphodon* (see pp. 224–5) which are both impressive. As inferred by its name, *Eudinoceras* was once thought to be a dinocerate, and both groups have been previously linked. These were among the last of the pantodonts, and were replaced by **ungulate** groups with more efficient teeth and digestive systems.

Tillodontia

The tillodonts are an extinct mammal lineage of uncertain affinities whose fossil record spans the Early Paleocene to Middle Eocene of Asia (Pakistan, Mongolia and China), North America (Canada, Mexico and the United States) and Europe (France and England). With only 20 genera so far documented, they were never highly diverse and are typically scarce as fossils, apart from the Late Paleocene and Early Eocene, with *Esthonyx* being an abundant **taxon**. The skulls of tillodonts exhibited a unique combination of features of various mammals including **ungulates** and carnivorans (see pp. 81, 83–4). Tillodonts had **hypselodont**, chisel-like incisors (similar to those of rodents, although tillodonts were definitely not rodent relatives) and reduced or absent canine teeth, although their **cheek teeth** exhibited V-shaped crests bearing a resemblance to those of pantodonts. Tillodonts started off as very small forms, with skull lengths of just five centimetres (two inches), and evolved into one of the earliest lineages of large mammals during the Paleogene. This pattern of increasing size peaked with the North American *Trogosus* (see pp. 206, 208). Tillodonts arose in Asia during the Early Paleocene, comprising small-bodied taxa including *Huananius*, *Benaius* and *Meiostylodon*. During the latest Paleocene/earliest Eocene (ca. 56 Ma), the tillodont *Esthonyx* was among the earliest to disperse to North America from Asia, where it became an abundant genus. The maximum species diversity of tillodonts is found in North America's fossil record, specifically from the Late Paleocene to Middle Eocene.

During the Early to Middle Eocene, at least four other tillodont genera lived in North America until the group's extinction there, with *Trogosus* being the last known representative. In Europe, tillodonts of the genera *Plesiesthonyx* and *Franchaius* emerged during the Early Eocene, but disappeared shortly afterwards. The rather reduced neocortex in the brain could have been a disadvantage for tillodonts such as *Trogosus* when facing competition with artiodactyls (see p. 104) for possibly similar food sources and when trying to avoid being preyed on by **primitive** carnivorans, the latter two groups which, during the Eocene, possessed a larger neocortex relative to the brain. These were possible causes of the extinction of tillodonts (Bertrand *et al.* 2024).

Taeniodonta

The enigmatic taeniodonts, likely basal placentals, were a group of "**archaic**" mammals characterised by robust builds, specialised, **hypselodont** teeth, and large claws. The name taeniodont means "banded teeth", presumably referring to ever-growing growth rings on the **cheek teeth**. The taeniodonts were never particularly diverse, with only nine confirmed genera so far known, although this could be because they lived in dry or arid climates not conducive to fossilisation. Taeniodonts are generally classified into two families: the small-bodied conoryctids (e.g. *Conoryctes* (see p. 180)), and the larger, more robust stylinodontids. The oldest known putative genus, *Schowalteria*, actually dates from the Late Cretaceous and was one of the largest mammals of its time. Undoubtedly, its burrowing habits helped taeniodonts survive the Cretaceous-Paleogene extinction event (see p. 169) as some Early Paleocene forms, such as the stylinodontids *Wortmania* and *Psittacotherium* (see pp. 178, 180) reached even larger sizes within the subsequent 500,000 years. Later stylinodontids (e.g. *Ectoganus*, *Stylinodon*) had evolved extreme **fossorial** features such as powerful forelimbs and large, wide claws. However, it is the bizarre, box-shaped skull with a deep jaw and large front teeth that taeniodonts are most noted for. While they had some resemblance to a rodent's chisel-like incisors, the largest teeth were in fact modified canine teeth. The trends of robust skulls and massive, tusk-like canine teeth were further developed in the Late Paleocene to Early Eocene taeniodont *Ectoganus*. These trends culminated with *Stylinodon* (see p. 206), a genus that appeared in the Eocene and persisted to the very end of that epoch. The taeniodonts became extinct by the Late Eocene possibly because rodents were better equipped for the niche of digging.

Dinocerata

During the Paleocene, an empty vacuum left by the demise of the dinosaurs led to mammals diversifying and flourishing. Many genera remained modest in size, while a few **herbivorous** groups grew to large sizes. Among these groups were the dinocerates (also known as uintatheres). The **primitive** dinocerates, such as *Prodinoceras* (see pp. 192–3) from the Late Paleocene, were pig-sized creatures with no horns, but large, tusk-like upper canine teeth. In the Early Eocene, the rhinoceros-sized *Bathyopsis* (see pp. 204, 207) continued the trend of large body size and huge upper canine teeth with large flanges of bone on the lower jaw to receive them. The largest dinocerates, *Uintatherium* and *Eobasileus* (see p. 204), were the size of elephants. The dinocerates were the largest mammals known from the Middle Eocene of North America and Asia.

Then by the late Middle and Late Eocene, brontotheriids (see pp. 135–6) took over as the largest land mammals, and dinocerates vanished. Even though they had elephant-sized bodies with broad, flat feet like elephants, dinocerate heads were truly weird. They had three pairs of knob-like horns on the top of their skull: one over their noses, one over their eyes and one on the back of their skulls. Even stranger were the huge, tusk-like upper canine which protruded downward below their lower jaw, which also had paired knobs along the bottom. However, the presence of head ornamentation was not uniform across this small group. *Gobiatherium* lacked the knob-like horns, fang-like upper canine teeth or mandibular flanges (Fig. 2.94). Instead, it had a highly flattened skull along with a set of arched nasal bones and an expanded septum that formed a nearly spherical snout. This structure may have been an **adaptation** for supporting a specialised growth of soft tissue. Lastly, the teeth of dinocerates were surprisingly small **brachydont** grinders, much too small for such a huge body, which would have needed a lot of food. These teeth could only chew soft, leafy vegetation, so presumably that is what they fed on in the jungles of North America and Asia during the Middle Eocene, and their disappearance by the Late Eocene was apparently related to the shrinking of the tropical forests as the Earth became cooler and drier. So where do dinocerates fit into the mammalian tree? Traditionally, the dinocerates were considered either part of the large **ungulate** supergroup or the obscure "cimolestans" (see pp. 77–8), but modern thinking has them possibly immediately outside the split between Perissodactyla (see p. 126) and Artiodactyla (see p. 104). On the other hand, they bizarrely may be related to the South American xenungulates.

Fig. 2.94. Reconstructed life appearance of the dinocerate *Gobiatherium mirificum*, from the Middle Eocene of Mongolia. Reconstructed length: 3 m.

Meridiungulata (South American Native Ungulates)

The South American native **ungulates** (SANUs) were a highly diverse group of peculiar, extinct placental mammals. After the Late Cretaceous, South America underwent a period of isolation from the rest of the world, so any proboscideans (see pp. 64–8), artiodactyls or **crown** perissodactyls (see p. 126) were unable to enter the continent until the Late Miocene or after. As a result, without competition from the ungulate lineages of Eurasia and North America, an enormous diversity of native hooved mammal lineages evolved in South America. A lot of these evolved into body forms **converging** on camels, pigs, rhinoceroses, horses and even hare-like mammals, filling those ecological roles in isolation from those lineages from other parts of the world. Sometimes, SANUs evolved unusual body forms unlike anything elsewhere around the world.

The SANUs include the Astrapotheria, Litopterna, Notoungulata, Pyrotheria and Xenungulata. Traditionally, they have all been grouped as Meridiungulata, as animals closer to each other than to other ungulates. As all groups of SANUs are extinct, it has long been controversial as to what their closest living relatives are, although recent analysis by Welker *et al.* (2015) of extracted protein sequences from remains of the notoungulate *Toxodon* and the litoptern *Macrauchenia* (see pp. 328–9) indicate that both groups may lie near the base of Perissodactyla. Since there are no organic molecules such as proteins left in the most recent astrapotherian or pyrotherian fossils to examine, their phylogenetic affinities continue to be mysterious. The SANUs are easily overlooked due to their lack of descendants, which is sad as the SANUs are an amazing example of convergent evolution, filling the roles of lineages absent from South America.

Xenungulata

The xenungulates form a small, enigmatic order of mammals which vaguely resembled tapirs in size and appearance, from the Late Paleocene to Middle Eocene of South America. The xenungulates, such as *Carodnia* (see pp. 227–8), are generally placed within the broader group of SANUs for showing similarities in dentition with the South American pyrotherians and astrapotherians (see p. 140). The first and second upper molars of *Carodnia* are **bilophodont**, a tooth pattern that also occurs in the pyrotherians. Some experts have linked xenungulates with the pyrotherians based on the shared bilophodont **cheek teeth**. However, bilophodonty may have evolved independently in both groups, since no clear **lophodont** pattern occurs in **primitive** pyrotherians. Alternatively, the xenungulates may be related to the dinocerates of Asia and North America on the basis of complex **lophate** third molars.

Pyrotheria

The pyrotherians were large, **digitigrade**, tapir-like mammals with relatively short, slender limbs and **pentadactyl** feet with broad, flat **phalanges**. The pyrotherians were never a diverse group, with only eight genera known, their fossils being restricted to the Paleocene through Oligocene deposits of Brazil, Peru and Argentina. Most of what is known about pyrotherians comes from a skull and related material from only *Pyrotherium romeroi* (see pp. 246–7).

Various features have linked pyrotherians with a diverse range of groups: pantodonts (see pp. 137–8), perissodactyls (see p. 126), xenungulates, embrithopods (see pp. 68–9) and the dinocerates, with most having no links to South American fauna. One controversial suggestion is that pyrotherians are basal members of the diverse notoungulates, as well as the considerably smaller Notostylopidae. This would make the hare-sized *Notostylops* one of the closer relatives of the rhinoceros-sized *Pyrotherium*! While very different in size and form, there are some similarities in skull morphology, although the notostylopids have more conservative features.

Astrapotheria

It was the astrapotherians which became the dominant trunked beasts of South America. A more inappropriate name could not have been chosen for astrapotherians—it means "lightning beasts". These **graviportal ungulates** were certainly not suited to a speedy life, having remarkably slender limbs and **plantigrade** hind feet. Actually, the name was coined as a link to the North American brontotheriids (see pp. 135–6), although the two groups are distantly related. In many respects, astrapotherian skulls looks similar to that of a modern tapir. A highly placed nasal cavity, retracted nasal bone and a domed forehead are distinctive features and very suggestive of a **proboscis**. Protruding from the mouth were four very large tusks which rubbed against each other, keeping the edges beveled sharp. Size differences have been noted between tusks in individuals for several genera, and this has been assumed to be a case of **sexual dimorphism**. The large body size and slim limbs have led many to assume astrapotherians led a **semi-aquatic** lifestyle, although they do not show any particular **adaptations** for swimming. Usually, astrapotherians are considered one of the weird SANUs, and distantly related to the perissodactyls. Recently, one study placed astrapotherians as being taxonomically closer to the Afrotheria (see p. 62), thus making them cousins to the proboscideans (see pp. 64–8), although this is not universally accepted. There are three families: eoastrapostylopids (containing only the **primitive** *Eoastrapostylops*), the trigonostylopids, and later the advanced astrapotheriids (the latter family which includes most genera) (Fig. 2.95).

With a more generalised body, *Trigonostylops* (see p. 201) did not closely resemble the much larger forms. It had smaller tusks, and less retracted nasals lacking any form of proboscis. *Trigonostylops* was one of the few genera which lived beyond South America, with specimens found in Antarctica. Fossils of later astrapotherians, *Antarctodon* (see pp. 200–1) were also found in similar sediments. Later astrapotherians grew to very large sizes. Known from almost a full skeleton was the modest-sized *Astrapotherium* which may have reached around 700 kilograms in weight. However, several other forms grew much larger. *Parastrapotherium* was a large, basal form, although there was some variation in size between the species. At around 1.5 tonnes, *G. snorki* was considerably larger than *Astrapotherium* (see pp. 265–6). Uniquely, the upper tusks in males grew to one metre long, and with horizontal lower tusks, the short-faced *Granastrapotherium* would have been unable to feed without an elongated proboscis (hence the species name *snorki*). Extensive wear along the full length of the tusks suggests that they were intensively used for feeding. The largest of the giant species was a massive Colombian species of *Hilarcotherium* (*H. miyou*). Based on some skull material, it could have weighed between four and six tonnes. A comparison with other **Neotropical** giants—the "gomphotheriid" *Notiomastodon* (up to 4.4 tonnes) (see pp. 65–6) and the giant ground sloth *Megatherium* (see pp. 328–9) (between 2.4 and four tonnes), potentially makes *Hilarcotherium* the largest known **terrestrial** mammal from South America. A disputed tusk was found from Late Miocene deposits in Brazil, making this potentially the latest known evidence of astrapotherians.

Notoungulata

The notoungulates ("southern **ungulates**") were by far the largest and most diverse **clade** of SANUs, which filled most of the **herbivore** niches in **Neotropical ecosystems** due to the absence of other herbivore groups that we are more familiar with. There are 14 families of notoungulates known, containing more than 150 genera and a few hundred

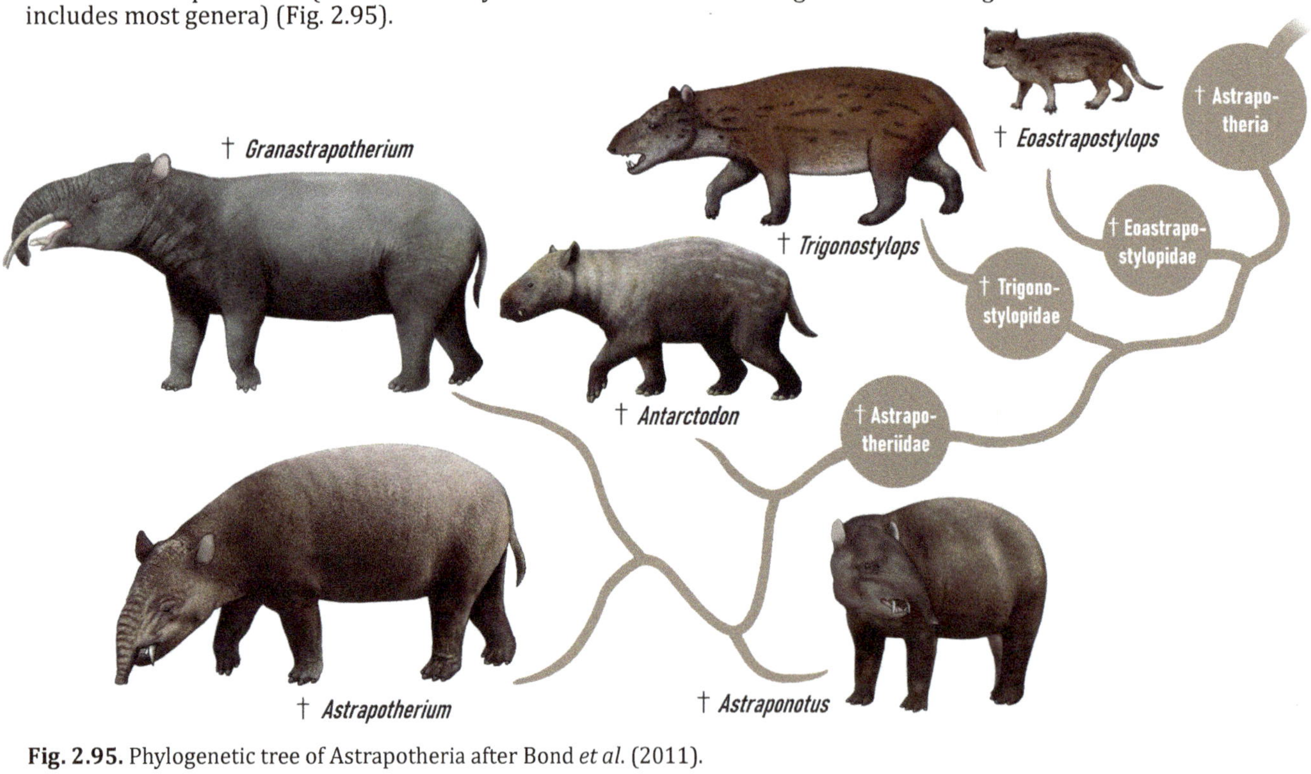

Fig. 2.95. Phylogenetic tree of Astrapotheria after Bond *et al.* (2011).

species. The earliest notoungulates date back to the Early Paleocene (ca. 61 Ma), possibly evolving from "condylarth" (see pp. 100–3) progenitors that had arrived from North America, and had survived right up to the Early Holocene. Roughly split into two groups of families, they include the large-bodied Toxodontia and the smaller-bodied Typotheria (Fig. 2.96). There is much uncertainty regarding the evolutionary affinities of many notoungulates. This is especially the case for notoungulates which branched before the divergence between toxodontians and typotherians. Among the most basal notoungulates were notostylopids, such as *Notostylops*, an unspecialised animal which vaguely resembled a wombat and is thought to have been a **browser** of low-lying vegetation. Notoungulate diversity declined following the Great American Biotic Interchange, which saw the connection between the two Americas (see pp. 252, 295). This decline has frequently been blamed on competition from North American immigrants, although climate change was likely a significant factor too. The evolutionary history of notoungulates is summarised by the following highlights.

Toxodontia

The notoungulates, a sister group to the odd-toed perissodactyls (see p. 126), filled the diverse roles of rabbits, horses and rhinoceroses. Toxodontians were the largest notoungulates, and lived from the Paleocene to the Holocene. Yes, some genera including the largest

Fig. 2.96. Phylogenetic tree of Notoungulata after Billet (2011).

species lived alongside modern humans. Among the toxodontians, the family Toxodontidae originated during the Late Oligocene with *Proadinotherium*. Among the possible descendants of *Proadinotherium* was *Nesodon* (see pp. 264–5), a medium-sized toxodontid from the Miocene. Toxodontids were restricted to South America, except for the giant *Mixotoxodon*, which successfully journeyed north into the Southern United States during the Pleistocene. Several genera completed the pseudo-rhinoceros feel by sporting a single horn on the forehead, such as the high headed *Hoffstetterius*, from the Middle to Late Miocene of Bolivia. The toxodontids culminated in the huge, **herbivorous** hippopotamus-like genus *Toxodon* (see pp. 328–9), which survived into the Early Holocene, becoming the last notoungulate to go extinct. Not all toxodontians had a hippopotamus-like appearance. They existed in a range of forms and sizes. The toxodontid *Trigodon* (Fig. 2.97) was built somewhat like a rhinoceros and even possessed a short, bony horn protruding from its forehead (but true rhinoceros horns consist of matted hair instead of bone), as did the sheep-sized *Adinotherium* (see pp. 264, 266). The homalodotheriids **converged** strikingly with the perissodactyl family Chalicotheriidae (see pp. 128–9). The notohippids such as *Rhynchippus* strongly resembled early horses in their body dimensions, and their dental anatomy was superficially horse-like.

Fig. 2.97. Reconstructed life appearance of the toxodontid *Trigodon gaudryi*, from the Late Miocene to Early Pliocene of Argentina and Brazil. Reconstructed length: 3 m.

The isotemnids, which are among the oldest and most **primitive** families of toxodontians (ranging from the Late Paleocene to Middle Miocene), were generally large animals with larger canine teeth than other early notoungulates, possibly for rooting around in soil. The isotemnids were herbivores and tended to have more **brachydont cheek teeth** than the notohippids. *Thomashuxleya* is the most well-known isotemnid, slightly larger than a sheep, possibly at around 1.5 metres (5 feet) in length and estimated to weigh 70 to 90 kilograms (150 to 200 pounds) (Croft, 2016). The Leontiniidae is a toxodontian family which lived from the Middle Eocene to Late Miocene and resembled flat-footed rhinoceroses. They share a few **derived** characters of the ankle and auditory region with notohippids and toxodontids. *Scarrittia*, the most well-known leontiniid, had **mesaxonic** limbs with reduced lateral toes.

Typotheria

One other lineage of notoungulates, the Typotheria, is characterised by a range of morphologies and **adaptations** just as diverse as those of Toxodontia. Although strictly considered '**ungulates**', few of the typotherians would resemble anything like a modern hooved mammal. Hooves were generally much smaller and claw-like, and many genera were notable diggers. All typotherian incisors were small compared to the large, flaring incisors of Toxodontia. The earliest typotherians are known from the Late Paleocene and grouped within the family Oldfieldthomasiidae. Oldfieldthomasiids had a more generalised dentition (i.e. having 44 teeth with no **diastema**), with the **cheek teeth** being **brachydont**. One group of typotherians called interatheriids were a very conservative family that were small-bodied and generally filled the niches of large rodents and rabbits. *Protypotherium* (see pp. 264–5), from the Late Oligocene to Late Miocene, is a fairly typical example of a rabbit-like genus, using sharp claws to create burrows. More specialised was *Miocochilius*, a larger genus from the Middle Miocene which had a reduced number of **digits** on the feet. Squirrel-like *Notopithecus* was once considered a Cretaceous primate (hence the *pithecus* part of the name), until it was realised that both the age and taxonomy were way out (Middle to Late Eocene). Interatheriids were the most successful notoungulates, currently with about two dozen named genera, and were some of the most abundant notoungulates within their **ecosystems**. Interatheriids were also the longest-ranging of all notoungulate families, living from the Early Eocene until the Late Miocene.

Uniquely among the notoungulates (and also all hooved animals), the mesotheriids had remarkable adaptations for a life of digging. All were heavily built, with reinforced bones in the elbow, sacrum and vertebrae to cope with the stresses on the body, and had deep claws. Mesotheriids share with rodents and lagomorphs (see pp. 148–9) enlarged, **hypselodont** incisors. In addition, the **hypsodont** molars were able to deal with chewing abrasive food (mostly tubers) and were also hypselodont in some species. Many also dug burrows for protection from weather and predators, like today's rabbits and wombats. The earliest mesotheriid fossils are potentially a species of *Trachytherus* (see pp. 246–7) known from the Late Eocene/Early Oligocene, but more definitive fossils are from the Late Oligocene onwards. However, the major mesotheriid diversity was during the Miocene. All genera are known from mid-latitude countries, being entirely absent from cold Patagonia and tropical regions. In particular, mesotheriids were common from the antiplano (high altitude grasslands) of Bolivia and northern Chile. Most genera were small- to medium-sized, with the largest genus being the late-living *Mesotherium*, which reached weights of up to 60 kilograms. It is the most recent known genus, living until the Middle Pleistocene. In the past, *Mesotherium* went by the name "*Typotherium*", hence why *Typotheria* is still used for the suborder.

The archaeohyracids ("ancient hyraxes") are completely unrelated to true hyraxes, neither did they look very similar. Archaeohyracids have historically been among the most obscure notoungulate families. The oldest archaeohyracids are known from the Late Paleocene and the family is among the oldest South American native ungulates to bear hypsodont molars. As in the interatheriids, archaeohyracids had a closed, **complete dentition**; all species had 44 teeth, with a number of **anterior** teeth being isolated by narrow diastemata. The hegetotheriids comprised small- to medium-sized notoungulates which varied from hare-sized to beaver-sized. Certain hegetotheriids had skulls and skeletons strikingly similar to those of rabbits (e.g. pachyrukhines), while others were built more like large rodents including the pacarana (*Dinomys branickii*). The hegetotheriids were somewhat **cursorial** notoungulates, and some of the later, smaller-bodied **taxa** were **fossorial**. Hegetotheriids were some of the most recent-splitting typotherian families. The fossil record of hegetotheriids dates back to the Late Oligocene, and the family persisted until the Early Pleistocene, seemingly vanishing shortly before the mesotheriids. Compared to all other notoungulates, hegetotheriids had the most simple teeth, with the upper pair in many species being devoid of any distinct **folds**. Given their hypselodont teeth and associated cranial features, hegetotheriids are widely believed to have consumed low-growing plants in open environments.

Litopterna

The litopterns are the second most diverse group of native **ungulates** on South America after the notoungulates (see pp. 140–3) and came in a variety of shapes and sizes. Litopterns consist of a number of South American families including the **primitive** sparnotheriodontids and proterotheriids, and the advanced, odd-nosed macraucheniids (Fig. 2.98). It is

Fig. 2.98. Phylogenetic tree of Litopterna after McGrath *et al.* (2018); (2020).

apparent that an important development in their evolutionary history was the reduction of toes per foot. From the earliest multi-toed forms, an increase in size and limb length led towards a more **cursorial** lifestyle, as in horse evolution. Today, all horses stand on a single hoof per limb. While unusual amongst vertebrates, it was not a unique event as a group of extinct South American native ungulates arrived at this evolutionary experiment well before the equids. Most groups are poorly known, although the proterotheriids are familiar due to some **convergent** features with horses. Early genera date back to the Late Paleocene, but it was during the Miocene that proterotheriids began to diversify. Typically, they had a reduced number of toes, usually one dominant **digit** and two side toes. *Thoatherium* shows the most extreme reduction, with only a single visible hoof and even lacking the side splint bones (remnants of other toes). Later genera were adapted to a drying environment. *Diplasiotherium* was a notably large genus from the Late Miocene to Early Pliocene, with teeth that were more **hypsodont** for chewing more abrasive vegetation. Possibly the most basal group were the megadolodines such as the Middle Miocene genus *Megadolodus*, unusual robust forms with self-sharpening tusks much like the similar-sized peccaries. Megadolodines had "primitive" teeth with very thick enamel and deep jaws to deal with tough items such as thick-rinded fruits and hard-shelled crustaceans. The last surviving proterotheriid was a species of *Neolicaphrium* that originated in the Late Pliocene and survived until the Late Pleistocene (possibly even into the earliest Holocene). Interestingly, it coexisted with ungulates of **Holarctic** origin (such as deer and horses) in open forest and **savannah**, although isotopic analysis shows that *Neolicaphrium* mostly ate fruit and some low-lying plants. The sparnotheriodontids are a particularly enigmatic extinct family of primitive litopterns known primarily from teeth. Along with the astrapotherians (see p. 140), the sparnotheriodontids were the only South American native ungulates known to have reached Antarctica. Specimens of the small Eocene *Victorlemoinea* (see p. 227) have been found in Antarctica and South America.

The most recognisable litopterns are the macraucheniids, which resembled various camelids (see pp. 104, 106–7). The hooves of macraucheniids were similar to those of rhinoceroses today, with a simple ankle joint and three digits on each foot. Thus, many macraucheniids may have been capable of rapid directional change when running away from predators. Early macraucheniids were very similar, although smaller and had less extreme retracted nostrils. Some such as the Late Oligocene to Early Miocene genus *Cramauchenia* had an appearance vaguely similar to that of a small llama or perhaps that of a stocky antelope. The skull of this animal was relatively elongated and provided with a slightly recessed nasal opening, which would indicate the presence of a strong, muscular lip, likely **prehensile**. In similar but larger and more recent forms, such as *Theosodon* (see pp. 265–6) and *Scalabrinitherium*, this lip gradually developed, eventually giving rise to a more prehensile structure with *Macrauchenia* (see pp. 328–9). *Scalabrinitherium* was a slightly robust, llama-like macraucheniid that had an elongated, low skull equipped with spatulate incisors and a nasal opening situated far back (intermediate between the position in *Theosodon* and *Macrauchenia*), suggesting the presence of a muscular upper lip. *Theosodon* probably gave rise to the large macraucheniids of the Pliocene and Pleistocene, including *Macrauchenia* and its sister genus *Xenorhinotherium*. *Macrauchenia* is the best-known and most recent macraucheniid. It became extinct during the Early Holocene. It was one of the few litopterns to survive the ecological pressures of the Great American Biotic Interchange, or GABI (see pp. 252, 295). One predator that *Macrauchenia* needed to evade were humans. Some bones have been discovered with cut marks indicative of butchery. It may instead have died out in the aftermath of the invasion of the Americas by human hunters. Once *Macrauchenia* had died out, so too did the Macraucheniidae and the entire litoptern order.

The Euarchontoglires: Rodents, Lagomorphs, Primates—And Humans

The Euarchontoglires

During the 1990s, DNA analyses of the phylogenetic affinities among mammals revealed that the rabbits and rodents constitute a **monophyletic** group (the Glires), although even more striking is that the primates are their nearest kin (Fig. 2.99). While this may be surprising to some people, our own nearest relatives after all the different kinds of primates, including lemurs, monkeys and apes, are actually the rabbits and rodents. Prior to the emergence of this molecular evidence, the prevailing view was that primates were all closely related, and that they could be close relatives of bats and elephant shrews as well as treeshrews and colugos (elephant shrews are now afrotherians (see p. 62), whereas bats are now laurasiatherians (see p. 74)). William King Gregory dubbed this grouping the "Archonta" in 1910. Historically, palaeontologists and zoologists suggested that the shared traits of small bodies and chisel-shaped incisors in rabbits and rodents resulted from **convergent evolution** from lineages with distinct origins. Nevertheless, in 1910, Gregory placed both gnawing lineages within the **taxon** Glires.

Glires

It comes as a surprise to many people that rabbits are not rodents. The distinctions between them were recognised as early as the 1850s, when scientists classified them into different orders. While both lineages are small **omnivores** or **herbivores** with chisel-shaped, **hypselodont** gnawing incisors, they differ greatly in various aspects of anatomy. The most important difference is that rodents only have two incisors in the upper jaw compared to four in lagomorphs, the latter which also possess more **cheek teeth**. They differ in many cranial and skeletal details as well. Lagomorphs are characterised by short tails and more elongated hindlimbs for **saltatorial** locomotion, whereas rodents are primarily climbers and runners (except jerboas, gerbils, kangaroo rats, kangaroo mice and several other species have independently developed the saltatorial locomotion of lagomorphs).

During most of the twentieth century, scientists considered lagomorphs and rodents to be unrelated lineages that had **convergently evolved** the gnawing habits in distinct ways. It was not until the 1980s and beyond that our views shifted in the opposite direction with new morphological studies and fossil discoveries, and eventually genetic data that was formerly unknown. Fossils of **"archaic"** ancestral Glires from the Paleocene and Eocene of Asia, which were close relatives of lagomorphs and rodents but did not belong to either lineage, provide the strongest evidence for the close affinities between both groups. These fossils, which go by the names "anagaloids" and

Fig. 2.99. Phylogenetic tree of Euarchontoglires after Esselstyn *et al.* (2017). The phylogenetic position of Arctostylopidae within Gliriformes follows Averianov (2020).

"eurymylids", are frequently categorised in their own **wastebasket taxon**, some, including *Gomphos* (see pp. 192–3) and *Mimotona* (see pp. 222–3), of which were more closely related to lagomorphs, whereas others, including *Rhombomylus* and *Tribosphenomys* (see pp. 192–3), were more closely related to rodents. Additionally, there were also the arctostylopids, which may be descendants of Asian gliriforms (Avarianov, 2020) and therefore also related to Glires, as indicated by the shape of their ankle bones. However, this has yet to be strongly supported. Arctostylopids were small, rabbit-sized mammals that are known mostly from jaw fragments and may have vaguely resembled modern hyraxes (Cifelli and Schaff, 1998). Certain scientists hypothesise that all these mammals originated from a lineage of **stem**-placentals from the Cretaceous of Mongolia called zalambdalestids.

In the Paleocene of Mongolia and China, there is an abundance of **taxa** from a lineage known as the pseudictopids, which are progressively more like lagomorphs and rodents. The pseudictopids were of similar size to lagomorphs and rodents, possessing most of the skeletal specialisations too, although they had yet to evolve the completely developed hypselodont gnawing incisors. Known from the Paleocene of China, *Mimotona* possessed the double pair of incisors in the upper jaw as in lagomorphs, although the molars retained a highly **primitive** crown pattern and were not quite lagomorph-like. Various other taxa from the Paleocene and Early Eocene of Mongolia, such as *Rhombomylus* and *Eurymylus*, possessed the single pair of gnawing incisors characteristic of rodents, although they lacked the molar specialisations which distinguish rodents from more primitive glires. As a result, the fossil record contains numerous "archaic" taxa that, beyond linking lagomorphs and rodents, enable us to trace their roots back to the Paleocene and Late Cretaceous.

Rodentia

Rodents are by far the most common and successful mammal order on Earth. Rodents are also extremely diverse (equating to more than 40% of all extant mammal species, or at least over 2,700 species in 518 genera) and disparate (ranging in body morphologies from the sheep-sized capybara, to **semi-aquatic** muskrats and beavers, to gliding, **fossorial** and **arboreal** squirrels, naked mole-rats, subterranean gophers and spiny porcupines, to hundreds of species of mice and rats). One need only consider the high reproduction rates in mice, rats or hamsters to understand why rodents are by far the most abundant mammals on Earth. The world would be overrun with rodents if predators were absent.

The defining characteristic of rodents is their chisel-like (**gliriform**) pair of upper and lower incisors, which function in gnawing tough food items and plants, and in certain lineages, to excavate burrows or fell trees. These incisors are **hypselodont** and rootless, requiring constant wear through gnawing and grinding against each other to maintain a sharp point. These form a "self-sharpening" system, where the softer **dentin** on the back edge wears away faster than the hard enamel band on the front. In case a rodent's incisors do not properly occlude and wear down, they can continue growing in a curve to the point of curling around or even penetrating the skull. **Posterior** to the incisors is a **diastema**, followed by a small number of premolars and molars ideal for crushing their diet of nuts, seeds and plants. This small-bodied niche of gnawing nuts, seeds and plants was highly successful, as evident from the multituberculates (see pp. 45–6), which filled this role for much of the Mesozoic, and numerous lineages of plesiadapiforms (see p. 150) from the Paleocene were rodent-like in build too. However, once rodents dispersed to Europe and North America from Asia during the Early Eocene, they started to replace the existing small mammal faunas, leading to the extinction of multituberculates and rodent-like plesiadapiforms by the Oligocene. The skulls of basal rodents had a highly **primitive** protrogomorph condition, in that the masseteric muscles which close the jaw were attached exclusively to a restricted area around the base of the **zygomatic arch** (seen in numerous other mammals). The rodents had radiated into several suborders by the Late Eocene. The sciuromorphs are only marginally more advanced than the basal protrogomorphs, with the key difference being a forward shift of the masseter muscle attachment onto the **rostrum**. Sciuromorphs comprise the squirrels and their closest kin. Sciuromorphs were plentiful during the Late Eocene, as they seem to have driven the primitive ischyromyids (a family of protrogomorphs that were larger relative to other early rodents, and lived in the Northern Hemisphere from the Late Paleocene to Early Oligocene) to extinction, and continued to radiate from the Oligocene onwards.

In the hystricomorph condition, the masseter muscle extends **anteriorly** and passes through a greatly enlarged infraorbital foramen (an opening for the passage of nerves) to attach to the snout. Among the most primitive hystricomorphs are certain African rodents, including a large diversification of **taxa** from the Eocene-Oligocene strata of the Fayûm Depression in central Egypt called Phiomorpha. The most astonishing of hystricomorphs are the enormous diversification of native South American rodents called caviomorphs. The caviomorphs likely colonised South America during the Late Eocene, possibly descended from African hystricomorphs which rafted there via a much narrower South Atlantic. (The New World monkeys also underwent the same journey.) The myomorph condition is the most specialised of all, characterised by a combination of features: a masseter muscle extending anteriorly to the zygomatic arch, and another part of the masseter passing through the infraorbital foramen. This condition occurs in most rodents (e.g. murids, cricetids). Historically, rodents were split exclusively into the three suborders Sciuromorpha, Hystricomorpha and Myomorpha—based largely around the development of muscles in the jaws. **Parallelism** and **convergence** are common in different groups and have muddied the taxonomic waters. Although molecular studies have been unable to resolve all disagreements, rodents now fall between five main suborders with two additions—the beaver-like Castorimorpha and the uniquely African Anomaluromorpha (Fig. 2.101). This latter group consists of two very divergent forms, the gliding scaly-tails, or anomalures, and the **terrestrial** springhares or springhaas. Most distinctively, springhares are known for their **saltatorial** hopping. Modern *Pedetes* are split between two populations in East and Southern Africa, and the small number of fossil species are largely known from the same distribution from the Miocene onwards. Most are known only from teeth and limited skeletal remains. They range from small forms roughly similar to jerboas in *Oldrichpedetes* and *Parapedetes* (the latter is placed in its own family) up to the relative giant *Megapedetes.* This larger form has species which range from East Africa north to Southern Europe. A migration of pedetids in Europe during the Miocene coincides with a warm period when African flora species were also found further north. Skeletal material of the Pliocene *Propedetes* are similar in size and saltorial behaviour to the modern species and is considered the ancestral form.

Among the suborder Castorimorpha, modern beavers are familiar animals, few people have seen these large secretive rodents. Synonymous with dam and lodge building, beavers are mostly active underwater or at night. It probably comes as a surprise to many that beavers have a long lineage dating right back to the Late Eocene and that the first castorids were predominantly burrowers. Some species created elaborate corkscrew burrows three metres deep. When first discovered, the origins of the burrows were unknown, and they were named as *Daemonelix/ Daimonhelix* fossils. The true identity of the burrows was only agreed upon after skeletons of *Palaeocastor* were found inside them (see p. 259). Early castorids are predominantly a North American group, but by the Miocene, aquatic species that resembled modern beavers showed up in Eurasia. Rodents typically dominate the niche of small-bodied mammals, although they can sometimes grow massive, including later species of beavers. The capybara (*Hydrochoerus hydrochaeris*), the largest extant rodent, weighs up to 50 kilograms, but *Castoroides ohioensis*, the largest beaver species, reached around 200 kilograms in weight, and lived in North American swamps until the end of the Pleistocene (Fig. 2.100). It resembled a modern beaver, living an aquatic lifestyle, although there are some noticeable differences: a slimmer tail, longer hind feet, broader teeth that were less efficient

Fig. 2.100. Reconstructed life appearance of the giant Pleistocene beaver *Castoroides ohioensis*. Reconstructed length: 2.5 m.

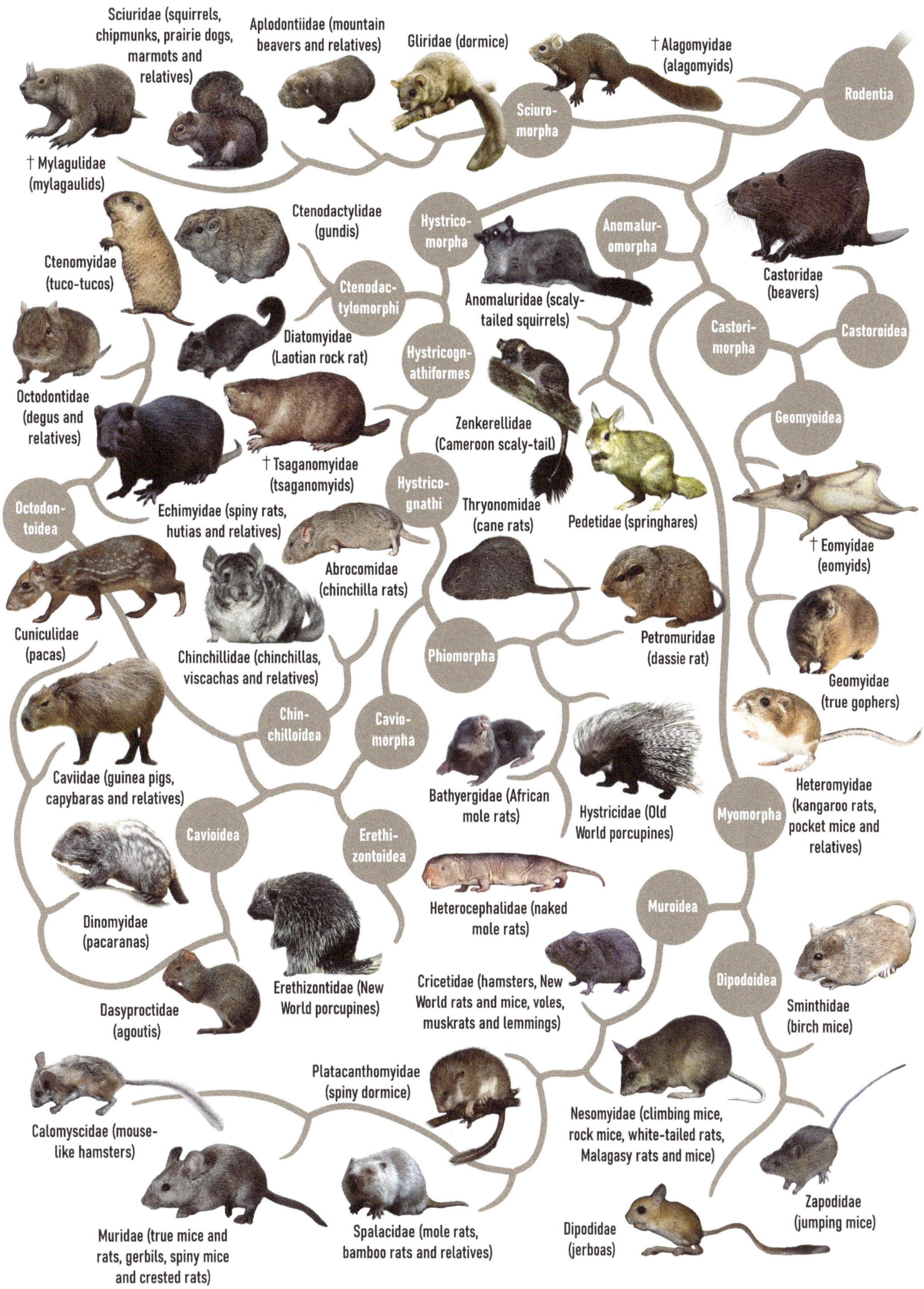

Fig. 2.101. Phylogenetic tree of Rodentia after Blanga-Kanfi *et al.* (2009), and Doronina *et al.* (2017). The phylogenetic position of Tsaganomyidae and Mylagaulidae follows that of Rodrigues *et al.* (2023), while that of Eomyidae follows Anemone *et al.* (2012), and that of Alagomyidae follows Asher *et al.* (2019).

at cutting trees, and a smaller brain, suggesting less complex social behaviour. It is assumed that they were not able to build lodges due to their large size. Among other giant rodents was the cow-sized Pleistocene capybara *Neochoerus*, weighing around 113 kilograms.

Meanwhile, South America's fossil record shows that rodents were capable of much larger sizes, with the largest being the gigantic *Josephoartigasia* (Fig. 2.102) at almost 500 kilograms. It was part of a lineage of super-sized rodents, split across two South American families of caviomorphs—the dinomyids and neoepiblemids. Only one extant species (a dinomyid) still exists—the pacarana (*Dinomys branickii*), which reaches a modest 15 kilograms. It was considered a cavioid (cavies, capybara, agouti), although more recently, pacarana have been shown to be an unusual chinchilloid, a group which includes the familiar chinchillas and viscachas. Dinomyids and neoepiblemids are regarded as separate families, although the relationship between the two is problematic. One suggestion is that neoepiblemids are a sister lineage to the true chinchillids, with the dinomyids being a more basal group to both. Another small extinct group, the cephalomids, may be even more closely linked to the neoepiblemids although they also show some similarities to the caviids. The extinct heptaxodontid giant hutias from the Caribbean islands were traditionally thought to be large versions of the extant hutias (octodontoids), but appear to be an additional lineage of large chinchilloids. Phylogenetic issues are due to poorly defined unique features within the extinct families (fragmentary remains), and thus there are little clear morphological similarities which unite all species.

In general, they all have heavily reinforced skulls (*Josephoartigasia* has a skull roughly the same as a large domestic cow!). The musculature and reinforced bones gave these rodents very high bite forces, equivalent to large felids (see pp. 86–9) and crocodylians (see pp. 26–30). Several suggestions have been put forward for this: digging with teeth, feeding on hard-shelled nuts, and defence against predators. The dinomyids and neoepiblemids date from the Early Miocene, with early forms probably resembling the much smaller potamarchines (e.g. *Potamarchus* (see pp. 270–1) and *Ferigolomys*, both from the Late Miocene). While dinomyids still survive with the pacarana, neoepiblemids did not make it beyond the Miocene. Amongst rodents, spines are present only in murids and three distantly related families of hystricognaths. Two families (hystricids and erithizontids) are especially well-endowed with a coat of long quills and are known as porcupines, although the two are not closely related. Hystricids are the typical large porcupines of terrestrial environments throughout Eurasia and Africa. All species have long spines on the body and enlarged nasal bones. Smaller forms have a bundle of flat spines at the end of a long tail, while the largest species have a short tail with hollow quills which rattles when they move. The earliest known genus, *Sivacanthion*, dates from the Middle Miocene, and appears to not be closely related to extant species. Relationships between the various fossil *Hystrix* species has seen species being lumped and split. Currently, the giant African *H. makapanensis* is thought to be separate from other large species that were present in Plio-Pleistocene Europe.

In comparison, the New World erethizontids were generally smaller and arboreal. An obvious difference with the Old World porcupines is that erethizontid quills are barbed and that there is noticeably more hair mixed among the spines. Modern genera are found across North, Central and northern South America, while fossil genera were very diverse in southern South America during the Miocene (e.g. *Bransamyopsis*, *Neosteiromys* (both from the Late Miocene of Argentina)). They included some of the largest known erithizontids and were more terrestrially adapted than extant genera. The earliest fossils (*Eopululo*, from the Late Eocene to Early Oligocene of Peru) include some of the largest rodents in Late Eocene habitats. *Erethizon* is the only genus to have migrated north into North America following the Great American Biotic Interchange, or GABI, during the Pliocene (see pp. 252, 295).

Lagomorpha

The rabbits, hares, and hamster-like pikas have always been grouped in their own order, the Lagomorpha. They comprise at least 109 extant species: 42 rabbit species in 10 genera, 33 hare species in one genus (*Lepus*), and 34 pika species in one genus (*Ochotona*). While many rodents are **omnivorous** or **carnivorous**, all modern lagomorphs are strictly **herbivorous**. Despite the familiarity, rabbits (and their relatives) have some very odd anatomical features. The nasal bones have a delicate,

Fig. 2.102. Reconstructed life appearance of the giant dinomyid rodent *Josphoartigasia monesi*, known from the Late Pliocene to Early Pleistocene of Uruguay. Reconstructed length: 3 m.

lattice-like texture (purpose unknown) and all teeth are **hypselodont**. In rodents, this only occurs with the incisors. Lagomorph teeth also have a distinctly unusual cusp morphology, making identification of fossil species more obvious. Rabbits and hares possess shorter tails than many rodents, and large ears that assist in detecting predators and to shed excess body heat. Finally, for a **quadrupedal** mammal, rabbits and hares have proportionally very long hindlimbs. As a result, they exhibit the recognisable bounding gait (other species show different gaits at slower speeds). There are just two families of lagomorphs that persist today: the Leporidae (rabbits and hares) and the Ochotonidae (pikas). Lagomorphs were more speciose in the past. The earliest known lagomorphs date from the Paleocene and Eocene of Asia and were **insectivorous**, unlike later species. During the Oligocene and Miocene, lagomorphs began to prosper and spread across Eurasia and North America as global conditions encouraged drier grasslands over forest **ecosystems**.

The earliest leporids are known from the Late Eocene of Asia, where genera including *Strenulalagus*, *Dituberolagus*, *Lushilagus* and *Shamolagus* represent highly **primitive** rabbits only marginally more derived than primitive Glires from the Paleocene to Early Eocene such as *Mimolagus*, *Mimotona* (see pp. 222–3) and *Gomphos* (see pp. 192–3). Leporids traversed the Bering Land Bridge (see p. 174) during the late Middle Eocene and started to radiate in North America too, developing into forms such as *Palaeolagus* (see pp. 235, 238–9) and *Megalagus*. The diversity of leporids remained low throughout the Miocene, potentially due to competition with ochotonids. When cool grasslands expanded during the Pliocene and Pleistocene, leporids experienced a significant evolutionary radiation. During the Late Miocene, they dispersed to Africa as well, and were among the immigrant **taxa** to cross into South America from North America via the Panama Isthmus during the Pliocene, where they radiated into new species (e.g. various native species of the cottontail rabbit genus *Sylvilagus*). A notable large Pleistocene form of the genus *Lepus* (hares) from the northern tundra (often called the Don hare) are known from mummified specimens complete with gut samples, found in the permafrost.

The oldest ochotonids are known from the Early Oligocene of East Asia, represented by genera such as *Desmatolagus* (see p. 250). They radiated in diversity during the Late Oligocene and Middle Miocene, with dozens of fossil genera known from Eurasia and North America, but the majority of these were larger than modern pikas and more rabbit-like in anatomy, existing in the lowlands where they outcompeted the leporids. Ochotonids were largely restricted to the Northern Hemisphere, although several species did reach Africa during the Miocene. There was a significant reduction in pika diversity by the Pliocene, and the previous ratio had reversed, resulting in only four ochotonid genera compared to 18 leporid genera. This may have been caused by a cooler, drier climate around the Miocene-Pliocene boundary, which led to reductions in **savannahs** and their vegetation, and the expansion of modern grassland plants favoured by leporids. It was not until later in their evolutionary history that pikas declined to a single genus, *Ochotona*, and became limited to high altitudes, after the ice sheets contracted at the end of the Pleistocene.

Additionally, several fossil species from both families had **adaptations** to island life, exhibiting **insular gigantism** and reduced mobility, including the leporid *Nuralagus* (Fig. 2.103), which was **endemic** to the Spanish island of Menorca in the Mediterranean Sea during the Pliocene (ca. 5–3 Ma), and the large island pikas *Gymnesicolagus* (from the Miocene of Mallorca and Menorca) and the late-surviving *Prolagus* (surviving in Corsica and Sardinia until sometime after 348 BCE). A prime example of insular gigantism, the lack of predators and competition in its isolated island habitat allowed *Nuralagus* to grow significantly larger than its continental relatives. In fact, reaching a shoulder height of 50 centimetres and weighing around 12 kilograms, *Nuralagus* is the largest lagomorph known to have ever lived. The animal also had a proportionally small head for its body size, and anatomical features suggesting reduced eyesight, hearing and ability to jump compared to other rabbits, probably since the lack of predators on Menorca made these abilities less important.

Fig. 2.103. Reconstructed life appearance of the giant leporid *Nuralagus rex*. Reconstructed length: 60 cm.

Euarchonta

As mentioned previously, William King Gregory originally coined the term Archonta ("rulers") in 1910 to group together treeshrews, bats, colugos, and primates, initially also including elephant shrews. That term became abandoned until the 1980s and 1990s, when it was revived in the anatomy-based phylogenies. During the late 1990s, it became adjusted when DNA studies revealed that treeshrews, colugos and primates are more closely related than previously thought (so the group was revised as "Euarchonta").

Scandentia

Treeshrews are rather squirrel-like in size and build, active on the ground and in the trees typically during the day (except the pen-tailed treeshrew (*Ptilocercus lowii*), which is **nocturnal**), and consuming a broad range of foods (seeds, fruit, leaves, insects and small vertebrates). The fossil record of treeshrews is scanty. The earliest known possible treeshrew is *Eodendrogale*, from the Middle Eocene of China, which is only represented by several teeth. More complete fossils are known from the Miocene of India, Pakistan, Thailand, and Yunnan, China, plus the Pliocene of India. A fossil species documented from the Early Oligocene of Yunnan is considered a close relative of the pen-tailed treeshrew, and shows that treeshrews have remained virtually unchanged since they originated.

Dermoptera

The strangest order of modern euarchontans is the Dermoptera, or colugos. They are occasionally called "flying lemurs", but this is inaccurate since they are not lemurs, neither are they capable of true flight. Colugos are, however, the closest relatives of primates. Colugos have a relatively sparse fossil record because they were seemingly limited to Asian rainforests, where fossilisation is unlikely. The earliest undisputed colugo known in the fossil record is *Dermotherium* from the Late Eocene to Early Oligocene of Thailand. The Plagiomenidae is an extinct mammal familiy which is commonly preserved in the Paleocene and Eocene strata of Eurasia and North America, represented only by teeth and jaws, and are considered possible relatives of colugos. Plagiomenidae is represented by two genera from the Late Paleocene to Late Eocene of North America, particularly from the Canadian Arctic and the Northern Rocky Mountains. After the Early Oligocene, colugos virtually disappear from the fossil record, with only two species surviving today—one in Indonesia and mainland Southeast Asia (*Galeopterus variegatus*), and one in the Philippines (*Cynocephalus volans*).

Primates

Historically, it was thought that small proto-primates scampered in the treetops above the heads of the last non-avian dinosaurs during the final days of the Cretaceous. Unfortunately, this was an incorrect assumption. Fossil remains of *Purgatorius* (see pp. 170–1) are now considered younger, dating from the earliest Paleocene, while any primate-like characters are possibly due to **convergence** (it may even fall outside **crown** placentals). However, fossils of true early relatives of primates, the plesiadapiforms, were present soon after, distributed across nine families. Resembling modern treeshrews and lemurs, plesiadapiforms are the earliest known mammals with finger nails, indicating a move towards grasping branches rather than using claws to grip. Intriguingly, a 1990 study found that postcranial features were similar enough to extant colugos that plesiadapiforms were also habitual gliders. There was little evidence to support it and current research indicates they are closer to primates. Confusion is hardly surprising as colugos, treeshrews and primates share a common ancestry.

While the majority of plesiadapiforms resembled squirrels or lemurs in build, certain genera had the size and form of a groundhog (e.g. *Platychoerops*, from the Late Paleocene to Early Eocene of Europe) or of a tarsier, with massive forward-facing **orbits** (e.g. *Carpolestes*, from the Late Paleocene of North America). Plesiadapiforms shared some molar features with true primates, and several of them possessed nails rather than claws, opposable thumbs, and tarsier- or lemur-like bodies. However, plesiadapiforms did not possess the key primate specialisations such as completely forward-directed orbits for stereoscopic vision, or a completely primate **cheek teeth** pattern. Additionally, the plesiadapiforms exhibited large, forward-projecting incisors separated from the cheek teeth by a large **diastema**. It is relatively more similar to the condition present in rodents, but the plesiadapiforms never evolved the complete chisel mechanism or **hypselodont** incisors present in all lagomorphs (see pp. 148–9) and rodents. Plesiadapiforms seem to have flourished in the Paleocene rainforests of Wyoming, Montana and Western Europe as indicated by the enormous diversity and abundance of their fossils. They may have fed on a range of vegetation, seeds and soft fruit.

The fossil record documets over 140 species within 11 families of plesiadapiforms. Plesiadapiforms exhibited their highest diversity in Europe and North America during the Paleocene, where **taxa** were so abundant and quickly radiating that they serve as time indicators for this epoch. The majority of plesiadapiforms went extinct during the Early Eocene with the arrival of true primates (omomyids and adapids), although several species survived until the late Middle Eocene (ca. 40 Ma). Given that many were **arboreal**, the extinction of plesiadapiforms during the Late Eocene may have been caused by the shrinking of rainforests, along with competition from true primates.

True primates exhibit various distinctive and unique specialisations, such as shortened snouts, forwardly-directed orbits promoting binocular vision, nails rather than claws on many of their **digits**, relatively large brains, and a suite of characters in the skull region and teeth. The modern true primates exhibit remarkable success, comprising 16 families, and at least 81 genera and 505 species of tarsiers, pottos, lorises, lemurs, monkeys and apes (Fig. 2.104). Additionally, hundreds of fossil genera and species are recognised, representing dozens of extinct families. Many primates exhibit a strong tendency towards an arboreal lifestyle, feeding primarily on seeds, fruit and leaves, whereas others (such as baboons, chimpanzees and humans) are largely or completely **terrestrial**.

Primates are broadly classified into two major lineages, the Strepsirrhini (formerly known as "prosimians"), which comprises galagos, lorises and lemurs; and the Haplorhini, which comprises the tarsiers and the simiiforms, or anthropoids (monkeys, apes, and humans). The strepsirrhine-haplorhine split occurred near the start of the **adaptive radiation** of primates, at about 55–90 Ma. Earlier splits are based on molecular estimates, whereas more recent splits are based on the limited fossil record. The oldest known true primate in the fossil record is the basal haplorhine *Altiatlasius*, dated to the Late Paleocene (ca. 57 Ma) of Morocco, and true primates were already diverse and had dispersed broadly across the northern continents by the earliest Eocene. The tarsier-like *Teilhardina*, a basal haplorhine from the Early Eocene, originated in Asia, before spreading to Europe then eventually North America within roughly 25,000 years. The earliest primates abruptly appear in the fossil record with no transitional fossils to indicate ancestry.

Strepsirrhini

The oldest and most **primitive** known strepsirrhines are the lemur-like adapiforms, which date back to the earliest Eocene and exhibit numerous morphological similarities with lemurs (such as relatively long-snouted skulls with small **orbits**). Adapiforms were a widespread lineage that radiated across Eurasia throughout the Eocene, with some arriving in North America (via Greenland when the landmasses were closer together). A few rare genera are also known from North Africa. The basalmost adapiforms include *Donrusselia* from Europe, and *Cantius* from both Europe and North America, the former which exhibits the most **ancestral** features. Adapiforms were very common and diverse in North America during the Early to Middle Eocene, but went extinct entirely by the Late Miocene (ca. 7 Ma) due to a cooling and drying global climate. The survivors of this early

Fig. 2.104A. Phylogenetic tree of Primates, particularly stepsirrhines, and haplorhines outside Platyrrhini and Catarrhini, after Orlando *et al.* (2008), Stevens *et al.* (2013), Silvestro *et al.* (2017), Seiffert *et al.* (2018), Rasmussen *et al.* (2019), and Tabuce *et al.* (2009).

radiation of strepsirrhines are the modern lemurs, or lemuriforms, **endemic** to Madagascar, plus the lorises and galagos of Asia and Africa, with over 100 species in 15 genera and five families. The split between lorisoids and lemuroids in Africa occurred during the Middle Paleocene (ca. 62 Ma) according to molecular clock estimates. Lorises diverged from the African galagos approximately 40 Ma, and subsequently immigrated to Asia. Lemurs most probably evolved from a very early branch of Asian adapiforms that, after immigrating to Africa, arrived in Madagascar during the Early Eocene (sometime between 47 and 54 Ma) via rafting. In addition, fossils of obscure close relatives of lemuriforms are known from the Eocene of Tunisia (*Djebelemur*) and Algeria (azibiids such as *Azibius* and *Algeripithecus*). Azibiids may be a sister group to the djebelemurids, and both groups, along with *Plesiopithecus* from the Late Eocene of Egypt, are possibly African **stem**-lemuriforms.

Since the Early Eocene, lemurs have evolved in isolation on Madagascar, whereas the galagos and lorises have had to face competition with other mammals such as monkeys and apes. Aye-ayes are thought to be the earliest diverging family of lemurs. Lemurs evolved into a diverse array of sizes and forms, from the diminutive Madame Berthe's mouse lemur (*Microcebus berthae*), weighing just 30 grams, right up to the nine kilogram indri (*Indri indri*), with many typical forms (such as ring-tailed lemurs (*Lemur catta*)) averaging two to three kilograms. These were completely dwarfed by 17 extinct species known collectively as giant lemurs, which died out during the Holocene Epoch. These included the gorilla-sized and mostly **arboreal** *Archaeoindris* (Fig. 2.105), a sloth lemur which weighed around 160 kilograms (350 pounds) and survived until about 412–199 BCE. The koala lemur *Megaladapis*, which measured 5 feet (1.5 metres) long, and possibly reached 50 kilograms (110 pounds) in weight, was a slow-moving climber which still inhabited the Madagascan forests until humans drove it to extinction by approximately 1280–1420 CE. The long-armed *Palaeopropithecus* was the most specialised of sloth lemurs, which is thought to have climbed in the canopy by suspension using long-curved fingers as in modern sloths; sloth lemurs disappeared by about 1620 CE. The archaeolemurids were the most **terrestrial** of all giant lemurs, **converging** on ground-dwelling monkeys such as macaques and surviving until about 1047–1280 CE.

Fig. 2.105. Reconstructed life appearance of the "gorilla lemur" *Archaeoindris fontoynontii*. Reconstructed height: 1.5 m.

Haplorhini

Among the oldest known haplorhines in the fossil record are *Archicebus* from the Early Eocene (55 Ma), and the Middle Eocene (45 Ma) *Xanthorhysis*, both discovered in China. At present, tarsiers are confined to several Southeast Asian islands, but have a fossil record dating back to the Middle Eocene of China and, controversially, from North Africa. Tarsiers have proportionately large eyes relative to their face, and have **nocturnal** habits. They are characterised by short snouts, small, hairless ears, large hands and feet with elongated **digits** bearing sticky pads for clinging and gripping, and long tails. Tarsiers are the only primates which are exclusively **carnivorous**, eating primarily invertebrates by jumping from branches; they feed on bats, birds, snakes and lizards too.

The simian primates (Simiiformes), or anthropoids, make up the remainder of primates. These comprise the Platyrrhini (New World monkeys), and the Catarrhini, the latter which encompasses the Cercopithecoidea (Old World monkeys) and the Hominoidea (apes and humans). Today, the New World monkeys live exclusively in South and Central America. These constitute over 170 species and 21 genera grouped into five families: the Pithecidae, Atelidae, Cebidae, Callitrichidae and Aotidae (Fig. 2. 104B). While many platyrrhines have tails, **prehensile** tails evolved independently in certain groups (e.g. atelids, capuchins). Additionally, they exhibit nostrils facing sideways ("platyrrhine" translates as "flat nosed") unlike the downward-facing nostrils of Old World monkeys ("catarrhine" translates as "down nosed").

The fossil record of Platyrrhini in South America is very rich, beginning with the oldest New World monkey fossils, known from a few individual teeth in Late Eocene strata (ca. 36 Ma) in eastern Peru. The earliest platyrrhine known from relatively complete fossil material is *Branisella* (see pp. 246–7), from around 26 Ma. It bears striking similarities to *Proteopithecus* from the Late Eocene of Africa, supporting the hypothesis that platyrrhines dispersed to South America from Africa on a raft of floating vegetation around 36 Ma, back when the Atlantic was probably 600 miles narrower than it currently is. (The same scenario holds true for the caviomorphs, or New World rodents, at roughly the same time.) The diversification of platyrrhines was well underway by the Miocene (ca. 15.5 Ma), consisting of at least 20 distinct genera reported from Colombia to Bolivia to Argentina.

Some of the platyrrhine species have been allocated to existing families, but the latest consensus is that they were early outgroups from the main platyrrhine families. They are quite varied in appearance, resembling more modern species such as sakis, titis and night monkeys. Modern families had replaced them by the Middle Miocene. The Oligocene *Canaanimico* (named from several teeth) and *Branisella* are both not closely related to the other species, but are off-shoots from the main Platyrrhini **clade**. A number of common features are seen in catarrhines, such as flat nails and a lack of prehensile tail. Several families evolved soon after the split between these two groups, and are referred to as **stem**-catarrhines. They include the Oligopithecidae (e.g. *Catopithecus*, *Oligopithecus*)—small genera which lived in Africa, were **insectivorous**, and lived during the Late Eocene to Early Oligocene.

Fig. 2.104B. Phylogenetic tree of Primates, particularly platyrrhines and catarrhines, after Orlando *et al.* (2008), Stevens *et al.* (2013), Silvestro *et al.* (2017), Seiffert *et al.* (2018), Rasmussen *et al.* (2019), and Tabuce *et al.* (2009).

The Propliopithecoidea includes larger species that resembled howler monkeys (*Alouatta*), were found in Africa, were slow moving, were **arboreal quadrupeds**, showed **sexual dimorphism**, and lived during the Early Oligocene. The Pliopithecoidea included medium to large species found in Europe and Asia. They resembled gibbons and spider monkeys, were suspensory climbers, had shorter faces, long limbs, a short tail, and lived during the Middle to Late Miocene. None of these three lineages left any modern descendants. Only two groups of catarrhines survive in modern times: the Cercopithecoidea (macaques, baboons, langurs, guenons) and the Hominoidea (apes and humans). Both are more closely related (as **crown** Catarrhini) to each other than the former groups.

Similar to the earliest tarsier, the earliest simiiforms lived during the Early Eocene, represented by fossils called *Eosimias* ("dawn monkey") from China (was the size of a pygmy marmoset (*Cebuella*), and was a tree dweller) and *Afrasia* from Myanmar, plus the larger Amphipithecidae from Pakistan, Thailand and Myanmar. Simiiforms had established a strong presence in Africa by the late Middle Eocene, represented by fossils including *Afrotarsius* (see pp. 242–3). They are even more well-known in the Late Eocene, particularly from the beds of the Fayûm Depression in central Egypt . Among these are squirrel-size forms including *Apidium* (see pp. 242–3), and the domestic cat-sized *Aegyptopithecus* (see pp. 242–3), *Parapithecus* and *Propliopithecus*, which may be regarded as stem-catarrhines. Old World primates disappeared from Eurasia by the Early Oligocene (with the exception of several Oligocene fossils discovered in Pakistan). The earliest known advanced catarrhines come from the Late Oligocene. Known collectively as Old World monkeys, many species of cercopithecids are instantly recognisable—baboons, macaques, langurs, vervets, colobus, **proboscis** monkey, mandrill, etc. However, cercopithecid dominance was not always the case. Old World primates or catarrhines consist of two extant groups, the cercopithecoids and the hominoids/ apes, and a number of fossil groups. A close relative of the common ancestor of cercopithecoids and hominoids is the gibbon-sized *Saadanius* (characterised by a longer face than modern catarrhines) from beds about 29 Ma, found near Mecca in Saudi Arabia. The next youngest fossil is *Kamoyapithecus* from beds dated at 24 Ma in Kenya. During the Early to Middle Miocene, it was the early hominoids which dominated forested and woodland environments in Africa.

Cercopithecids did not start to diversify until late in the Miocene. Although it shares many features in common, *Victoriapithecus* is the only member of the Victoriapithecidae, a closely related family to the Cercopithecidae. It was a small **terrestrial** primate found in East African woodland environments during the Middle Miocene. *Victoriapithecus* would have looked and behaved much like a modern vervet monkey (*Chlorocebus pygerythrus*). The oldest true cercopithecids also originated in Africa during the Middle to Late Miocene, with most known fossils resembling colobines, such as the diminutive *Microcolobus*. They were tree-dwellers, but were probably more **frugivorous** than later colobines. The balance between the catarrhines had shifted during the Late Miocene, and although we are not certain as to why, cercopithecids rapidly adapted to new niches. Colobines were common during the Pliocene and Pleistocene, with some environments supporting a range of genera up to 50 kilogram giants. Although modern African colobine sare largely arboreal, these ancestral genera (such as *Paracolobus*, *Rhinocolobus* and *Cercopithecoides*) showed **adaptations** to terrestrial life. The first non-African cercopithecid (the best known genus) was *Mesopithecus*, which ranged widely around Europe and West Asia. It behaved much like a Hanuman langur (*Semnopithecus entellus*), at home on the ground and in the trees. Colobines stretched further east, with the large and poorly known *Kanagawapithecus* reaching Pliocene Japan. Later lineages of colobines more closely related to extant langurs and leaf monkeys reached Southeast Asia during the Middle Pleistocene.

The diverse cercopithecid family is split into two subfamilies—the cercopithecines and the previously covered colobines. Cercopithecines are characterised by cheek pouches, and split into two tribes: cercopithecins (guenons) and papionins (baboons, mangabeys, geladas, macaques). The earliest cercopithecine fossils are known from the Late Miocene. Guenons are rarely found in the fossil record, a reflection of most being forest-dwellers (not the best location for fossil preservation), although several fossils have been found in the drier environments of the Arabian Peninsula and Kenya (*Nanopithecus*, from the Early Pliocene). The fossil record for the papionins is far better known. The earliest forms, such as the Late Miocene *Pliopapio*, were generalists that were equally active on the ground and in the trees. Some genera of baboons from the Plio-Pleistocene were massive, with the males of *Dinopithecus* (a true baboon) and *Theropithecus* (a gelada) estimated to get over 40 kilograms (even 77 kilograms has been suggested; compare that to the heaviest extant species—male mandrills can reach 32 kilograms).

Baboons and geladas evolved in different parts of Africa (Southern and East, respectively), and it was only during the Last Glacial Period (see p. 313) that geladas have been confined to the Ethiopian highlands. Baboons took advantage of the vacuum and now range across most of Africa (apart from rainforests and deserts). It was only the macaques and their closest relatives which have properly ventured and diversified outside Africa (one species of *Theropithecus* made it as far east as India (*T. oswaldi*)). An early form was the large, terrestrial *Paradolichopithecus*, from the Early Pliocene to Early Pleistocene of Europe, which is notable for several interesting features: it had an elongated muzzle like a baboon, and the ankle joints resembled those of *Australopithecus* (see pp. 156–8, 300–1) suggesting that it potentially spent time walking **bipedally**. Typical extant macaques are mostly confined to Asia, although the Barbary macaque (*Macaca sylvanus*) of North Africa is a notable outlier. During the Pliocene and Pleistocene, it also ranged further north into Europe. One isolated form, *M. majori*, was left stranded on the Mediterranean island of Sardinia during the Early Pleistocene, where it underwent shrinking in body size over time.

Hominoidea (Apes)

Today, apes have a global reach, but if you remove one upright, mostly hairless species, then the picture is less rosy. There are only a handful of non-human ape species, and these are restricted to Equatorial Africa and Southeast Asia. Meanwhile, their closest relatives are the cercopithecoid monkeys, which are more diverse and widely distributed throughout Africa, Asia and Europe (only just). Go back in the Early Miocene and the reverse was the case, with up to eight times the number of ape genera compared to

Fig. 2.106. Reconstructed life appearance of the basal hominoid *Oreopithecus bambolii*. Reconstructed height: 1.2 m.

the present. Among the features that distinguish apes from monkeys are the lack of tails and the greater range of movement in their shoulder joints, which enable certain species (e.g. gibbons and orangutans) to swing amongst branches using exclusively their arms (brachiation). Extant hominoid apes share a wide range of features, such as large brains, stiff back, mobile joints, long forearms, grasping ability and an absence of a tail. Apes also show a wide repertoire of locomotory styles with slow deliberate climbing in orangutans, acrobatic suspensory brachiation by gibbons and **bipedal** walking in humans. Whereas only a few species of apes survive today (two species of chimpanzee, along with gorillas, orangutans, and a few species of gibbons), ape diversity was significantly higher in prehistory. The earliest known apes are currently *Rukwapithecus* and *Nsungwepithecus*, from Late Oligocene beds (ca. 25.2 Ma) in Tanzania. Apes experienced a remarkable **adaptive radiation** in Africa and Eurasia throughout the Miocene, and they were more diverse and abundant than monkeys. More than 40 fossil genera and more than 100 species have been reported (with 14 genera from the African Miocene alone). They varied from domestic cat-sized (three kilograms) to gorilla-sized (80 kilograms), and had **omnivorous** or **herbivorous** diets.

The early hominoids from the Late Oligocene and Early Miocene bear little resemblance to their later relatives. Genera with good skeletal remains such as *Proconsul* reveal that they were primarily **quadrupedal** travelers on top of branches. Larger individuals of *Proconsul*, which were within the size range of chimpanzees, have been renamed as *Ekembo*. It would have resembled a large cercopithecoid monkey rather than an ape in overall anatomy, but had a coccyx (the small, triangular bone at the base of the spine) instead of a tail. It was only after the early apes such as *Afropithecus* migrated out of Africa (ca. 16.5 Ma) that they developed **adaptations** to thrive in new environments. These later genera are closer to modern hominoids and had thicker enamel for processing hard foods. Several genera diverged from the basic quadrupedal ape type. *Dendropithecus* was previously considered to be a close relative of gibbons, being a small, slender ape capable of brachiating. However, the two are not closely related as *Dendropithecus* shows various **primitive** features.

Far more puzzling was the last surviving ape in Europe (that is prior to the later migration of humans). Isolated during the Late Miocene on the Tuscano-Sardinia island, *Oreopithecus* (Fig. 2.106) was an island **endemic** with a host of odd features. The short pelvis is so similar to that seen in humans, that this small ape was believed to be a hominin and that it must have spent time walking bipedally. However, *Oreopithecus* was unlikely to be a runner due to the shape of its feet. Acting more like a steadying tripod when upright, the big toe (hallux) formed a unique 100 degree angle with the other toes. However, the structure of the spine was not consistent with bipedal walking, and the limbs were in similar proportions to other climbing apes. Other features include long **digits**, a short body, and a skull similar to that of a leaf-eating monkey. Confusingly, carbon isotope analysis of the teeth indicates that *Oreopithecus* was eating grasses and sedges, plants that do not grow in forests. It was possibly wading upright through wetlands to get to an unconventional food source. It died out when the landbridge was reformed to continental Italy.

The evolutionary split between gibbons (Hylobatidae) and the great apes (Hominidae, including humans) occurred about 17–18 Ma. After the initial split, the great ape lineage continued to diverge, with orangutans separating from the other great apes (members of the Homininae) around 12–14 Ma. Reinforcing the close relationship among all of the great apes, orangutans and their closest fossil relatives are a separate subfamily from the hominines: the pongines, which are also within the Hominidae. Pongine apes fall into three main groups: *Lufengopithecus*, sivapithecins and ponginins. While *Lufengopithecus* has generally been considered a primitive pongine, some scientists have noted features more reminiscent of hominines, such as postcranial remains indicating adaptations towards bipedalism. *Lufengopithecus* shows a far wider **sexual dimorphism** than seen in any other ape. Fossils previously referred to as "*Ramapithecus*" were once considered potential ancestors of humans. However, skull material was comparable to a previously named genus, *Sivapithecus*, which also closely resembled the modern orangutan. It lacked the adaptations for brachiating, preferring to move along the top of branches, and probably spent more time on the ground than modern orangutans. It may have been the first discovered, but is far from being the most famous sivapithecin. Although only known from teeth and lower jaws, the largest species of *Gigantopithecus* (*G. blacki*) had a body size larger than any other primate (Fig. 2.107). With so little material to work with, an

Fig. 2.107. Reconstructed life appearance of the pongine ape *Gigantopithecus blacki*, from the Early to Middle Pleistocene of China and Vietnam. Reconstructed height: 2 m.

actual size is speculative. Previous estimations of a standing head height at over three metres are exaggerated. In reality, it probably stood upright between 2.5 and 3 metres tall. The teeth were flat and robust, with thick enamel for crushing and grinding fibrous vegetation. In addition, evidence of plant remains amongst the teeth show that *Gigantopithecus* also regularly ate fruit. *Gigantopithecus* died out 300 ka during a period of a cooling climate shrinking its forest habitat, possibly alongside the hunting activity from the early human *Homo erectus* (see pp. 158–60). We do not know whether the two apes interacted but they were present in the same locations.

The modern orangutan, *Pongo*, is the only surviving genus of pongine ape and, while once more widely distributed, are now restricted to the Sunda archipelago of Southeast Asia. An earlier ponginine from mainland Asia was *Khoratpithecus* and it probably lived a similar lifestyle in the trees. Once thought to be integral to human evolution, dryopithecins are now considered an early side-branch of hominines, having split about 14 Ma. First dating from the Middle Miocene, they radiated throughout Southern and Central Europe (such as *Graecopithecus*, *Ouranopithecus*, *Pierolapithecus* and *Dryopithecus*) with several genera also in Africa (*Kenyapithecus* and *Otavipithecus*). The most familiar genus is *Dryopithecus*, a forest-dweller from various localities in Europe during the Middle to Late Miocene. It was largely a quadrupedal ape, although capable of suspensory behaviour. Once the climate began to cool and dry up by the end of the Miocene, the great ape radiation was decimated. Only a few survived into the Pliocene, while the Old World monkeys diversified instead. After the dryopithecins, the gorillas form the second lineage of hominines, having diverged about 8.9 Ma, although the two purported fossil genera (*Chororapithecus* and *Nakalipithecus*, from Ethiopia and Kenya respectively) are known only from fragmentary material. The jaws and teeth indicate large apes with adaptations to a **folivorous** diet. In regard to species diversity, the Hominini has been the most successful lineage of the Homininae. Extant species consist of the two chimpanzees and ourselves, with the two groups splitting sometime between 4 and 8 Ma.

Human Evolution

Humans and chimpanzees are grouped together within the tribe Hominini (Fig. 2.108), partly because we share more than 98% of our DNA. The earliest fossil widely recognised as belonging to the Hominini, or "hominins", was formally described in 2002 (other potential hominin fossils date back further). While the discoverers gave it the nickname "Toumai", its formal **binomial** name is *Sahelanthropus tchadensis*. The most well-preserved specimen is a nearly complete cranium from beds dated to roughly 6–7 Ma, discovered in the Sub-Saharan region of Sahel in Chad (hence the binomial, meaning "human from the Sahel of Chad"). *Sahelanthropus* would not have been dissimilar from modern chimpanzees in size and appearance (having prominent brow ridges and a small braincase), but it possessed strikingly human-like traits including reduced canine teeth, relatively small **cheek teeth** with wear patterns like those of later hominins, and a flattened facial region. While initially thought to be a habitual **biped** due to skull features, a recent report on femoral bones suggests that this was not the case.

A slightly later species is *Orrorin tugenensis*, from Late Miocene deposits of the Tugen Hills in Kenya, dated at 5.72–5.88 Ma. *Orrorin* is represented primarily by fragmentary fossils, including teeth with thick enamel characteristic of **primitive** hominins. Unlike *Sahelanthropus*, it appears that *Orrorin* did develop an early form of bipedalism as indicated by the femur. A little more recent than *Orrorin* are the fossils of *Ardipithecus kadabba*, discovered in Ethiopian strata dated at 5.54–5.77 Ma. *Ar. kadabba* is known from very limited material, although a single toe bone dated to 5.2 Ma suggests that hominins used the "toe off" mechanism during bipedal locomotion as far back as that time. By the end of the Miocene, fundamental human **adaptations** including a fully upright stance are believed to have emerged in hominins, although their brains remained small and they were still of comparable body size to the contemporaneous apes. The Pliocene witnessed a much richer diversity of hominins, with various primitive species temporally overlapping with the diversification of more derived forms.

The primitive forms included *Ar. ramidus*, discovered in 1992 from Ethiopian beds dated to about 4.4 Ma, and possessed human-like features including a U-shaped lower jaw (as opposed to the V-shaped lower jaw in apes) and reduced canine teeth. *Ar. ramidus* is represented by a partial skeleton, which is currently the earliest known and most complete hominin skeleton. *Ardipithecus* is possibly the earliest of "human-line" hominins. Having a chimp-like foot and elongated big toe allowed *Ardipithecus* to have a secure grip when climbing in the branches. Like chimps, the foot was also adapted for heel first walking on the ground. Despite walking upright, *Ardipithecus* was by no means a speedy long-distance walker. Additionally, Kenyan strata dating to roughly 3.5 Ma have yielded fossils of other "**archaic**" hominins including the poorly known *Kenyanthropus platyops*. The earliest species of the advanced genus *Australopithecus* had originated by about 4.2 Ma, which was the most diverse hominin genus during the Pliocene. Australopithecines are truly the perfect "halfway" step to humans. Depending on the species, they generally had tall faces with a prognathic jaw, small dentition, small braincase, curved fingers for **arboreal** activity, flat feet for bipedal walking, and

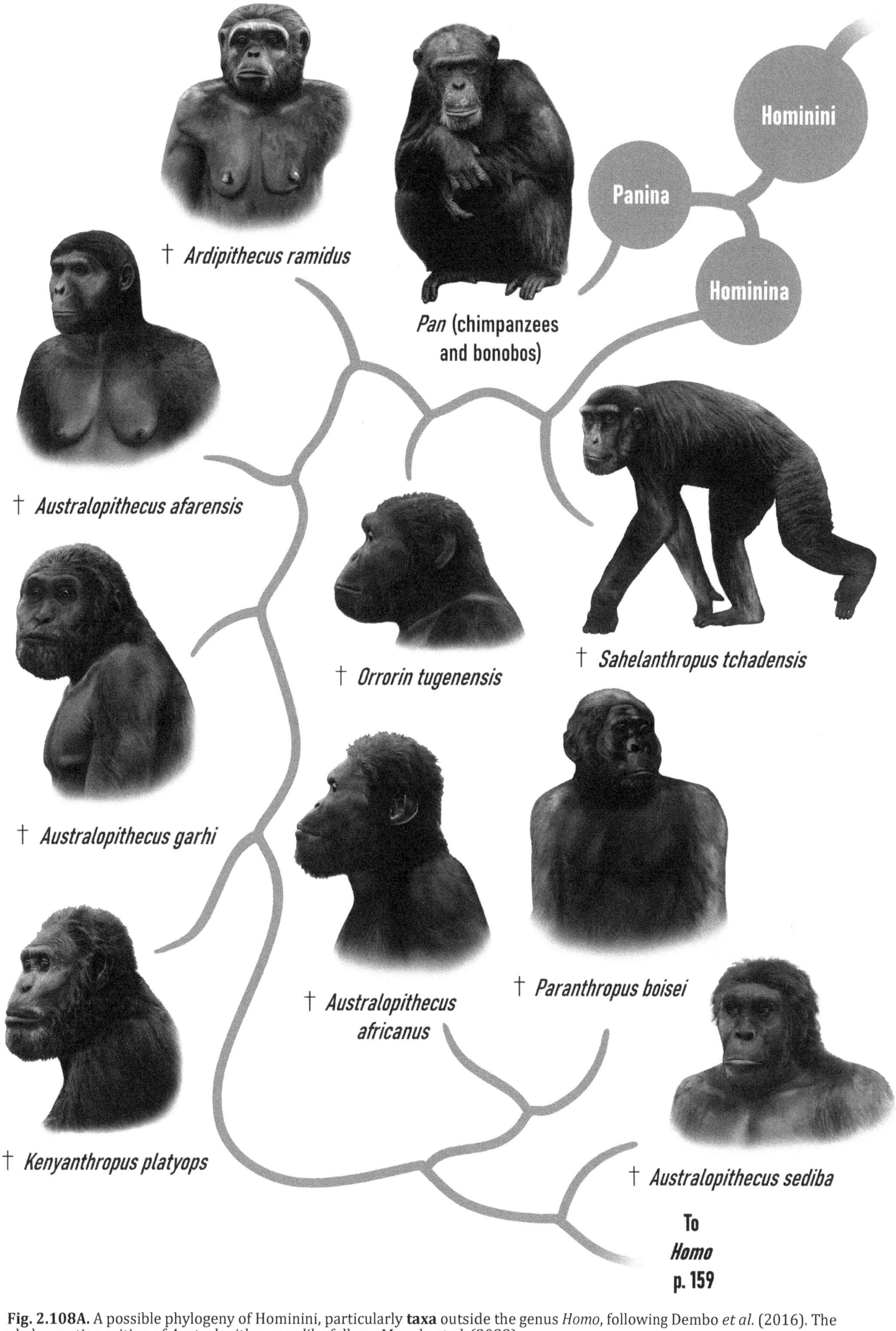

Fig. 2.108A. A possible phylogeny of Hominini, particularly **taxa** outside the genus *Homo*, following Dembo *et al.* (2016). The phylogenetic position of *Australopithecus sediba* follows Mongle *et al.* (2023).

pronounced **sexual dimorphism**. The earliest known australopithecine is *Au. anamensis*, from beds adjacent to Lake Turkana in Kenya dated between 2 and 3.9 Ma. In 2019, for the first time, a complete cranium of *Au. anamensis* was described, which improved our understanding of its anatomy and phylogenetic affinities. *Au. afarensis* is the best known of the primitive australopithecines, represented by remains from over 300 individuals including a partial (nearly 40% complete) skeleton yielded from 2.95–3.85 million year old beds close to Hadar, Ethiopia, and nicknamed "Lucy" by its discoverers Tim White and Donald Johanson. The specimen was named after the Beatles song "Lucy in the Sky with Diamonds", which was playing while those palaeoanthropologists celebrated their discovery. At the time of the specimen's discovery in 1974, *Au. afarensis* was the first primitive hominin to show evidence for bipedalism, although it was less upright compared to more recent hominins. (For further information on *Au. afarensis*, see pp. 300–1).

African hominin diversity had increased significantly by the Late Pliocene. As well as the "archaic" forms *Au. bahrelghazali* (from beds 3.5 Ma) and *Au. garhi* (from beds 2.5 Ma), these include one of the most well-known australopithecines, *Au. africanus*. *Au. africanus* was first described by Raymond Dart in 1924 on the basis of a juvenile skull (nicknamed the "Taung child"), and scientists initially doubted its ancestral link to modern humans. As higher-quality specimens were recovered from more South African caves by palaeontologists including Robert Broom (particularly an adult skull informally named "Mrs. Ples" in 1947), evidence solidified that *Au. africanus* was a small-brained, bipedal African hominin rather than a **quadrupedal** ape. *Au. africanus* was a relatively small, gracile hominin very similar to *Au. afarensis*, but possessed a larger jaw and cheek teeth, and lacked a **sagittal crest**. *Au. africanus* also had a brain volume of roughly 420–510 cubic centimetres. While *Au. africanus* was once regarded as a direct ancestor of *Homo* because of its gracile and remarkably human-like characteristics, current scientific consensus suggests that it is a side branch of the human evolutionary tree.

Besides *Au. africanus*, a variety of morphologically robust hominins are known from the Late Pliocene of Africa too. For a while, these were referred to the genus *Australopithecus*, assumed to be either separate species or robust males of *Au. africanus*. More recently, though, palaeoanthropologists have come to realise that these form a distinct robust side branch, belonging to the genus *Paranthropus*. The earliest known specimen pertaining to *Paranthropus* is the intriguing "Black Skull" (so named because its bones are dark-coloured), found in 1975 on the western shores of Lake Turkana in northern Kenya, from beds that date back to 2.5 Ma. While the braincase is small, the skull is robust with a prominent sagittal crest, broad, flat cheek teeth, and a dish-shaped facial region. The current scientific consensus is that the "Black Skull" belongs to the oldest species of *Paranthropus*, *P. aethiopicus*. After that came *P. boisei*, the most robust hominin ever, known from East African strata dating between 2.3 and 1.2 Ma. The first specimen of *P. boisei* discovered was informally named "Nutcracker Man" after its massive, thickly-enamelled molars, robust lower jaws, broadly flared cheekbones and prominent sagittal crest, indicating a diet of abrasive **C4 plants** such as grasses, with hard foodstuffs as fallback items. It was discovered by Mary Leakey in 1959 at Olduvai Gorge, Northern Tanzania, and initially described as "*Zinjanthropus boisei*" by her husband Louis Leakey.

P. robustus was the first discovered species of *Paranthropus* (found by schoolboy Gert Terblanche in 1938), known from South African strata dated at 1.6–1.9 Ma. It too possessed large lower jaws, broad cheek teeth, and a pronounced sagittal crest, although it was less robust than *P. boisei*. *P. robustus* coexisted with *Au. africanus* in Southern Africa, with fossils of both species found in the same cave systems. *P. robustus* was larger and more robust, with certain individuals weighing up to 120 pounds. Living around 1.9 Ma, the fossilised skeletons of *Au. sediba* discovered from Malapa cave in South Africa are so complete that scientists can see what entire skeletons looked like near the time when the genus *Homo* evolved. Details of the teeth, the length of the arms and legs, and the narrow upper chest resemble earlier australopithecines, while other dental traits and the broad lower chest resemble humans. It is currently believed that *Au. sediba* was a species that was on a separate evolutionary branch, sharing a common ancestor with the *Homo* lineage.

The Early Pleistocene also yields the earliest fossils of the modern genus *Homo*, which are easily differentiated from the contemporary *Australopithecus* and *Paranthropus* by a larger braincase, the lack of a sagittal crest, a reduced brow ridge, a flatter facial region, and smaller canine teeth and cheek teeth. The earliest known species of *Homo* is *H. habilis* (which translates as "handy man"), discovered by Louis and Mary Leakey at Olduvai Gorge in the 1960s, from **strata** dated to roughly 1.75 Ma. *H. habilis* had body proportions similar to australopithecines and possibly lived in large, multi-male, chimp-like groups. It was sexually dimorphic, and was both partially arboreal and capable of running. At first, all the oldest specimens of *Homo* were lumped together under the species *H. habilis*, although palaeoanthropologists now realise that these fossils are too disparate to be assigned to a single species, leading to the identification of multiple species. Among these are an anatomically more advanced skull now called *H. rudolfensis* (dated at ca. 1.9 Ma), first described by Richard Leakey, along with the short-lived albeit highly advanced *H. ergaster*, known from deposits dated at 1.6–1.8 Ma. The fossils of these species are also associated with early stone tools, particularly hand axes, choppers and scrapers of the "Oldowan" technology.

Some primitive Pliocene hominin species that survived into the Early Pleistocene (as late as 1.65 Ma) include *P. boisei* and *P. robustus*, *H. habilis* and *H. ergaster*. The most famous fossil of *H. ergaster* is a virtually complete skeleton of a male child whose age at death is estimated at about 7–11 years old, discovered on the western shores of Lake Turkana in 1984 by Alan Walker and his team. Informally named "Turkana Boy", it is estimated that he could have reached 1.85 metres tall if he had reached adulthood. Around 1.9 Ma, another species had arrived on the scene: *H. erectus*. In addition to being bipedal and standing erect (as indicated by the species name), this hominin was nearly the size of modern humans. The average brain capacity of *H. erectus* was around 1 litre (1000 cubic centimetres), a little less than modern humans. *H. erectus* produced hand axes and crude choppers ("Acheulean" technology), and was the earliest species to control fire. *H. erectus* was initially restricted to Africa, which was long home to all of our other early relatives. Fossil evidence suggests *H. erectus* left Africa around 1.8 Ma, as some Indonesian fossils (initially named "*Pithecanthropus erectus*" or "Java man") date back to that time.

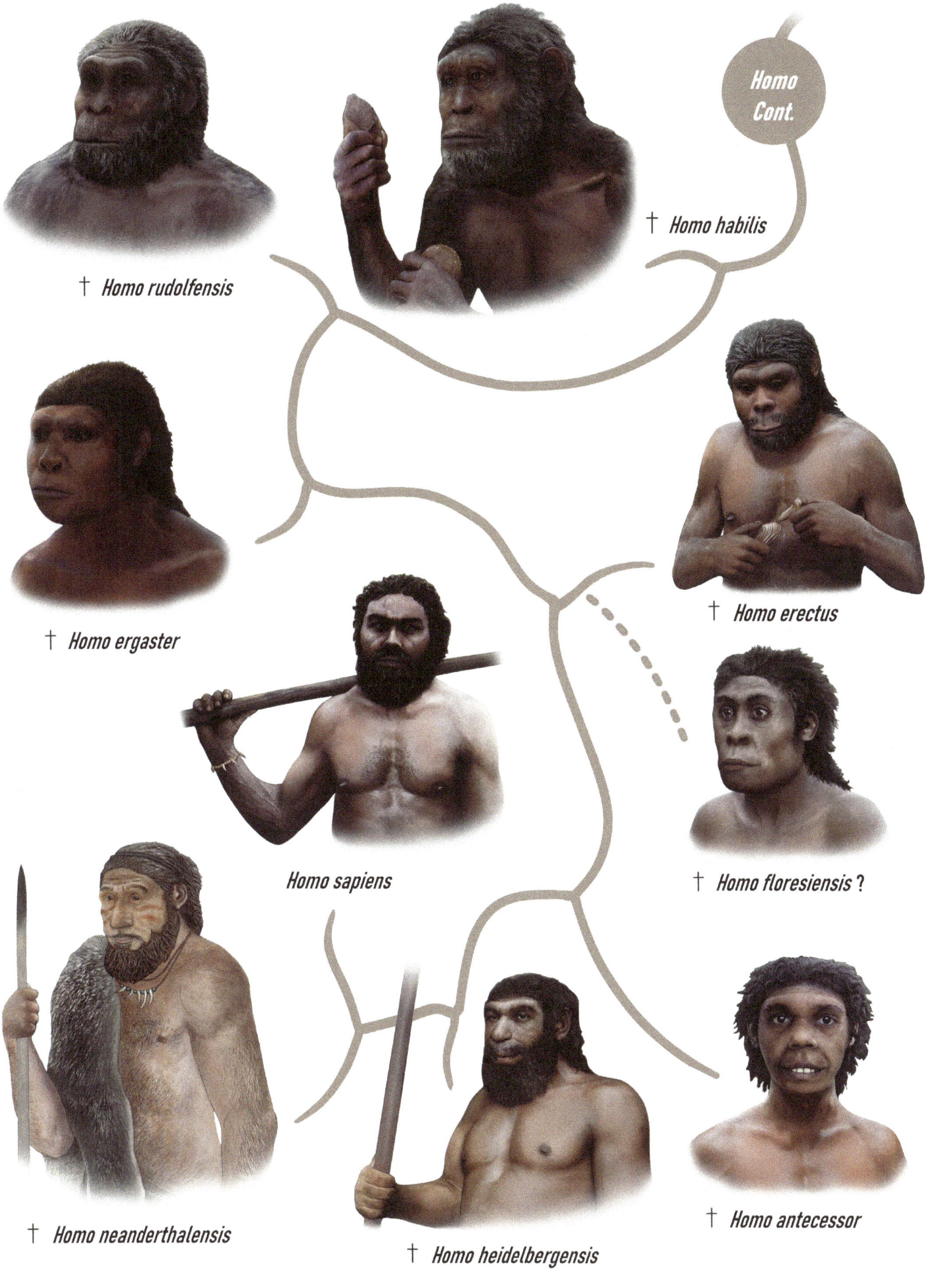

Fig. 2.108B. A possible phylogeny of Hominini, particularly among species of *Homo*, following Dembo *et al.* (2016).

Furthermore, fossils of nearly the same age are known from other parts of Eurasia, including Romania and the site of Dmanisi in Georgia. By about 500 ka, plenty of *H. erectus* fossils are found across Eurasia, such as the noted fossils from the Dragon Hill Bone site at Zhoukoudian, southwest of Beijing in China, initially dubbed "Peking Man" and dated to roughly 460 ka. According to the most recent dating, *H. erectus* possibly survived as late as 143 ka and potentially even as recently as 74 ka, contemporaneous with *H. sapiens*. *H. erectus* is considered the earliest widely distributed human species, and it was one of the longest surviving and most successful, existing for over 1.8 million years, from 1.9–0.143 Ma. While *H. erectus* had a more stable body plan and brain size than other hominins, there were changes and variations within the species over time. Living approximately 1.2–0.8 Ma, *H. antecessor* was the second earliest fossil hominin to have colonised Europe, known from just two sites in northern Spain. It had a very primitive skull, with low foreheads and brains only slightly larger than *H. ergaster*. Whether *H. antecessor* represents a primitive small-scale and unsuccessful colonisation of Europe or if it persisted and gave rise to later European hominins remains uncertain. Around 400 ka, another hominin had settled the Near East and Western Europe: *H. neanderthalensis*. In 1856, Neanderthals were the first humans to be recognised in the fossil record, but initially, their fragmentary bones were mistakenly believed to be those of diseased Cossacks who had met their deaths in caves. Some remains of an individual from a French cave called La Chapelle aux Saints, who died of rickets and old age, served as the basis of the earliest comprehensive description of Neanderthal skeletons. Following other discoveries during the early twentieth century, French palaeontologist Marcellin Boule fuelled the image that Neanderthals were slouching, club-wielding primitives. Boule's image of the Neanderthals was not significantly changed until the 1960s. Recent research has revealed that Neanderthals were very much unlike this outdated interpretation.

While their skulls differ from modern humans in exhibiting a sloping forehead, a pronounced browridge, larger jaws, a more robust mandible and a wider nose, their brain capacity, on average, was slightly larger than that of modern humans. While cave use is well-established, Neanderthals showed flexibility and regularly used open-air settlements while hunting. Their social structure and culture shows many similarities to modern humans. Neanderthals created sophisticated tools, wore clothing and jewellery, used strategy and teamwork to hunt, had complex speech, developed rudimentary medicine, cared for infirm members of their clan and performed specific burial rites with the dead. Compared to modern humans, neanderthals had stockier proportions with a wider chest, but proportionally shorter limbs, since they lived only in the cold environments of the glacial borders of Europe and the Near East, where their short, stocky physique would have been a good adaptation. They were also skilled toolmakers, known for producing Mousterian spear points, hand axes and other sophisticated objects, plus tools made from wood and bone. Some of these tools exhibit both simple and intricate carving, revealing their artistic abilities that no earlier hominin ever had. While previously classified as a subspecies of *Homo sapiens*, recent research and genetic evidence strongly suggest that Neanderthals were indeed a separate species. The Qafzeh and Es-Skhul caves on Mount Carmel in Israel contain the strongest fossil evidence for the existence of early anatomically modern humans in the Near East alongside Neanderthals, suggesting they were distinct populations. The first Neanderthal DNA sequences reported in 1997 reveal that they were genetically distinct from *H. sapiens*. However, the genetic evidence strongly suggests that interbreeding occurred between Neanderthals and *H. sapiens* in Eurasia after the latter migrated out of Africa. This is evident from the presence of Neanderthal DNA in the genomes of all non-African populations of *H. sapiens*. *H. neanderthalensis* was the sole extinct human species to have its DNA sequenced until 2010, when genetic research surprised the world with the discovery of yet another extinct human species that existed within the past 40,000 years.

In Denisova Cave, located in the Altai Mountains of Siberia close to the China-Mongolia border, Russian scientists excavated finger and toe bone fragments from a juvenile female hominin, along with several isolated teeth mingled with a bracelet and other artefacts. The occupation of Denisova Cave is estimated to have begun around 300 ka, with the youngest evidence dated at 52–76 ka. The analysis of mitochondrial DNA from the finger bone revealed that it had a distinct genetic sequence setting it apart from both Neanderthals and modern humans. It had unique nuclear DNA as well, but genetic evidence indicates that those humans were most closely related to Neanderthals. It is also possible that they interbred with modern humans, since roughly 3–5% of their DNA is found in the genomes of Australian Aboriginals and Melanesians. Based on mitochondrial DNA analysis, they diverged from the human line roughly 600 ka, and represent a distinct "out of Africa" dispersal event separate from the considerably older *H. erectus exodus* (1.8 Ma) and the more recent *H. rhodesiensis-H. heidelbergensis* migration to Eurasia (300 ka). These enigmatic humans with such unique DNA are now referred to as "Denisovans". Given the scarcity of fossils, we know very little about their physical appearance and lifestyle, or anything about them except that they carried DNA distinct from all other species of human. Indeed, scientists remain hesitant to assign a formal **binomial** name to the Denisovans due to the limited fossil evidence, which does not allow for a complete anatomical description of the species in any meaningful way. The Denisovans remain a mystery.

Rivalling the 2010 announcement of the Denisovans in its impact on our understanding of human evolution was the 2003 discovery of an "archaic" dwarfed species of *Homo* on the Indonesian island of Flores, called *H. floresiensis* (see pp. 336–7). Lastly, we find the earliest fossil skulls and skeletons appearing virtually identical to those of *H. sapiens*. A number of them, informally called "archaic *H. sapiens*" or formally, *H. heidelbergensis*, have been identified from several African sites dated at about 300 ka. Skulls discovered in Africa, dated at about 90 ka (including those from the Klasies River Caves in South Africa), exhibit a virtually modern human appearance and are identified as *H. sapiens*. Similar to *H. erectus*, early *H. sapiens* was confined to Africa for much of its history, before emigrating to Eurasia in multiple waves of migrations over a long period, from approximately 200 ka. In Eurasia, these humans encountered the Neanderthals, who they lived alongside for approximately 30,000 years. The Neanderthals went extinct by 40 ka, the reason which remains disputed. Theories include extinction by interbreeding with early modern human populations, inbreeding depression, competitive replacement, diseases passed on from modern humans, and climate change. Whatever occurred, modern *H. sapiens* soon came to dominate the globe.

3. Reconstructing Extinct Cenozoic Tetrapods

Reconstructions of extinct tetrapods are developed almost exclusively from fossilised teeth and bones. Soft tissues such as skin and muscle are almost always lost. Details including the colouration and patterning of the skin, feathers or fur usually decay prior to the fossilisation process. Furthermore, skeletons are frequently incomplete. A scientifically credible restoration of an extinct tetrapod's life appearance needs to be based on skeletal evidence, which in turn serves as a clear reference for the location and mass of muscles. External characteristics can only be added once the musculature is restored (Fig. 3.1). The reconstructions in this book generally follow this approach, apart from when the skeleton is unknown and therefore inferences need to be made from extant relatives or from other fossil relatives known from more complete material, or if the subject is almost identical to an extant relative.

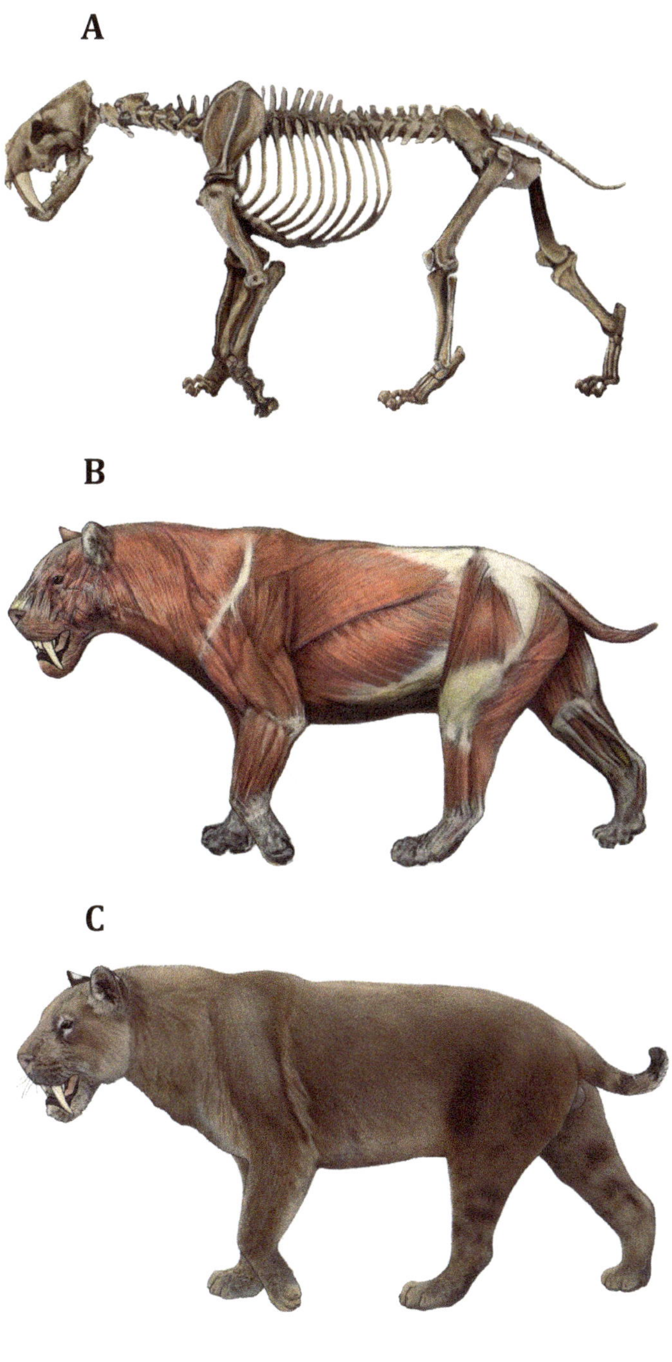

Fig. 3.1. Three stages of the reconstruction for an extinct sabre-toothed cat of the genus *Smilodon* are shown, working from (A) the skeleton, to (B) the muscles and cartilaginous structures (including the external ears (pinna) and nasal cartilages), to (C) the life appearance.

Skeleton, Musculature and Soft Tissues

Skeleton

Skeletal restorations are graphics that meticulously illustrate the skeletal components of a fossil animal. They take numerous lines of evidence into account in predicting skeletal ratios and articulation, and meticulously estimate the ratios of missing bones to bridge gaps in fossil evidence. Skeletal restorations are conjectures about the overall appearance of fossil species and, like all models of fossil species configuration, they are subject to change when new fossil evidence is discovered.

Skeletal restorations can be made easier upon the exceptional discoveries of complete, **articulated** skeletons, such as those preserved in the **Lagerstätten** of the Messel Formation (see pp. 210–7) or in the Ashfall Fossil Beds (see pp. 260–1). Unfortunately, for most fossil species, the skeleton has to be reconstructed from **disarticulated**, frequently isolated bones, which commonly involves assembling a skeleton from a few individuals to restore just one creature. This necessitates cross-scaling, which is accomplished by scaling skeletal components shared between two specimens to an equal size and presuming that uncommon parts can be scaled to an equal extent.

However, not every bone is an equally trustworthy size indicator. Small elements should be ignored since they are easily deformed by fossilisation and can be measured inaccurately, resulting in mistakes in a skeletal reconstruction. Larger bones such as the upper limb elements are more ideal, although not all are helpful since animals frequently grow with allometry, defined as the growth of body parts at various rates. Given the impacts of growing body weight on bones, allometry is highly prevalent—heftier animals require proportionately larger bones for body support. Cross-scaling can be hindered by allometry, which can also lead to significant inaccuracies; adult specimens should not be used as models for juveniles and vice versa. Skeletal restorations are executed in lateral aspect most often, although we may be able to restore extinct tetrapods in various perspectives if fossils make it possible. This greatly increases confidence when reconstructing a species three dimensionally or at an odd angle. The skeleton must also be articulated in a life-like, biologically plausible pose, which necessitates decisions regarding the articulation angles, and ranges of abduction and adduction, and flexion and extension. A reasonably "neutral" pose is typically better than a dynamic pose, since it is less likely to be deemed implausible.

Musculature

Much of the tetrapod skeleton is enveloped by muscle tissue, along with the associated tendon and ligament networks. The animal's overall size and bulk greatly influences muscle mass. Studying the attachment sites on bones, which are identified by ridges, crests or rough areas, is necessary for muscle restoration.

The position of jaw muscles around the skull's temporal bone is an anatomical feature common to all tetrapod heads. Jaw muscles normally run along the back of the jaw and are positioned within the skull bones, although in animals with strong bite forces, it is standard for insertion sites to extend beyond the outer skull surface. Facial regions of mammals have extra jaw muscles attached outside the dental row between the lower jaw's lateral side and the cheekbones. In fact, compared to other tetrapods, mammal facial regions often have significantly more muscle surrounding their jaws, enabling both jaw function and regulation over facial tissues. These are restorable by comparing the muscle arrangements of extant relatives with the scars and landmarks present on the skulls of extinct mammals. Scars and bony protuberances on the skull's **posterior** surface indicate the amount of muscle holding the head in place. If they are massive, they are unmistakable signs of bulky **anterior** neck musculature. Much of the neck's musculature consists of numerous individual muscles that attach to various points on the cervical vertebrae, skull, and portions of the anterior torso such as the pectoral girdle and scapula.

The muscles of the torso include those that extend along the vertebrae, encircle the ribcage, and support the pectoral girdle. They usually correspond to the underlying skeleton's outlines, although their coverage varies. While the vertebrae and ribs are frequently concealed, certain upper ribcage portions and the dorsal part of the limb girdles are often more protuberant. Among modern tetrapods, tail musculature differs considerably. The caudofemoralis (hindleg retractor muscles) attaching to both the hips and tail is a **primitive** trait among tetrapods. Most amphibians and reptiles exhibit a notable augmentation in the size of the caudofemoralis, since the **caudal** vertebrae are ventrally and laterally extended (through extra bones called chevrons) to support this occasionally massive muscle. Yet, mammals and birds abandoned this arrangement, repositioning all their main hindlimb muscles to the pelvis. This results in a rather sparsely muscled anterior portion of the tail and a distinct boundary between the tail and hindlimb musculature (Fig. 3.2).

The front thoracic vertebrae and pectoral girdle serve as attachment points for the upper forelimb musculature. The muscles that function the elbow surround the upper forelimb, the shoulder joint, and **proximal** part of the forearm, whereas the muscles that function the hand and wrist are primarily located immediately under the elbow, regulating the wrist and hand bones with ligaments. The hindlimb's musculature is largely similar to that of the forelimb, but because it serves as the main propellant limb in the majority of tetrapods, it frequently has deeper muscle tissue relative to the forelimb. The majority of the upper hindlimb muscles, apart from the caudofemoralis, are anchored by the pelvis, and the proportional sizes of the pelvic girdle can be used to estimate the distribution of these muscles. Additionally, a portion of the front pelvic musculature covers the anterior part of the knee to attach onto the lower leg. Along the femur are insertion points for extra knee-associated muscles. The majority of the ankle and foot muscles are located at the top of the

Terrestrial tetrapods and fliers

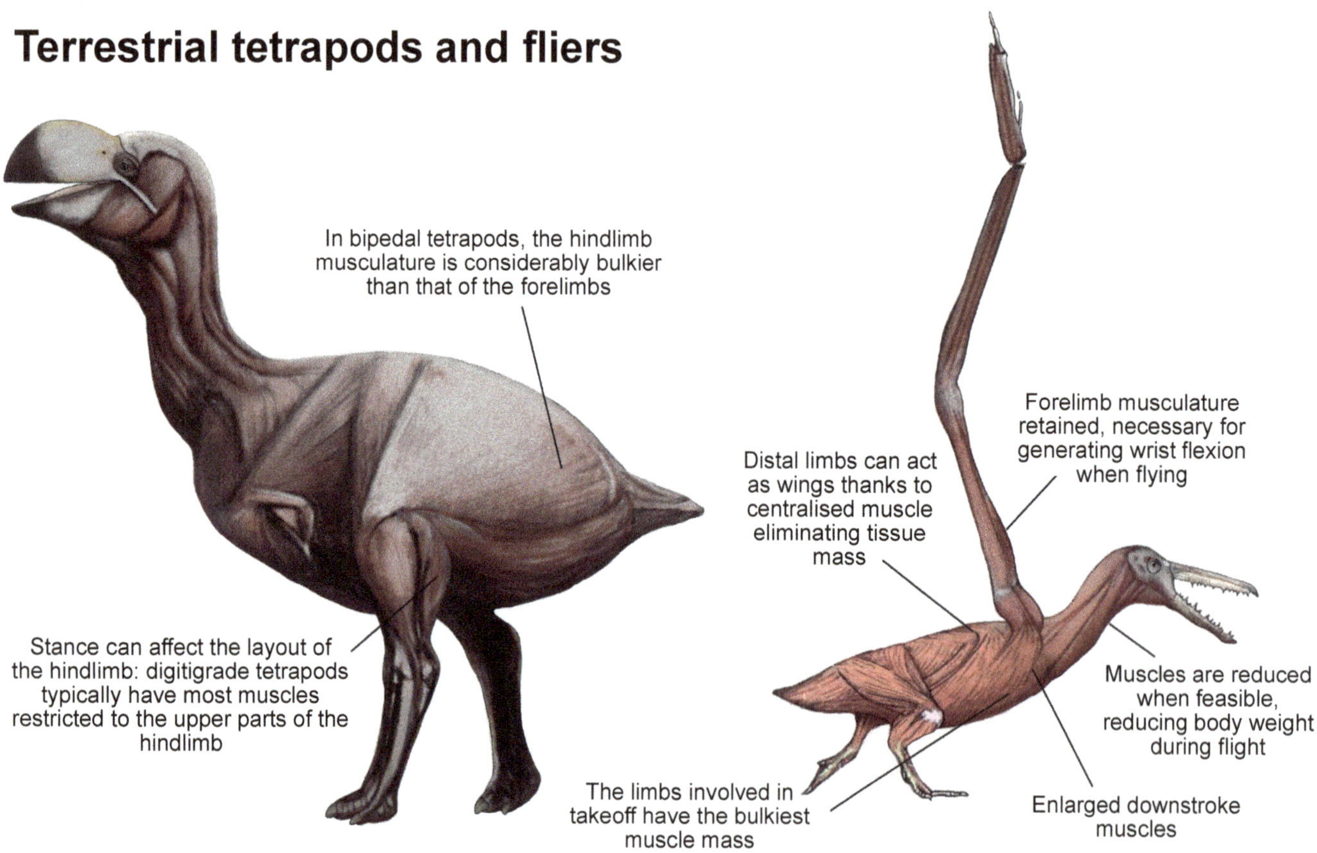

Fig. 3.2. Muscles in **bipedal terrestrial** animals and aerial fliers, exemplified by the large Miocene bird *Dromornis stirtoni* (left) and the extinct seabird *Pelagornis sandersi* (right), respectively.

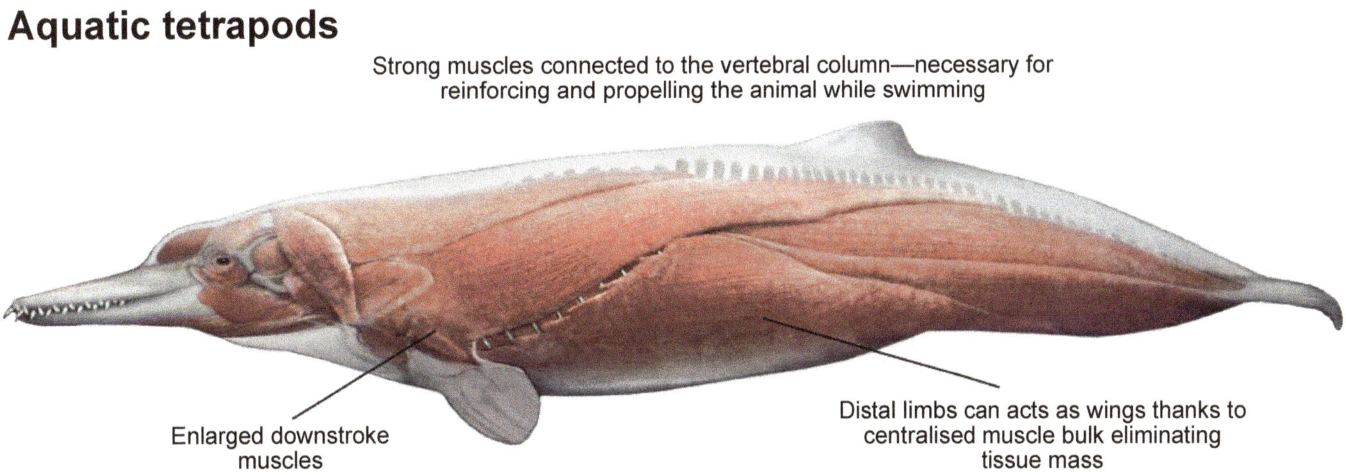

Fig. 3.3. The muscle layout and fat distribution of an aquatic animal, demonstrated by the cetacean (see pp. 121–6) *Squalodon whitmorei*. The bodies of extinct whales were almost certainly smoothed by fatty tissues and blubber as in extant species.

lower leg region, and they are linked to the bones of the **distal** limb by strong, elongated ligaments.

Adipose Tissue and Cartilage

Adipose tissues, or body fat, also play a vital role in determining tetrapod body shape, but are hardly ever fossilised. We can estimate where fatty tissues may be present in fossil tetrapods by looking at specific patterns in their distribution. In reptiles, fat is stored immediately **posterior** to their heads, all over their torsos, and significantly around their tails—occasionally to the extent where the adipose tissue in their tails bulges out. Birds and mammals typically store most of their adipose tissue all over their torsos rather than on long tails, but many mammals also have massive pads on their faces and necks (e.g. orangutans). Because of the significance of mass distribution to birds, fat is typically concentrated around the breast and abdomen in those animals, where it can be lifted most efficiently during flight. Many marine mammals including whales and seals use fatty deposits to help with thermoregulation, in which fat extends along their whole bodies to reduce heat loss (Fig. 3.3).

Additionally, the skeleton offers guidance for reconstructing additional soft tissues including cartilage, which is necessary for reconstructing ears and noses in mammals. In mammals, a highly retracted nasal aperture frequently corresponds to a **proboscis** of various lengths. Reconstructing the external ear (or pinna) is made easier by the fact that it is fixed in relation to the auditory meatus (ear canal) in the skull, although reconstructing the pinna's shape and size necessitate the use of functional extrapolation and phylogenetic inference (Antón, 2003).

Integument

Once all the musculature, body fat and connective tissues have been put in place, the following stage is to apply a suitable skin type. Understanding aspects of tetrapod skin anatomy is helpful, such as the structures it produces, how it interacts with skeletal elements and how it is fossilised. In animals, the integumentary system is the outermost layer of tissues, formed of an external layer of skin (epidermis), tissues lying just beneath it (dermis), and specialised derivatives that develop inside or on top of it including **osteoderms**, beaks, horns, nails, claws, scales and filaments including feathers and fur. While the integumentary system can leave a surprising quantity of evidence on bones, it is, frustratingly, mainly identifiable solely via preservation of soft tissues.

Epidermal Scales and Osteoderms

Epidermal scales are composed of heavily **keratinised**, locally thickened parts of the epidermis, and are mainly found in birds and reptiles. There are multiple differences between epidermal scales regarding form, size, spacing, and the extent of overlap. In order to allow the skin to flex, the epidermis thins between every scale, and to aid body movement, scales typically become smaller near regions of significant flexion, such as limb joints. While epidermal scales cover the entire body only in reptiles, they are also present on most bird hindlimbs, and even on the tails, feet and hands of some mammals. In birds, most hindlimb scales are now understood to be secondarily derived from feathers, while in mammals (e.g. pangolins), scales are convergent structures formed from modified hair. The greatest diversity in scale morphology is seen in reptiles, although a variety of scale-like structures are also present in birds and mammals. Scutes, a type of large, plate-like scale, are common in birds and reptiles, and usually arranged in rows along relatively immobile body regions, such as a bird's lower leg.

Compared to other non-mineralised integument types, epidermal scales are more likely to be preserved. Any scale patterns that are preserved well-enough in fossils is taxonomically important—placing the incorrect one on a fossil reptile would be like placing stripes on a leopard. While epidermal scales do not include a bony component, they possess osteological correlates when developing immediately above bone, in the form of shallow rugosities which can mimic the positioning of the superimposed scales (Figs. 3.4B, E). These could be large, distinct characteristics or slight irregularities, contingent upon the biological age of the tetrapod individual and the dimensions of its scales. Certain species have acute rugosities that precisely mirror the appearance of the skin, resulting in the formation of spikes, hornlets and other structures (Figs. 3.4A, 3.6B). On tetrapod skulls, scale correlates are frequently observed over the dorsal surface of the skull, on the

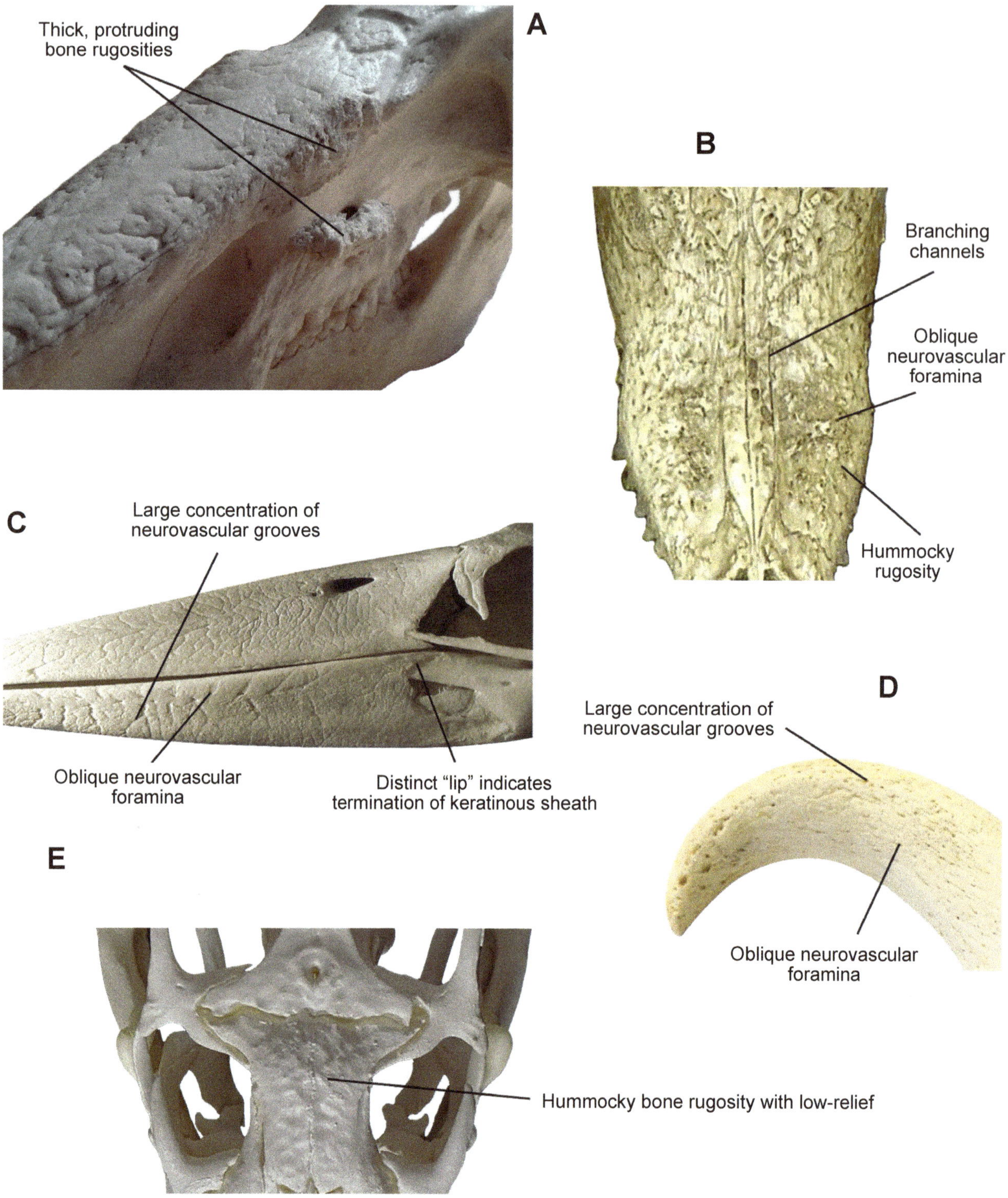

Fig. 3.4. Photographs of osteological features underlying certain epidermal tissues on modern animal skulls. (A) Armour-like dermis (bushpig **rostrum**); (B) Scale-like cornified epidermis (crocodile rostrum); (C) Cornified sheath (marabou stork rostrum) (D) Cornified sheath (water buffalo horn); (E) Epidermal scales (**frontal** area of komodo dragon skull).

facial region, and around the **orbits** and **temporal fenestrae**. **Osteoderms** are occasionally superimposed by reptile scutes. Osteoderms range from augmented sheets of thick bone assembled in rows (Fig. 3.5A), to minute bones scattered throughout a tetrapod's skin (Fig. 3.5B), to specialised bones fashioned like spikes or clubs. Among well-known examples of osteoderms are the shielded backs of armadillos and crocodylomorphs (see pp. 26–30), and the club-like tails in meiolaniid testudinatans (see pp. 15–7) and glyptodontines (see pp. 59–60). However, osteoderms are often dissociated from skeletons during fossilisation. Insights into the pattern of scales are provided by the arrangement of osteoderms. Apart from certain monitor lizards and all geckos, all extant reptiles have one scale covering each osteoderm, and it is fair to predict that certain extinct reptiles also had this trait. We should avoid depicting osteoderms as being too pronounced, since these structures are fixed within skin tissues rather than placed over them, and because the neighbouring skin can still be dense and augmented when bony armour is lacking.

Fig. 3.5. Osteoderms in extant reptiles. (A) Saltwater crocodile (*Crocodylus porosus*); (B) Gila monster (*Heloderma suspectum*).

Fur and Feathers

Keratinous strands that protrude from the skin comprise fibrous integuments, the only two forms which exist today being fur in mammals and feathers in birds.

Mammal fur is an adaptable structure that ranges widely in length and density. It has varied functions including thermoregulation (in which density and length vary to conserve heat and mitigate the effects of severe climates), protection (spiny fur, ear and nose hair, and eyelashes), visual signals (development of capes, tufts and manes) and, through whiskers, sensing surroundings. Among mammals, fur appears to develop in practically any place on the skin. In mammals with thick fur, all except the shortest coats conceals portions of the musculoskeletal outlines and thickens the overall body shape.

The insulating qualities of fur are more complicated than one may think, and they are of importance to fossil mammal restorations. Whereas a clear correlation exists between hair density, length, and the capacity to retain body heat, research on modern elephants demonstrates that short, sparse fur promotes heat convection off the body, which helps with cooling (Myhrvold, 2012). This raises the potential that large-bodied fossil mammals or those that existed in hot environments were not completely hairless. Fur can remain intact in exceptionally preserved specimens (as is the case in multiple mammal fossils from the Messel Formation). However, the fossil record of fur is extremely scanty prior to very recent deposits in which frozen and mummified mammal bodies are preserved. Remains of those mammals, such as steppe bison, woolly mammoths, steppe lions, cave bears and ground sloths, provide amazing details about their fur that enable fur density and length to be restored more confidently. Feathers, on the other hand, which today are found exclusively in birds, are the most advanced and versatile tetrapod integument ever. Bird feathers have the same fundamental purposes as mammal fur, including thermoregulation, protection, visual signalling and sensing, although their rigidness and ability to interlock enable them to produce both aerial surfaces and complex structures for display purposes.

Horns, Beaks, Claws and Nails

Among soft tissue horns, rhinoceros horns lack bony material and are composed solely of **keratin**. Soft tissue horns are never preserved except under special conditions (e.g. woolly rhinoceros remains in the Arctic permafrost). The presence of massive projections including rhinoceros horns is indicated by a halo of pronounced bony rugosities, which forms the contours of the horn base. The surface inside the halo may also be rugose, so it is crucial to determine the occurrence of unusually gnarly bone textures along the edges, to discern between armoured skin and the occurrence of a horn (Fig. 3.6A). Not all protruding skin structures can be predicted with certainty, since some (e.g. warthog facial warts) lack any osteological correlates.

The tissues of horn coverings, beaks, nails and claws are composed of tough, cornified (keratin-rich) sheaths (Figs. 3.4C–D). These are suited for hard, abrasive functions since they are inert, frequently dense epidermal tissues that are ever-growing. Fortunately, a few correlates are left on the underlying bone from the cornified sheaths, such as rather extensive vascular grooves, which are often branching systems on other bones, although can occasionally be single grooves on

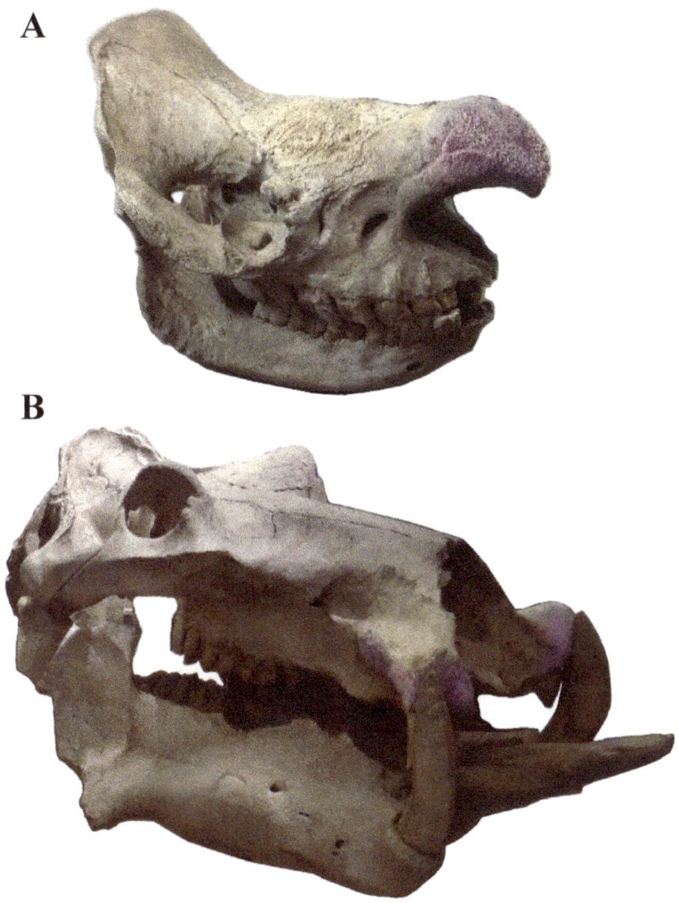

Fig. 3.6. Predicting epidermal proportions and armoured skin on extant species. (A) Skin protrusion (horns and crests)—ring-shaped rugosity distributex around the protrusion base (with supporting structure on rhinoceros skull); (B) Armoured skin—Rugosities dispersed around the front of the snout.

claws. These sheaths and their epidermal correlates may end suddenly, occasionally accompanied by a distinct "lip" on the bone, or they may grade into different tissues, resulting in reduced osteological scarring in a roughly gradational manner where this took place. This facilitates the accurate restoration of the sheathing's placement and distribution in fossil tetrapods. Among prime examples are fossil bird beaks and the bony horn cores of some extinct mammals such as arsinoitheriids (see pp. 68–9, 241–2). To verify or refute the occurrence of a claw sheath, one useful method is to search for characteristics including neurovascular foramina and vascular grooves on terminal **digits**.

Understanding the growth of cornified sheaths is helpful for estimating their presence and form in fossil tetrapods. Since it is made up of dead tissue, the cornified sheath has to develop at the junction between sheath and bone, with the addition of new keratin under older keratin in order to develop the sheath up from the inside out rather than by adding tissue to the exterior (Goss, 2012) (Fig. 3.7). Elongated soft tissue projections, including structures such as claws and horns, are formed when new layers of keratin displace the earlier layers of sheath from the bone. Once formed, sheaths are difficult to alter due to their inert nature, hence they are unable to cover body structures that get increasingly complicated with time. For example, a sheath cannot cover a bifurcating structure such as a deer's antler (which is bare bone once mature, with only temporary velvet tissue in early growth) or a pronghorn's horn (which uniquely sheds its keratin sheath annually, unlike true bovids) because bones cannot diverge beneath these hard, cornified tissues. Sheathed bones can only be reshaped gradually and in basic ways, with new internal layers of keratin displacing the sheath tissue and allowing for little changes in curvature or length. The skeleton can gain increased length and mass from the overlying cornified sheaths. Sheaths that develop quickly, including those on hooves, claws and beaks, generally augment the structure of the supporting bones into a similar but sharper shape. Massive grooves on the claws of giant, clawed mammals, such as ground sloths and chalicotheriids (see pp. 128–9), frequently serve as attachment sites for particularly large claw tissues. Yet, not every cornified sheath fits the contours of the underlying bone precisely. One case is the ridges of horns in bovids (see pp. 116, 118–9), as is the dramatic elongation of the lower jaw in skimming birds. Those peculiarities are produced by varying keratin accumulation rates and thicknesses, and are hard to estimate with certainty in fossil tetrapods.

Colouration

Once the skin, integument and horny structures have been restored, the colour pattern is usually the final element of a fossil tetrapod reconstruction to take into account. Sadly, these details are unknowable except for ultra rare instances in which they are preserved or can be reconstructed with reference to advanced analyses of fossilised pigments. Thus, colouration is among the most speculative components of our reconstructions, albeit one that cannot be avoided.

Taking colouration into account is assisted by some understanding of the ways tetrapods colour themselves, and how this is associated with habitat preferences and functional anatomy (Fig. 3.8). Biological pigmentation is the most prevalent method by which colour is produced. There are two main types of pigment present in animals. Melanin is the most abundant, and can be combined and diluted all over the integument and skin to produce shades of browns, dark reds, dull yellows, greys, and black. White colours and lighter shades are caused by little or no melanin. Camouflage requirements, along with a scarcity of carotenoids in **terrestrial** environments, was likely just as much the case in prehistory as it is today. The majority of bright colours present in terrestrial tetrapods (most vibrant in birds) have structural origins. Tetrapods without the ability to produce structural colour, such as the majority of

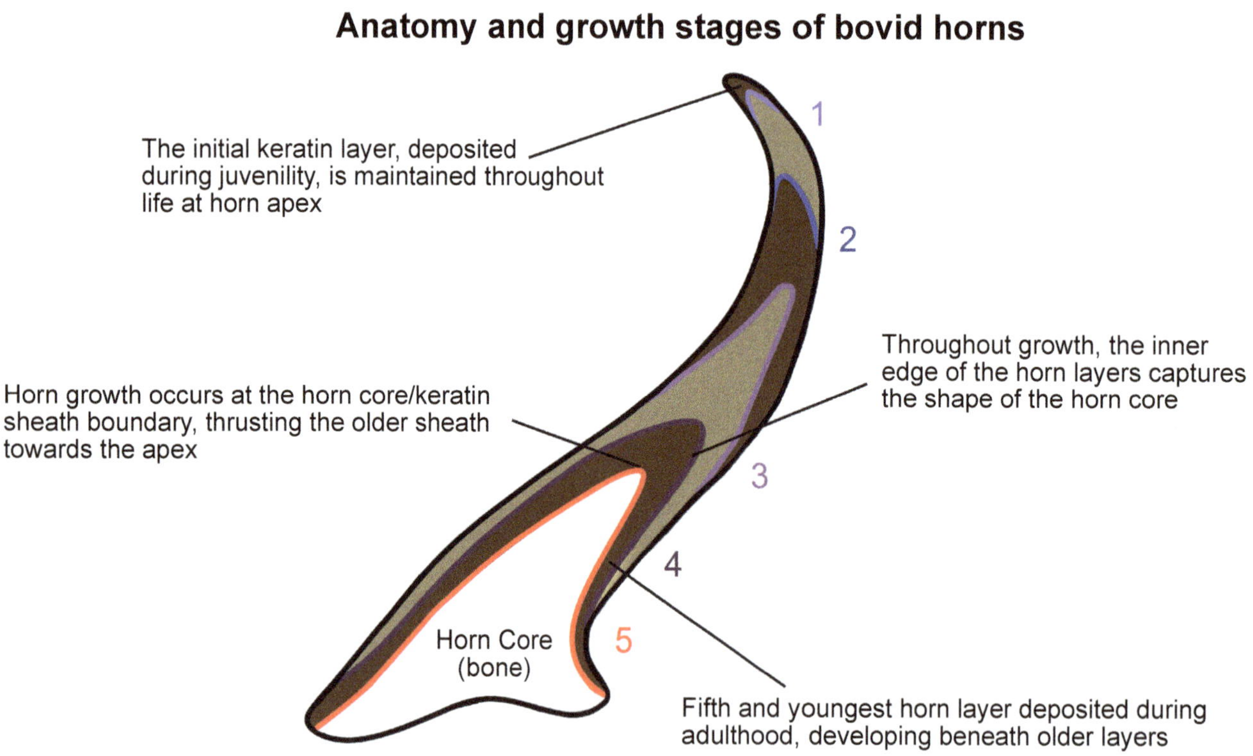

Fig. 3.7. A schematic overview of horn sheath growth in bovids.

mammals and likely most early humans, are therefore restored more credibly with melanin-compliant colours. Regarding feathered or furry tetrapods, shorter, more orderly filaments enable denser, more consistent patterning, whereas longer, denser pelts are more evenly coloured or irregularly blotched. In certain tetrapod groups, extra pigment types and structural colour can be added to melanin-based pigment patterns to create complex colour and shape configurations. The greatest influences on animal patterning and colouration include camouflage, which helps animals to blend in with their surroundings either to evade predators or conceal the approach to prey. This may explain the majority of colour diversity found in extant species. This may have also been the case for fossil tetrapods and, whenever pigments are fossilised and colouration can be identified, they typically follow modern camouflage strategies. Among these are countershading, in which the upper portion of the body is darker than the bottom, distributing shadows all over the body and decreasing the visibility of individuals from a distance (Vinther *et al.* 2016).

Fig. 3.8. The range and intricacy continuum of the colourations and patternings in modern tetrapods. (A) Vibrant colours, lack of earth tones; (B) Intricate and visually stunning patternings, multitude of earth tones; (C) Several earth tones used to generate basic patternings and regionalised colours; (D) Monochromatic colourations—generates visually stunning appearances with tight patternings and sharp contrast; (E) Uniform colours.

Among **ungulates**, species existing in open environments typically exhibit sharp distinctions between dark and paler areas, whereas more diffuse transitions occur in species existing in closed habitats such as forests. In an effort to blend in, tetrapods will frequently try to cover their eyes with dark stripes or patches to correspond to their inevitably dark pupils. It is possible that fossil tetrapods also employed all of these strategies.

Furthermore, pigmentation can mirror behavioural demands. **Diurnal** tetrapods are considerably more colourful than **nocturnal** ones since their complex colours are easily seen. Since dim light makes it more difficult to distinguish certain colours over the simple occurrence or absence of shadows, nocturnal tetrapods typically have neutral or subdued tones (whites, greys, blacks, browns). Whenever colours are present in nocturnal tetrapods, they mirror the needs for diurnal camouflage instead of being used in display. Other than making tetrapod bodies darker, melanin has numerous other purposes. Melanin is employed to reinforce tissues such as bird wings because it fortifies skin and integumentary structures against wear from both chemical and physical sources. Among other uses, melanin can absorb heat and light while simultaneously shielding tetrapod skin from ultraviolet radiation, which is ideal for **ectothermic** tetrapods that depend on surrounding heat to raise their body temperature. Tetrapods living in habitats exposed to intense sunlight, including deserts, typically have light-coloured, sparsely-pigmented skin that reflects light, and therefore heat, away from the body.

Selecting and implementing colour schemes can be somewhat challenging even when having all these things under consideration, but a carefully chosen colouration that takes the surrounding environment of the subject animal into account can make a reconstruction both memorable and very special. Additionally, patterning and colouration is very occasionally not lost by decomposition and fossilisation, the greatest examples being that of recently extinct Pleistocene megafauna uncovered from the Arctic permafrost (e.g. woolly mammoths, woolly rhinoceroses and steppe bison) or very dry caverns (e.g. giant ground sloths), in which mummified skin and fur can be preserved with the original colours retained. Restoring these mammals' colouration is further assisted by cave art of the same animals, where Palaeolithic painters recorded aspects of colour schemes which would normally be lost to time (Fig. 3.9).

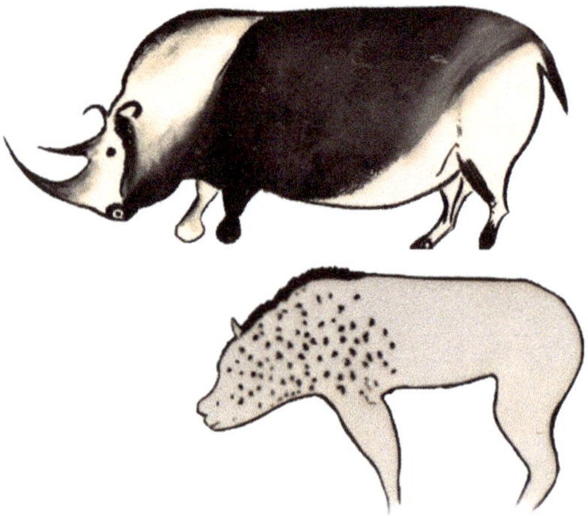

Fig. 3.9. Cave painting of a woolly rhinoceros (*Coelodonta antiquitatis*) showing a black transverse band across the middle of the body (top), and of a cave hyaena (*Crocuta crocuta spelaea*) (bottom), both from Chauvet Cave, southern France, dated at ca. 36 ka.

Behaviour

What we interpret as behaviour in extinct tetrapods is largely influenced by how modern relatives behave. Even without this assistance, inferences can still be drawn from the characteristics of a fossil. It is rarely easy to piece together the social behaviours of extinct tetrapods. While we can observe the behaviour of closely related species in their natural habitat, the majority of fossil tetrapods lack appropriate modern **analogues**. Thus, our hypotheses must be based on data from the fossils themselves. Our reconstructions can subsequently be guided by some general points drawn from our knowledge of modern tetrapods possessing similar features. In extant tetrapods, we can observe that specific body characteristics—such as form and size of structures including antlers and horns—played a part in social interactions between individuals of the same species. For example, modern elephant tusks—elongated upper incisors—are used by males in competitive shoving matches to gain access to mates. Likewise, deer and antelope employ their antlers and horns in contests to establish dominance within herds. By analogy, it is reasonable to infer that fossil mammals such as *Arsinoitherium* (see pp. 241–2), which lack close living relatives, may have used their horns in similar social or competitive behaviours.

It is not always necessary to restore fossil behaviour using comparisons to extant species—trace fossils (or groups of organisms fossilised together) provide insights into the lives and behaviours of extinct animals. These insights include the distance between animals moving in tandem, and how they sat and rested. Evidence pertaining to interactions between extinct species is predominantly biased towards indications of biting behaviours, including shed teeth and bite marks left in bones by **carnivorous** animals, or injuries caused by conspecifics during combat. Various other wounds, referred to as pathologies, indicate where animals suffered injury or contracted diseases. Pathologies are occasionally linked to particular diseases, which shed light on past events and are a fantastic source of inspiration for reconstructions. Tetrapod behaviour is nuanced and diverse, and while it sometimes involves aggression (Fig. 3.10), other times it is constructive, advanced, stationary and mundane. In fact, animals do spend most of their lives sleeping, bathing, grooming, feeding, and resting. In order to make our restorations scientifically credible, this is what our renderings of fossil tetrapods should depict.

Fig. 3.10. Life reconstruction of the head of the sabre-toothed cat *Smilodon populator* (see pp. 329–30), displaying his canine teeth, possibly to reinforce his existing status.

4. The Cenozoic

We now embark on a lively journey throughout the Cenozoic Era, the latest chapter in Earth's ongoing story of life and change. The beginning of the Cenozoic is marked by the Cretaceous-Paleogene (K-Pg) extinction event, the fifth major mass extinction to affect evolving life on Earth. This marked the end of over 150 million years of reptilian rule across the land and oceans throughout the preceding Mesozoic Era. Many species, including all the non-avian dinosaurs, along with various **terrestrial** and aquatic fauna and flora, became extinct in an event largely attributed to the impact of a large asteroid or other celestial body over six miles (10 kilometres) in diameter, crashing into a coastal area in what is now the Gulf of Mexico possibly at over 50,000 miles per hour (Fig. 4.1). With the energy of over one billion nuclear bombs, this asteroid left an impact crater (called the Chicxulub crater) in the Earth's crust more than 100 miles (160 kilometres) in diameter and more than 10 miles (16 kilometres) deep, in what is now the Yucatán Peninsula of southern Mexico. Seismic activity, tidal waves, volcanic eruptions and firestorms subsequently ravaged the world.

Fig. 4.1. Reconstruction of the Chicxulub impact in what is now the Yucatán Peninsula, Mexico, during the latest Cretaceous (ca. 66 Ma).

The asteroid hit an area of **carbonate** rock incorporating sulphur and combustible hydrocarbons, most of which was vapourised upon impact and thus, large amounts of sulphuric acid aerosols were ejected far up into the atmosphere to precipitate all over the world as rain acidic enough to kill terrestrial plants (Fig. 4.2A). This also caused ocean acidification, disintegrating plankton and other **calcareous** microorganisms in the oceanic surface waters and thus disrupting marine food chains. It is also possible that the roasting of the atmosphere by the collision itself led to the formation of nitrogen oxides which combined with water vapour to produce nitric acid rain, decimating the land and seas. Wildfires storming all over the Western Hemisphere were caused by hot debris thrown up by the impact that rained back down to the surface. This burning would have made the air even more contaminated by contributing soot and carbon dioxide to the atmosphere, heating up the planet via the greenhouse effect.

The amount of energy unleashed by the impact fed shockwaves that generated earthquakes, which in turn, triggered tsunamis that devastated coastal **ecosystems** and transported marine sand over enormous distances into continental interiors. The earthquakes also intensified volcanic activity, and they kept erupting for thousands of years. The dust cloud thrown up from the impact choked the atmosphere and blocked the sunlight, rendering the planet dark and cold for at least a decade (Fig. 4.2B). Globally, it is estimated that the land cooled by about 28°C and the oceans by about 11°C. Vegetation was inhibited from photosynthesising, woodlands collapsed, **herbivores** perished, and in turn, so did the predators. Whole ecosystems collapsed on both land and sea. Following the impact winter, the world underwent a phase of global warming caused by the evaporation of carbonates into carbon dioxide when the asteroid struck, whose prolonged presence in the atmosphere meant major warming would take place following the dissipation of more transient cooling gases. The warming temperatures hindered ecosystem recovery for several thousand years. The time it took for life to stabilise following the mass extinction varied depending on the location and groups of organisms. For instance, in the seas, the recovery of seafloor burrowing organisms took around 700,000 years, while it took millions of years for plankton to recover to anything close to their pre-extinction levels of biodiversity. On land, it took around four million years for biodiversity in South America to recover to healthy levels, while in North America it took around nine million years.

The asteroid was cataclysmic and permanently altered the trajectory of Earth's history. Incapable of surviving, approximately 75% of all plant and animal species became extinct. All terrestrial animals larger than a domestic dog were killed off. The most well-known casualties were dinosaurs, with solely a number of birds continuing their legacy to this day. In the seas, the Cretaceous-Paleogene extinction event wiped out almost all the large marine predatory reptiles (i.e. mosasaurs and plesiosaurs) and decimated sharks, bony fish, molluscs (including the total extinction of ammonites and belemnites), and numerous plankton species (which formed the base of the marine food chain). Among the animals that had survived were small mammals and lizards underground, small to medium-sized crocodylians, turtles and frogs underwater, and birds that had managed to fly to safer refuges. The mammals are commonly hailed as the mighty survivors in many accounts of the Cretaceous-Paleogene extinction event, the champions who took over the throne from the dinosaurs. This is accurate to some extent—mammals persisted, otherwise you would not be here to read this book, although recent studies reveal that this was a narrow escape, and their survival depended on what occurred during the months, years and centuries after the asteroid collision.

The asteroid was both a source of serious danger and a source of tremendous opportunity for the mammals. They were, rather unexpectedly, thriving during the latest Cretaceous, occupying a variety of ecological

Fig. 4.2. (A) Reconstruction of the asteroid impact during the Cretaceous-Paleogene extinction event approximately 66 Ma, from ground level. Any animal weighing above 10 to 25 kilograms (including non-avian dinosaurs such as the small herd of *Triceratops prorsus* (left), and pair of *Edmontosaurus regalis* (right), would die either during the impact or in the aftermath of it. **(B)** Reconstruction of the aftermath of the asteroid impact, in what is now North Dakota, United States, the devastation which had now undermined what was once the great kingdom of the dinosaurs. Yet, the dinosaurs were not the only large-bodied animals affected: in the skies, the last pterosaurs (top background) were now sealed to their fate, and in the oceans, all marine reptiles were now a memory. The debris of the shattered asteroid has ricocheted to various parts of the world, and burning pieces of rock start to crash down on regions including North America and Europe, causing widespread wildfires. The clouds of dust and ash from the catastrophe will also block out the sun for over a year, making it near impossible for plants to live. Yet, this desolate world retains survivors, including a small multituberculate of the genus *Mesodma*, patrolling the scorched landscape hunting for insects (bottom left). The millions of years that follow this hellscape will see a rampant recovery that will trigger the beginning of the Cenozoic Era. **(C)** Life reconstruction of *Purgatorius unio*, one of the closest fossils we have to the earliest placentals. Known mainly from teeth from the Early Paleocene of North America (66–63 Ma).

niches at the bottom of the dinosaur-ruled food web, such as **insectivores**, **florivores**, **omnivores** and bone crushers. Most of these animals were multituberculates (see pp. 45–6) or basal metatherians (see pp. 43, 46–7), while the eutherians (see pp. 43, 57) were relatively scarce. Over the last two million years of the Cretaceous, this situation remained continuous and there was no indication of any grave danger. Later, the circumstances suddenly changed. The sedimentary rocks that accumulated 66 Ma contain a thin line enriched with the element iridium, which is ubiquitous in comets, asteroids and meteorites, but scarce in the Earth's crust. This is the chemical blueprint of the asteroid, marking the transition from the Mesozoic to the Cenozoic Eras. There is no trace of non-avian dinosaur fossils above that line. Not only did the Cretaceous-Paleogene extinction event totally rearrange the food web, it also caused a mass extinction of most of the Cretaceous mammals that were around, and no mammal lineages were left unscathed.

Fossils from the US state of Montana show that by the earliest Paleogene, nearly every mammal that thrived in the region during the Cretaceous was extinct, with just seven species persevering. Various fossil sites in the region reveal what occurred during the following several hundred thousand years. There was a total number of approximately 23 mammal species in North America during this time. Out of all of these, just one was a metatherian—these marsupial progenitors underwent a near total extinction in the Northern Hemisphere after having been among the most diverse groups of mammals in the preceding Cretaceous. Taking into account the entire fossil record of Montana as well as additional data from Western North America, the overall picture is dire. Only seven per cent of terrestrial mammals survived the extinction (Longrich *et al.* 2016). The world which remained for these and other animals was ravaged and inhospitable, yet it is from this landscape that mammals rose to prominence. Within only 100,000 years of the mass extinction, local mammalian diversity had recovered and the number of species was approximately double that of the end-Cretaceous, then within 300,000 years, niche specialisation had started to take place.

The surviving mammals were smaller than the majority of those from the Cretaceous, and their dentition suggests that they had omnivorous feeding behaviours, whereas the casualties were larger-bodied and had more specialised feeding behaviours, either herbivorous or **carnivorous**. Although they were perfectly adapted to the most recent Cretaceous environment, their specialisations became weaknesses as the asteroid caused havoc. In contrast, the smaller generalists were better suited to consume whatever was available amid the turmoil that followed the asteroid impact. Eutherians, the placental precursors that previously played minor ecological roles in the Cretaceous, comprised most of the mammals that began to proliferate as ecosystems started to recover during the earliest Paleogene, having replaced the metatherians as the dominant mammals. Despite being broadly distributed across the Northern Hemisphere during the Late Cretaceous, over 50% of eutherian families went extinct completely. The precise number of placental lineages that persisted to leave descendants is unknown; the majority of estimates are about 10, but probably many more worldwide. Moreover, the anatomy of the surviving placental lineages is not directly known, since none are known from prior to the mass extinction. It can be assumed that they filled ecological niches as tiny, **nocturnal** insectivores.

No-one is so sure what enabled animals such as the oldest placental relatives to survive while numerous others died out. Certain elements of their life history may have assisted; small-bodied creatures require less food for survival, reproduce rapidly and have lots of young, a haphazard approach with a greater likelihood to be successful in an unpredictable environment. Early reproduction aids in adaptability, since less time is required for individuals to survive prior to replacing themselves numerous times over in the population. One factor that very likely played a role in the survival of numerous creatures was living in underground burrows, in which temperatures were more stable, thereby providing shelter from all of the intense heat, the fallout and the nuclear winter. There has been speculation that these mammals were also able to survive by exploiting the detrital food web. They were eating insects that could feed on detritus.

The earliest Cenozoic faunas consisted entirely of small-bodied mammals, including multituberculates, plesiadapiforms (see p. 150), basal carnivoramorphs (see p. 84), cimolestids (see p. 77) and **"archaic" ungulates** (see pp. 100–3). As can so far be deduced from their fragmentary fossils, none were larger than the largest mammals from the latest Cretaceous, being roughly the size of modern opossums or raccoons. Some of the earliest Cenozoic **communities** including those in Western North America lived in fern-dominated terrain, with several disaster **taxa** accounting for much of the biota, showing that the recovery of life was already underway 100,000 years after the asteroid strike. New eutherians emerged in what is now Montana and quickly became abundant. *Purgatorius* was a mouse-sized basal representative or close relative of the primate lineage, possessing rounded molar cusps for engaging in **frugivory**, and flexible tarsal joints for gripping onto and darting in trees (Fig. 4.2C). *Purgatorius*, or some other close relative, was our possible ancestor. Also known from Montana, one possible common ancestor to all the major **clades** of hooved mammals was *Protungulatum*, a small, insectivorous, tree-climbing eutherian that originated in the latest Cretaceous (Fig. 4.3A). Its phylogenetic affinities are controversial—many studies regard it as a placental according to its inner ear anatomy, whereas others consider it a **stem**-placental eutherian.

Among other enigmatic early placental genera scurrying through the vegetation in this earliest Paleocene Montana ecosystem were the terrestrial "arctocyonid" (see p. 101) *Baioconodon*, a coati-sized omnivore (Fig. 4.3B), and the cimolestid *Procerberus*, a slightly smaller animal resembling a large, predatory shrew (Fig. 4.3C). The earliest post-extinction mammal genera that were true herbivorous specialists included the fox terrier-sized periptychid (see pp. 180–1) *Mimatuta* (Fig. 4.3D). *Mimatuta* was low-slung and long-tailed, resembling a small badger but with a more domed skull, and had deeper lower jaws than *Baioconodon*, so this was a mammal accustomed to chewing. Immediately following the extinction of the non-avian dinosaurs, placentals such as *Mimatuta*, newly inflated in size thanks to an absence of large predators and an explosion in plant diversity, were the largest on land. This ecosystem was dominated by the minuscule ptilodontoid multituberculate *Mesodma* (Fig. 4.2B), comprising almost 75% of the community there. It possibly resembled a woodmouse, with prominent incisors, square jaws and a clambering locomotion,

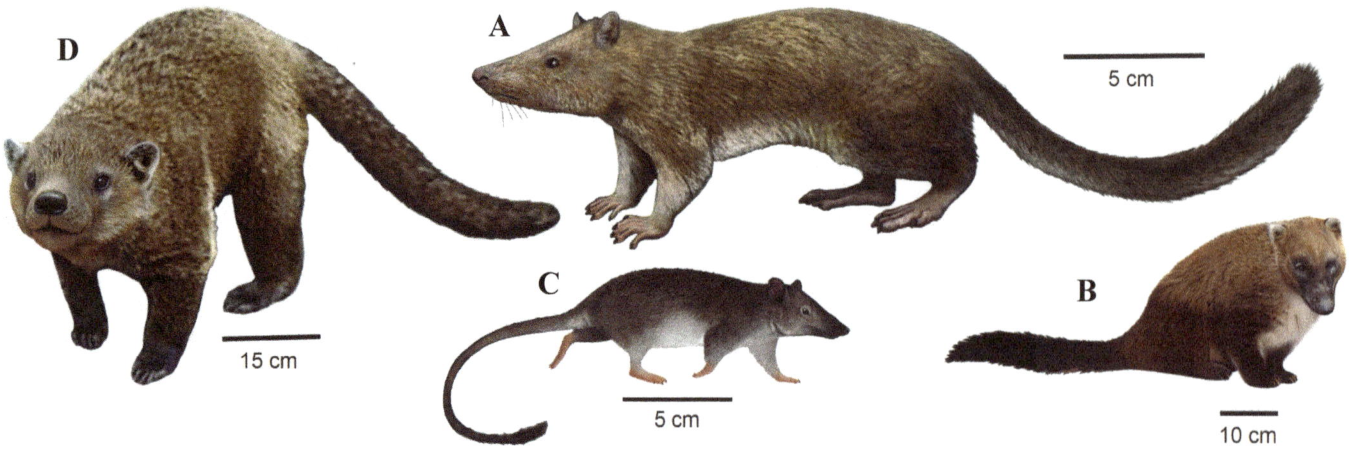

Fig. 4.3. Several examples of disaster taxa from the earliest Paleocene. (A) The possible early placental *Protungulatum coombsi*; (B) The "arctocyonid" *Baioconodon nordicus*; (C) The cimolestid *Procerberus formicarum*; (D) The periptychid *Mimatuta morgoth*.

but it had **plagiaulacoid** fourth premolars suited for pulverising tough woody stems. Thus, subsequent to a mass extinction, a turnover occurs, whereby new species emerge and the food web heals itself. It remains unclear precisely where the earliest placentals came from. Since they lack any clear Late Cretaceous forebears, this leaves questions regarding whether they merely evolved too quickly for the fossil record's frame rate to follow, or whether they originated from elsewhere, some place the fossil record has not documented, already headed in part towards an omnivorous diet, but obscured by the inconsistent preservation of environments over geological time. Perhaps their evolution happened off-camera, in some Cretaceous birthplace, diverging from each other and only ending up so different when a chance to broaden their geographic and ecological range presented itself? These are just several of the numerous unanswered questions concerning these enigmatic animals, and are frequently hard to answer conclusively.

Moreover, other major changes to the general ecosystem on land were in the works for tens of millions of years prior to the Cretaceous-Paleogene extinction event and include the Cretaceous Terrestrial Revolution (KTR). This event refers to a boost in the radiation of **angiosperms** and their coevolving pollinators, particularly insects, during the Middle to Late Cretaceous (ca. 125–80 Ma), which in turn fuelled the evolutionary and ecological divergence of mammalian lineages typical of Paleogene faunas. The various causes of the KTR included the continents drifting from one another during the Cretaceous following the fragmentation of the ancient supercontinent **Pangaea** in the preceding Jurassic, leading to the creation of new oceans and the expansion of temperate climates, stimulating angiosperm diversification. Increased tectonic activity created more montane habitats, which increased diversity in hot climates. The KTR was arguably still going on into the Early Paleogene with the continued increasing diversity of angiosperms, presumably up to 50 Ma, when they had attained ecological dominance. The KTR is regarded as a cause of explosions in mammalian ecological diversification by leading to new food sources (Grossnickle *et al.* 2019), driving radiations of insects that became food for insectivores and, especially during the Early Cenozoic, forming larger fruits and seeds that prompted additional diversifications in diet. Also, angiosperms are more productive than other plants, leading to a greater food availability overall. In fact, an escalation in niche partitioning among diverse lineages of mammals was facilitated by the general growth in complexity of angiosperm-ruled ecosystems during the Paleogene. It is throughout the Paleogene that diverse, angiosperm-dominated woodlands originated and the continued reorganising of terrestrial ecosystems took place (Fig. 4.4). The resilient survivors of the mass extinction proceeded to shape a new world whereby placentals predominated over all other mammals, and they started to attain unprecedented heights of anatomical disparities, giving rise to a host of species that inhabit the land, sea and sky. There was a planet to take over, broken ecological networks to mend, and while the surviving descendants of dinosaurs, the birds, would continue to be more prevalent (in overall mass) even to modern times, the mammals would typically occupy the highest position in the food chain. This new world, coextensive with the Cenozoic, is commonly referred to as the "Age of Mammals" since mammals were the animals that dominated large body size niches in both hemispheres on land. Yet, there are twice as many birds as mammals in regards to extant species. The Cenozoic is equally the age of birds, the age of bony fish, the age of codependent insects and angiosperms, and the age of grasslands. During this era, grasses played a highly significant role, forging the evolution of the mammals and birds that consumed them. In the Cenozoic, the diversity of modern vegetation started with the huge radiation of angiosperms, which was largely fuelled by the development of climatic zones around the world. Among bony fish, the spiny-finned teleosts underwent a massive radiation to become the world's most prevalent group of fish today.

The placentals, in particular, acquired new behaviours and traits, aiding them in adapting to the Early Paleocene havoc and exploiting vacant ecological niches. The development of numerous placental hallmarks occurred during this interval: advantages which contributed to shaping surviving generalists of the mass extinction into novel, disparate specialists. The success of placentals throughout the following 66 million years was driven by these features, which form the basis of our own human biology as well. The most important of these trademarks is the ability to produce well-developed offspring. Lengthened gestation gives placentals an advantage: soon after birth, newborns can frequently begin moving, socialising and obtaining food of their own. Larger young were able to develop into larger adults more easily, thus possibly allowing the earliest placentals to quickly grow in stature within several hundred thousand years of the extinction of non-avian dinosaurs, subsequent to 160 million years of being restricted to small sizes. It appears that the earliest placentals expanded in body size so quickly

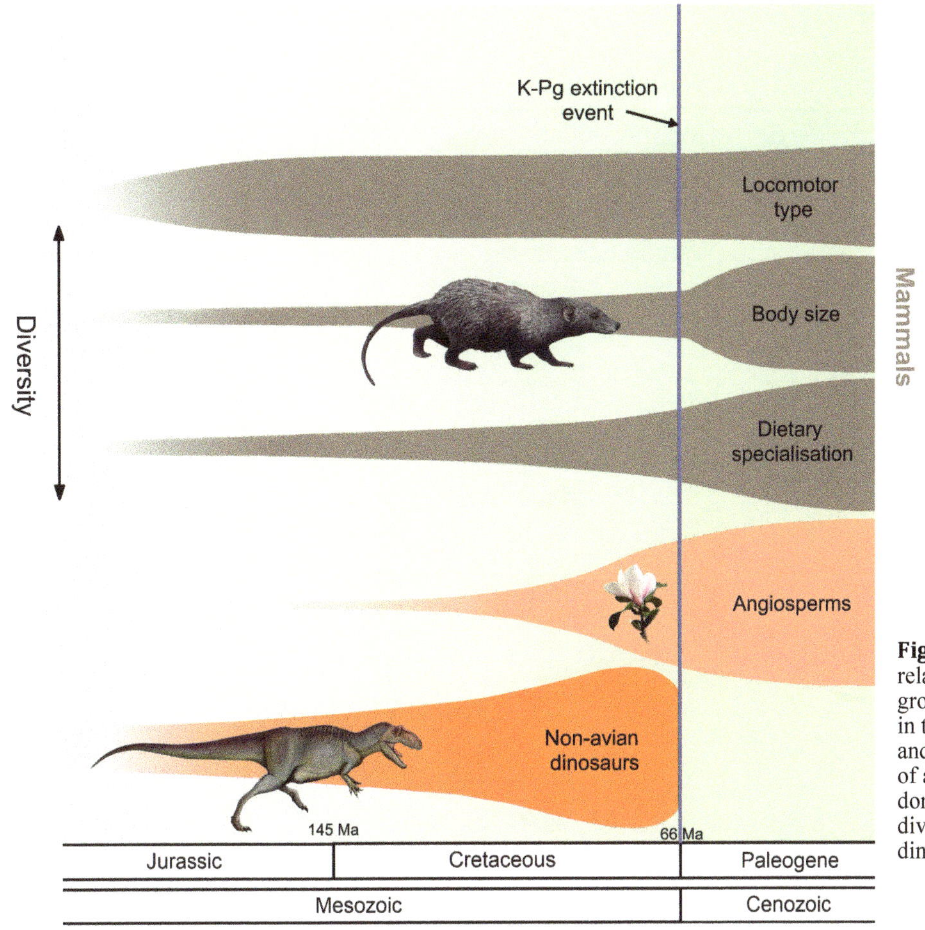

Fig. 4.4. Diagram displaying, relative to geological time, the growth of mammalian communities in terms of locomotor type, body size and dietary specialisation, the ascent of angiosperms to ecological dominance, and taxonomic diversification among non-avian dinosaurs. After Chen *et al.* (2019).

that their brains were, at first, unable to keep up. This discovery challenges the long-held belief that the absolute and relative sizes of mammal brains increased with time. At least initially, while there were plenty of vacant ecological niches to occupy, developing larger bodies was more vital than growing larger brains. In an unstable world of bountiful opportunities, large brains could have been disadvantageous due to being a metabolically expensive organ. In time, as ecosystems attained stability and the multiple newly evolved placentals experienced intensified competition, they developed larger brains. This expansion occurred largely in the neocortex, a part of the cerebrum concerned with greater sensory perception and cognitive functioning. However, that growth would not happen until the Eocene, when the "archaic" placentals gradually dwindled in diversity and the modern placental groups (e.g. bats, whales, horses) began dominating the world.

Finally, it is during the Cenozoic that the Earth itself gained its current look. The continents slowly moved to their modern positions in a process called plate tectonics, as Pangaea gradually broke up. Only a portion of Pangaea was separated by the end of the Mesozoic. The opening of the North and South Atlantic oceans date back to the Jurassic and Early Cretaceous Periods, respectively, although the split of Australia from Antarctica only commenced by the Middle Eocene. Throughout the Cenozoic, the Atlantic and Indian oceans continued to open, whereas contrary to the aforementioned oceans, the ancient Tethys Ocean separating Africa from Eurasia gradually closed. Antarctica remained centred on the South Pole, while the moving continents converged in a northward direction. The moving continents and the onset of glaciations in Antarctica led to changes in major ocean currents and weather patterns, which greatly affected the planet's climate. Whereas most of the Earth was relatively warm throughout the preceding Mesozoic, with a uniform greenhouse climate, the tendency during the Cenozoic is that the climates generally became cooler, drier and more variable, leading to the great glaciations of the Late Neogene and Quaternary Periods (Fig. 4.5). Despite this trend, there were separate events of global warming in which the climate was warmer than today's, most notably during the Early Eocene Climate Optimum (see p. 197), when Earth was even warmer than it had been during the Cretaceous, and later during the Pliocene. The locations of global landmasses also affected the dispersal of mammals between the continents, with mountains rising, and land bridges appearing and disappearing according to sea level changes and continental shift multiple times throughout the Cenozoic (Fig. 4.6). The Cenozoic saw the rise of many of Earth's great mountain ranges, including the Alps, the Himalayas, the Andes, and the Rocky Mountains of Western North America, as plates converged with one another. Life took modern shape throughout the Cenozoic, not only on land but in the oceans too, with all the major mammal lineages already existing by the beginning of the Neogene. Carnivores and herbivores evolved side-by-side, coping with competition and the ever-changing climates. A major event in the evolution of plant life was the expansion of grasslands, which began in the Late Paleogene Period (ca. 40 Ma) and accelerated during the Neogene Period, and life by the end of the Neogene was more diverse than ever. The large ice sheets that eventually covered much of the Northern Hemisphere in the Late Neogene then underwent a major expansion by the start of the Quaternary. Back then, the planet traversed a thermal threshold which triggered an interval of glacial-interglacial cycles (see pp. 311–3), a situation in which the planet remains locked at present. In comparison to other geological eras throughout Earth's history, the Cenozoic was relatively complex regarding geography and climate and thus, so was the development of its fauna.

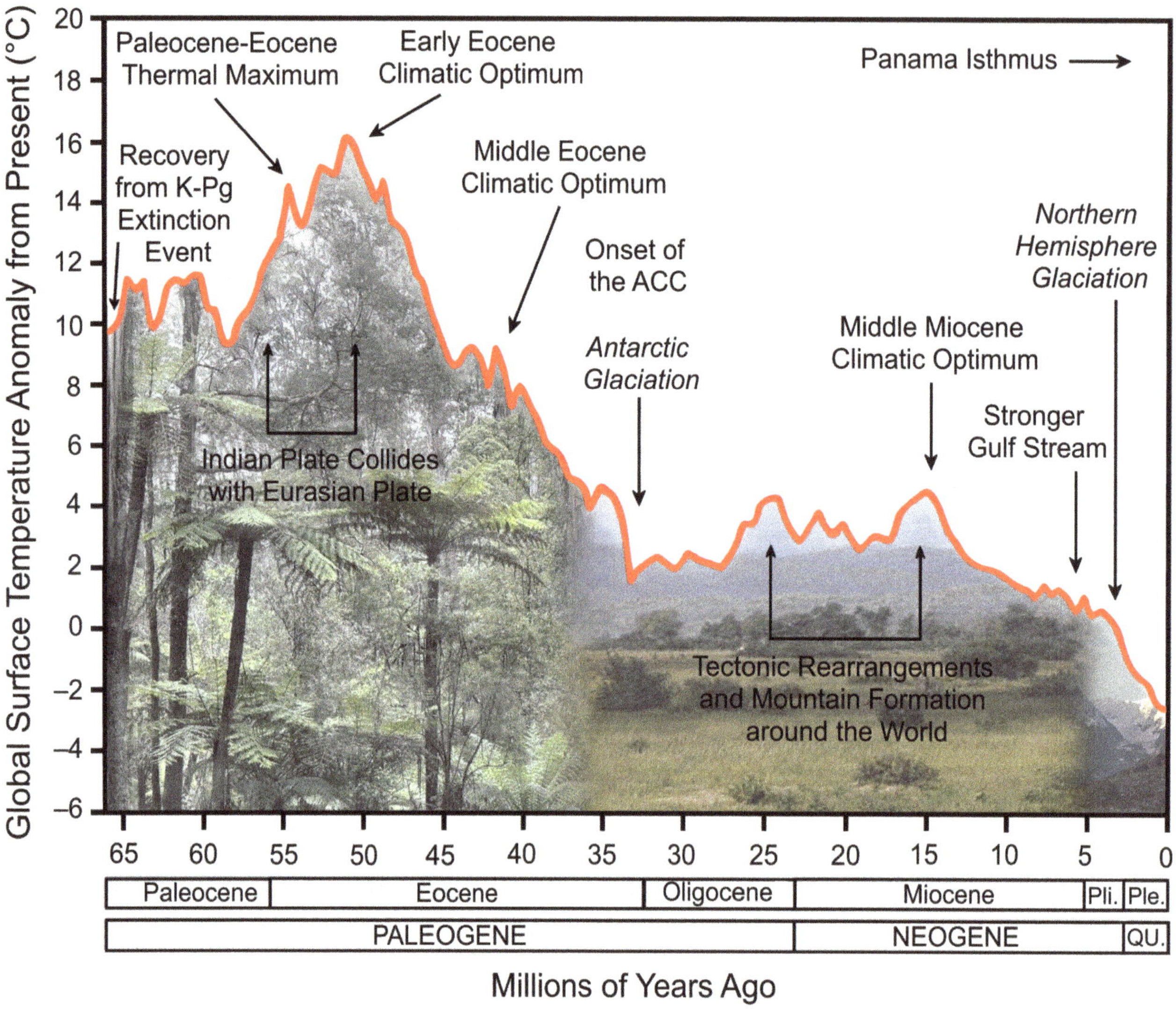

Fig. 4.5. Changes in the average global surface temperature, and major events in Earth's history, throughout the Cenozoic Era.

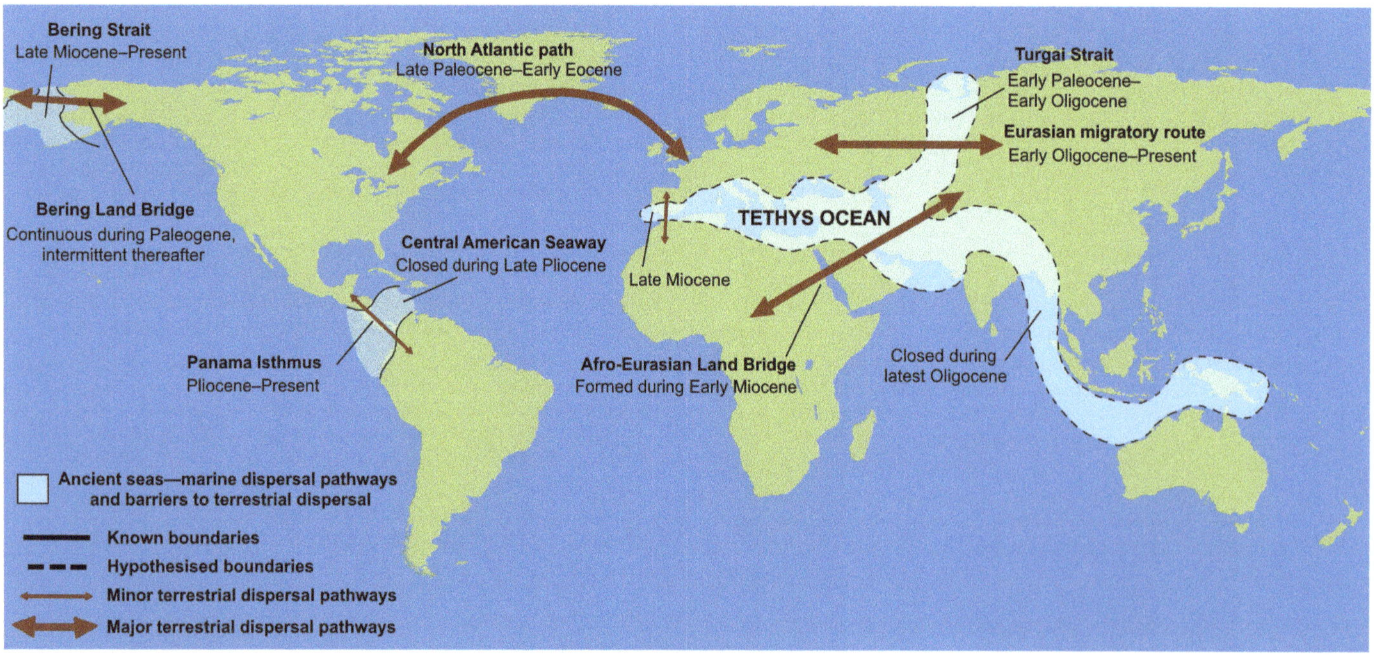

Fig. 4.6. The main faunal dispersal pathways and barriers within the Cenozoic Era.

5. The Paleogene
66–23.04 Ma

After the Cretaceous-Paleogene extinction event, myriad new forms of life emerged. At the beginning of the Paleogene Period, the mammals, reduced to tiny generalised **insectivores** or **omnivores** immediately following the extinction, began to thrive, quickly changing in order to take advantage of the ecological niches left vacant by the non-avian dinosaurs. These changes included an increase in taxonomic diversity and body size, the development of new locomotion styles, and a radiation in dietary diversity to incorporate plant material and other vertebrates. Within the Paleogene Period, the most diverse mammal groups throughout the Paleocene Epoch were **stem** members of extant **clades** (Grossnickle *et al.* 2019), but by the Early Eocene Epoch, the most diverse mammals were representatives of most modern mammalian orders (see p. 197). Also, during the Paleogene, reptiles had adopted more familiar forms and had changed little. The marine biota, hard hit by the mass extinction, gradually rebounded and had increased in abundance by the end of the Paleocene. From the Paleocene to Early Eocene, sharks dominated the oceans as the largest aquatic predatory vertebrates, to replace the extinct mosasaurs and plesiosaurs until fully aquatic whales evolved by the Middle Eocene. Compared to animal life, plant life was less affected by the mass extinction, with many plants persisting and continuing to radiate. One of the most significant innovations in Paleogene vegetation was the emergence of **closed-canopy forests** made up of giant **angiosperm** trees.

Throughout the roughly 42.97 million years of the Paleogene Period, the world underwent profound changes in climate, during which time there were several cool and warm cycles spanning about 10 million years. After declining at the end of the Cretaceous, global average temperatures rose again during the Early Paleogene, then significantly increased by about 5–8°C approximately 55.5 Ma, in an event called the Paleocene-Eocene Thermal Maximum, or PETM (see p. 197). Over a brief interval between only 10,000 and 30,000 years, high-latitude ocean surface temperatures increased by approximately 9°C, whereas those at low latitudes surged by 5°C. Tropical rainforests shrouded a large portion of the Earth as a result of higher temperatures combined with higher precipitation (Fig. 5.1). Even the polar swamps were sufficiently warm to support pond turtles, crocodylians, tapiroids (see p. 128), lemur-like primates, and numerous extinct Eocene lineages.

These conditions lasted for approximately 170,000 years, before temperatures swiftly decreased back to previous levels, then started to climb again for several million years during the late Early Eocene. Subsequent to this, a cooling trend started that would eventually lead to the beginning of major glacial episodes during the Quaternary about 50 million years in the future. This temperature decrease happened gradually and slowly until, around the Eocene-Oligocene transition, **mean annual temperatures** dropped considerably, by about 8°C over an interval of 400,000 years. Mean annual temperatures decreased from approximately 20°C to only 11°C. While the climate grew cooler, it grew drier as well. Grasses, which were better adapted to the cooler climates, evolved and dispersed (Fig. 5.2).

Fig. 5.1. A tropical forest **ecosystem** in North America during the Middle Eocene (Uintan NALMA (see p. 1), 46.2–42.0 Ma). *Centre*: the machaeroidine oxyaenid (see p. 82) *Apataelurus kayi* hunting the basal equid *Epihippus gracilis* (see pp. 132–3); *Top left*: a pair of the rodent *Paramys* sp. (see pp. 205, 209); *Centre background*: the brontotheriid (see pp. 135–6) *Metarhinus fluviatilis*, and immediately to the right, the gastornithid bird (see p. 36) *Gastornis (Diatryma) gigantea*; *Top right background*; adapiform primates (see p. 150) of the species *Notharctus tenebrosus* (see pp. 205, 209) leaping from tree to tree.

Throughout the Paleogene Period, the world's geography started to adopt a more familiar shape, as new oceans expanded the spaces between the continents, and new ocean currents formed. Africa and South America spread further apart, expanding the South Atlantic. The North Atlantic expanded too, and the progenitor of the Gulf Stream (a swift Atlantic current), started to get stronger. Among mountain-building events was the formation of the Rocky Mountains in Western North America. The Tethys Ocean (see p. 174) started to shrink as Africa started travelling northwards, uplifting the Alps. As continents broke up, the isolation of individual landmasses including South America and Australia took place, enabling their plants and animals to evolve independently. Europe and Asia were separated during the Paleocene and into much of the Eocene by a shallow sea that extended through the eastern margin of the Urals—the Turgai Strait (see p. 174)—and connected the Arctic Ocean to the Tethys Ocean. This limited the range of a number of animal populations.

Antarctica remained attached to the southern tip of South America until around 56 Ma, when the two continents started to drift apart. Then by roughly 40 Ma, Antarctica and Australia had become separated by an emerging expanse of water that still exists today, called the Tasmanian Passage. At some point between 36 and 23 Ma, a seaway called the Drake Passage formed off the tip of the Antarctic Peninsula. As time passed, the stage was set for a major new force in the Southern Ocean known as the Antarctic Circumpolar Current (ACC), which still circulates around Antarctica today and is the world's most powerful ocean current. This swirling current that is driven by winds and unhindered by land blocks warmer waters further north, preventing them from nearing the mainland. In addition, it dredges cold water from the seabed to the surface.

These two factors collaborated to induce a chilling effect on Antarctica. The ACC is believed to have formed between 41 and 23 Ma. However, it has been disputed as to exactly how the formation of this current affected the decrease in temperatures—and the increase in glaciation—on Antarctica. The decrease in temperatures was brought about by a decrease in atmospheric carbon dioxide levels probably because vast amounts of it were being buried in seafloor sediments or absorbed by marine plankton. This may have been partly responsible for the global cooling trend, and the onset of the ACC could have led to a further decrease in Antarctic and global temperatures.

From 45–32 Ma, carbon dioxide levels declined from between 1500 and 1000 parts per million to 720 parts per million, resulting in global cooling. For example, Lauretano *et al.* (2021) found that average annual temperatures in southeastern Australia gradually declined from 27 °C (±4.7 °C) during the Middle Eocene to 22–24 °C (±4.7 °C) during the Late Eocene, followed by a roughly 2.4 °C-step cooling across the Eocene-Oligocene boundary. This decrease in atmospheric carbon dioxide to below 750 parts per million and the resulting deep-ocean cooling by 4°C played a critical role in Antarctica glaciation (DeConto *et al.* 2008). Despite a few fluctuations, the overall Oligocene trend was towards cooler temperatures and increasing ice cover which correspond to decreasing levels of carbon dioxide in the atmosphere (Zachos *et al.* 2001). Cooling would have gradually lowered the snowline elevations, which would have slowly covered more of the continent's surface. After a point, a threshold was reached or climate feedback related to snow/ice-albedo and ice-sheet height/mass-balance would have initiated rapid ice sheet growth during times of favourable orbital periods (Pollard & DeConto, 2003).

The Paleogene ended with a relatively temperate global climate and favourable conditions for the spread and diversification of many life forms introduced earlier during this period.

Fig. 5.2. A wooded grassland ecosystem in North America during the Late Eocene (Chadronian NALMA (see p. 1), 37.2–33.9 Ma). *Centre middleground*: a flock of oreodonts (see pp. 107–8) of the species *Merycoidodon presidioensis* is being hunted by the large hyaenodontid (see pp. 82–3) *Hyaenodon (Neohyaenodon) megaloides*; *Bottom left foreground*: small oreodonts of the species *Merycoidodon gracilis*; *Centre background*: brontotheriids of the species *Megacerops coloradensis* (see pp. 135–6) forage in the background.

Paleocene
66–56 Ma

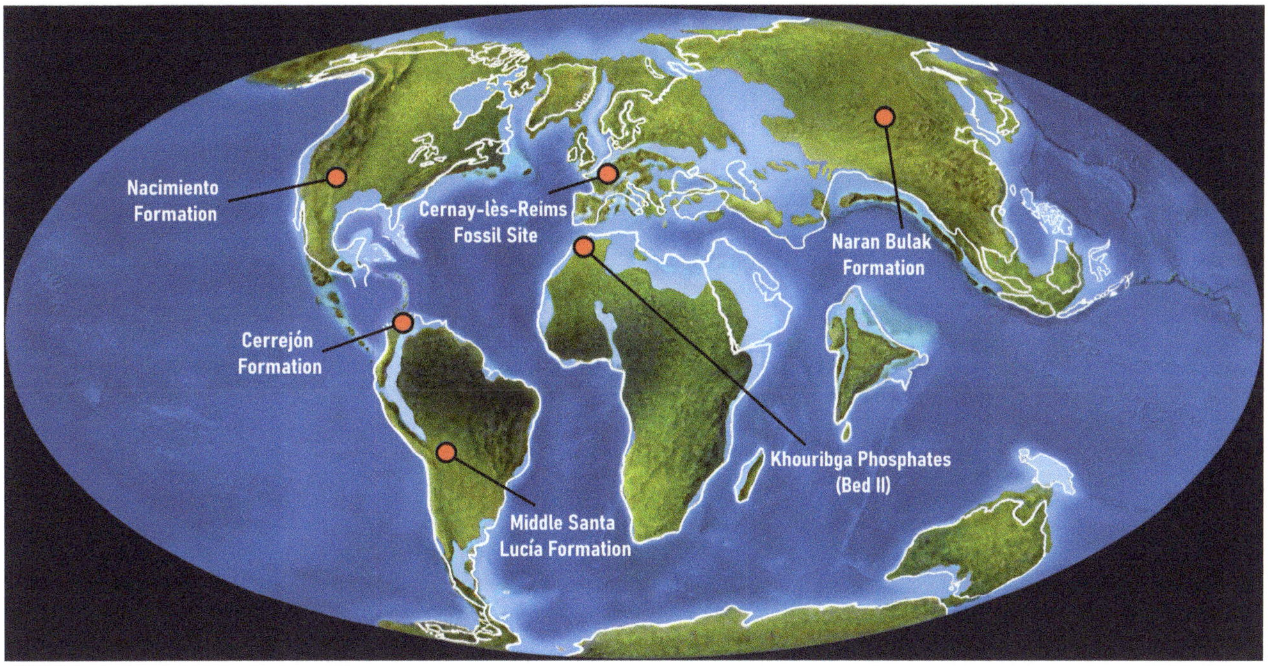

Fig. 5.3. Palaeogeographic map reconstruction of Earth during the Paleocene, with the position of the modern continents outlined and the approximate location of the fossil sites indicated.

In the aftermath of the Cretaceous-Paleogene extinction event (see p. 169), the full recovery of the Earth's natural cycles occurred during the first several million years of the Paleocene Epoch. The Paleocene climate was, much like in the preceding Cretaceous, tropical or subtropical, with global average temperatures being approximately 25°C. The poles remained temperate and lacked permanent ice sheets. Analysis suggests that throughout the Paleocene, global temperatures oscillated from cooler, to warmer and cooler again by the close of the epoch. A humid climate predominated and the prevalence of **arboreal** plants suggests that **closed-canopy forests** extended over much of the continents. There were also dry-wet variations as evident from large quantities of salts and gypsum that accumulated after ancient basins and floodplains, including the Tremp-Graus basin in the foothills of the Pyrenees, and the floodplains of what is now Tiupampa in Bolivia (see pp. 184–5), underwent desiccation.

During the Paleocene, the arrangement of the continents was already similar to that of today. Since the fragmentation of the ancient supercontinent **Pangaea** in the Early Triassic, various new ocean basins, including the Indian and Atlantic Oceans, had opened up throughout the Mesozoic. By the start of the Paleocene, the fragmentation of **Gondwana**, the southern portion of Pangaea, was well underway but not yet complete. South America and Australia remained connected to Antarctica (which was roughly in its present position). India and Africa, the other ancient Gondwanan "remnants", were a great distance from all these southern landmasses as well as from their current position near Eurasia. In the Paleocene, they drifted while remaining separated from the other continents, surrounded by the Tethys Ocean and the widening Indian Ocean. As a long, east-west trending maritime belt, the Tethys Ocean isolated Eurasia from India and Africa, and South America from North America. Thus, a continuous equatorial current, primarily driven by trade winds, flowed from east to west across the Pacific, Indian, Tethys and Atlantic Oceans, playing a crucial role in distributing heat and moisture around the globe. Europe and North America were still linked via Greenland, as supported by similarities in the Paleocene and Early Eocene fossil records of both continents. The Bering Land Bridge (see p. 174) linked North America to East Asia. The initial evolutionary diversification that resulted in Paleocene mammal diversity started long before the Cretaceous-Paleogene extinction event, and an assorted eutherian (see pp. 43, 57) mammal fauna had already evolved by the Late Cretaceous, comprising various shrew-sized **insectivores** and some **primitive** primate- and **ungulate**-like **taxa**. These eutherians joined other formerly prosperous mammal lineages including the metatherians (see pp. 43, 46–7) and multituberculates (see pp. 45–6). The metatherians were severely impacted by the Cretaceous-Paleogene extinction event possibly due to dietary specialisation, while the more generalised eutherians and multituberculates crossed this boundary with fewer losses (Pires *et al.* 2018). Throughout the Paleocene, mammals quickly radiated in body form and diversified into an unprecedented array of insectivores, **herbivores** and predators, although the largest were still no larger than a bear; the real mammalian giants would not evolve until the Middle Eocene. Alongside the new and hardier fauna of the Paleocene, new forms of vegetation other than palms and ferns became significant components of **terrestrial ecosystems**. The oldest trees of the walnut family and the oldest bean pods arose during the Early Paleocene—their nutritious seeds providing proteinaceous supplements to the diet of mammalian herbivores whose recent ancestors were largely insectivorous.

North America Nacimiento Formation, United States: Danian-Selandian, 65.7-61 Ma.

The Nacimiento Formation, located in the San Juan Basin of New Mexico, preserves one of the world's richest fossil records of mammal diversification from shortly after the Cretaceous-Paleogene extinction event (see p. 169). These fossils show that mammals increased in average body size throughout the Paleocene, with radiations into mostly **omnivorous** and **herbivorous** dietary niches. Most of the mammals are known largely from teeth and fragmentary fossils, but determined efforts to discover more complete skeletal material in recent years have significantly improved our understanding of the biology and ecology of these animals. The San Juan Basin was near sea level and covered with lush, densely-vegetated seasonal tropical forests receiving about two metres of rain per year. These were essentially rainforests, with rainfall probaby concentrated in the summers—monsoonal. The flora of the San Juan Basin, consisting primarily of **angiosperms**, was very diverse compared to contemporaneous floras described from higher latitudes. The **ecosystem** was dominated by **C3 vegetation**, suggestive of an environment with moderate sunlight intensity and temperatures, and high carbon dioxide concentrations, along with plentiful groundwater.

In the Late Cretaceous, mammals were probably fairly diverse, being mostly rat-like to badger-like in general body form, but by the Early Paleocene, they had morphologically diverse dental features and more specialised ecologies and diets, although most were not much larger than a marmot (*Marmota*). Only several small (opossum-sized) metatherians (see pp. 43, 46–7) are known from the Early Paleocene of North America, mainly placed in the genera *Swaindelphys*, "*Peradectes*" (see pp. 212, 216) and *Thylacodon*. The eutherians (see pp. 43, 57), particularly the placentals, were by far the most taxonomically and morphologically diverse mammals of the Nacimiento Formation. The Paleocene of North America saw numerous unusual "**archaic**" mammalian lineages which would fail to leave any modern descendants. These included many small mammals from the Early Paleocene, which have frequently been classified in the **polyphyletic** group "Insectivora" (see p. 75) since they share numerous **primitive** traits, including small body size and sharp, somewhat high-crowned teeth hardly different from their **insectivorous** Cretaceous forebears. Multiple genera are typically split among a few families including the Leptictidae (see p. 77), Pentacodontidae, Palaeoryctidae, and Cimolestidae (see p. 77).

The leptictid *Prodiacodon* likely resembled an elephant shrew, with an elongated skull and lower jaw. *Prodiacodon* was probably a small, agile insectivore, with clear **adaptations** for **saltatorial** locomotion. The forelimb morphology indicates that *Prodiacodon* may have also been adapted for digging. The medium-sized pentacodontid *Pentacodon*, about the size of a Solenodon (see pp. 75–6), is distinguished by considerably enlarged and nearly **molariform** lower fourth premolars, possibly used in crushing molluscs and arthropods. One "archaic" group which have uncertain affinities, arose in the Late Cretaceous and survived the mass extinction are the palaeoryctids, which had a generalised body plan and primitive tooth arrangement. The small, shrew-like *Palaeoryctes* (Fig. 5.4), which is known from an almost complete skull, had molars adorned with tall, pointed cusps ideal for puncturing insect **carapaces**, in contrast to the rodent-like cimolestid *Puercolestes*, a survivor of the mass extinction which had a dentition suited to a more generalist diet for crushing, puncturing and shearing.

The largely herbivorous pantodonts (see pp. 137–8) consisted of some of the earliest large mammals, represented by *Pantolambda* in North America. *P. bathmodon* is the smallest known pantolambdid, while *P. cavirictum* was considerably larger—at least 150 kilograms. *Pantolambda* was an omnivorous, wolverine-sized **plantigrade** that was likely **gregarious**, to afford protection from predators, and may have been **semi-aquatic**, an ecology that has been deduced for larger pantodont species from the Eocene. Ontogenetic studies of numerous individuals, along with dental wear analysis, suggests an unusually short lifespan (only seven years) for *P. bathmodon* in spite of its moderately large body size (around 17 kilograms) (Funston *et al.* 2021). Among other "archaic" placental groups were some rather interesting specialists. The taeniodonts (see p. 138) (Fig. 5.6) were a North American **endemic** lineage remarkable for their physical adaptations to a digging lifestyle, possibly first appearing in the Late Cretaceous. Their burrowing habits helped taeniodonts survive the mass extinction as some **primitive** forms such as the Early Paleocene *Onychodectes* and *Wortmania* reached larger sizes, but were still relatively generalised in body form. *Onychodectes*, which possibly resembled a large opossum, had a long, narrow skull with moderate-sized canine teeth. *Wortmania* reached weights of roughly 20 kilograms, possibly as soon as 500,000 years after the Cretaceous-Paleogene extinction event. *Wortmania* was a stylinodontid and possibly ancestral to the later (Middle Paleocene) *Psittacotherium*. It appears that all taeniodonts had adaptations for scratch-digging. These adaptations were more extreme in the stylinodontids, but the

Fig. 5.4. Reconstructed life appearance of the palaeoryctid *Palaeoryctes puercensis*. Reconstructed length: 12 cm.

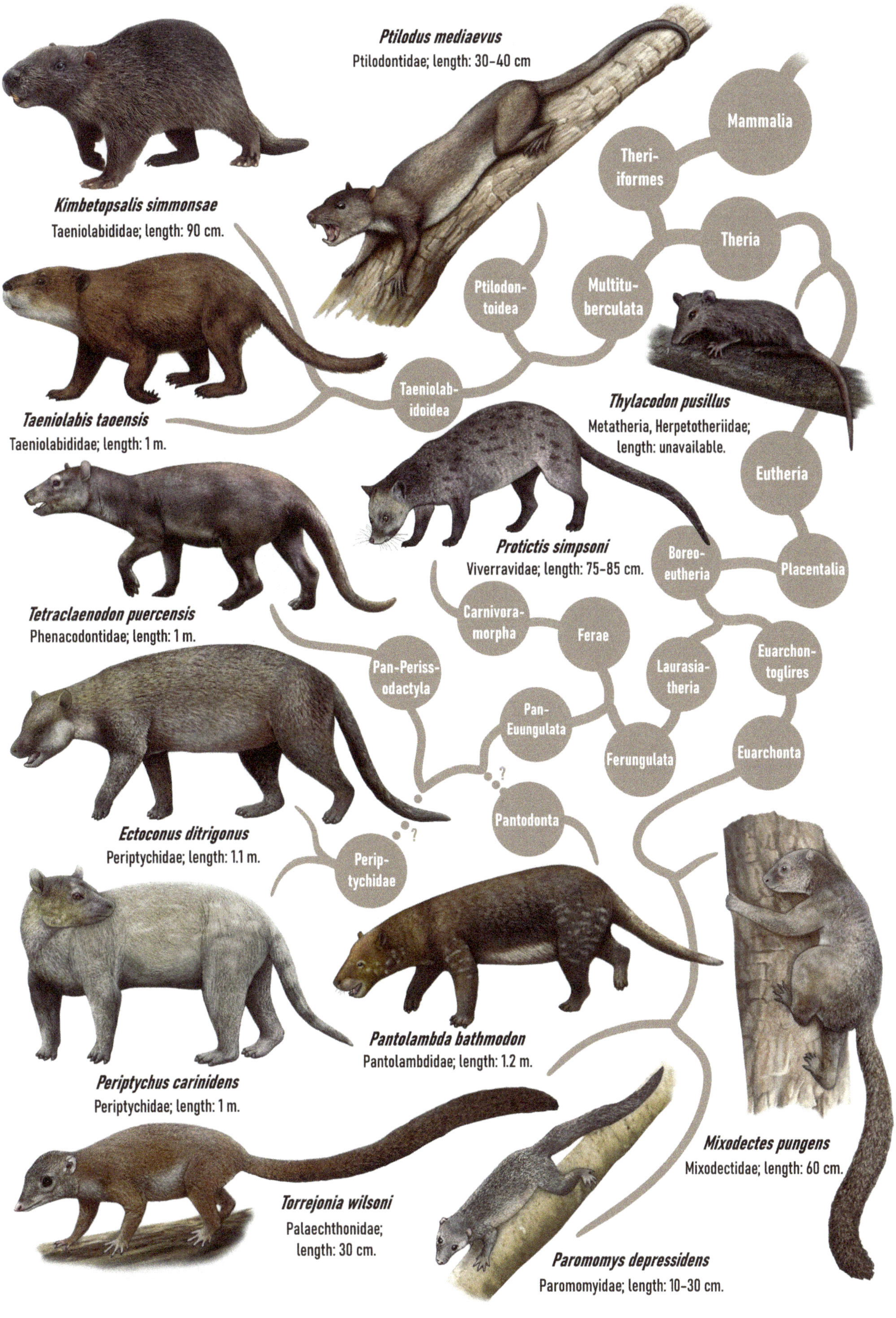

Fig. 5.5. Phylogenetic tree of a sampling of fossil mammal taxa, particularly multituberculates, metatherians, basal carnivoramorphs, herbivorous "archaic" **ungulates**, and euarchontans, from the Nacimiento Formation. The phylogenetic position of Periptychidae follows Puschel *et al.* (2019), that of Pantodonta follows Kort (2023) (the phylogeny of the study which indicates pantodonts are potential close relatives, or **stem** members, of Euungulata (p. 100)), and that of *Tetraclaenodon* follows Cooper *et al.* (2014) and Bai *et al.* (2021). The phylogeny within Euarchonta follows Chester *et al.* (2025).

North America Nacimiento Formation, United States: Danian-Selandian, 65.7–61 Ma, cont.

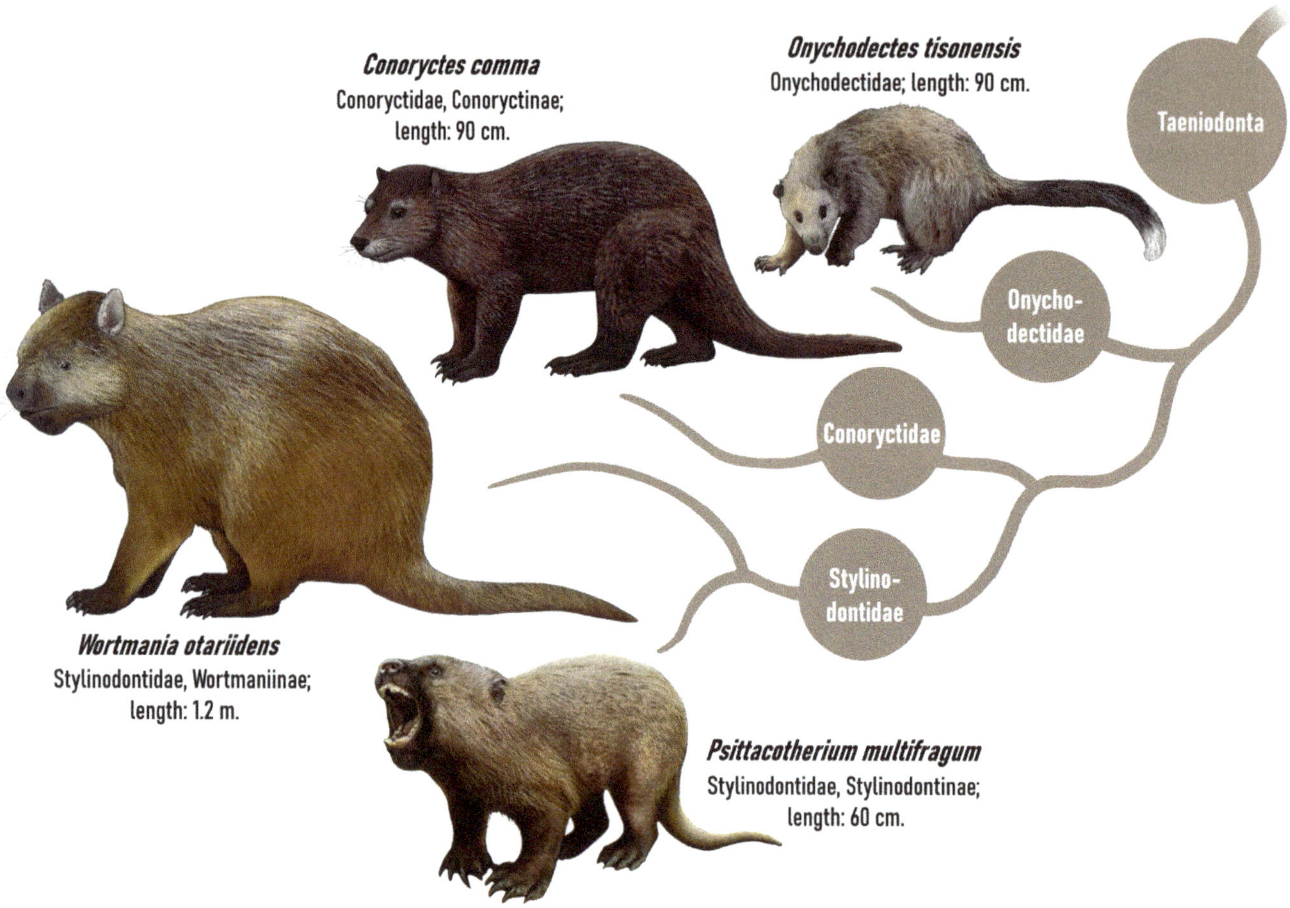

Fig. 5.6. The in-group phylogenetic relationships of taeniodonts from the Nacimiento Formation, after Williamson and Brusatte (2013).

Middle Paleocene *Conoryctes* had more specialised adaptations than were previously described (Kynigopoulou *et al.* 2024). *Conoryctes* had very powerful forelimbs, but also other specialisations found in extant scratch-digging mammals such as an expansive **distal** humerus for attachment of flexors and extensors of the forelimb, an ulna morphology that enabled better mechanical advantage, locking joints of the metacarpals to prevent the bones from overextending, and greatly enlarged and deep **unguals**. *Psittacotherium* had extreme **fossorial** adaptations (e.g. powerful forelimbs, large, deep claws). It also had a more highly **derived** dentition consisting of larger canine teeth and **cheek teeth** that became rootless, supported by strong, deep jaws, possibly a response to a more abrasive diet of tough vegetation, if the animals were feeding on underground roots and stems. *Wortmania* was intermediate in its fossorial adaptations between more gracile generalists such as *Onychodectes* and the larger (50 kilograms), more robust and more derived stylinodontids such as *Psittacotherium*.

Various lineages of "archaic" ungulates, which were previously grouped in the **wastebasket taxon** "Condylarthra" (see pp. 100–3), include the placental mammals most common as fossils, having typically retained **pentadactyl** hands and feet. The relationships of many of those lineages are not well-understood and they remain in flux. One lineage that is best represented in North America are the "arctocyonids" (see p. 101), which in Early Paleocene New Mexcio, ranged from poodle-sized (e.g. *Mimotricentes*, *Loxolophus*) to roughly wolf-sized (e.g. *Protogonodon*, *Deuterogonodon*, *Claenodon*). Some medium-sized **taxa** such as *Loxolophus* and *Chriacus* are known from nearly complete skeletons. *Loxolophus* (which likely appeared only around 300,000 years after the Cretaceous-Paleogene extinction event), was possibly capable of climbing trees. Similarly, *Chriacus* is also believed to have been **arboreal**, as it had a long, balancing tail and powerful limbs. It was an omnivorous "arctocyonid" that looked something like a South American coati in both size and appearance. Scans of the skull reveal some large **olfactory** bulbs relative to its size, showing an animal heavily reliant on the sense of smell, so possibly a primitive character among early ungulates. Certain derived characters of the inner ear also indicate that *Chriacus* may lie near the origins of the whole ungulate radiation, although it was not directly ancestral (Bertrand *et al.* 2020).

Perhaps the most peculiar family of Early Paleocene "archaic" ungulates are the periptychids. The majority of periptychids, including the conacodontines (e.g. *Militocodon*) and **paraphyletic** "anisonchines" (e.g. *Anisonchanus*, *Haploconus*) were small- to medium-sized omnivorous mammals with long, low bodies, resembling badgers or coatis in general body form. However, genera of the **monophyletic**, more herbivorous subfamily Periptychinae, such as *Periptychus* and *Ectoconus,* attained large body sizes. *Ectoconus* lived within 500,000 years of the Cretaceus-Paleogene extinction event, but *Periptychus*

Fig. 5.7. Phylogenetic tree of a sampling of "arctocyonids", "triisodontids" and mesonychids from the Nacimiento Formation, after Shelley *et al.* (2015).

lived slightly later (Middle Paleocene). *Carsioptychus*, from the Early Paleocene, was possibly ancestral to *Periptychus*. *Carsioptychus* and *Periptychus* have strongly striated teeth, but other periptychids, such as *Ectoconus*, often show weak crenulations. They are much more subtle in most periptychids. Not all periptychids were hard-object feeders, with analysis of tooth-wear not supporting hard object feeding even in *Periptychus*. All periptychids had primitive, elongated skulls with small braincases and large **sagittal crests**. Like the "arctocyonids", periptychids had plantigrade feet and a long heavy tail suggesting they were not fast runners.

If any group best represents the "archaic" ungulates, it would be the herbivorous phenacodontids (see p. 100), known from a wealth of well-preserved fossils from the Nacimiento Formation. The smallest and most primitive of these is the terrier-sized *Tetraclaenodon*, which is believed to have been **terrestrial** rather than arboreal according to tarsal morphology. As evident from the brain endocast and the size of the semicircular canals in the inner ear, *Tetraclaenodon* was somewhat agile, and had a relatively small brain relative to body size (Bertrand *et al.* 2022). *Tetraclaenodon* has reduced lateral and medial **digits** on the hind feet, ending in spade-shaped unguals. *Tetraclaenodon* also has a **brachydont** dentition with **bunodont** cusps, indicating that it fed primarily on leaves and fruit.

One group of "archaic" ungulates that includes the first placental mammals to evolve a distinctly **carnivorous** lifestyle were the mesonychians (see pp. 102–3), close to the ancestry of modern ungulates. In the Nacimiento Formation, mesonychians were largely represented by their likely **stem** relatives, the enigmatic "triisodontids". Among the adaptations for carnivory are narrow premolars exhibiting sharp cusps projecting backwards, and straight slicing blades on the molars formed by crests and cusps. The "triisodontids" reached sizes up to that of a modern day wolf (*Canis lupus*), and represented some of the largest North American carnivorous mammals throughout the Early Paleocene. The red fox-sized *Goniacodon*, estimated at about seven kilograms, is the most completely represented "triisodontid" in the fossil record so far, resembling a raccoon (*Procyon lotor*) with the

North America Nacimiento Formation, United States: Danian-Selandian, 65.7–61 Ma, cont.

robustness of a European badger (*Meles meles*). Attachment sites on the lower limbs indicate well-developed muscles associated with mobile hands and feet. The overall forelimb morphology indicates **pronation-supination** comparable to that of extant **scansorial** mammals. The elbow joint itself was mobile too, but also stable during forceful extension of the forelimb, suggesting that *Goniacodon* was a digger. The slightly larger (about 12 to 22.5 kilograms) *Eoconodon*, known mostly from near complete skulls, has a particularly large cavity above its teeth that housed a large nerve, inferring it had a very sensitive snout and had long, dense whiskers for sensing its surroundings. Bite force estimates indicate that *Eoconodon* (and "triisodontids" as a whole) possessed high bite forces up to 7363 Newtons, 1.65 times that of a spotted hyaena (*Crocuta crocuta*). This is likely to compensate for their dentition being less specialised than that of extant carnivorans (see pp. 81, 83–4). *Goniacodon* and *Eoconodon* are similar in possessing wide **zygomatic arches** and a distinct sagittal midline on the skull, for the attachment of powerful jaw muscles.

The largest known "triisodontid" is *Triisodon* (32 to 44 kilograms), a powerful, terrestrial animal with disproportionately large canine teeth relative to the rest of the dentition, probably used in gripping and subduing prey. *Triisodon* also had the most efficient carnivore dentition of all known "triisodontids", along with limb bones that are more robust relative to the animal's size than in any other "triisodontid". *Triisodon* had at least moderate digging abilities as evident from the thickness of the distal ulna, and the range of pronation-supination was limited as indicated by the humeroradial morphology (Toosey *et al.* 2024). Two valid species of *Triisodon* are known—*T. quivirensis* and *T. crassicuspis*. Features distinguishing the latter include its smaller size (about 20% smaller) and more elongated talonid (the grinding **posterior** portion) on the lower molars. In *Triisodon*, it is unknown exactly how large adults were and if they exhibited **sexual dimorphism** or regional variations in body size because fossils from this genus are often very incomplete. Compared to *Triisodon*, *Goniacodon* appears to have been more scansorially-adapted and had a more omnivorous diet (Grossnickle *et al.* 2016). The medium-sized *Eoconodon* was likely moving towards a carnivorous lifestyle with less efficient dentition compared to *Triisodon*. The mesonychids exhibited even more marked dental adaptations for carnivory than the "triisodontids", including narrower lower cheek teeth with deeply notched shearing blades for slicing flesh. Vaguely resembling a wolf with a proportionately

Fig. 5.8. A **closed-canopy forest** ecosystem in what is now New Mexico, Middle Paleocene. *Centre*: the mesonychid *Ankalagon saurognathus*; *Bottom left*: the leptictid *Prodiacodon crustulum*; *Bottom right*: the pentacodontid pantolestan (see pp. 77–8) *Pentacodon* sp.; *Right background*: the possible "triisodontid" *Oxyclaenus cuspidatus*, and immediately to the left, the "arctocyonid" *Mimotricentes subtrigonus*; *Left background*: the "anisonchine" periptychid *Anisonchanus sectorius*, and immediately above it, the mixodectid *Mixodectes pungens* vertically clings onto a tree trunk; *Top left background*: a paromomyid euarchontan (*Paromomys* sp.).

large skull, the bear-sized mesonychid *Ankalagon* was possibly the apex predator of its ecosystem (Fig. 5.8). A study by O'Leary *et al.* (2000) suggested that sexual dimorphism was present in *Ankalagon* based on robustness of the dentary, although comparisons were made between an adult with a subadult animal. Therefore, they were describing differences in ontogeny rather than sexual dimorphism. To document sexual dimorphism, aspects including canine tooth size would need to be compared, which would require large samples. Furthermore, some scientists have considered *Ankalagon* to be a large species of *Dissacus*, while others regard it as a sister genus. *Dissacus* is best represented in North America, where it underwent an increase in body mass from 8.5 to 21 kilograms throughout the Paleocene (Solé *et al.* 2021). The coyote-sized (12 kilograms) *D. navajovius* was a gracile and lightly-built mesonychid species, comparable in size to *D. europaeus* (see pp. 193–4), the former which is distinct in having the first molar significantly more reduced than the second molar (Zhou, 1995). *Ankalagon* is more derived than *Dissacus*, but more primitive than any other Late Paleocene or younger mesonychid.

Other carnivorous placentals include basal carnivoramorphs (see p. 84) such as *Protictis*, *Intyrictis* and *Bryanictis*. *Protictis* was a civet-like scansorial animal clearly focused on carnivory with its paired **carnassial** teeth. *Protictis* was a stem-carnivoran—not directly ancestral because it had already lost the third molar. The tall, pointed molar cusps indicate that *Protictis* was largely insectivorous. The location of the **orbits** implies that binocular vision was more poorly-developed than in extant carnivorans. The domestic cat-sized placental *Deltatherium* (Fig. 5.9) had elongated, sabre-like upper canine teeth, and powerful jaw muscles attached to the tall sagittal crest on the skull. Its phylogenetic affinities are controversial. Although its primitive dentition shares features with tillodonts (see p. 138) and pantodonts, the absence of the first premolar argues against the ancestral position of *Deltatherium* to both groups. Moreover, the large canine teeth and small incisors of *Deltatherium* are very different from the front teeth of tillodonts, in which the canine teeth are reduced, and the incisors are becoming more enlarged. *Deltatherium* is most probably related to the common ancestor of pantodonts and tillodonts.

Fig. 5.9. Reconstructed life appearance of the enigmatic "archaic" placental mammal *Deltatherium fundaminis*. Reconstructed length: 1 m.

Additionally, the diversity of euarchontans (see p. 149) was high, comprising paromomyids (e.g. *Paromomys*, *Plesiolestes*), and palaechthonids (e.g. *Anasazia*, *Palaecthon*, *Torrejonia*), both families which were probably plesiadapiforms (see p. 150). Paromomyids are characterised by a deep jaw and large, **procumbent** incisors for feeding on exudates, and an increased flexibility of joints that allowed for greater agility in trees. The pointed fourth premolars were possibly employed in opening hard seeds and fruit. Palaechthonids were small, agile climbers and behaved similarly to extant treeshrews (see p. 149). There were also the mixodectids such as *Mixodectes*, also probably plesiadapiforms, although relationships need to be better worked out. Although *Mixodectes* is well-known, it has been allied with poorly known taxa and placed in its own family because of its relatively derived teeth, which appear to be adapted for eating leaves. That is unusual among plesiadapiforms. Postcranially, it looks very much like *Plesiadapis* (see pp. 194–5) and other large plesiadapiforms.

While almost all of the mammals from the Nacimiento Formation are eutherians, the longest lived mammal group at the time, whose origins are rooted in the Mesozoic, were the multituberculates (see pp. 45–6), which were arguably already on the decline by the Middle Paleocene. Among the North American Paleocene forms are the squirrel-like ptilodontids of the genus *Ptilodus*, which took up a specialised arboreal lifestyle. This is evident from their sharp claws, grasping digits, a mobile foot that could be reversed backward to enable the animal to dash down tree trunks head-first, and a long, **prehensile** tail that functioned as an additional limb. The most distinctive characteristic of all is a **plagiaulacoid** lower premolar. *Ptilodus* was a typical medium-sized multituberculate, and is known from numerous well-preserved specimens from the Nacimiento Formation. The taeniolabidids, on the other hand, differ from the ptilodontids in having lego brick-like molars, a much heftier anatomy and more of a sprawling plantigrade stance. They approached the size of a small capybara (*Hydrochoerus hydrchaeris*) and likely had a completely terrestrial mode of life. One prime example is *Taeniolabis*, which is the largest known multituberculate, probably weighing 20 to 55 kilograms. It has a dentition superficially similar to modern rodents in possessing large, self-sharpening, gnawing incisors along with cheek teeth that have a complex, grinding surface. This, along with its bulky, box-shaped skull and short, rounded snout, indicates a specialised herbivorous diet (Williamson *et al.* 2016). One other beaver-sized taeniolabidid was *Kimbetopsalis*, with an estimated weight of 10 to 40 kilograms. It is clearly closely related to and possibly ancestral to *Taeniolabis*, distinct from the latter in possessing fewer cusps on the upper first molar and a double-rooted rather than single-rooted upper fourth premolar. Like *Taeniolabis*, *Kimbetopsalis* had prominent, chisel-like incisors adapted for slicing branches and leaves (Williamson *et al.* 2016).

This rich variety of mammals that filled in the ecological niches left vacant by the non-avian dinosaurs and more basal Cretaceous mammals, exploiting a wide variety of terrain, is a strong testament to the adaptability of mammals to thrive in ever-changing environments.

South America Middle Santa Lucía Formation, Bolivia: Danian, 64.5–64 Ma.

The deposits from the middle portion of the Santa Lucía Formation in Tiupampa, central Bolivia, are significant for yielding the earliest fossils of mammals from South America following the Cretaceous-Paleogene extinction event (see p. 169). This fossil site was likely to have been a tropical floodplain surrounded by the tributaries of a main river channel that abruptly overflowed during a flash flood approximately 64 Ma. The flash flood would have rapidly engulfed the area in a large quantity of fine-grained sediment and as a result, numerous **articulated** skeletons of various mammal species have been extraordinarily preserved. Some of these mammals would have been entombed in their burrows while others that were unfortunate enough to get caught up in the flood quickly drowned.

At the time of the flood, mammals in South America and across the world were very small and relatively unspecialised. In South America, the largest were no larger than a ground squirrel. Only as **ecosystems** recovered over the following 10 million years did mammals start to diversify and evolve larger body sizes. In addition to the mammal fauna, the deposits also preserve a rich diversity of other fossil vertebrates, including numerous freshwater **osteichthyan** fish, six snakes, three pleurodire turtles (see pp. 14–5), five crocodylomorphs (see pp. 26–30) and three amphibians (one caecilian and two anurans) (see pp. 12–3). These have given us a thorough understanding of the ecosystem in which those **primitive** mammals existed. **Terrestrial** habitats were dominated by hot, humid, low-lying tropical forests, scattered across an open **alluvial** plain, and bisected by meandering rivers. Tiupampa was likely dominated by a monsoonal climate, with a distinct wet and dry season, as evident from desiccation cracks, evaporite minerals and fossils of small lungfish from the extant South American genus *Lepidosiren*. To survive periods of drought, lungfish would/may have burrowed into the riverbed, enclosed themselves in a cocoon and lay dormant in a process called aestivation.

During the Cretaceous-Paleogene boundary, a strong connection between North and South America existed, allowing therian mammals, which may have first evolved in Asia (Williamson *et al.* 2014), to disperse to South America from the north (see p. 47), before the continent became isolated again for much of the Cenozoic. After the Cretaceous-Paleogene extinction event, the metatherians (see pp. 43, 46–7) soon became minor players in most ecosystems around the world (except in Australia), but maintained their dominance in South America through the Eocene. Some metatherians had arrived in South America by the end of the Cretaceous and there, they quickly became the most diverse lineage of mammals in South American Paleocene faunas such as that of Tiupampa. They included the **carnivorous** sparassodonts (see pp. 48–50) and **insectivorous** basal marsupialiforms called pucadelphyids. *Mayulestes*, which was about the size of a brown rat (*Rattus norvegicus*), is likely one of the oldest and most primitive known sparassodonts, as well as the largest carnivorous mammal at Tiupampa. It was likely **arboreal** and an agile climber, as indicated by the limb proportions and **prehensile** tail of the specimen (Argot, 2000). Its tail may have been less prehensile than those of extant opossums. *Mayulestes* had notched **cheek teeth** adapted for slicing meat and presumably fed on small vertebrates, performing an ecological role **analogous** to that of a marten. Other basal sparassodonts from Tiupampa include *Allqokirus*, which was of similar size to *Mayulestes* and also possessed dental adaptations for carnivory, but had a smaller, more gracile skull.

The most well-known metatherian at Tiupampa is *Pucadelphys*, a small, mouse-sized mammal known from over 30 individuals found congregated together in the same fossil burrow (Ladevèze *et al.* 2011). These represent adults, subadults and juveniles of both sexes. Unlike modern South American metatherians, *Pucadelphys* seems to have been **gregarious**, or was at least tolerant to regular social interactions. The skeletons show that males had particularly pronounced canine teeth, suggesting that they fought each other over females. *Pucadelphys* was a nimble generalist that was mostly terrestrial but also adept at climbing and digging. Because of its digging habits, *Pucadelphys* is most analogous to modern Australian dasyurid marsupials (e.g. quolls). *P. andinus* was likely insectivorous as indicated by its molars exhibiting small, acute cusps linked by narrow crests. One other basal metatherian from Tiupampa and a close relative of *Pucadelphys*, *Andinodelphys*, may have also been gregarious, as evident from the discovery of six articulated and jumbled skeletons. While *Andinodelphys* shares numerous similarities in dental and cranial anatomy with *Pucadelphys*, the former is distinct in having a larger skull and a longer snout, comparable in size to that of a mouse opossum (*Marmosa*) at 4.2 to 4.8 centimetres long. The metatherians at Tiupampa were also represented by true marsupials such as *Khasia*, a primitive microbiotherian (see p. 51) that is only represented by a single upper molar. The morphology of this tooth indicates that *Khasia* was a very small insectivore, like the modern monito del monte (*Dromiciops gliroides*).

Among the placental mammals at this site, *Alcidedorbignya* (Muizon and Marshall, 1992) is the most prevalent, which is believed to be a primitive pantodont (see pp. 137–8) as inferred by V-shaped ridges on the **occlusal** surface of the cheek teeth. Its occurrence in South America indicates that *Alcidedorbignya* was an early migrant into that continent from North America, and not that pantodonts originated in South America. It is known from multiple complete skeletons which show that the animal was tiny (roughly grey squirrel-sized, at around 500 grams) in comparison to most other pantodonts, the largest which weighed over half a tonne. A number of jaws from different-aged animals suggests that *Alcidedorbignya* lived in groups. It was a nimble **plantigrade omnivore** that may have been capable of both climbing trees (in a similar manner to a modern tree hyrax (*Dendrohyrax*)) and scampering in the undergrowth (Muizon *et al.* 2015). The fossils of kollpaniids, a family of small, primitive placental mammals formerly grouped as "condylarths" (see pp. 100–3), are also abundant at Tiupampa (e.g. *Tiuclaenus*, *Pucanodus*, *Molinodus*). However, these are known only from teeth and jaw fragments. They too likely descended from ancestors that arrived from North America. The Kollpaniidae has been tentatively grouped with the endemic litopterns (see pp. 143–4) on the basis of shared dental characters and has even been considered to lie near the origin of perissodactyls (see p. 126) (Chimento and Agnolin, 2020). The best-studied kollpaniid from Tiupampa is *Molinodus*, possibly a terrestrial, non-arboreal mammal with a diet of fruit and soft vegetation as evident from its **bunodont** molars.

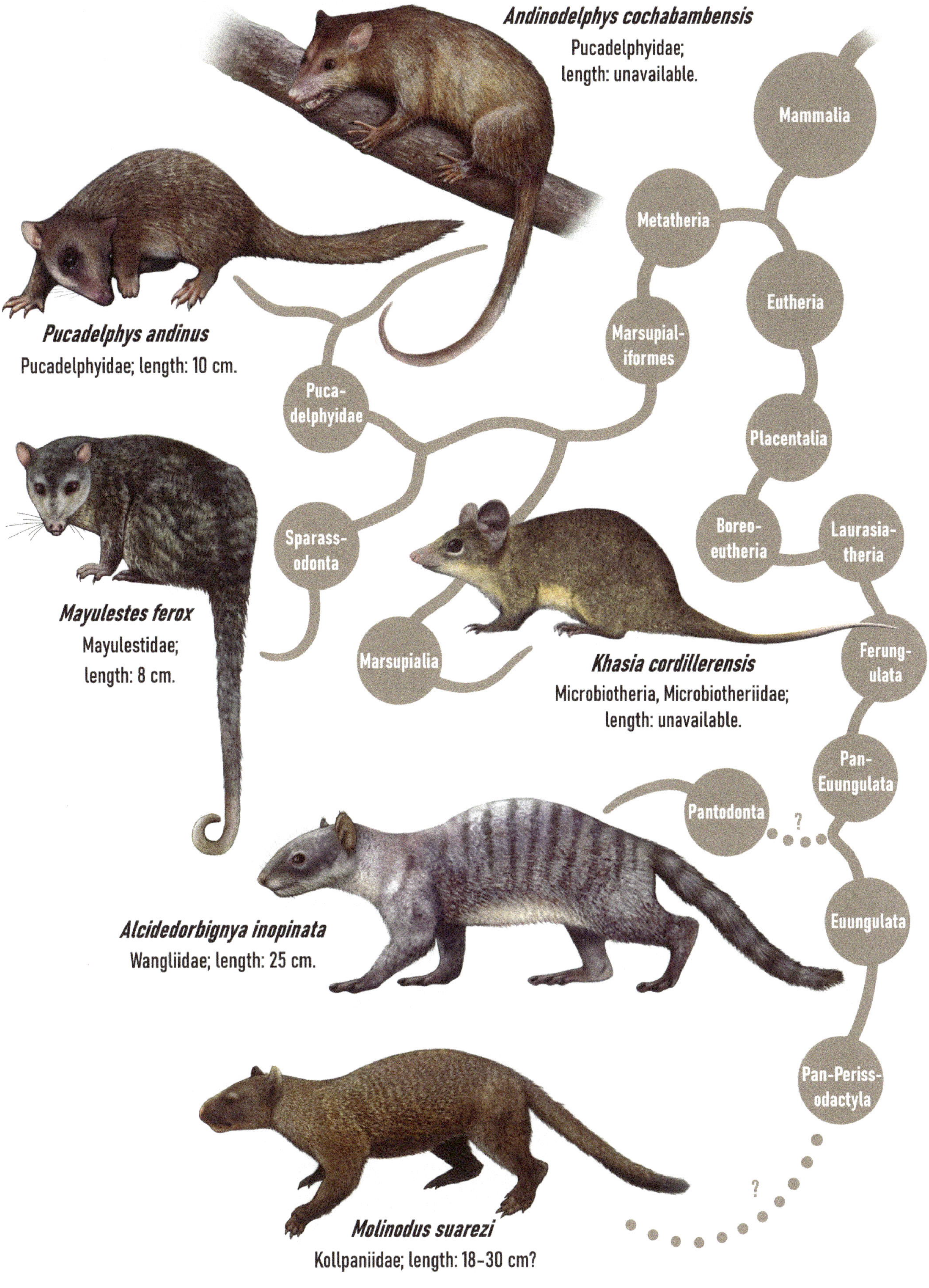

Fig. 5.10. Phylogenetic tree of a sampling of fossil mammal **taxa** from the Middle Santa Lucía Formation. The phylogenetic position of *Molinodus suarezi* follows Chimento and Agnolin (2020), while that of Pantodonta follows Kort (2023), the phylogeny in the latter study which suggests that pantodonts have plausible affinities with Pan-Euungulata (see p. 100).

South America Cerrejón Formation, Colombia: Selandian-Thanetian, 60-58 Ma.

Throughout the ten-million-year epoch following the extinction of the non-avian dinosaurs, the Earth remained, in certain places, home to a host of huge reptiles that were the largest animals alive during the Paleocene. The Cerrejón Formation, which is made up primarily of coal beds, has preserved an extraordinary environment in which giant snakes, crocodylomorphs (see pp. 26–30) and turtles dominated the landscape, and it is because of the plethora of these remarkable scaly beasts that this site in northern Colombia has received much attention. While the fauna is undoubtedly astonishing, the ecological significance of this Middle to Late Paleocene fossil site is frequently overlooked. The abundance of fossilised plants and leaves, along with signs of feeding damage from **herbivorous** insects, represents the earliest known fossil evidence of modern **Neotropical** rainforests (i.e. rainforests in Latin America). Plant families that have been discovered, which include a great variety of legumes and palms, still occur today in the Amazon Basin. However, the diversity of plants in contemporary rainforests is far greater than that of Paleocene forests, which may indicate that the Earth was undergoing a slow recovery from the Cretaceous-Paleogene extinction event (see p. 169) at the time the Cerrejón Formation accumulated. It could also suggest an early phase in the proliferation of Neotropical rainforests (Wing *et al.* 2009).

Estimates for the **mean annual temperature (MAT)** and and the precipitation levels vary. Certain studies have proposed balmy temperatures of over 28°C (~82°F), with an average annual rainfall of 2,500 millimetres (~8 feet), while others (Wing *et al.* 2009) propose a slightly cooler MAT of 24°C (~75°F) and greater rainfall levels of 3,240 millimetres (~11 feet) (Herrera *et al.* 2005). The more untoothed leaves there are in a certain environment, the higher the temperature is likely to be (Royer *et al.* 2012). **Angiosperms** with untoothed leaves comprise 74–78% of fossilised leaves yielded from Cerrejón, implying a high MAT for the surrounding environment. The vertebrate and plant fossils documented from the deposits, along with the geology of the Cerrejón Formation, indicates that the environment likely consisted of coastal wetland, fringed by humid tropical rainforests and criss-crossed by a complex network of rivers. The balmy temperatures and humid environment favoured cold-blooded reptiles, which grew to truly enormous sizes, partly due to the fact that any large predatory mammals or birds had yet to evolve in South America.

In 2009, one particular reptile species made headlines, breaking records as the longest, largest and heaviest snake of all time, aptly named *Titanoboa* (Head *et al.* 2009). Fossil material consisting mainly of ribs and highly robust, broad vertebrae from more than 28 individuals of *Titanoboa* suggest that the animal may have weighed in excess of 1,100 kilograms (2,500 pounds). It was a boid, a family of non-venomous snakes which typically use their muscle and bulk to subdue prey (see p. 24). This monster snake was originally assumed to be the apex predator in its wetland environment, specialising in hunting crocodylomorphs as is commonly depicted. However, the cranial features (e.g. the anatomy of the palate), along with the dental morphology and tooth count, suggest it was primarily **piscivorous**, an unusual diet for boids (Head *et al.* 2013). The crocodylomorphs that thrived in this Paleocene **ecosystem** belong to the family Dyrosauridae (see p. 30), three known species which lived in coastal and shallow, inland, freshwater environments. *Acherontisuchus* was a moderately large species, with a long, narrow snout indicative of a specialised piscivorous diet. *Acherontisuchus* was discovered with a bone present in many dyrosaurids and other crocodylomorphs called an ischial shaft, found in the pelvic region. This bone served as an anchor for the muscles that enabled the animal to maintain its pitch and buoyancy in the water, a fundamental task for living in the strong waves of coastal environments. In *Acherontisuchus*, however, this bone is considerably slenderer than in other dyrosaurids, suggesting a reduced ability and need to control its movement in rough water currents (Hastings *et al.* 2011). The animal's limbs could also support its entire weight, as in extant crocodylians. Unlike other dyrosaurids, *Acherontisuchus* was therefore able to navigate easily among the wetlands in which it lived, both in the water and on land.

In contrast to *Acherontisuchus*, *Cerrejonisuchus* was a miniature, short-skulled generalist with uncurved teeth, which probably fed on small reptiles, amphibians and mammals (Hastings *et al.* 2010). With compressed teeth, a cylindrical torso and longer limbs, *Cerrejonisuchus* is believed to have been more **terrestrially** adapted than other dyrosaurids. One other dyrosaurid with a short snout in proportion to the rest of its body is *Anthracosuchus*. Its snout is also very broad, which indicates that *Anthracosuchus* likely had a very strong bite force, since there is an enormous area for the attachment of strong and powerful jaw muscles. This, along with Its blunt, peg-like teeth, indicates a diet of tough-skinned prey such as small turtles. A difference in the muscle attachment on the ribs and less **osteoderms** allowed *Anthracosuchus* to be more flexible than modern crocodylians; it would have swam with a more undulating motion (Hastings *et al.* 2014).

While the gigantic snakes and dyrosaurids are fascinating and unique in their own right, no reptile from Cerrejón is more bizarre than *Carbonemys*, a pleurodire turtle of the family Podocnemididae (see pp. 14–5). *Carbonemys* has been described on the basis of a partial skull but a partial shell from a different individual may also belong to the same species (Cadena *et al.* 2012). It was a colossal animal with a streamlined, oval-shaped **carapace** as large as a Smart Car, a result of the hothouse climate that contributed to the emergence of the other giant reptiles with which it shared its habitat. The snout of *Carbonemys* is wide at the back and narrows abruptly towards the front, especially at the **premaxillary** region. Its diet is controversial, but it had a powerful bite sufficient to crush crocodile hatchlings. *Carbonemys* was most probably **omnivorous**, cropping vegetation, and preying on molluscs and small reptiles (both bountiful and diverse in its freshwater environment).

The warm, humid Paleocene environment did not persist for long and within several million years, the climate became slightly cooler and drier—these reptiles were unable to adapt. The cooler temperatures favoured smaller reptiles just as the hot climate had favoured larger ones. This, along with a change in habitat, led to the eventual extinction of these giant reptiles.

Fig. 5.11. Phylogenetic tree of the giant fossil reptile **taxa** from the Cerrejón Formation. The phylogeny within Dyrosauridae follows Hastings *et al.* (2014).

Africa Khouribga Phosphates (Bed II), Morocco: Danian–Thanetian, 66–56 Ma.

The phosphate-bearing strata of the Ouled Abdoun Basin in central Morocco surround the city of Khouribga. The phosphates are exploited in large industrial quantities for fertiliser and are among the very few fossil sites on Earth with large-scale deposition and fossils from shortly before and after the Cretaceous-Paleogene extinction event (see p. 169).

The rocks, which predominantly consist of phosphorite, with variable lithologies including **marl**, limestone, and claystone, had accumulated in a warm, shallow inland sea along the Atlantic coast of North Africa, at a time when sea levels were much higher than at present and when Northwestern Africa was at the intersection of the Atlantic Ocean and the ancient Tethys Ocean (see p. 174). Rarely do phosphate minerals occur in large concentrations in the geological record, requiring special conditions to form. Wind currents at mid-latitudes typically enable nutrient-rich waters to rise up from the deep sea, frequently resulting in greater biological productivity such as algal blooms. Additionally, the bottom of the water column needs to be **anoxic**, enabling the phosphorus absorbed by algae and other microorganisms to accumulate on the seabed subsequent to death. The phosphates were deposited in the marine basin over a prolonged period of time, from the Late Cretaceous (ca. 67 Ma) up until the Early Eocene (ca. 50 Ma). The marine **strata** are referred to as the "phosphate series" and are divided into several phosphate-rich beds known as, from bottom to top, Bed III, Bed IIb, BedIIa, Bed I and Bed 0. Bed III represents the Late Cretaceous (67–66 Ma), Bed IIb the latest Cretaceous (Maastrichtian) to late Selandian in the Middle Paleocene (66–61.66 Ma), Bed IIa the Late Selandian in the Middle Paleocene to Late Thanetian in the Late Paleocene (61.66–56 Ma), and Beds I and 0 the Early Eocene (56–50 Ma) (Kocsis *et al.* 2021).

The deposits are noted for their diverse fauna of aquatic vertebrates, dominated by **chondrichthyan** fish (sharks and rays), turtles and crocodylomorphs (see pp. 26–30) as evidenced by an abundance of **disarticulated** bone material. The fossils of Paleocene turtles include both cryptodires and pleurodires (see pp. 14–6). Among the cryptodires is *Euclastes*, a survivor of the mass extinction and a basal sea turtle, a **stem** relative of Cheloniidae (see pp. 14–5) which differs from geologically younger sea turtles in having a low, wide skull and a broad dentary (lower jaw) with a lengthened symphysis. The broad, flat jaws of *Euclastes* and its close relatives were **adaptations** for a **durophagous** diet, so they could crush shelled organisms including molluscs.

The pleurodires are represented exclusively by the extinct bothremydids (see pp. 14–5), a very diverse family that occupied numerous ecological niches and are represented by two lineages in the Khouribga phosphates—the taphrosphyins and bothremydins. The taphrosphyins were the most morphologically diverse bothremydids, with considerable variations in skull shape. *Tasphrosphys* is characterised by a snout that is expanded instead of narrow, with a **premaxilla** that is broader **anteriorly** than in other species of the genus, and a **maxilla** that is gently concave laterally. The **frontal** bone in the skull is reduced and present around the edge of the **orbit**. *Tasphrosphys* is also distinct in having a skull roof with a corrugated surface texture. *Rothonemys* is distinguished by a flattened skull with forward-facing and proportionally large nasal openings, the largest of any known bothremydid. In *Rothonemys*, the outer ridge on the palate is thick in cross-section with a slightly hollow inner surface and a broad, curved outer surface. In *Rothonemys*, the orbital margin has a tall, sharp rim (Gaffney and Tong, 2008). *Labrostachelys* is just as dramatic, with a long, narrow skull having small nasal openings, dorsally orientated nostrils and a shorter distance between the orbits compared to any other taphrosphyin. The front portion of the premaxilla extends to the front of the outer ledge of the palate, a feature not present in any other known turtle. On the other hand, the bothremydins such as *Bothremys* are distinguished by a lack of deep grooves on the **carapace**, producing an ornate pattern brought about by elevated polygons (Perez-García, 2021). Among other defining features are a near smooth external surface of the shell, a low carapace and a somewhat long overlap of the pectoral scutes on the **plastron** (Tong *et al.* 1998; Perez-García *et al.* 2012). In *Bothremys*, the orbits are rather small, protruding forwards and outwards.

Among other reptiles, the crocodylomorphs from the deposits belong to the successful dyrosaurid lineage (see p. 30), which in Africa are all represented by marine forms. Many of these possessed an elongated gharial-like skull, a long narrow snout and a uniform dentition as an adaptation to **piscivory**. Following the disappearance of Cretaceous marine reptiles including mosasaurs and plesiosaurs, they became the apex predators of the Paleocene oceans, until the emergence of the earliest fully aquatic whales by the Middle Eocene (see pp. 121–2). Two near complete skulls of *Chenanisuchus*, a species that first appeared in the Late Cretaceous and survived the mass extinction, are known. Analysis of the inner ear suggests that *Chenanisuchus* could walk on the seabed; this may have contributed to its survival, though the causal link is uncertain. *Chenanisuchus* is the most basal known dyrosaurid and is one notable exception in possessing the shortest snout in proportion to skull length of any known dyrosaurid. It has a flattened skull roof adorned with relatively deep pits isolated by rounded ridges (Hill *et al.* 2008). This animal also has laterally-facing orbits, hence the species name.

Arambourgisuchus is a dyrosaurid known from four specimens consisting of cranial material (Jouve *et al.* 2006). The skull is characterised by rounded orbits, massive **nares**, generally sharp, robust teeth that were not particularly slender, and a relatively narrow, slender snout. In contrast to the maxilla, which is laterally sculpted by deep furrows and ridges, the nasal bone is composed of one fused bone with very little ornamentation. There are also three ridges present between the orbits. *Atlantosuchus* is known from a partial skull with parts of the cervical vertebrae still attached. It too is characterised by an elongated, narrow and slender snout, but is smooth dorsally and anteriorly, and more adorned by furrows **posteriorly**. Furrows are present only on the lateral edges of the skull roof. The lateral edge of the external nares is higher relative to those of other dyrosaurids, and the teeth are conical and slender (Jouve *et al.* 2008). *Atlantosuchus* also had the longest snout relative to body length of any dyrosaurid (accounting for roughly 79% of the skull length).

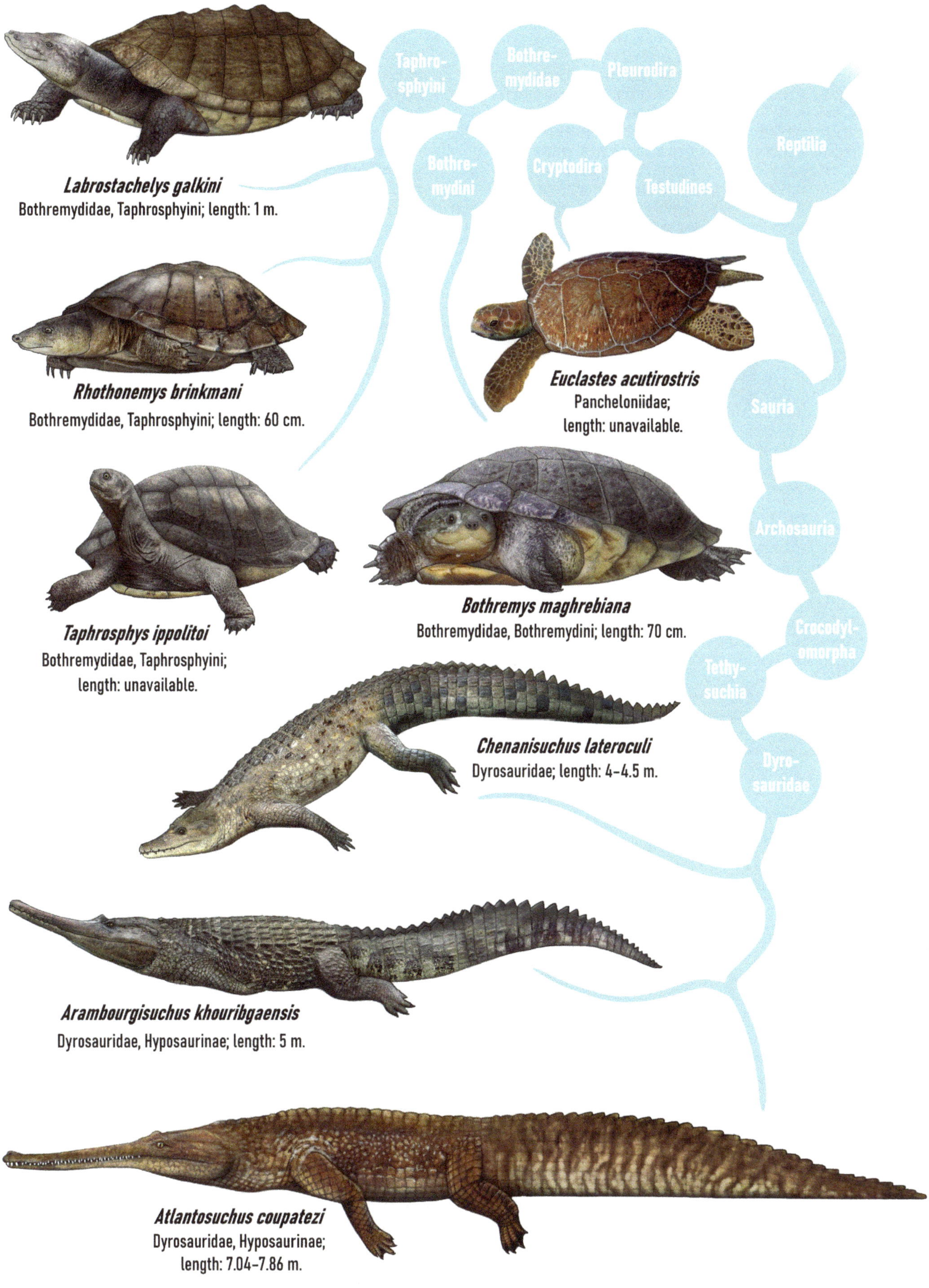

Fig. 5.12. Phylogenetic tree of a sampling of Paleocene reptile **taxa** from the Khouribga Phosphates. The phylogeny within Bothremydidae follows Carvalho *et al.* (2016), while the phylogeny within Dyrosauridae follows Salih *et al.* (2021).

Africa Khouribga Phosphates (Bed II), Morocco: Danian–Thanetian, 66–56 Ma, cont.

With the discovery of individual mammal remains in the 1990s came a greater focus of scientific interest to the area, which has remained to this day. The mammal fossils from Bed II, all belonging to **terrestrial** forms, have particularly made an important contribution to our understanding of the origins and early evolution of afrotherians (see p. 62). These fossils also include some of the earliest known placental mammals in Africa. They are, however, very scarce and rather fragmentary, known only from a few isolated occurrences; this contrasts with the large concentration of marine vertebrate material. The presence of fossils from over half a dozen Paleocene mammal genera in these marine deposits suggests that they must have undergone low-energy transportation from nearshore and terrestrial environments via rivers and estuaries prior to fossilisation. Some of the most basal afrotherians including *Ocepeia* and *Abdounodus* are known from the Khouribga Phosphates, both of which display **insectivore**-like and **primitive ungulate**-like morphological features. Further phylogenetic analysis is required to accurately assess where *Ocepeia* and *Abdounodus* fit into the mammalian evolutionary tree, but they provisionally appear to occupy a stem position relative to Paenungulata (Gheerbrant *et al.* 2016).

Ocepeia is represented by two species: *O. daouiensis*, from the Selandian, and the slightly younger (Thanetian) *O. grandis*, the latter which is distinguished by molars that are more massive and strongly-cusped (Gheerbrant *et al.* 2014). Based on dental material, *O. daouiensis* has a body mass estimate of 3.5 kilograms, whereas *O. grandis* is estimated at 10 to 12 kilograms. *Ocepeia* has **brachydont** molars exhibiting a morphology and wear pattern suggesting they functioned predominantly in shearing and not for grinding. Therefore, *Ocepeia* was likely to have been a specialised **folivore**. Some features that *Ocepeia* shares with primitive paenungulates include a broad nasal cavity and a very short bony projection in the orbits. Some peculiar anatomical characters of *Ocepeia* are features of the dentition and jawbones similar to those of simiiform primates (see pp. 152–6) (e.g. powerful jaws and a short **rostrum**), along with cranial bones consisting of numerous air cavities. The purpose of those air cavities is a mystery, but they were possibly used for enhancing sound transmissions.

Abdounodus is known from several upper and lower jaws and can be classified as relatively more derived according to its dental features. The teeth are more **bunodont**, with inflated crowns, less developed crests and cusps that are more bulbous and show substantial abrasive wear indicative of clear horizontal chewing movements. This abrasion also suggests a powerful vertical and grinding chewing mechanism in *Abdounodus*, and a diet of tough and abrasive plant material. This is evident from the robust and highly elongated molar roots (Gheerbrant *et al.* 2016). *Ab. hamdii* may have been largely **frugivorous** (Werdelin and Sanders, 2010). *Ocepeia* and *Abdounodus* are the first fossil taxa representing a morphological transition between the primitive tritubercular molar pattern (i.e. molars with three cusps, or tubercles, arranged in a triangle so that the upper molars interlock with the lower ones) and the advanced **bilophodont** molar pattern characteristic of paenungulates (Gheerbrant *et al.* 2016). The age and morphology of *Ocepeia* and *Abdounodus* support the **endemic** evolution of placentals in Africa, rather than afrotherians arising from taxa from **Laurasia**. One other contemporary basal afrotherian was *Hadrogeneios*, which is also the most recent known of these, and probably represents the most basal stem-paenungulate, as partly supported by its primitive tritubercular molar pattern. Despite being so basal, *Hadrogeneios* had a specialised and highly peculiar thickened chin, which is elevated well beyond the **cheek teeth** and sported reduced incisors. This dental apparatus has been suggested to have aided in pulling at tough fruits and leaves, with an advanced gripping ability (Gheerbrant, 2022).

Some of the most primitive undisputed proboscideans (see pp. 64–8) are also known from this site. These include *Eritherium*, documented by a single specimen consisting of the anterior portion of the skull and upper jaw (Gheerbrant, 2009). This exhibits primitive features including a short mandibular symphysis. Uniquely among proboscideans, *Eritherium* has a full set of teeth including canine teeth, with no **diastema** between the front and back teeth. *Eritherium* stood roughly 20 centimetres at the shoulder, weighed three to eight kilograms, and its molars are bunodont, with low, rounded cusps. The slightly younger *Phosphatherium* was approximately the size of a domestic cat, with a shoulder height of about 30 centimetres, and was nearly twice as heavy as *Eritherium*. The skull of *Phosphatherium* is fairly low and flat, but lacks the structures for a **proboscis** (Gheerbant *et al.* 1996). The snout of *Phosphatherium* is proportionally short, while the back end of its skull behind the orbits are elongated, which supported large, powerful jaw muscles. The dentition and jaw musculature are notably different between males and females, implying behavioural differences. The **heterodont** tooth morphology along with wear patterns on the teeth suggest that *Phosphatherium* may have had a broad diet consisting of rough vegetation and possibly animal matter. *Phosphatherium* was possibly a **semi-aquatic** animal, with a behaviour somewhat similar to that of extant tapirs *(Tapirus)* or pygmy hippopotamuses (*Choeropsis liberiensis*)—devoting much of the daytime lounging in mud or water, and coming onto land to forage during the night. *Phosphatherium* is the best known and most common fossil mammal from the deposits.

Aside from the afrotherians, the other major mammal lineage documented from the deposits are the **carnivorous** hyaenodonts (see pp. 82–3), which are, so far, the only known mammals of the Paleocene carnivore **guilds** in Africa. These include *Lahimia* from the Middle Paleocene and the slightly younger *Boualitomus*, both known from several lower jaws (Solé *et al.* 2009). The dentition of *Lahimia* and *Boualitomus* is clearly more specialised than other very primitive hyaenodonts which, in addition to the reduced front molars, exhibits a tendency towards a primarily cutting function. Among shared features between both genera are the reduction of the anterior cheek teeth (e.g. the disappearance of the first premolar), although this is more pronounced in *Lahimia*. The latter is further distinct from *Boualitomus* in features such as larger size and **carnassial** teeth that were specialised for shearing, despite being presumably older (Solé *et al.* 2009). Both genera suggest an early African radiation of hyaenodonts and presumably preyed on all the primitive afrotherians discussed above.

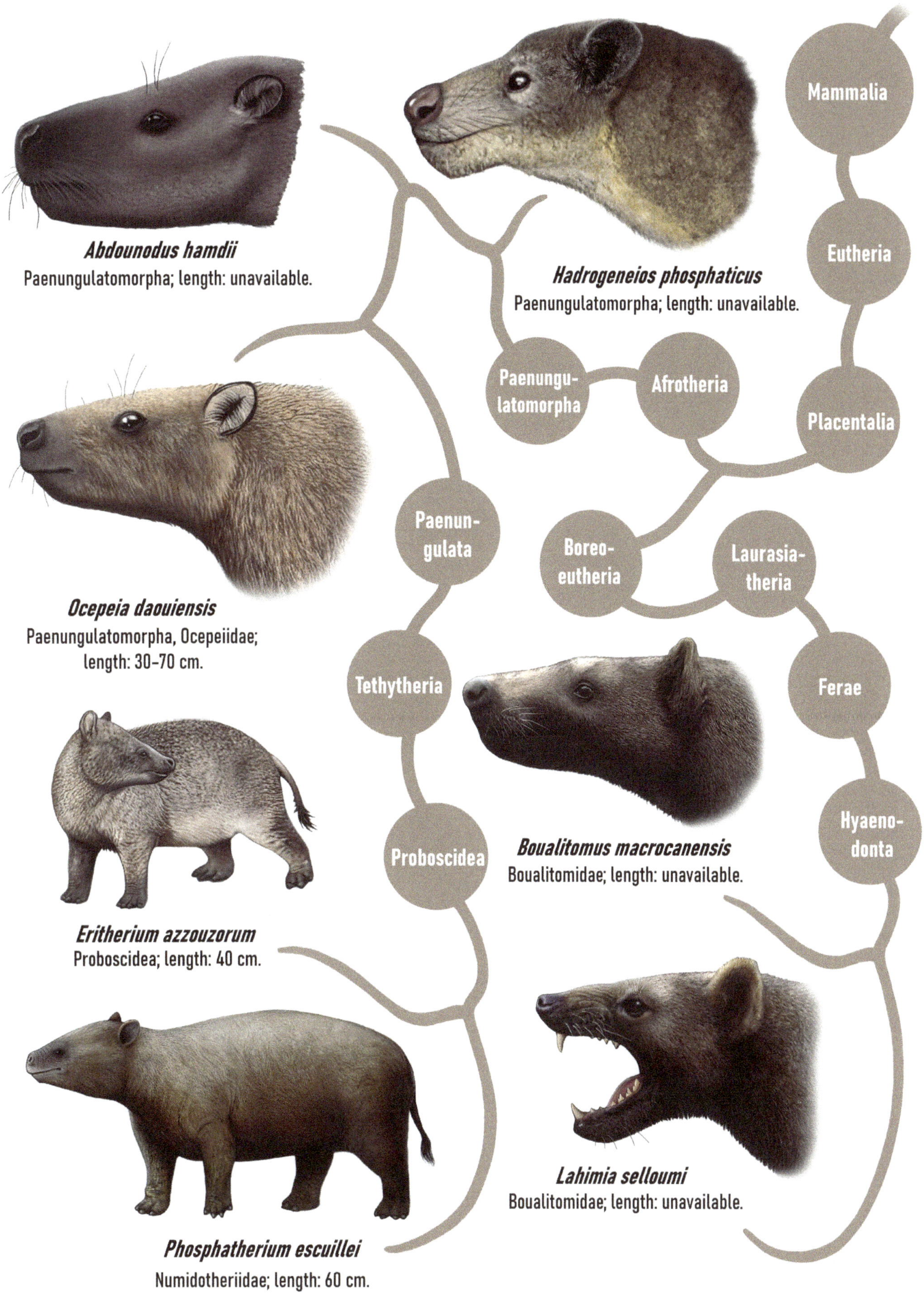

Fig. 5.13. Phylogenetic tree of a sampling of Paleocene mammal taxa from the Khouribga Phosphates.

Asia Naran Bulak Formation, Mongolia: Thanetian-Ypresian, 58.7–55.8 Ma.

From the Naran Bulak Formation of the Nemegt Basin in southern Mongolia, interpreted to have accumulated in **alluvial** and **lacustrine** settings (Dashzeveg, 1988; Yo *et al.* 2023), comes the most well-studied Paleocene mammal fauna from Asia, referred to as the Gashato Fauna and dated to the Late Paleocene. In recent decades, the record of Asian Paleocene mammals has advanced considerably, but our understanding still remains largely restricted given the fragmentary nature of most of the fossils. One glimpse at the Asian Paleocene fauna shows immediately how different it is from those of both North America and Europe. The pantodonts (see pp. 137–8) were particularly abundant, sharing the large mammalian niches with dinocerates (see p. 139) along with the **carnivorous** mesonychians (see pp. 102–3). Asian pantodonts were lighter-built species and smaller compared to their North American counterparts. *Archaeolambda planicanina*, known from a complete skeleton, was a gracile-bodied species that may have been **scansorial** given its low weight of just seven kilograms and the presence of claws (Lucas, 1998). However, this presumed scansorial lifestyle remains inconclusive since a detailed skeletal analysis has yet to be undertaken (Rose, 2006). The teeth are all roughly level, with even the canine teeth not exceeding the height of the other teeth, hence the species name.

Among the mesonychians, one distinct **clade** are the hapalodectids, represented by *Hapalodectes*. Given the previously proposed connections to cetaceans (see pp. 121–6), this **terrestrial** mesonychian has frequently been incorrectly depicted as rather otter-like. The hapalodectids had more slender jaws and highly compressed, elongated lower **cheek teeth** compared to other mesonychians, better adapted for shearing flesh. One unique feature of *H. dux* is the rear of the **orbit** being closed by a **postorbital bar**. This genus first appeared in Mongolia during the Late Paleocene and dispersed to North America via the Bering Land Bridge (see p. 174), surviving until the Early Eocene. The mesonychids include the species *Dissacus zanabazari*, characterised by a relatively short face, a prominent **sagittal crest**, and a highly reduced first metatarsal essentially devoid of function (Geisler and McKenna, 2007). In addition, we see the earliest and most basal dinocerates of the genus *Prodinoceras*. It was a relatively small and stocky animal, possibly living in marshes. The size of a Malayan tapir (*Tapirus indicus*), it was a large and ponderous animal for its **ecosystem**, but small compared to its later relatives. The 30 to 50 centimetre long skull of *Prodinoceras* is considered quite **primitive**, displaying a pronounced sagittal crest similar to the large "arctocyonids" (see p. 101) and had not yet evolved the remarkable knob-like horns typical of dinocerates. Like all dinocerates, however, it has massive upper canine teeth (supposedly used for fighting rival males) guarded by a pair of bony flanges on the lower jaw, and has the distinctive cheek teeth individually possessing two transverse crests.

The small mammal fauna was dominated by the extinct arctostylopids (see p. 145), primitive rodents, the extinct "eurymylids" and "anagaloids" (see pp. 144–5), and **stem**-lagomorphs (see pp. 144–5). The arctostylopids first appeared in East Asia during the Late Paleocene, where they are most diverse, and are represented solely by the genus *Arctostylops* at Naran Bulak. The arctostylopids, such as *A. macrodon*, have an unreduced dentition, comprising an evenly graded series lacking **diastemata**. The oldest and most primitive known Asian rodent, *Tribosphenomys*, appeared in the Late Paleocene, contemporaneous with the first North American rodents. The incisor enamel is double-layered, and highly primitive features of the known postcrania include a shallowly grooved **distal** tibia and distal astragalus (Rose, 2006). Among the "eurymylids", *Rhombomylus* is most primitive in possessing teeth composed of one layer of radial enamel (Martin, 2004; Rose, 2006). It was a generalised, predominantly terrestrial animal as evident from features of the hindlimb, and like most rodents, the manus was possibly used in manipulating food (Meng *et al.* 2003).

The "anagaloid" *Pseudictops* was a terrestrial mammal that shows evidence of **cursorial adaptations** such as a straight ulna, relatively distally elongated limb bones, and a stabilised tarsal joint facilitated by the calcaneofibular facet (Rose, 2006). The cheek teeth are similar to those of lagomorphs in being somewhat **hypsodont** and possessing crowns with many basins and transverse ridges, although they were rooted rather than being **hypselodont**. The oldest known stem-lagomorph, *Gomphos*, is represented by a specimen consisting of a nearly complete, **articulated** skeleton and skull. Like rabbits and hares, *Gomphos* has specialised, moderately long and thin ankle bones (astragalus and calcaneus) along with slender, elongated hindlimbs, which would have been ideal for jumping or running. However, unlike lagomorphs, the fibula is not fused with the tibia, and the calcaneus is not perforated with a canal (Asher *et al.* 2005). *Gomphos* is also more primitive in retaining a pair of incisors in the lower jaw, along with a moderately sized tail. There was also a bizarre sloth-like animal called *Ernanodon*, of the pangolin-related order Palaeanodonta (see pp. 79–80). It was a small, squat and massively built animal, with powerful and well-developed forelimbs armed with large claws. The massive and shortened skull of *Ernanodon*, equipped with a robust jaw, have simple, conical cheek teeth, which does not at all speak in favour of feeding on solid food. Furthermore, the incisors are completely absent. Therefore, this animal was most likely **insectivorous**.

Besides the fact that the abundance patterns of "anagaloids", mesonychians and pantodonts in Asia were notably different from those of other continents, the relative scarcity of some mammal groups that dominated in Europe and North America is striking. While **"archaic" ungulates** (see pp. 100–3) were highly abundant in both North America and Europe throughout the Paleocene, they were comparatively rare in Asia during this epoch, with pantodonts largely occupying their ecological roles. What this pattern indicates about the ecology of Paleocene Asia remains uncertain. The large quantity of ground-dwelling primitive relatives of lagomorphs and rodents, along with the relative scarcity of arboreal multituberculates (see pp. 45-6) and early primates, suggests that many continental habitats in Late Paleocene Asia may have been more open than those in Europe and North America, although direct evidence for vegetation structure remains limited. The Naran Bulak Formation demonstrates that Late Paleocene Asian mammal faunas became more diverse than their Early Paleocene counterparts. The Late Paleocene dispersal of these groups contributed to fundamentally reshaping the mammal faunas of North America and Europe as the Eocene unfolded.

Fig. 5.14. Phylogenetic tree of a sampling of fossil mammal **taxa** from the Naran Bulak Formation. The contentious phylogenetic position of *Prodinoceras martyr* within Pan-Euungulata (see p. 100) follows a recent phylogenetic analysis by Burger (2015), which places Dinocerata immediately basal to the divergence between Perissodactyla (see p. 126) and Artiodactyla (p. 104), while the phylogenetic position of Pantodonta potentially also within Pan-Euungulata follows Kort (2023), and that of Mesonychia follows Bast and Smith (2013).

Europe Cernay-lès-Reims, France: Thanetian, 59.2–56 Ma.

The record of Paleocene mammals is relatively limited in Europe compared to the rich fossiliferous strata of New Mexico in North America (see pp. 178–83), since much of the European continent was submerged by shallow seas for the majority of this epoch. By the Middle to Late Paleocene, areas of land arose in the region of what is now northern France, and Belgium. Sandstones and **marls** were formed in **fluvial** environments, which transported the remains of vertebrates that existed around the lakes and wetlands formed at the heart of geological basins including the Paris Basin. The best representation of the European Paleocene mammal fauna comes from the Cernay-lès-Reims fossil site in Marne, northeastern France.

The fauna essentially consisted of most of the same mammal lineages as those of North American fossil sites dated to the Late Paleocene, although there are substantial distinctions between the mammal **communities** of both continents, particularly in the proportion of taxonomic groups. The mammal faunas of Paleocene Europe were dominated mostly by "**archaic**" **ungulates** (see pp. 100–3), followed by multituberculates (see pp. 45–6) and plesiadapiforms (see p. 150). Among other relatively large **herbivores**, there were no pantodonts (see pp. 137–8) or taeniodonts (see p. 138) in Europe, which were so common in the North American faunas during the Middle to Late Paleocene. Additionally, the Late Paleocene mammal assemblages of Europe show a greater extent of geographic isolation and **endemism** relative to those of North America, contrary to the biotic interchanges that would occur by the start of the Eocene.

The "archaic" ungulates of Paleocene Europe included "arctocyonids" (see p. 101), which had radiated into a large number of genera including *Arctocyon*, *Arctocyonides* and *Landenodon*. *Arctocyon* was a predominantly **terrestrial** form about the size of a coyote (*Canis latrans*), albeit far more robust. It is known from a near complete skeleton (Argot, 2013), exhibiting the defining features of "arctocyonids", including: short limbs ending in clawed, **plantigrade** feet; an elongated tail; a moderately elongated skull with a pronounced **occipital** and **sagittal crest**; and tusk-like canine teeth. The lower canine teeth are longer and more robust than the upper canine teeth—the former were accommodated in a **diastema** and possibly used for rooting vegetation. It also has relatively simple, sharp premolars along with broad, **bunodont** and **brachydont** molars exhibiting a small number of accessory cusps that expanded the **occlusal** surface, inferring a bear-like **omnivorous** dietary niche. *Arctocyonides* is a smaller and more gracile close relative of *Arctocyon* from Cernay, while *Landenodon* has molars that are less bunodont than those of other "arctocyonids".

Another part of this community of "arctocyonids" was *Dissacus europaeus*, a jackal-sized representative of the family Mesonychidae (see p. 103) that had a meat-dominated omnivorous diet according to quantitative analysis of its molars (Chester *et al.* 2010). The genus *Dissacus* has a rather poor fossil record in Europe, consisting of a complete radius, some lower jaws and an incomplete humerus (Thewissen, 1991). While the limb proportions, flexible elbow joints, and **digitigrade** stance suggest **cursorial** abilities, the relatively rigid back and robust legs indicate that *D. europaeus* was not a fast runner by modern standards, though likely faster than many contemporaneous mammals. The other "archaic" ungulates were mostly specialised herbivores, which include the small hyopsodontids (see pp. 100–1) and the exclusively European pleuraspidotheriids (see p. 100). The teeth of both lineages were substantially more adapted for herbivory than those of "arctocyonids" and lack the impressive canine teeth. In Europe, the hyopsodontids attained their highest diversity during the Paleocene, represented by **taxa** including *Paschatherium* and *Tricuspiodon* at Cernay. The rat-sized *Paschatherium*, which was previously considered to be an early relative of afrotherians (see p. 62), was an **arboreal** form according to its limb and tarsal anatomy. *Tricuspiodon* was likely a more **fossorial** form as indicated by its well-developed shoulders and powerful forelimbs (Thewissen, 1991).

On the other hand, the pleuraspidotheriids exhibit features that are even more specialised, the best known genera which are *Pleuraspidotherium* (Fig. 5.16A) and *Orthaspidotherium* (Fig. 5.16B). The **cheek teeth** are incipiently **selenodont**, **analogous** to those of modern ruminants (see p. 110). *Pleuraspidotherium* and kin may have therefore been **browsers**, feeding mainly on leaves. A near complete skeleton attributed to *Pleuraspidotherium* suggests a generalised plantigrade animal and lacks the **adaptations** for cursorial locomotion found in later ungulates such as artiodactyls (see p. 104). It also may have been **scansorial** to some extent (Rose, 2006). The smaller *Orthaspidotherium* is slightly more derived than *Pleuraspidotherium* in basicranial anatomy and it is further distinct in having a double-rooted, unreduced upper second premolar, whereas that of the latter is single-rooted and reduced.

The lemur-like plesiadapiforms were a particularly common and important mammal group from the forests of Paleocene Europe. During the Paleocene, a mutualistic relationship with **angiosperms** (Sussman, 1991) and a change towards an increasingly omnivorous diet may have played an important part in the **adaptive radiation** of plesiadapiforms (Bloch *et al.* 2007). The smallest and most "archaic" plesiadapiforms from Cernay include representatives of the Microsyopidae, a family which shows diverse variation across size range and diet. These include *Berruvius*, which was approximately shrew-sized (about 20 to 60 grams) and possibly had a **folivorous** diet.

The plesiadapids show features in common with lemurs, with adaptations to increased herbivory. The most well-known and abundant genus is *Plesiadapis*, represented by several well-preserved skulls and near complete skeletons uncovered from Cernay. These skeletal remains show that it was an adept arboreal **quadruped** and somewhat squirrel-like in body structure, with similar climbing adaptations including sharp, laterally compressed claws, mobile ankle and elbow joints, short limb bones with prominent ridges and processes, and a long tail. The skull of *Plesiadapis* is rodent-like, consisting of a long, narrow snout equipped with large, gnawing lower incisors isolated from the molars by a large diastema, as in ptilodontoid multituberculates (see pp. 45, 183). The limb ratios and relatively robust skeleton indicates that *Plesiadapis* may not have been as agile

Fig. 5.15. Phylogenetic tree of a sampling of fossil mammal taxa from Cernay-lès-Reims. The phylogenetic position of Hyopsodontidae follows that of Orliac *et al.* (2012), while that of *Arctocyon primaevus* and *Dissacus europaeus* follows Bast and Smith (2013).

Europe Cernay-lès-Reims, France: Thanetian, 59.2–56 Ma, cont.

as later and more specialised arboreal mammals. *Chiromyoides* is one other plesiadapid known from Cernay. It was small (0.3 kilograms) and a potential descendant of *Plesiadapis* (Rose, 2006), characterised by a deep lower jaw, a short, robust snout, and cheek teeth reduced in size. Like *Plesiadapis*, *Chiromyoides* has sturdily built lower and upper central incisors, and was possibly an insect specialist much like an aye-aye (*Daubentonia madgascariensis*).

As well as the "archaic" ungulates and plesiadapiforms, other abundant mammal groups of this Paleocene **ecosystem** in Europe were the **insectivorous**, shrew-sized adapisoricids. Although known from isolated teeth and jaw fragments, adapisoricids attained high species diversity at the time, with taxa including the **primitive** *Bustylus* and *Afrodon*, characterised by simple, three-cusped upper molars, and the more derived *Adapisoriculus*, whose upper molars exhibit a trend towards developing two lambda (Λ)-shaped transverse ridges. Some archaic placental lineages had lower diversity and typically comprised one genus, such as the pantolestid (see p. 77) *Pagonomus*. The diversity of European multituberculates was bolstered during the Late Paleocene, with the emergence of the mouse-sized ptilodontoid taxa *Neoplagiaulax* and *Liotomus*. These were possibly vaguely squirrel-like, based on comparison to the better-known *Ptilodus* (see pp. 179, 183). *Neoplagiaulax* and relatives exhibit the charcteristic dental morphology of ptilodontids, including a **plagiaulacoid** lower premolar. During this time, *Neoplagiaulax* attained a rich diversity of species, comprising up to four species: *N. sylvani*, *N. eocaenus*, *N. copei* and *N. nicolai*. Additionally, multituberculates of the European family Kogaionidae thrived during the Middle to Late Paleocene. The best known kogaionid is *Hainina*, and was also the most successful. It was initially believed to be a ptilodontid due to its enlarged, plagiaulacoid lower premolar. *Hainina*, along with its Late Cretaceous forebear *Kogaionon*, are "archaic" multituberculates, possessing five upper premolars, along with rectangular molars exhibiting fewer cusps. This distinctive amalgamation of primitive and **derived** features suggests that during the Late Cretaceous, an enlarged plagiaulacoid premolar evolved independently in at least two occasions among the multituberculates.

Subsequent to various "experiments" in the evolution of rodent-like features, true rodents arrived in Europe by the Early Eocene, having dispersed from Central Asia. The earliest European rodents include ischyromyids (see p. 146): arboreal, squirrel-like taxa with simple limbs and dental morphologies. Yet, they already exhibit the defining features of rodents, including a pair of **hypselodont** incisors in the upper and lower jaws, isolated from the cheek teeth by a lengthy diastema. Rodents rapidly dispersed across Europe and predominated the small mammal faunas, while multituberculates declined in diversity and went extinct there by the Middle Eocene. By the close of the Paleocene, while various large mammals had already evolved in Asia and North America, Europe still had a mammal fauna essentially comprising incipient browsers (small- to medium-sized "archaic" ungulates), arboreal taxa (multituberculates, plesiadapiforms and certain "archaic" ungulates), small terrestrial taxa, and insectivores. Yet, the circumstances were to change in Europe too.

Fig. 5.16. Reconstructed life appearance of pleuraspidotheriids from Cernay-lès-Reims. (A) *Pleuraspidotherium aumonieri*; (B) *Orthaspidotherium edwardsi*. Scale bar: 10 cm.

Eocene
56–33.9 Ma

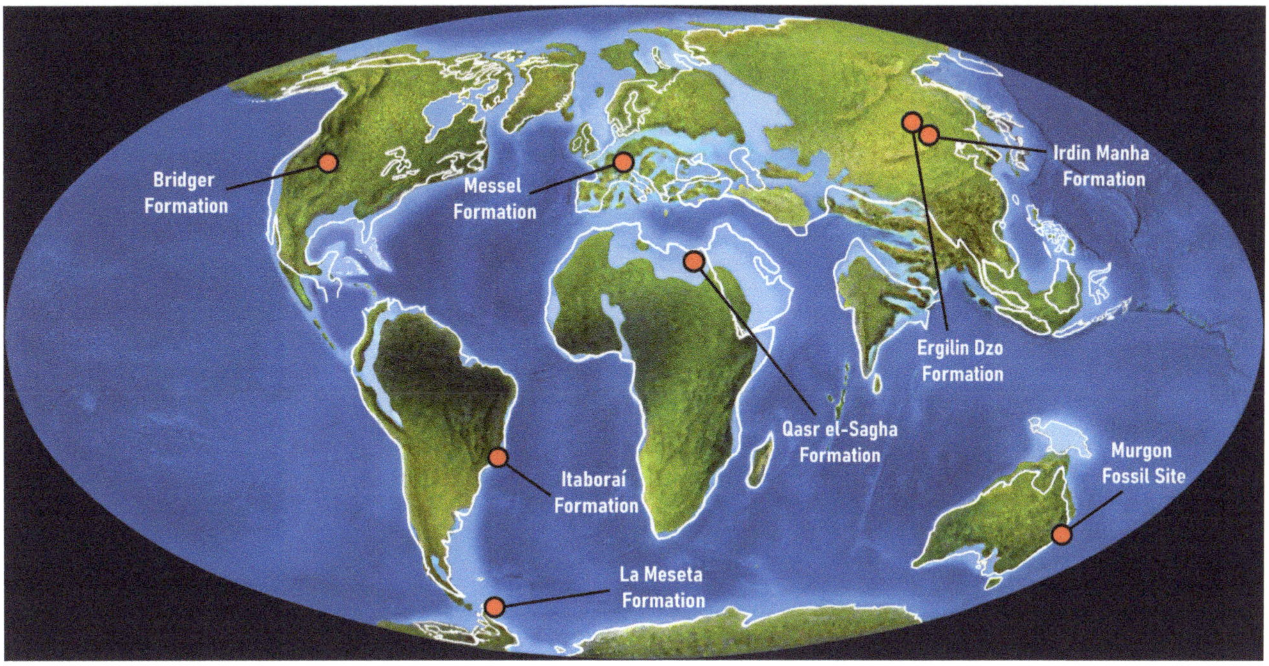

Fig. 5.17. Palaeogeographic map reconstruction of Earth during the Eocene, with the position of the modern continents outlined and the approximate location of the fossil sites indicated.

By the Eocene Epoch, the hothouse world got even hotter. The start of the Eocene is marked by the Paleocene–Eocene Thermal Maximum (PETM), a brief period of intense warming and ocean acidification caused by the release of carbon en masse into the atmosphere and ocean systems possibly as a result of extensive magmatism. This resulted in significant extinctions among deep marine foraminifera (single-celled amoeba-like microorganisms), which were adapted to the cooler Paleocene temperatures. Despite its substantial impacts on marine **ecosystems**, the PETM appears to have not had a major effect on Early Eocene **terrestrial** life, the overall evidence suggesting that it was relatively resilient to the rapid warming and environmental changes. The last pulse of warmth during this heating interval is called the Early Eocene Climatic Optimum (EECO), caused by additional carbon dioxide released by a string of intense hydrothermal events.

With average global temperatures of up to 30°C, the Early Eocene (56–48.07 Ma) is considered to have had the highest **mean annual temperatures** of the entire Cenozoic Era. The temperature gradients from pole to pole were relatively low, and precipitation was high in a world that remained largely free of ice. The increased humidity and warmer temperatures resulted in the advance of rainforests across the mid- and high-latitude regions, reaching southern England and even deep into the Canadian Arctic. These rainforests were more biodiverse and consisted of a greater diversity of tropical flora compared to Paleocene forests. The Early to Middle Eocene rainforests shared some similarities with modern East Asian paratropical forests in their structure and composition, particularly in their diverse evergreen broadleaf trees and dense, multilayered **arboreal** plant life including shrubs, vines and lianas. The increased warmth of the high latitudes, along with the ensuing northerly extension of rainforests, likely facilitated faunal exchanges across the land bridges that still joined North America to Europe and Asia. By the Early Eocene, there were over 30 orders of mammals, as disparate in size and shape as whales, rodents and bats. Most mammals were adapted to living in tropical forests (as in the preceding Paleocene), so the land faunas were dominated by tree-dwellers including lemur-like primates and **insectivorous taxa**, and "**archaic**" lineages of **ungulates** on the forest floor, many of which possessed dental **adaptations** for leaf-based diets. By the Middle Eocene (48.07–37.71 Ma), the separation of Australia from Antarctica opened up the Southern Ocean and formed the deep water Drake Passage between both continents. An isolated cold water passage was formed between both landmasses as warm equatorial currents were diverted away from Antarctica. A cooling trend began in the Antarctic region and the surrounding ocean started freezing, bringing cold water and ice floes northwards, exacerbating the cooling. However, decreasing carbon dioxide levels is considered to have been more important in Middle Eocene cooling. By the Late Eocene (37.71–33.9 Ma), the decrease in carbon dioxide concentrations had led to much lower mean annual temperatures, with increased seasonality and variability globally. The cooler temperatures and greater seasonality brought about an increase in mammalian body size and sparked a shift towards increasingly open **savannah**-like biomes, correlating with the thinning of forests. Many modern grasses originated during the Eocene, but were considerably limited in extent—today's grasslands were still far in the future. Throughout the Eocene, the continents continued drifting towards their modern positions. India began colliding with Asia around 55 Ma, folding and uplifting the Himalayas and Tibetan Plateau—a process continuing to this day.

Australia Murgon Fossil Site, Queensland: Ypresian, ca. 54.6 Ma.

Located in southeastern Queensland, the Murgon Fossil Site is notable for being the Australian palaeontological site with the widest variety of Eocene vertebrates, offering a unique window into the continent's biota during the Early Paleogene. The fossils of various animal groups are also significant for exemplifying the **Gondwanan** connections between Australia, Antarctica and South America, and for indicating large exchanges of species between the continents. While it is likely some species moved in both directions, the Antarctic land bridge facilitated a more prominent one-way movement, particularly from South America to Australia, via Antarctica, the latter continent which formed a lush, green connection (see pp. 200–3). The deposits consist predominantly of clays which settled in a shallow volcanic crater lake, shortly before Australia became an isolated continent. The surrounding habitat has not been determined, but the area surrounding Murgon was possibly a hot and humid swamp forest. There is no certainty about the geological formation of this site, but it probably corresponds to the Oakdale Sandstone Formation. In the earliest Eocene, Australia, Antarctica and South America were the final remnants of the Gondwanan supercontinent to be linked. The Australian marsupials reached Australia via Antarctica, the link between both continents which was by way of Tasmania.

The fossil fauna described from Murgon is known as the Tingamarra Local Fauna. The site has yielded fossils of the only known Australian salamanders and the oldest known songbird (which had a wingspan of 30 centimetres). The oldest known Australian frog fossil was also discovered there. The presence of a small madtsoiid snake (see pp. 23, 25) of the genus *Alamitophis*, also known from Patagonia, provides evidence of Australia's Gondwanan links with South America. *Alamitophis tingamarra*, a species estimated to have been 80 centimetres long, is known from some isolated vertebrae in addition to a fragmented dentary and rib. It is closely related to Argentine species (*A. argentinus*) from the Late Cretaceous. *Alamitophis* is characterised by relatively taller neural arches on the vertebrae than other small madtsoiids. It likely preyed on small mammals, frogs and lizards (Fig. 5.18). The possum-like *Djarthia* is the oldest and most **primitive** known Australidelphian marsupial (see p. 51), and may represent the ancestral form of modern Australian marsupials. *Djarthia* was slightly larger than a house mouse (*Mus musculus*). Skeletal material described include isolated molars, an incomplete cochlear, a tarsal bone, and upper and lower jaw fragments. It was an **insectivorous** generalist as evident from its dentition and had tarsal

Fig. 5.18. A rainforest **ecosystem** at the site of Murgon during the Early Eocene. *Centre*: The madtsoiid snake *Alamitophis tingamarra* stalks its prey: the basal australidelphian marsupial *Djarthia murgonensis*; *Bottom centre*: the carnivorous metatherian *Archaeonothos henkgodthelpi*; *Bottom right*: the extant frog genus *Lechriodus*; *Right background*: the small **omnivorous** marsupial *Thylacotinga bartholomaii*; *Top right*: a primitive songbird (Passeriformes indet.).

adaptations indicative of a grasping hind foot associated with **arboreal** habits. *Djarthia* lies near the base of the Australian marsupial radiation and indicates that the ancestors of South American microbiotherians (see p. 51) probably dispersed back across from Australia via Antarctica (Beck *et al.* 2008). Other notable metatherians (see pp. 43, 46–7) known from very fragmentary material include the unusual marsupial *Thylacotinga* and the **carnivorous** *Archaeonothos*. In *Thylacotinga*, the upper molar specimens show wear facets indicating a strong transverse masticatory stroke, a characteristic of diprotodontians (see p. 54), which may be linked to a dietary transition from insectivory to **herbivory**. *Archaeonothos* exhibits similarities to certain marsupialiforms from South America and Africa, and may have been an early dasyuromorph (see pp. 51–2). Its body mass is estimated at 40 to 80 grams.

Besides the marsupials, placental mammals already existed in Australia during this time. *Australonycteris* is the earliest known bat in the Southern Hemisphere and among the earliest known globally (Fig. 5.19). It is known from several postcranial bones, a partial ear bone, a single dentary bone and several isolated teeth. This "**archaic**" bat suggests the group had spread worldwide soon after it first showed up in the fossil record. *Australonycteris* could echolocate, unlike Early Eocene North American bat species, supporting the hypothesis that flight evolved before echolocation. *Australonycteris* shares numerous morphological features with extant microbats, although its evolutionary relationships to other bats are unclear. It was likely part of a primitive geographical dispersion from Asia (Hand *et al.* 1994), becoming isolated from the **crown** bat groups present on other landmasses. Additionally, the species *Tingamarra porterorum*, so far known from a single molar, an ear bone and an ankle bone, was possibly a placental, specifically a "condylarth" (see pp. 100–3). If true, then it would suggest that **terrestrial** mammals, including placentals, may have arrived in Australia alongside, or even before, the diversification of Australian marsupials. It would also have significant implications for arguments about "marsupial inferiority". However, more complete fossil material is needed to definitively confirm or refute the placental designation. It is thought that *T. porterorum* was a small, ground-dwelling animal, at about 20 centimetres long. The molars were evidently **bunodont** and generalised, and suggest a diet of leaves, fruit and insects.

Fig. 5.19. Reconstructed life appearance of the Murgon bat *Australonycteris clarkae*. Reconstructed wingspan: 20 cm.

The most common fossils at Murgon are of mekosuchine crocodylians (see p. 28) and giant trionychid turtles (see pp. 14–5), which have since become extinct in Australia. *Kambara* is the oldest known mekosuchine and was the largest animal in this ecosystem (Fig. 5.20). It had a generalised crocodylian bauplan, approaching the size of extant saltwater crocodiles (*Crocodylus porosus*). *Kambara* is intriguing in which the bite patterns vary between different species, with interlocking teeth in *K. implexidens* and non-interlocking teeth in *K. murgonensis*. Although the extent of torsion in the humerus is less than that of modern crocodylians, *Kambara* had shoulder adaptations that enabled greater motion and suggests more terrestrial activity then *Crocodylus*. However, *Kambara* may have been a **semi-aquatic ambush predator** because its skull is broad and flattened as in standard aquatic crocodylians. It likely hunted frogs, salamanders, freshwater turtles and small mammals such as *T. porterorum*.

Subsequent to the findings at Murgon, no Australian mammal fossils are known from the following 30 million years prior to the Late Oligocene deposits at Riversleigh (see pp. 244–5).

Fig. 5.20. Reconstructed life appearance of the mekosuchine crocodylian *Kambara implexidens*. Reconstructed length: 3 m.

Antarctica La Meseta Formation, Lutetian-Priabonian, 46.2-36 Ma.

The La Meseta Formation, located on Seymour Island (also known as Marambio Island), adjacent to the northern tip of the Antarctica Peninsula, consists of consolidated marine **siltstones** and sandstones that accumulated in a cool, temperate, deltaic-coastal environment during the Early Eocene. The sediments were specifically deposited in the littoral (nearshore) and shallow sublittoral zones where finer sediments can be transported and deposited by waves and currents. At this time, Antarctica was covered in dense temperate rainforests and very much unlike the continent we know today. The forests were dominated by *Araucaria* and *Nothofagus* trees, but podocarp foliage and palm trees flourished there also, along with **angiosperms** and even a diversity of South American native **ungulates**, or SANUs (see p. 139) and metatherians (see pp. 43, 46–7) (Fig. 5.21). Based on its geographical position, Antarctica may have served as a crucial migratory route for the ancestors of some of the most charismatic mammals of the Southern Hemisphere—Australian marsupials. The lush forested environment of Eocene Antarctica would, of course, ultimately transition into the freezing ice-covered landmass that it is today, isolated from the other continents by the strongest ocean currents on Earth. At La Meseta, sporomorphs—fossilised spores and pollen grains from ancient ferns and palm trees—have been discovered. Samples of pollen from other plants that typically live in tropical environments today have also been found. The remnants of these warm-climate plants can provide us with much information about what Antarctica was like at the time. Since these trees and other plants are not adapted to cold conditions, scientists believe that the coast during the Early Eocene of what is now the Antarctic Peninsula experienced pretty mild winters, with little or no frost. Some estimates for the **mean annual temperature** of that region of Antarctica was around 11 to 13 °C, with pronounced inter-seasonal variation including mean winter temperatures of -3 to 2°C. The surrounding ocean was a mild 12°C on average. By around the time the deposits were formed, the western part of Antarctica had just split from the tip of South America, but the eastern part was still mostly linked to Australia.

Isolated teeth and other fragmentary fossils of ancient metatherians inform us that a diversity of these small mammals thrived in West Antarctica. Judging from their dentition, it appears that some of them, including *Marambiotherium* and *Woodburnodon*, belonged to the Microbiotheria (see p. 51) and were small, **arboreal** or **scansorial insectivores**. The latter species weighed roughly one kilogram, over 20 times as much as the extant monito del monte (*Dromiciops gliroides*). The polydolopimorphian metatherian (see p. 48) *Antarctodolops* was **analogous** to, but not a close relative of, modern shrew opossums (see p. 51), and was likely a scansorial **omnivore**. Initially described in 1984, this opossum-like animal was the first land mammal ever identified in Antarctica's fossil record. Its ancestors likely crossed into the continent from

Fig. 5.21. A coniferous forest in Antarctica during the Middle Eocene. *Centre*: the basal astrapotheriid *Antarctodon sobrali*; *Top left*: the microbiotherian *Woodburnodon casei*; *Top right*: the polydolopimorphian *Antarctodolops dailyi*.

Fig. 5.22. Reconstructed life appearance of sparnotheriodontid litopterns from the La Meseta Formation. (A) *Notiolofos arquinotiensis*; (B) *Notiolofos regueroi*. Scale bar: 0.5 m.

South America, either over a series of closely-spaced islands or a continuous land bridge called the Weddellian Isthmus (Reguero *et al.* 2014). Other residents of Early Eocene Antarctica that came over from South America include medium- to large-sized **herbivores** such as the genus *Notiolofos*, a sparnotheriodontid litoptern (see pp. 143–4) **endemic** to Antarctica. The dentition of this animal indicates that it was a **browser**, feeding on leaves and twigs. Very few specimens have been found, but at least two species of *Notiolofos* are known to have existed. As determined by the dimensions of their molars, the larger species, *N. arquinotiensis* (Fig. 5.22A), was roughly the size of a muskox, weighing in excess of 200 kilograms, and was the largest **terrestrial** herbivore living in Antarctica at the time. Its smaller, sheep-sized relative, *N. regueroi* (Fig. 5.22B), only weighed around 25 to 58 kilograms. The fact that these two species differed greatly in size indicates that they may have both had specialised feeding ecologies, each consuming different types of vegetation so as to avoid direct competition (*N. arquinotiensis* may have included tough vegetation in its diet). This niche differentiation is also evident from dental wear facets.

Another large SANU documented from these Antarctic Eocene deposits is *Trigonostylops*, a peccary-sized (possibly 20 to 40 kilograms) **primitive** astrapotherian (see p. 140) (Fig. 5.23). *Trigonostylops* is represented only by isolated teeth from the La Meseta Formation and its limb anatomy is unknown, but a rather unusual, complete skull is known from South America (in Argentina). It likely lacked a **proboscis** for not having retracted nasals, but possessed a muscular and mobile snout for rooting. The skull includes a pair of well-developed, tusk-like, self-sharpening lower and upper canine teeth,

Fig. 5.23. Reconstructed life appearance of the basal astrapotherian *Trigonostylops wortmani*. Reconstructed length: 1.5 m.

a moderately large **diastema**, a small first premolar, and a battery of **cheek teeth** that are **lophed** and **brachydont**, suggesting that it fed primarily on soft vegetation. The upper incisors are absent in *Trigonostylops*, and the lower ones are so reduced that they likely served no important purpose. As in extant hyraxes, *Trigonostylops* may have therefore processed vegetation using only its cheek teeth. *Antarctodon* was an endemic basal astrapotheriid approximately the same size as *Trigonostylops* (at 18.5 to 38 kilograms) and is known exclusively from a distinct isolated tooth (most likely a lower fourth premolar), but more complete skeletons of other South American astrapotheriids show that this animal probably looked vaguely like a tapir with two prominent upper canine teeth. The known tooth is more brachydont and less **lophodont** than in later astrapotheriids (Bond *et al.* 2011). *Antarctodon* may represent an intermediate position between more basal astrapotherians such as *Trigonostylops* and the first advanced astrapotheriids.

La Meseta has also yielded material of different types of cetacean, including basal mysticetes of the genus *Llanocetus* (see pp. 123–4, Fig. 5.25), along with indeteterminate basilosaurids. *Llanocetus* had a long, narrow skull that retained small peg-like teeth instead of baleen; it would have preyed upon much larger prey, proportionally, than its modern relatives. *Llanocetus* apparently reached up to around 12 metres long and possibly resembled a minke whale (*Balaenoptera acutorostrata*). *Llanocetus* proves that mysticetes became giants before becoming filter feeders, and would have been among the most formidable predators of the Eocene seas. La Meseta is particularly famous for its penguin fossils, in which at least six genera and 14 species of well-established fossil penguins have been recognised from Seymour Island, known mostly from wing and foot bones. This is clearly the highest penguin diversity documented from any one small region, as a result of various **morphotypes** that may have evolved in an act of niche partitioning. However, it is difficult to compare each species as only isolated bones are known and proportions differ. *Anthropornis* is the largest penguin species known, which reached 1.8 metres in height. A defining feature is a bent joint in the wing, possibly a relic from flying ancestors. The slightly smaller genus *Palaeeudyptes*, known from an extensive quantity of material, was larger than all extant penguins, the smallest species (*P. gunnari*) being the size of an emperor penguin (*Aptenodytes forsteri*), at 1.2 metres tall (Fig. 5.24A), and the largest (*P. klekowskii*) being about 1.6 metres tall (Fig. 5.24B).

Antarctica La Meseta Formation, Lutetian–Priabonian, 46.2–36 Ma, cont.

On the whole, its osteological features appear to have been almost on a par with the enormous *Anthropornis* (Fig. 5.24C). The long, slender beaks of giant penguins like these can be incredibly long, up to two-thirds of the skull's length. Neck musculature was strongly developed to resist impacts and it is thought that they used the bill like a spear on larger prey. A larger body size and robust flippers would have also allowed the giant penguins to dive deeper and remain underwater for longer periods for hunting. One other morphotype corresponds to more medium-sized penguins with shorter beaks, resembling the Neogene species in configuration. *Archaeospheniscus* is among the oldest known penguins, measuring 75 to 85 centimetres in height. It retains a somewhat slender humerus, intermediate in morphology between that observed in standard bird wings and the dense structure present in extant penguins. Furthermore, the tarsometatarsus displays an unusual combination of features seen in primitive and modern species. It remains uncertain as to whether this indicates that *Archaeospheniscus* was ancestral to modern penguins or is an example of **parallel evolution**. *Delphinornis* was a slender-footed penguin about 90 centimetres tall that survived up until the Middle Miocene and had a more elongated tarsometarsus than the giant-sized penguins. It appears that Eocene Antarctic penguins attained higher diversity than their modern relatives in terms of body size.

Penguins are not the only fossil birds found in the La Meseta Formation. Recently described material has revealed the presence of truly colossal pelagornithid seabirds (see p. 36) such as *Dasornis* roaming the prehistoric polar skies (Fig. 5.25). This hints to an extended survival of these birds within the Southern Hemisphere, and their adaptability as pelagic predatory elements of coastal and marine **ecosystems** along with primitive penguins. A partial left dentary is of similar size to jaws of the largest known Neogene pelagornithids, suggesting an animal with a wingspan of up to six metres. Fragmentary material from the older La Meseta sediments are from similarly large individuals, along with more modest-sized forms with a 3.5 to 4.5 metre wingspan. This material shows that pelagornithids clearly reached large body sizes very early on in their history. Other seabirds included *Notoleptos giglii*, one of the earliest albatrosses (Diomedeidae). *N. giglii* was much smaller than its modern relatives, with a wingspan of only about 1.2 metres. It is reasonable to think that the contemporary predatory pelagornithids developed hyperspecialised flight earlier than albatrosses, relegating the first diomedeids to more generalist niches.

All these animals that shared this environment are highly significant to palaeontologists as Antarctica's fossil record is much less comprehensive than that of other continents. While Antarcticas's terrestrial mammals and other warm-adapted animals were still thriving, some drastic changes were just around the corner. As the climate cooled and glaciers became more prevalent across the continent by the Late Eocene around 36.5 Ma, many floral **communities** deteriorated. Analysis of fossil plant samples from the Antarctic Peninsula have shown that from the Late Paleocene to Middle Eocene, the continent's floral diversity declined by almost half. The tropical trees and ferns were gradually replaced by temperate forests dominated by southern beech trees. These trees are known to have lived on the continent since the Late Cretaceous Period, according to sporomorphs and fossil leaf impressions, and survived on Antarctica until as recently as 2.5 Ma. Today, Antarctica is a treeless landmass and a polar desert, without any terrestrial mammals, reptiles or amphibians and whose remaining vegetation consists primarily of hardy mosses, algae, lichens and grasses. The biodiversity of Antarctica was definitely hard hit after the Eocene while life continued to thrive on its two former neighbours.

Fig. 5.24. Reconstructed life appearance of palaeeudyptines ("giant penguins") from the La Meseta Formation. (A) *Palaeeudyptes gunnari*; (B) *Palaeeudyptes klekowskii*; (C) *Anthropornis nordenskjoeldi*. Scale bar: 1 m.

1. The small, primitive penguin *Archaeospheniscus wimani*, related to the giant penguins (Palaeeudyptinae).
2. A flock of the small, basal penguin *Delphinornis larseni* takes to the sea to feed.
3. *Llanocetus denticrenatus*, a toothed, basal mysticete cetacean.
4. *Dasornis* sp., a massive pelagornithid seabird.
5. Pelagornithidae indet.
6. A flock of the primitive albatross *Notoleptos giglii*.
7. *Antarctoboenus carlinii*, a **stem** relative of falcons.

Fig. 5.25. A coastal ecosystem in Northern Antarctica during the Middle Eocene.

North America Bridger Formation, United States: Ypresian–Lutetian, 49.5–46.5 Ma.

The Bridger Formation, confined to the Green River Basin in southwestern Wyoming, has preserved the most diverse assemblage of Eocene mammals from North America and has yielded some of the most amazing and unusual fossils. During the Middle Eocene, the Rocky Mountains were only a young mountain chain and the river basins amongst the hills were at considerably lower altitudes than at present. The basins were covered by dense, tropical forests in what is now Colorado, Utah, Wyoming and Oregon. The Green River Basin has preserved faunal successions extending roughly four million years. In the Early Eocene, a shallow lake was present in the basin that over time became infilled with pyroclastic deposits that in turn subsequently underwent **fluvial** dissection. Extensive layers of pyroclastic material are evident of intense volcanic activity. In a large wave, the ashes drifted downstream, which quickly transformed the entire structure of the valley. The mass deaths of freshwater turtles, which are represented by hundreds of skeletons from the Bridger Formation, have also been associated with these volcanic eruptions. The landscape of what is now the Western United States during the Middle Eocene must have been a spectacular sight from a bird's eye view. On the valley floor flowed a large, meandering fluvial system, with oxbow lakes formed in areas where the currents were reduced. In certain places, the forest would stretch to the edges of the river, while in areas where the flow was reduced, swampy plants proliferated. The lowlands were dominated by lush forests with a diverse array of tree species such as palms, maple, oak, birch, walnut and willow. Coniferous forests were distributed along the flanks of hilly terrain, and even higher towered volcanic landforms.

The first animals to catch one's attention in this environment would have been the giant, vaguely rhinoceros-like dinocerates (see p. 139), which were one of the mammal groups to peak in diversity in the Eocene after gaining a foothold during the Late Paleocene. They are represented by well-preserved skulls and postcranial elements, including the best known genera including the large *Uintatherium* and even larger *Eobasileus* (Fig. 5.26), the first truly gigantic mammals and the largest in North America at the time. They both possessed two pairs of bony **ossicones** on the head and another small pair of horns over the nose, perhaps used for mating displays. The more **primitive** *Bathyopsis* did not reach such gigantic dimensions. The skull of *Bathyopsis* was elongated (35 to 50 centimetres long) and depressed, with a relatively long **postorbital** region, and there were a pair of very short horns immediately behind the nostrils. A pair of long, blade-like teeth in all three dinocerates were presumably used for territorial fighting as in modern wild pigs. Most depictions of either species have the teeth exposed, but the presence of a bony flange on the lower jaw suggests the teeth were concealed when the mouth was closed. They fed on soft vegetation as indicated by their broad, simple, **brachydont cheek teeth** with V-shaped crests. As well as the dinocerates, perissodactyls (see p. 126) were radiating in the Middle Eocene too, represented by equids (see pp. 132–4) including the fox-sized *Orohippus*. These equids remained small and primitive, but had fewer outer toes than their earlier relatives such as the genus *Eohippus*. Early tapiroids (see p. 128) such as *Helaletes* were present too, which was relatively slender and long-limbed but started displaying the signs of the **proboscis** that characterises modern tapirs. Some of the earliest brontotheriid genera (see pp. 135–6), such as *Eotitanops* and the more advanced tapir-sized *Palaeosyops*, were also diverse, but had more primitive cranial proportions and lacked the characteristic horns of later genera. They did, however, possess large canine teeth which were probably used in **intraspecific competition**. *Palaeosyops* had a more complex tooth arrangement than *Eotitanops*.

Fig. 5.26. Reconstructed life appearance of the uintatheriid dinocerate *Eobasileus cornutus*. Reconstructed length: 4 m.

Artiodactyls (see p. 104) also quickly radiated subsequent to their initial dispersal around the globe during the Early Eocene. From the miniscule, chevrotain-like dichobunids (see p. 104), they diversified into over a dozen families, many of which resembled long-tailed deer to the non-specialist. In addition, some resembled pigs, such as advanced members of the Helohyidae which occupied the niche of large **omnivores**. The helohyid *Achaenodon*, known from partial skulls and lower jaws, was a massively-built, stocky animal with a short, robust snout and enlarged jaws that possessed prominent canine teeth and **bunodont** molars with inflated crowns and sometimes rugose enamel. The skull of *Achaenodon* has a pronounced **sagittal crest** and wide **zygomatic arches** to which powerful jaw muscles were attached, making the skull very broad. Because of these features and based on its massive teeth, *Achaenodon* was long classified as an entelodontid (see pp. 118, 120). Additionally, its forefeet and hind feet were **tetradactyl**, with unfused metapodials. Given its large size, *Achaenodon* would have been an uneasy target for Middle Eocene **carnivores** to prey on.

This suite of advanced large **herbivores**—dinocerates, perissodactyls and artiodactyls—abruptly displaced earlier members of the large herbivore **guild**. The "**archaic**" **ungulates** (see pp. 100–3), which were previously dominant during the Paleocene and Early Eocene, had largely vanished, with only the vaguely horse-like phenacodontids and the hyrax-like hyopsodontids (see pp. 100–1) persisting. *Hyopsodus* is known from well-preserved specimens from the Bridger Formation, whereas many other hyopsodontids are only represented by their unique molars, which are square-shaped and brachydont, possessing four prominent cusps. *Hyopsodus* was previously thought to have a sleek, elongated body and tail, giving it the appearance of a dachshund and the nickname "tube-sheep".

Fig. 5.27A. Phylogenetic tree of a sampling of fossil mammal **taxa**, particularly insectivorous mammals, euarchontoglires (see p. 144), and members of Ferae (see p. 74), from the Bridger Formation. The phylogenetic position of Leptictida within Placentalia follows Brady *et al.* (2024).

North America Bridger Formation, United States: Ypresian–Lutetian, 49.5–46.5 Ma, cont.

The recent discovery of complete skeletons have provided us with a better understanding of its proportions, and that rather it was more hyrax-like in build, with a thicker body and a more strongly arched spine. Given the wealth of fossil material, *Hyopsodus* may have thrived in an ecological role not filled by other primitive mammals, scampering through the dense undergrowth and dodging predators; it may also have been **fossorial**. Brain scans of skulls reveal that *Hyopsodus* had a good sense of smell and hearing, and was possibly using a basic form of echolocation for orientation. The sheep-sized *Phenacodus* is known from multiple complete skeletons. The skeletons are rather primitive and unspecialised, featuring a small skull with a proportionately small branicase, a robust, elongated body with an arched back, relatively straight limbs, and a long, heavy tail. The limb morphology and **digitigrade** stance suggests that *Phenacodus* was adapted for running, but less so than modern, highly specialised ungulates. *Phenacodus* had **pentadactyl** limbs, yet the hallux and thumb are reduced and starting to disappear (these would be lost entirely in nearly all advanced ungulates). These morphological characters suggest that the primary form of locomotion for phenacodontids including *Phenacodus* was walking and slow running.

The North American pantodonts (see pp. 137–8) survived until the Middle Eocene, and the scarce taeniodonts (see p. 138) and tillodonts (see p. 138) both disappeared by the end of the Eocene. *Stylinodon* represents the best known, largest (80 kilograms) and final member of the taeniodont group (Fig. 5.28). While it was naturally assumed that stylinodontids fed mostly on roots and tubers, the lack of abrasions on the teeth of *Stylinodon* suggests it was consuming different items. A polished surface is consistent with tough material such as branches, and it is possible that some taeniodonts were **browsers** of low vegetation, using the claws primarily for digging burrows. *Trogosus* is the only genus of tillodont represented by postcranial elements other than teeth and jaws (Fig. 5.29). It was bear-like in size and appearance, with a skeleton that was robust albeit relatively unspecialised, supposedly supporting a powerful animal estimated to have weighed around 150 kilograms (330 pounds). It had heavyset limb bones bearing prominent crests for anchoring powerful muscles, and large claws. This, along with the rodent-like, rootless second incisors, indicates a tendency of foraging for roots and tubers. The molars frequently exhibit signs of heavy wear, indicative of a diet high in abrasive plant material such as roots and tubers, although lateral grooves on the incisors indicate that they also plucked leaves from branches.

Like the large herbivores, the large carnivorous mammalian fauna was changing too. The "triisodontids" (see pp. 102–3) were long extinct and the dog-like, hooved mesonychids (see p. 103) were already much scarcer by the Middle Eocene. The limb joints of mesonychids were modified to limit rotational movements (**convergent** on those of extant ungulates), enabling them to achieve greater **cursorial** abilities over time (Rose, 2006). *Synoplotherium* and notably *Mesonyx* were slender mesonychids that intensified these trends in possessing somewhat more elongated and compact hind feet, helping facilitate efficient movement in a forward direction. *Synoplotherium* would have vaguely resembled a small wolf with a proportionately large head. The similar-looking *Mesonyx* had a relatively larger and more complex brain in comparison to those of "archaic" ungulates from earlier times, but less so than in modern carnivorans (see pp. 81, 83–4). It still would have been an intelligent mammal for its time and perhaps even an active hunter (Radinsky, 1977).

Harpagolestes was a bear-sized and more derived mesonychid best represented in North America (represented by *H. immanis* and *H. brevipes* at the Bridger Formation). It possessed heavily developed jaws, large teeth and a mandibular **condyle** morphology demonstrating a high degree of leverage for delivering a powerful bite. This, along with its robust limbs, indicates that *Harpagolestes* was a **scavenger** (Fig. 5.30), which was also capable of cracking bones as evident from wear on the molars. While both *H. immanis* and *H. brevipes* shared similarities in dental features and overall build, the latter species was larger overall and had a more **posterior** extension of the mandibular symphysis.

Fig. 5.28. Reconstructed life appearance of the stylinodontid taeniodont *Stylinodon mirus*. Reconstructed length: 1.7 m.

Fig. 5.27B. Phylogenetic tree of a sampling of fossil mammal taxa, particularly ungulates, from the Bridger Formation. The contentious phylogenetic position of Dinocerata within Pan-Euungulata (see p. 100), and immediately preceding the split between Perissodactyla and Artiodactyla, follows a recent phylogenetic analysis by Burger (2015). The phylogenetic position of Mesonychia follows Bast and Smith (2013), while that of *Hyopsodus* follows Orliac *et al.* (2012), and that of *Phenacodus* follows Cooper *et al.* (2014) and Bai *et al.* (2021). The phylogeny within Mesonychia follows Solé *et al.* (2023).

North America Bridger Formation, United States: Ypresian–Lutetian, 49.5–46.5 Ma, cont.

All three mesonychid genera went extinct in North America by the end of the Middle Eocene. During the Middle Eocene, the oxyaenids (see p. 82) and hyaenodonts (see pp. 82–3) were taking over from the mesonychians as the dominant carnivores. The stockier oxyaenids were among the top predators of their age and included the cougar-sized *Patriofelis*, which had a short, thick skull with prominent canine teeth and a small braincase. Its powerful forequarters, short limbs and bear-like feet suggest that *Patriofelis* was an **ambush predator** potentially with some short bursts of speed, but not for prolonged running or swimming. The limb proportions and inflexible vertebral column of *Patriofelis* suggest it was unlikely to be **semi-aquatic** or **arboreal**, and was instead a **terrestrial** predator adapted for hunting in dense vegetation. *Patriofelis* was probably a generalist carnivore, and the teeth were more robust than other oxyaenids, with heavy wear, so it likely fed on abrasive food items.

The cat-like *Oxyaena* had a flexible body, broad, low skull, long facial region, large lower jaw, and short limbs. It possesses a well-developed **carnassial** shearing dentition, adapted for crushing rather than shearing. However, it too was likely a generalist feeder, preying on eggs, insects and small- to medium-sized vertebrates. One specialised subfamily of oxyaenids includes the small (the size of a domestic cat) *Machaeroides*, which with its strong musculature and elongated, flattened upper canine teeth may have been capable of hunting prey much larger than themselves, such as early equids and dog-sized brontotheriids. These sabreteeth were protected by a distinct ventral flange at the front of the lower jaw, while the lower canine teeth were considerably shortened, nearly blending into the incisor row. The blade-like carnassials were more suited to **hypercarnivory** than those of other oxyaenids. These specialisations made machaeroidines the earliest known sabre-toothed predatory mammals. The more gracile hyaenodonts were more diverse than the oxyaenids and generally had proportionally larger skulls than other contemporary carnivores. Many of the hyaenodonts from the Bridger Formation do not fit major **clades**, but various side branches. They all show **adaptations** for forest-living, having proportionally short limbs to allow them to hunt in the undergrowth. One of the most well-known is *Sinopa*, a wide-ranging, fox-like and cursoriall hyaenodont characterised by more **derived** dental features and with species found across Eurasia and North America. *Tritemnodon* was a small, gracile and predominantly terrestrial hyaenodont that would have converged on modern civets in skeletal morphology, with an elongated body and short, powerful limbs. The marten-like *Prolimnocyon* was one of the smallest and most primitive hyaenodonts, characterised by a relatively long muzzle, a prominent sagittal crest, short, agile, semi-**plantigrade** limbs, and a very long tail. *Prolimnocyon* also had a relatively retrograde dentition, with well-developed premolars, specialised for feeding on small vertebrates and insects. It too was primarily a terrestrial predator.

Meanwhile, the “miacids” and viverravids, which were basal carnivoramorphs (see p. 84) that may have been the ancestral stock of modern carnivorans, remained small. The “archaic” oxyaenids and hyaenodonts were also on the decline and would, in time, be displaced by the true carnivorans. Near complete specimens of *Miacis* have shown that it resembled a pine marten (*Martes martes*), with a low-slung body, well-developed shoulders, mobile limb joints and an opposable hallux for grasping, making the animal an arboreal predator. Like numerous primitive placental mammals, but unlike modern carnivorans, *Miacis* has a **complete dentition** of 44 teeth, which it would have used for feeding on small mammals, birds, reptiles, eggs and fruit, particularly in the forest canopy. *Miacis* also has a relatively large braincase and, like modern felids (see pp. 86–9), had retractable claws. *Vulpavus* was a larger and rather lemur-like “miacid” with a locomotor behaviour comparable to that of modern coatis (*Nasua*), sharing the same arboreal adaptations with the genus *Miacis*. The slimmer viverravids such as *Viverravus*, a typical representative, differed from the “miacids” by lacking third molars and having an articulation between the calcaneus and fibula. In *Viverravus*, the ankle joint suggests it was better adapted to terrestrial locomotion than the “miacids”. The terminal **phalanges** of both “miacids” and viverravids were short and laterally compressed.

The small mammal fauna also underwent major developments. During the Early Eocene, the forest canopy was populated by the rodent-like multituberculates (see pp. 45–6) and the primitive, rodent-like plesiadapiforms (see p. 150). Among the rodent genera that were likely outcompeting the

Fig. 5.29. Reconstructed life appearance of the tillodont *Trogosus castoridens*. Reconstructed length: 1.5 m.

multituberculates by the Middle Eocene were *Paramys*. It was one of the earliest known rodents and a distant relative of squirrels. *Paramys* was a generalist capable of varied locomotion as indicated by its hindlimb anatomy. Its forelimbs resemble those of extant tree squirrels, whereas its powerful hindlimbs resemble those of extant ground squirrels. The plesiadapiforms were also being replaced by the diversification of rodents and two recent immigrant lineages of **crown** primates—primitive strepsirrhines of the extinct families Omomyidae and Adapidae (see p. 150). *Notharctus* and its relatives are the best known adapids, which were likely active arboreal animals and share many anatomical similarities with modern lemurs, but were less specialised in leaping and vertical clinging. Like modern lemurs, the metacarpals of *Notharctus* are short and the phalanges elongated, with the fourth **digit** of the manus being the longest as an adaptation for gripping. *Smilodectes* was a small, lightly built adapid with a small cranium and a moderately short snout, along with a rounded **frontal** bone (as in *Notharctus*) (Fig. 5.31). The larger visual cortex and relatively small **olfactory** bulbs indicate that *Smilodectes* had a **diurnal** lifestyle. Both genera have **selenodont** cheek teeth indicative of a **folivorous** diet.

The **insectivorous** mammal fauna continued to quickly transform, with certain groups such as the palaeanodonts (see pp. 79–80) diversifying into anteater-like genera. Among these are *Metacheiromys*, which likely used its powerful claws to dig into and dine on the termite and ant nests that were widespread in the Middle Eocene rainforests. *Metacheiromys*, known from numerous skeletons, is characterised by a cylindrical body with a long, low and narrow skull vaguely resembling that of modern pangolins and anteaters. Unlike those animals, however, *Metacheiromys* had robust canine teeth and a reduced number of cheek teeth, having simple horny pegs in its jaws that were used for grinding food. Its skeleton exhibits numerous fossorial adaptations such as a wide **distal** humerus, strong elbow joint and enlarged claw-like **unguals**. Other fossil insectivores from Bridger include leptictids (see p. 77) such as *Palaeictops*, which was relatively more omnivorous than other leptictids due to the more reduced and rounded molar crowns. *Palaeictops* may have moved mostly **quadrupedally**, but like extant elephant shrews, may have also engaged in **bipedal** leaping on its hindlimbs for speed and agility. Moreover, the forelimbs of *Palaeictops* have some digging adaptations, but it was not strictly fossorial.

The terrestrial vertebrate fauna of North America exhibits a number of remarkable trends during the Middle Eocene. Faunal diversity was similarly high as in the Early Eocene but then quickly declined during the Late Eocene, likely in response to a cooling climate.

Fig. 5.31. Reconstructed life appearance of the adapiform primate *Smilodectes gracilis*. Reconstructed length: 40 cm.

Fig. 5.30. The Green River Basin during the Middle Eocene. The large mesonychid *Harpagolestes brevipes* (right) and the artiodactyl *Achaenodon uintensis* (left) compete with each other for the carcass of a young brontotheriid (centre) that drowned the day before during a flood caused by a severe thunderstorm with prolonged rainfall.

Europe Messel Formation, Germany: Lutetian, ca. 47 Ma.

The Messel Formation, dated to the Middle Eocene, is a Konservat-**Lagerstätte**: a site of exceptional preservation and therefore one of the world's fossil "hotspots". Located in east-central Germany near Frankfurt, its geographical range is limited to the Messel Pit, a former Eocene lake of volcanic origin that was surrounded by dense subtropical rainforests as indicated by the fauna and flora. The Messel Formation is the richest European fossil site for understanding Eocene environments and **terrestrial communities**, having yielded thousands of exceptionally well-preserved fossils including complete **articulated** skeletons, wing membranes, feathers, hair and even stomach contents. During the Eocene, extensive areas of Europe, including parts of Germany, the Paris Basin and London, were submerged underwater as evidenced by marine fossils from those regions. Messel was located slightly further south to around the latitude of modern Sicily and Southern Spain. Messel was also bordered to the south by the ancestral Mediterranean Sea and being so surrounded by sea, was located on a large Central European island, and continental Europe continued to be an island archipelago.

The Messel Pit is believed to have formed above a maar (a broad, shallow volcanic crater). This maar was formed when magma moved upwards along steep faults that had formed previously in this area's geological history and reacted with the groundwater to trigger explosive eruptions. These eruptions gradually subsided, leaving behind a maar that was subsequently infilled with intruding groundwater and rainwater to form the Messel Lake. The base of the Messel Formation contains coarse, clastic sandstone and gravels, which was likely deposited before the lake environment stabilised and the Messel oil shale began to form. The coarse, clastic sediments are typical of deposits formed in a higher-energy environment, such as a riverbed or shoreline, and suggest a period of active erosion and deposition. Overlying those sediments is the Messel oil shale, the transition to that layer which indicates a shift to a low-energy, **lacustrine** environment. This oil shale, the main section of the **strata**, was formed by the gradual **anoxic** accumulation of mud and dead vegetation on the lakebed.

The anoxic conditions of the lakebed prevented aerobic microorganisms from decomposing organic material and also prevented **bioturbation**, as the lakebed was inhospitable to any organisms living on and within the lakebed. Therefore, complete skeletons are available and fully articulated due to the very low energy environment of the lake and the fine grained sediment of the oil shale. The soft tissues were exclusively decomposed by anaerobic bacteria, which were then fossilised themselves, leaving behind a "skin shadow" or outline of the original tissue. This outline was formed by the waste products of the bacteria as they decomposed the organic material. These excreted wastes included respiratory carbon dioxide, which reacted with iron in the lakebed sediment to form an iron **carbonate** called siderite. The accumulation of siderite formed a thin layer over the bacteria and surrounding soft tissues. This layer was subsequently filled with oily organic material. The combined effect of the siderite layer and the oily material formed dark, shadowy outlines around the soft tissues and integument. This unusual mode of fossilisation has provided a unique snapshot of a terrestrial **ecosystem** complete with leaves, branches, fruit and seeds, insects, fish, amphibians, turtles, crocodylomorphs (see pp. 26–30), snakes, lizards, birds and mammals.

A diverse fauna of recognisably modern freshwater fish including bowfins (*Cyclurus kehreri*) and one metre long gars (*Atractosteus messelensis*) inhabited the Messel waterways. The latter likely fed on smaller fish such as eels (*Anguilla ignota*) and perches (*Amphiperca multiformis* and *Palaeoperca proxima*). In turn, these **osteichthyans** and other aquatic vertebrates were likely preyed on by a large variety of crocodylians amongst the lake. The largest form was *"Asiatosuchus" germanicus*, a four metre long

Fig. 5.32. Reconstructed life appearance of a sampling of extinct crocodylians from Messel. (A) *"Asiatosuchus" germanicus*; (B) *Diplocynodon darwini*. Scale bar: 1 m.

Fig. 5.33. Reconstructed life appearance of the sebecian crocodylomorph *Bergisuchus dietrichbergi*. Reconstructed length: 1.5 m.

crocodylid (see pp. 27–8) that used its robust tail for propulsion in the water as its modern relatives do (Fig. 5.32A). The most abundant and successful genus was *Diplocynodon*, a 1.5 metre long caiman-like member of the alligator **clade**, a group which reached smaller sizes but was more species-rich (Fig. 5.32B). *Diplocynodon* mostly ate flesh as evident from its short snout, and was an opportunistic predator. The neck, back and underside were covered in **osteoderms**, but the tail was almost devoid of them. Contrary to the aforementioned crocodylians, the 1.5 metre long sebecian *Bergisuchus* was a terrestrial, **cursorial** predator with long limbs held directly underneath the body (Fig. 5.33). Unlike modern crocodylians, *Bergisuchus* had separate nasal openings and laterally facing **orbits**, as indicated by the snout shape.

Large boid snakes (see p. 24) such as the 6 foot long *Eoconstrictor fischeri*), ancestral pythons such as the 3.2 foot long *Messelopython freyi*, and small lizards such as the early amphisbaenian (see pp. 21–2) *Cryptolacerta* (Fig. 5.34), often visited these waters as they do in the rivers of modern tropical forests. *E. fischeri* was likely a generalist, and gut contents reveal that it hunted small crocodylians and large lizards. It also possessed heat-sensing organs in its upper lip area, although since it hunted cold-blooded vertebrates, heat sensing possibly arose as an **adaptation** for better sensing the environment. Three other smaller fossil booids (see p. 23) are known from Messel, one of which was **arboreal** (*Messelophis variatus*), and two of which were terrestrial (*Rageryx schmidi* and *Rieppelophis ermannorum*), although all the smaller snakes lacked heat-sensing pits. *Cryptolacerta* is represented by a single specimen including an articulated and almost complete skeleton, lacking only the end of the tail. It has relatively small limbs and a thick skull with a morphology further supporting the amphisbaenian-lacertid molecular relationship and suggesting that this particular group of headfirst burrowing reptiles evolved reinforced skulls prior to snake-like bodies. Also, Messel has the honour of being the site of the only known fossils of vertebrates mating, which constitute nine male-female pairs of the small, aquatic pig-nosed turtle *Allaeochelys crassesculpta*. These turtles had scaleless skin, enabling oxygen absorption from the water, but it made them vulnerable to Messel's lethal volcanic gases too.

Fig. 5.34. Reconstructed life appearance of the lacertoid lizard *Cryptolacerta hassiaca*. Reconstructed length: 7.5 cm.

The avifauna is more diverse at Messel than any other major vertebrate group, with over 70 species known to date and many modern arboreal families including swifts, hummingbirds, nightjars, parrots, trogons, mousebirds, rollers and hoopoes being well-documented, as well as various waterbirds including **primitive** ibises, flamingos, kingfishers and rails, and **stem**-gannets. *Messelastur* was a parrot relative that occupied the niche of a falcon or hawk. The most predominant arboreal bird is the unspecialised and **insectivorous** *Messelirrisor*, the earliest known member of the hoopoe family (Upupidae) that resembled its modern relatives but was smaller in size. Its bill was slightly curved and its fossils show black bars on the tail feathers. Among water birds, *Eocoracias* was an early relative of kingfishers that is noteworthy for a melanosome (colouration) analysis that determined it was either grey or had non-iridescent structural colour. *Masillastega* was an early relative of gannets and boobies, and had a similar-shaped conical beak. Contrary to those birds, *Masillastega* was bound to sub-tropical freshwater lakes rather than marine habitats, and the beak was probably unfit for plunge diving. *Masillaraptor* was a long-legged, stem-falconiform (see p. 35) that possibly specialised in preying on snakes, and likely had well-developed flight capabilities.

Several ground-dwelling birds are known too, such as *Palaeotis* (see p. 33), a one metre tall, basal palaeognath (see pp. 32–4) that was flightless as indicated by the lack of a keel. *Palaeotis* is characterised by a narrow beak and wings that were relatively large compared to modern ostriches, but smaller than those of flying birds. *Messelornis*, a generalised, long-legged, moorhen-sized early relative of rails, is by far the most commonly found fossil bird, representing two thirds of all bird fossils from Messel, and is known from hundreds of skeletons. It may have hunted near the waters' edge looking for seeds, insects

Europe Messel Formation, Germany: Lutetian, ca. 47 Ma, cont.

and fish. One specimen found has even shown a fish still in the birds' throat. The largest and tallest bird of the ecosystem was *Gastornis geiselensis*, which is closely related to waterfowl (Fig. 5.35). It was previously assumed that these giant birds dominated Early Paleogene ecosystems as the apex predators, in the "absence" of large mammals. Flightless gastornithids (see p. 36) were frequently depicted as running down unfortunate mammalian victims, commonly early horses, which were dispatched with a blow from their deadly beak. For various reasons, this story is a complete fabrication. Standing as tall as an average male human, there is no doubt that *G. geiselensis* would have been an intimidating bird. A large head on a strong neck with a deep hatchet-shaped beak looks like it could do some serious damage. Yet, the beak had straight edges and lacked a hooked tip, as one would typically see in **carnivorous** birds. Short lower legs that were not built for sprinting and blunt hoof-like claws complete an unlikely predator. Deep muscles anchoring the lower jaw were certainly capable of generating a powerful bite, but it was not for tearing into a carcass. The beak was built to crack hard fruits and tough seeds, and subsequent calcium isotope analyses have confirmed that this bird was a **herbivore**.

Fig. 5.35. Reconstructed life appearance of the gastornithid bird *Gastornis geiselensis*. Reconstructed height: 1.75 m.

There is an interesting pattern regarding the mammals at Messel: ancient lineages such as the "cimolestans" (see pp. 77–8) and hyaenodonts (see pp. 82–3) played more specialised roles in the ecosystem, whereas modern lineages such as the eulipotyphlans (see pp. 74–7), rodents, primates, perissodactyls (see p. 126) and artiodactyls (see p. 104) were more generalised and all still largely resembled the ancestral mammal body plan. This suggests that the generalised lineages, with their broader adaptability, may have had a longer evolutionary trajectory, diversifying into the modern mammalian orders we see today. Considering the incredible diversity of insects, especially beetles, that populated the Messel rainforests, there is no surprise that many of the famous mammals from this site include small, insectivorous genera such as *Macrocranion*, a primitive eulipotyphlan of the extinct family Amphilemuridae. The large ears, small eyes and a long, dexterous snout indicate that *Macrocranion* may have been a terrestrial and mostly **nocturnal** animal, leaping **quadrupedally** on the forest floor with its long, powerful hindlimbs hunting for insects and fruit. It also hunted fish according to fossilised stomach contents. Its teeth were similar to those of modern erinaceids (hedgehogs), but unlike them, it did not exhibit the typical spiny fur. More hedgehog-like and sturdily-built was the amphilemurid *Pholidocercus*, which bore a dense covering of relatively long bristles on its back. Uniquely, it had a helmet of thick scaly skin on its head and a thick, elongated tail that was scaly. These features indicate that *Pholidocercus* was a ground-dweller, using its horny cephalic shield as an adaptation for digging. The leptictidans (see p. 77) at Messel include *Leptictidium*, the best known form given the number of perfectly preserved specimens, some with stomach contents suggestive of an **omnivorous** diet, consuming insects, leaves and small vertebrates. A major source of argument concerns whether *Leptictidium* hopped or ran in movement. It was clearly **bipedal** as the very shortened forelimbs were in strong contrast to the remarkably elongated and powerful hindlimbs balanced by a long tail (with 40+ vertebrae!). Studies on bone structure concluded that the leg articulation was too weak for jumping but the long feet were perfectly adapted for hopping. So pretty inconclusive.

The most complete specimens of the tiny metatherian (see pp. 43, 46–7) "*Peradectes*" are known from this site, a genus also known from the Americas that shows substantial similarities with today's South American opossums including relatively large canine teeth, simple premolars and slender jaws. Like most extant opossums, "*Peradectes*" had a long, **prehensile** tail as inferred by the structure of the **caudal** vertebrae, an adaptation for an arboreal lifestyle (Fig. 5.39). Like their modern counterparts, it too may have had an insectivorous or omnivorous diet. *Amphiperatherium* was a larger mammal compared to "*Peradectes*", but had a somewhat shorter tail, suggesting that the former was more adapted to a terrestrial lifestyle than an arboreal one. The insectivorous apatemyids (see p. 78) were also present at Messel. *Heterohyus* was a small, arboreal form known from three skeletons, which had a long tail, protruding incisors and a large skull in proportion to the body, along with super-elongated third and fourth fingers (Fig. 5.36).

Fig. 5.36. Reconstructed life appearance of the apatemyid *Heterohyus nanus*. Reconstructed length: 25 cm.

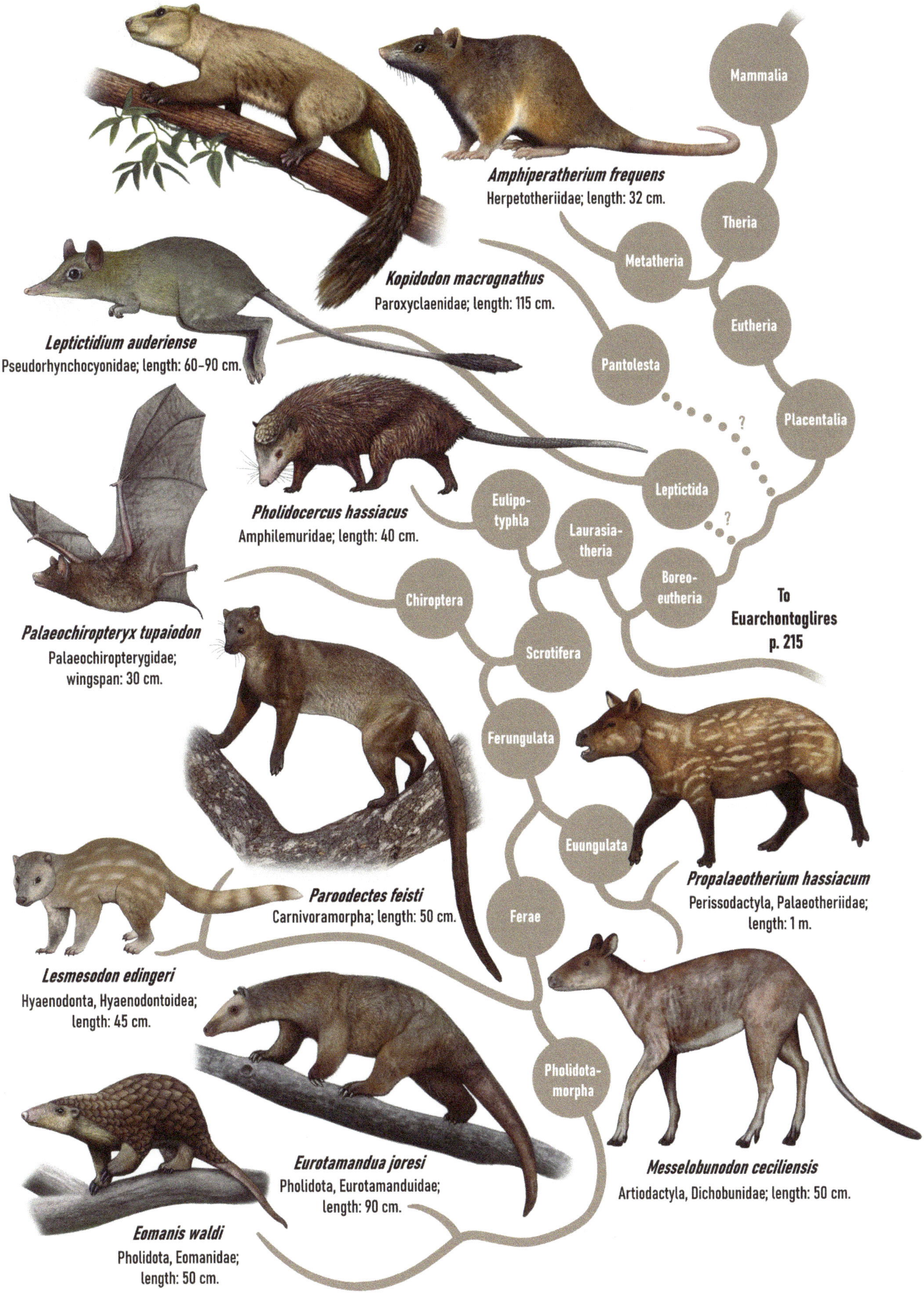

Fig. 5.37A. Phylogenetic tree of a sampling of fossil mammal **taxa**, particularly metatherians, and eutherians (see pp. 43, 57) outside Euarchontoglires (see p. 144), from the Messel Formation. The phylogenetic position of Leptictida and Pantolesta (see pp. 77–8) within Placentalia follows Brady *et al.* (2024), and Solé *et al.* (2024), respectively.

Europe Messel Formation, Germany: Lutetian, ca. 47 Ma, cont.

Heterohyus possibly resembled and behaved very much like the modern aye-aye (*Daubentonia madagascariensis*), using its elongated fingers for extracting grubs from bark it ripped open with its gnawing incisors, an example of **convergent evolution**.

At Messel, the most abundant insectivorous mammals preserved as fossils are bats, with hundreds of well-preserved skeletons recovered. The best known form is the small and highly specialised *Palaeochiropteryx*, which had moderately broad wings indicative of low wing-loading, enabling slow but manoeuvrable flight through the dense forest cover and close to the water's surface. This mode of flight would have made the bats vulnerable to toxic volcanic gases emitted by the lake sediments, which may explain why a large number of complete skeletons were fossilised at the site, showing exquisite preservation of the fur, wing membranes and outlining of external ears. While several features of the skull and skeleton are primitive, the enlarged cochlea suggests that this bat was already capable of echolocation. Like most extant bat species, it was reddish-brown in colour as indicated by chemical analysis of fossilised melanosomes preserved in the hairs of various fossil specimens (Colleary *et al.* 2015). Among other Messel bats, *Archaeonycteris* was a medium-sized bat that lived in the overstory, and *Hassianycteris* was a large, fast-flying bat preying on speedy, tough-shelled insects in the open spaces above the canopy. As well as the insectivores, the otter-like pantolestid "cimolestans" (see pp. 77–8) are also part of the Messel fauna.

The exceptional skeletons of *Buxolestes* prove the **semi-aquatic** habits of these ancient "cimolestans", with their well-developed hindlimbs, short, robust humerus and a long, muscular tail consisting of vertebrae with wide, transverse processes. Many other skeletal traits including massive, bony claws are similar to those of **fossorial** mammals, indicating that *Buxolestes* was also able to forage underground and dig shelters like modern otters. This combination of semi-aquatic and semi-terrestrial habits is further evident from the stomach contents of several fossil skeletons, indicating a diet primarily consisting of fish, as well as some seeds and fruits (Fig. 5.40). *Kopidodon* was a squirrel-like paroxyclaenid "cimolestan" that was well-adapted to an arboreal lifestyle, as evidenced by the mobile elbow joints, flexible ankles, a long, bushy tail and the presence of large, laterally-compressed claws along with opposable thumbs. It possessed long canine teeth likely used for defence and, like other paroxyclaenids, its dentition suggests that *Kopidodon* was omnivorous. The skeleton was raccoon-like in proportions, albeit more robust, and *Kopidodon* may have filled a similar ecological role.

At Messel, the rodents and primates essentially consisted of the same families that lived in Europe during the Early Eocene. The most abundant fossil rodent at Messel is the squirrel-like ischyromyid (see p. 146) *Ailuravus*, an almost marmot-sized rodent represented by several skeletons from the site, preserving fur impressions that reveal an elongated, bushy tail, along with gut contents composed primarily of leaves. The forelimbs were shorter than the hindlimbs, and the pointed claws resembled those of tree squirrels, suggesting arboreal habits. Additionally, the enamel on the molars of *Ailuravus* was highly crenulated. *Masillamys*, represented by several skeletons, was a smaller, squirrel-sized ischyromyid, with short limbs as in digging rodents such as voles, although it lacked fossorial limb adaptations. Contrary to fossorial animals, its tail was somewhat elongated, although not bushy as in the case of *Ailuravus*. Its lifestyle is uncertain, but it may have been an arboreal generalist (Fig. 5.40).

Among the primates, there are three genera known from Messel, all lemur-like adapiform strepsirrhines (see p. 150): *Godinotia*, *Europolemur* and *Darwinius*. *Godinotia* had massive orbits and a short snout, indicating that it was nocturnal. The dentition indicates that it was **frugivorous** and partly insectivorous, which is the case for *Europolemur* as well. In *Europolemur*, the skull was relatively short, with small orbits and a reduced **sagittal crest**. *Europolemur* had a grasping hallux (big toe) and the hand was large relative to the forearm, a feature possibly associated with vertical climbing. The former two genera are known only from partial remains, while the exquisitely preserved holotype of *Darwinius*, named "Ida", obtained an inordinate amount of media attention in 2009 (Franzen *et al.* 2009), lauded as a revolutionary scientific find. *Darwinius* is perhaps the most famous fossil ever found at Messel and shows arboreal characteristics such as binocular vision, grasping hands, and nails instead of claws, like *Godinotia* or *Europolemur*, but was more robust in build. The manus of *Darwinius* was considerably stockier than that of *Godinotia* or *Europolemur*, despite there being no great difference between the lengths of the metacarpals and basal **phalanges**. It was also possibly nocturnal as indicated by the large orbits. The dental morphology and preserved gut contents confirm that *Darwinius* fed on leaves and seeds.

Among the truly carnivorous mammals at Messel, which are very rare, were the hyaenodont *Lesmesodon*, and basal carnivoramorphs (see p. 84) represented by the genera *Paroodectes* and *Messelogale*. *Lesmesodon* was likely a generalised terrestrial animal, although the shape of the scapula suggests it was not cursorial. It is also characterised by unfissured terminal phalanges, unlike other hyaenodonts. Other features that can be gleaned from the intact skeleton of a juvenile (which includes fur) are short whiskers, small ears and a bushy tail comparable to that of a fox or a squirrel. *Paroodectes* was a **plantigrade**, civet-like predator that may have used its long tail for balance when climbing between branches. The well-developed clavicles characteristic of basal carnivoramorphs, but vestigial in the

Fig. 5.38. Reconstructed life appearance of the palaeotheriid *Eurohippus messelensis*. Reconstructed length: 60 cm.

majority of extant carnivorans (see pp. 81, 83–4), also suggest it was a nimble climber, likely hunting small, arboreal animals. *Messelogale* was also an arboreal predator (Fig. 5.39), but is known only from juvenile specimens which lack the upper third molars, possibly because they have not yet erupted. Features of *Messelogale* include a narrow facial region, a short basicranium region, and a wide, flat socket for the shoulder joint. Some of the most unusual faunal elements of Messel are the carnivoran-related pholidotans (see pp. 79–80) such as the scaled *Eomanis*. This ancestral pangolin was very much similar to its modern counterparts, although the scales were not covering the tail and limbs, and it had a lesser capability to roll itself up. One pholidotan was originally described as the first known non-South American anteater. *Eurotamandua* looked very similar to modern tamanduas and possessed no scales. A change of classification followed when it was determined that they also lacked the distinctive fused joints of xenarthans (see p. 58). It is now considered a furry stem-pangolin. By the Middle Eocene, perissodactyls had taken over from the "**archaic**" **ungulates** (see pp. 100–3)(which Messel is completely devoid of) as the most abundant ungulates of terrestrial ecosystems, represented by palaeotheriids (see pp. 134–5) of the genera *Eurohippus* (Fig. 5.38) and *Propalaeotherium*, and basal rhinocerotoids (see p. 129) of the genus *Hyrachyus*. One specimen of *Eurohippus* preserves a foetus inside, along with parts of the uterus, revealing that reproduction in early equoids was very similar to that of modern horses, despite the large differences in size and morphology. The limbs of *Eurohippus* were long and slender, and the premolars were non-**molariform**. *Propalaeotherium*

Fig. 5.37B. Phylogenetic tree of a sampling of fossil mammal taxa, particularly euarchontoglires, from the Messel Formation. The phylogeny within Adapiformes (see p. 150) follows Seiffert *et al.* (2009), while the phylogeny within Rodentia follows Vianey-Liaud and Laurent (2021).

Europe Messel Formation, Germany: Lutetian, ca. 47 Ma, cont.

was larger and more robust in build than *Eurohippus*, and the skull was less wedge-shaped. Complete articulated skeletons of *Propalaeotherium* show that it was a small, stocky animal, with a fairly short neck and short forelimbs, suggesting that it was not a fast animal. It also had a short tail ending in a broad tassel. Several specimens have had stomach contents preserved which indicate that they **selectively fed** on leaf matter and soft fruit (evidenced from grape seeds) on the forest floor. *Hyrachyus* was a small to medium sized (1.5 metres) rhinocerotoid (Fig. 5.40) with a long, narrow skull devoid of the characteristic horns of modern rhinoceroses. It had a **complete dentition**, with spatula-shaped incisors, small non-molariform premolars, and molars that were starting to develop the characteristic pi (π)-shaped **occlusal** pattern of rhinocerotoids. It bore four toes on the forefeet and three on the hind feet, like many other early perissodactyls including *Propalaeotherium*. As opposed to most other contemporaneous ungulates, *Hyrachyus* was a somewhat cursorial animal. The artiodactyls, on the other hand, are comparatively rare, represented by the greyhound-sized dichobunid (see p. 104) *Messelobunodon*. Its characteristic features include a very long tail, an advanced limb arrangement and vestigial lateral toes, along with elongated, powerful hind limbs indicating a more cursorial lifestyle compared to some of its relatives. Additionally, it had a wedge-shaped skull with an elongated **rostrum**, a distinctly arched back and long, slender, **pentadactyl** forelimbs, the last-mentioned feature which no longer occurs in modern artiodactyls.

1. *Eoconstrictor fischeri*, a primitive species of boa.
2. *Masillatrogon pumilio*, a primitive species of trogon, suggesting that the family arose in the Old World.
3. *Titanomyrma gigantea*, largest known ant (queens measuring up to 7 cm), sting was absent, lacked a crop so was possibly carnivorous.
4. Stag beetle (Lucanidae indet.).
5. "*Peradectes*" sp., an opossum-like peradectid metatherian (see pp. 47–8).
6. *Quasisyndactylus longibrachis*, a possible relative of extant kingfishers.
7. *Parargornis messelensis*, an early relative of treeswifts and hummingbirds, characterised by elongated tail feathers and short wings.
8. *Plesiocathartes kelleri*, a primitive relative of the extant cuckoo-roller. (*Leptosomus discolour*) of Madagascar.
9. *Messelogale kessleri*, a civet-like basal carnivoramorph.
10. *Masillaraptor parvunguis*, a long-legged, possibly basal relative of falcons.

Fig. 5.39. A rainforest canopy ecosystem at Messel, Germany, during the Middle Eocene.

1. *Messelirrisor halcyrostris*, a small, basal relative of hoopoes.
2. Jewel beetle (Buprestidae indet.).
3. *Masillastega rectirostris*, a possible early sulid (the bird family including gannets and boobies), the size of a common shag (*Gulosus aristotelis*).
4. *Eocoracias brachyptera*, a primitive species of roller, in which studies of the plumage suggest non-iridescent structural colouration.
5. *Rhynchaites messelensis*, a primitive ibis, otherwise called the "snipe-rail" for its morphological resemblances to rails and painted snipes.
6. *Allaeochelys crassesculpta*, a relative of the modern pig-nosed turtle (*Carettochelys insculpta*).
7. *Buxolestes piscator*, a superficially otter-like pantolestan mammal up to 81 cm long.
8. *Atractosteus messelensis*, the "Messel gar".
9. *Macrocranion tupaiodon*, a small (30 cm long) amphilemurid eulipotyphlan (see pp. 75, 77).
10. *Messelornis cristata*, the "Messel rail", a likely terrestrial bird that, due to its ts relatively short wings, was probably capable of only short bursts of flight.
11. *Hyrachyus minimus*, a basal rhinocerotoid.
12. *Masillamys parvus*, an ischyromyid rodent.
13. *Messelastur gratulator*, a bird whose ecology remains poorly known, but is generally agreed to lie somewhere near the origin of parrots, passerines and falcons.

Fig. 5.40. A rainforest floor ecosystem at Messel, Germany, during the Middle Eocene.

Africa Qasr el-Sagha Formation, Egypt: Priabonian, 37.2–33.9 Ma.

The Qasr el-Sagha Formation forms part of the Late Eocene **strata** in the Fayûm Depression in central Egypt, about 80 kilometres southwest of Cairo, and west of the River Nile. The shales and sandstones of the Qasr el-Sagha Formation were deposited in a deltaic to nearshore marine environment in the Tethys Ocean (see p. 174), and on the then gradually subsiding, unstable passive continental margin of North Africa. Throughout the Eocene (and into the Oligocene), continental sediments slowly advanced to the west and north with a simultaneous gradual marine regression (i.e. a seaward movement of the shoreline caused by falling sea levels or uplifting land). As a result, continental and shallow marine strata overlap deeper marine deposits across much of Northern Egypt. Erosional troughs within the rock sequences preserve concentrations of fossils of both **terrestrial** and marine origin. Marine transgressive phases (i.e. a landward shift of the shoreline caused by rising sea levels or subsiding land) in the rock sequence are evident from transported, fragmented mollusc shells. Assemblages of various fossil types are present in both the regressive and transgressive sequences of the Qasr el-Sagha Formation. Although fossils of terrestrial mammals are uncommon, the remains of cetaceans (see pp. 121–6), **stem**-sirenians (see pp. 69–71), snakes, crocodiles, turtles, **osteichthyan** fish bones, **chondrichthyan** (shark, skate and ray) teeth, bivalve and gastropod shell fragments, corals, polychaete worm tubes and silicified wood are all abundant at Qasr el-Sagha. The silicified wood, most of which is waterworn, is a testament to the area's lush past in which the coastlines were dominated by swamps and mangrove forests. The proportion of marine mammals to terretrial mammals is comparatively higher in the nearshore marine sequences than in the **fluvial** deposits, likely due to taphonomic and facies differences when the remains had accumulated (Holroyd *et al.* 1996).

The fossils of "archaeocete" cetaceans from the basilosaurid lineage (see pp. 122–3) are particularly noteworthy, ranging from single bones to complete skeletons and representing several hundred individuals, concentrated in a single extensive locality called Wādī al-Ḥītān (meaning "Valley of the Whales" in Egyptian Arabic). This locality is important in demonstrating the morphology and lifestyle of cetaceans during their evolution from a terrestrial existence to a marine one. Whereas modern cetaceans also have a distinctly **homodont** dentition, the basilosaurids retained the well-defined range of morphologically varied teeth present in other mammal groups used for processing their prey. Their diet consisted mainly of fish, sharks, squid and other marine mammals. Basilosaurids are characteristically long, serpentine animals due to an elongation of the lumbar vertebrae. This facilitated a movement via tail propulsion **analogous** to that seen in modern cetaceans.

Basilosaurus is believed to have been abundant and widespread across the Tethys Ocean, acting as the apex predator of its environment in a similar manner to modern killer whales (*Orcinus orca*) (Fig. 5.41). In addition to being constantly oversized—*B. cetioides* reached 17 to 20 metres long, while the related *B. isis* was smaller, at 15 to 18 metres long—most depictions show the early cetacean to be very slim, with a shrink-wrapped skull. Looking at any cetacean's skeleton which consists of a skull, vertebrae, ribs and pectoral bones, one might be

Fig. 5.41. A scene set in the shallow seas off the coast of Northern Egypt during the Late Eocene. *Centre*: a large male of the basilosaurid cetacean *Basilosaurus isis*; *Bottom right*: a pair of the early dugongid sirenian *Eotheroides sandersi*.

under the impression of a slim swimmer. However, the body mass is largely made up of extensive organs, musculature and blubber. Although *Basilosaurus* was obviously very elongated, its body would conform with modern cetaceans. Without a suitable covering of flesh, the skull looked decidely reptilian (in keeping with the sea serpent vibe) and the exposed teeth were prominent, whereas the teeth in *Basilosaurus* were hidden behind fleshy cheeks and lips. Most early depictions do not show hind limbs or a dorsal fin. A 16 metre long specimen of (most likely) *B. isis* had preserved 35 centimetre long hindlimbs which are not connected to the vertebral column and limited **articulation**. It is thought they may been used during mating as stabilisers. While there is no definite proof of a dorsal fin or its actual position on the body, it is probable that they possessed one. The vertebrae was similarly-sized along the full length, which indicates **anguilliform** movement mostly in a vertical plane. Filled with buoyant marrow (modern cetaceans have solid vertebrae) and surrounded by weak axial muscles, *Basilosaurus* was probably a shallow-water swimmer that relied on small bursts of speed.

Basilosaurus is distinguished from other basilosaurids by its larger body size and more elongated lumbar, **anterior caudal**, and **posterior** thoracic vertebrae. Its front flippers retained an articulated elbow joint, a remnant of their terrestrial ancestry, allowing for limited flexion and rotation within the forelimb. The pelvis in *Basilosaurus* was fully isolated from the vertebral column, and the hindlimbs were extremely reduced and vestigial. These skeletal features are less pronounced in other basilosaurids, such as *Dorudon* (Fig. 5.42), which had a more compact body and shorter vertebrae. Due to their morphological similarity but smaller size, dorudontines were initially misidentified as juvenile *Basilosaurus* individuals. However, the discovery of juvenile *Dorudon* specimens has confirmed that they represent a distinct genus. As in other basilosaurids, the nostrils were positioned midway between the snout and the top of the head.

Fig. 5.42. Reconstructed life appearance of the basilosaurid cetacean *Dorudon atrox*. Reconstructed length: 5 m.

Fig. 5.43. Seagrass meadows in a nearshore marine environment off the coast of Northern Egypt during the Late Eocene. *Centre*: the early dugongid sirenian *Eosiren libyca*; *Right background*: a sea turtle (Cheloniidae indet.); *Left background*: a small requiem shark of the extinct genus *Physogaleus*; *Top left background*: the basilosaurid cetacean *Saghacetus osiris*.

Africa Qasr el-Sagha Formation, Egypt: Priabonian, 37.2–33.9 Ma, cont.

Furthermore, many juvenile and newborn *Dorudon* fossils have been found with unhealed bite marks on their skulls consistent with predation from the larger *B. isis*. One possible scenario is that a pod of *Dorudon* had come to a bay to give birth, only for their young to fall prey to some ravenous *B. isis*. This predator-prey relationship is further supported by the stomach contents of an old male *B. isis*. (Voss *et al.* 2019). Other basilosaurids from Qasr el-Sagha include *Saghacetus*, which is distinguished from other basilosaurids by its smaller body size (about four metres long), moderately elongated lumbar and caudal vertebrae, and more **primitive** skeletal features, reflecting a basal position within the basilosaurid lineage (Martínez-Cáceres *et al.* 2017).

Among other marine mammals from Qasr el-Sagha, *Eotheroides* and *Eosiren* represent early fully aquatic members of the sirenian lineage. *Eotheroides* is characterised by a gently curved snout, relatively small tusks, and a lower jaw with a comparatively unspecialised dentition, contrasting with the more reduced and specialised teeth of modern sirenians adapted for grazing on seagrass. *Eosiren* is more derived than earlier stem-sirenians in being larger and possessing **pachyostotic** ribs, a more robust skull and mandibles, and a broader, flatter skull roof. While its overall anatomy was already quite similar to that of modern **crown** sirenians (Fig. 5.43), *Eosiren* retained primitive features such as canine teeth, thigh bones, and a relatively larger pelvis. The close association of its thigh bones with the pelvis reflects an advanced stage in hindlimb reduction and a clear **adaptation** to a fully aquatic lifestyle (Zalmout and Gingerich, 2012).

In the coastal swamps of Egypt during the Late Eocene, the early proboscideans (see pp. 64–8) *Moeritherium* (Fig. 5.44A) and *Barytherium* (Fig. 5.44B) coexisted, likely not competing with each other as their adaptations and feeding ecologies were markedly different. Isotopic data supports a **semi-aquatic** lifestyle for both species. The feeding ecology of *Moeritherium* was likely more strongly linked to a semi-aquatic lifestyle and a diet of freshwater vegetation, while the diet of *Barytherium*, though still connected to freshwater sources, may have been more diverse and potentially included terrestrial elements (Liu *et al.* 2008). Stable isotope analysis also suggests that *Barytherium* may have been slightly more terrestrial than *Moeritherium*, as further indicated by the forefoot structure. The more familiar and pig-sized *Moeritherium* was initially thought to be a direct ancestor of modern elephants, but is now considered an extinct side branch of Proboscidea. Known from relatively complete skeletal remains, *Moeritherium* had a highly elongated torso and short, stout limbs for moving in and out of water, and high-placed **orbits** for better visibility while submerged. It also had an elongated skull and enlarged, tusk-like second incisors. Being low to the ground, and having the nasal openings close to the front of the skull, a trunk would have been a highly unlikely characteristic. *Barytherium* was among the first large-sized proboscideans, a **graviportal** species standing two metres at the shoulder. *Barytherium* had eight short, stubby tusks, with four in both the upper and lower jaws. The upper tusks were vertical and pointed outwards, while the lower tusks projected forwards. The elevated nasal region suggests that *Barytherium* may have had a well-developed tapir-like **proboscis**, though evidence of the exact structure is lacking.

The coastal **ecosystem** was also home to a small, very successful and distinctive hyaenodont (see pp. 82–3) of the genus *Apterodon* (Fig. 5.45), which is represented by dental and limited postcranial material but has some rather interesting features. It lacked the **carnassial** teeth seen in other hyaenodonts and the short jaws were capable of heavy chewing, possibly specialising on shellfish and crustaceans. A flattened tail and a degree of flexion in the joints comparable to that of otters implies that apterodontines spent a considerable amount of time in and around water (Grohé *et al.* 2012).

Fig. 5.44. Reconstructed life appearance of basal proboscideans from the Qasr-el Sagha Formation. (A) *Moeritherium lyonsi*; (B) *Barytherium grave*. Scale bar: 1 m.

Some of the largest snakes are also known from Qasr el-Sagha, with a small quantity of fossils (mostly vertebrae) from the madtsoiid (see pp. 23, 25) *Gigantophis* having been recovered (Fig. 5.46). Reaching up to seven metres (with some estimates at 10 metres), it was only exceeded by the earlier Paleocene boid (see p. 24) *Titanoboa* (see pp. 186–7). Some earlier material (Early Paleocene) from Pakistan has also been described as *Gigantophis*, suggesting that dispersal across seas was not a barrier (Rage *et al.* 2013). Although *Gigantophis* has been recovered from coastal deposits, the skeletal material does not show any specific adaptations to aquatic life. Its large size and comparison with giant snakes including *Titanoboa* and anacondas have also indicated that *Gigantophis* may have preyed on large terrestrial animals such as *Moeritherium*. Like extant constrictor snakes, next to the cloaca, it had a small pair of claw-like pelvic spurs, which are modified femurs **articulated** to the hips and are external remnants of the hindlimbs. These are comparable to the vestigial hindlimbs of basilosaurids and may have had a similar function (used for gripping during mating rituals).

Fig. 5.45. Reconstructed life appearance of the apterodontine hyaenodont *Apterodon saghensis* in a coastal swamp environment in what is now Egypt, Late Eocene. Reconstructed length: 1 m.

Fig. 5.46. Reconstructed life appearance of the madtsoiid snake *Gigantophis garstini*. Reconstructed length: 10 m.

Asia Irdin Manha Formation, China: Ypresian–Priabonian, 48.6–37.2 Ma.

The Irdin Manha Formation is dated to the Middle Eocene and situated in what is now the eastern Erlian Basin in Northern China's Inner Mongolia province, just south of the Mongolian border. The Irdin Manha Formation consists of two units: the upper unit predominantly consists of sandy **conglomerates**, whereas the lower unit is composed of coarse sandstones and muddy sandstones, deposited in a **fluvio**-deltaic environment. The Irdin Manha Formation is highly fossiliferous in both units, with the lower unit having the greatest concentration of mammal fossils (Ni *et al.* 2010). During the timespan the deposits were accumulated, there was a gradual shift from densely forested habitats with abundant soft-leaved vegetation to more open, shrubby landscapes. This likely influenced dietary adaptations in herbivorous mammals, although true hypsodont teeth would not become widespread until later epochs. The Irdin Manha Formation exemplifies well the transitional period for some mammal groups that have existed for a long time but only during the Middle Eocene were starting to become large (e.g. brontotheriids (see pp. 135–6) and hyaenodonts (see pp. 82–3)). The Middle Eocene mammal faunas of Asia share numerous similarities with those of North America, reflecting a history of faunal interchange across the Bering Land Bridge (see p. 174). However, by this time, the free interchange characteristic of the Early Eocene had significantly declined, with only limited exchanges occurring thereafter. The "anagaloids" (see pp. 144–5), from which the lagomorphs (see pp. 148–9) and rodents had already evolved, survived in China for much of the Middle Eocene. **Stem**-lagomorphs also persisted, such as *Mimolagus*, first regarded as a basal lagomorph due to its ankle morphology (Bleefeld and McKenna, 1985; Dashzeveg and Russell, 1988), but it is now placed outside Lagomorpha given the absence of the notched upper front incisor (Rose, 2006).

Many other lineages that had declined elsewhere survived in East Asia well into the Middle Eocene, such as pantodonts (see pp. 137–8) and the bear-sized oxyaenid (see p. 82) *Sarkastodon*. It is the largest and most impressive species of oxyaenid, possibly acting as an apex predator in East Asia. Known only from one large skull, there has been much debate as to the actual total size of *Sarkastodon*. Some estimates come in at an impressive 800 kilograms, but these figures are way out, as oxyaenids had large heads compared to their body. When based on skeletal remains of the related *Patriofelis*, the true size of *Sarkastodon* was probably between 200 and 300 kilograms (well within Siberian/Amur tiger range). The hyaenodonts are represented by *Propterodon*, a primitive hyaenodontid that is known only from lower jaw fragments and is the closest known relative to the genus *Hyaenodon* (see pp. 82–3) (Borths *et al.* 2016). The long, blade-like **carnassials** and fewer accessory cusps formed a more advanced shearing surface in *Propterodon* than in more basal hyaenodonts.

The **carnivorous** hooved **ungulates** of the mesonychid lineage (see p. 103) persisted in Asia long after they went extinct elsewhere, represented by the widespread genus *Harpagolestes* at Irdin Manha. It is represented by the Asian species *H. leei*, in which the cheek teeth are similar in size to that of *H. brevipes*, but smaller than that of *H. immanis* (see pp. 206–7, 209). *H. leei* exhibits the typical strong, curved canine teeth and heavy molar wear of the genus, but there are subtle differences in tooth morphology (e.g. an unusually elongated lower fourth premolar). Originally classified as a giant mesonychid, *Andrewsarchus* has only recently (and controversially) been linked to entelodontids (see pp. 118, 120). Known from only one huge 83 centimetre long skull, *Andrewsarchus* became renowned as the largest terrestrial mammalian carnivore ever known. It was initially depicted as a wolf-like predator (up to six metres long!) minus the claws, but this view is outdated, and the skull is very similar to entelodontids suggesting that *Andrewsarchus* may be their closest relative. What little we do know of *Andrewsarchus* is that the snout was long and powered by muscles anchored to broad cheeks, but the jaws were relatively weak. The eyes were set low down and widely apart. Otherwise, we have no idea about its appearance without finding more skeletal material, but I have assumed similar body proportions to entelodontids (Fig. 5.47).

Besides all these Early Eocene relic lineages, Middle Eocene Asia also witnessed the emergence of more advanced lineages, such as some of the most primitive haplorhines (see pp. 152–6). *Tarkops* was a small, early, marmoset-like omomyid primate that possessed **bunodont** cheek teeth so characteristic of Eocene primates, suggesting a diet of fruit and soft vegetation. In *Tarkops*, the lower jaw was deep in proportion to the height of the molar crowns (Ni *et al.* 2010). In Asia, the perissodactyls (see p. 126) consisted of a diverse range of brontotheriids and tapiroids (see p. 128) by the Middle Eocene, along with advanced rhinocerotoid lineages including the long-limbed hyracodontids and paraceratheriids (see pp. 129–30). *Protitan* is a poorly-known brontotheriid that approached the size of a modern rhinoceros. On the **frontal** and nasal bones was a pair of low, bony protuberances. The skull was saddled along the forehead, with the occiput (the posterior part of the skull) slightly elongated to the rear, typical of horn-bearing brontotheriids. The nasal bone was very broad and was orientated downwards slightly.

Lophialetes is one example of the strange, early tapiroids known from Irdin Manha. It had a physique more akin to a horse or a deer than its pig-like modern cousins, standing about 0.5 metres at the withers. It also had slender, highly-elongated limbs with **tridactyl** hooved feet each bearing much of its weight on the central **digit** as an **adaptation** for high-speed running in more open environments, which was essential for surviving the increase in aridity. Its skull had an expanded nasal region that may have supported a small trunk comparable to both modern saiga antelope (*Saiga tatarica*) and tapirs (*Tapirus*), but not as well-developed. The hyracodontid *Triplopus* was an agile animal as indicated by its long, slender limbs. *Triplopus*, as its name indicates, already had tridactyl feet and, compared to any other rhinocerotoid, the radius was proportionately longer. The premolars had a primitive morphology in being subtriangular shaped with two transverse ridges. *Forstercooperia* was a small, basal paraceratheriid known from many cranial fossils but some limited postcranial material. *Forstercooperia* is more primitive than later paraceratheriids in having unspecialised, conical incisors, non-**molariform** premolars and a shorter **diastema**. Other features include three incisors, lower first premolars and lower canine teeth, primitive among rhinocerotoids, and the nasal incision terminated slightly behind the canine teeth.

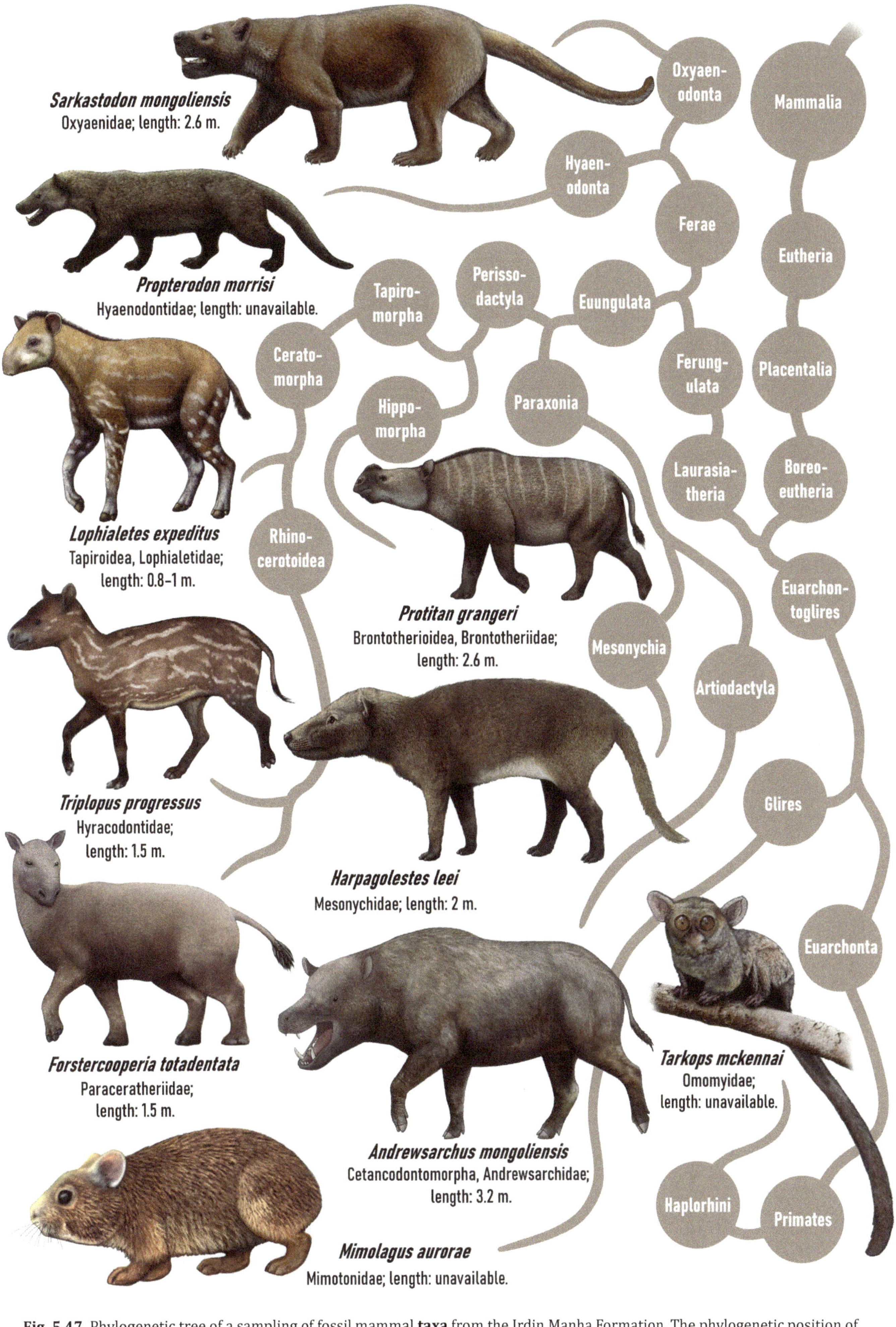

Fig. 5.47. Phylogenetic tree of a sampling of fossil mammal **taxa** from the Irdin Manha Formation. The phylogenetic position of Mesonychia follows Bast and Smith (2013), while that of *Andrewsarchus* follows Yu *et al.* (2023).

Asia Ergilin Dzo Formation, Mongolia: Priabonian-Rupelian, 35.7-33.7 Ma.

The Late Eocene Ergilin Dzo Formation, distributed in what is now Southeastern Mongolia, has yielded many **terrestrial** mammal fossils in which a diverse range of perissodactyls (see p. 126) predominate. The palaeoenvironment in which this formation accumulated is inferred to have been warm and humid, with somewhat closed habitats, as indicated by the wealth of **herbivorous** mammals with **brachydont** dentitions. On the other hand, the lack of primates and the scarcity of crocodylomorphs (see pp. 26–30) from this site indicate that the region had some more open habitats and was drier compared to other Eurasian faunas from the same time (Tsubamoto *et al.* 2022). The geology implies that the Ergilin Dzo Formation was formed in a floodplain setting characterised by a braided river system (Saneyoshi *et al.* 2010). By the Late Eocene, the Asian brontotheriids (see pp. 135–6) such as *Embolotherium* possibly roamed the plains and open forests in large herds. *E. andrewsi* is known from complete skulls, several jaws and miscellaneous skeletal elements, but no complete skeletons have yet been found. The nasal horn of this species was greatly elevated and protruded forward like a single, flattened battering ram. However, the internal construction was more lightweight and fragile than seen in the North American **clade**, and the rams do not appear to differ between sexes. Rather than for physical bouts, an enlarged nasal cavity has led scientists to consider that the horn acted as a resonating chamber by amplifying their calls during signalling.

Other perissodactyls were at their zenith in Asia during the Late Eocene, including deperetellid tapiroids (see p. 128) such as *Teleolophus*, **primitive** chalicotheriids (see p. 128) such as the schizotheriine *Schizotherium,* and many amynodontid rhinocerotoids (see p. 129) such as the tapir-like *Cadurcodon*. Features of *Teleolophus* include **hypsodont**, highly **bilophodont cheek teeth** and an inverted U on the upper molar crowns, typical of deperetellids. The asymmetrical femoral trochlea as well as the elongated, slender limbs suggest that *Teleolophus* preferred relatively open habitats and was consequently adapted to fast running. *Schizotherium* may have been ancestral to numerous forms of schizotheriine chalicotheriids, which dispersed to Europe, North America and Africa. *Schizotherium* had started to develop higher-crowned cheek teeth, although they were more specialised and less hypsodont than those of more recent schizotheriines. W-shaped crests were prevalent on the upper molars. Unlike similar but later chalicotheriines (e.g. *Moropus* (see pp. 255–6) and *Ancylotherium* (see pp. 302–3)), the metatarsals were longer than they were wide. *Cadurcodon* had a short, yet tall skull with reduced nasal bones and an expanded nasal incision, suggesting the presence of a tapir-like **proboscis**. Its stocky, hippopotamus-like body indicates that *Cadurcodon* spent at least part of its time in water, sensing out and harvesting waterside vegetation with its short proboscis. Other rhinocerotoids included a diversity of long-limbed hyracodontids and eggysodontids (see p. 129). *Ardynia* was a medium-sized hyracodontid distinguished by a slender lower jaw, and molars that are relatively more hypsodont than in other hyracodontids. It also had a highly reduced or absent lower second premolar and a relatively shortened premolar series. The eggysodontid *Allacerops* was a small, terrestrial **browser** characterised by features including **heterodont** upper premolars, and robust, near-vertically attached canine teeth. The fossils of pantodonts (see pp. 137–8) from Ergilin Dzo are largely limited to skulls, but the massive *Hypercoryphodon* is impressive, the last and largest known representative of its order. *Hypercoryphodon* must rank as one of the largest terrestrial Eocene mammals alongside the dinocerate *Eobasileus* (see p. 204) and the brontotheriids. *Hypercoryphodon* is characterised by traits such as a narrow skull roof, and upper molars that are bilophodont and square-shaped. It is generally considered a possible descendant of *Coryphodon* (see pp. 137–8).

Among the artiodactyls (see p. 104), the hippopotamus-like "anthracotheriids" (known from indeterminate material) (see pp. 120–1), and pig-like primitive entelodontids (see pp. 118, 120) such as *Brachyhyops*, were quite abundant, along with a variety of deer-like basal ruminants (see p. 110), including the lophiomerycids and archaeomerycids (see p. 105). *Brachyhyops* had a short snout, along with somewhat weakly developed jugal flanges on the skull, and the mandibular tubercles so characteristic of entelodontids were absent. Like entelodontids, however, the **zygomatic arches** were extended and deepened **posterior** and ventral to the **orbits**. The lophiomerycids such as *Lophiomeryx* were similar to modern chevrotains in build, were some of the most dentally primitive ruminants, and retained **tetradactyl** feet with the third and fourth metapodials being unfused. The archaeomerycids such as *Miomeryx* are distinguished by features including a short facial region, a low, narrow cerebral area in the skull, and **pentadactyl** forelimbs and tetradactyl hindlimbs. *Miomeryx* itself was small and is characterised by cheek teeth that are both brachydont and **selenodont**. A myriad of lagomorphs (see pp. 148–9) and rodents, mainly **taxa endemic** to Asia, were quickly taking over as the antique "anagaloids" (see pp. 144–5) vanished.

The predatory mammals included a combination of the primitive hyaenodonts, some of the last mesonychids (see p. 103), represented by indeterminate fossils from Ergilin Dzo, and the more advanced true carnivorans (see pp. 81, 83–4), including primitive amphicyonids (see pp. 90–1), also represented by indeterminate material, and the "stenoplesictids" (see p. 84). The "stenoplesictid" *Asiavorator* was very similar to modern genets and civets in skeletal morphology, and was predominantly terrestrial, although it was possibly still capable of climbing trees. The hyaenodonts are represented by various species of the very successful genus *Hyaenodon* (see pp. 82–3). The largest species is *H. gigas*, estimated to be slightly larger than a lion (*Panthera leo*). It had a very strong, muscular neck that supported a heavy and relatively short skull measuring around 50 centimetres long. *H. gigas* was relatively long-bodied and had slender limbs. *H. gigas* was more **digitigrade** than other contemporaneous hyaenodonts, being adapted to active movement on hard ground. *H. gigas* may not have been able to run long distances though. It was likely an **ambush predator**, briefly but actively pursuing prey, like many of today's felids (see pp. 86–9). By the end of the Eocene in Asia, perissodactyl-rich faunas adapted to warm, humid conditions were largely replaced by faunas dominated by lagomorphs and rodents, which were adapted to cooler and drier climates. This faunal turnover, which corresponds to the Grande Coupure in Europe (see p. 230), is referred to as the "Mongolian Remodelling" (Meng and McKenna, 1998; Kratz and Geisler, 2010).

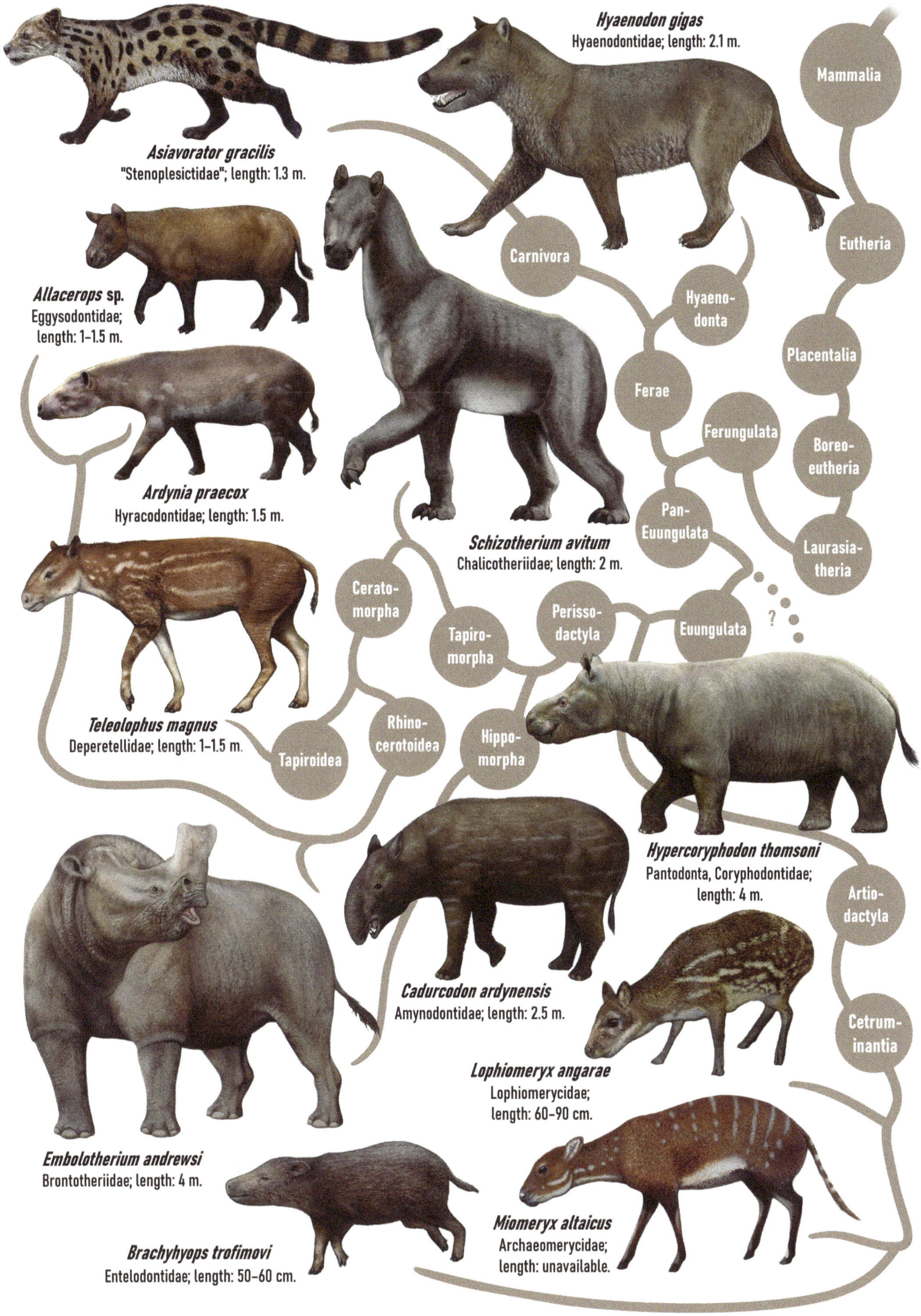

Fig. 5.48. Phylogenetic tree of a sampling of fossil mammal taxa from the Ergilin Dzo Formation. The tentative phylogenetic position of Pantodonta follows Kort (2023), which recovers pantodonts as **stem**-ungulates (members of Pan-Euungulata (see p. 100)).

South America Itaboraí Formation, Brazil: Ypresian, 53–50 Ma.

The Itaboraí Formation, whose geographic range is limited to the São José de Itaboraí Basin in Southeastern Brazil, near the town of Rio de Janeiro, is composed predominantly of limestone sequences that were deposited in **alluvial-lacustrine** settings during the Early Eocene. It is highly fossiliferous, having preserved a diverse fossil assemblage rich in gastropods and vertebrates, among which the mammals are most abundant and significant as fossils. The remains of these organisms were preserved in naturally-formed fissures in the limestone, in which sediments rich in teeth and bones from the neighbouring regions had gradually accumulated. Since these specimens cannot be directly dated as a result of this fossilisation process, it is hard to calculate their exact ages. This process has also meant that the bones are fragmented, **disarticulated** and belong to various species jumbled together, so teeth and jaw fragments have been used to identify most of the named fossil mammals from this site more easily.

The habitat of the Itaboraí Formation was essentially a dense tropical forest as indicated by the plant and pollen fossils, which show that the flora consisted of tropical trees such as palms. Although grasses were present, they were not particularly abundant. The climate was hot and humid, and temperatures varied little with the seasons. However, there may have been seasonal variation in precipitation, and heavy rainfall was likely accountable for transporting faunal and floral remains into the limestone fissures. The deposits are inferred to have formed during an interval when global temperatures were extremely high, referred to as the Early Eocene Climatic Optimum (EECO) (see p. 197).

During the Early Eocene, the marsupials and related metatherians (see pp. 43, 46–7) still dominated the mammalian fauna of South America. It was not until the Middle Eocene that placentals surpassed the marsupials as the most taxonomically diverse and abundant mammal group, having always retained this dominant role ever since. Nonetheless, numerous species of South American marsupials persisted as small, **nocturnal frugivores** and **omnivores**, partly as a result of the continent's isolation. Today, these ecological roles are filled by primates including tarsiers and lorises in Asia, and galagos in Africa. The first primates that arrived in South America by the Late Eocene were **diurnal** and there are few species of nocturnal primates in South America at present. Therefore, marsupials and placentals could not have competed directly for this nocturnal way of life. According to the morphology of isolated postcranial elements from Itaboraí, both **arboreal** and **terrestrial** metatherians were part of this fauna (Szalay and Sargis, 2001). The most common metatherian from Itaboraí is the vaguely opossum-like, **primitive** polydolopimorphian (see p. 48) *Epidolops*, and among the few whose cranial morphology is well-represented. *Epidolops* may have been a nocturnal or **crepuscular** animal living an arboreal lifestyle. Unlike any modern animal, *Epidolops* had a **plagiaulacoid** third premolar in the lower and upper jaws, specialised for shearing. Among extant mammals, plagiaulacoid teeth are present exclusively in several Australian marsupials such as bettongs (*Bettongia*) and pygmy possums (family Burramyidae). *Epidolops* had reduced molars with low, rounded cusps, which are **analogous** to those of the aye-aye (*Daubentonia madagascariensis*) and striped opossum (*Dactylopsila trivirgata*). It appears that its specific dentition was suited to an omnivorous diet, feeding on fruit and grubs.

The most abundant South American native **ungulate** (see p. 139) from Itaboraí is the possum-sized *Colbertia*, which was part of an early radiation of small- to medium-sized and relatively unspecialised notoungulates (see pp. 140–3) of the family Oldfieldthomasiidae. *Colbertia* had a moderately elongated skull and **brachydont** molars with a **lophodont** morphology, consisting of cusps that are structured together to form elongated ridges called **lophs**. This dentition suggests that it was a **browser**, like most other mammals from the site (López *et al.* 2011). The tarsal bones suggest that *Colbertia* still retained **plantigrade** feet, an **ancestral** condition for notoungulates, and had generalised locomotor behaviour.

Protolipterna, along with other protolipternids, is believed to be among the most basal litopterns (see pp. 143–4), the second most diverse and common group of South American native ungulates (see p. 139) from Itaboraí after the notoungulates. *Protolipterna* had a **digitigrade** stance, unlike primitive notoungulates such as *Colbertia*, and thus enabling faster and more efficient **cursorial** locomotion. *Protolipterna* was perhaps the most agile terrestrial mammal at the site, with a running or jumping gait comparable to that of a hare or a chevrotain. *Protolipterna* had brachydont, lophed molars for shearing foliage, like *Colbertia*, but also a pair of long, fang-like upper canine teeth. Other primitive litopterns from Itaboraí include

Fig. 5.49. Speculative reconstructed life appearance of the metatherian *Gaylordia macrocynodonta*. Reconstructed length: 30 cm.

Fig. 5.50. Phylogenetic tree of a sampling of fossil mammal **taxa** from the Itaboraí Formation. The phylogenetic position of Notoungulata and Litopterna follows Welker *et al.* (2015)

Victorlemoinea, which is estimated to have weighed 25 to 58 kilograms (Vizcaíno *et al.* 2012) and is best represented there. Although traditionally regarded as a macraucheniid, several studies have suggested that it may instead belong within Sparnotheriodontidae (Francis et al. 2006). Like other early litopterns, *Victorlemoinea* had **selenodont** molars (Rose, 2006).

Carodnia is the best known xenungulate (see p. 139) and the largest known South American mammal during the Eocene, being approximately five to six feet long. It was likely a browser feeding on fruit and soft vegetation, as indicated by analysis of the enamel microstructure and pattern of mastication (i.e. horizontal wear facets that are **posteriorly** directed in the lower **cheek teeth**) (Bergqvist and von Koenigswald, 2017). It is one of the rarest fossil mammals from Itaboraí, known from associated cranial, dental and postcranial material including a complete lower jaw, several fragmented limb bones and numerous vertebrae. It had sturdy limbs, short, **pentadactyl**, plantigrade feet, and **digits** ending in flat, wide and unfissured hoof-like unguals.

Some intriguing placental mammals from Itaboraí include *Riostegotherium*, one of the oldest known armadillos. It is known from limb elements along with isolated **osteoderms** that resemble those of modern dasypodid armadillos (see pp. 58–60). From these remains, it is apparent that armoured armadillos existed in South America as far back as the Early Eocene (Oliveira and Bergqvist, 1998). It had two types of osteoderms: one comprised a fused mass which in modern armadillos form rigid front and back shields

South America Itaboraí Formation, Brazil: Ypresian, 53–50 Ma, cont.

called bucklers, whereas the other consisted of isolated but movable elements in mobile bands. It is possible that the bucklers of *Riostegotherium* were composed of loosely connected elements and separated by an unknown number of mobile bands, probably more numerous than in extant armadillos (Bergqvist *et al.* 2019).

Among other mammals, there are also some fossilised limb elements strongly resembling those of a gliding mammal, including long, thin bones, flexible ankles characteristic of tree-climbing mammals, knees with jumping **adaptations**, and locking elbow joints. No extant gliding South American mammals exist. These fossils have not been given a formal scientific name since they are likely to belong to a previously recognised species. According to its size, the best identity guess is currently *Gaylordia*, a marsupialiform metatherian around 30 centimetres long which has grinding molars (possibly for crushing shelled invertebrates) and large canine teeth (Fig. 5.49). In addition to mammals, the fossil vertebrate fauna of Itaboraí includes an array of lizards, snakes, amphibians and crocodylians, along with four species of birds (Fig. 5.51). The crocodylians include the caiman *Eocaiman*, which was very similar to its modern relatives but distinguished by its relatively robust and well-developed fourth **maxillary** tooth. The fossil birds, which are particularly uncommon mostly because of their delicate, **pneumatised** bones, include early relatives of cuckoos (*Eutreptodactylus*) and *Palaeopsilopterus*. The latter was a primitive phorusrhacid (see pp. 41–2), more heavily built than *Psilopterus* (see p. 262).

Itaboravis was a small, terrestrial, **carnivorous** cariamid bird (see p. 42) closely related to modern seriemas; it is known only from postcranial elements consisting of the left coracoid and humerus, sharing features of the latter with tinamous (Mayr *et al.* 2011). *Diogenornis*, a likely ancestor of ratites (see p. 32), is known from a good number of well-preserved specimens. Phylogenetic analyses suggest that it is most closely related to Australian ratites (emus and cassowaries). *Diogenornis* had a somewhat narrow beak for probing seeds, fruit and invertebrates as in modern cassowaries and tinamous, along with relatively broad wings. These rather generalised features are most likely a relic from a flying ancestor.

Fig. 5.51. A wooded lakeshore at the sie of Itaboraí, Early Eocene. Clockwise from bottom right: the medium-sized xenungulate mammal *Carodnia vieirai*, a pair of the seriema relative *Itaboravis elaphrocnemoides*, the large, primitive ratite *Diogenornis fragilis*, the primitive phorusrhacid ("terror bird") *Palaeopsilopterus itaboraiensis*, the primitive caiman *Eocaiman itaboraiensis*, a large flock of the cuckoo relative *Eutreptodactylus itaboraienis*, and a wading bird (Gruiformes indet.).

Oligocene
33.9–23.04 Ma

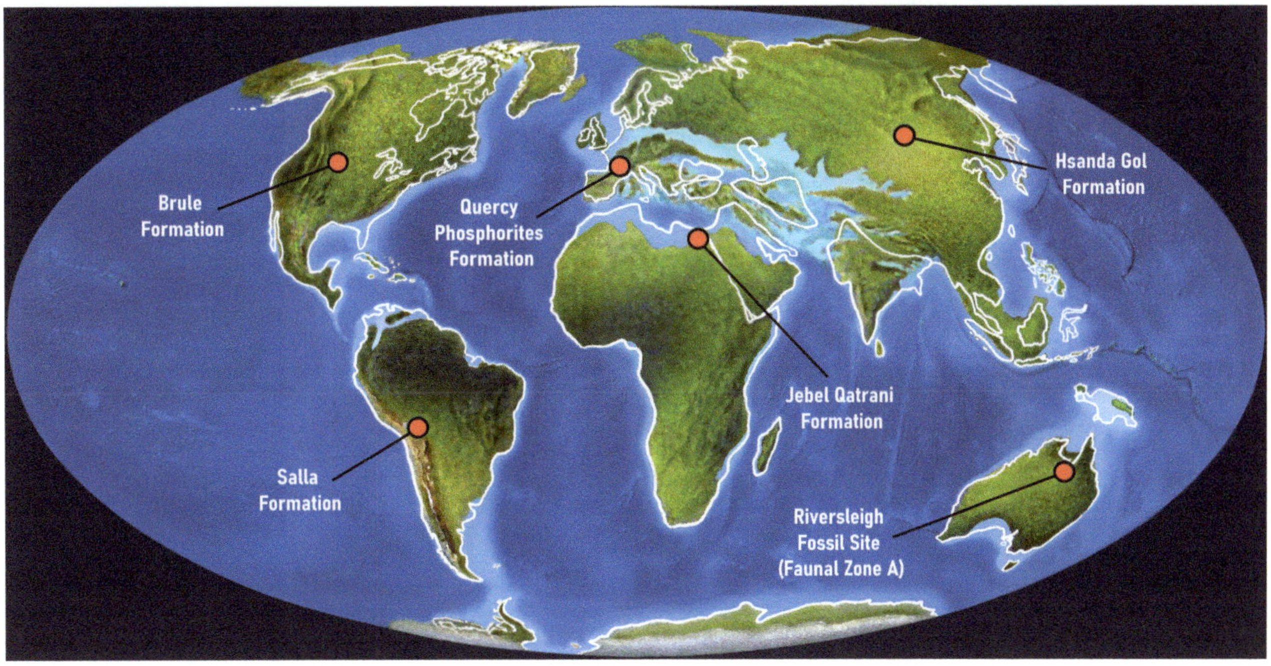

Fig. 5.52. Palaeogeographic map reconstruction of Earth during the Oligocene, with the position of the modern continents outlined and the approximate location of the fossil sites indicated.

The Oligocene marked the beginning of the transition from a hothouse to an icehouse. The beginning of the Oligocene is marked by a minor extinction event and faunal turnover in Eurasia called the Grande Coupure (see p. 230). Also, by the start of the Oligocene, the splitting of Australia from Antarctica was a significant factor in initiating a circumpolar current, which limited the exchange of heat between Antarctic waters and the warmer equatorial seas. An abrupt and significant cooling known as the Eocene-Oligocene Transition (EOT) occurred, leading to a drop in average temperatures in the high latitudes by approximately 5°C, although the impacts were considerably more marked further inland, such as in the North American interior, where **mean annual temperatures** decreased by about 8°C. Global climates experienced further increases in seasonality and variability as land and ocean temperatures decreased. This was the most dramatic and lasting change in temperature since the Cretaceous-Paleogene extinction event (see p. 169).

In large areas of Antarctica, a positive snow balance had occurred, where the annual snowfall exceeded the annual snowmelt and thus, a new glaciation had dawned. The ice sheets grew and engulfed significant portions of the continent during this Early Oligocene glaciation (possibly spanning several hundred thousand years). The initial growth of the Antarctic ice sheets, around 34 Ma, was primarily in East Antarctica, before the ice gradually expanded westwards, reaching the coast of West Antarctica at least seven million years later. This glaciation is primarily supported by the occurrence of ice floe-transported detritus in the Indian sector of the Southern Ocean. As the ice sheets expanded, a large amount of water was drawn from the oceans and stored in the ice. This caused a significant drop in global sea level, potentially by about 75 metres, with the East Antarctic Ice Sheet being the primary driver. This global cooling and sea level fall contributed to major turnovers and extinctions in marine **taxa**, particularly among plankton. Warm-water, surface-dwelling tropical and subtropical plankton species were notably affected. The Early Oligocene (33.9–27.3 Ma) was characterised by a multitude of various events including a decrease in taxonomic diversity in forests of the mid- and high-latitude regions as climates got cooler, precipitation got scarcer, rainforests retreated to the Equator and grasses expanded from water-bank habitats into open tracts. Many of the "**archaic**" forest-dwelling groups of mammals vanished, such as many of the lemur-like primates and "archaic" **ungulates** (see pp. 100–3) that once dominated the Eocene forests. In their places were **primitive** relatives of modern lineages including dogs, rhinoceroses, horses and camels. The Early Oligocene was marked by major faunal turnovers, including a relatively free exchange of animals across the northern continents, as demonstrated by the likeness of their vertebrate faunas. The Northern Hemisphere's higher latitudes saw a transition in vegetation from tropical evergreen broadleaf forests, as had dominated the Eocene, to temperate **deciduous** woodland composed of both broadleaf and evergreen trees. The Late Oligocene (27.3–23.04 Ma) was typified by the spread of grasslands and prairies, which was closely associated with the proliferation of **grazing** mammals. Globally, grasses became increasingly abundant, and a cooling phase with seasonal rainfall persisted. Mammals continued to grow larger as the mammalian radiation continued and landscapes became more open. During this time, numerous lineages of ungulates began to become more **cursorial**, in order to adapt to the plains that were opening up as the Eocene rainforests retreated.

Europe Quercy Phosphorites Formation, France: Bartonian-Chattian, 38-25 Ma.

In Europe, the best picture that we have of **terrestrial** vertebrate faunas during the Oligocene comes from the karstified phosphate deposits in the region of Occitanie, southern France. The deposits demonstrate changes in mammalian diversity from the Late Eocene to Early Oligocene. For most of the preceding Eocene, Europe was a tropical archipelago, but towards the Eocene-Oligocene boundary, the continent was approaching its modern configuration as sea levels fell and the emergence of solid ground ensued, forming corridors of dry land that connected former islands and opening up migration pathways for terrestrial mammals. The Turgai Strait (see p. 174) closed as land bridges merged Europe with the larger Asian landmass, and the wave of new immigrants from the east had a catastrophic effect on the faunas that had evolved in geographical isolation on the European Archipelago. The turnover of the European mammalian faunas was so intense that this event has been termed "La Grande Coupure", meaning "The Great Break". This turnover coincided with the pattern towards a more temperate climate, as the lush Eocene rainforests were replaced by drier and more open woodlands. The fact that the invading mammals had evolved in an environment that was distinctly continental, and therefore already arid and seasonal, meant they had an advantage over the **endemic** European mammals.

Late Eocene faunas remained dominated by endemic lineages of the European archipelago such as the palaeotheriids (see pp. 134–5), dozens of endemic families of deer-like artiodactyls including the anoplotheriids and xiphodontids (see p. 104), many endemic rodents, and the last of the omomyid and adapid primates (see p. 150). By contrast, the Oligocene faunas were strikingly different. The new replacement lineages dominated by immigrants from Asia included predatory mammals such as the leftover hyaenodonts (new species of *Hyaenodon* (see pp. 82–3)) along with some of the first ursids (see pp. 92, 94–5), mustelids (see p. 99) and "stenoplesictids" (see p. 84), as well as amphicyonids (see pp. 91–2) and nimravids (see p. 85).

Cephalogale (also present in Asia and North America) was a **primitive** member of the ursid subfamily Hemicyoninae. It lacked the large body proportions of modern ursids, but was a meduim-sized **omnivore** with a dentition adapted mostly for grinding, which retained some shearing features. The molars were significantly expanded, which each had a wide **posterior** surface forming a substantial portion of the tooth. The small, **scansorial** carnivorans (see pp. 81, 83–4) included the early amphicyonid *Cynodictis*, which resembled a large, slender marten. Compared to its "miacid" precursors (see p. 84), *Cynodictis* had developed a more specialised dentition—the last molars had disappeared and the cusps were less sharp. The body was elongated and the limbs relatively short, but longer than those of its "miacid" precursors. *Cynodictis* was a typical carnivoran of the European faunas in the Late Eocene, but during the Early Oligocene, gradually gave way to forms such as *Amphicynodon*, a **stem**-pinnipedimorph (see pp. 91, 94) that was no larger than a domestic cat and resembled a raccoon or genet. *Amphicynodon* had a long, low skull that differed from that of *Cynodictis* in its dentition, with robust canine teeth and molars that were more **bunodont**, suggesting it was more omnivorous. *Amphicynodon* also had an unspecialised skeleton with an elongated, flexible spine and **plantigrade** feet that allowed for a mobile articulation, enabling a hind foot reversal movement, potentially 180 degrees, which is similar to modern climbing carnivorans. While it may have hunted on the forest floor at least occasionally, *Amphicynodon* was capable of climbing trees headfirst either to evade larger predators (such as hyaenodonts) or in search of food. *Amphicynodon* was an **ambush predator** as indicated by its body proportions, likely preying on small, nimble vertebrates (Fig. 5.57).

Additionally, the oldest Oligocene members of the nimravid family belonged to the leopard-sized species *Eusmilus bidentatus* (Fig. 5.53), distinguished by advanced **hypercarnivorous** features. Known from good cranial and mandibular material, this dirk-toothed cat developed highly elongated upper canine teeth (longer than in any other nimravid), which accommodated in a mandibular sheath when closing

Fig. 5.53. Life reconstruction of the nimravid *Eusmilus bidentatus*, in the woodlands of Early Oligocene France. Reconstructed length: 1.5 m.

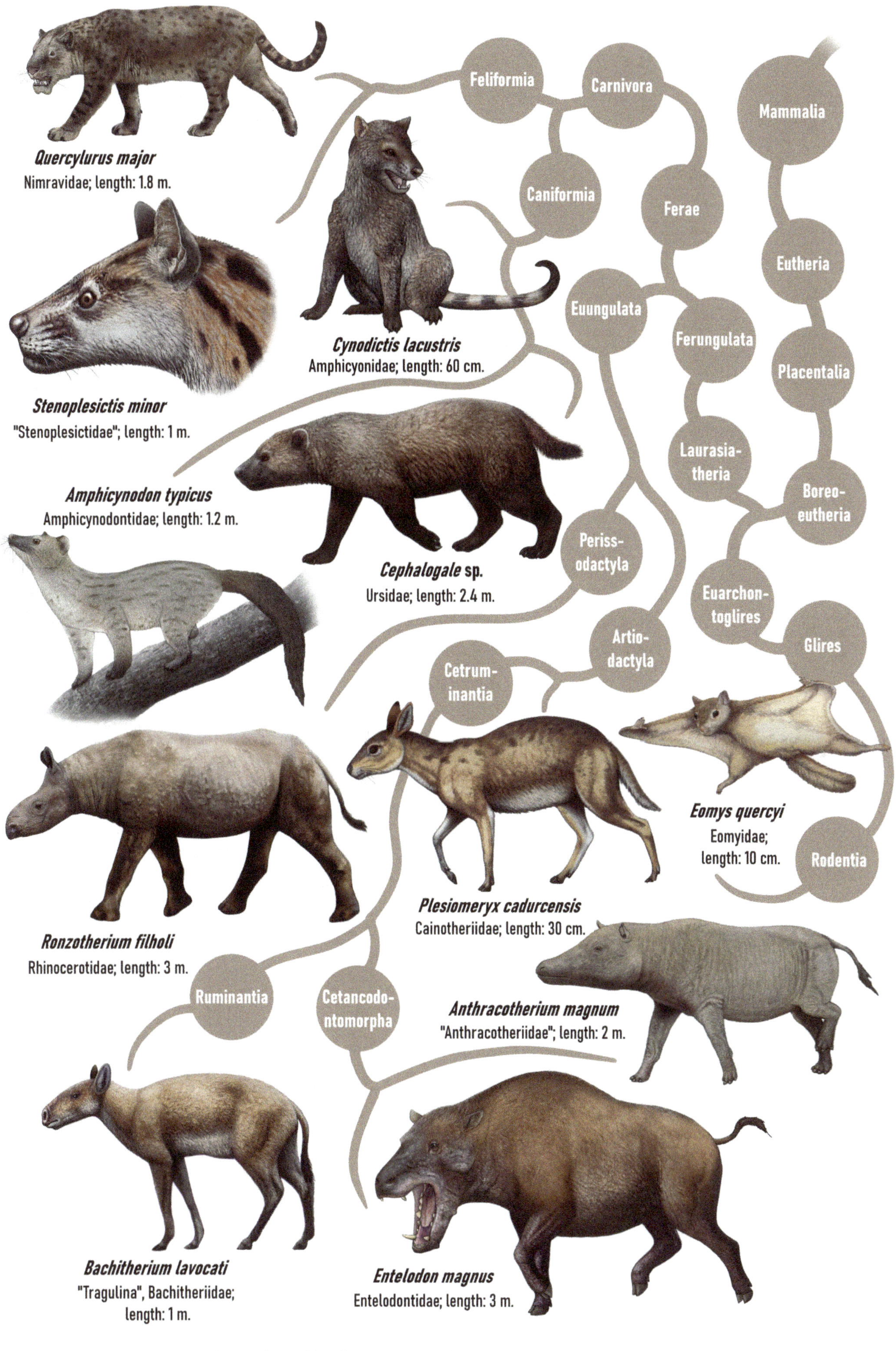

Fig. 5.54. Phylogenetic tree of a sampling of fossil mammal **taxa** from the Quercy Phosphorites Formation.

Europe Quercy Phosphorites Formation, France: Bartonian–Chattian, 38–25 Ma, cont.

the jaws. One defining feature of *E. bidentatus* is the markedly reduced dentition, which has 26 teeth rather than the 44 typically present in carnivorans, in order to make space for the huge canine teeth. Its jaws were structured in such a way that they could open at a gape of 90 degrees. The **adaptations** of this predator were not suited for preying on giant, tough-skinned **herbivores**, but instead for the fast and efficient dispatching of small- to medium-sized vertebrates. Although only known from limited remains, *Quercylurus* was a giant among the nimravids and the largest predator of its time, approaching the size of a brown bear (*Ursus arctos*) and weighing in at about 200 kilograms. It was a plantigrade **carnivore** with sturdy limbs more similar to those of an ursid than a felid (see pp. 86–9), but how it hunted remains unknown.

As well as diverse and advanced nimravids, Early Oligocene Europe also documents a variety of small, primitive, **polyphyletic** feliforms (see pp. 84–5) called "stenoplesictids", which looked like modern genets and civets in dimensions and anatomy. *Stenoplesictis* was the most abundant genus at the time, which had high-crowned **carnassials** with well-differentiated cusps. Additionally, the teeth of "stenoplesictids" showed trends in the reduction of the second molar, the third molar which had already been lost. In numerous ways, *Stenoplesictis* and its close relative *Palaeoprionodon* were most **analogous** to modern palm civets, with **hypocarnivorous** to **mesocarnivorous** diets. Among the perissodactyls (see p. 126), the last of the palaeotheriids survived alongside rhinocerotids (see pp. 130–2) and amynodontids (see p. 129). The Early Oligocene documents the first arrival of rhinocerotids in Europe, which included the basal form *Ronzotherium*, a gracile, long-limbed mammal with an elongated femur and humerus as **cursorial** adaptations to the increasingly drier climate. The skull of *Ronzotherium*, which lacked a nasal horn, possessed two pairs of upper incisors that were much reduced and one pair of long, tusk-like, forward-facing lower incisors. It was a **browser**, as evident from its **brachydont** teeth and retracted nasal bones indicative of a **prehensile** upper lip.

While the endemic perissodactyls experienced more complete extinction events, the artiodactyls primarily experienced a decline in diversity with some species persisting despite the turnover. Some Late Eocene taxa had lingered on into the Early Oligocene, including the antique dichobunids (see p. 104) such as the genera *Dichobune* and *Tapirulus*. Contrarily, a variety of artiodactyl families that were previously successful during the Late Eocene, such as the anopolotheriids and xiphodontids, had disappeared due to competition with immigrant hippopotamus-like "anthracotheriids" (see pp. 120–1), pig-like entelodontids (see pp. 118, 120) and a diversity of advanced basal ruminants (see p. 110). Yet, the small-bodied cainotheriids (see p. 104) were notably resilient to the arrival of these ruminants, having increased in abundance in the faunas of Oligocene Europe after surviving the Grande Coupure. Among these cainotheriids were the hare-sized primitive form *Plesiomeryx* which, like all cainotheriids, had a **complete dentition** with a **selenodont** morphology.

The ruminants had radiated into several genera such as *Bachitherium*, which had quite an unusual appearance due to its limb proportions comparable to those of moschids (see p. 115), its short, robust neck, and long, massive skull. The forelimbs were shorter relative to the hindlimbs, although the difference in length between them is smaller compared to tragulids (see p. 110). This suggests an adaptation for manoeuvring through dense scrub, dodging predators with powerful jumps and abrupt lateral swerves. The jaws were equipped with tusk-like teeth that were in fact **caniniform** first premolars, with the lower tusks behind the upper ones. The ruminant *Prodremotherium* (Fig. 5.55) is regarded as a possible ancestor of Giraffidae (see pp. 112–4) according to the development of the neck vertebrae, the lengthened and fused metapodials, and reduced canine teeth (Danowitz *et al.* 2015).

Fig. 5.55. Reconstructed life appearance of the ruminant *Prodremotherium elongatum*. Reconstructed length: 1 m.

The "anthracotheriids" were the most abundant bunodont artiodactyls, prevailing in a diversity of taxa. These included *Anthracotherium*, a medium-sized form (pygmy hippopotamus-sized) with enlarged canine teeth and robust jaws, whereas others such as *Elomeryx* (Fig. 5.56) and *Bothriodon* were slightly smaller pig-like forms with longer, narrower skulls and more gracile limbs. The canine teeth were relatively smaller in both genera compared to *Anthracotherium*. *Elomeryx* displayed **sexual dimorphism**, in which the canine teeth of males were more elongated than those of females and had serrated posterior edges. In all these "anthracotheriids", the canine teeth, along with the spatulate incisors, were used for uprooting aquatic vegetation. Broadly spreading toes (five on the forefeet and four on the hind feet) indicates that they were comfortable walking on wet, marshy ground.

Fig. 5.56. Reconstructed life appearance of the "anthracotheriid" *Elomeryx crispus*. Reconstructed length: 1.5 m.

Besides the abundance and diversity of "anthracotheriids", some of the most majestic Asian immigrant taxa were the massive entelodontids such as the widely distributed Eurasian genus *Entelodon*. Standing roughly 1.35 metres at its high-humped shoulder (exhibiting long neural spines on the thoracic vertebrae), it was the size of a really large wild boar (*Sus scrofa*). The forelimbs were more powerful and sturdy than the hindlimbs, and the femur was considerably longer than the **distal** hindlimb elements, such as the tibia, suggesting that *Entelodon* was not a fast runner. However, the fact that it had only two toes on each foot (the lateral **digits** which had disappeared) along with its highly elongated metapodials indicate that *Entelodon* and related genera were still capable of running efficiently in open habitats. The powerful forequarters and very short neck supported a long, low and wide pig-like skull with expanded **zygomatic arches** protruding prominently from both sides.

The first arrival of lagomorphs (see pp. 148–9) to Europe was among the most important events in the history of the continent's small mammal faunas. Eulipotyphlans (see pp. 74–7) including moles (*Myxomygale*) became prevalent following the extinction of "**archaic**" placental families such as the nyctitheriids. The "archaic" Eocene rodents, such as the ischyromyids (see p. 146), had all disappeared by the Early Oligocene, and were superseded by immigrant **clades** including the aplodontids (which today includes the mountain beaver), sciurids (squirrels), glirids (dormice), cricetids (hamsters) and various advanced extinct lineages, as well as lagomorphs. Additionally, the small non-marsupial metatherians (see pp. 46–7) did not suffer major losses during the Grande Coupure, with lineages of *Peratherium* (see p. 240) and *Amphiperatherium* (see pp. 212–3) persisting throughout the Oligocene.

The extinct theridomyids, which were endemic to Europe and originated during the Early Eocene, remained the most abundant rodents. After having peaked during the Eocene, the diversity of theridomyids had decreased by the Early Oligocene (and likely have no extant relatives). The theridomyids exhibited an evolutionary trend towards higher-crowned molars (and thus a more abrasive diet) and predominantly included **lophodont** taxa such as the rat-sized *Pseudoltinomys*. As indicated by its body proportions, with shortened forelimbs and elongated rear limbs, *Pseudoltinomys* had a **saltatorial** mode of locomotion in a fashion similar to modern springhares and gerbils, in semi-arid terrain and open woodlands (Fig. 5.57). The exact phylogenetic position of theridomyids within the order Rodentia remains highly uncertain. The aplodontids were a prosperous lineage of browsing rodents during the Oligocene, having developed upper molars that were increasingly selenodont. The earliest sciurids to disperse to Europe are typically associated with the genus *Palaeosciurus*. They possibly resembled modern squirrels, in spite of several minor differences in the teeth. The earliest species were likely ground squirrels as indicated by the limb ratios. The humerus and femur were long relative to the radius and tibia, like extant sciurids.

In the Early Oligocene, the cricetids were the most distinctive immigrants, becoming a highly successful family subsequent to their first arrival after the Grande Coupure. The glirids, which arose during the Middle Eocene, diverged into multiple genera with varying ecomorphologies and sizes, such as *Gliravus*. The earliest European cricetids, such as *Eucricetodon*, were small rodents possessing brachydont molars consisting of prominent bunodont cusps linked by low ridges. These rodents are also characterised by the presence of three molars and the absence of the last premolar in each jaw. One extinct family of rodents that originated in Europe during the Early Oligocene were the squirrel-like eomyids, characterised by a brachydont dentition similar in structure to those of the contemporary cricetids, but were distinct in consisting of three molars and one premolar per tooth row. *Eomys* is the only eomyid known from more than just teeth and from a complete skeleton, preserved with a furry outline including a skin membrane called a **patagium** linking the front and hindlimbs. As in modern flying squirrels, this membrane was supported by a bony spur projecting from the elbow, so eomyids are the earliest known gliding rodents.

Fig. 5.57. A scene from the site of Quercy, southern France, during the Early Oligocene, with the stem-pinnipedimorph *Amphicynodon leptorhynchus* hunting a rodent of the species *Pseudoltinomys gaillardi*.

North America Brule Formation, United States: Priabonian–Rupelian, 34–30 Ma.

In North America, the best record of **terrestrial** mammals from the Oligocene comes from the Rocky Mountains region in the Western United States. Upon continuous erosion, the mountains had become much reduced, and the entire region had become a vast plain roughly 700 to 1000 metres above sea level, in which fine-grained sediment continued to accumulate in slow-flowing **fluvial** systems. An exceptional quantity of mammal fossils have been unearthed from the White River Group, which forms widespread badlands covering areas of Wyoming, Colorado, Nebraska and South Dakota. The lower subunit includes the Chadron Formation, which is of Late Eocene age and consists of light grey claystone beds, deposited when the forests had lost their lushness but retained lots of tropical and subtropical trees. At the top of the Chadron Formation, these sediments transition into the Brule Formation which dates from the latest Eocene to Early Oligocene, and is mostly composed of **mudstones** and **siltstones** with some sparse channel **conglomerates**. During this time, the land may have been covered by more open woodlands interspersed with expanses of grassland consisting of shrub-like vegetation. Overall, the rocks of the White River Group were deposited in an **alluvial** setting, with the predominance of clays and sandstones suggestive of a mature river valley with a relatively flat **topography** and meandering fluvial systems, punctuated by swamps and lakes.

Many of the mammal groups that were around before the large climatic deterioration showed no change at all, apart from a dwarfing event in the oreodont (see pp. 107–8) genus *Miniochoerus*. Most of the lineages that existed in the Late Eocene were, apparently, already tolerant of the drier, more sparsely wooded environments, so the changes in flora did not have that much impact. In the Early Oligocene, the mammalian fauna that existed in what is now the Big Badlands was very similar to that of the Late Eocene, but without the brontotheriids (see pp. 135–6) and camel-related oromerycids (see pp. 104–5). By far, the most abundant mammals were oreodonts, with the genus *Merycoidodon* represented by hundreds of skulls from nearly every level of the section. This abundance of material indicates that *Merycoidodon* lived in vast herds for protection like modern zebras on the Serengeti plain. It was a typical pig-sized oreodont with an elongated body and short limbs. The upper and lower hindlimb bones were roughly equal in length, suggesting that it was not adept at running. *Merycoidodon* had a **complete dentition** of 44 teeth in the jaw, contrary to modern ruminants (see p. 110). A pit in front of the **orbits** infers the presence of a scent gland for territorial marking as in modern deer.

More generalised in form were the eporeodontines, with species of *Eporeodon* being among the largest oreodonts, which are significantly rarer as fossils presumably because they did not live in large herds (Fig. 5.58A). *Eporeodon* had a stocky, triangular skull (wider at the rear) and long, robust, **caniniform** upper incisors. Unlike many oreodonts, the molars of *Eporeodon* remained **brachydont**. *Leptauchenia*, as large as a medium-sized dog, had a very short skull, with a short muzzle that tapered towards the front (Fig. 5.58B). An enlarged **auditory bulla** suggests that *Leptauchenia* had large ears and good hearing. The limb ratios suggest that *Leptauchenia* was possibly an open habitat rock climber similar to modern rock hyraxes (*Procavia capensis*) (Wilhelm, 1993; Lander, 1998). The even smaller *Miniochoerus* (small dog-sized) also had a short skull but, contrary to other oreodonts, had reduced canine teeth that were of little use for defence (Fig. 5.58C). *Agriochoerus* (Fig. 5.58D) was a medium-sized close

Fig. 5.58. A sampling of oreodonts from the Brule Formation. (A) the eporeodontine merycoidodontid *Eporeodon major*; (B) the leptaucheniin merycoidodontid *Leptauchenia decora;* (C) the miniochoerine merycoidodontid *Miniochoerus gracilis;* (D) the agriochoerid *Agriochoerus antiquus*. Scale bar: 0.5 m.

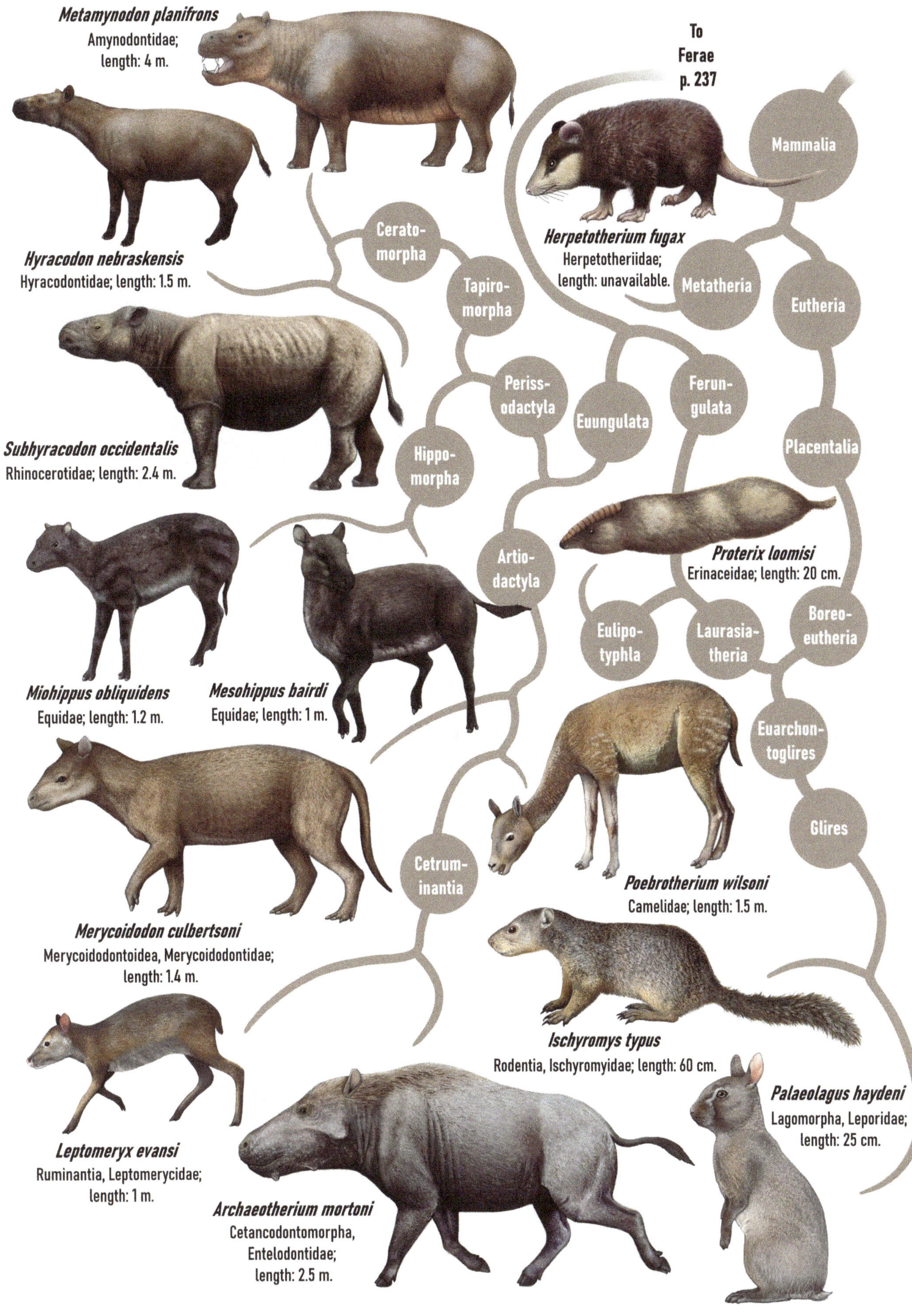

Fig. 5.59A. Phylogenetic tree of a sampling of fossil mammal **taxa**, particularly metatherians (see pp. 43, 46–7), and eutherians (see pp. 43, 57) outside Ferae (p. 74), from the Brule Formation.

North America Brule Formation, United States: Priabonian-Rupelian, 34–30 Ma, cont.

relative of oreodonts unusual for having claw-like hooves, although the skeleton shows no specialisations for climbing or digging (modern tree-kangaroos (*Dendrolagus*) are considered a good **analogue**). Its limbs were longer and more gracile than those of oreodonts, with five **digits** on the front (with a vestigial thumb) and four on the back.

In addition to oreodonts and their relatives, **primitive** camelids (see pp. 104, 106–7), which likely lacked humps and were the size of small- to medium-sized deer, were also abundant, along with the giant pig-like entelodontid (see pp. 118, 120) *Archaeotherium* and small, antlerless basal ruminants including the leptomerycids and hypertragulids (see p. 105). *Leptomeryx*. was one of the most primitive ruminants and resembled modern chevrotains. It was also distantly related to chevrotains as indicated by its delicate stature and enlarged, tusk-like upper canine teeth present in males. The first premolar was caniniform and the molars were typically brachydont, with a **selenodont** tooth pattern indicating that it was a **browser**. The possibly **frugivorous** *Hypertragulus* (Fig. 5.60) was similar to *Leptomeryx* in appearance, but had a narrower **rostrum** and the molars were higher-crowned. On the forearm, the ulna and radius were fused, unlike the condition in *Leptomeryx*. The orbits of *Hypertragulus* were also smaller and less prominently bulging.

Fig. 5.60. Reconstructed life appearance of the basal ruminant *Hypertragulus calcaratus*. Reconstructed length: 1 m.

In many ways, the camelid *Poebrotherium* closely resembled a small llama, with a narrow skull, long neck and gracile body, and had a reduced number of digits compared to other basal artiodactyls (see p. 104). It had an **unguligrade** foot stance with appressed digits, and lacked the pads typical of later camelids. *Poebrotherium* was most probably a **mixed feeder**, according to its generalised dentition, and adapted to a wide range of environments. *Archaeotherium*, the North American equivalent of the Eurasian genus *Entelodon* (see pp. 231–3), lived in forested environments, with footprint evidence of an individual stalking and chasing a small proto-rhinoceros through a wet marsh. Lots of *Archaeotherium* skulls have been found with bite marks evident of **intraspecific competition**. Like many entelodontids, *Archaeotherium* exhibits heavy dental wear on the tips of its canine teeth and on its blunt molar cusps, demonstrating that its diet consisted of tough, abrasive foods. It was evidently a **scavenger** and **omnivore**, and even a bone-crusher, consuming nearly anything it could including some plant material, although the front teeth do not display soil scratches from rooting in the ground. One locality in Wyoming with numerous fossil skeletons of *Poebrotherium* provides proof the animals were killed and partially eaten by *Archaeotherium*, indicating that this predator likely even hoarded carcasses in a meat cache. The bodies of *Poebrotherium* generally consist of rib cages with **articulated** forelimbs, indicating that the pelvis and hindlimbs were consumed. Dense stacking of bones suggests a rapid burial of a single event.

Among perissodactyls (see p. 126), the early equids (see pp. 132–4) *Mesohippus* and *Miohippus* were abundant (both the size of German Shepherds), as were rhinocerotoids (see p. 129) including the hippopotamus-like amynodontids wallowing in the waterways, the lightly-built, pony-sized hyracodontid *Hyracodon* galloping across the open forests, and the cow-sized rhinocerotid (see pp. 130–2) *Subhyracodon*. *Mesohippus* is the oldest known equid with **tridactyl** forelimbs, and had a noticeably longer skull (with a gap appearing behind the front teeth in

Fig. 5.61. Life reconstruction of the hyaenodont *Hyaenodon horridus* in an open grassland in what is now the Western United States, Early Oligocene.

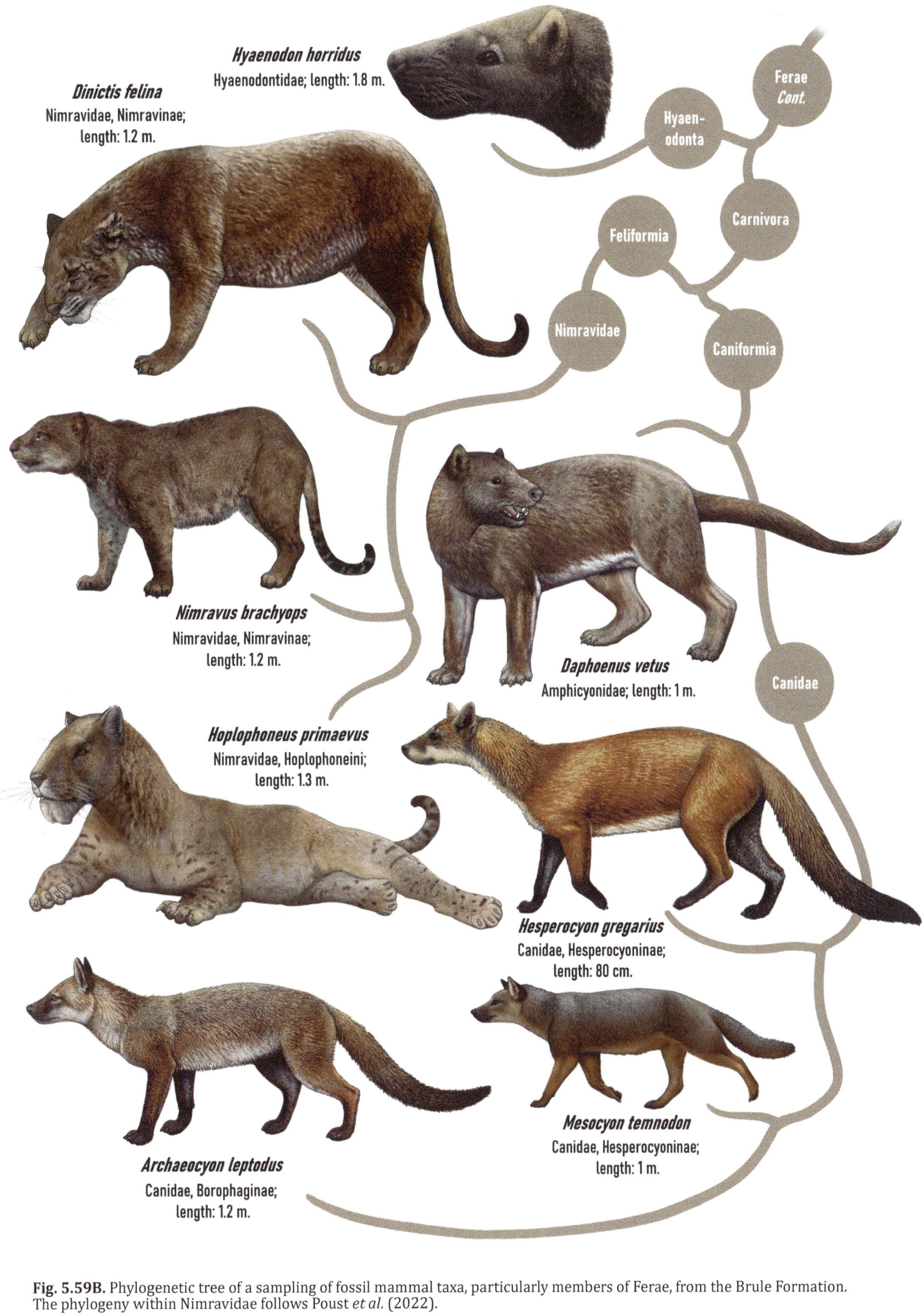

Fig. 5.59B. Phylogenetic tree of a sampling of fossil mammal taxa, particularly members of Ferae, from the Brule Formation. The phylogeny within Nimravidae follows Poust *et al.* (2022).

North America Brule Formation, United States: Priabonian–Rupelian, 34–30 Ma, cont.

contrast to earlier equids) and larger braincase than its precursors. It also had rounder orbits that were placed farther back and wider apart. As indicated by its brachydont **cheek teeth** and dental wear, *Mesohippus* was a browser, presumably feeding on fruit, leaves and twigs. *Miohippus* was more advanced than *Mesohippus* in possessing a slightly longer skull, a deeper facial **fossa,** and an additional crest on the upper molars that enabled the animal to grind vegetation more effectively.

Hyracodon was the last hyracodontid known from North America, which had very long, slender, tridactyl limbs (like the contemporary primitive equids), a short, wide muzzle and simple, rhinoceros-like, brachydont molars adapted for browsing low vegetation. Studies of its limb bones suggest that its running capabilities were no better than a modern forest hog. *Subhyracodon* was a hornless, browsing rhinocerotid. The skull was elongated and depressed, and the dentition was highly specialised compared to earlier forms—the upper canine teeth and upper third incisors were lost, and the upper second premolars were **molariform**. *Subhyracodon* relied on speed to avoid predators as indicated by its relatively long limbs, which unlike many "**archaic**" rhinocerotids, were tridactyl instead of **tetradactyl**. If any amynodontid is familiar it would be *Metamynodon*, the stockily-built genus which contributed to the idea that the whole family was amphibious. It is now thought that only the metamynodontines spent considerable time in water and show specific **adaptations** to an aquatic lifestyle: high placed orbits and nasal openings, a wide muzzle, short limbs, and a broad rib cage that would have assisted with buoyancy. The dentition indicates a **grazing** diet of tough vegetation and males may have used their large canines for uprooting plants along riverbanks or for intraspecific competition.

Preying upon the **herbivores** were the last of the ancient hyaenodonts (see pp. 82–3) of the genus *Hyaenodon* (see pp. 82–3), along with a much richer diversity of true carnivorans (see pp. 81, 83–4) such as mongoose-like primitive canids (see pp. 92–3), abundant nimravids (see p. 85) and early amphicyonids (see pp. 91–2). The canid *Hesperocyon* has the characteristic dog-like ear structure, encased in bone instead of cartilage. In *Hesperocyon*, the tooth count was reduced to 42, with the last molar absent from both sides of the upper jaw. Its hands and feet retained five digits each, contrary to the disappearance of the first finger and large toe in more advanced canids. The digits became increasingly compact, signifying a more upright and **digitigrade** stance. Fossil coprolites suggest that its diet consisted of small rodents and lagomorphs (see pp. 148–9) (Wang and Tedford, 2008). Similar to *Hesperocyon* in size and body proportions was the more omnivorous *Archaeocyon*, which was among the oldest known borophagines. *Archaeocyon* had a generalised skeleton and retained a **plantigrade** stance. It was more advanced than *Hesperocyon* in its dentition, with modifications such as the addition of an accessory cusp on the upper molars. *Mesocyon* was a small, coyote-sized hesperocyonine canid that was one of the first canids to attain moderate sizes and **hypercarnivorous** specialisations. *Mesocyon* is characterised by an unusually robust skull and very short muzzle.

Many skeletons of the well-known North American hyaenodont *Hyaenodon horribilis* (Fig. 5.61), the largest known *Hyaenodon* species on the continent (estimated at 40 to 60 kilograms), were found around a waterhole where it was able to **ambush** prey. Other **carnivores** were unsafe; an individual of the nimravid *Dinictis* was found with fatal *H. horribilis* tooth puncture marks on its skull. Differences in dentition altered the way nimravids hunted. The long-lived genus *Dinictis* had shorter, thicker and more conical canine teeth which were capable of delivering a greater puncture bite compared to other nimravids. Its short, semi-plantigrade feet indicate that it would not have been able to run long distances in pursuit of prey. *Dinictis* had the short-limbed/long backed form of a fossa (*Cryptoprocta ferox*), and was presumably semi-**arboreal**.

The less agile nimravid *Hoplophoneus* is known from several specimens that have been found with damage caused by conspecifics usually aimed towards the back of the skull and the orbits, and this may give a clue to how the longer toothed nimravids dispatched prey. The limbs of *Hoplophoneus* were more robust than those of *Dinictis*, and the cervical vertebrae were larger and with advanced processes for muscle attachments. *Nimravus* had long, appressed metapodials, like a modern felid (see pp. 86–9) but unlike other nimravids, indicating that it hunted by pursuit instead of ambush. Its upper canine teeth were also more poorly-developed than those of other nimravids, reaching a similar relative size of those in modern felids. The discovery of a *Nimravus* skull with its upper canine teeth lodged into the humerus of another *Nimravus* signals a fatal case of intraspecific competition. *Daphoenus* was a coyote-sized, early amphicyonid characterised by a long tail, elongated body and rather short limbs that show it was an ambush predator. It had a plantigrade stance as evident from fossil footprints. *Daphoenus* excavated burrows for sheltering their offspring and for evading larger predators. Fossil coprolites indicate that *Daphoenus* hunted and masticated its food in a similar manner to modern coyotes (*Canis latrans*).

In the small mammal **communities**, rodents (particularly modern lineages such as beavers, pocket gophers, hamsters and sciuromorphs (see pp. 146–7) including the extinct *Ischyromys*) predominated, along with mouse-like metatherians, common lagomorphs and a whole array of **insectivorous** mammals. *Ischyromys* was rather squirrel-like in body proportions and generally resembled a prairie dog (*Cynomys*), although the tail was more robust and elongated, possibly more so than in the genus *Paramys* (see pp. 205, 209). *I. typus* already possessed well-developed incisors like those of modern rodents. *Palaeolagus* was an ancestral rabbit that had shorter

Fig. 5.62. Reconstructed life appearance of the leptictid *Leptictis acutidens*. Reconstructed length: 30 cm.

hindlimbs than its modern counterparts and therefore was probably not good at bounding **quadrupedally** but rather at scampering like squirrels. Its skull and dentition also remained much more primitive than in many later rabbits. Some specimens show specialisations in the inner ear region suggesting acute hearing. *Proterix* was an armour-headed, mole-like relative of hedgehogs, presumably using its forehead in a spade-like fashion to hunt for burrowing invertebrates. The lumbar vertebrae appear to be very well-developed for strong, undulating bodily movements. This, along with the strongly developed nasal and neck musculature, the auditory opening in the ear region being deflected downwards, and limited vertebral movements indicate a **fossorial** lifestyle. This remains inconclusive until more complete postcrania is found.

The leptictid (see p. 77) *Leptictis* (Fig. 5.62) is known from well-preserved skeletons and, despite similarities with *Leptictidium* (see pp. 212–3), there were obvious physical differences in the skeleton which prevented *Leptictis* from moving in the same manner. The hindlimbs of *Leptictis* were less proportionately elongated relative to the forelimbs, showing that it could undergo quadrupedal movement, a more basal condition among leptictids. *Herpetotherium* was an omnivorous non-marsupial metatherian known from many well-preserved skeletons that show it was similar in appearance and size to modern opossums. Limb bone adaptations reveal that *Herpetotherium* was a terrestrial quadruped, unlike modern opossums. Studies of petrosal morphology in the skull have concluded that the North American *Herpetotherium* was more closely related to marsupials than the European herpetotheriid genera (see pp. 47–8) *Peratherium* (see p. 240) and *Amphiperatherium* (see pp. 212–3).

Perhaps the best known fossil bird from the Brule Formation is *Bathornis*. Originally known as "*Neocarthartes*", it was described as a terrestrial cathartid (New World vultures and condors). Multiple species of *Bathornis* are known, ranging in size between one and two metres tall (Fig. 5.63A). The largest species (*B. geographicus*) and those of *Paracrax* (Fig. 5.63B) were significant predators, coexisting alongside large predatory mammals during the Early Oligocene. Some differentiation of habitat occurred between coexisting species, with *Bathornis* found in moister environments and *Paracrax* found in more arid regions. Among the reptiles, the genus *Stylemys* (Fig. 5.64) was a herbivorous tortoise with many species ranging in size from as small as five centimetres to as large as one metre. Although *Stylemys* displays the same neck anatomy as extant tortoises, it had more primitive jaw muscles and the limbs did not have the burrowing specialisations of later genera.

Fig. 5.64. Reconstructed life appearance of *Stylemys nebrascensis*. Reconstructed length: 1 m.

In summary, many of the herbivorous mammals described above had reasonably primitive, brachydont teeth for feeding on shrubs and soft leaves. The lack of arboreal mammals such as multituberculates (see pp. 45–6) and primates also demonstrate that the thick forests disappeared as nearly all the small mammals seem to have been well-adapted to a terrestrial lifestyle. From the Late Eocene until well into the Miocene, the White River Fauna was well-established and stable, having changed very little.

Fig. 5.63. Reconstructed life appearance of bathornithids (see pp. 35, 42) from the White River Group. (A) *Bathornis grallator*; (B) *Paracrax gigantea*. Scale bar: 1 m.

Africa Jebel Qatrani Formation, Egypt: Priabonian–Rupelian, 34–30 Ma.

The Jebel Qatrani Formation is located in the Fayûm Depression in central Egypt, where it overlies the Qasr el-Sagha Formation (see pp. 218–21). The rocks, mainly sandstones, were deposited in meandering rivers and ephemeral lakes during the Early Oligocene, when Africa remained isolated from the other continents and was largely stable tectonically, with no significant uplift yet occurring in the eastern or southern regions. At the time, the Sahara Desert had not yet formed, and Africa lacked much of the vegetative zonation that would arise from the establishment of a generalised horseshoe of progressively arid habitats surrounding the equatorial rainforests of the central and western regions of the continent. Instead, much of Africa was likely covered in a mosaic of humid tropical and subtropical rainforests, interspersed with more arid or seasonally dry habitats in localised areas. Fossil evidence from the Jebel Qatrani Formation supports this interpretation, and reveals that a lowland floodplain with swamp forests, lakes and slow-flowing rivers existed in what is now Northern Egypt—all comprising a rich **ecosystem** capable of sustaining a wide variety of plants and animals (Fig. 5.65).

Africa also continued to evolve a strange **endemic** fauna. The foremost components were proboscideans (see pp. 64–8) of the moderately large genera *Palaeomastodon* and *Phiomia*. These were both similar to later proboscideans in having the external nasal cavity retracted to above the molars, and had one pair of incisors enlarged as tusks in the upper and lower jaws. The upper tusks were curved downwards in both genera, although the lower tusks were scoop-shaped in *Palaeomastodon* whereas they were straight and flat in *Phiomia*. Both genera had **brachydont cheek teeth**, but those of *Palaeomastodon* exhibited a **lophodont** pattern, whereas those of *Phiomia* were shorter and conveniently **bunodont**, with cusps that were more rounded. In both genera, a large **diastema** was present between the cheek teeth and tusks in the lower jaw, and contrary to later proboscideans, the molars erupted vertically instead of horizontally. The postcrania was robust, consisting of columnar limbs designed for efficient locomotion as in later proboscideans. Certain species of *Palaeomastodon* (*P. beadnelli*) were comparable to the modern Asian elephant (*Elephas maximus*) in size, while *Phiomia* was slightly smaller.

The most abundant mammals were the hyracoids (see pp. 70, 72–3), which back in the Oligocene of Africa, were the most diverse and dominant **herbivore** group. The largest known genus is *Titanohyrax* (the size of a small rhinoceros), with species ranging in size from 23 kilograms up to 1.3 tonnes. *T. ultimus* was the largest species, represented solely by teeth. *Titanohyrax* was a **cursorial** genus, and its specialised **selenodont** dentition with **molariform** premolars and strongly **lophed** upper molars is reminiscent of that of numerous perissodactyls (see p. 126) and indicates that it was a **browser**. The medium-sized gazelle-like titanohyracid *Antilohyrax* had a brachydont dentition **convergent** with bovid **ungulates** (see pp. 116, 118–9), suggesting that it was a **folivorous** browser.

Fig. 5.65. Palaeoenvironmental reconstruction of the Jebel Qatrani Formation. Early Oligocene of Egypt. Clockwise from bottom right: a congregation of the side-necked turtle *Neochelys fajumensis*, the eusuchian crocodylomorph (see pp. 26–30) *Eogavialis africanum*, a monitor lizard (*Varanus* sp.) attacking the small metatherian (see pp. 43, 46–7) *Peratherium africanum*, the shoebill relative *Goliathia andrewsi*, the distant flamingo relative *Palaeolodus* sp., the embrithopod *Arsinoitherium zitteli*, a flock of egret-like birds (Ardeidae indet.), the hyaenodontid *Bastetodon syrtos*, the large stork *Palaeohippiorhynchus dietrichi* perched on a bush, the heron-like bird *Xenerodiops micter* perched on a log, the hyracoid *Saghatherium antiquum*, a pair of jacana birds (*Javipes nymphaeobates*) perched atop waterlilies, the dugongid sirenian (see pp. 69–71) *Eosiren imenti* surfacing above the water, and the "anthracotheriid" *Bothriogenys gerringei* swimming in the river.

Fig. 5.66A. Phylogenetic tree of a sampling of fossil mammal **taxa**, particularly placentals outside Boreoeutheria (see p. 9), from the Jebel Qatrani Formation.

Africa Jebel Qatrani Formation, Egypt: Priabonian-Rupelian, 34–30 Ma, cont.

Like other hyracoids, its upper incisors were enlarged as small tusks. One unusual feature is the **pectinate** incisors, strangely shared with the gliding colugo, and were probably used for grooming fur. *Geniohyus* is named after its morphological similarities to pigs (e.g. bunodont molars), with early research intially attributing the genus to Suidae (see pp. 108–10). The skull of *Geniohyus* was very elongated **anteriorly**, equipped with a long narrow snout and an unusually short **postorbital** region. Other features include a strong **sagittal crest**, an exceptionally narrow palate, and brachydont cheek teeth. *Megalohyrax* is was an abundant hyracoid, with robust limbs and a massive body. Unlike modern hyraxes, the skull was low and elongated, with the **orbits** not positioned anteriorly. Contrary to other contemporaneous hyracoids, the dentition of *Megalohyrax* was more generalised, with cheek teeth that were **bunoselenodont**, while like most other **primitive** hyraxes, the premolars show a tendency towards molarisation. *Saghatherium*, which resembled modern hyraxes but was roughly twice as large, had deep, long jaws and likely fed primarily on grasses. Additionally, species within this genus exhibit marked **sexual dimorphism** (a more massive jaw in males) and possibly had complex mating behaviours. *Saghatherium* is more primitive than modern hyraxes in which the manual **digits** had cleft, claw-like terminal **phalanges**.

Even more extraordinary than all these hyracoids was *Arsinoitherium*, an embrithopod (see pp. 68–9) resembling a rhinoceros in size and overall appearance. It possessed a pair of massive, hollow, conical, bony horn cores projecting from the nasal region of the skull. Each of these had a single, small, knob-like horn placed immediately behind. These were likely coated in a horny sheath consisting of **keratinised** skin as in modern cattle, contrary to the matted fibrous structures with no bony core that comprises the horns of modern rhinoceroses. The limbs were pillar-like and terminated in **pentadactyl** feet, being generally **analogous** to those of elephants. *Arsinoitherium* had a **complete dentition** of 44 teeth, with features inferring that it was a selective browser. The **graviportal** limbs, dental wear and details of bone chemistry indicate fully **terrestrial** habits. The only fossil artiodactyls (see p. 104) from Jebel Qatrani are the ubiquitous **semi-aquatic** "anthracotheriids" (see pp. 120–1) such as the small-bodied and vaguely pig-like genus *Bothriogenys*. It had the characteristic features of "anthracotheriids" including a complete dentition, a considerably elongated muzzle, and a low skull. In *Bothriogenys*, the canine teeth remained reduced and there was no marked sexual dimorphism as in other "anthracotheriid" genera. Other distinctive features include upper molar crowns possessing five cusps.

Besides all these large herbivores, the small mammal **communities** included a variety of **insectivores**, and the possible afrotherian (see p. 62) *Ptolemaia*, which became a gracile, coyote-sized **carnivore** or **omnivore**. One partial skull is known, which is long, low and has a relatively small braincase. *Ptolemaia* had large canine teeth and robust premolars, the upper **posterior** premolars which were larger than the molars. The molars became increasingly smaller posteriorly, the crowns which exhibited low cusps that were rounded to somewhat sharp, with dental wear indicating a diet of abrasive or tough food items. The rodents all belonged to endemic families, such as the extinct Phiomyidae, grouped within the order Hystricomorpha (see pp. 146–7). There were no lagomorphs (see pp. 148–9) such as were widespread across the Northern Hemisphere by the Early Oligocene. Even more remarkable are the predators—there were no true carnivorans (see pp. 81, 83–4) living in Africa during the Oligocene and only the widespread hyaenodonts (see pp. 82–3) persisted. *Bastetodon*, a recently identified genus of hyainailourine (Al-Ashqar *et al.* 2025), was a leopard-sized (27 kilograms) apex predator with a short, **rostrum** and a more reduced dentition compared to other hyainailourines. Its cranial anatomy, particularly its robust zygomatic arches and its reduced dentition with pronounced shearing **adaptations**, suggests attachment surfaces for powerful jaw muscles and a **hypercarnivorous** diet, respectively. The short rostrum contributed to giving *Bastetodon* a strong bite force. The discovery of *Bastetodon* strengthens the hypothesis that the Hyainailourinae diversified in Africa and then dispersed into Eurasia in multiple waves during the Miocene. *Masrasector* was a relatively small (skunk-sized) teratodontid hyaenodont. The limb bone morphology, particularly of the **distal** humerus, indicates a stable elbow joint and that it was a fast running, primarily terrestrial predator, possibly preying on the local rodents and small hyracoids. On the other hand, the rather blunt cusps of the premolars could also indicate a predilection for hard-shelled molluscs (**durophagy**) (Borths *et al.* 2017).

Jebel Qatrani is particularly renowned for its diverse primate fauna, with the most abundant fossil record of primitive simiiforms (see pp. 152–6) provided by the Fayûm deposits. Some taxa were part of the early catarrhine radiation while others such as the afrotarsiids directly preceded the platyrrhine-catarrhine divergence, placed outside **crown** Simiiformes. *Afrotarsius* is only known from a single partial lower jaw with three complete teeth, meaning its phylogenetic relationships have been challenging to determine. It was initially classified as a tarsiid, although the most recent phylogenetic analyses have found it to be a **stem**-haplorhine (Chaimanee *et al.* 2012; Jaeger *et al.* 2020). The similarities in molar morphology between *Afrotarsius* and a stem-haplorhine from the Middle Eocene of Myanmar called *Afrasia* suggests that simiiforms colonised Africa from Asia by dispersing across the Tethys Ocean (see p. 174) sometime during the Middle Eocene, just before *Afrotarsius* shows up in the African fossil record. The early catarrhines included parapithecids such as *Apidium*, among the most well-known and abundant Fayûm primates. Some distinctive anatomical features of parapithecids include a relatively deep distal femoral articulation distinct from other primates. Overall, the limb anatomy suggests that *Apidium* was an active **arboreal quadruped** capable of jumping from horizontal supports. The most recent family of Fayûm simiiforms to show up in the fossil record are the propliopithecids, which includes *Aegyptopithecus*. They are also the largest Fayûm primates, with which also have the most **derived** anatomical features. *Aegyptopithecus* is believed to have been a slow-moving arboreal quadruped as evident from the distal articular surface of the femur, deeper than in modern catarrhines. *Aegyptopithecus* resembled the howler monkey (*Alouatta*) in size and in various details of the forelimb long bones, but according to overall femur anatomy, was more heavily-built. As in *Apidium*, *Aegyptopithecus* had relatively small orbits suggestive of a **diurnal** lifestyle.

Fig. 5.66B. Phylogenetic tree of a sampling of fossil mammal taxa, particularly placentals within Boreoeutheria, from the Jebel Qatrani Formation.

Australia Riversleigh Fossil Site (Faunal Zone A), Queensland: Rupelian–Aquitanian, ca. 28.4–23.03 Ma.

Located in the basin of the Gregory River in northwestern Queensland is Riversleigh, which is Australia's most famous fossil site and given world heritage status since 1994 due to its unparalleled richness, the exceptional quality of the fossil material found and the stretch of time encompassed by its record, with intervals of deposition spanning from the Late Oligocene to the Early Holocene. It is from the fossil fauna preserved in a complex mosaic of karstic limestone distributed across over two hundred individual localities (collectively known as the Riversleigh Fauna) that we now have a very good picture of Australian **ecosystems** throughout the Late Paleogene and Neogene Periods. The Riversleigh fossils are noteworthy in that they have been preserved in soft freshwater limestone. Consequently, in contrast to the majority of fossil sites, the animal bones were preserved in their original three-dimensional form instead of being crushed. The high calcium **carbonate** content also contributed to the exceptional preservation of the fossils. The fossil localities represent areas where caves and fissures were once active, having developed through the dissolution of limestone bedrock by water, a process characteristic of karst landscapes. This generated natural traps and feeding grounds for **carnivores** that at regular intervals, and possibly abruptly, became infilled and preserved. The Riversleigh fossils are now exposed on the surface because the areas where caves and fissures once existed have been uplifted and eroded. The huge assemblages of marsupials and other vertebrates, which record over 20 milion years of faunal evolution on the isolated continent, are a result of these special conditions. The Oligo-Miocene fossil assemblages at Riversleigh are divided into four "Faunal Zones" (FZ) according to time: FZ A, spanning the Late Oligocene (28.4–23.03 Ma), FZ B, spanning the Early Miocene (23.03–15.97 Ma), FZ C, spanning the Middle Miocene (15.97–11.608 Ma) and FZ D, spanning the Late Miocene (11.608–5.332 Ma) (Myers *et al.* 2017).

From the Late Oligocene to Middle Miocene, the area over the limestone karst terrain was covered with extensive tracts of lush, tropical forests (not too different from those found in Borneo today), and the weather was characterised by intense rainfall and mild temperatures. More open areas, with karst soil, occurred at forest edges or around inland freshwater streams and lakes. Even during these greenhouse conditions in Australia, it is possible that climatic gradients existed, potentially with marginally drier conditions at the continent's centre. Except for bats and several monotremes (see p. 44), all the Riversleigh mammals were marsupials, many of which were occupying niches no longer seen in the much drier habitats of modern Australia. It has been estimated that during the Late Oligocene (and into the Early Miocene), marsupials formed ~85% of species-level diversity among Australia's mammals (Long *et al.* 2002). The following is a sample of genera from Faunal Zone A:

The macropodiforms (see pp. 54–6) were all small-bodied and included numerous structurally **primitive** forms (*Wururoo, Ganawamaya, Nambaroo, Galanarla*) belonging to Balbaridae, an extinct family characterised by features such as **brachydont** and **lophodont cheek teeth**, lateral constriction of the cranium behind the **orbits**, and large sinuses. Skeletons of the genus *Nambaroo* indicate that balbarids may have been primarily **quadrupedal** but could most likely hop like modern kangaroos. *Wururoo* is characterised by large, robust upper third premolars adapted for crushing hard food items. *Ganawamaya*, which was about the size of a bettong (*Bettongia*), had a short tibia-fibula contact and reduced lateral foot movement, indicating a shift towards ground-based locomotion (Den Boer *et al.* 2019). Both *Wururoo* and *Ganawamaya* were **browsers**, feeding on fruit and soft vegetation. *Nambaroo*, which was roughly the size of a quokka (*Setonix brachyurus*), is characterised by reduced third premolars along with molars exhibiting well-developed **lophs**, and was possibly **omnivorous**. *Nambaroo* also had large toes and mobile feet, and the skeleton shows flexibility in locomotion, with some hopping and some **scansorial** (semi-**arboreal**) behaviour. *Galanarla* is a poorly represented balbarid. It is relatively small, similar to *Ganawamaya* in size and morphology. Given the limited fossil material, how *Galanarla* is related to other balbarids currently cannot be resolved. *Galanarla* is thought to have been a **mixed feeder** like modern pademelons (*Thylogale*). Additionally, there were also primitive potoroids such as *Bulungamaya*, a genus with **bunolophodont** molars and **plagiaulacoid** lower third premolars. The basal macropodid *Gumardee* is distinguished by bunolophodont cheek teeth, 11 sharp ridges on the upper third premolar and a highly reduced upper last molar. There was **sexual dimorphism** in *Gumardee* as evident from variation in tooth size, particularly in the third premolars. Species of *Gumardee* became progressively **bilophodont** over time, and among macropodiforms, a completely bilophodont condition was only attained by balbarids and macropodids. *Bulungamaya* and *Gumardee* were likely browsers.

Among the vombatiforms were the **herbivorous** wombat-like wynyardids and ilariids (see p. 56), both browsers. *Namilamadeta* was a **terrestrial** wynyardid roughly the size of a beagle, characterised by a proportionately deep snout and relatively robust molars, the upper ones which exhibit two transverse series of three cusps (Rich and Archer, 1979; Long *et al.* 2002). The small-bodied ilariid *Kuterintja* is characterised by large lower incisors protruding horizontally from the lower jaw, cheek teeth with no transverse connecting crests, and conical-cusped lower molars. *Ngapakaldia*, a sheep-sized, ground-dwelling herbivore, is among the oldest and most primitive known diprotodontids (see p. 56). *Ngapakaldia* is characterised by a **posteriorly** raised skull, a flattened snout, broad first incisors, and second and third **digits** slenderer than the fourth and fifth digits on the hind foot. *Ngapakaldia* possibly lived on the shores of lakes and fed on riparian vegetation, although its teeth seem adapted for slicing leaves. The sheep-sized *Marada* is the sole known member of an extinct family of herbivorous wombat-like vombatiforms called Maradidae, distinguished by a unique amalgamation of primitive and specialised features. *Marada* appears to be most closely related to true wombats (Vombatidae) (Crichton *et al.* 2023). Key features of *Marada* include a short, broad **rostrum**, large, robust incisors, and a steep decrease in cheek tooth size from front to back, suggesting **adaptations** for processing hard plant matter. The predatory marsupials included primitive thylacinids (see p. 52). *Badjcinus* is the oldest and most basal known thylacinid, at 1.7 to 3.1 kilograms. It had a more **ancestral** cranial anatomy than later thylacinids, with features similar to dasyurids (see p. 52). It was likely an arboreal generalist carnivore, occupying the niche of a quoll (*Dasyurus*), and eating insects and small vertebrates.

Fig. 5.67. Phylogenetic tree of a sampling of fossil mammal **taxa** from Faunal Zone A of the Riversleigh deposits. The phylogeny within Balbaridae and of the potoroids follows Butler *et al.* (2018).

South America Salla Formation, Bolivia: Chattian, 27–26 Ma.

The Salla Formation consists of fossiliferous, multicoloured **siltstones** and claystones distributed across the hills and valleys of what is now western Bolivia, some 3,500 metres above sea level. The beds accumulated on a **fluvial** plain during the Late Oligocene, when the area was at a considerably lower altitude than at present, possibly less than half as high as it is today. Much of the South American landscape became drier in the Oligocene than in the Eocene, with continuous volcanism associated with the formation of the Andes mountain range affecting the regional climates. The cooler climate led to the widespread extension of grassland biomes such as **savannah**, and the presence of **carbonate** nodules suggests that the region had a distinct dry season. The mammals most likely lived in partially open environments including open woodland. Most species fed on gritty, low-growing plant material, while others were **frugivores** and **browsers**. **Scansorial** and **arboreal** mammals also occurred, but were much less common and diverse than **terrestrial** species.

Some of the oldest South American monkey fossils and rodent skulls are known from this site. The early platyrrhine (see pp. 152–3) *Branisella* is known only from isolated teeth and cranial elements. *Branisella* was possibly more terrestrial than any modern platyrrhine since its teeth are more high-crowned. However, there is heavy wear on the molars suggesting that *Branisella* fed on abrasive foods, with poorly-developed shearing crests indicative of a frugivorous diet (Kay *et al.* 2001). Like most modern South American primates, *Branisella* was possibly **diurnal**. Its exact phylogenetic position is still debated, with some studies placing it as sister to all non-pitheciid platyrrhines, while others suggest a closer relationship with marmosets and tamarins. Fossils of various rodents (such as early relatives of modern chinchillas and agoutis) are more plentiful than at earlier South American fossil sites, mirroring this **clade's** increasing significance in the continent's **ecosystems**. The remains of xenarthrans (see p. 58) comprise some of the oldest well-preserved material from this lineage, such as the **primitive** genus of glyptodontine (see pp. 59–60) *Clypeotherium*, with a possible body mass of over 500 kilograms. At Salla, the notoungulates (see pp. 140–3) highlight the pattern of increasingly **hypsodont** teeth that began in the Early Oligocene, the majority of species which had **hypselodont cheek teeth**. During the Oligocene, this **adaptation** was developed by the notoungulates well ahead of their ecological equivalents on other landmasses. The notoungulates comprised numerous primitive species of families such as Mesotheriidae and Toxodontidae, the latter which became the dominant family during the Miocene.

Trachytherus, the most completely represented Oligocene mesotheriid and typotherian, had enlarged and hypselodont first upper incisors, with enamel exclusively on the front surface, like most other typotherians. The molars were arched and almost hypselodont, with a thick outer layer of **cementum**, suggesting a diet of gritty/abrasive foodstuffs such as low-lying plant material or roots and tubers. It was terrestrial and **fossorial**, but was possibly less adapted for digging than geologically younger mesotheriids. The archaeohyracids were small, scampering generalists in the woodland underbrush. *Archaeohyrax* is characterised by a skull with a tall, blunt snout and a robust lower jaw. The first pair of upper incisors were enlarged and implanted obliquely, whereas the first two pairs of lower incisors pointed forward and upwards. The cheek teeth were very hypsodont, suited for grinding tough, fibrous vegetation such as grasses, a diet evident from **mesowear** analysis (Croft and Weinstein, 2008). While skulls and jaws are well-known, the postcranial skeleton remains largely unknown. The notohippid *Rhynchippus* possessed a long skull, compact body and elongated limbs characteristic of **grazing**, plains-adapted mammals such as horses. *Rhynchippus* was in fact related to *Toxodon* (see pp. 328–9), having evolved a continuous horse-like dentition adapted for grazing via **convergent evolution**. The canine teeth did not form tusks as in other toxodontians, but had the same cropping structure as the incisors. Convoluted enamel on the molars enabled *Rhynchippus* to chew tough vegetation. The middle ear region was also highly **pneumatised**, resembling that of other primitive toxodontians, and had a particularly well-developed tympanic crest (a bony ridge/projection that aids in the overall protection of the eardrum). The cow-sized leontiniid *Anayatherium* was a heavyset animal with an elongated neck and body, as well as sturdy limbs with hooved **tridactyl** feet, and a small tail. The skull had an inflated, tusk-like pair of first incisors that functioned as canine teeth. True canine teeth were absent, resulting in a shortened muzzle. The cheek teeth were **brachydont** or subtly hypsodont, indicating that it was a **folivorous** browser feeding in areas with plentiful trees and scrub.

Besides the increasing hypsodonty, one other notable trend in Late Oligocene South American ecosystems is the development of large body size in several lineages, many of which vanished soon afterwards. Lion-sized sparassodonts (see pp. 48–50) from the family Proborhyaenidae are the most striking examples of this trend. Often depicted as a large bear, the large sparassodont *Paraborhyaena* had a large skull somewhat resembling that of a mustelid (see p. 99) in being short-snouted. *Paraborhyaena* is known only from fragmentary cranial material (including a deep lower jaw) that indicates a skull length of approximately 40 centimetres. Its powerful teeth mean that it may have been able to crunch bones, so it is possible that they were more **scavengers** than large hunters, converging on modern hyaenas. The Oligocene was also a time of gigantic pyrotherians (see pp. 139–40). In *Pyrotherium*, the long skull's most notable feature is a large nasal opening between the **orbits**, with muscle attachment sites indicative of a **proboscis**. Forward-facing incisors were chisel-shaped, with four in the upper jaw and two in the lower jaw, although it is the cheek teeth which are the most distinctive. The brachydont molars had thicker, wrinkled enamel in a unique pattern not present in any other South American native **ungulate** (see p. 139). Very dense bones and short, slender limbs are suggestive of a massive, **graviportal** beast that walked on **plantigrade** feet. *Pyrotherium* is estimated to have weighed between 3 and 3.5 tonnes. Despite often thought of as a **semi-aquatic** species, the fact that its fossils are restricted to dry environments prone to droughts suggests that a good modern **analogy** for *Pyrotherium* would be a browser like the black rhinoceros (*Diceros bicornis*), able to strip vegetation using the front teeth. It is uncertain what led to the evolution of large size in multiple groups independently during this time. Global climatic trends possibly had an important influence, as similar patterns have been observed on other landmasses.

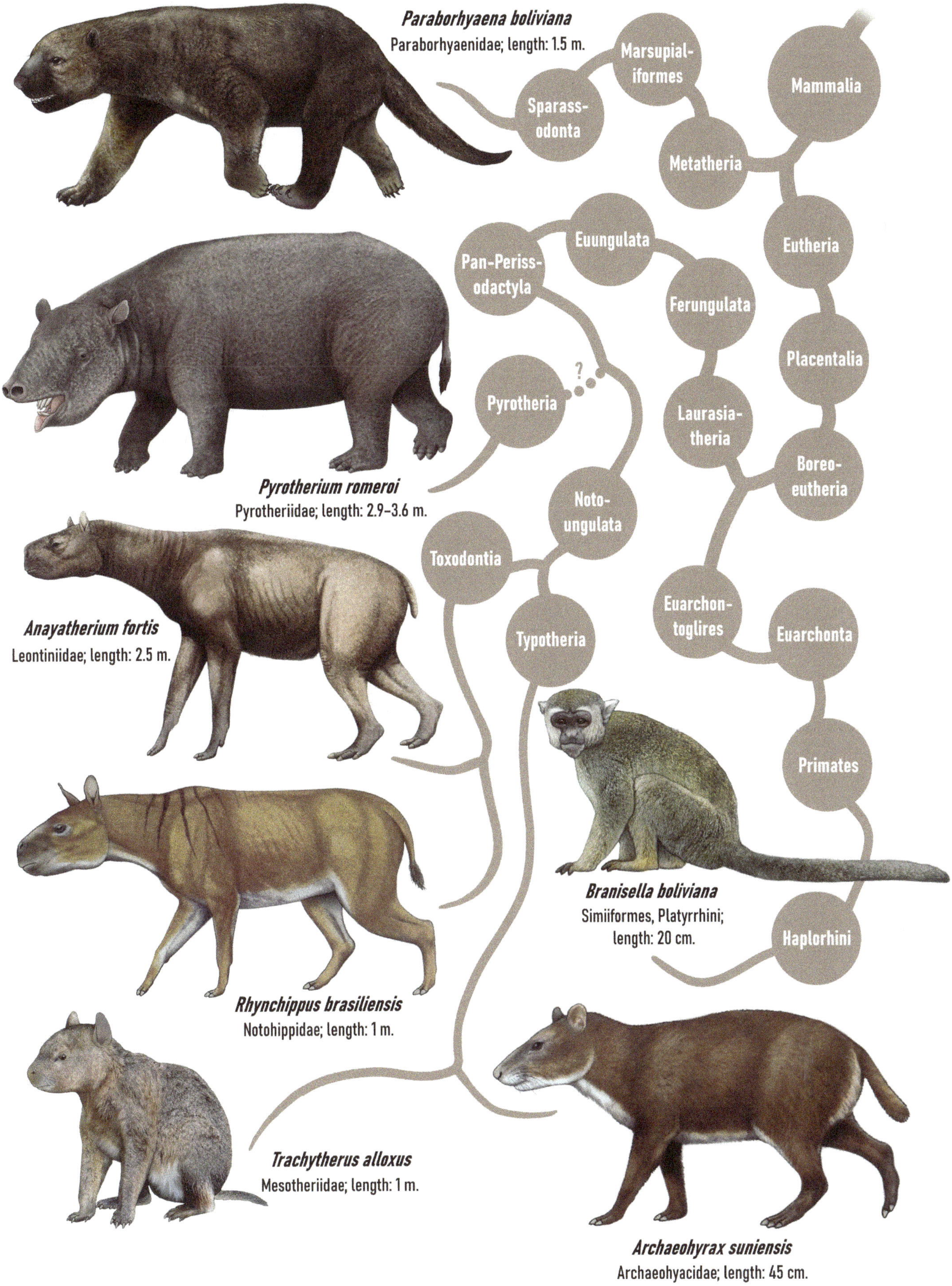

Fig. 5.68. Phylogenetic tree of a sampling of fossil mammal **taxa** from the Salla Formation. The phylogenetic position of Notoungulata follows Welker *et al.* (2015). The tentative phylogenetic position of *Pyrotherium* as a **stem**-perissodactyl follows Kramarz and MacPhee (2023), which supported the placement of South American native ungulates, including pyrotherians, within Laurasiatheria.

Asia Hsanda Gol Formation, Mongolia: Rupelian, 33.4–31 Ma.

The Hsanda Gol Formation, situated in what is now the Tsagan Nor Basin of southern Mongolia, was deposited in a continental semi-desert with ephemeral rivers during the Early Oligocene. Palynological analyses from contemporaneous sites in Mongolia and China have concluded that much of the area back then was covered by a woody scrubland in which arid-adapted plants such as saltbush, Mormon tea and *Nitraria* bushes predominated (Leopold *et al.* 1992). The sparse trees that dotted the landscape, concentrated close to groundwater (Leopold *et al.* 1992), were **deciduous** and broad-leaved, relatives of modern birch, oak and elm. The fauna from Hsanda Gol differed substantially from the Asian faunas of the Late Eocene as well as those from elsewhere around the world during the Oligocene. The brontotheriids (see pp. 135–6), amynodontids (see p. 129), entelodontids (see pp. 118, 120), the last of the mesonychians (see pp. 102–3) and the tapiroids (see p. 128) that thrived during the Late Eocene, had all disappeared from this region. However, the "**archaic**" didymoconid leptictidans (see p. 77) lingered on, as did the hyaenodonts (see pp. 82–3) and paraceratheriids (see pp. 129–30). The paraceratheriids and eggysodontids (represented by *Allacerops* (see pp. 222–3)) were both abundant, along with a diversity of deer-like ruminants (see p. 110) including small, long-limbed gelocids (an extinct family of hornless ruminants from the Eocene to Oligocene and whose exact placement within Ruminantia is somewhat unclear) adapted for **grazing** and running. Various new mammal groups appeared for the first time, such as several advanced rodents and a whole new array of advanced but generally small carnivorans (see pp. 81, 83–4) including mustelids (see p. 99), amphicynodontids (see pp. 91, 94) and some of the earliest true felids (see pp. 86–9), sharing the **carnivorous** niche with mustelid-like basal carnivorans such as *Palaeogale* (along with the cat-like nimravids (see p. 85) and the dog-like hyaenodonts, having already been well-established).

Fig. 5.69. Life reconstruction of the hystricomorph rodent *Tsaganomys altaicus* in the confines of its burrow. Early Oligocene of Mongolia. Reconstructed length: 20 cm.

Known from a partial skull and lower jaw with associated postcranial elements, *Didymoconus* was a leptictidan with no ecological **analogues**, although it had an inflated mastoid region (a bony projecton which plays a role in middle ear pressure regulation and protects the inner ear) in the skull **converging** on modern arid-adapted elephant shrews and kangaroo rats (Gingerich, 1981; Mellett, 1968). The lagomorphs (see pp. 148–9) and rodents were by far the most diverse and abundant **herbivores** of this fauna, most of which possessed **hypsodont** teeth for feeding on abrasive plants. *Tsaganomys* was a **fossorial** rodent analogous to mole-rats (using its incisor teeth for digging in a similar way) but is most closely related to hystricognaths (porcupines and kin), and is an indicator of sandy soil in arid areas (Fig. 5.69).

Amphicticeps was a small, **hypercarnivorous** amphicynodontid with a robust skull relative to its body size, consisting of a short, wide **rostrum** and medium-sized canine teeth. Some **primitive** features include a reduced upper second molar, an upper third molar either highly reduced or completely absent and a **carnassial** notch on the upper fourth premolar. The marten-sized *Palaeogale* was a hypercarnivore whose phylogenetic affinities remain controversial. The dental anatomy is mixed, consisting of both feliform (specialised carnassials) and mustelid (reduced lower second molars) characteristics. It also has an ossified, single-chambered **auditory bulla** unlike feliforms (see pp. 84–5). However, *Palaeogale* is generally considered a basal feliform (Fig. 5.70), and it was possibly semi-fossorial.

Among the felids, fragmentary fossils of a lower jaw have been tentatively assigned to *Proailurus*, the most basal known member of the family. As is evident from more complete remains found elsewhere in the Northern Hemisphere, it was a small and compact, ocelot-sized carnivoran with large **orbits**, a long tail, and semi-**plantigrade** feet ending in sharp, semi-retractable claws. Its limb morphology and locomotor behaviour resembled that of modern fossas (*Cryptoprocta ferox*), and it was possibly **scansorial**. Although both the skull shape and teeth morphology are largely similar to those of today's felids, the dentition had not yet undergone the drastic reduction of molars and premolars typical of modern felids.

Among the hyaenodonts (the largest predators), five species of *Hyaenodon* are known from Hsanda Gol, with *H. pervagus* being the least specialised. It is very poorly known and not so representative of the genus *Hyaenodon* as a whole (which is possibly a **wastebasket taxon**), which primarily consisted of predators as large as wolves or smaller. Many of these species are likewise poorly known and classed within this genus mainly based on their characteristic shearing teeth. *H. pervagus* is differentiated from the other Asian species of *Hyaenodon* in being medium-sized (smaller than *H. gigas* (see pp. 224–5)), and by features including premolars that are relatively larger and more sharp-edged, and a very short carnassial notch on the upper second molar (Lange-Badre and Dashzeveg, 1989).

In Asia, the diversity of Early Oligocene large mammals was much lower, with the large body size niche being dominated by rhinocerotoids (see p. 129). The Hsanda Gol Formation is particularly well-known for yielding fossils of the second largest, tallest and heaviest known **terrestrial** mammal that has ever existed (only surpassed by the elephant *Palaeoloxodon namadicus* (see pp. 67–8)): the paraceratheriid rhinocerotoids of the species *Paraceratherium transouralicum* (Fig. 5.71). The lack of complete skeletons makes it difficult to definitively determine the exact size of *P. transouralicum*, but it is estimated to have reached 4.8 metres (15.7 feet) at the shoulder and weighed 15 to 20 tonnes. As in sauropod dinosaurs, *P. transouralicum* had hollow openings in the neck and back vertebrae to lighten the skeleton without compromising its structural integrity, which was crucial for supporting its massive size.

The lower limbs and toe bones of *P. transouralicum* were elongated and relatively slender, in contrast to the shorter, thicker and more robust lower limbs and stumpy toes found in other gigantic, **graviportal** mammals such as elephants. Although it approached enormous sizes and weighed twice as much as a bull African bush elephant (*Loxodonta africana*), it never abandoned the long-toed blueprint of its **cursorial** rhinocerotoid ancestors. *P. transouralicum* would have been unable to run or move fast due to its great size, but the long limbs allowed for a larger stride, making locomotion more energy-efficient over long distances. Travelling long distances would have been essential in habitats where resources were scarce. Thus, they possibly had extensive territories and underwent migration. The unusually wide atlas and axis vertebrae, and the nearly horizontal **posterior** articular face of the axis, indicate that *P. transouralicum* had a more flexible neck than other rhinocerotoids, enabling it to **browse** foliage at greater heights. *P. transouralicum* had a colossal skull over 1.5 metres (5 feet) long, with a **prehensile** upper lip (as indicated by the deep nasal incisions), large, conical, tusk-like incisors for stripping leaves from branches (the lower incisors pointed forwards and the upper ones were larger and pointed downwards) and simple, **brachydont cheek teeth** adapted solely for browsing on soft vegetation (the upper molars have a pi (π)-shaped pattern on the **occlusal** surface like most rhinocerotoids). This rhinocerotoid had such a long neck and its head was elevated so high above the ground that it likely browsed off the treetops, occupying an ecological niche similar to the giraffe (*Giraffa camelopardalis*). The skull had a characteristic arched upper profile and, as in all other primitive rhinocerotoids, lacked any rugosities for the attachment of horns. Even though names such as *Indricotherium* and *Baluchitherium* have been given to various fossils from these beasts, they are all now believed to be synonymous with the genus *Paraceratherium*, the first name given to this animal. *P. transouralicum* was considerably larger than any Oligocene predator, so it was resistant to predation apart from when calves were small and vulnerable.

These giant rhinocerotoids were the sole large species at Hsanda Gol, and medium-sized species were almost absent. The Hsanda Gol Formation includes a poor sample of the hugely diverse perissodactyls (see p. 126) and artiodactyls (see p. 104) that were so common elsewhere in the Northern Hemisphere during the Early Oligocene. It is apparent that certain **ungulates** (equids (see pp. 132–4), rhinocerotids (see pp. 130–2), chalicotheriids (see pp. 128–9), entelodontids, "anthracotheriids" (see pp. 120–1) and tragulids (see p. 110)) were mostly browsing forms that may have been unable to cross into the semi-arid Hsanda Gol region. The general faunal composition is

Fig. 5.70. Phylogenetic tree of a sampling of fossil carnivorans from the Hsanda Gol Formation. The phylogenetic position of *Palaeogale* follows Wang and Zhang (2015).

Asia Hsanda Gol Formation, Mongolia: Rupelian, 33.4–31 Ma, cont.

further indicative of an arid environment, with not many tall trees to support the **arboreal** mammals, although some gallery forests possibly existed along rivers. Virtually all the terrestrial mammals were adapted either for living near the ground and within the restricted brush cover (many of the deer-like ruminants and their predators, along with the enormous diversity of fossorial and ground-dwelling lagomorphs and rodents) or for browsing on the treetops (such as the paraceratheriids) (Meng and McKenna, 1988). In summary, while the mammal fauna at Hsanda Gol shares many genera and families with Europe (particularly the artiodactyls and many of the same carnivorans), it was much more arid-adapted than those of Europe at the time. A much greater quantity of mammals in Asia bore hypsodont dentitions and there is abundant floral evidence for scrubbier plant life and more abrasive diets. Even though most of these Asian mammal **taxa** dispersed to Europe during the Grande Coupure (see p. 230) to outcompete the **endemic** faunas, Europe remained a considerably milder and wetter continent than both North America and Asia during the Oligocene.

Fig. 5.71. Palaeoenvironmental reconstruction of a semi-desert **ecosystem** in Mongolia during the Early Oligocene. In the foreground (from left to right): a hyaenodont of the species *Hyaenodon pervagus* rests underneath the shade of a tree, and the leptictidan *Didymoconus colgatei*. In the middleground (from left to right): the basal ruminant *Eumeryx culminis*, and the lagomorph *Desmatolagus gobiensis*. In the background is a family of the paraceratheriid rhinocerotoid *Paraceratherium transouralicum*.

6. The Neogene

23.04–2.58 Ma

Although the break between the Paleogene and Neogene Periods was not heralded by a dramatic extinction event, it was marked by a change in the fossil record, with many "**archaic**" mammal groups having disappeared and most life forms from this time closely resembling their modern relatives. The Neogene Period also heralds the beginning of a genuinely modern world, with broadly spaced continents and a steep temperature gradient from chilly high-latitude regions to warm equatorial areas. Once the Antarctic Circumpolar Current (see p. 176) was firmly established by the Early Neogene, Antarctica became truly isolated as a cold zone. A change from **carbonate**-rich to silica-rich ocean floor sediments corresponds to the change from warm to colder waters.

Key aspects of modern global geography were also coming about throughout the Neogene. India kept on nudging into Asia, thrusting up the Himalayas ever-so higher and contributing to the closure of the Tethys Ocean (see p. 174). Since the Arabian, African and Australian plates drifted northwards, Italy collided with Switzerland and France to thrust up the Alps, Spain with France to form the Pyrenees, and later Australia with the Eurasian plate to create Indonesia. During these mountain-forming processes, the continental crust underwent horizontal compression while rocks of about the same density were pushed together by colliding plates. Sea levels decreased on average as a result of a slight reduction in the area occupied by the continents and an increase in the extent of oceanic basins. It is possible that all these events aided in the cooling of the planet, which remodelled its **ecosystems**. Grasslands spread worldwide, albeit at different rates in different regions. They appear to have been established in East Asia concurrently with North America during the Oligocene, before expanding during the Miocene. Mixed grasslands and scrublands were present by the Early Miocene, and by the Late Miocene, the landscape very much resembled today's African **savannahs** (Fig. 6.1). Savannah **communities** of bovids (see pp. 116, 118–9), giraffids (see pp. 112–4), rhinocerotids (see pp. 130–2) and equids (see pp. 132–4) were prevalent across Europe and West Asia by the Late Miocene. Grasslands were also established in Africa by this time.

Meanwhile, many large and small **terrestrial** mammalian **herbivores** and their predators had to adapt to a demanding life on the open plains. As grasslands replaced forests in mid-latitude regions, **ungulates** developed morphological changes including **hypsodont** teeth adapted for **grazing** and long limbs for running away from predators. A number of South American native ungulates (e.g. notoungulates (see pp. 140–3)) developed hypsodonty well before their counterparts in North America, although this was to cope with ingesting volcanic ash from the uplifting Andes mountain range, rather than grass, which only dispersed all over the temperate regions of South America by the Miocene-Pliocene boundary. The Andes mountains were folded up by the subduction of a denser oceanic plate, the Nazca plate, beneath continental crust, the South American plate, which led to compression of the western edge of the latter plate.

Despite the shifting climate, atmospheric carbon dioxide was relatively stable throughout much of the Neogene. From the Middle Miocene onwards, carbon dioxide levels began to decline significantly. The precise reason for this decline is unclear, but it may be associated with the uplift of the Himalayas and Tibetan Plateau. It has been suggested that this new phase of mountain uplift, which intensified the weathering of silicates, removed a significant amount of carbon dioxide from the atmosphere as the gas reacted with the decomposing rock minerals to generate bicarbonates. These bicarbonates are soluble, which are carried by **fluvial** systems and eventually settle as sediment on the ocean floor. Whatever the cause, it had profound impacts.

Fig. 6.1. A view of a grassland ecosystem at Venta del Moro, eastern Spain, during the Late Miocene (Turolian ELMA (see p. 9), ca. 9.0–5.3 Ma). From left to right: the beaver *Dipoides*, the bovine bovid *Miotragocerus*, the hipparionin equid (see pp. 132–4) *Hipparion*, the hyaenid (see pp. 88–90) *Thalassictis*, the proboscidean (see pp. 64–8) *Tetralophodon*, the ursid (see pp. 92, 94–5) *Agriotherium*, the camelid (see pp. 104, 106–7) *Paracamelus* and the cervid (see pp. 115–7) *Dicrocerus*.

When the Mediterranean Sea largely dried up into an enormous salt pan during the Late Miocene (see p. 253), the removal of all this salt from the sea had triggering effects on worldwide climate, since the Pliocene climate was noticeably cooler compared to that of the Late Miocene, and also considerably drier on land (leading to the disappearance of numerous Miocene terrestrial mammal groups). Some of this cooling was possibly also a result of the continued fall in atmospheric carbon dioxide as the Himalayas and Tibetan Plateau continued to rise, which accelerated the weathering of rocks into carbon dioxide-absorbing soils.

A major aspect of the Neogene was drier conditions in many regions of the planet as well as worldwide cooling. This drying pattern was especially pronounced in Australia, but it took place in the Americas and Asia as well. It was intensified by the formation of rain shadows associated with the throwing up of the Himalayas, the Rocky Mountains and the Andes. The major group of plants that successfully exploited these new dry environments were grasses. During the Neogene, new open habitats were formed as a result of drier climates, possibly in conjunction with greater pressure from grazing mammals and a greater frequency of wildfires.

The Neogene plant life took on a more modern shape as the climate and geography became more like that of today's. Throughout the Neogene, warmth-adapted vegetation became progressively limited to the tropics and subtropics due to the cooler climates. North of the subtropics, regions with more strongly seasonal climates saw the emergence of oaks, maples, and beeches as the dominant tree species. Temperate woodland expanded across the Northern Hemisphere during the Neogene. The generally similar composition of these woodlands in different regions suggests they may have originated from a common ancestral flora that dispersed via land bridges across the North Atlantic and Bering Strait during the warmer Paleogene Period. Since there is considerably less land at equivalent latitudes, temperate woodlands have developed at a considerably lesser extent in the Southern Hemisphere.

The massive Antarctic ice sheets of the Oligocene, which had contracted during the Early and Middle Miocene, expanded to almost their modern size by the Late Miocene (ca. 14 Ma). Increased snowfall on Antarctica brought about by increased water circulation in the atmosphere, derived from the Atlantic Ocean, resulted in the ice sheet expanding to its current extent by that time. By the end of the Miocene, the Antarctic ice sheet expanded enormously, becoming even more extensive than it is at present. Two episodes of glacial expansion, lasting roughly 15,000 years, took place 5.2 and 4.8 Ma, separated by a relatively warm interval. Yet, at the North Pole, ice had not begun to form all year round, which would not happen until the Late Pliocene, around 2.8 Ma (see p. 295). The Earth had entered an Ice Age by the time the Neogene Period ended.

Fig. 6.2. In what is called the Great American Biotic Interchange (GABI), mammals dispersed between the Americas, starting during the Late Miocene (over 10 Ma), and especially during the Late Pliocene (ca. 3 Ma) when the Panama Isthmus was fully established. Some South American native groups managed to disperse north, while most migrants were North American native groups travelling south.

Miocene
23.04–5.333 Ma

Fig. 6.3. Palaeogeographic map reconstruction of Earth during the Miocene, with the position of the modern continents outlined and the approximate location of the fossil sites indicated.

The Miocene represented a time of tremendous change. Global climate went through numerous fluctuations, from a severe icehouse at the end of the Oligocene to an interval of gradual warming throughout the Early Miocene. Dry, **savannah**-type plant life had evolved in dry, low-latitude regions by around 16 Ma, during the Middle Miocene Climatic Optimum (MMCO), the warmest period on average since the abrupt cooling at the end of the Eocene. The MMCO is one of the most mysterious Cenozoic climatic events, though changes in rates of tectonic degassing (the release of gases, including carbon dioxide, from the Earth's mantle as tectonic plates interact and move) may have been a primary driver. Other factors such as changes in ocean circulation, volcanic activity, and even variations in the Earth's orbit probably also contributed to the MMCO. Then between 14 and 12 Ma, global temperatures had plummeted during the Middle Miocene Climate Transition, associated with a decline in atmospheric carbon dioxide levels. The East Antarctic Ice Sheet expanded significantly, and permanent ice sheets finally draped the continent.

While ocean temperatures abruptly decreased in the polar regions, they remained relatively warm in low latitudes, averaging around 22–24°C (72–75°F). Extensive ice sheets possibly started to form on Greenland as early as 7–8 Ma, yet the Arctic generally remained warm enough to sustain trees there until the Late Pliocene. This Early to Middle Miocene warming, followed by a sharp cooling, is thought to have resulted in the receding of tropical **ecosystems** to lower latitudes, the spread of northern coniferous woodland, and greater seasonality. These changes led to the radiation of modern sedges and grasses. Throughout the Miocene, and globally, there were higher levels of primary production, (i.e. plant growth), which in turn led to a higher net diversity of **herbivore clades**, and an ensuing higher net diversity of predators. Additionally, kelp forests first evolved in the Miocene oceans, quickly becoming one of the planet's most productive ecosystems. In the Early Miocene (ca. 19 Ma), the Arabian Peninsula collided with Eurasia, cutting off the connection between the Indian Ocean and the Mediterranean Sea, and allowing a faunal interchange to occur between Eurasia and Africa. This led to the closure of the Tethys Ocean (see p. 174), leaving behind remnant seas including the Mediterranean, Red, Black and Caspian Seas. This closure, and the associated changes in water circulation and salinity, contributed to increased aridity in Southern Europe. During the Late Miocene (ca. 5.7 Ma), the Mediterranean Sea almost completely dried up because the cooling climate reduced sea levels, disconnecting the Atlantic from the Mediterranean, and evaporation rates exceeded inflow from rivers, in an event called the Messinian Salinity Crisis. What was left behind were vast, white expanses of salt on the former sea floor from which former islands towered like mountain chains. By around the Miocene–Pliocene boundary, the Strait of Gibraltar later reopened and the Mediterranean catastrophically refilled within a few months or years, in an event called the Zanclean Flood. Similar to the preceding Oligocene, the general trend of environmental change during the Miocene was one of continuously decreasing global temperatures and increased aridity, which accelerated the spread of grasslands at the expense of forests. The expansion of grasslands prompted the appearance of new mammalian groups and the extinction of some older ones. For example, the even-toed **ungulates** (see p. 104), especially ruminants (see p. 110), expanded at the expense of their odd-toed cousins (see p. 126) as they were better adapted to feeding on the tough, poorly-nutritious grasses due to their highly complex stomachs.

North America Agate Fossil Beds, United States: Burdigalian, ca. 20 Ma.

The Early Miocene witnessed significant mammalian dispersals between continents, continuing patterns established since the Early Eocene. The relatively warm climate and abundant vegetation extending toward the Bering Strait likely facilitated faunal exchanges between Eurasia and North America. While tectonic activity influenced the region's long-term geography, the accessibility of the Bering Land Bridge (see p. 174) was primarily controlled by sea-level fluctuations. Regardless of the specific drivers, the land bridge served as a vital corridor that allowed various Eurasian mammal lineages to enter North America and vice versa. Early waves of immigration are evident in the earliest Miocene faunas. Among the most notable fossil sites of this age is the Agate Fossil Beds National Monument in the badlands of northwestern Nebraska, which has yielded hundreds of fossil mammal skeletons representing some of the best Miocene mammal specimens. The beds were formed in a **fluvial** channel and sandbars, where many thousands of bones were swept into. In the Miocene of North America, the rain shadow formed by the Rocky Mountains, along with changes in global climate, was causing a significant decline in wood cover, and the Great Plains developed increasingly open **savannah**-like vegetation, with forest cover confined to sheltered river valleys.

The wooded savannahs were bisected with rivers, streams, oxbow lakes and ephemeral waterholes. These environments preserve a rich **ecosystem** of migratory waterbirds, crustaceans, fish, ampibians and reptiles as well as mammals. Permanent inhabitants comprised small alligators, **stem**-pinnipedimorphs (see pp. 94, 96) and the **semi-aquatic** oreodont (see pp. 107–8) genus *Promerycochoerus*. **Carnivores** such as *Daphoenodon* constructed dens along riverbanks to take advantage of tree cover and abundant prey. These valuable water sources expanded and contracted, providing for a diverse **community** of plants and animals during the dry season. The fierce summer droughts frequently led to mass die-offs, giving rise to the amazing fossils discovered at Agate Springs, including the Waterhole site at Carnegie Hill, which preserves large numbers of the **cursorial** herding rhinoceros *Menoceras*, and at least 17 individuals of the strange schizotheriine chalicotheriid (see pp. 128–9) *Moropus*. Intense rainfall during the wet season provided relief and would flood the lower plains to produce temporary ponds.

The most abundant mammal from the Agate Fauna was *Menoceras*, an Old World immigrant. *Menoceras* lived in **polygynous** herds. It was pig-sized, but sported a pair of small, parallel horns on the front of the snout only in males. Thus, they were most likely used in sparring as well as for defence against predators. Like modern rhinoceroses, these horns were composed of **keratin**, so were not fossilised, although the location of their attachment is indicated by a pair of rugosities on the nasal bones. Other distinctive features include the second incisors being absent, well-defined ridges and cusps on the upper molars, and a shortened cranium. *Diceratherium* also had paired horns, although unlike its cousin *Menoceras*, the shape of the skull was low and elongated, and the teeth consisted of simple molars without crests. Moreover, in *Diceratherium*, the horns were present in both sexes. *Diceratherium* specialised more towards **browsing** while *Menoceras* was a **grazer**.

Among other Early Miocene immigrants were several small, weasel-like mustelids (see p. 99) such as *Zodiolestes* and *Oligobunis*, the wolverine-like mustelid *Megalictis*, and amphicyonids (see pp. 91–2) including the genus *Daphoenodon*. *Zodiolestes* occupied the same ecological niche as modern ferrets, and is distinguished by features such as a fairly elongated skull along with sharp **carnassial** teeth a notch on the upper fourth premolars. *Oligobunis* (Fig. 6.4) was the size of an American badger (*Taxidea taxus*), although it had a relatively small skull, an elongated body and tail, and forelimbs longer than the hindlimbs. *Oligobunis* was possibly an active predator, occupying the ecological niche of small felids (see pp. 86–9) which had yet to arrive in North America.

Megalictis is possibly the largest **terrestrial** mustelid to have lived (jaguar-sized), with a skull nearly as broad as that of an American black bear (*Ursus americanus*). *Megalictis* was well-adapted to engage in bone-crushing as indicated by its strong bite force and blunt, robust **cheek teeth**. It was likely a generalist predator. The forelimb joints and proportions indicate that *Megalictis* was an **ambush predator**. Numerous fossils are known from the Bear Dog Hill locality, indicating that *Megalictis* possibly occupied the abandoned dens of *Daphoenodon*. *Daphoenodon* represents a midpoint in amphicyonid diversity. This carnivore exhibits a mosaic of anatomical traits, combining the lithe forelimbs of cursorial cheetahs and wolves with the muscular shoulders and hindlimbs of bears and large felids. It also possessed long metapodials which increased stride length. It is the most abundant genus of fossil carnivoran (see pp. 81, 83–4) from Agate, and it is possible that these animals specialised in chasing their prey on open plains. The specimens show

Fig. 6.4. Reconstructed life appearance of the oligobunine mustelid *Oligobunis crassivultus*. Reconstructed length: 30 cm.

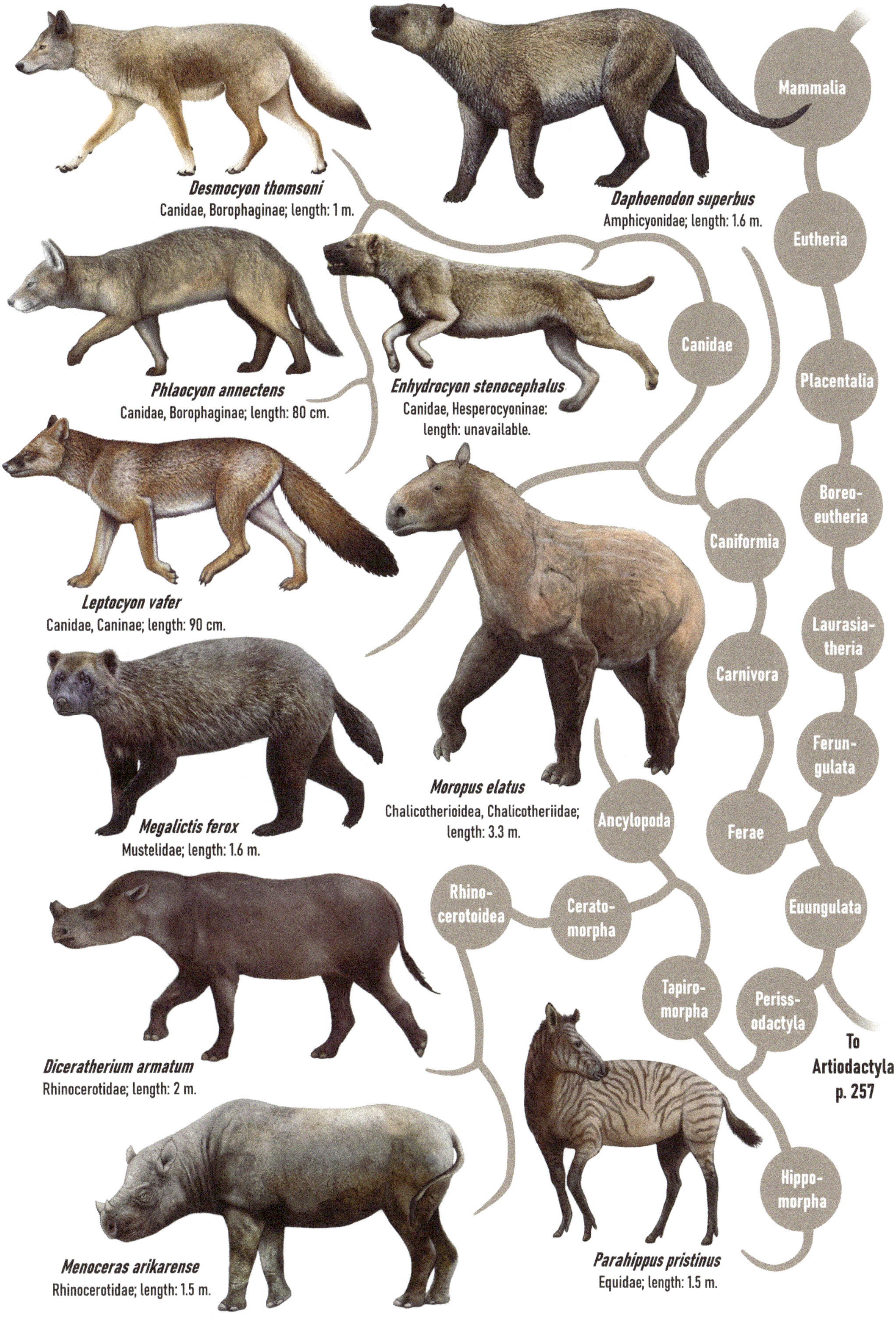

Fig. 6.5A. Phylogenetic tree of a sampling of fossil mammal **taxa**, particularly members of Ferae (see p. 74), and perissodactyls (see p. 126), from the Agate Fossil Beds.

North America Agate Fossil Beds, United States: Burdigalian, ca. 20 Ma, cont.

pronounced **sexual dimorphism** in *Daphoenodon*, with the males being larger, more robust and having longer jaws equipped with longer premolars than the females. Gnawed **ungulate** remains from Nebraskan den sites indicate a bone-crushing diet. These burrows exhibit remarkable similarities to the bone accumulations of hyaenas, suggesting *Daphoenodon* was **gregarious** like modern hyaenas and African wild dogs (*Lycaon pictus*). This is further evident from fossils of young, mature and old individuals.

Another immigrant was *Moropus*, which was one of the largest forms, reaching up to five metres into the canopy when rearing up on its strong hindlimbs. It resembled a massive, muscular horse with three heavy claws on each foot. The claws were used for defence as well as feeding. The claws on the forelimbs were larger and also retractable to keep the tips from wearing, a feature **convergent** with felids. *Moropus* tended to have low numbers of scratches on teeth (indicating softer leaves) but higher gouging (caused by twigs) with a lot of grit thrown in. *Moropus* had a **digitigrade** stance. *Moropus* also exhibited sexual dimorphism, with males having considerably thicker skulls and limbs than females. Thus, males possibly led herds which they defended from competitors in a similar manner to modern horses. Other important immigrant groups were pecoran ruminants (see pp. 110–8). One group were the moschids (see p. 115), which included *Blastomeryx*, an "**archaic**" form that did not have antlers (as in modern musk deer (*Moschus*)), and the tusk-like canine teeth were longer in males than in females, suggesting that they must have been used primarily for sexual display. It may have resided in the undergrowth of the woodlands, evading larger predators.

The Agate Fossil Beds also yield the largest and last surviving entelodontids (see pp. 118, 120) of the genus *Daeodon* (formerly *Dinohyus*), which was as large as an American bison (*Bison bison*), had a massive 90 centimetre long skull and was evidently eating meat as fossils show fracture wear on molars commonly seen in species that are able to crack bones. *Daeodon* is distinguished from other entelodontids by a set of unique dental features (e.g. square upper third premolar and an upper fourth premolar with two cusps (Prothero, 2007)), the shape and comparatively small size of the cheekbone flanges (smaller and less flaring than in *Archaeotherium* (see pp. 235–6)), and the small chin tubercles, along with the fused lower hindlimb bones. *Daeodon* was likely territorial, preferring to live solitary lives browsing, hunting and driving other predators off their kills. Males would have been highly aggressive, engaging in **intraspecific competition** including face biting, as evident from tooth marks on their skulls. Males may also have used their flanges for signalling their health to mates and competitors.

Additionally, the fauna consisted of a diversity of antelope-like and gazelle-like camelids (see pp. 104, 106–7), which remained present only in North America. *Oxydactylus* was an early member of the aepycameline lineage of camelids that was able to avoid competition with other Miocene camelids and browsers by reaching higher into the trees. It was evidently a swift runner, as it still had pointed hooves. The metapodials were long but only partially fused and not splayed as in the condition of modern camels. While adapted for life on the grassland, *Stenomylus* and its relatives had incredibly **hypsodont** last molars, taking up most of the jaw, although the very slender muzzle shows that they were likely **mixed feeders**, like the modern pronghorn (*Antilocapra americana*). Confirmation of herding behaviour in this animal comes from mass fossil assemblages consisting of dozens of complete skeletons, as a result of explosive volcanic activity. *Stenomylus* was a highly specialised camelid, with features including highly reduced upper second and third premolars, retracted nasal bones and elongated limbs. The protoceratids (see p. 111) included the unusual *Syndyoceras*, which had a V-shaped pair of horns on its snout and long horns above the **orbits** that curved upwards and inwards. The canine teeth were long enough in this genus to be used in rooting through the soil, as in other protoceratids. Genera with more elaborate horns were sexually dimorphic, females being either hornless or in the case of *Protoceras* having only one pair of small horns. These horns were likely used for sexual display or intimidating rivals.

The unbranched and non-**deciduous** appendages above the orbits, typical of dromomerycids (see pp. 114–5), may have been characteristic of males but appear to have been present in both sexes of *Aletomeryx*. At least the males possessed long, recurved upper canine teeth. *Aletomeryx* was the size of a roe deer (*Capreolus capreolus*), with very long and slender limbs. In *Barbouromeryx*, the skull was equipped with short, vertical horns above the orbits, with a bulbous top. There was another bulbous structure which protruded from the nape region at the back of the skull. *Barbouromeryx* had a short **diastema** between the canine teeth and the premolars, and relatively elongated limbs. Oreodonts (see pp. 107–8) were at their zenith, with many forms possibly descending from the **primitive**, sheep-sized genus *Merycoidodon* (see pp. 234–5). These included various forms with hypsodont teeth—some such as *Merychyus*, characterised by a small, shortened skull, were fleet-footed and possessed limbs that were longer than earlier oreodonts, but no longer than any modern suid (see pp. 108–110), while others such as the promerycochoerines had pig-like bodies and short limbs, so they had an apparently amphibious lifestyle like tapirs. *Promerycochoerus* had distinctive, wide-flaring cheekbones often showing damage indicative of intraspecific competition. Nasal shortening is notable in the merycochoerines such as *Merycochoerus*, suggesting that they had a mobile upper lip. It also had enlarged, tusk-like canine teeth probably used in defence. Both **clades** were closely related (in fact, the two often get confused) and include some of the more robust oreodonts.

These shared the woodlands and streams with the peccaries, along with some of the last "anthracotheriids" (see pp. 120–1). The small "anthracotheriid" *Arretotherium* is characterised by a relatively short snout without a diastema, along with a more **selenodont** dentition and more robust lower jaw than other "anthracotheriids". **Tridactyl** equids (see pp. 132–4) were radiating in diversity, with the last of the small, primitive *Miohippus* (see pp. 235, 238) coexisting with more advanced equids exhibiting very slightly higher-crowned teeth (those teeth remained **brachydont**), as well as several anchitheriines such as *Parahippus*. *Parahippus* had a longer facial region than *Miohippus*, reflecting better

Fig. 6.5B. Phylogenetic tree of a sampling of fossil mammal **taxa**, particularly artiodactyls (see p. 104), from the Agate Fossil Beds.

North America Agate Fossil Beds, United States: Burdigalian, ca. 20 Ma, cont.

grazing abilities. The radius and ulna of *Parahippus* were fused, likely a more **derived** condition than in more primitive forms such as the genus *Miohippus*.

In addition to the immigrant carnivorans already mentioned, there was a large radiation of native canids (see pp. 92–3). *Leptocyon* was one of the most primitive members of the modern subfamily Caninae. It was a miniscule, kit fox-like animal that had long, narrow jaws equipped with simpler, more widely spaced premolars than the borophagines. The skull and teeth were well-suited for capturing small, speedy vertebrates, and *Leptocyon* was likely **omnivorous**. Once thought to be synonymous with the fox genus *Vulpes*, *Leptocyon* is distinct in having a more pronounced **postorbital** constriction of the skull along with lower limb bones that were more robust and had larger **distal** articulation surfaces. Some primitive borophagines included *Phlaocyon* and *Desmocyon*. *Phlaocyon* was a small to medium-sized canid that somewhat resembled a raccoon (*Procyon lotor*). Its **bunodont** teeth, with no meat shearing specialisations, indicate that *Phlaocyon* was omnivorous. However, two large genera (15 kilograms), reversed this trend to become **hypercarnivorous** with robust jaws. *Desmocyon* is characterised by a wide **rostrum** and palate, along with the presence of a **frontal** sinus. **Hypocarnivorous** features are absent in its dentition (Wang *et al.* 1999). *Enhydrocyon* was among the last surviving hesperocyonines. It was also the first canid to act as a dominant predator in its ecosystem given its robust skull and tooth morphology, as well as among the first to attain large body size. *Enhydrocyon* is characterised by a particularly short skull, somewhat resembling that of an otter, and the dentition has specialisations towards a hyaena-like diet, including unusually large premolars.

While the oak savannahs and open woodland of the Agate Fossil Beds were desirable habitats for many animals (such as *Moropus* and *Daeodon*, its two most popular denizens), many other aquatic and semi-aquatic animals from the wetlands in addition to *Arretotherium*, *Promerycochoerus* and *Metamynodon* included a variety of birds such as members of the extinct duck genus *Dendrochen* (Fig. 6.6). *Dendrochen* included species known from North America and Eurasia from throughout the Miocene. *Dendrochen* was very similar to extant stiff-tailed ducks (*Oxyura*), to which it was either a close relative or part of the same group. Like modern ducks, it was semi-aquatic and was capable of flight. Other birds included crowned cranes of the modern genus *Balearica*, one of the oldest representatives of the crane family (Gruidae) and are presently limited to Africa. However, *Balearica* is known from the Early Miocene of Nebraska, France and Germany. An ulna and radius from Agate possibly belong to this genus, or may be closer to sandhill cranes (*Antigone*). Like its modern relatives, *Balearica* was possibly an omnivore that foraged in wetlands and grasslands. *Parictiornis* was a primitive member of the pratincole lineage, a group of small birds which, like sparrows, hunt flying insects on the wing. Despite their hunting method, they are close relatives of wading birds including plovers.

Among reptiles, *Alligator mcgrewi* was a basal member of the extant genus *Alligator*, but considerably smaller. Its skull was only about 15 centimetres long and 10 centimetres wide, and from that the body mass of this reptile has been estimated at just three kilograms (although it seems like an underestimate, as that is under half the mass an adult Cuvier's dwarf caiman (*Paleosuchus palpebrosus*)). *Al. mcgrewi* likely hunted crustaceans and small vertebrates in the Nebraskan wetlands, and it is most closely related to Chinese alligators (*A. sinensis*). With a slender, streamlined body, the 1.5 metre long stem-pinnipedimorph *Potamotherium* (see p. 91) was a proficient swimmer, and analysis of its fossils suggests it had good sight and hearing, albeit a poor sense of smell. Although broadly resembling a giant otter (*Pteronura brasiliensis*) in size and form, this

Fig. 6.6. A riparian wetland habitat in what is now the Agate Fossil Beds National Monument, Early Miocene. At the bottom left is the crocodylian *Alligator mcgrewi*; at the bottom right is the stem-pinnipedimorph *Potamotherium valletoni*; the partly submerged hippopotamus-like animal on the left is the "anthracotheriid" *Arretotherium leptodus*, and immediately behind it are three ducks of the genus *Dendrochen* floating on the water's surface. In the background, partly submerged in the water, are the oreodonts *Promerycochoerus carrikeri*; flying in the air is a pair of the extinct crane *Balearica exigua*.

semi-aquatic animal is among the most primitive known relatives of seals and sea lions. *Potamotherium* was fairly generalist, and spent its time hunting anything from aquatic crustaceans, to fish, to waterfowl in the waterways. Finally, lagomorphs (see pp. 148–9) and rodents continued to diversify. The Agate Springs yield a diversity of mylagaulids, cricetids, heteromyids (see p. 147), jumping mice, squirrels, pocket gophers and beavers. Like many early beavers, *Palaeocastor fossor* was mainly a terrestrial and **fossorial** animal rather than semi-aquatic. Its skull was short, low and wide (which is typical of other fossorial rodents), the forelimbs were broad and powerful, and the tail was shortened. *P. fossor* is particularly famous for having created elaborate corkscrew burrows three metres deep (Fig. 6.7). When first discovered, the origins of the burrows were unknown, and they were named as *Daemonelix/ Daimonhelix* (the "Devil's Corkscrew") fossils. The true identity of the burrows was only agreed upon after skeletons of *P. fossor* were found within the chamber. Burrows were excavated using incisors rather than claws. The helical shape maintained consistent temperature and humidity in a dry environment (similarly shaped burrows are used by lizards and spiders). Evidence of 200 burrows close together suggests that *P. fossor* lived in colonies in a similar manner to modern prairie dogs (*Cynomys*). There were off-shoot chambers used for sleeping, nurseries, latrines and food storage. Fossil skeletons of other occasional inhabitants including the entoptychine pocket gopher *Gregorymys*, the larger beaver *P. magnus* and the carnivorous mustelid *Zodiolestes* have also been found in the burrows. Thus, *Zodiolestes* was a major predator of *P. fossor*, the long and slim body of the mustelid being ideal for entering the beavers' burrows so it can kill them in their own homes. A similar predator-prey relationship is seen today between the black-footed ferret (*Mustela nigripes*) and the prairie dog.

Fig. 6.7. "Devil's corkscrew" (*Daemonelix*) fossil burrows constructed by the primitive beaver *Palaeocastor fossor*, an individual which is hiding underground in its oddly-shaped burrow system from the mustelid *Zodiolestes daimonelixensis*. Early Miocene of Nebraska.

North America Ashfall Fossil Beds, United States: Serravalian, ca. 12 Ma.

Located in Nebraska, the Ashfall Fossil Beds represent a unique and extensive deposit of volcanic ash several metres thick, deposited during the Late Miocene. This ash originated more than a thousand kilometres away from the Bruneau-Jarbidge caldera in present-day Idaho—a large volcanic caldera formed when the volcano collapsed following a major eruption around 12 Ma. The Bruneau-Jarbidge caldera is part of the Yellowstone hotspot, a relatively stationary mantle plume responsible for volcanic activity as the North American Plate moved westward over it for millions of years. The resulting ash plume drifted eastward across the Great Plains, depositing volcanic ash that caused a mass mortality event and preserved an exceptional fossil record at the site. The animals living there had died in the volcanic eruption from lung failure caused by inhaling volcanic glass when approaching a waterhole seeking relief, preserved in the exact pose they were in when they perished. This mass-death assemblage provides an excellent reconstruction of the **ecosystem** during that time: a diverse **ungulate** fauna and a mosaic of **savannah**-woodlands in a warm, equable climate with high primary productivity.

One particular locality in northeastern Nebraska, dubbed the "Rhino Barn" and with a museum built surrounding the excavated site, records a huge number of the aceratheriin rhinocerotid (see pp. 130–2) *Teleoceras* (over one hundred complete skeletons), small, **grazing tridactyl** and **monodactyl** equids (see pp. 132–4), camelids (see pp. 104, 106–7), and various birds and freshwater turtles, simultaneously. Many animals were preserved three-dimensionally with their bones **articulated**; some *Teleoceras* individuals retained the contents of their last meal in their mouths, while one was found with an unborn foetus. *Teleoceras* had a long, low body, barrel-shaped chest, short, thickset limbs, and **hypsodont cheek teeth** adapted for grinding grass, as well as a single, small horn on the snout. *Teleoceras* most likely had a **semi-aquatic** lifestyle like hippopotamuses, wallowing in the lakes and rivers during the daytime for protection and coming out at night to graze, a lifestyle further supported by oxygen isotope studies on the teeth.

The equids occupied the antelope niche in this ancient ecosystem. All equids from Ashfall were exclusively New World genera, except for *Cormohipparion*. One common tridactyl equid was the hipparionin *Cormohipparion*, which is characterised by a highly complex enamel pattern on its cheek teeth, long, narrow muzzle and teeth relatively more **brachydont** than other Late Miocene equids. *Cormohipparion* most probably fed on a combination of browse and grasses. The genus was ancestral to Eurasian species of *Hipparion*, as *Cormohipparion* is the earliest equid known from Eurasia, and then *Hipparion* appeared. The smallest and most common equid was *Pseudipparion*. It was tridactyl, exhibited a shallow depression on its skull above the teeth, and was a grass eater. The medium-sized *Neohipparion* represents the culmination of **adaptation** towards a pure grass diet among North American hipparionin equids. The donkey-sized equinin *Protohippus* had a smooth facial region without depressions in front of the **orbits** and was tridactyl but functionally single-toed, like all hipparionins. The incisors were arranged in a straight row, which enabled the animal to crop grasses at ground level more easily, a feature absent in the modern genus *Equus*. The equinin *Pliohippus*, standing up to 1.25 metres at the shoulder, retained side toes, but they were variable in size and vestigial in some individuals. *Pliohippus* is distinguished by deep, complex facial **fossae** in the skull (the purpose which is not fully understood), a characteristic absent in the modern genus *Equus*. Furthermore, it had strongly curved cheek teeth in comparison to those of *Equus*.

The artiodactyls (see p. 104) included several genera of camelids including *Aepycamelus*, which had a relatively long neck, achieved by giraffe-like lengthening of the cervical vertebrae, and elongated, stilt-like limbs, both features which enabled **high browsing** in the tree canopy above the height of the other contemporary animals. *Procamelus* is the earliest genus that can truly be recognised as a camel, from its skull shape to tooth arrangement to foot bone morphology. In *Procamelus*, the working incisors were limited to the lower jaw, acting against a pad on the upper. The batteries of massive, hypsodont cheek teeth were suited for chewing abrasive plant material. The wrist bones of the two functioning **digits** are merged into one cannon bone and the toe tips are spread out, indicating that it possessed weight-bearing foot pads as in modern camels, similar to *Aepycamelus* and contrary to many earlier camelids. One interesting artiodactyl is the long-snouted moschid (see p. 115) *Longirostromeryx*. It was the last surviving blastomerycine and had distinctive gazelle-like features, but it was not a specialised runner, being found in more closed forest environments. One notable change in the faunal **community** was the decline in diversity of **browsing herbivores** with brachydont teeth, as grazers with hypsodont teeth gradually took over. Such browsers included the moschids, oreodonts (see pp. 107–8), dromomerycids (see pp. 114–5), protoceratids (see p. 111) and the anchitheriine equids.

Among carnivorans (see pp. 81, 83–4), the last of the amphicyonids (see pp. 91–2) were disappearing, their place being taken by canids (see pp. 92–3). The **hypercarnivorous** canid *Aelurodon* was among the earliest borophagines to develop jaws sufficiently powerful and teeth adequately heavy to be capable of bone-crushing. It resembled the grey wolf (*Canis lupus*) in body dimensions, but had slightly shorter **distal** limb bones and a shorter neck. *Aelurodon* may have been an **ambush predator** according to analysis of the elbow joint morphology (Figueirido *et al.* 2015). The borophagine *Epicyon* is perhaps the largest canid to have ever lived (the size of a brown bear (*Ursus arctos*)), and possessed a short, massive skull with strong jaws, making its head shaped more like that of a lion (*Panthera leo*) than a wolf. Like the spotted hyaena (*Crocuta crocuta*), *Epicyon* had inflated lower fourth premolars and it too was adept at crushing bone. This adaptation would have allowed *Epicyon* to **scavenge** as well as hunt. Also noteworthy is the presence of the beaver *Dipoides* (see p. 251) and smaller rodents such as *Ceratogaulus* and silky pocket mice (*Perognathus*), along with several species of shrews, moles and bats. Commonly called the horned gopher, *Ceratogaulus* was not closely related to pocket gophers, but instead was a sciuromorph rodent and thus closer to squirrels. *Ceratogaulus* had paired horns on its head, the purpose of which is unclear, with theories for that having been suggested such as **intraspecific competition**, attracting mates or digging, but it is known that both genders had them.

Fig. 6.8. Phylogenetic tree of a sampling of fossil mammal **taxa** from the Ashfall Fossil Beds. The phylogeny within Equidae, including the basalmost position of *Pseudhipparion*, follows Famoso and Pagnac (2011).

South America Santa Cruz Formation, Argentina: Burdigalian, 18–16 Ma.

Most of what is understood about extinct South American mammals comes from the impressive Early Miocene fossils from the **fluvial** sandstones and **mudstones** of the Santa Cruz Formation in southern Patagonia, being almost incomparable in their completeness and state of preservation. Near complete skeletons, with the skull preserved, have been recovered for over 80 mammal genera from more than 30 families, almost all of which have no modern descendants. The fossil fauna and flora from Santa Cruz suggests that during the Early Miocene, the region had a seasonal climate, moderated by the adjacent Atlantic Ocean. Its winters were cool and wet (typically above freezing), while its summers were hot and dry. The landscape was possibly similar to areas of present day central Chile. The region received adequate rainfall to support forests, although these were relatively sparse and confined around lakes and streams. In between the forested areas were more open **savannahs** with low-growing vegetation such as grasses, resulting in an array of habitats able to support a broad variety of organisms.

Fossils of reptiles and amphibians are known from the Santa Cruz Formation, but this site is among South America's richest fossil bird deposits, with at least 23 species recognised. The phorusrhacids (see pp. 41–2) have the highest diversity. The first phorusrhacid fossil to be obtained from Santa Cruz was a jaw fragment, but was initially not identified as a bird because of its massive size. The genus it belonged to was *Phorusrhacos* (Fig. 6.9A), which stood about 1.6 metres tall, had a skull over 60 centimetres long, had a body mass similar to that of a male ostrich (at under 130 kilograms) and was among the dominant **terrestrial** predators in its **ecosystem**. It had an elongated and strong tarsometatarsus (considerably longer than that of other phorusrhacids), suggesting that it could run at high speeds. *Patagornis* was a smaller (medium-sized) phorusrhacid with a slenderer, more lightly built body and a long, narrow, gently curved mandibular symphysis with an uncurved beak tip (Fig. 6.9B). Studies of its inner ear anatomy have shown that it had a relatively low hearing range compared to other phorushracids (Degrange *et al.* 2015). The smallest known phorusrhacid (70 to 80 centimetres tall) was the gracile-bodied *Psilopterus* (Fig. 6.10), which was similar to modern seriemas in size and appearance but more heavily built, and

Fig. 6.10. Life reconstruction of the small phorusrhacid *Psilopterus bachmanni*. Early Miocene of Argentina. *Centre background*: a flock of the **primitive** ibis *Protibis cnemalis*; *Top right background*: the falcon *Thegornis musculosus*.

Fig. 6.9. Reconstructed life appearance of phorusrhacids from the Santa Cruz Formatiuon. (A) *Phorusrhacos longissimus*; (B) *Patagornis marshi*. Scale bar: 1 m.

Fig. 6.11A. Phylogenetic tree of a sampling of fossil mammal **taxa**, particularly sparassodonts and xenarthrans, from the Santa Cruz Formation. The phylogeny within Folivora (see pp. 61–2) follows Toledo *et al.* (2015).

South America Santa Cruz Formation, Argentina: Burdigalian, 18–16 Ma, cont.

the wings were much more reduced. The skull of *Psilopterus* had a deep **maxilla**, indicating a loss of cranial kinesis (flexibility) and an increased rigidity. It also had a hooked **rostrum** extending beyond half the length of the skull.

At Santa Cruz, the highly diverse mammals encompass species from almost all mammal families that existed on the continent at the time. The mammals even consist of members of lineages that are typically rarely preserved as fossils, such as anteaters and New World monkeys. Metatherians (see pp. 43, 46–7), particularly the sparassodonts (see pp. 48–50), continued performing the roles of predatory mammals, from the small, mongoose-like *Cladosictis* to the significantly larger, hyaena-like *Borhyaena* and bear-like *Arctodictis*. *Cladosictis* is known from skeletons of at least three individuals. This small sparassodont was an active predator that may have been mostly terrestrial, using its low stance for hiding in the underbrush, but was able to climb trees. *Cladosictis* had an elongated skull with a dentition resembling that of a **carnivorous** placental, although it had three pairs of **carnassials** (with all molars functioning sequentially as the carnassials) instead of one (formed of the upper fourth premolar and first lower molar in placentals). It possibly raided the nests of the contemporary phorusrhacids. *Acyon* is distinguished from the closely related *Cladosictis* in being larger, having relatively longer limbs, a shallower lower jaw, premolars with slicing **adaptations** that were more pronounced, and wider premolar **diastemata** in adults.

Some basal borhyaenoids were relatively small predators and occasionally **arboreal**. *Lycopsis* is known from good fossils that reveal a strange amalgamation of anatomical features in its limb bones, with adaptations for both running (e.g. straight limbs and semi-**digitigrade** feet) and climbing. It was probably a terrestrial **ambush predator**, hunting small vertebrates such as rodents, but could also grasp and climb trees as indicated by its semi-opposable thumbs. With short limbs, grasping feet and a long muscular tail, *Prothylacinus* was probably even more agile in trees, able to carry large weights in the strong jaws, an **analogue** for the modern fossa (*Cryptoprocta ferox*) and clouded leopard (*Neofelis nebulosa*).

The borhyaenid *Borhyaena* had a large, sturdy skull, short muzzle, blade-like **cheek teeth** and a long, powerful neck resembling that of modern hyaenas. In contrast to proborhyaenids, *Borhyaena* possessed massive canine teeth that did not grow continuously throughout life. Its limb joints were well-adapted for stability and quick movements, and it had proportionally more elongated bones in the forefeet compared to other sparassodonts. Being a heavily-built mammal, it would have hunted its prey by ambush. One specimen consists of a scar on the lower jaw, indicating that **intraspecific competition** took place between individuals, demonstrating aggressive behaviours analogous to those observed in extant Tasmanian devils (*Sarcophilus harrisii*). *Arctodictis* is the largest known borhyaenid, regarded as a more generalised carnivore according to its dentition and was a ground-dweller capable of hunting large prey including ground sloths, litopterns (see pp. 143–4) and toxodontians. It also had a possible **plantigrade** stance and had relatively fewer **cursorial** adaptations than *Borhyaena*. The cheek teeth were adapted for cracking bones, as evident from wear on the molars.

Most spectacular of all, though, was the diversification of South American native **ungulates** (see p. 139). The notoungulates (see pp. 140–3) consisted of toxodontids such as *Nesodon*, homalodotheriids, and the vaguely rabbit-like interatheriids and hegetotheriids.

Adinotherium was a primitive toxodontid that resembled a scaled down form of its later rhinoceros-like kin (Fig. 6.12A). Its forelimbs were more elongated than those of other toxodontids, so that the shoulders were of similar height to the pelvis, making *Adinotherium* better adapted at running. It was probably a **herbivore** that moved in flocks, **grazing** grasses and foliage on the plains. The skull roof was adorned with a small horn that may have been used for sexual display. The more advanced toxodontid *Nesodon* had a bulky, rhinoceros-like body, but the skull was enormous and very tall, almost disproportionate to the rest of the body and the limbs, which were relatively thin. *Nesodon* was initially thought to have been a grazer given its highly **hypsodont** cheek teeth. However, **microwear** analysis on the enamel suggests that it may have been a **browser** that ate soft leaves, possibly supplementing its diet with fruit or bark. The skull morphology shows that it was a **mixed feeder**, and the molars had an **occlusal** surface sporting a complex pattern of enamel **folds** which comprise a sequence of low blades adapted for shearing all kinds of vegetation.

Homalodotherium had strongly-curved claws instead of hooves and longer forelimbs than hindlimbs. The forelimbs were digitigrade whereas the hind feet were plantigrade. *Homalodotherium* had highly versatile forelimbs, with flexible wrist joints and insertion points for powerful muscles. Like its ecological equivalent, the chalicotheriids (see pp. 128–9), *Homalodotherium* possibly stood **bipedally** when feeding, using its claws to pull down branches and leaves. *Protypotherium* was a typical interatheriid with a rodent-like skull that tapered to a pointed snout, and included a full complement of 44 teeth, which were relatively **derived** in being **hypselodont** but were highly simplified. It had a long tail and short neck, and the elongated hare-like body was supported by short but strong limbs, equipped with clawed, digitigrade feet. These robust claws indicate that it was capable of digging burrows or modifying those disused by other mammals.

The interatheriid *Interatherium* had a more elongated and weasel-like body with short, squat limbs (contrary to the cat-like body proportions usually present in interatheriids) that have led some scientists to hypothesise that it lived in underground dens, although the hindlimbs do not exhibit any specific **fossorial** adaptations. Others have proposed that *Interatherium* was **semi-aquatic**, favouring wetlands or marshes. *Interatherium* had a short but deep skull with a powerful jaw. *Interatherium* also had a **complete dentition** of 44 teeth (the **ancestral** condition in notoungulates) as in other interatheriids, but differed in having a diastema between the incisors and premolars. Among hegetotheriids, *Pachyrukhos* had a rabbit-like skull with a dentition suited for feeding on nuts and abrasive vegetation, and its elongated hind feet indicate a hopping

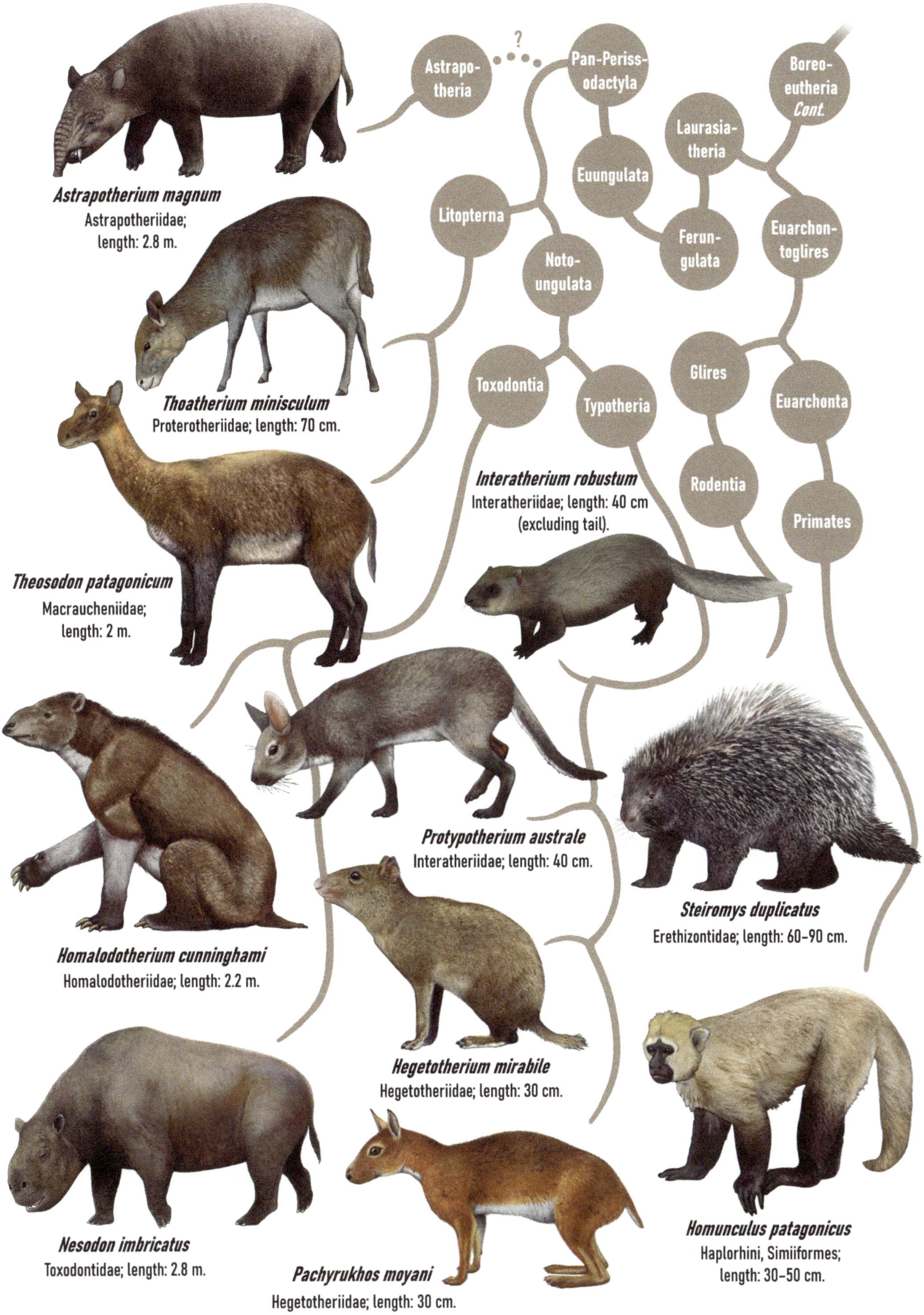

Fig. 6.11B. Phylogenetic tree of a sampling of fossil mammal taxa, particularly boreoeutherians, from the Santa Cruz Formation. The phylogenetic position of Notoungulata and Litopterna follows Welker *et al.* (2015). The tentative phylogenetic position of Astrapotheria as a **stem**-perissodactyl follows Kramarz and MacPhee (2023), which supported the placement of South American native ungulates, including astrapotherians, within Laurasiatheria.

South America Santa Cruz Formation, Argentina: Burdigalian, 18–16 Ma, cont.

locomotion style. The complexity of the auditory region indicates a good sense of hearing and that it bore a pair of long ears. *Pachyrukhos* was possibly **nocturnal** as evident from its large **orbits**. These similarities with rabbits can be attributed to **convergent evolution**, as *Pachyrukhos* occupied the same ecological niche, although unrelated. *Hegetotherium* reached the size of an American badger (*Taxidea taxus*), and had a high skull, orbits that were almost closed **posteriorly** and **tetradactyl** limbs (with the third toe being the longest). *Hegetotherium* was more robust in overall morphology compared to other hegetotheriids. Other distinctive features of *Hegetotherium* among hegetotheriids were a short muzzle, deep **zygomatic arches**, molars with high, sharply pointed edges and large, forward-protruding hypselodont incisors (with enamel on the outside). The incisors are obliquely implanted and somewhat chisel-like, different from the more rodent-like incisors of mesotheriids and interatheriids (in some respects). *He. mirabile* occupied a cursorial, **diurnal** niche, likely feeding on abrasive vegetation.

Among the litopterns were the small, gracile, horse-like proterotheriids *Thoatherium* and *Diadiaphorus* (Fig. 6.12B), and multiple species of the larger, long-necked, guanaco-like macraucheniid genus *Theosodon*. *Thoatherium* was a fast runner, according to its long limbs. The toes were also very much reduced to only a single, horse-like hoof. The side toes were even more atrophied in *Thoatherium* than in modern horses, showing that **monodactyly** had developed to a greater extent and evolved much earlier in litopterns. *Thoatherium* had generalised, **brachydont** cheek teeth adapted for browsing leaves. The great difference in overall size between *Thoatherium* (weighing only around 15 kilograms) and modern horses (500 to 600 kilograms on average) probably reflects different ecological niches. *Diadiaphorus* was a sheep-sized, cursorial proterotheriid that had three **digits** on each foot, with only the central one bearing the animal's weight. This toe had a large hoof, while the two lateral toes were vestigial, comparable to the morphology of primitive equids such as *Merychippus* (see pp. 132–3), and not completely lost as in the smaller litoptern *Thoatherium*. In contrast to equids, the lower limb bones of *Diadiaphorus* were not fused together. *Diadiaphorus* had a short skull with quite a large braincase, along with brachydont cheek teeth adapted for browsing on leaves. As in most extinct equids, the feet of *Theosodon* were **tridactyl** instead of monodactyl. Strikingly, instead of having nasal openings at the front of its skull, *Theosodon* had them placed on top, midway between the forehead and the end of the snout. The nasal openings were also directed upwards instead of forwards, probably as a specialisation for feeding on thorny vegetation. Certain extant ungulates such as the gerenuk (*Litocranius walleri*) and giraffe (*Giraffa camelopardalis*) have similar feeding behaviours.

There were also relic groups from the Paleogene, such as astrapotherians (see p. 140). Known from an almost complete skeleton was the eponymous *Astrapotherium*, which had high, retracted nasal bones on the skull that suggest the presence of a short trunk. *Astrapotherium* had short plantigrade feet, and the forelimbs were significantly stronger than the hindlimbs. It had large, tusk-like upper and lower canine teeth in its jaws, the upper ones sharpening themselves by grinding against the lower ones, as in modern hippopotamuses. The upper incisors were absent and the lower ones projected forward to function against a horny pad on the roof of the upper jaw, similar to the grass-gathering feeding style in modern cattle. *Necrolestes* was an enigmatic mole-like mammal that was the last member of a line of ancient Mesozoic mammals called meridiolestidans (see p. 46), unrelated to any modern group (Fig. 6.13). It had an upturned snout, which may have supported a pad of thick, toughened skin, used like a shovel for digging underground. *Necrolestes* shares a strange characteristic with golden moles called a third upper limb bone, which is absent in any other known mammal. This "limb bone" is in fact an ossified tendon of a flexor muscle. Since *Necrolestes* and golden moles are not closely related, it is a case of convergent evolution given the similar fossorial adaptations of both mammals.

The xenarthrans (see p. 58) included eight families represented by anteaters, ground sloths and armadillos. The basal anteater *Protamandua* had a **prehensile** tail and thus, was very likely arboreal (Fig. 6.14). It foraged on social insects such as ants and termites. *Hapalops* was a modest sized sloth with a cylindrical body that likely housed an extensive gut for fermenting its leaf-rich meals. *Hapalops* had a long, slender skull with a highly narrow snout, inferring that it was a **selective feeder**. *Hapalops*

Fig. 6.12. Reconstructed life appearance of some well-represented examples of South American native ungulates from the Santa Cruz Formation. (A) the toxodontid notoungulate *Adinotherium ovinum*; (B) the proterotheriid litoptern *Diadiaphorus majusculus*. Scale bar: 0.5 m.

Fig. 6.13. Reconstructed life appearance of the meridiolestidan *Necrolestes patagonicus*. Reconstructed length: 15 cm.

possessed few teeth, with only four small pairs of cheek teeth and no incisors. The limb anatomy suggests that it was probably capable of digging, and may have been adapted to life both in the trees and on the ground. Phylogenetic studies suggest that *Hapalops* represents an early branch of the sloth family tree prior to the divergence of the lineages Megatheriidae, Nothrotheriidae and Mylodontidae (see pp. 61–2); it is currently not assigned to any specific family. Some megatheriids from Santa Cruz include the medium-sized *Prepotherium*, which is distinguished by a less pronounced convex edge of the lower jaw and was primarily a terrestrial **folivore**. The basal megatherioid *Schismotherium* was a small, gracile, arboreally-adapted animal as indicated by a hip joint that enabled high mobility and well-abducted stances, and possessed primitive features such as **bilophodont** cheek teeth. *Eucholoeops* differed from other megalonychids (see p. 61) in possessing a very deep lower jaw, and transversely expanded upper second and third molars. The muzzle was short, wide and prominent, but less so than in more derived megalonychids. It was slightly more agile than *Prepotherium*, and is considered arboreal or semi-arboreal.

Peltephilus was a large fossorial armadillo with a pair of "horns" on its snout, which were formed by pointed **osteodermal** armour plates in front of the orbits. There is some uncertainty to their purpose, but a more plausible hypothesis is for defence than for display. *Peltephilus* had a stout, wide skull with U-shaped batteries of large, triangular teeth, which led to the opinion that it was carnivorous. However, a reanalysis of the jaw musculature and limb anatomy indicates that it fed on tough underground tubers. *Stegotherium* was a dasypodid (see pp. 58–60) with an unusual, elongated and nearly toothless skull that represents a **myrmecophagous** specialisation, a similar lifestyle to the modern giant armadillo (*Priodontes maximus*).

Fig. 6.14. Reconstructed life appearance of the primitive anteater *Protamandua rothi*. Reconstructed length: 1 m.

Stegotherium was more toothless than modern armadillos. The near complete lack of teeth in *Stegotherium* is a unique feature among armadillos. As in anteaters, the dentition was much reduced, formed of several pairs of weak, cylindrical pegs in the posterior region of the jaw. The shell was similar to modern armadillos in consisting of mobile plates arranged in bands along the back, although the osteoderms are distinguished by a more compact bone structure and granular appearance. *Propalaehoplophorus* is the earliest glyptodontine (see pp. 59–60) known from a partial skeleton. Although it has defining characteristics of the group including a high, hemispherical **carapace** with almost all the osteoderms solidly connected to each other, *Propalaehoplophorus* was anatomically less specialised than Late Cenozoic glyptodontines. Its osteoderms are also considerably thicker, with a flower-like arrangement of grooves on their surface. In Early Miocene glyptodontines such as *Propalaehoplophorus*, only the back teeth are lobed, whereas in later glyptodontines, all teeth are lobed.

Among rodents, modern species of New World porcupines (erethizontids (see pp. 147–8)) are found across central and northern South America (with only one species, *Erethizon dorsatum*, found in the United States and Canada), while species of the fossil genus *Steiromys* were very diverse in the southern half of South America during the Miocene. They include some of the largest known erethizontids and were more terrestrially adapted than extant species. *Steiromys* may have bore a closer resemblance to North American porcupines than extant South American ones. The limbs of *Steiromys* lacked the adaptations for gripping and climbing present in **Neotropical** porcupines, and whether it had a prehensile tail remains unknown. Compared to modern New World porcupines, *Steiromys* had more massive masticatory muscles, more robust incisors and thicker molar enamel, suggesting that it was mostly terrestrial, feeding on abrasive foods. The primates are represented by the arboreal **quadruped** *Homunculus*, which lacked a **sagittal crest** and had a relatively short face, along with medium-sized orbits, indicative of diurnal behaviour. Based on dental morphology, *Homunculus* had a **frugivorous** diet, supplemented with leaves. Some phylogenetic studies place *Homunculus* as a stem-platyrrhine excluded from any modern group, while others consider it a **crown** platyrrhine.

This site's astonishing species diversity is perhaps a result of its age, since its fossils had accumulated during the Middle Miocene Climatic Optimum (MMCO) (see p. 253). Once this warm phase ended around 15 Ma, numerous mammal groups from Santa Cruz became limited to lower latitude regions and would never again inhabit the southernmost parts of South America.

South America Urumaco Formation, Venezuela: Tortonian–Messinian, 10–5.3 Ma.

The Urumaco Formation, located in northern Venezuela, was formed in a deltaic environment, where the rocks were deposited in both **fluvial** and marine settings, and in which faunal remains were washed into from inland sources. During the Middle to Late Miocene, the region was composed of tropical wetlands with an interconnected series of grasslands, rivers, lakes and gallery forests—conditions **analogous** to those of the modern Pantanal. This habitat was inhabited by an incredibly diverse assemblage of land animals of all sizes that lived in marine, brackish and continental waters. The wetlands provided favourable conditions for the native reptilian fauna, with several lineages of crocodylians reaching enormous sizes and radiating into various ecologies during this time. The Urumaco Formation has the largest number of fossil reptiles in all of northern South America and is currently identified as the site with the highest diversity of fossil crocodylian species in the world, with at least 14 species coexisting together.

The long-bodied and robust *Gryposuchus* is the largest known representative of the Gavialidae (see p. 29), which lived further inland, with one species, *G. croizati*, measuring about 10 metres long (Fig. 6.15A). *Gryposuchus* was very likely the most successful gavialid, with an existence possibly spanning the majority of the Miocene, being distributed between Argentina and Venezuela. This domination may be attributed to the fact that *Gryposuchus* was one of just two freshwater adapted gryposuchine genera (the other being *Hesperogavialis*), while the others (such as the shorter snouted forms *Aktiogavialis* and *Ikanogavialis*) were either predominantly marine-, coastal- or estuarine-based predators. *Charactosuchus* had a gharial-like appearance with a long, narrow snout but bore no relation to them, being more closely related to modern crocodylids (see pp. 27–8) than to gavialids (Fig. 6.15B). The fossil caiman diversity at Urumaco is unparalleled, which includes some of the largest known crocodylians ever. *Purussaurus* (Fig. 6.16) was the Miocene king, which used its decidedly powerful bite force (5.3 to 7 tonnes!) on a variety of prey species including massive pleurodire turtles and large mammals, facing no real competition from sympatric, smaller predators. It had a hugely stout and robust skull containing teeth that were edged with small ridges to help puncture the flesh of large vertebrates, and those towards the rear of the mouth were conical, ideal for crunching into bone. Living alongside this giant was a close relative which hunted in a very different fashion. *Mourasuchus* (Fig. 6.17) had a U-shaped lower jaw and a long, flattened snout housing many small teeth used for skimming small prey, in a similar manner to pelicans and baleen whales. This way it would have avoided competition with its fearsome cousin. *Mourasuchus* had a long and slender humerus, possibly suggestive of weakened forelimbs, potentially associated with a primarily aquatic lifestyle. One group of small caimans that included *Globidentosuchus* were **durophagous** and dedicated bivalve hunters, possessing short, rounded snouts and blunt teeth (Fig. 6.15C).

Fig. 6.15. Reconstructed heads of a sampling of crocodylians from the Urumaco Formation. (A) the gavialid *Gryposuchus croizati*; (B) the crocodylid *Charactosuchus mendesi*; (C) the alligatorid (see pp. 28–9) *Globidentosuchus brachyrostris*. Not drawn to scale.

Some of the giant crocodylians may have played a role in the evolution of the large body size of the appropriately named pleurodire (see pp. 14–5) *Stupendemys*, putting pressure on the animal through predation. *Stupendemys* was the largest freshwater turtle to have ever lived, with a truly stupendous total **carapace** length of over three metres, which had a low-arched profile. Its closest extant relative is the much smaller Arrau turtle (*Podocnemis expansa*) from the Amazon region, which it most closely resembles. Some features of the approximately triangular-shaped skull of *Stupendemys* (e.g. a robust, broad jaw structure) suggest that it was a generalist feeder—a wide dietary range would have aided this animal in maintaining its giant body size. The **anterior** margin of the carapace was marked by a deep indentation flanked by prominent spikes in males. These spikes were likely covered by a horny sheath as indicated by deep grooves; they may have been used in **intraspecific competition** over mates and territory as in modern snapping turtles. In overall shape, the carapace of *Stupendemys* was longer than it was wide, measuring 1.8 metres (6 feet) in width (Fig. 6.18).

Urumaco has also preserved a rich diversity of mammals including giant ground sloths, glyptodontines (see pp. 59–60) such as the medium-sized (about two metres long) genus *Boreostemma*, toxodontids (see pp. 141–2), sirenians (see pp. 69–71) of the genus *Nanosiren*, and the giant, **herbivorous** capybara-like rodent *Phoberomys*. Ground sloths of the families Mylodontidae and Megatheriidae (see pp. 61–2) inhabited the coasts of Urumaco. *Eionaletherium* is a very peculiar ground sloth as it had proportionally long legs when compared to other mylodontids. It shows similarities to the marine genus *Thalassocnus* (see pp. 61–2), despite not being fully aquatic-adapted as its bones were not particularly dense. This sloth likely fed on aquatic vegetation, and it seems to have filled a niche

Fig. 6.16. Reconstructed life appearance of the giant caiman *Purussaurus mirandai*. Reconstructed length: 10 m.

Fig. 6.17. A riverbank near a gallery forest in what is now northern Venezuela, Late Miocene. *Centre*: the giant filter-feeding caiman *Mourasuchus arendsi* basks on the riverbank, and immediately to the right, a mata mata turtle (*Chelus* sp.); *Bottom right*: the gavialid *Ikanogavialis gameroi*; *Bottom left*: a cormorant (Phalacrocoracidae indet.); *Top left*: a flock of pelicans (Pelecanidae indet.); *Top right*: a pair of ibises (Threskiornithidae indet.).

similar to that of the marsh deer (*Blastocerus dichotomus*) and moose (*Alces alces*), which also possess long legs that help with traversing waterways (Fig. 6.19). As indicated by its small, conical tusks, *N. sanchezi* fed in shallower waters than its dugong kin, possibly feeding on smaller seagrasses near coastlines. *Phoberomys* is one of the largest rodents to have lived, weighing about 700 kilograms in weight and being comparable in size to a buffalo (Fig. 6.20A). Contrary to popular belief, the closest relatives of *Phoberomys* are not capybaras (*Hydrochoerus hydrochaeris*), but rather chinchillas (*Chicnhcilla*). *Phoberomys* was a common resident of the area, where it would have fed on abrasive vegetation that grew on the brackish and coastal waters of the delta. *Phoberomys* had robust hindlimbs in contrast to the shorter and more slender forelimbs, which suggests that it would have been able to sit on its haunches and use its hands to manipulate food, like many other rodents. It is not known how aquatic these animals were, but it is unlikely they were adept swimmers and instead, seem to have been efficient runners.

Other rodents in the **ecosystem** included hystricognaths of the extinct families Neoepiblemidae and Dinomyidae (see pp. 147–8). A fossil skull from *Neoepiblema* (Fig. 6.20B) shows that the neoepiblemids were not particularly brainy, but possessed a good

South America Urumaco Formation, Venezuela: Tortonian–Messinian, 10–5.3 Ma, cont.

sense of smell. The small dinomyid *Potamarchus* (Fig. 6.20C), widely distributed in South America during the Late Miocene, is characterised by strong crenulations on the **occlusal** surface of the **cheek teeth**.

Avian remains from Urumaco are rare. The genus *Pelagornis* of the family Pelagornithidae (see p. 36) would have been the largest of the birds in the area, where it would have coexisted with flamingos, ibises, frigate birds and pelicans. The remains of *Pelagornis* (mostly well-preserved fragments of the **maxilla**) are known from earlier strata of the region, although remains from neighbouring formations from within the same time as the Urumaco Formation are a good indication that it was likely present in the area as well. Like many seabirds today, it may have used mangrove forests as a safe place for nesting.

The constantly changing environment and increasingly drier climate would have contributed to the disappearance of this ancient delta and may have reduced the long-term survival of the giant reptiles, favouring smaller species more resilient to ecological shifts. The giant reptiles were overspecialised and were unable to survive when their habitat changed, contrary to the smaller caiman species.

Fig. 6.18. A river **ecosystem** in which the Urumaco Formation was deposited, Late Miocene. *Centre*: the giant pleurodire turtle *Stupendemys geographicus*, with two arrau turtles (*Podocnemis expansa*) feeding on alge that has accumulated on the former's carapace; *Bottom right*: the mata mata turtle relative *Chelus columbianus*, and immediately above it, a shoal of the **osteichthyan** fish cf. *Megaleporinus* sp.; *Bottom left*: a pair of catfish (Pimelodidae indet.), and immediately behind them, a sawfish (*Pristis pectinata*); *Top left*: a cichlid (Cichlidae indet.), and immediately behind it, a pair of rays (*Rhinoptera* sp.); *Top right background*: the dugongid sirenian *Nanosiren sanchezi*.

1. A pair of the piranha-like osteichthyan fish *Colossoma* sp.
2. A shoal of the osteichthyan fish cf. *Megaleporinus* sp.
3. A Suckermouth catfish (Hypostominae indet.).
4. *Eionaletherium tanycnemius*, a giant ground sloth of the family Mylodontidae.
5. A surfacing river dolphin of the extinct genus *Saurocetes*.
6. A small alethinophidian snake of the genus *Colombophis*.
7. A spoonbill (Plataleinae indet.).
8. A small typotherian notoungulate (Miocochilidae indet.).
9. *Aktiogavialis puertoricensis*, a gavialid.
10. A pair of modern capybaras (*Hydrochoerus hydrochaeris*).
11. *Boreostemma* sp., a glyptodontine.
12. *Gyrinodon* sp., a toxodontian notoungulate of the family Toxodontidae.

Fig. 6.19. A freshwater environment in which the Urumaco Formation was deposited, Late Miocene.

Fig. 6.20. Reconstructed life appearance of giant rodents from the Urumaco Formation. (A) the neoepiblemid *Phoberomys pattersoni* (B) the neoepiblemid *Neoepiblema* sp.; (C) the dinomyid *Potamarchus* sp. Scale bar: 0.5 m.

South America Santa María Group (Chiquimil and Andalhualá formations), Argentina: Messinian, 7–5.5 Ma.

The **conglomerates** and sandstones of the Chiquimil and Andalhualá Formations (part of the Late Miocene to Early Pliocene Santa María Group (Moya *et al.* 2018)), located in the northwestern Argentinian province of Catamarca, are among the richest South American palaeontological sites for Late Miocene mammals. The beds straddle the southeastern margin of an arid high plateau called the Atacama Plateau, that landform which is characterised by a diverse **topography** consisting of a mix of basins and small mountain chains. The region has preserved a rich fossil record as each basin functioned as a reservoir for sediments weathered from neighbouring mountains that entombed and fossilised the remains of various vertebrates including birds and mammals. Since this process occurred almost continuously, the deposits provide a comprehensive account of changing faunas, biomes and climates within an interval that has been termed the "Age of the Southern Plains". The climate was typically colder and more seasonal than in the Early to Middle Miocene, resulting in more open and arid environments in many regions, especially Patagonia. However, low-lying marshlands, dry-mixed grasslands and highland forests occurred in northwestern Argentina, due to the region's varied topography and proximity to the Equator.

Rodents and xenarthrans (see p. 58) comprised the majority (more than 60%) of mammals existing in the region at the time. Among xenarthrans, the ground sloth *Pronothrotherium* was an early nothrotheriid (see p. 62), which is known from an almost complete skeleton and had a narrow, elongated muzzle that shows it was a **selective feeder**. Its tooth count (18) was identical to many other sloths, although the **caniniform** teeth were situated midway between the front teeth and the **molariform cheek teeth**, a typical feature of that lineage. The armadillos included *Vassallia*, a pampatheriid (see pp. 59–60) similar in dimensions to the modern giant armadillo (*Priodontes maximus*). Among pampatheriids, *Vassallia* had a notably robust skull and the **osteoderms** had a simpler ornamentation pattern. It was a **terrestrial herbivore** that consumed low-growing plants. The glyptodontine (see pp. 59–60) *Eleutherocercus* is regarded as a probable ancestor (or at least near the origin) of *Doedicurus* (see pp. 328–9), with a tail anatomy somewhat anticipating that of the latter. While *Eleutherocercus* resembled *Doedicurus* in overall appearance, it differed in various details, such as the carapace plates being more united and the **distal** end of the tail being spatula-shaped, slightly depressed and ending in a rounded contour.

The fossils of notoungulates (see pp. 140–3) are also particularly common, suggesting that these mammals played an important role in this bygone **ecosystem**. Other notable mammal species included *Cyonasua*, a procyonid (see pp. 98–9) twice as large as a modern raccoon (*Procyon lotor*) that is the earliest known carnivoran (see pp. 81, 83–4) from South America. *Cyonasua* colonised the continent when ecosystems there were still ruled by predatory metatherians (see pp. 43, 46–7), specifically **carnivorous** opossums along with sparassodonts (see pp. 48–50). Many of the last sparassodonts occupied **hypercarnivorous** dietary niches, whereas *Cyonasua* was an opportunistic **omnivore**, and were therefore unlikely to have competed for resources. This may be one possible reason why *Cyonasua* and its kin were able to colonise the continent long before other carnivorans. Like modern coatis (*Nasua*) and raccoons, *Cyonasua* was a generalised, terrestrial animal with some **fossorial** and climbing abilities as indicated by its limb bones. *Chapalmalania* was a procyonid that **converged** on the giant panda (*Ailuropoda melanoleuca*) in body form. The dentition of *Chapalmalania* is suggestive of omnivory and the animal apparently had a bone crunching bite not unlike that of a hyaena. *Chapalmalania* may have opportunistically fed on **carrion** as supported by a glyptodontine skeleton with bite marks corresponding to the teeth of *Chapalmalania*. The carnivorous metatherians included *Thylacosmilus*, which is the most well-known thylacosmilid, a lineage of sparassodonts that superficially resembled sabre-toothed cats through convergent evolution. Unlike any sabre-toothed placental, the convex skull housed a pair of **hypselodont** upper canine teeth. These were longer than in any conventional sabre-toothed cat and fitted against long flanges on the lower jaw when the mouth was closed, possibly for protection. The roots of the sabreteeth arched upwards above the **maxilla** and atop the **orbits**. It is presumed that the strong neck muscles would have helped power the sabreteeth into prey. *Thylacosmilus* most probably dispatched prey in the same way as placental sabre-tooths, since the nape, **carnassial** area and **premaxillary** region are all comparably well-developed, which are suspected to have, respectively, bolster neck-driven head pulling, increase gape and withstand the high impact when attacking large prey. There may even have been a **keratinous** structure or some other soft tissue covering the top of the snout and head, as indicated by pits and grooves on the maxilla (Janis *et al.* 2020).

Among the birds was *Andalgalornis*, a small, slender phorusrhacid (see pp. 41–2) that had a larger beak relative to its body size than any other member of that lineage. Biomechanical analysis has suggested that this beak was relatively weak and not well-adapted for grappling large prey. *Andalgalornis* most probably killed its prey with strikes from its legs or in a series of swift, repeated blows with its hatchet-like beak. One study of its cervical vertebrae has shown that they were suited for vertical striking. The teratorn (see p. 39) *Argentavis* may have been the heaviest flying bird to have lived, possessing one of the largest wingspans. Its humerus was roughly the length of a whole human arm! Most estimates suggest it weighed about 70 kilograms (154 pounds). Given that *Argentavis* resembled a vulture, the best comparison is with the largest extant aerial land bird, the Andean condor (*Vultur gryphus*). It has been hypothesised that *Argentavis* would have required headwinds to propel itself off the ground, although its strong hindlimbs possibly gave it a running or leaping start. However, its broad wings would have been incapable of flapping when the animal stood, and it would have required a strong takeoff before being able to execute a complete downward stroke. Its huge wings would have made its wing loading sufficient for powered flight.

The carbon isotopes preserved in fossil mammal teeth and sediments from Catamarca demonstrate that a significant change occurred in the continent's ecosystems towards the close of the Miocene. In South America, the spread of **C4 vegetation** at the expense of **C3 vegetation** may have opened up new niche opportunities, possibly reducing resource competition amongst **endemic taxa**.

Fig. 6.21. Phylogenetic tree of a sampling of fossil mammal and bird taxa from the Miocene deposits of Catamarca.

Europe Sansan Fossil Site, France: Langhian, ca. 15 Ma.

The classic Middle Miocene fossil site of Sansan, in southwestern France, has been studied extensively since it was first discovered in 1834, with nearly 200 plant and animal species documented in the process. The deposits were formed in the meander of a river or in an oxbow lake fringed by warm, seasonal subtropical forests, which transitioned into more open wooded regions further from the waterways. The flow would be restricted during the dry season, but during seasonal flooding, water would gush into the oxbow lake or meander from the main river, drawing in all kinds of detritus such as mammal remains. As the waters ebbed, those remains were stranded and eventually buried in the mud-rich sediments, which have preserved a broad variety of vertebrates from a very rich avifauna, including owls, pheasants and waterfowl, to giant proboscideans (see pp. 64–8). The barbourofelids (see pp. 85–6) and amphicyonids (see pp. 91–2) persisted as the most prominent members of the **carnivore guild**. The **primitive** barbourofelid *Sansanosmilus* is the animal whose name comes from this site, which would have been relatively felid-like given its cranial proportions with a shortened muzzle. As well as long, laterally-compressed upper canine teeth, *Sansanosmilus* already displayed advanced features typical of barbourofelids, such as an enlarged mandibular flange and a small mandibular coronoid process (the projection at the front of the ramus (the vertical part) of the lower jaw). *Sansanosmilus* is distinguished from its more derived relative *Barbourofelis* in being smaller (leopard-sized), possessing shorter upper canine teeth, less blade-like **carnassials**, open **orbits**, a longer tail, and foot bones showing a more **plantigrade** stance.

The Middle Miocene records the first occurrence of one of the largest and most prosperous amphicyonids. *Amphicyon major* was lion-sized and had an elongated skull, with a relatively long muzzle and rather long, robust canine teeth. *A. major* had a relatively **complete dentition** as in other amphicyonids, with three molars and four premolars (a primitive condition in carnivorans (see pp. 81, 83–4)), the **posterior** molars which were enlarged. The limb proportions were generally bear-like, including shortened metapodials, which are typical of plantigrade animals. All of these characteristics suggest that *A. major* was an active predator that hunted through **ambush** or a combination of ambush and pursuit. The diet of *A. major* has been compared to that of brown bears (*Ursus arctos*), suggesting a similar opportunistic feeding behaviour. Since the Early Miocene, the amphicyonids faced competition from the arrival of the earliest true ursids (see pp. 92, 94–5) including the specialised hemicyonine genus *Hemicyon*. Unlike modern ursids, this animal had a more elongated back and longer, straight, and parallel metapodials, suggesting a **digitigrade** stance. Such body proportions are actually more similar to those of a big cat, and *Hemicyon* may have been a fast runner and an active predator, possibly engaging in group hunting. The dentition of *Hemicyon* was definitely not felid-like, but exhibited more carnivorous specialisations than modern ursids. *Plithocyon* was even more specialised and exhibited greater **cursorial adaptations**, with more felid-like and gracile limb bones. Furthermore, its massive carnassials suggest a greater inclination towards **hypercarnivory**.

These gracile hemicyonines were not the sole predators that confronted the hypercarnivorous barbourofelids and amphicyonids in occupying the large carnivore guild. The several biotic interchanges that occurred during the Early Miocene resulted in the barbourofelids being in direct competition with true felids (see pp. 86–9) of genera including *Pseudaelurus*, which dispersed widely across Europe at the time. The largest species was the slenderly built and leopard-sized (around 30 kilograms) *P. quadridentatus*, which showed a tendency towards relatively high-crowned, flattened upper canine teeth. Thus, *P. quadridentatus* is a strong candidate for the ancestral stock of machairodontine felids. Compared to the contemporary barbourofelid *Sansanosmilus*, *P. quadridentatus* was relatively longer-limbed and more digitigrade. Generally considered more derived than *P. quadridentatus* was the smaller (wild-cat sized), ocelot-like feline *Styriofelis*, which survived longer (into the Late Miocene). This is possibly due to greater ecological flexibility and adaptability to diverse prey types and habitats—traits that may have conferred a survival advantage amidst shifting environmental conditions and increasing competition. *Styriofelis* had body proportions that were broadly similar to those of modern felines of similar size, but had shorter and more robust forelimbs and metapodials, suggesting a primarily **terrestrial** lifestyle.

A major faunal shift in the **herbivore community** in Europe during the Early to Middle Miocene, besides the radiation of rodents such as glirids and cricetids (see p. 147), was the emergence and dispersal of the earliest ruminants (see p. 110) sporting cranial appendages. The cervids (see pp. 115–7) maintained their high diversity in this part of Western Europe. The forest-dwelling *Dicrocerus* is the oldest known cervid to have antlers with a distinct permanent base. The antlers themselves were shed and regrown yearly but, unlike most modern deer, both males and females had antlers, with those of the latter being smaller and more slender. In *Dicrocerus*, the pedicle was generally longer than the antler itself. *Heteroprox* was a modest-sized cervid with protruding tusks and simple antlers that consisted of a long stem ending in a two-pronged fork. The top portion of the antler was possibly shed periodically, but unlike modern cervids, there was no clear divide between the permanent and non-permanent portions. *Heteroprox* may have been a **semi-aquatic browser** as evident from its limb ratios.

The diversity of bovids (see pp. 116, 118–9) remained relatively low in Western Europe. The earliest known bovid, *Eotragus*, was a small ruminant similar in size and proportions to modern duikers or the pygmy antelope (*Nesotragus batesi*), but possibly related to the boselaphins (e.g. nilgai (*Boselaphus tragocamelus*)). It was a forest species that weighed around 18 kilograms (40 pounds) and ranged across Europe, West Asia and North Africa. The horn cores were conical, short, and oval in cross-section, positioned slightly posterior to the orbits. One successful family of horned ruminants were the palaeomerycids (see p. 114), represented by *Ampelomeryx*. *Ampelomeryx* sported three non-**deciduous** appendages on the skull, with two above the orbits and the third located at the back of the skull, on the **occipital** region. However, the two **frontal** appendages were broad, flattened, and laterally projected over the orbits rather than vertically, helping shade the eyes. The most distinctive characteristic of *Ampelomeryx* is its single, flattened occipital appendage, forming a Y-shape.

Fig. 6.22A. Phylogenetic tree of a sampling of fossil mammal **taxa**, particularly proboscideans, and laurasiatherians (see p. 74) outside Euungulata (see p. 100), from the Sansan Fossil Site.

Europe Sansan Fossil Site, France: Langhian, ca. 15 Ma, cont.

Among other artiodactyls (see p. 104), the suids (see pp. 108–10) peaked in diversity in Central and Western Europe in much of the Middle Miocene, represented by the listriodontines and tetraconodontines at Sansan. These were the dominant suids in Western Europe at the time. *Listriodon* was a **lophodont** listriodontine, having replaced more primitive, semi-lophodont members of that subfamily such as the genus *Bunolistriodon*. The lophodont teeth along with dental **microwear** analysis indicates that *Listriodon* had a strictly herbivorous diet. This suggests that it browsed on tough vegetation instead of rooting like generalised suids, so as to avoid food competition. *Listriodon* was possibly adapted to living in more open environments as inferred from its elongated **distal** limb elements. The tetraconodontine *Conohyus* likely resembled a small boar in body and cranial proportions, but was distinct in its dentition. The premolars were equipped with conical cusps showing dental wear similar to that of hyaenas, suggesting a diet of hard foodstuffs such as seeds. Throughout the Middle Miocene, the European tetraconodontines decreased in body size, from the 70 kilogram *Conohyus* to its offshoot genera such as the 40 kilogram *Parachleuastochoerus*.

Dorcatherium, the most diverse tragulid (see p. 110) genus ever, is an indicator of the persistence of humid environments in Central and Western Europe. It exhibited marked **sexual dimorphism** unlike any other known tragulid. Males had a more robust and elaborate skull roof with well-developed bulges and ridges, along with sabre-like upper canine teeth, whereas the females had weakly developed cranial ridges and reduced upper canine teeth. In males, these structures were likely used for protection during **intraspecific competition** and for sexual display. One family of perissodactyls (see p. 126) that entered Europe from North America were the equids (see pp. 132–4). *Anchitherium* was one of very few anchitheriines to have entered Europe, despite the great diversity of this lineage in North America. The forelimbs of *Anchitherium* were short and the hindlimbs comparatively elongated, reminiscent of the limb ratios of basal equids and reflective of an adaptation to movement on soft substrates. These limb ratios, along with a somewhat elongated neck and body, and a small skull, made *Anchitherium* vaguely resemble a deer. Its muzzle was shorter than in modern horses, possibly because of the **brachydont** teeth, which were smaller than those of **grazing** equids but well-adapted to a **folivorous** diet. *Anchitherium* was not such a specialised cursorial equid like many others of its time.

In the Early to Middle Miocene, the rhinocerotids (see pp. 130–2) underwent a substantial turnover. The ancestral aceratheriins were replaced by more advanced species of the hornless genus *Brachypotherium*, prevalent in Central and Western Europe, but also present in Africa and Asia. It had hippopotamus-like body dimensions, short limbs, and a relatively short, wedge-shaped skull that largely retained a **hypsodont** dentition. The orbits were located somewhat far forward and above the second molar. One other perissodactyl family that arrived in Europe during this time were the chalicotheriids (see pp. 128–9), which also underwent a major turnover as the primitive schizotheriines were generally replaced by the larger and more derived chalicotheriines such as *Anisodon*. This genus is distinguished by a considerably shorter and taller skull. Very long forelimbs contrasting with short hindlimbs gave chalicotheriines a distinctive sloped back and upright walk. Like anteaters and ground sloths, they were able to protect their long, curved claws by walking on their thickened knuckles. Extra weight-bearing pads on their hips and hindlimbs suggest that they also spent considerable time sitting vertically while feeding. Those claws inevitably doubled as formidable defensive weapons.

Among the first proboscideans to disperse to Europe were the "gomphotheriids". *Gomphotherium angustidens* had a low, elongated (**longirostrine**) skull with an upper pair of long, downward-curving tusks and relatively shorter spatulate tusks in the lower jaw. The upper tusks were covered by a layer of enamel, contrary to modern elephants. The purpose of these tusks was presumably to dig or scrape up soft aquatic vegetation. The molars were **bunodont** and brachydont, with the third molars distinctively exhibiting three pairs of cusps. During the Middle Miocene, deinotheriids of the species *Prodeinotherium bavaricum* and *Deinotherium giganteum* successfully colonised Europe not too long after the "gomphotheriids". *D. giganteum* was among the tallest and largest animals during the Miocene. Other than larger size, the genus *Deinotherium* differed from *Prodeinotherium* in having a shorter and narrower skull roof, a less upturned articulation between the skull and cervical vertebrae, and a **rostrum** that was not orientated downwards parallel to the downturned mandibular symphysis. Deinotheriid trunks are usually depicted as being very short like a tapir, with the broad muscle attachment base at the skull indicating a wider, shorter **proboscis** compared to modern elephants. The proboscis would have most probably been of intermediate length, since deinotheriids would otherwise have had some difficulty drinking from waterholes with those long limbs. The deinotheriids may have processed their food in a similar manner to modern tapirs, with the second and third molars functioning in a powerful vertical shearing motion, with minimal lateral motion. This feeding method contrasts to that of "gomphotheriids", which chewed by lateral grinding.

The small mammal fauna consisted of a high diversity of rodent species. The theridomyids (see p. 233) had already disappeared, during the earliest Miocene. Many rodent lineages persisting into the Middle Miocene, such as the sciurids, had remnant members from the Oligocene. There was great stability among Early to Middle Miocene sciurids, with the continuation of the conservative Oligocene genus *Palaeosciurus*. The beavers such as *Steneofiber*, which arrived in Europe following the Grande Coupure (see p. 230), are strongly evident of humid conditions. *Steneofiber* was half as large as the modern beaver (*Castor*), but may have had a semi-aquatic lifestyle similar to that of modern water voles. *Steneofiber* was more terrestrially adapted than extant beavers, and could dig burrows. Moreover, ochotonid lagomorphs (see pp. 148–9) became abundant in the small mammal faunas, having previously been scarce during the Oligocene. Among eulipotyphlans (see pp. 74–7), larger than modern shrews and armed with elongated incisors was *Dinosorex*. The teeth were reinforced by iron particles embedded into the enamel, which allowed *Dinosorex* to hunt larger prey. It is thought that as environmental conditions became more changeable at the end of the Miocene, the larger-bodied terror shrews could not adapt to a reduced food supply in the winter.

Fig. 6.22B. Phylogenetic tree of a sampling of fossil mammal taxa, particularly **ungulates**, from the Sansan Fossil Site.

Europe Cerro de los Batallones, Spain: Tortonian, 10–9 Ma.

The Cerro de los Batallones site complex, located in the Madrid basin of central Spain, is a sequence of pseudo-karstic cavities divided into ten individual sites. They were formed during the Late Miocene, far from the edges of a giant freshwater lake. The cavities retained some standing water, even during intense droughts. Therefore, they functioned as natural traps for **ungulates** including moschids (see p. 115), several indeterminate species of boselaphin bovids (see pp. 116, 118–9), suids (see pp. 108–10), giraffids (see pp. 112–4), equids (see pp. 132–4) and rhinocerotids (see pp. 130–2), as well as proboscideans (see pp. 64–8). These animals would have been drawn to the cavities by the moisture, crossed into them intentionally or fell in by accident, and then be incapable of escaping because of the texture, depth and steepness of the walls. In a number of the cavities, the bodies of the trapped **herbivores** served as bait that lured local predators (which at several sites account for the vast majority of remains, at over 90%), such as the machairodontine felids (see pp. 86–7) *Promegantereon* and *Machairodus aphanistus*, small conical-toothed cats, amphicyonids (see pp. 91–2), ursids (see pp. 92, 94–5), hyaenids (see pp. 88–90) and ailurids (see p. 98), along with a diversity of mustelids (see p. 99). The sediments that covered the remains must have flooded in rather rapidly, otherwise the skeletons would not be so intact and **articulated**.

The Late Miocene witnessed continuously receding broadleaved forests in Europe giving way to open woodlands interspersed with larger grassy areas, similar to those observed today in certain regions of India where monsoonal weather predominates (Fig. 6.24). From the east, more mammals that had evolved **adaptations** for living in such open environments entered Europe, having often joined the already established groups without a clear or rapid replacement of potential competitors already occupying similar ecological niches. This suggests a period of coexistence rather than complete displacement. The newcomers included hipparionin equids and machairodontine felids, and bovids became widespread, as did the giraffids and the rhinocerotids. The barbourofelids (see pp. 85–6) had vanished from Eurasia, while the sabre-toothed true felids became dominant. The giraffid *Decennatherium* had a rather short **frontal** pair of **ossicones** and a more prominent **posterior** pair, the latter which had curved, flattened and flaring ossicones, with ridged edges. *Decennatherium* also had a highly elongated and slender facial region. *Decennatherium* was a **primitive** sivatheriine, with longer and slenderer limbs than later representatives.

The tiny *Micromeryx* was a close relative of the musk deer (*Moschus*) and the last surviving member of the European moschids. Evidence of predation on the leg bones of some fossils reveals that mustelids were the likely culprit (a similar predator-prey interaction occurs with modern moschids). Whereas modern moschids feed largely on lichens, the teeth of fossil species show a much wider dietary range, such as leaves and fruit in the case of *Micromeryx*. In addition, the great diversity of suids abruptly declined by the Late Miocene, with several distinctive lineages including the **browsing** listriodontines (see pp. 276–7) having disappeared. The suines persisted and became more diverse, with the arrival of eastern immigrants such as *Microstonyx*. It was a massive pig (about 300 kilograms) that generally resembled modern suids, with muscle attachment sites on the skull indicating a mobile, disc-like snout used for rooting, although it also differed in possessing very small canine teeth. Instead of well-developed canine teeth, the alveolar ridges above the upper canine teeth (especially in males) were very large and resembled a pair of bumps on the muzzle.

The rhinocerotids occupied much of the large browser **guild** during this time, the most abundant which was *Aceratherium*, a modest-sized, hornless aceratheriin which had a **brachydont** dentition along with elongated limbs, and retained a functioning fifth metacarpal. *Aceratherium* may have had a tapir-like lifestyle as indicated by the similar limb ratios. Males possessed a pair of robust, tusk-like lower incisors that were much larger than those of females, an adaptation for browsing dense plant growth. Despite the length of the incisors, it is thought that they were concealed by muscular lips in life, as in extant Asian rhinoceroses (which possess large incisors too). Like the bovids, the hipparionin equids peaked in diversity in Eurasia during the Late Miocene. The medium-sized, largely forest-dwelling form *Hippotherium* persisted, less adapted to **grazing** than modern horses because of lower-crowned teeth. Compared to other hipparionins, the **rostrum** of *Hippotherium* was relatively short. In *Hippotherium*, the **preorbital fossa**, characteristic of Neogene equids, was large and triangular-shaped, extending deep into the upper jaw. The metapodials were generally broad and short, and thus considerably more robust than in *Cormohipparion* (see pp. 260–1) (Bernor *et al.* 2021). *Tetralophodon* was a four-tusked proboscidean like the "gomphotheriids", to which it was not closely related and which it had replaced. *Tetralophodon* was also larger than "gomphotheriids" and its molars were more **hypsodont**. Furthermore, it had a shorter, more elephant-like skull and lies near the origin of modern elephants. *Tetralophodon* is distinguished mostly by the ridges of the molars (in fact, its name means "four-ridged tooth") and were a response to the aridification of the climate, since it needed these molars to effectively grind tough, fibrous plants.

Among the **carnivore** guild, *Promegantereon* is one of the earliest known sabre-toothed felids, comparable in size and overall proportions to a small leopard, but with a shorter tail and more elongated neck. Interestingly, most fossils are of young adult individuals. They were at an important developmental stage, being forced to depart from their mothers' territories, and given that pressure and inexperience, they were more prone to take risks than older individuals. This bias is prevalent in fossil carnivore remains from trap sites. There are also high incidences of canine tooth breakages present among *Promegantereon* samples at Batallones. This suggests that they dispatched prey with their upper canine teeth as in modern felids, a much riskier hunting practice given the absence of protruding incisors. This is also the case for *Machairodus aphanistus*, the apex predator at Batallones. *Machairodus* was the most widespread genus of machairodontine in Eurasia during the Late Miocene, distributed between the Iberian Peninsula and North America. Several near complete skeletons from this site have shown that *M. aphanistus* was a generalised machairodontine with tiger-like skeletal proportions and elongated hind limbs well-suited for leaping. However, compared to modern big cats, *M. aphanistus* had a longer and narrower skull, with smaller **orbits**. Its upper canine teeth were moderately long,

Fig. 6.23A. Phylogenetic tree of a sampling of fossil mammal **taxa**, particularly carnivorans, from Cerro de los Batallones.

Europe Cerro de los Batallones, Spain: Tortonian, 10–9 Ma, cont.

laterally compressed and coarsely serrated on both sides. The neck was stronger and more elongated, as indicated by the well-developed temporal crests on the lower jaw, which provided an increased attachment surface area for the temporalis muscle, crucial for jaw closing and neck movement. *M. aphanistus* was most probably an **ambush predator**, using its relatively elongated hindlimbs to thrust the animal's body forward in an abrupt, short sprint, while its powerful forelimbs, with large claws, could undergo significant lateral rotation that allowed the animal to grip and subdue prey. *M. aphanistus* evidently may have been susceptible to niche partitioning, with bear-sized amphicyonids such as the genus *Magericyon*, the former preferring densely vegetated habitats while the latter hunted in more open environments. Amphicyonids and ursids potentially competed with *Machairodus* for food, whereas the smaller predators may have been less likely rivals.

Magericyon was about the size of a large leopard (91 kilograms). Contrary to many other amphicyonids, *Magericyon* had a **hypercarnivorous** dentition, consisting of upper second molars smaller than the first, absent second premolars, single-rooted lower third premolars and laterally compressed canine teeth (Siliceo *et al.* 2015). Morphofunctional studies have concluded that *Magericyon* had strong neck and jaw musculature that assisted in stabilising and extending the head and jaws when biting (Siliceo *et al.* 2015). Compared to *Amphicyon* (see pp. 274–5), *Magericyon* had developed a more specialised dentition for killing and consuming large prey efficiently. The emergence of the earliest large ursids in Europe was an important event among this lineage. *Indarctos* was one of the earliest members of a subfamily of large-bodied ursids called Ailuropodinae, which were significant components of the Late Miocene **community** of carnivorans (see pp. 81, 83–4) and today consists only of the strictly herbivorous giant panda (*Ailuropoda melanoleuca*). *Indarctos* had brachydont molars, with the posterior grinding part of the upper second molar similarly well-developed to that of modern brown bears (*Ursus arctos*) and spectacled bears (*Tremarctos ornatus*). The lower jaw of *Indarctos* was distinctly slender, with a pronounced chin that only occurred in mature individuals. Known largely from numerous lower jaws, the relatively long-limbed *Eomellivora* was closely related to the honey-badger (*Mellivora capensis*) and was much larger than modern **terrestrial** mustelids, weighing

Fig. 6.24. Palaeoenvironmental reconstruction of Cerro de los Batallones during the Late Miocene. *Foreground*: the primitive hyaenid *Protictitherium crassum* (left) and the sabre-toothed cat *Machairodus aphanistus* resting underneath the shade of a tree (right). Behind the latter is the moschid *Micromeryx flourensianus*. *Background*: the proboscidean *Tetralophodon longirostris* (left). *Far background*: from left to right is the suid *Microstonyx major*, a pair of indeterminate boselaphin bovids, and the giraffid *Decennatherium rex*. *Top left background*; a bird-of-prey of the genus *Haliaeetus*.

about 35 kilograms. *Eomellivora* likely lived in much the same manner as modern honey-badgers and wolverines. The dentition of *Eomellivora* indicates that it was more hypercarnivorous than wolverines.

Protictitherium was a small, civet-like, primitive hyaenid which had semi-retractile claws indicating that it was largely **scansorial**, possibly to elude larger predators. According to its generalised dentition, *Protictitherium* had a varied diet including insects and **carrion**, like extant terrestrial civets. Its fossils are a highly frequent occurrence at Batallones, which is unsurprising considering its adaptations for **scavenging**. The nimble, unobtrusive *Protictitherium* was capable of scurrying between the feet of the larger predators and swiping its meal from the scraps of their kills. *Simocyon* was a leopard-sized early relative of the red panda (*Ailurus fulgens*) according to the morphology of its teeth, ear and tarsal bones. Although the red panda is predominantly herbivorous, the skull and dentition of *Simocyon* suggests that it was not only hypercarnivorous, but possibly also adept at cracking bones, similar to modern hyaenas. According to its skeletal anatomy, *Simocyon* was scansorial like red pandas, although it possibly spent most of its life on the ground. Like red pandas, *Simocyon* possessed a radial sesamoid, a peculiar wrist bone that functions as a "false thumb".

Unlike *Styriofelis* (see pp. 274–5), the primitive feline *Leptofelis* possessed limb bones adapted for fast, efficient terrestrial locomotion. This feature is also present in the European wildcat (*Felis silvestris*), but is likely a result of **convergent evolution**, since the specific dental characters preclude *Leptofelis* from being ancestral to extant felines. The skeleton of *Leptofelis* consisted of a mixture of features, such as the attachment of a strongly developed quadratus plantae muscle on the tarsal bones, showing that it was more adept at climbing than many modern felids, both to capture small prey in trees and to evade larger predators. *Leptofelis* may have used its jumping capabilities to catch small vertebrates while hunting on the ground, in a similar manner to the caracal (*Caracal caracal*) and serval (*Leptailurus serval*).

Fig. 6.23B. Phylogenetic tree of a sampling of fossil mammal taxa, particularly ungulates, from Cerro de los Batallones.

Asia Liushu Formation, China: Tortonian-Messinian, 11-6 Ma.

The Liushu Formation, located in the Linxia Basin of Gansu province, northwestern China, consists of **carbonate**-cemented **siltstones**, with abundant fossils being found in an intercalated reddish matrix of **mudstones** referred to as the "Hipparion red clays/ Hipparion Beds". Mammal fossils are preserved in concentrations called "bone nests", which often correspond to depressions in the floodplain of ancient river systems where the remains would accumulate subsequent to being transported by floodwaters. The **herbivore community** was dominated by **grazing** species, with woodland **browsers** being scarce. This, along with other evidence including pollen analysis, suggests that mostly semi-arid steppe/grassland habitats prevailed. These were affected by a stronger East Asian monsoon than today, with monsoonal rains potentially providing water sources for the fossil sandgrouse *Linxiavis*, a group of birds known for their water-carrying abilities (Li *et al.* 2020). A diverse fauna of large and unique mammals including **primitive** giraffids (see pp. 112–4), cervids (see pp. 115–7), bovids (see pp. 116, 118–9), chalicotheriids (see pp. 128–9) of the genus *Ancylotherium* (see pp. 302–3)), rhinocerotids (see pp. 130–2), **tridactyl** hipparionin equids (see pp. 132–4), and proboscideans (see pp. 64–8) of the genus *Tetralophodon* (see pp. 278, 280)), existed in the region.

The giraffid *Palaeotragus* was a browser resembling a very large okapi (*Okapia johnstoni*) in size and body proportions, with a long, thick neck and two simple **ossicones** growing upwards and backwards from above the **orbits**. Unlike the modern okapi, it had a limb and foot bone structure adapted for running in the open plains. The bovids included forms such as the large, goat-like caprine *Protoryx*, characterised by semi-**hypsodont** molars, a high, narrow snout, and upright-inserted horn cores that were slightly curved backwards, divergent and laterally compressed. *Protoryx* was a **mixed feeder**, feeding on both soft and abrasive vegetation. *Sinotragus* is distinguished by a short cranium, a perpendicular, high **frontal** bone, and horn cores with distinct backward and outward curves, oval in cross-section at their base. Its phylogenetic position is controversial, probably belonging to the subfamily Hippotraginae or Caprinae. It is also possible that it was close to the origin of the Ovibovini (e.g. muskox) because of features including a narrow skull, short, thick horns and a pronounced cranio-facial flexure, although its dental anatomy and horn-core structure is different

Among perissodactyls (see p. 126), the slender, medium-sized equid genus *Hipparion* reached a very high diversity by the Late Miocene, being most successful in Asia and North America and probably living in large herds. The hypsodont **cheek teeth** (although lower and broader than those of *Equus*) with elaborate patterns of enamel were the key adaption that *Hipparion* had for living in open habitats. *Hipparion* is represented by *H. coelophyes* at Liushu, characterised by features such as a single **preorbital fossa** situated relatively far in front of the orbit and a nasal slit situated above the upper second premolar (Li *et al.* 2017). The rhinocerotids included a wide variety of elasmotheriines and aceratheriins. One early elasmotheriine, *Iranotherium*, provides evidence of variation between genders, with larger rugosities present in males suggesting more musculature around the jaws and overall robustness. It also possessed a domed boss, but it was located on the **anterior** end of the snout. The more derived *Sinotherium* is unique among elasmotheriines in having the horn situated between the nasal and frontal bone, a transitional stage between the **ancestral** nasal position, and the advanced frontal position in the genus *Elasmotherium* (see pp. 131–2). In *Sinotherium*, the presence of a rugosity **posterior** to the nasofrontal horn suggests that the animal was tandem-horned. Among the aceratheriins was the successful *Chilotherium*, known from all over Eurasia. *Chilotherium* was hornless and possessed a pair of immense, curved tusks on the lower jaw for display and defence. One particular fossil consists of a skull and jaw bone from an adult female bearing a deep injury on its forehead. The tusks were recurved too much to inflict an injury so high on the animal's skull, and the only predator that was capable of inflicting the wound was the massive, hyaena-like percrocutid (see p. 90) *Dinocrocuta*, with the puncture mark perfectly corresponding to the shape of the **carnivore's** canine teeth. The occurrence of just one puncture mark and ensuing bone growth around the injury suggests that the rhinocerotid had managed to survive the predator's attack and recover.

Dinocrocuta was the largest and strongest known **terrestrial** carnivoran (see pp. 81, 83–4) from Liushu, which reached an upper weight of 300 kilograms and had a 43 centimetre long skull (roughly twice the length of a spotted hyaena (*Crocuta crocuta*)), dwarfing the contemporaneous hyaenids (see pp. 88–90) such as *Thalassictis* and felids (see pp. 86–9) including *Amphimachairodus*. The skull and dentition of *Dinocrocuta* were very powerful for generating a bone-crushing bite. The same series of teeth capable of dismembering carcasses were also adapted for tearing into the hides and bones of some of the largest steppe herbivores. Whether *Dinocrocuta* was social or solitary is currently uncertain, but it may have been both an active hunter and a **scavenger**. The more slenderly built hyaenid *Thalassictis* had a dentition somewhat resembling that of canids (see pp. 92–3) but with a better ability to crush bone. The skeletal mophology suggests that *Thalassictis* was only moderately adapted for **cursorial** locomotion. Nevertheless, it was possibly an active hunter, adapted for hunting in more open woodland habitats than more primitive hyaenids such as *Protictitherium* (see pp. 279–81).

The best preserved fossil material of the lion-sized homotherin felid *Amphimachairodus giganteus* are known from the famous "Hipparion Beds". It is distinct from *Machairodus* (see pp. 278–80) in possessing more blade-like **carnassials**, comparatively smaller lower canine teeth and larger, forward-facing incisors. *A. giganteus* had a long, highly muscular neck, rather elongated limb bones and a longer tail than many other large homotherins. The metailurin felid *Metailurus major* is characterised by a shortened facial region and the absence of the upper second premolar. It was larger than a male cougar (*Puma concolor*), having similar body proportions but with relatively more elongated hindlimbs indicative of powerful leaping. The relative gracility of the limb elements is comparable to that of the snow leopard (*Panthera uncia*). Among the ursids (see pp. 92, 94–5) was the primitive species *Ursavus tedfordi*, weighing over 90 kilograms. The muzzle was short and broad relative to the back of the skull, and *U. tedfordi* started showing the general features present in modern ursids, such as a large grinding dentition suited to a broad, **omnivorous** diet.

Fig. 6.25. Phylogenetic tree of a sampling of fossil mammal **taxa** from the Liushu Formation.

Africa Nawata Formation, Kenya: Tortonian–Messinian, 9–5.5 Ma.

In the Turkana Basin, near the southwestern corner of Lake Turkana in northern Kenya, lies the Lothagam Group, a rock unit that is composed mostly of fine-grained volcanic sediments. The sedimentary environment corresponds to a large, meandering **fluvial** system, and the fauna unearthed is amazingly rich. Several depositional episodes have been documented at the Lothagam Group, dating from the Middle Miocene to Early Pleistocene Epochs, but the most fossiliferous deposits are found in the Nawata Formation, which accumulated during the Late Miocene. Back then, the river's course was perennial, but prone to pronounced seasonal fluctuations, with a dry season lasting roughly four months according to faunal, floral and geological evidence. An abundance of oysters, **osteichthyan** fish, turtles, crocodylians and hippopotamuses (the most commonly preserved mammals), is evident of extensive bodies of water. These indicators, along with the overall mammalian composition, signals the presence of lush gallery forests that, further from the rivers, transitioned into wooded **savannah** with thorny scrub (Fig. 6.29). The Great Rift Valley was still just a young geological feature, and because this valley had an effect on the distribution of humidity and winds, combined with disparities in global climate through time, the grasslands and savannah were considerably more limited in extent than today, with a greater proportion of Africa being shrouded by woodland. Africa was no longer an isolated continent and thus, many of its mammals were similar to those of Eurasia.

The crocodylians included particularly large forms such as the long-snouted *Euthecodon*, estimated to have been around 10 metres long (Fig. 6.26). *E. brumpti* had a remarkably elongated and slender **rostrum**, unlike its closest extant relative, the West African dwarf crocodile (*Osteolaemus tetraspis*). *Euthecodon* had only four **premaxillary** teeth, whereas other crocodylians typically have five. The sharp, double-keeled teeth of *Euthecodon* are believed to have been adapted for **piscivorous** feeding habits and would have been useless for holding onto mammalian prey such as antelope. Additionally, the earliest known occurrence of the Nile crocodile (*Crocodylus niloticus*) and slender-snouted crocodile (*C. cataphractus*) have been documented from these deposits.

Among the most spectacular mammalian finds are three exceptionally well-preserved skeletons of fossil carnivorans (see pp. 81, 83–4), representing the giant, leopard-sized mustelid (see p. 99) *Ekorus*, the hyaenid (see pp. 88–90) *Ictitherium* and a small scimitar-toothed cat of the genus *Lokotunjailurus*. The oldest fossils of the dirk-toothed cat genus *Dinofelis* are also known from Nawata. Alongside the felids (see pp. 86–9), the hyaenids arrived in Africa during the Late Miocene, an important wave of immigration from Eurasia even if *Dinofelis* likely first evolved in Africa. Weighing nearly 44 kilograms (97 pounds), *Ekorus* was significantly larger than the honey-badger (*Mellivora capensis*), which is the largest modern **terrestrial** mustelid in Africa, but also exhibited functional morphological differences. *Ekorus* had proportionately longer limbs. The forelimb bones reveal that the muscles involved in digging, advanced in the honey-badger, were not as powerful and that it had a more upright stance, like that of a hyaenid or a felid, instead of squatting like a honey-badger. The ensuing picture of *Ekorus* is a large, **cursorial** predator that would have actively chased down its prey in relatively more open environments. It was **hypercarnivorous** as indicated by its felid-like dental morphology.

With its gracile body, short limbs and a dentition adapted for a diet of both plants and small- to medium-sized mammals, *Ictitherium* more closely resembled a civet than any modern hyaena. It also had not yet developed the powerful bone-crushing bite of modern hyaenas. *Ictitherium* was a highly common and successful genus of hyaenid, being found across Eurasia as well as in Africa. The concentration of their fossils implies that this early hyaenid may have hunted in packs and had an

Fig. 6.26. Life reconstruction of the crocodylid (see pp. 27–8) *Euthecodon brumpti* on a grassy riverbank. Late Miocene of Kenya.

Fig. 6.27. Phylogenetic tree of a sampling of fossil mammal **taxa** from the Nawata Formation.

Africa Nawata Formation, Kenya: Tortonian-Messinian, 9–5.5 Ma, cont.

advanced social hierarchy, like its extant relatives. *Lokotunjailurus* was about the same shoulder height as a small lion, albeit relatively more slender in build, with a somewhat smaller skull and longer metapodials. The claws of the second to fourth **digits** were also more reduced. The most distinctive feature of this animal was a massive dewclaw on the first digit of the front paws, which would have been a conspicuous feature in life, even when concealed by skin and fur. This large dewclaw was likely an **adaptation** for grappling onto and immobilising prey. *Lokotunjailurus* had the most **derived** dental anatomy among homotherins, and had a much more elongated lumbar region in the vertebral column than in the genus *Homotherium* (see pp. 323–4, 343–4). One other predator was *Viverra leakeyi*, similar in physical appearance to modern terrestrial Asian civets of the genus *Viverra* but considerably larger, around the size of a small leopard and weighing approximately 40 kilograms, making it the largest known viverrid (see p. 88). The teeth of *V. leakeyi* were more adapted for slicing meat than those of extant species of *Viverra*, or its closest modern relative the African civet (*Civettictis civetta*). This characteristic, along with its large size, indicates that *V. leakeyi* was an active hunter of small- to medium-sized vertebrates.

The large carnivorans would have had a wide range of prey to choose from among the various species of hipparionin equids (see pp. 132–4), aardvarks, antelopes, suids (see pp. 108–10), giraffids (see pp. 112–4), and cercopithecid monkeys (see p. 154), and they may have even hunted young hippopotamuses, rhinocerotids (see pp. 130–2) and proboscideans (see pp. 64–8). The **primitive** form of the modern white rhinoceros genus *Ceratotherium*, has been documented from Nawata, possibly a recent immigrant from the Arabian Peninsula. During the Late Miocene, *Ceratotherium* showed modifications in skull structure, which became more elongated and enhanced the animal's ability to **graze** on short grass, characteristics somewhat replicated in hipparionin equids belonging to the genus *Eurygnathohippus*. This equid appears to have combined higher-crowned **cheek teeth** with additional major adaptations of its feeding apparatus such as a long muzzle and modified incisors to help it obtain greater access to short, hard grasses.

Among artiodactyls (see p. 104), giraffids made their first entry, as clear immigrants from Asia. From roughly 6.5 Ma, the tragelaphin, reduncine and hippotragine antelopes ascended to prominence, alongside impalas, and possibly relatively later the alcelaphines. The suids were very diverse, with representatives such as *Nyanzachoerus*, a possible early relative of modern warthogs and part of the extinct subfamily Tetraconodontinae. Beautifully preserved crania of *N. tulotus* show that they sported a variety of knob-like protrusions on the top surface, plus flaring cheekbones. These prominent lumps would have been covered with an additional layer of skin, although the tusks were only medium-sized. The males were larger the heads were more ornamented compared to those of the females, a case of marked **sexual dimorphism**. It is possible that the ornaments were used in courtship. One of the earliest hippopotamids (see p. 121) is *Archaeopotamus* (Fig. 6.28), which was of gracile build and larger than the modern pygmy hippopotamus (*Choeropsis liberiensis*). *Archaeopotamus* differed in having a narrower skull, with the **orbits** not raised above the level of the forehead as in modern hippopotamuses. *Archaeopotamus* was possibly already amphibious in nature. It possessed three pairs of incisors and a very long, narrow mandibular symphysis. For this reason, the term "narrow-muzzled hippopotamuses" has been dubbed for *Archaeopotamus* and its relatives. The genus *Archaeopotamus* is represented by two species at Nawata, both previously assigned to *Hexaprotodon*: *A. lothagamensis* and *A. harvardi*. *A. harvardi* was smaller than *A. lothagamensis*, with femurs roughly the same size as those of modern hippopotamuses. *A. harvardi* also had a more riparian lifestyle.

Fig. 6.28. Reconstructed life appearance of the hippopotamid *Archaeopotamus harvardi*. Reconstructed length: 3 m.

Known from a fragmentary skeleton, *Leptorycteropus* was a primitive aardvark that was considerably smaller and of lighter build than the modern species (*Orycteropus afer*). Like the extant aardvark, the vertebral column was arched, although the limbs were less robust and more elongated. As evident from the length of the tooth row, the snout would have been somewhat shorter. These features indicate that *Leptorycteropus* had a less specialised diet and perhaps less specialised digging abilities.

Among proboscideans, the primitive elephantid *Stegotetrabelodon* had a pair of straightened upper tusks and a long, slender mandibular symphysis that supported a pair of long lower tusks protruding forwards and downwards, reminiscent of the related "gomphotheriids". This animal would have typically carried its head fairly elevated as indicated by the low-set **occipital condyles** along with the notable downwards slant of the symphysis. Despite the **brachydont** molars, their **lamellar** structure suggests a primarily grass-based diet. As the Miocene ended, the **topographic** alterations (i.e. uplift of the Great Rift Valley) in East Africa were adequate to begin having a significant effect on the regional climate. Temperatures drop with altitude, and mountains force air to rise, causing it to cool and release moisture. This leads to conditions favourable for rainforest on one side and a rain shadow on the other. The evolution of Africa's modern fauna regarding composition and structure was taking shape during the Late Miocene, with immigration and extinction—as well as in-situ evolution of new species—being vital factors. The fossils from Nawata are of great importance in understanding this trend and the transition of forest habitats to the more open forest-savannah mosaics that would distinguish East Africa during most of the Pliocene and Pleistocene.

1. *Eurygnathohippus turkanense*, a hipparionin equid.
2. *Nyanzachoerus tulotus*, a tetraconodontine suid.
3. *Tragelaphus kyaloae*, an extinct species of the extant spiral-horned antelope genus *Tragelaphus*.
4. The extant impala (*Aepyceros melampus*).
5. The extant Nile crocodile (*Crocodylus niloticus*).
6. *Stegotetrabelodon syrticus*, an elephantid.

Fig. 6.29. A scene at Lothagam, Kenya, during the Late Miocene.

Australia Riversleigh Fossil Site (Faunal Zones B-D), Queensland: Aquitanian-Zanclean, ca. 23.03–5.332 Ma.

By far the richest and most diverse fossil assemblages at the Riversleigh world heritage site (see pp. 244–5) are from the Miocene deposits. Among placental mammals, over 35 fossil bat species have been found from these deposits, which is a world record, and consist of microbats, along with relatives of modern ghost bats and leaf-nosed bats. The monotremes (see p. 44) are represented by a single family, the platypuses (Ornithorhynchidae). The marsupials are represented by a total of over 150 species divided into five orders: the Dasyuromorphia (including **primitive** thylacinids), Notoryctemorphia (marsupial moles), Peramelemorphia (bandicoots), Diprotodontia and the extinct Yalkaparidontia. The Diprotodontia are further subdivided into three suborders: the macropodiforms (including bettongs and a great variety of "**archaic**" forms, some of which were rat-sized, others **carnivorous**), the vombatiforms (including over a dozen families, such as medium- to large-sized diprotodontids and palorchestids, primitive koalas and thylacoleonids) and phalangeriforms (including a diversity of possums and cuscuses) (see pp. 51–7). It is outside the scope of this book to review all these species in detail, so I will focus on some of the better known, more spectacular and more unusual **taxa** instead.

Three species of the Late Oligocene to Late Miocene monotreme genus *Obdurodon*, which strongly resembles the extant platypus *Ornithorhynchus*, were present and were easily separated by size. They kept teeth into adulthood, unlike the extant platypus, which solely possesses teeth as a juvenile. This indicates *Obdurodon* may have fed on different prey to *Or. anatinus* (the large *Ob. tharalkooschild* is believed to have crunched down freshwater turtles!). The differences in skull and bill morphology between *Obdurodon* and *Ornithorhynchus* indicates that the former fed in the water column or surface, instead of on the riverbed like the latter. Among the dasyuromorphs, the thylacinids included Middle Miocene genera more advanced than *Badjcinus* (see pp. 244–5), which they had replaced. *Muribacinus* was a **quadrupedal**, fox terrier-sized thylacinid (1.5 kilograms), somewhat resembling a civet. Some distinguishing features include a relatively short **rostrum** (Wroe, 1996). The crests and cusps of the molars were reduced or lengthened to form a shearing blade, specialisations that were also present in *Mutpuracinus*. *Mutpuracinus* is the smallest known thylacinid (quoll-like in size and appearance) with a body mass of roughly 1.1 kilograms. *Mutpuracinus* is distinguished from other thylacinids by features such as premolar crowns that were more **posteriorly**-directed. *Muribacinus* was a generalist carnivore, while *Mut. archibaldi* was **hypercarnivorous**.

Nimbacinus was a thylacinid similar in size and appearance to a fox, weighing around 5.3 kilograms. *Nimbacinus* had a more specialised ear region than its earlier relatives such as *Badjcinus*. Contrary to other early thylacinids, the jaws of *Nimbacinus* were likely powerful enough to subdue prey larger than itself. However, it was less specialised for carnivory than the modern thylacine (*Thylacinus cynocpehalus*). A well-preserved skull has been used to determine that like the modern thylacine, it was likely a swift runner and relied on endurance rather than speed to capture prey. *Nimbacinus* was likely an opportunistic hunter and not hypercarnivorous. The Middle Miocene hypercarnivorous dasyurid *Whollydooleya*, known only from an isolated lower second or third molar, was a distant relative of the Tasmanian devil (*Sarcophilus harrisii*). From this single tooth, *Whollydooleya* is estimated to have been two to three times larger than a possum, weighed 20 to 25 kilograms (twice as heavy as *S. harrisii*) (Archer *et al.* 2016). With a dentition unique among known mammals (a massive, dome-like upper third premolar), the ferret-sized dasyuromorph *Malleodectes*, from the Middle to Late Miocene, presumably held a niche unlike that held by any other mammal, extinct or extant. In fact, its teeth most closely resembled the modern slender bluetongue skinks (*Cyclodomorphus*) of Australia, which are mostly specialised for eating snails. As such, *Malleodectes* is commonly proposed to have fed largely on snails, although the size and strength of its jaws and teeth suggests it also dined on other prey (such as small vertebrates) at least occasionally.

The marsupial moles are represented by the Early Miocene *Naraboryctes*, which document the mosaic evolution of skeletal and dental characters of the modern genus (*Notoryctes*) from a progenitor that mostly lived above the ground. Unlike the modern genus, primitive marsupial moles such as *Naraboryctes* were adapted to living in humid, forested environments, which is where they likely originated (Archer *et al.* 2011). Among peramelemorphs, *Crash bandicoot*, from the Middle Miocene, is regarded as the oldest known peramelid according to its dental characters. It is known only from a fragmentary right **maxilla** with the first three molars complete. It must not have had a dissimilar appearance to modern bandicoots and is estimated to have weighed about one kilogram. The Oligo-Miocene *Yarala* is the smallest known peramelemorph, which would have hunted for invertebrates in the forest leaf litter (Fig. 6.34). Some cranial features of *Yarala* present in bandicoots include nasal bones that are narrow across their whole length, and a Y-shaped jugal bone (part of the **zygomatic arch**) that connects to the maxilla. The snout of *Yarala* was relatively shorter than other bandicoots. *Yarala* possibly had an **insectivorous**-carnivorous diet according to its dental structure, intermediate between that of dasyurids and more specialised bandicoots.

Among diprotodontians, the macropodiforms included primitive macropodids such as the Late Oligocene to Early Miocene *Cookeroo*. *Cookeroo* was small, about the size of a Tammar wallaby (*Macropus eugenii*), and primarily moved around quadrupedally on the rainforest floor, possibly bounding like tree-kangaroos (*Dendrolagus*). The Middle to Late Miocene *Ganguroo* was anatomically similar to both macropodids (e.g. **bilophodont** molars) and potoroids (e.g. long, narrow rostrum), but most likely belongs to the latter family. Over time, *Ganguroo* species increased in body size and had an increasingly reduced number of premolar cusps, likely in response to shifts in climate and plant life during the Middle Miocene (Cooke *et al.* 2015). The Early to Middle Miocene balbarid (see p. 244) *Balbaroo* (Fig. 6.31), about the size of a domestic cat, had unusually large canine teeth forming prominent "fangs", which were possibly used for display and **intraspecific competition** as in some modern **ungulates** such as water deer (*Hydropotes inermis*). Its overall limb proportions suggest that its primary mode of locomotion was quadrupedal bounding and slower-speed hopping.

Fig. 6.30A. Phylogenetic tree of a sampling of fossil mammal taxa, specifically monotremes, and marsupials outside Diprotodontia, from Faunal Zones B–D of the Riversleigh deposits. The phylogeny within Thylacinidae follows Rovinsky *et al.* (2019).

Australia Riversleigh Fossil Site (Faunal Zones B–D), Queensland: Aquitanian–Zanclean, ca. 23.03–5.332 Ma, cont.

The Late Oligocene to Miocene genus *Ekaltadeta* was a giant (5 to 10 kilograms) early relative of the musky rat-kangaroo (*Hypsiprymnodon moschatus*) and was possibly **omnivorous**, with a preference for meat. This inference is largely based on its **plagiaulacoid** third premolars. The lower first incisors were elongated, suggesting a stabbing or gripping function, potentially for capturing and handling prey. Unlike the premolars, the molars had relatively large, horizontal **occlusal** surfaces with a **bunodont** morphology, characteristic of omnivores. *Ekaltadeta* can therefore be viewed as an opportunistic feeder, ecologically categorised somewhere between foxes and wild boars. The vombatiforms included a variety of primitive, rainforest-adapted koalas, all from the Early to Middle Miocene. *Nimiokoala* was one third the size of the extant species (*Phascolarctos stirtoni*). *Nimiokoala* likely had more generalised feeding habits than extant koalas, and had a more elongated, possum-like snout. *Nimiokoala* had a **selenodont** dentition, with a large number of cusps and accessory shearing edges. It had the most intricate molars of any known fossil koala, and may have included more fibrous vegetation in its diet than other contemporary koalas such as the similar-sized genus *Litokoala*.

Litokoala seems to be more closely related to the extant koala than *Nimiokoala* was, and exhibits many cranial similarities to the extant species indicating a face that was koala-like instead of possum-like. Both *Nimiokoala* and *Litokoala* appear to have been substantially more active and agile tree climbers than modern koalas, as evident from the small body size and proportionately large **orbits**, which suggest excellent night vision (Black *et al.* 2014). Even larger but more basal than either genus was *Priscakoala*, distinguished from all other koalas by the more **brachydont** teeth and absence of crenulations on the enamel surface. The Late Oligocene to Miocene *Nimbadon* was a koala-like and sheep-sized zygomaturine diprotodontid represented by 24 complete **articulated** skeletons (Black *et al.* 2012). It was a proficient climber as indicated by its flexible limb girdles, proportionately long arms, and semi-opposable **digits**. It is also the largest known **arboreal** marsupial from Australia, weighing up to 70 kilograms. As indicated by its **lophodont cheek teeth**, *Nimbadon* was a **browser**, consuming primarily leaves and stems. *Nimbadon* possessed elongated feet with large, retractable claws that were larger and more robust than those of koalas, suggesting it was even more adapted for climbing and gripping. The zygomaturine *Neohelos wortmani*, represented by numerous specimens, was typical of the earliest diprotodontid species from the Late Oligocene to Early Miocene, being barrel-bodied and with a passing resemblance to a non-burrowing wombat. Some of the distinguishing features include upper first molars with a square-shaped occlusal profile and four-cusped upper third premolars. The Early to Middle Miocene *Silvabestius* was a sheep-sized browsing zygomaturine represented by two well-preserved skulls discovered closely together, likely a mother and a juvenile. Some researchers believe that this is particularly a foetus in the womb of its mother who died before birth. *Silvabestius* is distinguished from other zygomaturines by the retention of small upper canine teeth. Unlike later zygomaturines, the maxilla of *Silvabestius* lacked a constricted **diastema** (Black and Archer, 1997).

The primitive palorchestid *Propalorchestes* was a ground-dwelling **herbivore** that had already developed the highly retracted nasal morphology typical of palorchestids. *Propalorchestes* is distinguished from the genus *Palorchestes* (see pp. 333–4) in being smaller and possessing features such as a narrower neurocranium, a shorter and more robust lower jaw, a shorter, deeper and broader maxilla, and unspecialised forelimbs (Black, 2006). The true wombats (vombatids) were represented by at least two genera from Riversleigh. The primitive genus *Rhizophascolonus*, known from the Late Oligocene to Early Miocene, was a medium- to large-sized vombatid which had developed **hypselodont** upper incisors by the Early Miocene. However, its cheek teeth retained roots, a feature not seen in later vombatids. This makes it an intermediate form in the evolution of wombat dentition.

The most primitive thylacoleonids, of Early Miocene age, are also known from Riversleigh. *Microleo* is the smallest and most basal known thylacoleonid, characterised by features including the shape and morphology of the upper third premolars (e.g. a protuberant posterior cusp), and weakly-crenulated molars (Gillespie *et al.* 2016). *Lekaneleo* was a slightly larger thylacoleonid distinguished by features such as a reduced **sagittal crest** and a relatively wide outer ear canal (Gillespie *et al.* 2019). *Microleo* may have competed with larger thylacoleonids such as *Lekaneleo* for similar prey, but the smaller size of the former probably allowed access to higher parts of

Fig. 6.31. Reconstructed life appearance of the balbarid macropodiform *Balbaroo fangaroo*. Reconstructed length: 60 cm (excluding tail).

Fig. 6.30B. Phylogenetic tree of a sampling of fossil mammal taxa, particularly marsupials within Diprotodontia, from Faunal Zones B–D of the Riversleigh deposits. The phylogeny within Phascolarctidae and Thylacoleonidae follows Gillespie *et al.* (2016).

Australia Riversleigh Fossil Site (Faunal Zones B–D), Queensland: Aquitanian–Zanclean, ca. 23.03–5.332 Ma, cont.

the forest canopy. *Microleo* and *Lekaneleo* possibly hunted a variety of insects, birds and lizards found in the forest canopy, while remaining out of reach of the larger, closely-related predators such as the genus *Wakaleo*. The largest known predatory mammal genus at Riversleigh, *Wakaleo* was a dingo-sized thylacoleonid characterised by the loss of the first premolar and lower canine teeth. Its sturdy skull was shorter and broader than that of *Lekaneleo*, but longer and narrower than the later *Thylacoleo* (see pp. 296–7, 333, 335). *Wakaleo* probably lived as **scavengers** on the forest edges.

Among phalangeriforms was the primitive ektopodontid *Chunia*. Ektopodontids had massive, forward-facing orbits (possibly for **nocturnal** vision), short snouts, and the most distinctive and elaborate teeth of all known marsupials (e.g. molars with transverse **lophs** divided into patterns of struts or ribs distributed longitudinally or radially). Compared to other ektopodontids, *Chunia* had simpler molars that resembled those of phalangeroid possums. Quite possibly the strangest marsupial from Riversleigh is the possum-like Oligo-Miocene species *Yalkaparidon*, nicknamed "Thingodonta" (Fig. 6.35). The exact phylogenetic position of *Yalkaparidon* is uncertain given its unusual morphology: it has a highly primitive basicranium in the skull resembling that of bandicoots, molars that are morphologically similar to those of marsupial moles, and massive, hypselodont incisors. Although the incisors are diprotodontian-like, no other characteristics provide compelling evidence for those affinities. Thus, *Yalkaparidon* is presently grouped in its own order (Yalkaparidontia). *Yalkaparidon* is suspected to have eaten earthworms and probably caterpillars according to molar characters shared with Madagascan worm-eating tenrecs. In addition to mammals, most of the other vertebrates still living in Australia today were present in the Early to Middle Miocene rainforests at Riversleigh,

Fig. 6.32. Reconstructed life appearance of the madtsoiid snake *Yurlunggur camfieldensis*. Reconstructed length: 6 m.

consisting of several types of flightless birds such as primitive ratites (see p. 32), along with cockatoos, bowerbirds, lyrebirds and a myriad of passerines. The ratite *Emuarius* is sometimes known as the "emuwary" for exhibiting anatomical similarities to both emus (in the feet and lower leg bones) and cassowaries (in the skull and femur) (Fig. 6.33A). Thus, it has a phylogenetic position between emus and cassowaries. The avifauna also included the extinct dromornithids (see p. 36) which likely had better-developed vision and a broader range of dietary preferences than ratites. The fleet-footed *Barawertornis* (80 to 95 kilograms) is the smallest known species, roughly the size of a cassowary (Fig. 6.34). *Dromornis*, reaching an impressive three metres tall and weighing in around 500 kilograms, had a particularly large skull and deep beak, unlike those seen in ratites (Fig. 6.33B). In fact, its cranium was wider and higher than it was long. *Dromornis*

Fig. 6.33. Reconstructed life appearance of a sampling of large, flightless birds from the Riversleigh deposits. (A) the ratite *Emuarius gidju*, from the Late Oligocene to Early Miocene; (B) the dromornithid *Dromornis stirtoni*, from the Late Miocene. Scale bar: 1 m.

1. *Gawinga aranaea*, an ancient ringtail possum.
2. *Archerops annectens*, an extinct genus of leaf-nosed bat.
3. *Trichosurus* sp., the extant brushtail possum.
4. A gecko (Gekkonidae indet.).
5. *Barawertornis tedfordi*, a small, primitive dromornithid.
6. *Cookeroo hortusensis*, a small, basal macropodid kangaroo.
7. *Australlus disneyi*, an extinct member of the rail family (Rallidae).
8. *Yarala burchfieldi*, a bandicoot-like marsupial of the extinct family Yaralidae.
9. *Burramys brutyi*, an extinct relative of the extant mountain pygmy possum (*Burramys parvus*).
10. The extant Australian tree frog genus *Litoria*.
11. *Neohelos wortmani*, a primitive diprotodontid.
12. A pair of logrunners (passerine birds **endemic** to Australasia) of the genus *Orthonyx* perched on a liana.
13. *Corvitulasoides grandiculus*, a large, extinct songbird, intermediate in size between lyrebirds and ravens.
14. *Nimiokoala greystanesi*, a primitive koala.

Fig. 6.34. A rainforest floor **ecosystem** at Riversleigh during the Early Miocene.

Fig. 6.35. Reconstructed life appearance of the "Cohen's Thingodonta" *Yalkaparidon coheni*. Reconstructed length: 20 cm.

Australia Riversleigh Fossil Site (Faunal Zones B–D), Queensland: Aquitanian–Zanclean, ca. 23.03–5.332 Ma, cont.

also had larger, more muscular hindlimbs, built for strength and stability to support its large body weight.

The lower vertebrates included goannas, madtsoiid snakes (see pp. 23, 25), skinks, crocodylians and pleurodire turtles (see pp. 14–5), along with plentiful frogs. *Yurlunggur* (Fig. 6.32), named after a serpent in aboriginal mythology, was adapted to living both in burrows and possibly **semi-aquatic** environments. Known from a complete skull and lower jaw, its anatomy indicates that it did not evolve from small **fossorial** lizards but from ancestors that resembled monitor lizards (Palci *et al.* 2018). The crocodylians were represented by an early radiation of mekosuchines (see p. 28) on the continent (Fig. 6.36). The largest mekosuchine was *Baru*, otherwise known as the "cleaver-headed crocodile", which was a more generalised form distinguished by a very broad skull, deep jaws and a highly emarginated tooth row. Its teeth were long, blade-like and slightly recurved, adapted for a diet of large vertebrates. One skull was found with thylacoleonid bones in its mouth, so *Baru* may have been the apex predator of its ecosystem. Its lifestyle remains largely unknown until postcranial material is described. *Mekosuchus* and *Trilophosuchus* fitted the small predatory niches that would be taken later by monitor lizards. The goanna-like *Mekosuchus* had large orbits, specialised rear teeth for crushing shellfish and a limb bone structure that may suggest an arboreal lifestyle, although there is no evidence for climbing abilities in *Trilophosuchus*, which was most probably a **terrestrial** hunter. In *Trilophosuchus*, the neck musculature, and the inclination of the posterior skull, indicates that the animal raised its head high, a posture more characteristic of terrestrial animals. *Trilophosuchus* was an unusual mekosuchine, with a short, deep skull, a blunt snout, large, laterally-directed orbits, and three distinct crests extending along the length of the skull roof.

Throughout the Early Miocene, Australia was in one of its lushest greenhouse intervals, which persisted until the latest Middle Miocene, with the final members of relatively "archaic" lineages (e.g. balbarids, wynyardids) coexisting with the earliest members of numerous modern lineages. After the Middle Miocene, Australia gradually became drier as the continent drifted northwards towards a zone of low rainfall. From around 15 Ma, Australia experienced a significant cooling trend (McGowan and Li, 1994), with a steady decrease in both temperatures and rainfall, plunging Australia into the onset of one of its most sudden and profound cool phases. This cooling trend was punctuated by a minor greenhouse interval during the Early Pliocene, but otherwise continued towards the modern climate. The rising aridity led to a reduction of the northern and central rainforests, with grasslands becoming more dominant, although truly arid habitats that characterise much of Central and Western Australia today did not fully develop until the Plio-Pleistocene. Riversleigh shows that by the Miocene-Pliocene boundary, the marsupial composition of Australia remained broadly similar, but with new taxa (and very few left from Faunal Zones B and C), and shifts in the ecological roles of some groups. Overall, this cooling trend significantly impacted the evolution of most mammal lineages, with some forest-adapted ones going extinct only to be replaced by others that diversified in these more arid environments (e.g. kangaroos and wombats). Over the last 15 million years, many mammal lineages, along with some birds and reptiles, underwent gigantism, possibly because larger herbivores could consume more of the lower quality vegetation which, in turn, may have selected for larger carnivores that could more effectively hunt them, eventually culminating in the iconic megafauna of Pleistocene Australia (see pp. 332–5).

Fig. 6.36. Reconstructed life appearance of various mekosuchine crocodylians from Riversleigh. (A) The "cleaver-headed crocodile" *Baru darrowi*; (B) *Mekosuchus sanderi*; (C) *Trilophosuchus rackhami*.

Pliocene
5.333–2.58 Ma

Fig. 6.37. Palaeogeographic map reconstruction of Earth during the Pliocene, with the position of the modern continents outlined and the approximate location of the fossil sites indicated.

While the Pliocene Epoch saw significant cooling and the beginnings of widespread Northern Hemisphere glaciation, it represented a transition towards the full icehouse state that became dominant in the Pleistocene. The onset of the Pliocene is often marked by a period known as the "Pliocene Climatic Optimum" (PCO), a time characterised by globally warmer sea surface temperatures than today, and a reduced temperature gradient between the poles and the Equator. This period (ca. 4.4–4 Ma) saw average global temperatures potentially as high as 4°C warmer than at present. Then around 3.2 Ma, the Northern Hemisphere experienced a significant cooling event, possibly involving the expansion of the Greenland ice sheet. This led to a stronger latitudinal temperature gradient and the Sahara's initial aridification pulse. Following this initial glacial episode was a rise in average global temperature (about 2–3°C higher than today) occurring roughly 3.1–3 Ma called the mid-Pliocene Warm Period, partly caused by increased greenhouse gas concentrations.

After this interval, at around 2.6 Ma, the Earth's climate ultimately shifted from the generally warm and humid conditions that characterised most of the Cenozoic Era as the entire globe, especially the Northern Hemisphere, plunged into a new era of frequent, large-scale glaciations that drastically altered landscapes and **ecosystems**. Unlike earlier episodes of Pliocene cooling, a distinct glacial-interglacial dynamic (see p. 313) had become secured from about 3 Ma, with glacial-interglacial cycles occurring in 41,000 year cycles. The Pliocene's initial glacial pulses, though less extensive than later Pleistocene ice sheets, significantly influenced middle and northern latitude climates, particularly through increased seasonality and aridification. Thus, the initial glaciation at the end of the Pliocene brought about the initial shift in vegetation from woodlands to tundra-like open landscapes in much of Northern Eurasia. These climatic changes necessitated new survival strategies, including hibernation in rodents, and seasonal migrations and herd behaviour in **ungulates**. Additionally, at around 2.8 Ma, the connection between North and South America was finally completed, forming the Panama Isthmus and closing the Central American Seaway (see p. 174), a process that some studies (Vallejo-Hincapié *et al.* 2024) have estimated as having started during the Middle Miocene (ca. 15–13 Ma). This closing affected the world's climate by changing the circulation of ocean currents. Warm equatorial ocean currents were cut off, which initiated an Atlantic cooling cycle, with cold Arctic and Antarctic waters dropping temperatures in the now-isolated Atlantic Ocean. The blockage of Panama provided the Arctic with moisture, which is naturally cold because of the low amount of solar input it receives year-round. All the North Pole requires to form a floating ice cap is a source of moist clouds capable of precipitating snow and ice. Therefore, it seems that the formation of the Panama Isthmus contributed to the growth of the Arctic ice cap. The formation of the isthmus also caused fauna and flora that were once separated to immigrate freely between North and South America in an event called the Great American Biotic Interchange (GABI), significantly impacting local ecosystems, especially in the south. Animals such as cats, dogs, tapirs, deer and peccaries dispersed southwards, while opossums, anteaters, armadillos, sloths, porcupines, caviomorph rodents (see pp. 146–8) and hummingbirds moved northwards (Fig. 6.2). Additionally, the Pliocene witnessed various other global climatic changes including the expansion of **savannahs** (particularly in Africa and Asia), semi-deserts in Central Asia, and monsoons in India. By the end of the Pliocene, desert had likely developed in the Western Sahara due to the intense Northern Hemisphere glaciation.

Australia Allingham Formation, Queensland: Zanclean, 5.3-3.6 Ma.

The Allingham Formation is located at the Bluff Downs fossil site on the banks of Allingham Creek in the Cape York Peninsula, northern Queensland, and is among Australia's most significant Pliocene fossil sites partly because of its distinctive fauna. In the Pliocene, the palaeoenvironment at Bluff Downs consisted of extensive waterways surrounded by riparian forests. The presence of lake and river sediments, along with the fossils of darters and pygmy geese, is evident of large permanent water bodies. A shallow, turbid lagoon was part of the ancient **topography** as indicated by the presence of long-necked testudinids and short-necked chelids (see pp. 14–5). The fossils of macroalgae (*Chara*) and **osteichthyan** fish indicate the presence of persistent freshwater in the area, and the existence of complex rivers and creeks with diverse aquatic life is supported by fossils of short-necked chelids. Flamingos, which are now extinct in Australia likely due to the gradual disappearance of permanent water bodies, waded in the shallows for food. There were high levels of humidity and rainfall in this region, and the overall environment was comparable to Kakadu National Park today. There were pockets of more closed forest as evident from fossils of mammals adapted for both **terrestrial** and **arboreal** locomotion, including ringtail possums. Those forests would have constituted a small part of the **ecosystem**, as indicated by the scarcity of possum fossils.

Following a trend that was to peak during the Pleistocene, many mammals and reptiles of Pliocene Australia grew to massive sizes. Among marsupials is the horse-sized (roughly 500 kilograms) diprotodontid (see p. 56) *Euryzygoma*, characterised by a deep, narrow snout and unusual, laterally expanded cheek bones. The tips of those extensions are not rugose, indicating that *Euryzygoma* lacked horny growths. The cheekbones are better-developed in some fossil skulls but less so than in others, a possible indicator of **sexual dimorphism**, with only the males having broad, flaring cheekbones. Those extensions were most probably used for sexual display. Numerous characteristics in the skull and teeth, such as rugose molar enamel and incipiently horseshoe-shaped crowns on the upper third premolars, indicate close phylogenetic affinities and potential ancestry to the genus *Diprotodon* (see pp. 333–5). *Euryzygoma* was a **browser**, eating primarily leaves.

The giant vombatid (see p. 56) *Phascolonus* is the largest known **fossorial** marsupial. In addition to overall size, *Phascolonus* is distinguished by a relatively large skull with a distinct inward depression on the cranial roof. The limbs were stout and powerful as well as short. Compared to extant wombats, the **distal** end of the humerus in *Phascolonus* was relatively broader, indicating that this huge wombat was just as adept at burrowing as extant species. As in other wombats, the molars of *Phascolonus* were **hypselodont**, bi-lobed and curved, although the incisors were considerably broad other than being long and curved. It had strap-shaped upper incisors, a unique feature among wombats. *Phascolonus* most probably fed primarily on leaves and grasses, whereas extant wombats typically consume leaves, roots and shoots. *Koobor* is an enigmatic vombatiform whose phylogenetic position has been a subject of debate due to the limited fossil material, but has been placed within Phascolarctidae (see pp. 56–7) (e.g. Black *et al.* (2014)). Known only from three molars and a partial upper jaw, *Koobor* is characterised by an amalgamation of features including narrow, elongated molars with a **selenodont** condition, indicating a browsing diet. Early relatives of modern kangaroos and bandicoots thrived here too, as did *Thylacoleo crassidentatus*, a precursor of the leopard-sized Pleistocene thylacoleonid (see p. 57) *T. carnifex* (see pp. 333, 335). The macropodids (see pp. 54–6) are only represented by fragmentary jaws and isolated teeth at Bluff Downs, represented by species such as the moderately large *Macropus dryas*, distinguished by elongated upper third premolars with a near straight longitudinal crest, and **hypsodont** molars indicating that it had an abrasive diet.

Perameles allinghamensis, known only from an isolated upper molar, was a very large (about 2.19 kilograms) member of the modern bandicoot family Peramelidae (see pp. 52–4). It may have been an **omnivore**, possibly occupying a niche more similar to the northern brown bandicoot (*Isoodon macrourus*) than any other species of *Perameles*, largely foraging for insects from the forest soil. *T. crassidentatus*, known only from teeth and fragmentary jaws, was a medium-sized thylacoleonid (roughly beagle-sized), distinguished from *T. carnifex* by features such as the **plagiaulacoid** last premolar being less **carnassial**-like, and a considerably wider mandibular ramus in the area of **cheek tooth** implantation, as well as smaller body size. *T. crassidentatus* was possibly at least partially arboreal and fed on smaller prey than *T. carnifex*. The teeth of *T. crassidentatus* was adapted for shearing meat, although they were possibly less specialised than those of *T. carnifex*.

A variety of kangaroos would certainly have fallen prey to formidable **semi-aquatic** crocodylians of the genus *Paludirex*. *Paludirex* was a medium-sized mekosuchine (see p. 28) characterised by large, conical teeth, upwards-directed nostrils and a proportionately extremely broad snout (broader than in other mekosuchines), which suggest that it was a specialised **ambush predator** in shallow water like many modern crocodylians. The skull of *Paludirex* was flatter and less robust than in the genus *Baru* (see p. 294) and the contemporary *Quinkana* (see pp. 334–5). As in *Baru*, the **premaxilla** of *Paludirex* was slightly sloped and did not terminate in an almost vertical line. The teeth of *Paludirex* are characterised by minor crenulations that were likely not true serrations, unlike the condition in *Quinkana*.

Known solely from isolated vertebrae, *Liasis dubudingala*, commonly referred to as the Bluff Downs Giant Python, would have made short work of relatively large **herbivores** such as juvenile diprotodontids, as indicated by its body size as well as the palaeoenvironment and favourable climate. *L. dubudingala* was possibly the largest Australian snake ever, rivalling modern reticulated pythons (*Malayopython reticulatus*) and anacondas (*Eunectes*) in size. *L. dubudingala* is placed within the extant genus *Liasis* based on the extraordinarily high neural spines of the vertebrae, a feature also present in *L. mackloti* (the water python) and *L. olivaceus* (the olive python). The tall neural spines (which provide a larger surface area for muscle attachment, especially those involved in climbing) and overall vertebral proportions are a likely indicator that *L. dubudingala* was partially arboreal. Like all Australian pythons, *L. dubudingala* likely incubated its eggs by coiling around them before they hatched.

Fig. 6.38. Phylogenetic tree of a sampling of fossil mammal **taxa** from the Allingham Formation. The phylogenetic position of *Koobor jimbarrati* follows Black *et al.* (2014).

South America Chapadmalal Formation, Argentina: Zanclean-Piacenzian, 4.5–3.3 Ma.

The rocks of the Chapadmalal Formation, situated along the coast of eastern Argentina, were deposited when the region was a warm, somewhat dry **savannah**, with rodents and xenarthrans (see p. 58) comprising the majority of mammal species (almost two thirds) at the site. Some of those species are still extant today, such as the chinchillid *Lagostomus* (viscachas), the caviid *Microcavia* (mountain cavies) and the big hairy armadillo (*Chaetophractus villosus*). These include the geologically earliest fossils of any extant South American **terrestrial** mammal. In general, the mammal fossils are preserved in beds of fine-grained, **aeolian** sediment called **loess**, which are commonplace in the Late Neogene fossil sites of Patagonia. These loess-derived deposits have also preserved fossil burrows, some of which contain bones of their supposed residents. **Primitive** ctenomyid rodents (which today include tuco-tucos), small hegetotheriid notoungulates (see pp. 140–3) of the genus *Paedotherium* and highly specialised metatherians (see pp. 43, 46–7) of the genus *Argyrolagus* were some of the most abundant burrowers, and various kinds of small rodents opportunistically inhabited these burrows or alternatively built their own. Some giant terrestrial rodents from Chapadmalal included dinomyids (see pp. 147–8) of the cow-sized genus *Telicomys*, most closely related to the pacarana (*Dinomys branickii*). *Telicomys* had unusually wide incisors, a very flat **occipital** and deep masseteric **fossae** (muscle attachment **adaptation** for strong bite forces).

Paedotherium was clearly an adept digger as most of its fossils have been discovered in ancient burrows. Yet, it was capable of moving really fast, employing a specialised hare-like mode of locomotion called the half-bound gait. As in all hegetotheriids, *Paedotherium* had very simple, **hypselodont cheek teeth** with specialisations for shearing plant material, possibly in both wooded and open habitats. However, it remains unclear how the lifestyle of *Paedotherium* was different from those of similar-sized caviomorph rodents such as chinchillids and caviids (see pp. 146–8) that also possessed hypselodont dentitions. Whether *Argyrolagus* was a polydolopimorphian (see p. 48) or a paucituberculatan (see p. 51) remains to be resolved, due to its unusual teeth, being highly simplified and hypselodont, along with the highly specialised skull and postcrania. Another issue is that the known skulls do not preserve the ear region/basicranium, which is often very useful for inferring phylogenetic relationships. *Argyrolagus* was a desert-adapted **herbivore** with reduced forelimbs and elongated hindlimbs, making it superficially jerboa-like. *Argyrolagus* had **didactyl** feet, a long, pointed snout, and large **orbits** and ears that suggest it was possibly **nocturnal**. Its elongated, **saltatorial** hindlimbs were likely well-suited for **bipedal** leaping, and functional analysis (Abello *et al.* 2020) suggests that it had a **fossorial** lifestyle. Its dental morphology indicates that it fed on arid-adapted plants.

The glyptodontine (see pp. 59–60) genus *Panochthus* was a 1.5 tonne herbivore with a dome-shaped **carapace** consisting of hundreds of rounded scutes. The skull was relatively deeper and shorter than other glyptodontines, with a retracted nasal opening. The stubby, wedge-shaped tail was composed of osseous bands armed with relatively short spikes for defence. In one specimen, the tracheal rings were preserved, which are somewhat flattened ventrally and dorsally, and most **analogous** to those of the California sea lion (*Zalophus californianus*) among extant mammals (Zamorano, 2020).

Important modifications occurred in the **carnivore guild** of South America from the Pliocene onwards. Once the sparassodonts (see pp. 48–50) had vanished during this interval, placental carnivorans (see pp. 81, 83–4) invaded to occupy the continent's terrestrial predatory niches. This shift served as an opportunity for other mammal groups to take up a predatory way of life, such as the large armadillo *Macroeuphractus* and an extinct lineage of opossums within the modern Didelphidae (see p. 51) called the Sparassocynini. *Macroeuphractus* had a sturdy skull with well-developed jaw muscles, and large, conical **caniniform** teeth indicative of a predatory lifestyle. *Macroeuphractus* was fossorial like many armadillos and thus was possibly able to penetrate into the burrows of hegetotheriids and caviomorph rodents to prey on defenceless young. There are no armadillos today specialising in carnivory, although the majority are generalist **omnivores**, and certain species such as the six-banded armadillo (*Euphractus sexcinctus*) have the ability to kill and feed on small vertebrates. Recent phylogenetic analyses on armadillos have confirmed that *Macroeuphractus* has closer relations with pampatheriids (see pp. 59–60) and glyptodontines than with modern armadillos. If true, these predatory behaviours would have evolved numerous times among armadillos.

Sparassocynus was particularly small for an opossum, and its dentition was more suited for cutting flesh and less for grinding other foodstuffs compared to other opossums. *Sparassocynus* had a broader skull and shorter muzzle than modern opossums. A rather large tympanic bulla is one distinctive characteristic of *Sparassocynus*. Various lineages of rodents and tiny marsupials such as argyrolagids developed bullae the same size as or larger than that of *Sparassocynus*, possibly as a specialisation for perceiving sounds of specific frequencies. This characteristic assists small mammals in detecting and eluding predators. The lifestyle of *Sparassocynus* may have been analogous to that of small dasyurid marsupials (see p. 52) from Australia called mulgaras (*Dasycerus*), preying on insects, other arthropods and small vertebrates, and constructing burrows for refuge. Among other predators, the phorusrhacids (see pp. 41–2) such as the mesembriornithine *Llallawavis* persisted, among the last in South America. *Llallawavis* is known from an exceptionally preserved and near complete **articulated** skeleton preserving eye bones, tracheal rings and even ossified tendons (Degrange *et al.* 2015). CT scans reveal that the inner ear was fine tuned to hear relatively low frequencies, and that *Llallawavis* possibly produced low-pitched sounds, seemingly for intraspecific communication (e.g. during courtship) or sensing swift and nimble prey. Contrary to modern birds, the cranial bones were fused together, which is regarded as an adaptation for withstanding stresses on the skull when dispatching larger-bodied prey such as small notoungulates. Additionally, four species of small mammals from the Chapadmalal Formation were part of a significant modern lineage that had dispersed to South America while the phorusrhacids immigrated northwards—the sigmodontine rodents, referred to as New World rats and mice. These comprise the most diverse subfamily of the Cricetidae (see p. 147), and were some of the first North American immigrants as part of the Great American Biotic Interchange, or GABI (see pp. 252, 295).

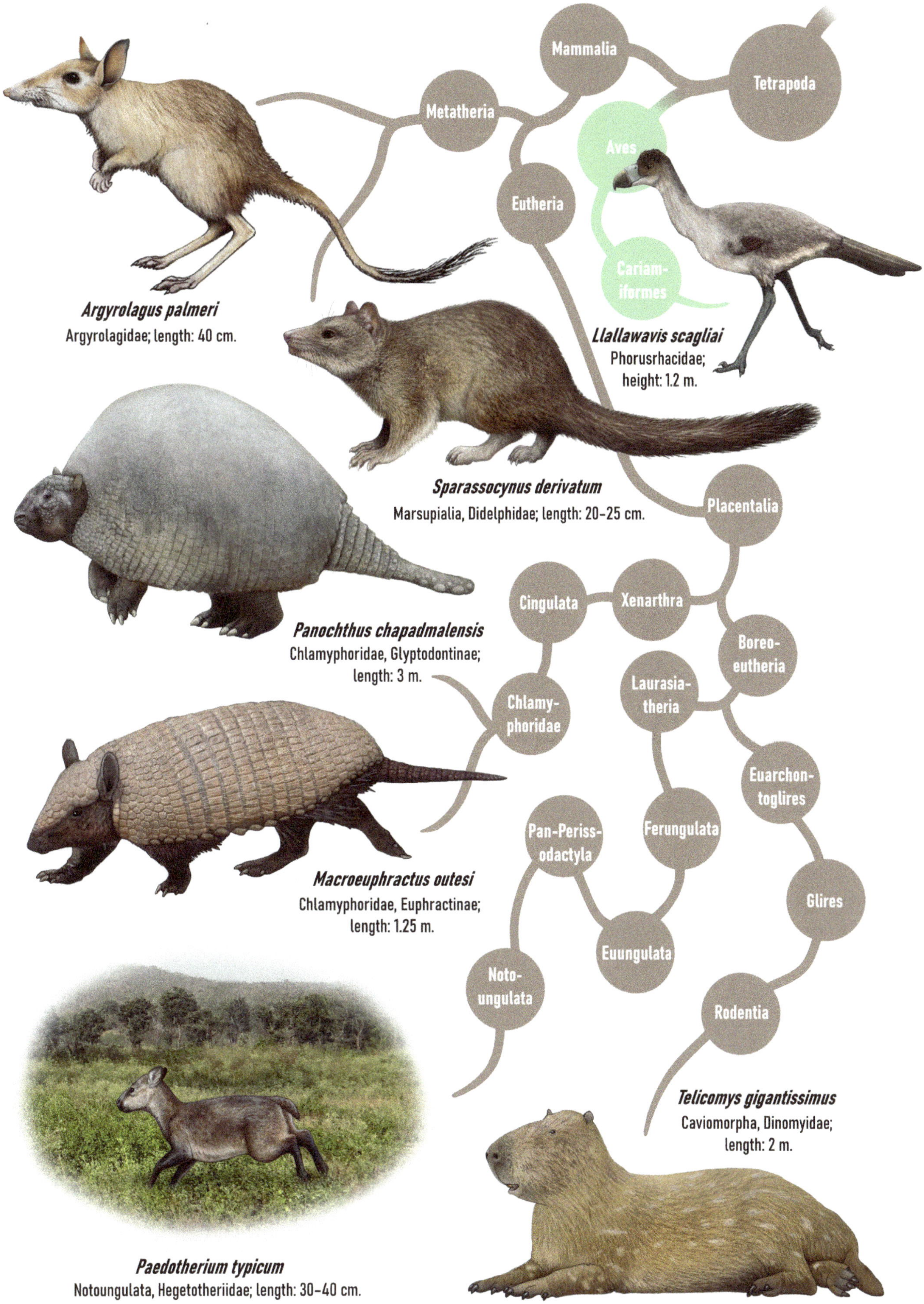

Fig. 6.39. Phylogenetic tree of a sampling of fossil mammal **taxa** from the Chapadmalal Formation. The phylogenetic position of Notoungulata follows Welker *et al.* (2015), while that of *Sparassocynus* within Didelphidae follows Beck and Taglioretti (2020).

Africa Laetoli Beds, Tanzania: Zanclean, 4.36–3.63 Ma.

The Laetoli Beds are situated south of Olduvai Gorge and near Lake Eyasi in Northern Tanzania. This site is among Africa's most significant palaeontological and palaeoanthropological fossil sites. It is special in that the deposits there, contrary to those at other fossil sites in East Africa, formed in a dry **savannah** far from bodies of water and thus shed a potentially different light on Middle Pliocene habitats and **ecosystems** in the region. The variety of flora and fauna represented is suggestive of a mosaic of arid to semi-arid grassland with scattered bush and tree cover, and patches of acacia woodland, similar to the modern-day local setting.

Measured directly from the tooth enamel composition of **grazing herbivores**, **C4 vegetation** became more widespread across Africa by the Pliocene. Due to the lower nutritional value of C4 vegetation, herbivorous mammals have had to modify their feeding behaviour in response to the shifting plant life (Cerling *et al.* 2015). The mix of scrub and open, dry bushland enabled various mammals to specialise in feeding on various kinds of vegetation. Many herbivores had not yet completely adapted to the proliferation of C4 vegetation. Many species from Laetoli, a greater number than in most, if not all, present day African ecosystems, were **mixed feeders**. These include the muskox relative *Makapania* and the **primitive** alcelaphine (see pp. 116, 118) *Damalacra*, along with *Simatherium*, a possible ancestor

Fig. 6.40. Life reconstruction of the primitive hominin *Australopithecus afarensis* (a female) walking around on beds of volcanic ash in what is now Laetoli, Tanzania, during the Late Pliocene (ca. 3.7 Ma).

of the modern African buffalo (*Syncerus caffer*) (Fig. 6.43). During the Pliocene, the bovids continued to diversify, surpassing pigs in terms of the number of species. Throughout the Pliocene of East Africa, the suids (see pp. 108–10) exhibited a pattern of significant dental development, several species which had evolved larger teeth and taller crowns, with the rate of change increasing during this epoch. The **browsers** included *Giraffa jumae*, a possible ancestor of modern giraffes that had the distinctive size and body proportions, but retained primitive features in the skull similar to the more ancient palaeotragine giraffids (see pp. 113–4). Impala (*Aepyceros*) and **tridactyl** hipparionin equids (see pp. 132–3) (e.g. *Eurygnathohippus*) grazed in more open habitats amongst the trees, alongside giant warty pigs, flocks of guineafowl and ostrich chicks, and a diverse array of proboscideans (see pp. 64–8).

The proboscideans included *Elephas ekorensis*, closely related to mammoths and Asian elephants, and *Loxodonta adaurora*, a close relative of and hardly distinguishable from modern African elephants. Living alongside the earliest members of *Elephas* and *Loxodonta* was *Deinotherium bozasi* (Fig. 6.41), which was **endemic** to Africa and the last *Deinotherium* species. It is differentiated from other species of the genus by a narrower and taller skull, a shorter mandibular symphysis and a smaller but more elevated nasal aperture. Like modern elephants, many of these proboscideans were browsers, although *L. adaurora* was a grazer. It is uncertain as to why *Loxodonta* species shifted to a C4 diet and then back again, although this could be due to the large number of proboscidean species competing with each other. As trees became less dense and the savannah spread out to its complete extent, the sole surviving proboscidean in Africa would ultimately be the modern descendant of the extinct grazing species *L. adaurora*.

The Laetoli Beds are also special in that the huge assemblage of tetrapod species and abundant remains of a hominin (see pp. 156–60) identified as the gracile-bodied species *Australopithecus afarensis* are accompanied by thousands of fossil footprints. These were produced in wet volcanic ash that solidified to preserve a collection of trackways (Fig. 6.40). The ash layer was cemented to tuff by soft rain, with the trackways not being destroyed in the process. These were gradually covered by other ash deposits over time. The trackways include **ungulate** hoofprints as well as the imprints of a **bipedal** mammal tentatively concluded to have been *Au. afarensis*. The latter was very different from modern humans in being considerably smaller (over one metre tall), and retained ape-like facial proportions (e.g. tall face, flat nose and massive, overhanging lower jaw) along with large canine teeth. *Au. afarensis* had a braincase typically under 500 cubic centimetres, slightly larger (about 20%) than that of a similar-sized ape (e.g. chimpanzees), but one third the size of a modern human's. Males were substantially larger than females, with a much greater size difference than in modern humans. Males probably stood about 4 feet 11 inches (1.5 metres) in height and weighed up to 42 kilograms (92 pounds), whereas females may have been around 3 feet 5 inches (1 metre) tall and weighed 29 kilograms (64 pounds). This pronounced **sexual dimorphism** suggests that *Au. afarensis* probably had a **polygynous** society, although it is very hard to accurately determine the group dynamics of primitive hominins. *Au. afarensis* also had long upper limbs comparable to those of orangutans, which has sometimes been regarded as an **adaptation** for partial **arboreality**, but is more likely just a basal feature acquired from the last common ancestor of chimpanzees and humans, with no adaptive function.

The body hair remained long, with hair loss believed to be linked to humans' subsequent adaptation to long distance running. *Au. afarensis* was clearly a competent biped, as evident from the skull being squarely balanced over the vertebral column, the pelvis atop the hips, the angle of the femoral head and a knee morphology that enabled better balance when walking on two legs. Further evidence for walking upright include the absence of knuckle impressions amongst the footprints. Unlike apes, the feet lacked the mobile big toe; rather, they possessed an arch (the bending of the foot sole) present in modern humans. Computer simulations based on data from *Au. afarensis* limb bones and the distance between the footprints from Laetoli suggests that, on average, the hominins were walking at one metre per second or faster, a similar walking speed to modern humans. The Laetoli footprints show that their feet were relatively less efficient at walking than modern humans, but had a slightly inturned stance, suggesting significant capability of climbing and moving around in trees. The Laetoli footprints also support the view that bipedalism preceded the increase in brain size in human evolution.

Regarding diet, it is possible that *Au. afarensis* was a generalist **omnivore** which, according to carbon isotope studies of the teeth, had a primarily plant based-diet, consuming both C4 savannah vegetation and **C3 forest vegetation**. Likewise, it seems that *Au. afarensis* could take advantage of a wide range of food sources in a variety of environments. The molar shape and size suggests that *Au. afarensis* could feed on hard, brittle foodstuffs as well, although dental **microwear** analysis indicates they were consumed only occasionally, possibly as "fallback" items when fruits were unavailable. In summary, *Au. afarensis* was among the earliest and anatomically most primitive of the australopithecines and thus, is regarded as a good general ancestor of all later hominins.

As well as *Au. afarensis*, other prints reveal the occurrence of at least twenty different animal species, including baboons (*Parapapio ado*), machairodontine felids (see pp. 86–7) of the genera *Dinofelis*, *Megantereon* and *Homotherium*, hyaenas, bushpigs, gazelles, several types of antelope including the modern genus *Tragelaphus*, giraffes (*Giraffa*), buffaloes, rhinoceroses (ancestors of the modern white rhinoceros species, *Ceratotherium simum*), hipparionin equids, chalicotheriids (see pp. 128–9), elephant relatives (including the extinct *D. bozasi*), hares and miscellaneous birds. A lot of these animals in addition to the australopithecines are represented by body fossils from Laetoli. *P. ado*, for example, was among the later species of baboons, being agile in trees and on the ground, while other baboon species were restricted to predominantly **terrestrial** lifestyles due to their larger size (Fig. 6.43). A small number of trackways are superimposed, suggesting that they were covered up quickly. The imprints of proboscideans, equids and other water-dependent **taxa** show up exclusively in the upper unit, indicating the arrival of the rains in the dry grasslands (Hay and Leakey, 1982). One extraordinary series of hipparionin footprints has been inferred to represent a foal walking ahead of its mother, a form of behaviour prevalent in modern equids such as zebras. The gait employed by these hipparionins is known as "running walk", a lateral-sequence movement that gives significant stability when traversing across slippery

Fig. 6.41. Reconstructed life appearance of the deinotheriid proboscidean *Deinotherium bosazi*. Reconstructed shoulder height: 3 m.

Africa Laetoli Beds, Tanzania: Zanclean, 4.36–3.63 Ma, cont.

surfaces including wet volcanic ash. Two unique footprints are associated with a tridactyl, clawed, approximately horse-sized ungulate, and meticulous studies of these imprints have revealed that they correspond to the foot morphology of *Ancylotherium*, a schizotheriine chalicotheriid that lived in Eurasia and Africa during the Late Miocene to Early Pleistocene (Fig. 6.43). *Ancylotherium* moved about on the ends of its metapodials, and the massive claws were typically retracted so that solely the tips contacted the ground. *Ancylotherium* may have had a more elongated muzzle than other known chalicotheriids, and the dentition was more **hypsodont**, indicating a diet of tougher vegetation. The known body fossils of the species *An. hennigi* are rather fragmentary.

The major predators of hominins in the ecosystem at this time were most probably leopards (*Panthera pardus*) and the fellow machairodontines *Dinofelis barlowi* and *Megantereon cultridens* (Fig. 6.42A–B), the latter two whose carbon isotope ratios show a high indication of preying on hominins. Additionally, most mustelids (see p. 99) in Africa had declined because of increased competition from new predators including felids and canids (see pp. 92–3). *D. barlowi* had a skull that was jaguar-like in size and proportions, but with moderately flattened upper canine teeth (intermediate between the long, flattened upper canine teeth of true sabre-toothed cats and the conical, round-section teeth of modern felids), along with relatively larger **carnassials**. The forelimbs were proportioned like those of a tiger or small lion, albeit more robust, whereas the hindlimbs were comparable to those of leopards (Werdelin and Lewis, 2001). The stocky build and powerful forequarters of *D. barlowi* reveal that it hunted by **ambush** rather than pursuit, pouncing on its prey and wrestling it to the ground with its strong forelimbs while it dispatched using its large canine teeth. *M. cultridens* was more specialised than *D. barlowi* but it too was an ambush predator, distinguished by short, sturdy limbs, a shortened, stiffened lumbar region of the spine, an elongated and muscular neck, and well-developed, blade-like upper canine teeth supported by a short mandibular flange. Its size and physique suggests that *M. cultridens* was probably capable of climbing trees like modern leopards, but its rigid lumbar region may have limited its manoeuvrability. That rigidity may have enhanced the strength for subduing prey while the massive neck muscles were built to generate a strong "canine shear-bite". One other extinct felid from Laetoli that may have hunted, at least occasionally, our hominin ancestors is the long known but only recently described (Hemmer, 2023) pantherine *Panthera principialis* (Fig. 6.42C). This lion-sized (120 to 200 kilograms) felid is generally accepted as close to the common ancestor of lions and leopards, or at least predating their full divergence.

The diversity of mammal species that evolved in Africa during the Pliocene have survived in larger numbers down to the present day. Thus, Africa had started to look modern at a substantially earlier time than the Americas or Eurasia, while still retaining a larger proportion of its older variety of large-bodied mammals.

Fig. 6.42. Reconstructed life appearance of various extinct felids from Laetoli. (A) *Dinofelis barlowi*; (B) *Megantereon cultridens*; (C) *Panthera principialis*. Scale bar: 1 m.

1. *Ancylotherium hennigi*, a schizotheriine chalicotheriid.
2. *Makapania broomi*, a massive caprine bovid with laterally positioned horns.
3. *Loxodonta adaurora*, a precursor of modern African elephants.
4. A hipparionin equid (*Hipparion* sp.).
5. *Damalacra acalla*, an alcelaphine bovid.
6. *Parapapio ado*, a baboon that resembled mangabeys in appearance.
7. The extant white rhinoceros (*Ceratotherium simum*).
8. *Giraffa jumae*, regarded as a potential ancestor of the modern giraffe (*Giraffa camelopardalis*).
9. *Simatherium kohllarseni*, a bovine bovid somewhat similar to the modern African buffalo (*Syncerus caffer*) in size and appearance.
10. A small flock of the extant helmeted guineafowl (*Numida meleagris*).

Fig. 6.43. A savannah ecosystem near Laetoli, in Tanzania, during the Late Pliocene.

North America Blanco Formation, United States: Zanclean–Calabrian, 4.75–1.6 Ma.

The Blanco Formation, of Early Pliocene to Early Pleistocene age, consists of gravels and sandstones deposited by winds and east-flowing rivers. The eroded and weathered **calcareous** material from the underlying bedrock resulted in the characteristic white colour and composition of the Blanco Formation. This is how the Blanco (white) Canyon in northern Texas got its name, which in turn became the type locality for the Blanco Formation. The Blanco Formation's fossil record, along with the presence of other evidence, suggests that the prevailing environment was semi-arid grassland, likely with narrow tree-lined areas along watercourses and in valleys. The lake deposits and **aeolian** silt indicates fluctuating moisture conditions in the region during the Pliocene. Fossils of reeds and aquatic plants indicate the presence of seasonal ponds, which may have been important water sources for the fauna.

In North America, most of the mammals that thrived in the Late Miocene grasslands, including dromomerycids (see pp. 114–5), protoceratids (see p. 111), rhinocerotids (see pp. 130–2) and horned mylagaulid rodents (see pp. 147, 258–9), had disappeared by the beginning of the Pliocene. Many other mammal families, including equids (see pp. 132–4), camelids (see pp. 104, 106–7) and antilocaprids (see pp. 111–2), also suffered declines in diversity. As evident from palaeobotanical data, North America experienced a marginally warmer climate during the Pliocene than during the Late Miocene. The Pliocene climate was relatively drier in the Rocky Mountains and Great Plains regions, with the development of vast modern steppe-like environments or grasslands. A mix of dense woodlands and semi-arid grasslands characterised the Pacific Northwest, with grasslands becoming more prevalent during drier periods. The Gulf Coast, which served as a wooded haven for the last of the moschids (see p. 115), palaeomerycids (see p. 114), protoceratids, large flying squirrels, woodland-adapted peccaries and several native equids during the Late Miocene, had become a subtropical **savannah**, and these **taxa** had become extinct on the continent. These were replaced by a new mammalian assemblage that inhabited the North American grasslands.

While the diversity of North American mammal families was significantly reduced, the remaining families had radiated into new genera and species. Equids were once again abundant, with numerous equid species of both the hipparionin (e.g. *Nannippus*) and equinin (*Calippus*) lineages. Among the earliest species of the modern genus *Equus* was *E. simplicidens*, also known as the Hagerman Horse, often considered to be the oldest species of zebra (Fig. 6.44A). These widespread **grazers** possibly contributed to the proliferation of grasslands by overgrazing and decimating all other types of plants apart from the hardiest grasses that can quickly rejuvenate (Webb and Opdyke, 1995). *E. simplicidens* had a somewhat stocky appearance with a thick neck and straight shoulder, similar to zebras, and a short, slender, donkey-like skull. While almost all the equids had developed much larger body sizes throughout the preceding Miocene, the genera *Nannippus* and *Calippus* reversed that trend. Both genera were grazers and dwarfed by their contemporary relatives. The hipparionin *Nannippus* was the size of a domestic sheep (75 to 90 kilograms) and is characterised by small and very **hypsodont cheek teeth** relative to other hipparionins, strongly curved **posteriorly** and possessing a thickened cement covering (Fig. 6.44B). The hand and foot bones were more elongated and slender than in most other hipparionins (Morgan *et al.* 2008). Weighing 49 to 132 kilograms, *Calippus* was a **tridactyl** equinin characterised by a broad, quadrangular muzzle and a relatively short postcanine **diastema** (Fig. 6.44C).

Fig. 6.44. Reconstructed life appearance of various extinct equids from the Blanco Formation. (A) *Equus simplicidens*; (B) *Nannippus peninsulatus*; (C) *Calippus regulus*. Scale bar: 1 m.

The camelids were also diverse, including forms such as *Hemiauchenia blancoensis* (Fig. 6.45), a North American relative of llamas that persisted until the latest Pleistocene, and the genus *Titanotylopus* (Fig. 6.46). *H. blancoensis* was about twice as heavy as any modern llama, with shorter limbs and large, **caniniform** upper first premolars. It was a hypsodont **mixed feeder** with a tendency towards **browsing**. *Hemiauchenia* would later migrate to South America during the Pleistocene where it developed into new species, perhaps even counting modern llamas as its descendants. *Titanotylopus* had particularly long, massive limbs compared to its body size and shows evidence of **sexual dimorphism**. Males were about 20% larger than females, and sported long canine teeth. As evident from the elongated neural arches on the back vertebrae, *Titanotylopus* had a hump for storing fat as in modern camels. *Titanotylopus* also had relatively wide second **digits**, indicating that like modern camelids, it had true padded toes (Kurtén and Anderson, 1980).

Fig. 6.45. Life reconstruction of the camelid *Hemiauchenia blancoensis*. Reconstructed length: 2.4 m.

Fig. 6.46. Life reconstruction of the giant camelid *Titanotylopus spatulus* browsing on a tree branch. Late Pliocene of the Southern United States.

The antilocaprids were reduced to several genera, while cervids (see pp. 115–7) emigrated from Eurasia to North America by the earliest Pliocene, having replaced the dromomerycids. There were several cervid species present in North America by the Pliocene, with some fossils corresponding to the modern white-tailed deer (*Odocoileus virginianus*). There was a great diversity of peccaries, including the genera *Platygonus* and *Mylohyus* (see pp. 318, 320–1).

Among proboscideans (see pp. 64–8), the "gomphotheriids" were still common, including forms such as *Stegomastodon* (Fig. 6.47). As in modern elephants, but contrary to its close relatives, *Stegomastodon* only possessed two upper tusks, which curved upwards and measured over three metres long. The deep skull and short, robust lower jaw indicates powerful vertical jaw movements. Its enamel-covered cheek teeth are characterised by an intricate pattern of ridges and knob-like protuberances, providing a broad grinding surface that enabled the animal to feed on grass (Lucas *et al.* 2011). Additionally, the diversity of rodents remained higher than it was during the Miocene, with numerous species of marmots, squirrels, beavers, jumping mice, pocket mice and pocket gophers, and a wide range of cricetids (the most significant components being voles and lemmings).

Feeding on all these **herbivores** was a diversity of carnivorans (see pp. 81, 83–4), comprising various species of mephitids (see p. 99) (the modern spotted skunk *Spilogale*), procyonids (see pp. 98–9), canids (see pp. 92–3) (including *Canis lepophagus* (Fig. 6.48A) and the last surviving bone-crushing borophagines such as *Borophagus* (Fig. 6.48B)), ursids (see pp. 92, 94–5) and the final North American hyaenids of the genus *Chasmaporthetes*. Felids (see pp. 86–9) were also prevalent, with a range of machairodontines (*Homotherium* and *Dinofelis*), cougars and even some of the earliest American "cheetahs" of the genus *Miracinonyx* (see pp. 324, 326).

C. lepophagus is the oldest known species of the genus, and was potentially ancestral to coyotes (*C. latrans*) and grey wolves (*C. lupus*). Key features of *C. lepophagus* include a more slender skull and skeleton compared to modern coyotes, smaller overall body size, a more pronounced **sagittal crest** and a less expanded braincase. *Borophagus* was a more efficient bone crusher than *Epicyon* (see pp. 260–1), with the enlarged lower fourth premolars being even more powerful. To aid in bone crushing, *Borophagus* had a robust jaw and a domed forehead that assisted in reinforcing the jaw apparatus so as to sustain the enormous bite force necessary for cracking bones. Perhaps this was an **adaptation** to pack hunting where full utilisation of a carcass was preferred. The bone-cracking abilities of *Borophagus* are further evident from coprolites, which also suggest that it filled a niche that has not survived in modern North American **ecosystems** (Wang *et al.* 2018).

Chasmaporthetes, a genus widely distributed across Eurasia and Africa, is the sole hyaenid that managed to cross the Bering Land Bridge (see p. 174) over to North America. *Chasmaporthetes* was never adequately successful to be a major element of the New World carnivoran **guild**. *Chasmaporthetes* was clearly adapted to running in open areas as its limb bones were long and gracile with cheetah-like proportions, reflective of an active hunting lifestyle (Fig. 6.49). Its cheek teeth were sharp-edged and narrow like those of felids. The premolar intercuspid notches suggest that

North America Blanco Formation, United States: Zanclean–Calabrian, 4.75–1.6 Ma, cont.

Chasmaporthetes was possibly **hypercarnivorous** instead of **durophagous** like its modern relatives are (except the aardwolf (*Proteles cristata*)). *Chasmaporthetes* and the much more abundant canid *Borophagus* may have competed with each other for resources (Wang and Tedford, 2008).

Among the reptiles were familiar lizards and snakes, along with alligators, softshell turtles (*Trionyx*), mud turtles (*Kinosternon*) and even giant tortoises twice as large as the Galápagos tortoise (*Chelonoidis niger*). The genus *Hesperotestudo* (Fig. 6.50) was the largest North American tortoise ever, with a shell that reached 1.25 metres in length, was almost twice as broad as high, and was heavily composed of large dermal **ossicles** that in life were sheathed in **keratin**. Its closest extant relative is the gopher tortoise (*Gopherus*). All these reptiles further support the diverse nature of the ecosystem and are good indicators of the warming temperatures in the Great Plains during the Pliocene. In addition to the Eurasian immigrants (e.g. the cervids and hyaenids already mentioned, felids, voles and lemmings), by the Late Pliocene, the South American immigrants consisted of a diversity of xenarthrans (see p. 58) including the mylodontid ground sloth (see p. 61) genus *Glossotherium* (which was one of the first ground sloths to have migrated to North America), plus the glyptodontines (see pp. 59–60). As evident from its broad snout, *Glossotherium* was likely a **bulk feeder**, consuming low-growing vegetation. Modern bulk feeders tend to be grazers, and similar habits have been inferred for certain mylodontids including *Glossotherium*. Although it had a predominantly grass-based diet as evident from isotope analysis on the teeth, *Glossotherium* may have also fed on a variety of leaves and would be best described as a "browser-grazer" (Naples, 1989). Its ear anatomy reveals that *Glossotherium* could sense low frequency sounds and possibly produced them with its large nasal cavity, using them for long range communication like modern elephants (Blanco and Rinderknecht, 2008).

Fig. 6.47. Reconstructed life appearance of the "gomphotheriid" proboscidean *Stegomastodon mirificus*. Reconstructed shoulder height: 2.8 m.

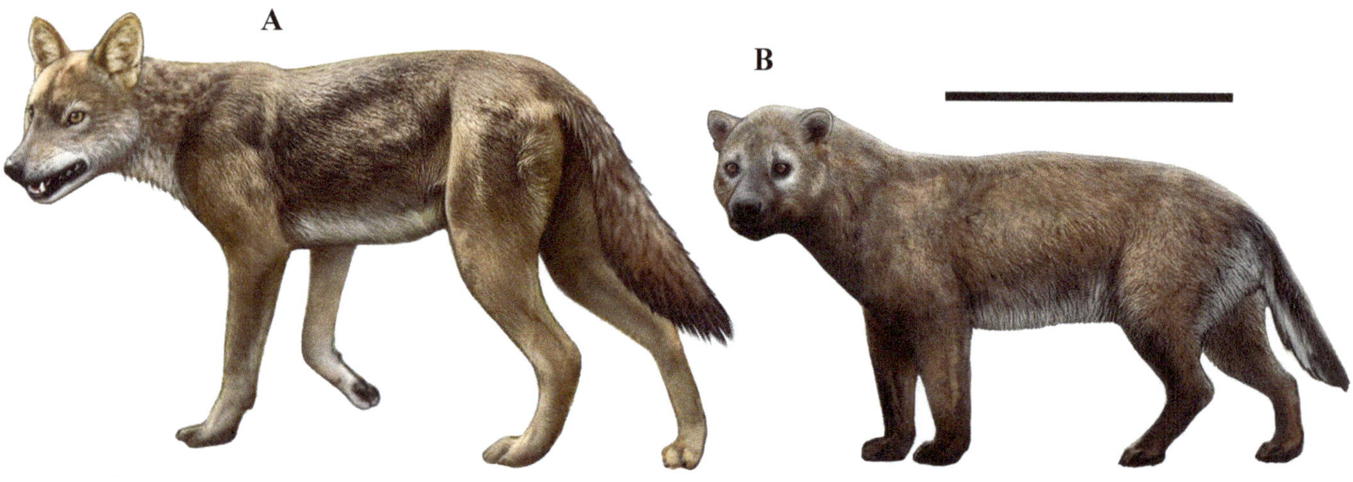

Fig. 6.48. Reconstructed life appearance of canids from the Blanco Formation. (A) *Canis lepophagus*; (B) *Borophagus diversidens*. Scale bar: 0.5 m.

Fig. 6.49. A scene in the Southern United States during the Late Pliocene, in which a pack of the hyaenid *Chasmaporthetes ossifragus* are hunting down a white-tailed deer (*Odocoileus virginianus*). In the background is the "gomphotheriid" proboscidean *Stegomastodon mirificus*.

Fig. 6.50. Life reconstruction of the mylodontid ground sloth *Glossotherium chapadmalense* (left background) looking back at a giant tortoise of the species *Hesperotestudo turgida* (right foreground). Late Pliocene of the Southern United States.

Europe Milia Fossil Site, Greece: Piacenzian, 3.6–2.58 Ma.

Dated from the Middle Pliocene to Early Pleistocene, the fossils of the Milia site complex are distributed in the unconsolidated **fluvial** sand deposits (Tsoukala, 1834) adjacent to the Haliacmon River, in northern Greece, into eleven main localities (MIL 1–MIL 11). The fossils are quite dissociated and spread across a distance of more than 20 kilometres, near the village of Milia. The fossil assemblage of the Milia site complex illustrates a Pliocene eden in the Southern Balkans, one that vanished with the advent of the extreme climatic shifts of the Early Quaternary. The climate in Milia and elsewhere in Europe was generally warm, humid and non-seasonal throughout the Pliocene. The **terrestrial** vegetation was composed of evergreens and temperate mixed forest in which cypresses predominated. In this environment, the Middle Pliocene fauna primarily consisted of mammal **taxa** also found in Europe during the latest Miocene, but there was also significant faunal change driven by environmental shifts and the arrival of new groups from other continents. The proboscideans (see pp. 64–8) included the basal elephantoid *Anancus* (Fig. 6.51A) and the more **primitive** Eurasian Borson's mastodon, named *Mammut/Zygolophodon borsoni*, which were the largest **herbivores** of the Pliocene woodlands. There is some confusion as to which genus the latter animal should belong to, although it is probably a Eurasian lineage of *Mammut* (Fig. 6.51B).

Anancus was possibly derived from an Asian species of *Tetralophodon* (or at least a *Tetralophodon*-like ancestor), and a number of important characteristics sets it apart from its close progenitors. Although *Anancus* had a dentition barely different from that of earlier elephantoids, it had a considerably taller and shorter skull (with an elevated dome) compared to the lengthened (**longirostrine**) skull of *Tetralophodon*, superficially resembling that of modern elephants. It also had a significantly shortened (brevirostrine) lower jaw, and lacked the lower tusks present in Early to Middle Miocene "gomphotheriids". The upper tusks were unusually elongated, and slender and outwardly divergent from each other, measuring over three metres long. The modern body dimensions of elephants appear in *Anancus*, although the limbs were shorter. Unlike true elephants, the molars did not consist of **lamellae**, but had hill-like cusps, similar to tapir and pig molars. *"M". borsoni* is known from several partial skeletons, including an impressive pair of complete tusks with a slight upward curvature, an absence of enamel bands and length of up to five metres, the longest known amongst proboscideans. The lower jaw (especially the mandibular symphysis) was shorter than that of more primitive mammutids such as the genus *Zygolophodon*. Unlike more primitive mammutids, permanent premolars seem to have been absent in the jaws of *"M". borsoni*. In *"M". borsoni*, the hills of the molar cusps were transformed into sharp ridges, with more clearly defined shearing edges. In addition to the elongated upper tusks, *"M". borsoni* developed features resembling those of *Anancus* via **parallel evolution**, including the shortened face, and the reduction and disappearance of the mandibular tusks. According to dental **mesowear** analysis, *Anancus* was a browse-dominated **mixed feeder**, whereas *"M". borsoni* seems to have been a pure **browser** (Saarinen and Lister, 2016). The upper tusks of both proboscideans would have made impressive structures for sexual display, digging for food or water and as weapons.

The ruminants (see p. 110) were dominated by gazelles of the genus *Gazella*, which had become common in Europe during the Pliocene. Large bovids boosted their diversity with the new taxa *Alephis* and *Grevenobos*, the former which was a lightly-built bovine more slender than cattle and bison (Fig. 6.53). *Alephis* resembled tragelaphines such as the nyala (*Tragelaphus angasii*), and just like them it is very likely that it had marked **sexual dimorphism** in the horns. The formidable, elongated horns extended upwards and were either slightly divergent (*A. tigneresi*) or curved outwards (*A. lyrix*), unlike those of earlier, closely related genera such as the smaller *Parabos*, which had short, straight horns. *Alephis* had shorter and more robust limbs than its earlier relatives, with lengthened metapodials built for sprinting over hard substrates, and the **cheek teeth** were **hypsodont**, specialised for **grazing**. *Grevenobos* is an extremely obscure animal, but it appears to also fall within the **clade** Bovini. *Grevenobos* had long, unbranched horns and possibly resembled a cross between an oryx and a domestic cow. Among suids (see pp. 108–10), the extinction of the giant form *Microstonyx* (see p. 278) was made up for by the emergence of the earliest modern suids of the species *Sus arvernensis*, possibly ancestral to extant wild boars (*Sus scrofa*).

In this warm world, however, several short episodes of cooling occurred in the Northern Hemisphere

Fig. 6.51. Reconstructed life appearance of proboscideans from Milia. (A) *Anancus arvernensis*; (B) *"Mammut" borsoni*. Scale bar: 1 m.

before a major glaciation event around 3.2 Ma. That event resulted in a considerable turnover within the Eurasian large mammal faunas, and some components whose origins were rooted in the Miocene, including the final European giraffids (see pp. 112–4), went extinct. Large bovids (see pp. 116, 118–9) of the genus *Alephis* had gone extinct before being replaced by *Leptobos* across Europe, a giant, lightly-built bovine similar to the modern genus *Bos* in various aspects of anatomy (see p. 119). Additionally, the climatic cooling had a major effect on the cervids (see pp. 115–7), which underwent a major radiation. The Late Miocene to Early Pleistocene *Croizetoceros* (Fig. 6.52), one of the oldest known cervines, developed into larger species with increasingly intricate antlers such as *Croizetoceros*, which had up to five tines. The antlers were long and lyre-shaped, and the tines branched off tangentially from the main antler. Furthermore, new cervid taxa related to the modern genus *Cervus* arose, such as *Praeelaphus*, a small cervine distinguished by advanced, four-tined antlers (Croitor, 2017).

Fig. 6.52. Reconstructed life appearance of the cervid *Croizetoceros ramosus*. Reconstructed length: 1.2 m.

The diversity of perissodactyls (see p. 126), especially rhinocerotids (see pp. 130–2), had declined substantially since the Miocene-Pliocene boundary. Once the aceratheriins had vanished, and the range of dicerotins became limited to Africa, the only lineage that persisted in Europe were the dicerorhinins, which by the Late Pliocene, began with the large tandem-horned genus *Stephanorhinus*. Within the **guild** of large browsers was the earliest species of *Stephanorhinus*, *S. jeanvireti* (Fig. 6.54), which is distinguished by features including an elongated mandibular symphysis, a massive, highly elongated facial region, relatively large nasals reaching far forwards and slender limbs, in addition to massive body dimensions (Guerin, 2013).

The hipparionin equids (see pp. 132–4), just like the rhinocerotids, also experienced a decline in diversity throughout the Pliocene. *Hipparion crassum*, a large species with robust and massive metapodials, defined the genus in Europe. The tapirs seem to have been the sole perissodactyls to have thrived in Europe past the Miocene-Pliocene transition, and were largely successful during the Pliocene. *Tapirus arvernensis*, most closely related to the Malayan tapir (*Tapirus indicus*) was an apparently abundant tapir species that was fairly small in size (about two metres long) and looked very similar to modern species. Therefore, its behaviour was likely **analogous** too, possibly feeding on aquatic vegetation using its **proboscis** and living in wooded areas with lots of water around.

Among rodents, the occurrence of a porcupine (*Hystrix* sp.) can be deduced from several fossils and some gnawing marks (porcupines have a propensity of gnawing on bones for calcium intake essential for the growth of their quills). As a result of shifts in plant and herbivore **communities**, the climatic transition occurring around 3.2 Ma had a significant impact on the **carnivore** faunas. Many of the modern-looking carnivorans (see pp. 81, 83–4) that had arisen in the Early to Middle Pliocene lingered on, occasionally consisting of new taxa such as the immigrant ursid (see pp. 92, 94–5) *Ursus etruscus* (Fig. 6.53). Comparable to the brown bear (*Ursus arctos*) in size, it had a complete set of premolars, a feature carried on from *Ursavus* (see pp. 282–3), although the humerus was still of primitive structure. The massive, sturdily built *Agriotherium* (see pp. 94–5) survived in Europe up until the Late Pliocene, while *Indarctos* (see pp. 279–80) did not persist beyond the Miocene. Among felids (see pp. 86–9), the machairodontines underwent an explosion in diversity, with the first species of *Homotherium*, including *H. latidens* (see pp. 343–4), having dispersed and become part of the expanded big cat guild.

The proliferation of grazers, the gracile morphology of various **ungulates** and the radiation of pursuit predators indicates that an expansion of steppe and grassland habitats occurred in Europe in response to the glaciation event of around 3.2 Ma (Figs. 6.53, 6.54). Two major mammalian dispersals to Europe were linked to the onset of periodic extensive glaciations that began around 2.6 Ma, although these events did not drastically reorganise entire **ecosystems**.

The first major dispersal event was the arrival of some of the earliest true elephants belonging to the genus *Mammuthus* (*M. meridionalis*), which following an initial coexistence, superseded the final basal elephantoids of the genus *Anancus*, with the mastodont *"M". borsoni* having gone extinct prior. The second event, not long after the arrival of the earliest *Mammuthus*, were the earliest modern horses of the genus *Equus* displacing the last *Hipparion* species. Besides the local disappearance of their respective Pliocene guild counterparts *Anancus* and *Hipparion*, the arrival of *Mammuthus* and *Equus* was associated with the disappearance of a "warm forest assemblage" and the arrival of species adapted to more open, parkland-like environments. The extinction of certain woodland-adapted species such as the tapirs (*Tapirus arvernensis*) and the primitive wild boars (*Sus arvernensis*) can be regarded as a major consequence of the glaciation that struck the Northern Hemisphere around 2.6 Ma. Meanwhile, this glaciation had a significant impact on the large reptile faunas, including massive tortoises such as *Titanochelon*, which could not adapt to the pronounced seasonal fluctuations in temperature

Europe Milia Fossil Site, Greece: Piacenzian, 3.6–2.58 Ma, cont.

linked to the glacial-interglacial cycles. On the other hand, their smaller kin of the modern genus *Testudo* were able to easily adapt, and still persists, given their **adaptations** to burrow underground and hibernate during cold weather. The Plio-Pleistocene faunal turnover in Europe, while affecting a limited amount of mammal species, was significant due to the key taxa involved and the broader ecological implications of their replacement. This turnover saw the initial entry of several large mammal groups that would dominate Europe during much of the Pleistocene, such as hyaenas, wolves and pantherine felids.

Fig. 6.53. Some representatives of the fauna found at the site of Milia during the Late Pliocene. *Left foreground*: the tapir *Tapirus arvernensis*; *Right foreground*: a small group of the bovine bovid *Alephis* sp.; *Centre*: the bovine bovid *Grevenobos antiquus*. *Left background*: the ursid *Ursus etruscus*, and immediately to the right, two other individuals of *Alephis* sp.; *Right background*: a lone proboscidean of the genus *Anancus*.

Fig. 6.54. Some representatives of the Late Pliocene to Early Pleistocene fauna found in Southern Europe. Wading in the river are a pair of the extinct European hippopotamus (*Hippopotamus antiquus*) (while not known from Milia, the earliest hippopotamus fossils currently known from Europe date to the Early Pleistocene (ca. 2.1 Ma), of Italy (Coste San Giacomo and Chiusi Basin) and Greece (the site of Elis)). In the background is the narrow-nosed rhinoceros *Stephanorhinus jeanvireti* (right), and two male proboscideans of the species *Anancus arvernensis* engaging in a bout of pushing and shoving with their tusks to establish their status (left). In the far background is a herd of the cervid *Croizetoceros ramosus*. Flying in the air is a flock of herons (*Ardolea* sp.).

7. The Quaternary
2.58 Ma–present

The Quaternary Period encompasses an ongoing Ice Age and is the geological period that we are presently in, characterised by the periodic expansion and contraction of continental ice sheets caused by cycles in the Earth's rotational axis and in Earth-Sun relations, which in turn are brought about by variations in the Earth's orbit known as Milankovitch Cycles. Changes in atmospheric carbon dioxide and the amount of solar radiation that enters the atmosphere are influenced by these cycles. The effects of Milankovitch Cycles influence climate with regular cycles of cooling and warming. The three Milankovitch Cycles comprise: variations in the shape of Earth's orbit around the sun called eccentricity, the angle of Earth's axis known as obliquity, and the wobbling of Earth's axis known as precession. Together, these three Milankovitch Cycles change the angle at which solar energy reaches the Earth's surface near the poles and the amount of solar energy (insolation) that the planet receives. The periodic fluctuations in the solar input to the Earth were expressed as a cooler and warmer climate cycle, therefore the glacial-interglacial cycles.

This pattern of alternating cool and warm intervals persisted throughout the Pleistocene Epoch, right up to the warm Holocene Epoch in which we live today (see pp. 345–6). However, while these cycles were repeating, they varied in their duration and constancy of conditions. Over the past 2.58 million years, the global climate has fluctuated wildly, and especially throughout the past 600,000 years. The last Ice Age, known technically as the Last Glacial Period, was the most severe of all, which ended around 11.7 ka. To cope with the glacial conditions, many large mammals developed dense, protective coverings of fur as an **adaptation** to the cold, such as woolly mammoths, woolly rhinoceroses and bison. These were preyed on by sabre-toothed cats, lions, canids (see pp. 92–3) and our own ancestors—large-brained hominins that had started to disperse all over the world.

During the coldest part of the Last Glacial Period, known as the Last Glacial Maximum (ca. 26–20 ka), a vast area of the Northern Hemisphere was shrouded by ice sheets and montane glaciers. Since ice sheets locked away a huge portion of the planet's water, sea level falls led to the exposure of continental shelves (and smaller insular shelves) all across the planet (Fig. 7.3). One prime example is the Bering Land Bridge (see p. 174) between Alaska and Eastern Siberia, which was created when a large portion of seafloor was exposed between both regions. As an ecological zone, Beringia, as it is called, spanned from the Taymyr Peninsula in north-central Russia to the Yukon Territory, Canada. The biome it would have supported is the mammoth steppe,

Fig. 7.1. Late Pleistocene steppe **ecosystem** of Northern Eurasia during a glacial period. In the foreground is a group of cave/steppe lionesses (*Panthera spelaea*) chasing down a female and two calves separated from a herd of steppe bison (*Bison priscus*). In the background, from left to right, are a herd of reindeer (*Rangifer tarandus*), a pair of woolly rhinoceroses (*Coelodonta antiquitatis*), and herds of woolly mammoths (*Mammuthus primigenius*), saiga (*Saiga tatarica*) and horses (*Equus ferus*) on the move.

which consisted of a specific combination of low bushes, grasses and other herbaceous vegetation (Fig. 7.1). The **periglacial** zones along the southern margins of the ice sheets would have supported similar vegetation. As the largest landmass, Eurasia had the greatest diversity of ecological conditions. Although the majority of Eurasia's mid- and high-latitude regions were devoid of ice during this time, this was not due to higher temperatures than that of Northern North America, but rather to a relative lack of precipitation. Indeed, almost the whole of Northern Eurasia was covered by either treeless steppe or extreme polar desert.

The last glacial and interglacial periods saw significant changes in the mammalian fauna, culminating in the demise of the Pleistocene megafauna by the Late Pleistocene and Early Holocene. Of course, impressively large **terrestrial** vertebrates are still with us today, think Komodo dragons, ostriches, hippopotamuses, rhinoceroses and elephants. However, in comparison to 50 ka, there are few large ones left (Fig. 7.2). In North America, for instance, there are approximately a dozen extant species which are megafaunal; before the extinctions started around 13 ka, at least 30 more existed. Australia and South America experienced similar large-scale continental losses. It is estimated that more than 100 mammal species, some 75% of which weighed over 45 kilograms (100 pounds) vanished from North and South America combined at the close of the Pleistocene. Interestingly, disappearances in Africa, South Asia and Southeast Asia were much fewer and staggered over prolonged intervals. Losses in Northern Eurasia (occurring at around the same time as those of the Americas) fall between these extremes and they contributed another dozen megafaunal species to the extinction list, as did those happening relatively earlier in Australia (see pp. 332–5). All these megafaunal extinctions were most likely caused by varying combinations of factors, including climatic changes that led to the drying out of many habitats, and pressure imposed by the dispersal of *Homo sapiens*, who hunted various megafaunal species for food and mastered the control of fire for purposes such as cooking, making more sophisticated hunting tools and clearing vegetation.

Throughout the Quaternary, climate has had a substantial influence on the Earth's continents and oceans. As India carries on ploughing into Asia and intensifying the Himalayan uplift, Arabia and Africa into Eurasia, and Australia into Indonesia, with corresponding and frequently devastating earthquakes, sea levels have risen and fallen dramatically. While the ice sheets extended from north and south, a considerable amount of the water that would have otherwise been restored to the oceans was locked in the frozen landmasses and sea levels decreased significantly. Ever since human civilisation began during the Early Holocene, our impact on the landscape has become increasingly significant and we have become more dominant in our surroundings, often to the detriment of many other extant species. Over the past several thousand years, and especially in the last few centuries, the ecosystems that have grown up on land have come under increasing pressure from human populations. The geography of the world as it exists today may only be temporary. Anthropogenic activities such as deforestation and burning fossil fuels, which emit greenhouse gases, appear to be triggering global warming to an extent that could potentially flood many low-lying terrestrial regions around the world due to melting glaciers. It is not yet possible to predict with certainty whether the Quaternary's natural cycles of glacial advance and retreat will persist and if so, whether current human influence on global temperatures will be overcome by the next glacial period. If this does happen, the geographical area exposed above sea level will increase in size as the planet returns to a windy, frozen state once again.

Fig. 7.2. Vanished (and vanishing) megafauna: Within the last 50,000 years, large-bodied mammals were present in almost all habitable parts of the world, even encompassing the majority of islands. Most of these are extinct now; a select few representative **taxa** are featured on the map as black and white illustrations. With the exception of extant elephants and rhinoceroses (depicted in colour), every genuinely large or supermegafaunal species (weighing more than 2,000 kilograms/ 4,400 pounds) has vanished entirely.

Pleistocene
2.58–0.0117 Ma

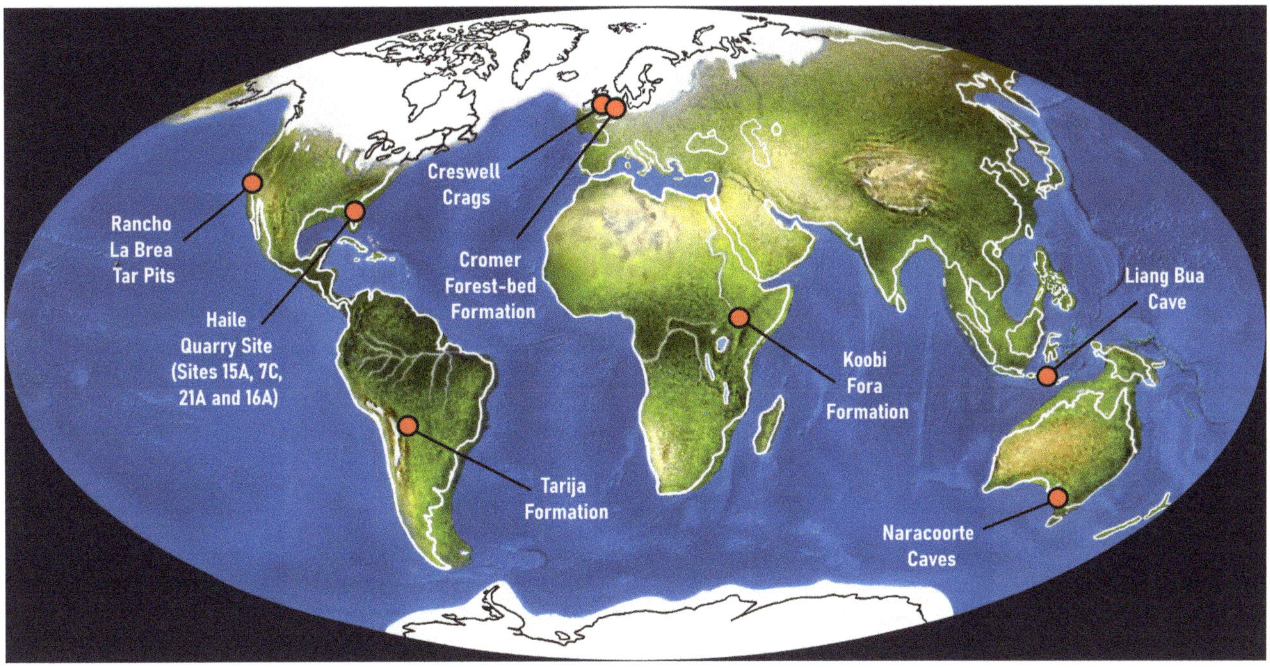

Fig. 7.3. Palaeogeographic map reconstruction of Earth during the Pleistocene, with the position of the modern continents outlined and the approximate location of the fossil sites indicated.

The Pleistocene spans the most recent periods of global cooling, otherwise known as Ice Ages. Within the glacial-interglacial cycles, the glacials were represented by cold temperatures and the advancement of ice sheets over vast areas of the Northern Hemisphere. The interglacials were characterised by global temperatures similar to today's and sometimes even warmer, which resulted in ice sheets decreasing in extent and therefore higher sea levels. Some interglacials were also cooler than today. During the glacials, woodlands and forests were scarce in northern areas, where vast grasslands with huge amounts of plant biomass (mostly mosses, lichens, sagebrush and short grass) existed to support all the **grazing megaherbivores** such as mammoths and horses. During the warmer interglacials, however, northern habitats were dominated by open woodlands, which were composed of both coniferous tree species such as pine and **deciduous** species such as oak and elm.

During the Middle Pleistocene Transition, 1.2–0.8 Ma, the Earth underwent a new considerable decrease in temperature and the climate cyclicity increased in duration, with the cyclicity of glacial cycles changing from 41,000 year cycles (which began in the Late Pliocene) to asymmetric 100,000 year cycles, which coincided with changes in the Earth's orbital eccentricity and made climate variation more extreme. The interglacials of the Middle and Late Pleistocene represented relatively shorter periods of time in comparison to the glacials, and typically lasted 10,000 to 15,000 years. The differences in the duration of glacials and interglacials is due to timings related to different orbital cycles. The reason why the changes during glacial-interglacial cycles are asymmetric (where transitions into glacial periods are slow and transitions into interglacial periods are rapid) is because it is faster to melt ice/decrease the planet's albedo than it is to build up ice/increase the planet's albedo. The Eemian Interglacial (130–115 ka) was the last major interglacial period. It is often found to be one of the warmest interglacials of the last 500,000 years, during which time the climate was very temperate, considerably warmer than that of the Holocene, by about 1–2°C on average. The Late Pleistocene (129–11.7 ka) is distinguished by a predominantly cool climate interspersed with geologically brief temperate stages called interstadials. The Last Glacial Period (115–11.7 ka) was a major glacial pulse that marked the end of the Pleistocene. The ice sheets blanketing North America and Eurasia peaked in extension between about 30–19 ka, after which they receded. For example, the onset of retreat of the British-Irish Ice Sheet is asynchronous but is definitely underway from 22 ka; maximum extent was between 30–22 ka. Final deglaciation was by climate warming from around 19 ka. The ensuing sea level rises involved tremendous alterations to the geography of Northern Europe and numerous other regions around the globe. There were some minor and brief glacial readvances during the 'Younger Dryas' cooling circa 13–12 ka (see p. 327). Throughout the Pleistocene, mammals abounded beyond the margins of the ice sheets. Compared to small animals, large animals were more adapted to conserving body heat and withstanding severe cold. The distributions of numerous animals grew and shrank in response to shifts in plant life and climate. For instance, hippopotamuses spread northwards from Africa to England (as far north as Stockton-on-Tees in County Durham) during the interglacial periods, while on the other hand, reindeer spread southwards from the Arctic to Southern Europe during the glacial periods. During the Pleistocene, continental and marine faunas were both essentially modern, except that giant **terrestrial** mammals, or megafauna, were considerably more diverse around the world than today.

Africa Koobi Fora Formation, Kenya: Zanclean-Calabrian, 4–1 Ma.

Located along the eastern shores of Lake Turkana, in northern Kenya, the Plio-Pleistocene Koobi Fora Formation is composed of **lacustrine** and **fluvial** sediments, intercalated with hydrated volcanic tuffs that serve as the foundation for correlation and dating. Many **terrestrial** mammals, including over 90 large ones and over 80 small ones, have been preserved as fossils, providing a good snapshot of life in East Africa during much of the Pleistocene. The Koobi Fora Formation is best known for its Early Pleistocene specimens of early members of the genus *Homo* (including the oldest known fossils of *Ho. habilis*), but those of the genera *Australopithecus* and *Paranthropus* have also been discovered (see pp. 157–8). *Homo* and *Australopithecus* appear to have coexisted in East Africa for a few hundred thousand years, one reason which may be avoidance of competition through exploitation of different foodstuffs (Leakey and Lewin, 1978).

Throughout the Pleistocene, the African climate experienced repeated oscillations in rainfall and temperature, alternating between arid-cool intervals and humid-warm intervals associated with the glacial-interglacial cycles. Those oscillations increased in duration and intensity from about 2.8–1 Ma, during the Late Pliocene to Middle Pleistocene (deMenocal, 2004). Throughout the Pleistocene, the glacial-interglacial cycles caused Africa's biomes to repeatedly expand and contract, which had a marked effect on the distribution of plants and animals. During arid-cool intervals, grasslands and **savannahs** with arid-adapted species spread into lowland forests, while the Sahara expanded into semi-desert regions. During humid-warm intervals, forests expanded into the montane shrublands south of the Sahara, and into the savannahs in equatorial Africa (Dupont, 2011). These oscillations would have had significant impacts on the size and composition of large mammal populations. For instance, the interval between 2.8 and 1.8 Ma was a time of accelerated faunal turnover in East Africa at least partially caused by climate change (Bobe and Behrensmeyer, 2004; Ferraro *et al.* 2013). This turnover resulted in major evolutionary changes in which dispersions, **speciation**, extinctions, and modifications within various lineages played a role.

Fig. 7.4. Reconstructed life appearance of the baboon *Theropithecus brumpti*. Reconstructed shoulder height: 90 cm.

Unlike Eurasia and the Americas, Africa hosted a Pleistocene mammalian megafauna not much different from that of today. In addition to the aforementioned hominins, among other fossil primates from Koobi Fora are cercopithecids (see p. 154) including the baboon *Theropithecus brumpti* (Fig. 7.4), known from beautifully preserved skulls and lower jaws. As with modern geladas (*T. gelada*), *T. brumpti* had powerful masticatory muscles and broad **cheek teeth** suited for grinding fibrous plants, although its snout was more elongated. *T. brumpti* also exhibited significant **sexual dimorphism**, with the males having exceptionally large canine teeth likely used for display. The well-developed masseter muscles in the snout may have allowed *T. brumpti* to crack open and feed on nuts.

Africa remained inhabited by a diversity of ancient proboscideans (see pp. 64–8) during the Early Pleistocene, including *Deinotherium bozasi* (see pp. 300–1), and **primitive** species of the modern African and Asian elephants. After approximately 2.3 Ma, the extinct *Palaeoloxodon* lineage, represented by *P. recki* (Fig. 7.5), emerged as the dominant proboscideans, while the ancestors of modern African elephants (the genus *Loxodonta*) experienced a decline. Known from an impressive, near complete skeleton, *P. recki* had very different body proportions compared to extant elephants, with its disproportionately reduced skull and lengthened forelimbs that contributed towards a rather slanted back. The animal possessed rather small tusks relative to its body size, and the cranium, with its low-set **orbits** and tall, highly domed forehead, contributed to this elephant's distinctive appearance.

There was also a greater diversity of hippopotamuses, with *Hippopotamus gorgops* strongly resembling and being larger than its extant relative, *Hi. amphibius* (Fig. 7.7). The main feature distinguishing it from *Hi. amphibius* was the position of its orbits. Extant hippopotamuses have orbits set high on the skull, but *Hi. gorgops* had them even further elevated, essentially giving it a periscopic view, and the more vertical **occipital** bone at the back of the skull suggests that the animal could raise its muzzle at a larger angle. These characteristics indicate that *Hi. gorgops* had a more aquatic lifestyle than extant hippopotamuses. *Hi. gorgops* died out by the Middle Pleistocene likely due to environmental shifts reducing its habitat and food sources.

With a hint of moose about it due to the broad, sweeping palmate **ossicones**, the sturdy, thick-necked giraffid (see pp. 112–4) *Sivatherium maurusium* is one of the largest known terrestrial artiodactyls (Fig. 7.6). It also had a short pair of conical ossicones above the orbits, and its shoulders were very strong to support the neck muscles necessary for lifting the massive head. Its forelimbs were more robust than those of other giraffids and this is attributed to the presence of the heavy cranial ossicones. Like the contemporary modern long-necked giraffes (*Giraffa*), it is believed that *S. maurusium* had soft, mobile lips and possibly a long tongue for **browsing**. In *S. maurusium*, however, there is a pattern towards progressively shortened metacarpals in proportion to the metatarsals, which could correspond with the increasingly **grazing** lifestyle of advanced sivatheriines.

The most taxonomically diverse large mammals were the antelope and cattle, which have remained vital parts of Africa's fauna to this day. Many are representatives of modern genera, but the diversity of extinct **taxa** was much higher. From about 3–2 Ma, an important transformation occurred in the antelope **guild**, in which immigration of antilopines and caprines (see pp. 116, 119), speciation and extinction played a role. While the climate became cooler, more bovid species emerged by approximately 2.7 Ma, during the Late Pliocene, in order to keep up with the changing environments. By the Early Pleistocene (ca. 2.5 Ma), the new species consisted of those adapted to open environments, a notable shift which was the increased prominence of larger, open-country bovids, most notably among the alcelaphines (see pp. 116, 119).

A particularly large alcelaphine was the giant hartebeest *Megalotragus*, a genus that resembled a tall hartebeest with wildebeest-like body proportions and very long, curved horns (Fig. 7.6). Changes in the dentition (e.g. a shift to **hypsodonty**) of later alcelaphines such as the modern wildebeest (*Connochaetes*) reflect the increasing adaptability to thrive on the resources found in open savannahs. Among extinct cattle was *Pelorovis oldowayensis* (Fig. 7.7), which was similar in size to the modern African buffalo (*Syncerus caffer*), although the limb bones,

Fig. 7.5. Reconstructed life appearance of the elephantid proboscidean *Palaeoloxodon recki*. Reconstructed shoulder height: 4.27 m.

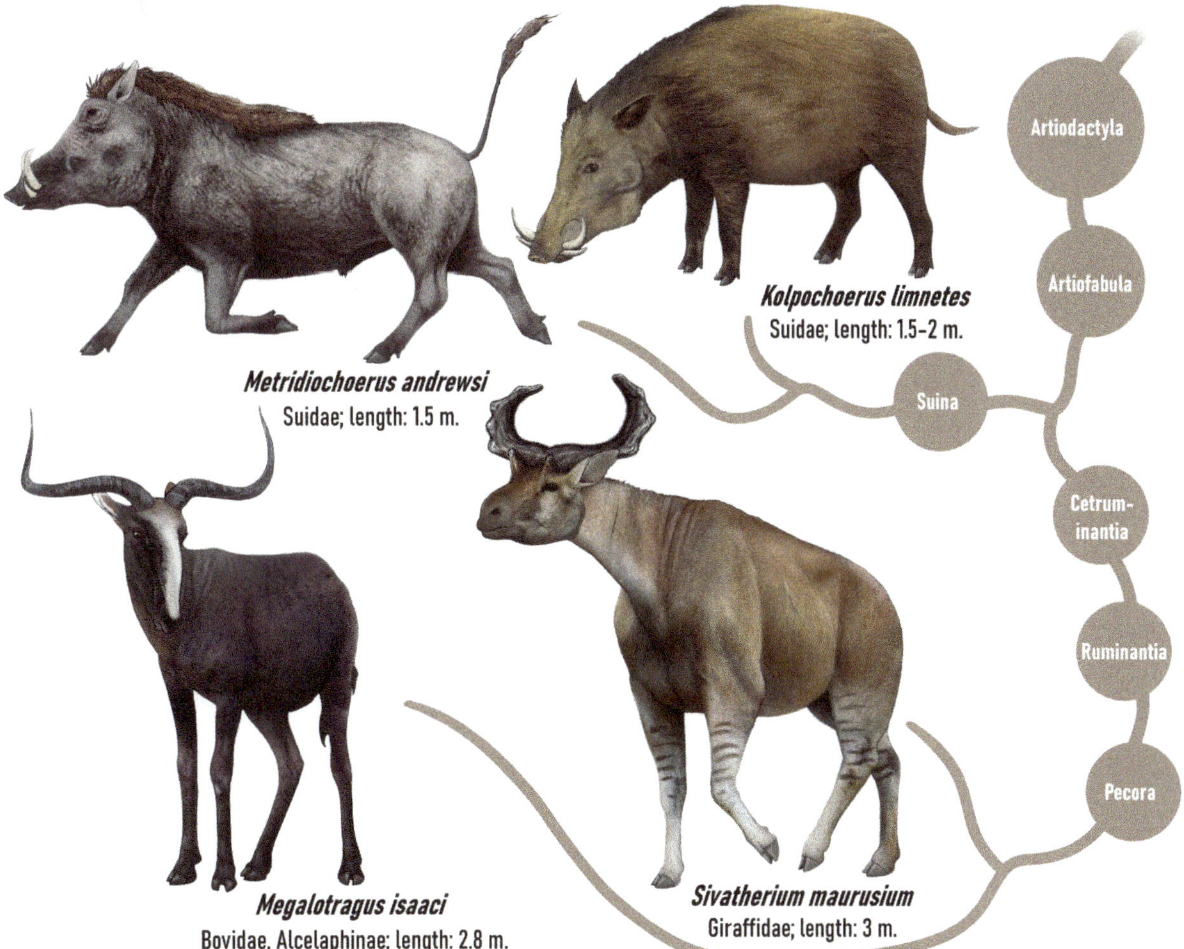

Fig. 7.6. Phylogenetic tree of a sampling of fossil artiodactyls from the Koobi Fora Formation.

Fig. 7.7 (overleaf). A semi-arid savannah at the site of Koobi Fora, Kenya, during the Early Pleistocene. *Right foreground*: a pair of *Hippopotamus gorgops*; *Middleground*: from left to right, a pair of the scimitar-toothed cat *Homotherium* sp. resting underneath the shade of a tree, a group of the African wild cattle *Pelorovis oldowayensis*, a pair of the zebra *Equus koobiforensis*, a pair of the elephantid *Palaeoloxodon recki*, and a leopard (*Panthera pardus*) resting on a tree branch; *Background*: from left to right, a modern giraffe (*Giraffa camelopardalis*), a group of gazelle (*Gazella* sp.), a family of the African bush elephant precursor *Loxodonta adaurora* (see pp. 300, 303), a reduncine bovid (*Kobus sigmoidalis*), the extant suid *Hylochoerus* sp., and a lechwe (*Kobus leche*).

Africa Koobi Fora Formation, Kenya: Zanclean–Calabrian, 4–1 Ma, cont.

especially the **distal** elements, were much more elongated, giving the bovine a taller appearance. The most notable distinctions are in the skull, which was significantly longer in a manner similar to those of alcelaphine antelopes, a trait inferring a similar lifestyle of grazing low-growing vegetation. The horns, which were colossal in males, are unique in protruding backwards, outwards and lastly forwards. As the **keratinised** sheaths do not fossilise, the span of the horns must have been up to three metres. Equids (see pp. 132–4) belonging to the modern genus *Equus*, including primitive ancestors of zebras (e.g. *E. koobiforensis*), entered Africa some 2.3 Ma. The native equids of the extinct hipparionin lineage lived alongside the newcomers for a while and exhibited a distinctive pattern of tooth development, being similar to that of suids (see pp. 108–10) and elephants from the Pliocene. In addition, there was a significant turnover in the suid guild during the Early Pleistocene between about 2 and 1.6 Ma, with the disappearance of tetraconodontines (the *Nyanzachoerus* lineage (see pp. 285–6)), and the emergence of *Kolpochoerus* and various species of *Metridiochoerus* (Fig. 7.6). The latter genus resembled modern warthogs, but was over twice as large and sported two massive pairs of tusks that curved upwards and outwards. The cheek teeth were hypsodont and exhibited an intricate, knobbly enamel pattern paralleling that of specialised grazers such as equids. *Kolpochoerus* was characterised by well-developed and expanded **zygomatic arches** along with spectacular recurved tusks. The African forest hog and bushpig are regarded as possible descendants for exhibiting these features, but to a more moderate extent. The suids were a highly significant component of the overall African Plio-Pleistocene faunal diversity, before declining to the three living genera—*Phacochoerus* (modern warthogs), *Hylochoerus* (modern forest hogs) and *Potamochoerus* (modern bushpigs and river hogs)—and were likely just as significant to the predators as prey. The transformations in the framework of the bovid and suid guilds would have had a direct impact on the large **carnivore community**.

Many of the modern large carnivores including lions, leopards and cheetahs lived alongside some of the more primitive sabre-toothed cats of the genera *Dinofelis*, *Megantereon* and the more open-country *Homotherium* up until around 1.5 Ma. *Dinofelis* seems to have persisted in Africa until a later time than the other machairodontines (ca. 1.2 Ma). One reason was that, being less specialised than the sabretooths

Megantereon or *Homotherium* (near obligate predators of large **ungulates**), *Dinofelis* was more adaptable and more capable of adjusting to changes in environmental conditions and availability of prey. The somewhat specialised *D. piveteaui*, the latest known species of *Dinofelis*, was larger than *Megantereon* and therefore more formidable, but simultaneously retained more advanced climbing **adaptations** than *Homotherium*. It could withdraw into forests and subsist on smaller prey, since its comparatively shorter upper canine teeth were less likely to break when contacting bone than those of the "classic" sabretooths. Evolutionary changes in predator and prey elements of the African mammal community around 1.7 Ma are linked to shifts in vegetation (such as more open grasslands) due to climate change. In Africa, the extinction of the three aforementioned machairodontine genera, and possibly three species of hyaenid (see pp. 88–90) afterwards, highlight the severity of the Quaternary climatic changes. *Dinofelis* and *Megantereon*, as **ambush predators**, would have been significantly impacted by reduced vegetation cover, shifts in prey size and behaviour (particularly suids and antelope) that impacted hunting success, and potentially increased competition with open-terrain predators such as lions, hyaenas and *Homotherium*. It is likely that *Homotherium* was severely affected by reductions in the size and availability of their prey as dry-season droughts intensified. The eventual disappearance of all these machairodontines, due to combined changes in climate, vegetation, and prey structure and composition, left the modern big cats as the only felids of the large carnivore guild, which had therefore adopted its modern composition and structure—a major transformation.

Subsequent to that, the African fauna in general started to reflect this modern stamp. The giraffids and suids were eventually whittled down in diversity and populations to the modern limited variety of species while the deinotheriids, chalicotheriids (see pp. 128–9) and **tridactyl** hipparionin equids were all going extinct. The turnovers of the mammal groups in Africa that happened during the Early Pleistocene may be considered the final major stages in shaping the continent's modern fauna. The megafaunal extinctions that took place in Africa during the Late Pleistocene to Early Holocene included at least 24 total extinctions of species, such as the last species of *Megalotragus*, which went extinct around 5.5 ka probably because of a reduction in grasslands following the Last Glacial Period (see p. 313), and the giant buffalo *Syncerus antiquus* (see p. 119), which was abundant in the southern, eastern and northern regions of Africa until over 4 ka, possibly due to aridification and competition with domestic cattle for resources. However, localised extinctions were much more prevalent, with mammals such as black wildebeest and white rhinoceroses vanishing from the Cape region, and giraffes and zebras from Northwestern Africa, for example.

North America Haile Quarry Site (Sites 15A, 7C, 21A and 16A), United States: Piacenzian-Calabrian, ca. 2.6–1 Ma.

The Early Pleistocene deposits of the Miocene- to Pleistocene-age quarry site of Haile in Alachua County, northern Florida, is an extensive karst deposit mostly made up of limestones and unconsolidated clays. The **terrestrial** fauna from Haile indicates that, during the Early Pleistocene, Florida was covered in subtropical forests interspersed with great expanses of **savannah**. Certain mammals that first appeared in North America by this time include *Smilodon*, jaguars and various rodents such as voles and lemmings. Following the arrival of all these new Eurasian immigrants, various **taxa** characteristic of the Miocene and Pliocene had vanished, including the last mega camels, "gomphotheriids" (see p. 66), borophagine canids (see pp. 92–3) and North American hyaenids (see pp. 89, 305), and the final hipparionin equids (see pp. 132–4) were declining, with the modern single-toed genus *Equus* persisting. The deposits at some localities, such as Haile 15A, about 2.6–2.2 Ma in age, had accumulated in a spring surrounded by forest according to the presence of tapirids (see pp. 127–8), cervids (see pp. 115–7), beavers and flying squirrels. The most abundant mammal from Haile 15A is *Equus*, which in addition to the llama *Hemiauchenia* and the cotton rat (*Sigmodon*), are indicators of neighbouring grassland environments. The episodes of clay accumulation suggest instances of a tranquil water environment, either a lake, pond, or slow-flowing river. This is further evident from plentiful fossils of amphibians and freshwater fish, and pleurodire turtles (see pp. 14–5).

There are four equid species known from Haile 15A, two hipparionins (*Nannippus* and *Cormohipparion*), which are both rare as fossils, and two species of *Equus* (*E. simplicidens* and a smaller species known from insufficient material to confirm identification). The most abundant artiodactyl (see p. 104) was the modern white-tailed deer (*Odocoileus virginianus*), whose fossils from Haile 15A show that it has existed in Florida during the last 2 to 2.5 Ma without considerable changes. An early record is also preserved at Haile 15A for the llama *Hemiauchenia macrocephala*, which survived in Florida until the close of the Pleistocene before it went extinct (Webb, 1974). Fossils of the the **omnivorous** long-nosed peccary, *Mylohyus* (Fig. 7.8), from Haile 15A show it was larger and had a more elongated mandibular symphysis than later Pleistocene members of the genus. *Mylohyus* is rare in the Haile deposits, possibly because it lived in small social groups. A **diastema** between the canine teeth and front premolars gave *Mylohyus* a highly elongated snout.

Fig. 7.8. Reconstructed life appearance of the long-nosed peccary *Mylohyus nasutus*. Reconstructed height: 0.75 m.

Mylohyus had forelimbs somewhat longer than the hindlimbs, and very tall neural arches on the back vertebrae to support the massive head. The carnivorans (see pp. 81, 83–4) from Haile 15A include *S. gracilis*, the oldest and smallest *Smilodon* species, weighing about 55 to 100 kilograms (Fig. 7.9). It was a stocky, leopard-sized felid (see pp. 86–9) that had a more **derived** dentition (e.g. larger incisors) and cranial features (e.g. larger mastoid region, which is the bony prominence below and behind the ear) than *Megantereon cultridens* (see p. 302), and was slightly larger-bodied, but had a considerably smaller mandibular flange. *S. gracilis* was likely ancestral to both *S. fatalis* and *S. populator*, which it would later be replaced by at around 500 ka. Other localities such as Haile 7C, about 2.2–1.9 Ma in age, had formed in a freshwater lake, as determined by alternating sands and clays, and a plethora of freshwater and **semi-aquatic** vertebrates including **osteichthyan** fish, salamanders, frogs, pleurodire turtles, crocodylians and tapirs of the extinct species *Tapirus lundeliusi*, similar in dimensions and appearance to the modern mountain tapir (*T. pinchaque*). The terrestrial fauna indcates that the lake was largely surrounded by woodland or forest. By the Early Pleistocene, there were many new immigrant New World porcupines (erethizontids) and xenarthrans (see p. 58) from South America, the latter group comprising armadillos and the huge megatheriid ground sloth (see pp. 61–2) *Eremotherium*, known from Haile 7C (Fig. 7.10A). It was a generalist which dispersed as far north as the Southern United States by around 2.5 Ma.

Like *Nothrotheriops*, but unlike *Megalonyx* (see pp. 323, 327), *Eremotherium* did not expand its range further north, probably due to an intolerance of cooler temperatures. *Eremotherium* had a relatively slender skull more lightly built than that of *Megatherium* (see pp. 328–9), and the dentition consisted of uniformly shaped, **hypsodont cheek teeth**. *Eremotherium* had relatively long limbs that terminated in hands with three **digits**, unlike the four-fingered hands of the closely related *Megatherium*. The large **occlusal** surface of the teeth along with the moderately broad snout indicate that *Eremotherium* was possibly a **mixed feeder**. Haile 7C has also yielded the remains of the glyptodontine genus *Glyptotherium*, which was similar to *Glyptodon* (see pp. 328–9) in having a deep jaw with hypsodont teeth along with a dome-like **carapace** consisting of hundreds of small hexagonal scutes, but smaller in body size. *Glyptotherium* is distinguished from *Glyptodon* by numerous differences including, but not limited to, a slender zygoma (i.e. cheek bone), a much more arch-shaped carapace with a saddle-like overhang above the tail, a considerably more rectangular and less rounded ventral edge of the carapace, and somewhat longer tail armour. The somewhat smaller pampatheriids (see pp. 59–60) of the genus *Holmesina* are known from both Haile 15A and Haile 7C. *Holmesina* is characterised by a long, low skull as well as flexible armour, was up to two metres long and weighed around 227 kilograms, four times larger than the modern giant armadillo (*Priodontes maximus*) (Fig. 7.10B). The large dimensions and slow habits of larger pampatheriids including *Holmesina* made them unable to survive on an **insectivorous** diet, so they switched to **grazing** tough plant material in the savannahs of the Americas using their **primitive**, peg-like teeth.

Fig. 7.9. Two sabre-toothed cats (*Smilodon gracilis*) confronting the glyptodontine *Glyptotherium* sp., the latter which is crouched in a defensive position. Early Pleistocene of Florida (ca. 1 Ma).

Fig. 7.10. Reconstructed life appearance of fossil xenarthrans from Haile. (A) The megatheriid ground sloth *Eremotherium eomigrans*; (B) The pampatheriid armadillo *Holmesina floridanus*. Scale bar: 1 m.

North America Haile Quarry Site (Sites 15A, 7C, 21A and 16A), United States: Piacenzian–Calabrian, ca. 2.6–1 Ma, cont.

In one limestone cave locality, Haile 21A, about 1.6–1 Ma in age, near complete skeletons of two individuals of the stocky homotherin felid *Xenosmilus* have been discovered (Martin *et al.* 2000) (Fig. 7.11), along with prevalent fossils of the flat-headed peccary *Platygonus cumberlandensis* and some fragmentary material from *Smilodon gracilis*. The locality was a probable den for the big cats, and many if not all of the peccary remains would have been brought back by these **carnivores** (Martin *et al.* 2011). *Xenosmilus* had the sabre-like, coarsely serrated upper canine teeth typical of homotherins, but were relatively shorter and broader. *Xenosmilus* is further distinguished by a narrow forehead, and some pronounced constriction **posterior** to the **orbits**. The upper canine tooth and third incisor were closely spaced, allowing the upper canine teeth and incisors to function together as a single unit more effectively when biting. This trait has given *Xenosmilus* the name "cookie-cutter cat" to interpret this specific craniodental anatomy, and infers a different method of dispatching prey, by hunting through a "bite and retreat" method in which its front teeth were used to inflict deep wounds. Its short and sturdy **distal** limb bones indicate that *Xenosmilus* was an **ambush predator** (Fig. 7.12), hunting in a similar fashion to *Smilodon* and thus, probably faced competition. Why *Xenosmilus* went extinct by the Middle Pleistocene remains uncertain.

P. cumberlandensis had a pig-like muzzle and enlarged tusks that may have been used in defence. It was **herbivorous** and possessed relatively large, hypsodont molars that bore transverse crests. Thus, it most probably consumed coarse plant material, including roots and tree pods. *P. cumberlandensis* possibly had a complex digestive tract similar to that of modern ruminants (see p. 110). In life, it would have most closely resembled the extant Chacoan peccary (*Catagonus wagneri*), its closest modern ecological and taxonomic **analogue**. The abundant fossils of *P. cumberlandensis* indicate that it was **gregarious** like all modern peccaries. Compared to *P. cumberlandensis, Mylohyus* had a proportionately larger skull, more elongated, gracile limbs and a shorter tail. As evident from dental **microwear** analysis, *Mylohyus* fed on even harder foodstuffs than *P. cumberlandensis*, including woody browse such as twigs, attaining niche partitioning. Carbon isotope analysis shows that the diet of both peccaries, along with those of other contemporary **ungulates** such as the llama *Hemiauchenia* (see p. 305), varied over time, probably due to warming climates and/or drier environments.

Other lineages of predators were diverse, known from other localities of similar age to Haile 21A, including Haile 16A. These predators included canids (see pp. 92–3) of the medium-sized species *C. edwardii* and the Armbruster's wolf. The former is probably most closely related to the modern coyote (*C. latrans*), whereas the latter may potentially be an early species of the genus *Aenocyon* (see pp. 323–4), called *Ae. armbrusteri* (Bartolini-Lucenti *et al.* 2024). There were also medium-sized tremarctine ursids (see pp. 92, 94–5) of the species *Arctodus pristinus* (lesser short-faced bear) (Fig. 7.13). *Ar. pristinus* was the smaller of the two species of the short-faced bear genus *Arctodus*. *Ar. pristinus* possibly weighed over 500 kilograms, and when standing on its hindlimbs, it reached three metres—these figures make this bear comparable to a large brown bear (*Ursus arctos*) in size. Compared to *Ar. simus* (see pp. 322–3), *A. pristinus* was shorter-limbed, and had premolars and first molars that were relatively smaller and more broadly spaced.

The North American mammal fauna remained relatively stable for much of the Pleistocene until around 135 to 195 ka, with an increased exchange of mammals between Eurasia and North America. The immigrant species from Eurasia included bison, and new species of deer and rodents. Many of these were inhabitants of the grassland faunas of the mammoth steppe **ecosystem** (see pp. 311–2) that bordered the ice sheets. Those immigrant species likely competed with or displaced some native ones, and could have influenced vegetation patterns, leading to shifts in plant **communities** and the overall structure of North America's ecosystems.

Fig. 7.11. Reconstructed life appearance of the machairodontine felid *Xenosmilus hodsonae*. Reconstructed length: 2 m.

Fig. 7.12. A scene depicting *Xenosmilus hodsonae* preying on a flat-headed peccary (*Platygonus cumberlandensis*). Early Pleistocene of Florida (ca. 1 Ma).

Fig. 7.13. Life reconstruction of a lesser short-faced bear (*Arctodus pristinus*) mother and cub. Middle Pleistocene (ca. 500 ka) of Florida.

North America

Rancho La Brea Tar Pits, United States: Upper Pleistocene-Greenlandian, 0.05–0.011 Ma.

Located in Los Angeles of Southern California, United States, the Rancho La Brea Tar Pits comprise a series of over 100 individual pits, which have resulted from excavations at the site since 1901. The sediments (primarily sands) were deposited after having been carried by **alluvial** fans formed in the neighbouring Santa Monica area. The natural asphalt that seeped up from the ground then saturated the sands, producing an adhesive and possibly shallow mass. The site is particularly famous for yielding myriad mammalian fossils dated to the Last Glacial Period (see p. 313), in which the asphalt often produced a deposit sufficiently viscous for trapping animals. A thin layer of water, dust, or leaves would conceal the deposits, and animals would saunter in for a drink, get stuck, and eventually die. As evident from thorough analysis of fossils obtained from one of the pits (Pit 91), palaeontologists have deduced that the skeletons of trapped animals were buried quite rapidly, though not before predators successfully snatched away most of the limb elements from the carcasses' exposed sides (Spencer *et al.* 2003). A number of those **carnivores** were unable to escape with their food and instead became trapped themselves, having become extra bait. Consequently, an abnormally large proportion of carnivore remains have been preserved at La Brea, which encompass over 90% of all the vertebrate fossils. Within the deposits, secondary movement of the bones because of limited flow has resulted in thorough mingling of the bones in all the pits, making it extremely difficult to put together a skeleton with every bone belonging to the same individual. Rancho La Brea's fossil fauna is highly diverse, encompassing fossils of large carnivorans (including ursids (see pp. 92, 94–5), felids (see pp. 86–9) and canids (see pp. 92–3)) and a range of mammalian **herbivores** (including bovids (see pp. 116, 118–9), tayassuids (see p. 110), camelids (see pp. 104, 106–7), cervids (see pp. 115–7), equids (see pp. 132–4), giant ground sloths and mammoths), along with a variety of birds and smaller animals. In combination with fossil plants and pollen analysis, this fauna signifies a temperate climate, marginally cooler and moister than at present, and a mix of greenery that was dominated by buckwheat, sagebrush and pines.

Among ursids, the giant short-faced bear *Arctodus simus* was the apex predator of the continent and the largest carnivoran of all time, standing up to 12 feet tall on its rear limbs. Once considered to be **hypercarnivorous**, the giant tremarctines including *Ar. simus* lacked any specific **adaptations** for pursuing prey. Although they were long limbed, sheer bulk probably meant they were unable to chase prey over any distance. Instead, they were probably **omnivorous** with **carrion** being an important part of their diet. One suggestion is that *Ar. simus* was a **kleptoparasite** of machairodontine felid/large canid kills (Fig. 7.14). The felids were diverse, from the very popular sabre-toothed cat *Smilodon fatalis* and the scimitar-toothed cat *Homotherium*, to lions (larger than modern lions), jaguars (*Panthera onca*), cougars (*Puma concolor*), bobcats (*Lynx rufus*) and the so-called American "cheetah" *Miracinonyx trumani*. Among other carnivorans, modern ringtails (*Bassariscus astutus*) and raccoons (*Procyon lotor*) were also prevalent, while the mustelids (see p. 99) consisted of weasels (*Neogale frenata*) and American badgers (*Taxidea taxus*). Intermediate in size between *S. gracilis* (see pp. 318–9) and *S. populator* (see pp. 329–30), *S. fatalis* is the best known species, thanks to the vast collection of fossils from La Brea comprising almost one million bones spanning all ages. It was a massive machairodontine, lion-like in bodily dimensions but with very muscular, sturdy limbs, inferring a much greater body mass than that

Fig. 7.14. A wooded grassland **ecosystem** adjacent to one of the tar pits at the site of Rancho La Brea, Late Pleistocene. *Foreground*: a giant short-faced bear of the species *Arctodus simus* (left) confronting a *Smilodon fatalis* (right) over its kill (a Western horse (*Equus occidentalis*)); *Right background*: at the tar pit, from left to right, is a pack of dire "wolves" (*Aenocyon dirus*), a carcass of the ancient bison (*Bison antiquus*), a Western horse, and the carcass of a Columbian mammoth (*Mammuthus columbi*) **scavenged** by more dire "wolves". *Bottom left*: a striped skunk (*Mephitis mephitis*); *Right middleground*: a caracara (*Caracara plancus prelutosus*), and behind it, three asphalt storks (*Ciconia maltha*); *Top right background*: a turkey vulture (*Cathartes aurora*); *Right far background*: a herd of Columbian mammoths; *Left far background*: two other asphalt storks, and behind them, a herd of ancient bison.

Fig. 7.15A. Phylogenetic tree of a sampling of fossil mammal **taxa**, particularly proboscideans (see pp. 64–8), xenarthrans (see p. 58) and carnivorans (see pp. 81, 83–4), from the Rancho La Brea Tar Pits.

North America Rancho La Brea Tar Pits, United States: Upper Pleistocene-Greenlandian, 0.05–0.011 Ma, cont.

of a tiger (*Panthera tigris*) or a lion (*Panthera leo*), at 160 to 280 kilograms. *S. fatalis* had a stubby tail, a shortened back and a long, powerful neck, as in *Megantereon cultridens* (see p. 302), but with a larger skull possessing sabre teeth that were more hypertrophied and recurved, and serrated on both sides. *S. fatalis* also exhibited a highly reduced mental protuberance (bony flange) on the lower jaw. A lot of bones from *S. fatalis* exhibit signs of injury or disease, indicative of fighting and a hard lifestyle. Some of those bones also show signs of healing, possibly suggesting tolerance of wounded or ill members of a group, inferring social behaviour. Male and female *S. fatalis* were about the same size, and exhibit small to absent **sexual dimorphism** smaller than that of modern lions, but similar to that of modern wolves. This kind of sexual dimorphism indicates that rather than having a lion-like harem system, *S. fatalis* had a familial pack structure comparable to the extended family groups of canids.

Homotherium serum, which is significantly rarer at La Brea than *S. fatalis* perhaps due to a more solitary lifestyle, had shorter upper canine teeth than sabre-toothed cats. *H. serum* was a North American species of *Homotherium* distinguished by a narrower **diastema** between the upper canine teeth and third incisors, a broader forehead, and upper canine teeth that were not particularly enlarged, but highly flattened. The forelimbs were also longer in proportion to the hindlimbs, giving this species a somewhat sloped appearance. *H. serum* was more of a **cursorial** predator, favouring open habitats and mainly preying on **grazers**. The American lion (*Panthera atrox*) may have resulted from a population of steppe lions (*Panthera spelaea*) (see pp. 343–4) becoming isolated south of the encroaching continental ice sheets around 200 ka. Despite regular cycles of glacial and interglacial periods, American lions never intermixed with their Alaskan brethren, as both probably avoided the dense boreal forests. Evidence from La Brea suggests that American lions were more solitary than their modern counterparts due to very few fossil skeletons compared to other large predators.

Fig. 7.16. A scene in which a Pleistocene North American jaguar (*Panthera onca augusta*) has just killed a California tapir (*Tapirus californicus*) for food.

The Pleistocene North American jaguar (*P. onca augusta*) was relatively longer-limbed, with metapodials that were more elongated, so it was clearly able to hunt in more open areas than modern South American jaguars (Fig. 7.16). It was also larger than modern jaguars, by about 15 to 20% (Seymour, 1989), as populations further north are larger than those in warmer equatorial areas. *M. trumani* was a close relative of the cougar characterised by a proportionately small head with a foreshortened face, along with canine teeth that were short and conical, unlike those of many contemporary predators. Given its flexible vertebral column, heavy steering tail and long limbs, *M. trumani* has often been imagined as an equivalent to the modern cheetah (*Acinonyx jubatus*). However, recent research indicates it was not as specialised in running as cheetahs because it retained retractile claws that would have restricted its ability to run very fast. It seems *M. trumani* was ecologically very versatile, and at least some inhabited mountains and steep, rocky slopes, denning in caves and hunting large bighorn sheep (*Ovis canadensis*) and mountain goats (*Oreamnos americanus*) in dangerous cliff faces, somewhat like modern snow leopards (*Panthera uncia*) in the Himalayas. Nonetheless, many studies suggest *M. trumani* is the reason why modern pronghorns (*Antilocapra americana*) are able to run at speeds so much beyond that of coexistent extant carnivores—it evolved alongside this now extinct predator (Barnett *et al.* 2005; Higgins *et al.* 2023) (Fig. 7.17).

The canids included modern wolves, coyotes and foxes, which had immigrated from Eurasia during the Early Pleistocene. The most famous and common canid is the extinct dire "wolf" (*Aenocyon dirus*), a short-limbed, robust carnivore and scavenger represented by many thousands of specimens from La Brea. Its dental and cranial morphology closely resembled that of the modern grey wolf (*Canis lupus*) due to **convergent evolution**, although its teeth were more massive and exhibited more advanced shearing adaptations. *Ae. dirus* also had a more powerful bite force than any species of *Canis* (possibly about 2,300 Newtons). As per recent genetic studies (Perri *et al.* 2021), the genus *Aenocyon* was in fact unrelated to any modern canid lineage, and was part of a distant South American lineage that evolved independently from wolves, coyotes and jackals, as supported by the lack of evidence for hybridisation with other canids. Carbon isotope studies show that *Ae. dirus* favoured prey occupying open environments such as **savannah**, in contrast to *S. fatalis*' predilection for woodland-dwelling prey including deer, tapirs and woodland-dwelling bison (DeSantis *et al.* 2019).

The variety of herbivores that these predators fed on included a great diversity of immigrant Eurasian species. Among artiodactyls (see p. 104), the deer included modern white-tailed deer (*Odocoileus virginianus*) and the peccaries are represented by *Platygonus compressus*, the latter which was better suited to consuming grasses than earlier species of the genus (see pp. 320–1). The bovids encompassed caprines, or goat-antelopes, a familiar group that diversified into quite different forms, some of which do not particularly resemble domesticated sheep and goats. Restricted to the bleak northern tundra today are muskox (*Ovibos moschatus*), the sole extant species of a group which once ranged across Eurasia

Fig. 7.15B. Phylogenetic tree of a sampling of fossil mammal taxa, particularly **ungulates**, from the Rancho La Brea Tar Pits.

North America Rancho La Brea Tar Pits, United States: Upper Pleistocene-Greenlandian, 0.05–0.011 Ma, cont.

and North America in far more temperate environments. The helmeted muskox *Bootherium* had longer limbs than the modern species and lived in open woodland from Alaska to Texas. *Bootherium* also had a thicker cranium, a significantly longer muzzle, and higher, downward curving horns that were merged along the midline of the cranium. The shrub-ox *Euceratherium*, which appears to have favoured hilly terrain, was of similar build and appearance, but had horns that were more powerfully-built, and curved upwards and forwards. *Euceratherium* was a **browser** that fed on shrubs and tree leaves as determined by fossilised dung pellets.

With a pair of truly impressive horns (spanning up to two metres across!) was the much larger bovine *Bison latifrons*, the long-horned or giant bison. With a potential weight of nearly two tonnes, it was probably the largest bovid that ever lived. In keeping with larger body sizes, long-horned bison would have had shorter fur around the head compared to the modern species. It lived on plains and in woodlands (though those horns may have caused manoeuvring difficulties in dense forest), throughout much of North America, reaching northern Mexico. It seems to have gone extinct roughly 20 ka, giving way to the smaller migratory species *B. antiquus*, also known as the ancient bison. The latter is the most frequently unearthed large herbivorous mammal from La Brea, suggesting that it was likely a swamp- or woodland-dwelling animal. It is unknown whether *B. antiquus* lived in vast herds like modern prairie bison, since forest bovids typically do not live in groups that large. *B. antiquus* had larger bones than modern bison, along with extremely tall neural arches. It became progressively widespread in parts of Western North America from about 18–10 ka, after which it was replaced by its direct descendant the modern American bison (*B. bison*). The remains of two species of antilocaprid (see pp. 111–2) have also been recovered: the modern pronghorn and the diminutive *Capromeryx*, the latter which is distinguished by having short, pointed, non-branching horns. *Capromeryx*, over 50% smaller than the modern pronghorn, was typically more gracile than other antilocaprids given its reduced size and weight. It was one of the few antilocaprids with dual horn cores (i.e. they were actually four-horned), with a larger back pair and a shorter front pair. The neck and back vertebrae of the camelid *Camelops* are characterised by prominent neural arches like that of the dromedary (*Camelus dromedarius*), so *Camelops* may have had a hump. Some vegetational remnants were discovered between the teeth of various specimens, demonstrating that *Camelops* was a generalist herbivore eating any accessible plant material. There is also direct evidence of hunting by the first humans to cross over the Bering Land Bridge (see p. 174).

Among perissodactyls (see p. 126), **monodactyl** equids (see pp. 132–4) of the modern genus *Equus* were quite diverse, the most commonly found species from La Brea being the Western horse, *E. occidentalis*, which occurred all over the Pacific coast from Canada to Mexico. It is comparable to modern zebras in anatomy and proportions, although not closely related. The teeth of *E. occidentalis* had less intricate enamel **folds** and ridges than modern equids. Isotope studies on the teeth suggests that *E. occidentalis* occupied a distinct dietary niche from modern equids, probably living as a **mixed feeder** ingesting a large quantity of woody browse from shrubs and other vegetation. Tapirs of the extinct species *Tapirus californicus*, although relatively scarce, favoured the more wooded areas of the landscape. As in most modern tapirs, *T. californicus* is thought to have been predominantly solitary. Dominating the landscape were the Columbian mammoths (*Mammuthus columbi*), which stood four metres tall at the shoulder. They are characterised by a tall, single-domed skull,

Fig. 7.17. A scene depicting the extinct American "cheetah" (*Miracinonyx trumani*) hunting a modern pronghorn (*Antilocapra americana*) on the prairies of the Western United States during the Late Pleistocene.

a sloped back and a pair of curved tusks up to 4.2 metres long. The molars had separated enamel ridges, which were covered in "prisms" orientated towards the grinding surface. Carbon isotope analysis indicates that *M. columbi* was not limited to browsing or grazing. *M. columbi* likely had a reduced covering of hair compared to the Eurasian species, due to the warmer climate in North America. It did not live in the Canadian Arctic areas, which were instead inhabited by woolly mammoths (*M. primigenius*). The distributions of both species possibly overlapped, and molecular studies indicate that they interbred.

Among ground sloths, *Nothrotheriops* is the most familiar species of nothrotheriid (see p. 62), better known as the Shasta ground sloth. It was the smallest Pleistocene North American species and found only throughout the dry environments of the Western United States and Mexico. Complete skeletons, including preserved hair, internal organs and tendons, have given us an accurate image of their appearance. Frequently regarded as comparable to an American black bear (*Ursus americanus*) in size, *Nothrotheriops* is implied to have weighed 250 kilograms on average, smaller than the contemporary ground sloths including *Paramylodon* and *Megalonyx*, both of which weighed over a tonne. Like other species of North American sloths, *Nothrotheriops* is believed to have had a less specialised and more **primitive** cranial anatomy in comparison to South American nothrotheriids. The **premaxilla** was tapering and narrow, although less so than in its South American kin. These skull features, along with the rather small size of *Nothrotheriops* and given that all sloths are believed to be foregut fermenters, suggests that it was a low browsing generalist, consuming low trees and shrubs, and sometimes grasses.

The more heavily built *Megalonyx* is the best known and largest species of the Megalonychidae (see p. 61), which lived up to the end of the Late Pleistocene as far north as Alaska and Canada. Like *Nothrotheriops*, it was a browsing animal, characterised by broad, peg-like teeth, massive jaws, and a short, blunt snout. *Megalonyx* also had **plantigrade** feet, contrary to the majority of sloths, enabling it to stand up on its hindquarters in a semi-erect stance to reach high for vegetation while using its stout tail for balance. It likely inhabited a range of habitats, dwelling primarily in mixed conifer-hardwood forests dominated by spruce. *Paramylodon*, also known as the Harlan's ground sloth, was a mylodontid (see p. 61) morphologically similar to the genus *Glossotherium* (see pp. 306–7). *Paramylodon* was distinct from other North American ground sloths in being a mixed feeder with a preference for grasses. Skull studies have revealed that there are two different skull **morphotypes**, suggesting that sexual dimorphism may have been present, although it remains uncertain as to which is male or female. The forelimb anatomy implies that *Paramylodon* was possibly capable of burrowing.

As the sister group to the New World vultures, the teratorns (see p. 39) such as *Teratornis*, the largest, most famous and most abundant fossil bird from La Brea, have been traditionally depicted as bald-headed, oversized condors scavenging carcasses. Closer inspection of their skeletons suggests that this is probably incorrect. It had a wingspan up to 3.8 metres (12.5 feet), with a wing area of about 17.5 square metres and weighed about 15 kilograms (33 pounds), a third larger than the largest living extant bird today, the Andean condor (*Vultur gryphus*), and twice the size of the modern California condor (*Gymnogyps californianus*). As the largest bird at La Brea, its powerful bill could have ripped holes in large carcasses and opened them up for smaller scavengers to reach the edible parts (Fig. 7.18). *Teratornis* was one of the more **terrestrially**-adapted teratorns (strong legs for walking) and would have likely hunted small prey (a large beak and gape).

According to one interesting recent study by O'Keefe *et al.* (2023), using pollen, geochemistry and sedimentary data, the ecosystem in Southern California began transitioning from woodland to shrubland around 14.7 ka as the region warmed and rainfall decreased. At the same time, human populations, which depended on hunting large mammals, were on the increase. As evident from prevalent charcoal in **lacustrine** sediments in Southern California, our fires occasionally burned uncontrollably, further degrading the dry, stressed landscape. Climate change combined with the growing impacts of humans may have ultimately resulted in the extinction of large, specialised mammals in the region around 13 ka before they died out elsewhere in North America. These include giant ground sloths, camels, horses, ancient bison, sabre-toothed cats, American lions and dire "wolves". The ecosystem of Southern California had completely transformed into chaparral shrubland by around 12.9 ka, with the only remaining large predators being the highly adaptable coyotes and cougars in addition to humans. Why many of the aforementioned megafauna became extinct in North America as a whole by the start of the Holocene continues to be highly contentious. The Younger Dryas, an abrupt cooling event that occurred in much of the Northern Hemisphere about 12.9–11.7 ka, is frequently cited as one of the contributing factors, characterised by increased atmospheric carbon dioxide, increased seasonality, and substantial shifts in vegetation. This event disrupted the prevailing warming trend following the Last Glacial Period and may have dramatically reduced the habitable space for numerous species. Indeed, one study by Stewart *et al.* (2021) has found that decreases in North American megafaunal populations correlate with declines in global temperature.

Fig. 7.18. Life reconstruction of the large, scavenging teratorn, *Teratornis merriami*, perched on the carcass of a giant short-faced bear (*Arctodus simus*) stuck in one of the tar pits at La Brea, during the Late Pleistocene.

South America Tarija Formation, Bolivia: Calabrian–Upper Pleistocene, 1.1–0.1 Ma.

Among the richest South American palaeontological sites for mammals is the Tarija Formation of southern Bolivia, from which thousands of fossils have been obtained. The majority of those specimens are of xenarthrans (see p. 58), although fossils of immigrant groups from North America are adequately represented too. From the Tarija Formation, there is a greater diversity of mammal species within groups that entered South America during the Great American Biotic Interchange, or GABI (see pp. 252, 295) (including sabre-toothed cats, canids (see pp. 92–3), ursids (see pp. 92, 94–5), camelids (see pp. 104, 106–7), equids (see pp. 132–4), and "gomphotheriid" proboscideans (see p. 66)), than within the relic native groups, most of the immigrant **taxa** of which are **carnivores**. Therefore, the Tarija Formation approximates a modern South American mammal assemblage more closely than any of the other South American fossil sites featured in this book. Yet, this mammal assemblage was markedly different from one found in present day South America. Over one third of the mammal families present at this site are now extinct on the continent and comprise the most abundant mammals, including notoungulates (see pp. 140–3), litopterns (see pp. 143–4), ground sloths, glyptodontines and "gomphotheriids". The fossils were preserved within **fluvial** sediments in grassland at an elevation of about 6,000 feet above sea level, and provide a glimpse of life in South America prior to human arrival around 15 ka.

The notoungulates included the massive, hippopotamus-like *Toxodon*, which had a rather peculiar gait, with its forelimbs shorter than the hindlimbs. The steeply-sloped rear of the skull suggests that *Toxodon* held its head low for consuming huge amounts of low-growing plants. *Toxodon* was a **bulk feeder** as evident from its wide, **hypselodont** incisors, although isotopic analysis suggests that it did not feed only on grasses. It also had **hypsodont cheek teeth** (with curved, bow-shaped crowns) that grew during the animal's lifetime to compensate for wear induced by the mixture of grit in its food. It had short, muscular limbs with **plantigrade** feet terminating in three short, functional **digits**, with much of the body weight being borne by the central digit. *Toxodon* possibly galloped in order to evade predators, although as in modern rhinoceroses, its size likely served as its primary protection. The short limbs, dense bones and relatively high-positioned **orbits** and nasal openings infer that *Toxodon* was at least **semi-aquatic**.

The large, long-necked, camel-like litoptern *Macrauchenia* is among the most iconic of South American native **ungulates** (see p. 139). The most distinctive feature is a long, narrow skull which had a large, retracted nasal opening between the eyes, popularly interpreted as housing a trunk in the living animal. Trunks are useful tools for animals with short necks, and this puts the idea at odds with *Macrauchenia*. A snout fully enclosed by bone also indicates that a trunk is a very unlikely feature. Evidence suggests that the nose was possibly swollen in a similar manner to the moose (*Alces alces*) or saiga (*Saiga tatarica*). The retracted nostrils were possibly a protection from being impaled on tough spiny branches when feeding (giraffes are a less pronounced example). Hypsodont teeth are suggestive of chewing on grass, and thus *Macrauchenia* was possibly an opportunistic feeder of high and low vegetation. Another unusual feature is the proportions of the robust limbs, which were not especially suited for fast running. Instead, they were adapted for manoeuvrability, quick turning probably over broken terrain (Fariña *et al.* 2005).

The megatheriid ground sloth *Megatherium* is the largest known xenarthran, exceeded in size only by several proboscideans and the extinct rhinocerotoid *Paraceratherium* (see pp. 249–50) among land mammals. Like all other megatheriids, *Megatherium* had a robust skeleton, being able to balance using a broad tail into a tripodal position to reach up into the canopy, probably occupying a giraffe-like niche. Surprisingly, fossil trackways show that the sloth was one of the few species capable of also walking **bipedally** (at almost four miles per hour). Unlike the broad-nosed species from other families, *Megatherium* possessed a particularly narrow muzzle and **prehensile** lips to selectively **browse** in a similar fashion to modern giraffes (Fig. 7.20). A deep jaw and hypsodont teeth suggest a rough fibrous diet. A combination of the teeth showing strong, vertical biting forces, and muscle attachment to the elbow which parallels those seen in carnivores, led to the fanciful suggestion that *Megatherium* **scavenged** from carcasses. This has since been disproved by carbon isotope analysis on bones showing that the species was definitely an obligate **herbivore**. The mylodontid ground sloth *Scelidotherium* had a narrow, elongated muzzle that suggests it was a **selective feeder** (Fig. 7.21). This feeding style was assisted by its particularly long and thin predental spout. This is a toothless projection at the tip of the lower jaw found in all sloths, although it varies between species due to differences in feeding behaviour. The lower lips, which worked together with the upper lips and tongue during feeding, were supported by the predental spout. *Scelidotherium* also possessed limbs with digging **adaptations**, and it possibly consumed underground stems such as tubers, along with other kinds of vegetation.

Among the glyptodontines was *Glyptodon*, characterised by a rounded, highly elongated **carapace**, along with a cone-shaped tail consisting of mobile **caudal** rings formed by large, conical **osteoderms**. Its robust limbs may have enabled *Glyptodon* to stand bipedally for feeding, mating, defence and/or observing its surroundings, as in extant giant armadillos (*Priodontes maximus*). The most striking feature of *Glyptodon* is its very specialised, telescoped cranium, compressed to such a great degree from front to rear that the brain cavity is located above rather than **posterior** to the **cheek teeth**. *Glyptodon* was particularly well-armoured, even possessing conical osteoderms lining the front and ventral sides of its carapace for protection of the neck and stomach. The narrow snout of *Glyptodon* suggests it was a selective feeder. *Doedicurus* had a formidable tail that was surrounded by a sequence of smooth caudal rings at the base for mobility and terminated in a club of solid bone at the tip, which had shallow depressions that supported blunt spikes. This feature would have been used primarily in territorial fighting (Blanco *et al.* 2019), supported by evidence of damage on the armour of some specimens. Biomechanical studies are further evident that the tail was strong enough to damage a rival's carapace. *Doedicurus* had a fatty cushion underneath the carapace for lowering the force of impacts during fighting, and probably also as a fat storage for leaner times. A secondary hump on top of the shell is thought to have been a fat reserve as well. *Doedicurus* was a bulk feeder as indicated by its relatively wide snout.

Fig. 7.19. Phylogenetic tree of a sampling of fossil mammal taxa from the Tarija Formation. The phylogenetic position of *Toxodon platensis* and *Macrauchenia patachonica* follows Welker *et al.* (2015).

South America Tarija Formation, Bolivia: Calabrian–Upper Pleistocene, 1.1–0.1 Ma, cont.

The major predators of the region were the machairodontine felid *Smilodon populator*, the Argentinian short-faced bear *Arctotherium angustidens* and large canids of the wolf-like genus *Protocyon*. *S. populator* was the largest and most robust species of *Smilodon*, with an upper weight estimate of over 400 kilograms. Other than size (about 50% larger) and bulk, the animal is distinct from the North American *S. fatalis* (see pp. 322–4) in features including comparatively shorter metapodials, higher placed nasal bones, a straighter upper profile in the skull, and forelimbs slightly more elongated in proportion to the hindlimbs. *S. populator* lived across much of South America, between Venezuela in the north and Argentina in the south, where it preyed on large mammals including *Macrauchenia* and evidently *Doedicurus*. *A. angustidens* may have evolved its large size **convergently** from its northern counterpart *Arctodus* (see pp. 320–3), probably for dominance over other predators when competing over the largest carcasses (Mitchell *et al.* 2016) (Fig. 7.22). The closest living relative of *A. angustidens* is the much smaller spectacled bear (*Tremarctos ornatus*) from the mountainous regions of South America. *Protocyon* was an important predatory component. It was large (weighing 16 to 25 kilograms) and related to the extant South American canids (cerdocyonines such as the maned wolf). Unlike these, *Protocyon* was a **hypercarnivore** according to its dental morphology. *Protocyon* probably hunted medium-sized **grazers**, although it would not have missed the chance of consuming **carrion** as well (remains of "gomphotheriid" bones with bite marks attributed to *Protocyon* have been found (Dominato *et al.* 2011)). Following the extinction of *Protocyon* , the bush dog (*Speothos venaticus*) was left as the sole surviving hypercarnivorous canid in South America, a medium-sized species that preys on rodents and other vertebrates in lowland tropical **ecosystems**.

Among immigrant herbivores, *Palaeolama* (Fig. 7.21) was a much larger relative of modern llamas, at around 300 kilograms. It had a slender skull consisting of a shallow, robust lower jaw and a long **rostrum**. Its **brachydont** dentition indicates that it was a **mixed feeder**, with a diet composed predominantly of fruit and leaves, supplemented with grass. *Palaeolama* also specialised in forest browsing. *Palaeolama* was relatively thickset in build as evident from its limb bones including short, robust metapodials. Limbs such as these are often linked to moving across rough and uneven ground, which is indicative of being well-suited to evading predators in wooded regions as well (Dompierre, 1995; Scherer, 2013). The equids included *Equus insulatus*, which was very abundant at this site in comparison to the strange *Hippidion*, with *H. principale* being less common than its close relative *H. devillei*. The most peculiar feature of *Hippidion* is an elongated, arched nasal bone, assumed to have housed enlarged tubular nasal passages as in the modern saiga. This would have been an adaptation for moistening the air it inhaled from the cool, dry environment in which it lived. The highly elongated nasal notch along with the muscle attachments on the skull infer that *Hippidion* had a prehensile upper lip for consuming tree leaves and shrubby plants, a lifestyle further supported by its less hypsodont dentition. All three equid species had relatively short, thickset limbs as an adaptation for surviving in mountainous terrain.

Fig. 7.20. Life reconstrution of the giant megatheriid ground sloth *Megatherium americanum* browsing on the treetops, while a social group of modern capybaras (*Hydrochoerus hydrochaeris*) saunter by in the foreground. Late Pleistocene of South America.

All proboscidean fossils from Tarija belong to the "gomphotheriid" *Cuvieronius*, which had a lower and more elongated skull relative to modern elephants, along with twisted tusks that had a spiral-shaped band of enamel extending their whole length, similar to the tusks of narwhals (*Monodon monoceros*). In contrast to modern elephants, *Cuvieronius* had simpler, brachydont molars, showing that it was a browser consuming leaves, bark and twigs, although carbon isotope analysis has shown that it also fed on grasses and other kinds of vegetation. The proportions of grass and other vegetation consumed by different individuals varied, and their diet possibly varied with the seasons. *Cuvieronius* was adapted to cool, high-altitude environments associated with the Andes mountain range. During the Pleistocene megafaunal extinction around 11.7 ka, a broad range of mammals became extinct in South America. Every mammal species weighing over a tonne and about four fifths of those weighing more than 45 kilograms had disappeared from the continent. The largest surviving South American mammal is the Baird's tapir (*Tapirus bairdii*), which is only under 10% as large as *Megatherium*. Additionally, the extinction wiped out the last surviving representatives of nine native South American mammal lineages, which were litopterns, notoungulates, and seven xenarthran groups including glyptodontines, pampatheriids and five sloth families (Megalonychidae, Megatheriidae, Nothrotheriidae, Mylodontidae and Scelidotheriidae) (see pp. 59–62).

Fig. 7.21. A scene in which the scelidotheriid ground sloth *Scelidotherium tarijense* is browsing on a bush, while a group of the massive llama *Palaeolama weddelli* forage in the background. Late Pleistocene of South America.

Fig. 7.22. A scene in which the carcass of the toxodontid notoungulate *Toxodon platensis* has been washed up onto a riverbank by a flood and is being occupied by the tremarctine ursid *Arctotherium angustidens*, seeing off a passing jaguar (*Panthera onca*) with a languid look. Late Pleistocene of South America.

Australia Naracoorte Caves, South Australia: Chibanian–present, 0.5–0 Ma.

The vertebrate fossil deposits of the Naracoorte Caves world heritage site have preserved a record of faunal diversity for the southeastern part of the South Australia state, spanning the Middle to Late Pleistocene. The limestone in the region was produced by coral and other marine organisms approximately 20 Ma, when southern Australia was submerged under shallow sea. After a fall in sea level, which rendered the landscape a dry limestone plain, parts of the limestone were dissolved by groundwater, resulting in the formation of caves and sinkholes. For about half a million years, the caves served as feeding spots for predators and pitfall traps in which mammals and other **terrestrial** vertebrates fell into sinkholes and were then unable to escape, resulting in an extraordinary collection of fossils. The fossils were preserved in fans of sediment that blew and washed in from the surface. The sediments were dispersed across the cave floor as the cone enlarged, ultimately accumulating up to the ceiling and closing the entrance around 15 ka, sealing the caverns and its treasures until discovery. The fossil record spans several glacial periods and the arrival of humans to Australia. The fossil traps are particularly noteworthy for the extinct Australian megafauna, with nearly all groups of the continent's native marsupial mammals attaining large body sizes, including giant macropodid kangaroos (see pp. 54–6) up to 10 feet (3 metres) tall, along with diprotodontids (see p. 56) and the bizarre palorchestids (see p. 56).

Among the familiar macropodids, there were giant short-faced kangaroos (a completely different subfamily to the macropodines), giant protemnodons and relatively giant, **omnivorous** rat-kangaroos. Safe to say things were often larger in the past! The largest marsupial today is the male red kangaroo (*Osphranter rufus*) at around 60 to 70 kilograms, which can look a 1.8 metre tall human in the eye. This is pretty small fry when you consider that the largest ever marsupial was four metres long and weighed in at 2.5 tonnes. Probably the best known of the fossil macropodids are the short-faced sthenurines. Only a handful of genera such as *Sthenurus* (72 to 173 kilograms) and *Simosthenurus* (around 118 kilograms) are known but they were relatively large in size, with the smallest species (*Procoptodon gilli*) weighing around 50 kilograms. Up until the discovery of the giant *Macropus titan*, the sthenurines held the title of largest known macropodid—*Procoptodon goliah*. They were primarily **browsing** species, using their height and long forelimbs to reach up to low branches. Unlike other macropodids, sthenurines possessed only one toe on the hind foot, which ended in a blunt, hoof-like toe. An inflexible spine and robust hindlimbs also altered sthenurine locomotion. Unlike other macropodids, they moved by **bipedal** striding, in a similar manner to theropod dinosaurs. *Sthenurus* had **hypsodont cheek teeth** and a more elongated snout than other sthenurines, so its bite force at the incisors was likely less powerful than other sthenurines. In comparison to other sthenurines with broader incisors, that would have enabled the plucking of a larger amount of small leaves with each bite, probably eating small-leaved shrubs.

Simosthenurus and *P. goliah* were very short-faced, and had large masticatory muscles and small incisors, enabling a powerful bite. These features, along with the large surface of their cheek teeth (no larger than in other sthenurines), would have enabled them to feed on tough plant material, but not on abrasive vegetation such as grasses. They are thought to have fed on dryland browse including shrubs and trees, specialising in manipulating their food with their hands in a koala-like manner. *Simosthenurus* is thought to have lived in more coastal regions than *Sthenurus*, possibly in open forests and woodlands. Both *P. goliah* and *Simosthenurus* were similar in having small incisors within a very short, robust skull. The small incisors indicate they were not **grazers**, as they would have been limited to small bites and it is hard to gather grass with the forepaws. *P. goliah* may have had difficulty getting enough to feed on if it grazed on grass. It has been suggested that it was possibly a **mixed feeder**, consuming both browse and abrasive vegetation, potentially overlapping with the diet of the giant extinct wombat *Diprotodon*, but using its long, powerful arms and upright stance for grasping and pulling down branches (Prideaux, 2004).

The highly diverse macropodid genus *Protemnodon*, believed to be a browser, is characterised by a somewhat flattened skull with a long snout. Unique to large species of *Protemnodon*, the molars were highly complex, with numerous fine ridglets emanating all over the crowns that strengthened the teeth, unlike in any extant grazing macropodid. The large-bodied species of *Protemnodon* (*P. anak*) had very short hind feet and much more robust arms, whereas those from New Guinea, such as *P. tumbana*, were morphologically more like the short-footed tree-kangaroos (*Dendrolagus*). *P. anak* (around 131 kilograms) had very odd body proportions, with a long tibia, long arms and a very long neck. The **proximal** and **distal** humerus morphology suggests a a greater capacity for weight-bearing and a more stable joint compared to extant macropodids (Jones *et al.* 2022), thus suggesting it was mainly **quadrupedal** (Jones and Janis, 2024). *Propleopus* was a large relative of modern musky rat-kangaroos (*Hypsiprymnodon moschatus*), weighing up to

Fig. 7.23. Reconstructed life appearance of the giant monitor lizard *Varanus (Megalania) priscus*. Reconstructed length: 5.5 m.

Fig. 7.24. Phylogenetic tree of a sampling of fossil mammal **taxa** from the Naracoorte Caves. The phylogeny within Macropodidae follows Richardson *et al.* (2024).

Australia Naracoorte Caves, South Australia: Chibanian-present, 0.5-0 Ma, cont.

70 kilograms. It had a dentition equipped with enlarged third premolars, transformed into oblong blades, but the molars were tuberculate (i.e. having small, rounded protuberances/projections), which made it possible to grind both plant and animal matter.

The largest marsupials of all were the diprotodontids and palochestids, wombat-like marsupials that were closely related to true wombats. *Palorchestes* had muscular forelimbs with a peculiar, limited range of motion at the elbow, along with massive, koala-like claws. These may have been **adaptations** for grasping food more easily (Richards *et al.* 2019). *Palorchestes* was a **selective feeder** as evident from dental **microwear**. *Palorchestes* possibly also had a protrusible, giraffe-like tongue as indicated by the elongated mandibular symphysis. The retracted nasal bones have led to the assumption that *Palorchestes* had a short, tapir-like **proboscis**, giving it the common name "marsupial tapir", although recent analysis on the skull morphology suggests the presence of muscular, **prehensile** lips instead. *Zygomaturus* was a late surviving zygomaturine diprotodontid living in wetter areas of southern Australia and thus was possibly **semi-aquatic**, scooping up water plants with its shovel-like incisors. It had laterally prominent zygomatic arches and a curiously arched muzzle, with the nostrils raised as in modern hippopotamuses, and with bony nasal projections that have sometimes been interpreted as small horns. *Zygomaturus* had lower-crowned cheek teeth with wider crests compared to those of the diprotodontid *Diprotodon*, thus the former presumably fed on coarse, woody vegetation including bark roots and stems, with its short face and powerful jaws.

Resembling a rhinoceros in wombat's clothing, *Diprotodon* was comfortably the largest marsupial to ever walk the earth (at 2.5 to 3.5 tonnes). *Diprotodon* is characterised by an enormous but lightweight skull, a pair of chisel-like upper incisors and a pair of elongated, forward-pointing lower incisors, and sturdy, pillar-like limbs, with the upper limb bones being more elongated than the lowers. The

Fig. 7.25. Reconstructed life appearance of the mekosuchine crocodylian *Quinkana fortirostrum*. Reconstructed length: 3.5 m.

Fig. 7.26. Life reconstruction of the large madtsoiid snake *Wonambi naracoortensis* feeding on an Australian Tasmanian devil (*Sarcophilus harrisii*). Late Pleistocene of Australia.

development of this skull was likely linked to the need for accommodating larger teeth and muscles to process coarse vegetation efficiently. While incredibly large, the food requirements of *Diprotodon* were proportionally lower than expected for such a massive animal, due to its lower metabolic requirements than smaller Australian marsupials. Hundreds of skeletons are known of *Diprotodon* across the continent, giving a very good insight into the habits of an extinct species. *Diprotodon* may have walked relatively slowly, since it had unusual feet with large ankles and wrists, but very small toes. Multiple species were once named, but a thorough reexamination found that there was only one species that showed dramatic **sexual dimorphism**. Sexual dimorphism is common in **polygynous** breeding animals, such as elephants. Males probably led solitary lives, only coming together during the breeding season to fight with other males for dominance. Females travelled in multiple family herds between waterholes in the dry season, making long treks following the same routes. The earliest fossils of *Diprotodon* date back to 1.65 Ma but they only encountered humans by about 65 ka.

Preying on all these marsupials was the modern thylacine species *Thylacinus cynocephalus* (see pp. 52–3), and *Thylacoleo carnifex*. The latter was a leopard-sized thylacoleonid (see p. 57) that had a rounded skull equipped with massive, **plagiaulacoid** third premolars that functioned as **carnassials**, similar to those found in felids (see pp. 86–9). It also had blunt, **caniniform** incisors that were likely used for gripping prey rather than delivering a killing bite, and fewer molars than earlier thylacoleonids to accommodate the enlarged carnassials, a condition linked to a highly specialised, **hypercarnivorous** diet. *T. carnifex* also had powerful forelimbs and a highly manoeuvrable elbow joint, along with enormous hooked claws on the large, semi-opposable thumbs for grasping, manipulating and possibly even disemboweling prey. Scratch marks in one steep-sided cave locality indicate that *T. carnifex* also had no difficulties climbing back-and-forth. Most of the scratches were attributed to juveniles, suggesting the cave was probably used for rearing young. Weighing no more than 100 kilograms, *T. carnifex* was the thylacoleonid which best fitted the title of "marsupial lion". *T. carnifex* was the apex terrestrial mammalian predator across the continent.

During the Pleistocene, Australia was also home to a giant monitor lizard, *Varanus priscus*, which is the largest known terrestrial lizard (Fig. 7.23). *Megalania* was originally given its own genus, but was subsumed under *Varanus* as its closest relatives are the extant Komodo dragon (*V. komodoensis*) and Lace monitor (*V. varius*). Size estimation by researchers has plagued *Megalania*. Anything between 3.5 and nearly 8 metres long has been suggested. However, when using Komodo dragons as a model, large individuals come in around seven metres long. Although they undoubtedly overlapped in time with early modern Australian aboriginals, which must have been a frightening experience, we do not have any evidence of interaction. Among other reptiles, the large mekosuchine crocodylian (see p. 28) *Quinkana* (Fig. 7.25) was a dominant predator of terrestrial environments during the Pleistocene, with long limbs for lifting the body off the ground and a broad, short snout equipped with flattened, serrated, sharp teeth. A bite from *Quinkana* would cause massive blood loss. There is also no current evidence that this species interacted with modern aboriginal humans, although there was suitable overlap of time.

Although not known from complete skeletons, *Wonambi* was a madtsoiid snake (see pp. 23, 25) with more generalist features. This reflects the cooler, drier conditions of southern Australia during the Plio-Pleistocene where *Wonambi* lived compared to the warm Miocene rainforests (see pp. 288–94). Its skull was disproportionately small and the jaw could not be completely **disarticulated**, indicating that it consumed medium-sized prey (Fig. 7.26). *Wonambi* was also the last surviving madtsoiid. In addition, Australia harboured huge, flightless dromornithid birds (see p. 36), or "mihirungs". The last mihirung, *Genyornis* (Fig. 7.27), survived alongside the famous Late Pleistocene Australian fauna of giant short-faced kangaroos, *Diprotodon*, *T. carnifex* and "*Megalania*". *Genyornis* was a generalist compared to its relatives, being large-bodied and ostrich-sized but with a smaller, wedge-shaped skull. It also had very distinctive feet almost like that of an ostrich (*Struthio camelus*), with hooved, compact toes. Paintings on cave walls of large birds (distinct from emus) and carved footprints are proof of a coexistence between dromornithids and the first humans to Australia.

While it is possible that humans were responsible for the extinction of *Diprotodon* and much of the other Australian megafauna that died out around 44 ka, either directly through hunting or changes to land management through firestick burning, other factors such as increasing aridification of the Australian environment were probably more significant.

Fig. 7.27. Reconstructed life appearance of the dromornithid bird *Genyornis newtoni*. Reconstructed height: 2.3 m.

Asia Liang Bua Cave, Indonesia: Chibanian–present, 0.19–0 Ma.

Liang Bua is a limestone cave situated in the Manggarai District some 13 kilometres northwest of Ruteng in western Flores, an island forming part of the Lesser Sundas archipelago situated east of Java. While certain Indonesian islands have previously been connected to Australia or Asia via land bridges, others are separated by deep oceanic trenches and have undergone isolation with limited dispersal of species between them. Flores was formed relatively recently, sometime within the last 15 million years, and was home to adept but very strange **endemic** species during the Pleistocene, which have undergone **insular dwarfism** or **insular gigantism**.

The dominant **grazers** in Pleistocene Flores were stegodontid proboscideans (see p. 67) of the genus *Stegodon*, distinguished by a short, deep skull, **brachydont** molars exhibiting parallel ridges, and no permanent premolars. One highly unusual feature are tusks often growing so close together that there was insufficient space for the trunk to fit in between. Older individuals possibly had to rest their trunks sideways over either tusk. Flores has been populated in successive waves of dwarfed *Stegodon* (*S. sondaari*, *S. florensis insularis* and *S. sumbaensis*) as well as ancestral large species (such as *S. florensis*), which were able to island-hop through Indonesia by swimming. *S. sondaari* lived during the Early Pleistocene (ca. 0.9 Ma) and reached weights of roughly 490 kilograms. *S. sondaari* weighed about 15% of the probable ancestral *Stegodon* found on the mainland (*S. elephantoides*). During the Middle Pleistocene, *S. sondaari* seems to have been replaced by the relatively massive (about 1.7 tonnes) *S. florensis*, the latter which gave rise to the 570 kilogram subspecies *S. f. insularis* by the Late Pleistocene. *S. sumbaensis*, also from the Late Pleistocene, reached the smallest dimensions—possibly at 250 kilograms. These pygmy stegodontids were likely major prey of the Komodo dragon (*Varanus komodoensis*), which occupied the apex predator niche on Flores and continues to do so on the island's edges. It is thought that these large lizards were able to **ambush** *Stegodon* in a similar manner to how they currently hunt buffalo and deer. *V. hooijeri*, a considerably smaller extinct species of monitor lizard, was also native to Flores. *V. hooijeri* occupied a substantially different niche from the Komodo dragon, exhibiting morphological similarities to the Gray's monitor (*V. olivaceus*) and Nile monitor (*V. niloticus*)—both **omnivores**. Like the Gray's monitor, *V. hooijeri* probably fed mostly on fruit.

Several extinct bird species are also known from Liang Bua. The largest was *Leptoptilos robustus*, a giant stork closely related to the Marabou stork (*L. crumenifer*) and adjutants (*L. javanicus*, *L. dubius*). *L. robustus* was possibly **carnivorous**, surviving on **carrion** and small vertebrates. *L. robustus* had a slender build typical of extant *Leptoptilos* species, but unusually heavy, thick-walled bones, and thus was formerly presumed to have been functionally flightless. Recent studies of wing bones revealed no proportional reduction in wing length, indicating it was completely flight-capable. Additionally, vulture fossils of the genus *Trigonoceps* have been recovered from Liang Bua. Today, this genus is represented solely by the white-headed vulture (*T. occipitalis*) of sub-Saharan Africa. It remains unclear as to whether those vulture fossils represent an endemic species or subspecies. Fossil data from Liang Bua indicates that mice and rats were the most common vertebrates on Flores, constituting roughly 80% of all fossil discoveries from the cave, but this is potentially due to accumulation biases from predators. Flores was once home to several species of giant rat. The only extant species is the Flores giant rat, *Papagomys armandvillei*, which has a close extinct relative called *P. theodorverhoeveni*. The locomotion of these species is **terrestrial** and **arboreal**, respectively. The largest Flores murid is *P. armandvillei*, at 1.2 to 2.5 kilograms, whereas the omnivorous *P. theodorverhoeveni* weighed 0.6 to 1.6 kilograms. Other extinct giant rats include *Spelaeomys florensis*, a terrestrial species thought to have been **herbivorous**.

Hominins (see pp. 156–60) have a long history in Flores, with the oldest known evidence of tool use dated to at least 1 Ma. It remains uncertain as to whether these original setters descended from *Homo erectus* or the more **primitive** *Homo habilis* (see pp. 158–9), but regardless of which hominin initially reached Flores, they later evolved into a whole population of dwarfed humans called *Homo floresiensis*, characterised by features including robust legs and unusually flat feet. The oldest fossils of *H. floresiensis* are dated to 0.7 Ma. These dwarf humans have been informally named the "hobbits" due to their short stature. They had considerably smaller brains than one might expect for their size (380 cubic centimetres), probably an **adaptation** for conserving energy in an island environment with restricted resources, as brains are metabolically expensive organs. However, the region of their brains linked to higher cognition was roughly equal in size to that of modern humans, thus they did not necessarily have lower intelligence—stone artefacts and cut marks on pygmy elephant remains indicate they hunted cooperatively. However, butchery marks on *Stegodon* bones are scarce, and competition with other predators over carcasses would have been fierce, so some authors have proposed that the main prey of *H. floresiensis* were giant rats. *H. floresiensis* was presumably an endemic species and no other hominins are known from the Lesser Sundas before modern humans (*H. sapiens*) arrived some 50–46 ka.

The extinction of *L. robustus*, and *Trigonoceps* from Flores, occurred simultaneously with that of *Stegodon* and *H. floresiensis*, and was most probably triggered by the loss of large carrion, which likely formed the main dietary component of both bird species. Regarding the extinction of all megafauna on Flores, except the Komodo dragon, three hypotheses have been proposed: a volcanic eruption, climate change, and the arrival of modern humans. Although a volcanic eruption cannot be ruled out as a contributing factor, it seems highly improbable to be the causal trigger of the extinctions, partly because multiple layers of volcanic tephra are known from Liang Bua and other sites on Flores. There is so far little substantial evidence to confirm dramatic climatic changes on Flores when the extinctions occurred. Modern humans were the most probable root cause of the extinctions, about 50–46 ka. As the extinction event seems to be limited exclusively to *Stegodon* and its **scavengers**, it appears that overhunting was the most likely cause of extinction due to a lack of evidence for habitat deterioration or invasive species during this time. It is possible that *H. sapiens* displaced *H. floresiensis* via competition. The Holocene would witness further extinctions, including *V. hooijeri* and several giant rats (possibly due to invasive species), and introduce most of the fauna associated with Flores today.

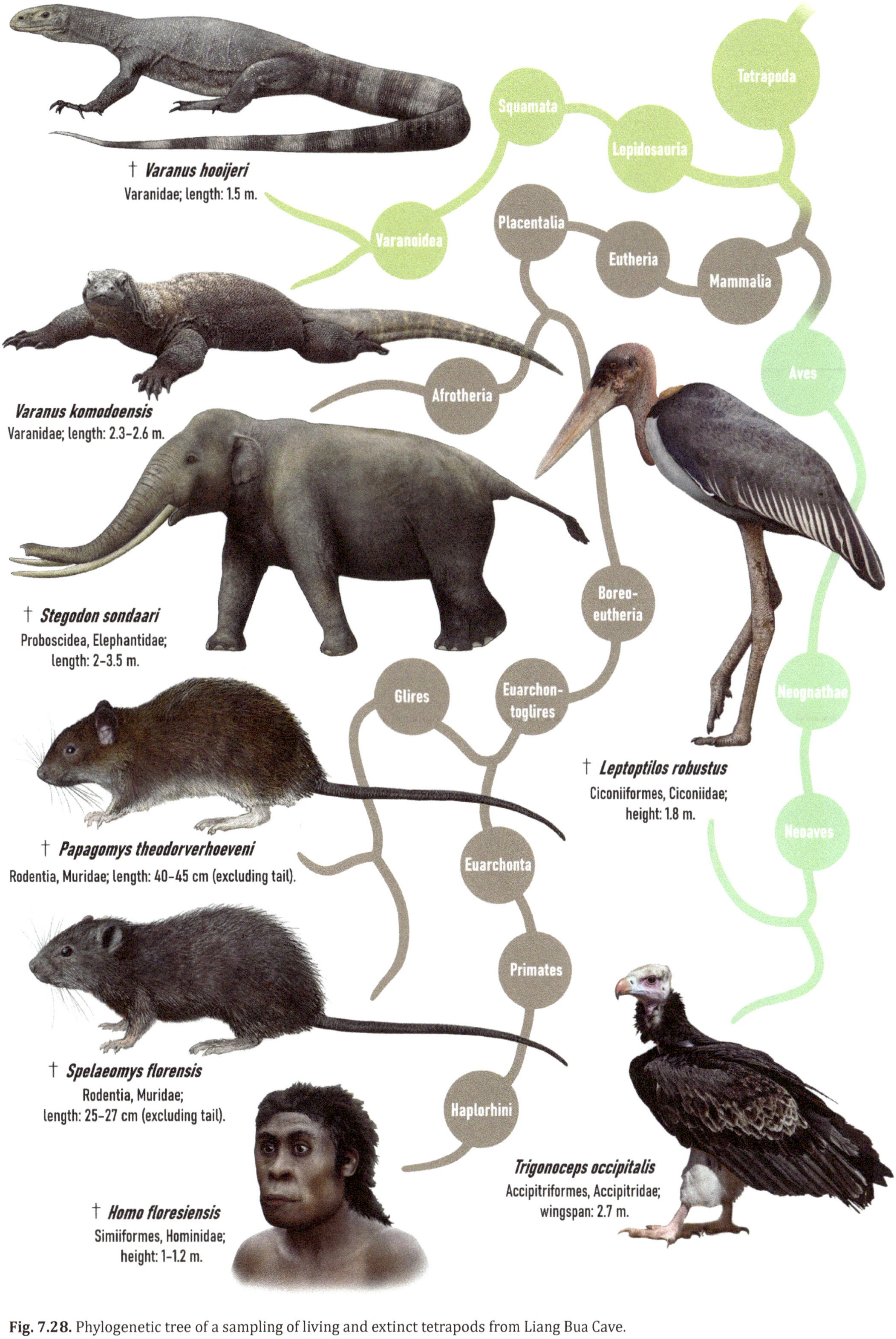

Fig. 7.28. Phylogenetic tree of a sampling of living and extinct tetrapods from Liang Bua Cave.

Europe Cromer Forest-bed Formation, England: Gelasian–Chibanian, 2–0.5 Ma.

The Cromer Forest-bed Formation (CFbF) represents a sequence of Early to Middle Pleistocene marine, estuarine and freshwater deposits located along the northern Norfolk coast in eastern England. The CFbF also outcrops further south along the Norfolk coast and into Suffolk, but its type area is indeed west of Cromer. Accumulation occurred at the base of a low energy, shallow **fluvial** channel as indicated by the fine-grained sediment consisting of detritus muds and organic rich silts, which has yielded rich vertebrate material. This lowland fluvial channel meandered past banks fringed by vegetation; sparse mixed **deciduous** forest covered the land, with open grasslands maintained by herds of **grazing** animals. The Early to Middle Pleistocene small mammal fauna from the CFbF includes species typical of riverside habitats such as beavers, desmans and water shrews, along with vole species of the genus *Microtus*, which indicate local herb-dominated vegetation on the floodplain. The rapid spread of voles throughout Eurasia is a good indicator of the expansion of the steppe and cooler conditions by the start of the Pleistocene. The largest rodent from the CFbF is the extinct giant beaver *Trogontherium* (Fig. 7.29), which strongly resembled modern beavers, but had a tail lacking the characteristic transverse processes of the vertebrae, implying that the tail was cylindrical in cross section rather than flat. The incisors of *Trogontherium* were finely grooved and the lower ones were especially elongated, while the morphology of the **anterior** part of the jaw and its rough surface suggests the presence of a muscular upper lip. Its lifestyle was possibly **analogous** to that of a coypu (*Myocastor coypus*) instead of a beaver (*Castor*), the former which today inhabits the waterways of temperate South America, burrowing in river banks, spending a lot of time in water and feeding mostly on aquatic plants.

Fig. 7.29. Reconstructed life appearance of the beaver *Trogontherium cuvieri*. Reconstructed length: 1.5 m.

As in North America, the Early Pleistocene in Europe was dominated mainly by mammal lineages that still persist today, although represented primarily by extinct species. Among these were "**archaic**" mammoths (*Mammuthus meridionalis*) (Fig. 7.30A) and the two-horned, steppe-adapted Etruscan rhinoceros (*Stephanorhinus etruscus*) (Fig. 7.31A). Bovids (see pp. 116, 118–9) and cervids (see pp. 115–7) were diverse, in addition to a variety of equid species (see pp. 132–4) of the modern genus *Equus*, including *E. stenonis* (Fig. 7.31B). *Mammuthus meridionalis* was a large mammoth that stood four metres tall at the shoulder, and is characterised by a distinct but relatively low dome on the forehead and a markedly sloped back. *M. meridionalis* also sported long, robust, twisted tusks that diverged from the upper jaw. It had **brachydont** molars with relatively few (~ 13) thick enamel ridges (much less than in geologically younger mammoths), adapted to a **browsing** diet of leaves and shrubs in woodlands. This suggests it flourished in relatively temperate climates and thus, lacked a dense covering of fur.

The Late Pliocene relic *S. etruscus* was a rather large rhinoceros (up to four metres long and two metres at the shoulder), with elongated, gracile limbs well-suited for **cursorial** locomotion. The brachydont teeth, the horizontal position of the head and the shape of the skull suggest that *S. etruscus* was a browser, much like the black rhinoceros (*Diceros bicornis*). *S. etruscus* quickly spread from Eastern Europe, becoming one of the most abundant rhinoceros species in the continent's biomes during the Late Pliocene to Early Pleistocene, although its specific dispersal routes and the timing of its appearance in different regions are still being debated. *E. stenonis* was a rather robust equid, standing between 1.45 and 1.55 metres at the shoulders. It differed in appearance from all modern equids in having a more elongated and

Fig. 7.30. Proboscideans from the Cromer Forest-bed Formation. (A) the southern mammoth (*Mammuthus meridionalis*); (B) the straight-tusked elephant (*Palaeoloxodon antiquus*); (C) the steppe mammoth (*Mammuthus trogontherii*). Scale bar: 1 m.

Fig. 7.31. (A) Life reconstruction of the narrow-nosed rhinoceros *Stephanorhinus etruscus*; (B) Life reconstruction of the equid *Equus (Allohippus) stenonis*. Early Pleistocene of England.

proportionately larger head, with a long snout, an undulated dorsal profile and a deep nasal notch. *E. stenonis* had relatively large and stocky limbs, with proportionately elongated metapodials. In general, *Equus stenonis* was morphologically close to zebras.

The **carnivores** encompassed **primitive** ursids (see pp. 92, 94–5), a diversity of felids (see pp. 86–9) including scimitar-toothed cats of the genus *Homotherium*, pantherine felids and lynxes, numerous types of foxes and wolves, and a whole range of mustelids (see p. 99), including weasels, stoats, martens, wolverines and otters. The bear *Ursus etruscus* (see pp. 309–10), of Late Pliocene to Early Pleistocene age, gave rise to a lineage that led to increasingly **herbivorous** and anatomically bulky populations. *U. deningeri*, the earliest member of this branch, had brachydont **cheek teeth** with a broad **occlusal** surface. *U. deningeri* is characterised by an amalgamation of primitive and **derived** features: the lower jaw was slenderer than that of *U. etruscus* and contemporary modern brown bears (*U. arctos*), but had features that would be more developed in the later cave bear *U. spelaeus* (see pp. 343–5), such as an elongated neurocranium with a straight dorsal profile (Fig. 7.32A). The sturdy limbs also suggest digging abilities since, like *U. spelaeus*, *U. deningeri* was a hibernating species as evident from abundant fossils in cave deposits elsewhere across Europe.

During the Early Pleistocene and into the Middle Pleistocene, the diversity of wolf-sized canids (see pp. 92–3) underwent a notable decline in Eurasia, transitioning from multiple species to just two main types: the small, short-limbed **hypercarnivorous** Mosbach wolf (*C. mosbachensis*) and the larger, long-limbed *Canis (Xenocyon) lycaonoides* (Fig. 7.33). This shift likely occurred due to factors such as climate change, competition, and the evolution of larger wolves. *C. lycaonoides* (also hypercarnivorous) had a robust skull and dimensions similar to those of modern grey wolves (*Canis lupus*) of northern populations. *C. lycaonoides* is distinguished by a short neurocranium in proportion to its overall skull size, along with the presence of lower third molars and a lack of cusps on the anterior portion of the premolars. Some regard *C. lycaonoides* as an early relative of dholes (*Cuon alpinus*) and African wild dogs (*Lycaon pictus*) due to similarities in forelimb morphology (Rook, 1994).

Fig. 7.32. (A) Life reconstruction of the cave bear *Ursus deningeri*; (B) Life reconstruction of the bovid *Bison schoetensacki*. Early Pleistocene of England.

Therefore, one taxonomic affinity has suggested that *C. lycaonoides* belongs in the genus *Lycaon*. The earliest known fossils of the genus *Bison* in Europe (*B. (Eobison) georgicus*) come from Daminisi, Georgia, at around 1.77 Ma, but *Bison* had become more

Europe Cromer Forest-bed Formation, England: Gelasian-Chibanian, 2-0.5 Ma, cont.

common in Europe from about 1.5 Ma (long before they arrived in North America), which replaced the final *Leptobos* of Pliocene ancestry. *B. schoetensacki* was a small, woodland-adapted bison species which had relatively small and gently upward-recurved horns, along with limb bones and metapodials that were more gracile than in later species (Fig. 7.32B). One major event was the dispersal of giant megacerine cervids alongside the already giant *Cervalces*. The megacerine cervids included *Praemegaceros verticornis*, which was one of the largest deer of its time and bore heavy antlers that protruded laterally from the skull as in other megacerines, but was distinct in having antlers with a less palmate and more cylindrical shape (Fig. 7.33). It is believed to not be ancestral to *Megaloceros*, but instead a separate lineage that evolved into a large form independently. It was probably adapted to mixed grassland and forest environments.

The broad-fronted moose (*Cervalces latifrons*) is the largest known cervid, standing up to 2.5 metres at the shoulder and weighing about 1.2 tonnes! It had antlers with a relatively elongated beam, longer than in the modern moose (*Alces alces*) and comprising about half of the antler's length (Fig. 7.34). The antlers of the males had large, slightly concave palmate lobes each with seven large tines on average, and may have been used for display purposes instead of fighting, since these cervids are thought to have been solitary like its modern relatives. Judging by the huge span of its horns (also up to 2.5 metres), *C. latifrons* likely favoured open terrain. *C. latifrons* resembled the modern moose in possessing long limbs, a short neck, and a similarly elongated facial region and narrow jaw. The molars of *C. latifrons* indicate a similar diet of woody browse, although the less specialised structure of the facial region suggests a broader diet, likely including a significant proportion of fairly tough herbaceous vegetation. Other than these Asian immigrants to Europe, there were also some African immigrants, the most impressive being the massive *Hippopotamus antiquus*, a hippopotamus weighing 3.5 to 4.2 tonnes. The anatomy of *H. antiquus* was more poorly adapted than that of *H. amphibius* to **terrestrial** locomotion, with a much larger body mass and shortened metapodials, while several other features were better adapted to an aquatic lifestyle (e.g. **orbits** more elevated). Like *H. amphibius*, *H. antiquus* sported four pairs of incisors

Fig. 7.34. Reconstructed life appearance of the cervid *Cervalces latifrons*. Reconstructed length: 3 m.

Fig. 7.33. A scene in which a pack of the canid *Canis (Xenocyon) lycaonoides* is hunting down a cervid of the genus *Praemegaceros*. Middle Pleistocene of England.

and fed largely on aquatic plants rather than the terrestrial grasses favoured by extant hippopotamuses. Other African immigrants included the spotted hyaena (*Crocuta crocuta*) and the early human *Homo antecessor* (see pp. 159–60)—a set of fossil footprints from the locality of Happisburgh that corresponds to the latter species are the earliest known human footprints outside Africa, dated to over 800 ka.

Among the pantherine felids, the robustly built European "jaguar" (*Panthera gombaszoegensis*), also believed to have African ancestry, was larger on average than the modern jaguar (*P. onca*), but smaller than modern lions (*P. leo*) and tigers (*P. tigris*) (Fig. 7.35). The multiple known intact lower jaws of *P. gombaszoegensis* have jaguar-like characteristics and dentition, although the limited postcranial material shows affinities to the lion. Recent phylogenetic analysis based on cranial characters shows that the closest living relative of *P. gombaszoegensis* is the tiger rather than the jaguar, forming a sister **taxon** relationship with *P. tigris* (Chatar *et al.* 2022). However, it should be noted that even though *P. gombaszoegensis* may be among the basal taxa within the tiger lineage, it was not a true tiger nor was it ancestral to the modern tiger. Given this close relationship, *P. gombaszoegensis* probably led a solitary lifestyle. *P. gombaszoegensis* became extinct shortly after the arrival of lions (*P. leo*) and leopards (*P. pardus*) in Europe by the Middle Pleistocene.

Fig. 7.35. Reconstructed life appearance of the pantherine felid *Panthera gombaszoegensis*. Reconstructed length: 1.9 m.

Europe throughout the Early Pleistocene was cool and dry relative to the Pliocene, but there were warmer and wetter periods too. By the late Early Pleistocene, milder glacial pulses alternated with warmer interglacials. Vegetation types would have shifted continuously between interglacial-glacial cycles even in the more muted cycles of the Early Pleistocene. Long-limbed equids derived from *Equus stenonis* were diverse by the late Early Pleistocene, including the relatively small and gracile *E. altidens*, a close extinct relative of wild asses. Among rhinoceroses, the primitive *Stephanorhinus etruscus* gave rise to more gracile species adapted to open, drier environments, such as *S. hundseimensis*, which had longer horns and more **hypsodont** molars for grazing grasses. The emergence of macaques (*Macaca sylvanus*) and wild boar (*Sus scrofa*) is suggestive of the more temperate climate of these woodlands. Additionally, red deer (*Cervus elaphus*), fallow deer (*Dama*) and roe deer (*Capreolus capreolus*) made their initial appearance in the woodlands. The primitive fallow deer relatives (e.g. *Dama roberti*) and the modern species (*D. dama*) share the trait of having palmate antlers, although the palmation is narrower in the former. By the start of the Middle Pleistocene, the glacials had intensified significantly, with sagebrush-dominated steppes south of the ice boundaries and temperate deciduous woodlands arriving during the interglacials. The majority of Early Pleistocene mammals survived, although the **ecosystems** were dominated by many more cold-adapted steppe taxa including equids, bovids and mammoths. Some new cervids included *Megaloceros savini*, an advanced species of megacerine abundant in Western Europe during the Middle Pleistocene. *M. savini* possessed large antlers spanning two metres across, although it was smaller-bodied compared to its Late Pleistocene kin (see pp. 343–4). The basal tines of the antlers are well-known to diverge horizontally from the burr. The antlers were relatively flattened, although they lacked the distinctive palmation found on the antlers of the other *Megaloceros* species.

Among the proboscideans (see pp. 64–8), the straight-tusked elephant (*Palaeoloxodon antiquus*) dominated the interglacials (Fig. 7.30B), whereas the glacial episodes were dominated by the steppe mammoth (*Mammuthus trogontherii*) (Fig. 7.30C), both species which had replaced the southern mammoth (*M. meridionalis*). *Palaeoloxodon antiquus* was a large relative of the modern African forest elephant (*Loxodonta cyclotis*), the former which was up to 4.2 metres tall at the shoulder and possessed a pair of downward-curving tusks about three metres long. It also had a prominent bony crest on the skull roof, known as the parieto-**occipital** crest, which anchored powerful neck muscles to support the heavy head. The **lamellar** structure of the molars is considered intermediate between the *Loxodonta* and *Mammuthus* lineages due to the number and extent of the lamellae. The dental pattern suggests that *P. antiquus* was a **mixed feeder**, a dietary habit that evolved to avoid competition with the grazing mammoths and attain niche partitioning. Similar to modern elephants, *P. antiquus* likely had a very sparse fur covering and large ears given that it lived in temperate conditions.

Represented by a near complete skeleton found at the locality of West Runton, *M. trogontherii* was the largest mammoth to occupy Europe, with a shoulder height of up to 4.5 metres. Its tusks could reach lengths of five metres, characterised by a pronounced upward curvature and recurved tips. The high-domed skull was shorter, the lower jaw was smaller and the molars had a more elaborate design compared to *M. meridionalis*, with a larger number of straight and folded lamellae. Thus, it was better adapted to a diet of tough steppe grass. *M. trogontherii* had a relatively less slanted back than the woolly mammoth (*Mammuthus primigenius*) (see pp. 342–3), partly because its tusks were less curved and its hindlimbs were more elongated. As partly evident from its short tail, *M. trogontherii* was the first mammoth species to adapt to life in cold regions, and it likely developed a thin coat of fur. Like *P. antiquus* and other large mammoths, *M. trogontherii* possessed lengthened forelimbs perhaps as a result of a scaling factor associated with the requirement to maintain and balance the massive-tusked skulls (Larramendi *et al.* 2017).

Europe Creswell Crags, England: Upper Pleistocene-Greenlandian, 0.05–0.01 Ma.

The Creswell Crags world heritage site consists of a 450 metre long limestone gorge concentrated on the border of Nottinghamshire and Derbyshire in central England (Currant and Jacobi, 1997), the walls of that gorge which incorporate a complex network of caves, shaped by a vertically moving fault scarp straddling the site's western margin, in which Late Pleistocene and Early Holocene sediments were deposited (Jacobi *et al.* 1998; Stewart and Jacobi, 2015). There are approximately 23 caves, many of which have preserved fossils representing a rich fauna of large Pleistocene mammals (Currant and Jacobi, 2001). Some of the caves and rock shelters served as predator dens (especially among spotted hyaenas (*Crocuta crocuta*) over the last 120,000 years) and were occupied by early modern humans during the Last Glacial Period (see p. 313). The Late Pleistocene in Europe comprises the last glacial-interglacial cycle, commencing 129 ka. The ultimate Late Pleistocene glaciation, the Last Glacial Period, was particularly intense, with almost the whole of Northern Europe blanketed by ice sheets; sea levels lowered by over 120 metres, leading to the emergence of land bridges between Britain and Europe. Even areas of the North Sea were periodically exposed and formed open grasslands, known as Doggerland. Yet, the Last Glacial Period was possibly not the most intense Quaternary glaciation. The 'Anglian' glaciation of circa 450 ka was arguably of greater extent in the UK and is often said to be the most extreme glaciation during the Quaternary. During the Anglian, the British ice sheet reached further south and diverted the course of the River Thames.

Most of the temperate elements that had persisted in Europe during milder glacial-interglacial cycles had become extinct, such as the narrow-nosed rhinoceros, *Stephanorhinus etruscus*, and the straight-tusked elephant, *Palaeoloxodon antiquus*. The European hippopotamus (*Hippopotamus antiquus*) had gone extinct and was replaced by *H. amphibius* during the Middle Pleistocene (by around 400 ka), much earlier than the Last Glacial Period. Populations of *H. amphibius* probably went regionally extinct during the Last Glacial Period, though. Various temperate genera including *Stephanorhinus* and *Palaeoloxodon* were replaced by **taxa** suited to the cool sagebrush steppes and snowfields that bordered the ice sheets, including woolly mammoths (*Mammuthus primigenius*), woolly rhinoceroses (*Coelodonta antiquitatis*), steppe bison (*Bison priscus*), horses (*Equus* sp.), saiga (*Saiga tatarica*), ibexes (*Capra ibex*), and the last and largest of the megacerine cervids (see pp. 115–7) (*Megaloceros giganteus*), along with those mammals associated with the modern Arctic tundra such as muskox (*Ovibos moschatus*) and reindeer (*Rangifer tarandus*), all represented at Creswell Crags. While aurochs (*Bos primigenius*) did survive in warmer parts of the Last Glacial Period, they are often regarded as a warm woodland indicator, being a **grazer** preferring warmer woodland mosaic habitats.

Mammuthus primigenius was well-adapted to the cold steppe-tundra biome, characterised by a high, single-domed forehead along with a thick coat comprising an outer layer of elongated guard hairs and a shorter undercoat insulating its body against the frigid temperatures. Its tail and ears were short to minimise heat loss, like the ancestral *M. trogontherii* (see pp. 338, 341). *M. primigenius* was smaller than its Middle Pleistocene forebears (*M. trogontherii*), measuring up to 3.5 metres tall at the shoulder. *M. primigenius* sported long, strongly-recurved tusks, which were used in **intraspecific competition** and for digging up underground stems from the icy permafrost of the tundra. The molars had a greater quantity of thinner **lamellae** (up to 30 on the third molars), having a significantly larger grinding surface in comparison to that of *M. trogontherii*, an excellent **adaptation** to the poor, fibrous sedges and grasses of the steppe-tundra **ecosystem**. In fact, *M. primigenius* had the most complex molars of any known elephant. The feet and limbs had rather short hair, an adaptation for walking on ground that is typically frozen and periodically buried in snow. The animal's external appearance is well-understood thanks to the numerous, well-preserved frozen carcasses discovered in the Siberian permafrost, in which the low temperatures have remarkably preserved the soft tissues.

Coelodonta antiquitatis was a tandem-horned rhinoceros of the dicerorhinin lineage, considerably better adapted to the cool and dry steppe-tundra environment than the genus *Stephanorhinus*. Like *M. primigenius*, *C. antiquitatis* also had long, thick fur that allowed it to withstand the permafrost and exceptionally cold temperatures of the last glacial pulse. Unlike many depictions, it probably avoided heavy snow, although wear on the lower surface of the massive, sabre-like front horn indicates that these may have been used for sweeping snow away from the ground while foraging. Furthermore, *C. antiquitatis* had **hypsodont cheek teeth** that were well-suited to the tough, herbaceous steppe vegetation. *C. antiquitatis* had a more elongated head and body, and shorter limbs, than other rhinoceroses. The elongated skull gave the head a deep, downward-directed sloping position which assisted grazing on short vegetation. Its shoulder formed a prominent hump, which accommodated a fat reserve to assist survival through harsh winters. As in the woolly mammoth, the feet and limbs of *C. antiquitatis* were coated in short fur too.

Bison priscus was a robustly built bovid that was similar to modern bison in body dimensions, but had a more prominent shoulder hump, a trait associated with the low position of the skull, and had a pair of relatively longer horns that could sometimes span 1.2 metres across. Several external characteristics, such as fat deposits, fur length and even colouration (a reddish-dark brown), are known to us from many portrayals in cave paintings across Western Europe (including the Pin Hole Cave locality at Creswell Crags) and from mummified specimens discovered in the Arctic permafrost, the most well-known being "Blue Babe", found in Alaska. Aurochs, the extinct wild ancestor of all domestic cattle, were prevalent in the Late Pleistocene landscapes, with the males exhibiting massive horns that were recurved upwards and outwards from the base, and inwards at the ends. The females exhibited horns that were slenderer and solely recurved upwards. Compared to most modern cattle breeds, the skull was much more massive and elongated, and the limbs slenderer. Aurochs had an athletic body shape similar to various wild bovines, with a notable emphasis on the neck and shoulder muscles, particularly in bulls. Aurochs lived in small herds primarily during winter, although throughout the summer, they generally lived solitarily or in smaller social groups. Like modern cattle, aurochs would have likely attained social status via combat and displays, with both bulls and cows participating.

Fig. 7.36. Phylogenetic tree of a sampling of extinct mammals from Creswell Crags.

Europe Creswell Crags, England: Upper Pleistocene-Greenlandian, 0.05-0.01 Ma, cont.

As one of the largest Eurasian **herbivores**, the aurochs were impressive and formidable beasts that were held in high importance to early modern humans. They were readily portrayed in Palaeolithic cave paintings, including at Creswell Crags. In Europe, the musk ox survived as the only ovibovin bovid during the Late Pleistocene, which today is native to the tundra of Greenland, Northern Canada, and Alaska, well-adapted to the harsh Arctic winter given its double-layered fur. The saiga is a similar example, which today survives only in the steppe habitats of Central Asia. This species has a short **proboscis**, which is significantly more prominent in the males and gives these animals an unusual look. The reindeer was also an abundant occupant of the cool, steppe habitats during the Late Pleistocene, which at present is a migratory species occupying the cool tundra and boreal forests of the high Eurasian latitudes. Other contemporary species from the Late Pleistocene that persisted in Europe to this day include the moose (*Alces alces*) and ibex. This combination of tundra and steppe components is found in the small mammal fauna too, as shown by the co-occurrence of lemmings with susliks and marmots, forming **disharmonious faunas**. The Late Pleistocene steppe-tundra is therefore unlike any modern ecosystem.

The megacerine cervids of the genus *Megaloceros* lingered on during the Last Glacial Period, when it reached its largest dimensions (two metres in shoulder height). *Megaloceros giganteus* has been called the "Irish Elk" for being so frequently discovered in Irish bog deposits, although it was not an elk nor was it solely Irish. It was in fact an enormous cervid related to fallow deer (*Dama*) and ranged widely throughout much of Northern Eurasia, from Western Europe to China. *M. giganteus* was more robustly built than the modern moose/elk and, possibly because of a dissimilar social structure, had more pronounced **sexual dimorphism**. Since antler size is related to body size, mature males of *M. giganteus* sported a pair of enormous palmate antlers that spanned over 11 feet from end to end, weighed 35 to 45 kilograms and had distinct backward-pointing tines. These massive antlers were supported by a heavily-augmented skull and neck (Lister, 1994), and a sturdy body over two metres at the shoulder. A **cursorial** lifestyle is essential to understanding how the body size and antler spread of *Megaloceros* developed. The largest cervid will also be the fittest and fastest, and antler size is an accurate reflection of stag health. These factors could have aided in the development of massive antlers in a large cervid living in open habitats (Geist, 1999). While the huge antlers would have precluded *M. giganteus* from inhabiting the dense woodlands where extant cervids live, they most probably had a high adaptive value as display ornaments, precluding energy loss and other harmful effects of seasonal male fights.

In addition, there were many large and cold-adapted predators; they are typically preserved in cave deposits, although they likely did not spend much of their time in caves. These include the versatile spotted hyaena, which was represented by its own cave subspecies, *Crocuta crocuta spelaea*, the Eurasian cave or steppe lion, *Panthera spelaea* (the largest felid in its ecosystem, measuring about three metres long), the scimitar-toothed cat *Homotherium latidens* and the enormous, distinctly dome-headed cave bear *Ursus spelaeus* (also three metres long). Many species of carnivorans (see pp. 81, 83–4) from the Middle Pleistocene persisted into the Late Pleistocene, but evolved into larger and more robust forms regarded as distinct species or subspecies. The cave hyaena ranged from Iberia to East Asia, and was larger and more robust than its extant African relative, but its metatarsals and metacarpals were shorter and thicker, while the humerus and femur were more elongated, therefore suggesting different habitat preferences. Their use of caves as dens is widely accepted, although sites in open environments are also known. Cave hyaenas would not have had the complex social structure of its modern relatives. During the Pleistocene, there was a considerably higher biomass available for scavenging from and thus, cave hyaenas possibly did not need to hunt frequently, which would have had chain reactions for sociality. Cave hyaenas probably would have lived in smaller groups and had smaller territories, as food was more plentiful. The cause of its extinction roughly 31 ka remains uncertain, but was probably due to a combination of factors including climate change (i.e. decline in grasslands) and competition with other predators (wolves and humans).

Panthera spelaea is the most famous of the extinct lions. It was clearly different from modern lions (*Panthera leo*) in a number of physical features—a mane reduced to a ruff in male lions, a longer and narrower muzzle, shorter and more robust limbs, lighter fur, and pronounced dark markings on the face, shoulder and side. Cave paintings also indicate that cave lions hunted in small prides, in much the same manner as modern lions in India and Namibia do today. Preserved bones were first found in caves, hence the name, but the lions actually lived on the open northern steppes across Eurasia, from Western Europe to eastern Russia and into Alaska (Beringia). It was in Alaska that the last of the cave lions died out around 13 ka. *Homotherium latidens* was a Eurasian species of scimitar-toothed cat, characterised by slightly longer forelimbs than that of a lion, along with a more elongated neck, and a shorter lumbar region and tail. It had powerful forelimbs for tackling large prey as in all machairodontines, but it also had marked adaptations for sustained running on open terrain, such as smaller claws with reduced retractability. *H. latidens* was a rather lightly-built scimitar-toothed cat, substantially lighter than the Pleistocene lions it coexisted with in the biomes of Middle to Late Pleistocene Europe. *H. latidens* is represented only by an isolated upper canine tooth from Creswell Crags (specifically the locality of Robin Hood Cave) which, along with a 28,000 year old lower jaw from the North Sea (Reumer *et al.* 2003) confirm it survived into the Late Pleistocene. This species has a relatively continuous fossil record in Eurasia between the Late Pliocene (ca. 3 Ma) and the Middle Pleistocene (ca. 400 ka). Maybe it lived in low densities between 400 and 28 ka, or perhaps it went regionally extinct and then immigrated from a different population. Hopefully new discoveries will be made to fill this gap.

Of all the Eurasian Late Pleistocene carnivorans, the cave bear *Ursus spelaeus* represents the most extreme case in the development of greater body size and robustness, taking the trends seen in *U. deningeri* (see p. 339) to their utmost. *U. spelaeus* was mostly herbivorous like its Middle Pleistocene progenitor, possessing even larger molars that were suited to grinding tough vegetation, although unlike modern brown bears (*U. arctos*), abrasive foodstuffs such as tubers were not a major dietary component as evident from dental **microwear** analysis. *U. spelaeus* had a large, short skull with a raised front as well as a body more thickset than modern brown bears. *U. spelaeus* also had a shorter snout, relatively retracted nasals and a robust lower jaw, traits associated with its strong masticatory system. As its name implies, cave bear fossils are generally discovered in caves, and frequently in large numbers, where the species hibernated during the severe winters of the Last Glacial Period. There is evidence to suggest that *U. spelaeus* preferred to hibernate exclusively in caves rather than in other locations including thickets, unlike the more versatile brown bear. Cave bears that were unable to find suitable caves would have suffered a high winter mortality rate as a result of this specialised hibernation habit. Thus, the cave bear faced a declining pool of available caves as human populations gradually grew, ultimately leading to the species' extinction by about 24 ka.

All these mammals lived alongside the neanderthals during the earlier part of the Late Pleistocene (ca. 60–40 ka), the first human occupants of the Creswell Crags caves. The Neanderthals visited the caves with the seasons in order to track herds of mammoths, horses, bison and reindeer, which were a vital source of food. The Neanderthals left behind flint hand axes and scrapers, antler and bone tools, and animal bones bearing traces of butchery marks, although no Neanderthal fossils have been discovered at Creswell Crags. The oldest known fossils of *Homo sapiens* from the site date to 28 ka, and are of Cro-Magnons, early modern humans associated with the Upper Palaeolithic period (40–10 ka) in Europe. There are earlier occurences of *H. sapiens* in Great Britain (e.g. the Kents Cavern **maxilla** dated at >40 ka, from Devon). *H. sapiens* used the natural shelters at Creswell Crags as seasonal camps while they tracked migrating mammoths and left more sophisticated flints than the Neanderthals, and shaped bone and antler artefacts.

The Holocene and Beyond

The Holocene Epoch (11.7 ka–present)—an interglacial we regard as our time—is one of the most climatically stable intervals of the Quaternary. The end of the final Pleistocene cold phase (the Younger Dryas cooling event (see pp. 313, 327)) and a rapid rise in global temperatures signalled the start of the Holocene. Following a peak in Late Pleistocene glaciation during the Last Glacial Maxium (see p. 311), circa 20 ka, global sea levels rose by about 125 metres up to the present day as glaciers and ice sheets retreated, though the rise was uneven. The highest rates of sea level rise occurred between about 16 and 8 ka, during so-called meltwater pulses, with sea-level rise sometimes exceeding 10 to 15 millimetres per year. Both the Antarctic Ice Sheet and the Laurentide Ice Sheet in North America contributed significant meltwater. After about 6–7 ka, sea levels approached modern values and remained relatively stable until the modern era, when human-driven warming has already caused an exceptional rise of more than 0.2 metres in just the past century.

Throughout most of the Holocene, global average temperatures varied by less than about 1°C until the onset of recent anthropogenic warming. This stability was not absolute, as regional climate variations occurred due to factors such as latitudinal temperature differences and local ocean–atmosphere interactions. Although far more stable than the abrupt, large-amplitude fluctuations of the Last Glacial Period (see p. 313), natural variability on millennial and centennial timescales persisted, leaving evidence in marine sediments, lake cores, and tree rings. One notable warm spell was the Mediaeval Warm Period (ca. 950–1250 CE), likely driven by a combination of increased solar radiation, reduced volcanic activity, and changes in ocean and atmospheric circulation that brought warm water northwards from the tropics. Regional in character, it was a prosperous interval in European history, coinciding with Norse voyages to the North Atlantic and beyond. The cold spells included the Little Ice Age (ca. 1300–1850 CE) in Europe, driven by reduced solar output and a reversal of the North Atlantic Oscillation (an atmospheric circulation pattern over the North Atlantic and surrounding regions). This cooling, especially pronounced in Europe, contributed to famines, social unrest, and revolutions.

In the early Holocene, thriving biodiversity and strong seasonal cycles provided stable environmental conditions that humanity could exploit. Earth's seasons, driven by axial tilt, had existed long before the Holocene, but within this relatively stable climatic framework, they proved essential for the rise of agriculture. Almost as soon as these conditions emerged, communities in the Middle East—then wetter and more fertile than today—began transitioning from hunter-gatherer lifestyles to farming. Agriculture emerged gradually through a complex, uncoordinated process shaped more by chance events than by foresight. Farming arose independently in at least 11 different parts of the globe at different times, progressively developing cultivated strains of a huge variety of crops, comprising familiar types such as wheat, rice, maize, potatoes and sugarcane. Alongside plant cultivation, humans domesticated a range of animals. Dogs had already been domesticated in the Late Pleistocene, while cats, sheep, goats, cattle, pigs, horses, and chickens were gradually brought under human influence during the Holocene through selective breeding of **morphotypes** and varieties most suited to human needs. This process allowed many of these species to expand their ranges, creating a form of mutualism in which both humans and the animals benefited.

Agriculture radically transformed the relationship between humanity and the natural world. The arable lands owned by various families inevitably differed in water availability and soil type, thus certain crops and livestock were more productive compared to others. Subsequent to feeding their families, the early farmers could use any excess food for trading. Communities of farmers congregated at open markets to barter their produce. They started exchanging produce for other goods and for skills. The farmers required oil, stone and twine. They sought the products of toolmakers, carpenters and masons who now, for the first time, could trade for food instead of having to grow it themselves. As the amount of trades grew, the markets evolved into towns and later cities across most of the fertile river valleys in the Middle East. Civilisation had dawned. It gained momentum with every generation and technological advancement. Hydropower, steam power and electrification were developed and refined, laying the foundation for the modern technologies we rely on today. Every generation, in these increasingly intricate societies, could develop and advance because the living world remained stable and could be counted on to provide the resources and conditions for our survival.

Fast-forwarding to the present day, the Earth is undeniably a human world. This has not always been the case, and may not remain so, but for now our species exerts an influence on the biosphere unmatched by any other organism in the planet's history. With the advent of the Industrial Revolution in the late eighteenth century, modern humans started emitting into the atmosphere vast amounts of carbon dioxide and other greenhouse gases by burning fossil fuels that had remained locked up in continental and marine reservoirs for millions of years. The major consequence of these greenhouse gas emissions is a gradual warming of the world, and average global temperatures have already risen by slightly over 1°C since 1880. This warming will, in turn, lead to an increase in the frequency of droughts, wildfires, and a shift in the distributions of plants and animals. By making comparisons to our current atmosphere with those of the past, we can project how future global climates may appear. The current atmospheric composition is comparable to that of the Oligocene, that transition from a hothouse to an icehouse. According to the Intergovernmental Panel on Climate Change (IPCC), high-emissions scenarios could see atmospheric carbon dioxide concentrations approach levels last experienced during the Eocene (around 800 ppm) by the end of the twenty-first century. Such concentrations would likely be accompanied by global temperatures comparable to those of that epoch. Avoiding this trajectory would require far more rapid reductions in carbon emissions than are currently planned.

Meanwhile, the world's human population has risen at a staggering rate over the last century. As of this book's release, there are over eight billion people on Earth, making humans a dominant force in almost all **ecosystems**. Much of this growth has resulted from declining child mortality due to advances in healthcare. Concerns about overpopulation are therefore common, yet demographic trends suggest that this issue may ease over time. For decades, fertility rates have been decreasing across the planet, and the global population is projected to peak in the latter half of the twenty-first century, alongside increasing urbanisation and expanded access to education for women. By sheer amount, we are exceedingly abundant. Humans currently account for 36% of the mass of all mammals. Another 60% of the mass are domesticated mammals. A mere 4% of Earth's mammal mass is wild. For birds, it is slightly better. Some 30% of Earth's bird mass is wild, whereas the remaining 70% comprise domesticated species, almost exclusively chickens. By around 2020, the total mass of materials produced by humans had grown to roughly equal that of all living biomass on Earth, and it continues to increase. If the planet were examined today in the same way palaeontologists interpret the fossil record—by analysing the distribution of skeletal remains—it would suggest that something highly unusual was occurring: a small number of species would appear to account for the majority of vertebrate biomass. Such a pattern would typically be interpreted as evidence of severe environmental disruption, consistent with a mass extinction event.

Additionally, we are instigating new ecosystems. Quite possibly the modern analogue to the post-extinction landscapes is the city, largely stripped of trees and megafauna, with low productivity. Numerous species are unable to persist under such conditions, and those that do survive need to adapt, down to even their most simple habits. Since the Last Glacial Period, the largest tetrapods have either disappeared from every continent, or are at high risk of extinction (Fig. 7.2). The Earth is starting to look like a post-extinction planet, with the man-made ecosystem a refuge for disaster species. Those adapted to our own environments, generalist species able to survive on refuse such as rats, opossums, raccoons, foxes, gulls or storks, as well as those that we breed for our own needs including dogs, pigs and chickens, are flourishing. The ecosystems that have developed since the Cretaceous-Paleogene extinction event, 66 Ma (see p. 169), have experienced increasing strain especially in the last century. This is due to various anthropogenic factors including habitat loss, fragmentation and degradation, pollution, the widespread use of pesticides, herbicides and other chemicals, hunting and overexploitation, and the effects of invasive species, all of which are increasingly compounded by rapid climate change. If these stresses continue to accumulate, the consequences for both the biosphere and human societies could be severe. Yet, there is also cause for cautious optimism. Our capacity to reflect on the state of the planet, coupled with the analytical tools that allow us to reconstruct past environmental upheavals and recognise their consequences, provides a basis for informed action. By understanding how life has responded to periods of profound disruption in the past, we can better anticipate potential futures and identify pathways that help avert disaster.

When it comes to environmental change, complacency carries real risk. A 'business-as-usual' trajectory—in which greenhouse-gas emissions and ecosystem degradation continue unchecked—would lead to climatic conditions without precedent in human history, unfolding within our own lifetimes. At the same time, narratives focused solely on inevitable catastrophe are equally unhelpful. Outcomes in conservation and climate mitigation are not binary. Claims that humanity has six or eleven years to avert dangerous climate change are not absolute deadlines, but indicators of narrowing windows in which action can meaningfully reduce harm. Acting promptly does not promise a return to past environmental states, just as inaction does not imply immediate or total collapse. Ecosystems that existed during the early twentieth century and before have already been irreversibly altered, and the consequences of those changes continue to accumulate. What remains within our control is the scale and severity of future damage. Preventing current environmental pressures from precipitating a major mass extinction will require fundamental changes in how we live, consume, and extract resources. Earth cannot sustain lifestyles as resource-intensive as those typical of economically developed societies, nor can it support such demands while allowing wild species to flourish. Meaningful mitigation therefore depends on reducing consumption and rapidly ending reliance on fossil fuels. If insufficient action is taken at local, national, and global scales, the consequences for human societies will continue to intensify over the coming decades. Our long-term well-being—both as individuals and as a species—depends on fostering a more reciprocal relationship with the planet's natural systems. Only by doing so can we preserve both the diversity of life on Earth and our place within it. Change is inevitable, as it always has been in the history of life, but we can allow the Earth to progress at its own pace, while letting the passage of geological time gradually lead us into future worlds and faunas.

Further Reading

Chapter 1: Introduction

Beck, R. M., Bininda-Emonds, O. R., Cardillo, M., Liu, F. G. R., & Purvis, A. (2006). A higher-level MRP supertree of placental mammals. *BMC Evolutionary Biology*, 6, 1–14.

Bininda-Emonds, O. R., Cardillo, M., Jones, K. E., MacPhee, R. D., Beck, R. M., Grenyer, R., Price, S. A., Vos, R. A., Gittleman, J. L., & Purvis, A. (2007). The delayed rise of present-day mammals. *Nature*, 446(7135), 507–512.

Churakov, G., Kriegs, J. O., Baertsch, R., Zemann, A., Brosius, J., & Schmitz, J. (2009). Mosaic retroposon insertion patterns in placental mammals. *Genome Research*, 19(5), 868–875.

Donovan, S. K., & Paul, C. R. C. (1998). *The Adequacy of the Fossil Record*. Wiley, New York.

Hallström, B. M., & Janke, A. (2008). Resolution among major placental mammal interordinal relationships with genome data imply that speciation influenced their earliest radiations. *BMC evolutionary biology*, 8, 1–13.

Hallström, B. M., Kullberg, M., Nilsson, M. A., & Janke, A. (2007). Phylogenomic data analyses provide evidence that Xenarthra and Afrotheria are sister groups. *Molecular biology and evolution*, 24(9), 2059–2068.

Hennig, W. (1966). *Phylogenetic Systematics*. University Illinois Press, Urbana, Illinois.

Kriegs, J. O., Churakov, G., Kiefmann, M., Jordan, U., Brosius, J., & Schmitz, J. (2006). Retroposed elements as archives for the evolutionary history of placental mammals. *PLoS Biology*, 4(4), e91.

Laurin, M., & Anderson, J. S. (2004). Meaning of the name Tetrapoda in the scientific literature: an exchange. *Systematic Biology*, 53(1), 68–80.

Liu, F. G. R., Miyamoto, M. M., Freire, N. P., Ong, P. Q., Tennant, M. R., Young, T. S., & Gugel, K. F. (2001). Molecular and morphological supertrees for eutherian (placental) mammals. *Science*, 291(5509), 1786–1789.

McCormack, J. E., Faircloth, B. C., Crawford, N. G., Gowaty, P. A., Brumfield, R. T., & Glenn, T. C. (2012). Ultraconserved elements are novel phylogenomic markers that resolve placental mammal phylogeny when combined with species-tree analysis. *Genome research*, 22(4), 746–754.

Meredith, R. W., Janečka, J. E., Gatesy, J., Ryder, O. A., Fisher, C. A., Teeling, E. C., ... & Murphy, W. J. (2011). Impacts of the Cretaceous Terrestrial Revolution and KPg extinction on mammal diversification. *Science*, 334(6055), 521–524.

Murphy, W. J., Eizirik, E., O'Brien, S. J., Madsen, O., Scally, M., Douady, C. J., ... & Springer, M. S. (2001). Resolution of the early placental mammal radiation using Bayesian phylogenetics. *Science*, 294(5550), 2348–2351.

Murphy, W. J., Pringle, T. H., Crider, T. A., Springer, M. S., & Miller, W. (2007). Using genomic data to unravel the root of the placental mammal phylogeny. *Genome Research*, 17(4), 413–421.

Nishihara, H., Okada, N., & Hasegawa, M. (2007). Rooting the eutherian tree: the power and pitfalls of phylogenomics. *Genome Biology*, 8, 1–10.

Novacek, M. J. (1992). Fossils, topologies, missing data, and the higher level phylogeny of eutherian mammals. *Systematic Biology*, 41(1), 58–73.

O'Leary, M. A., Bloch, J. I., Flynn, J. J., Gaudin, T. J., Giallombardo, A., Giannini, N. P., ... & Cirranello, A. L. (2013). The placental mammal ancestor and the post-K-Pg radiation of placentals. *Science*, 339(6120), 662–667.

Prothero, D. R., & Schoch, R. M. (1994). *Major Features of Vertebrate Evolution*. Paleontological Society Short Courses in Paleontology, no. 7. Paleontological Society, Lawrence, Kansa. 270 pp.

Shipman, P. (1981). *Life History of a Fossil*. Harvard University Press, Cambridge, Massachusetts.

Simpson, G. G. (1945). *The principles of classification and a classification of mammals*. American Museum of Natural History.

Song, S., Liu, L., Edwards, S. V., & Wu, S. (2012). Resolving conflict in eutherian mammal phylogeny using phylogenomics and the multispecies coalescent model. *Proceedings of the National Academy of Sciences*, 109(37), 14942–14947.

Waddell, P. J., Kishino, H., & Ota, R. (2001). A phylogenetic foundation for comparative mammalian genomics. *Genome Informatics*, 12, 141–154.

Waddell, P. J., & Shelley, S. (2003). Evaluating placental inter-ordinal phylogenies with novel sequences including RAG1, γ-fibrinogen, ND6, and mt-tRNA, plus MCMC-driven nucleotide, amino acid, and codon models. *Molecular Phylogenetics and Evolution*, 28(2), 197–224.

Chapter 2: The Evolution of Cenozoic Tetrapods

Intro

Benton, M. J. (2024). *Vertebrate Palaeontology, Fifth Edition*. John Wiley & Sons.

Coates, M. I., Ruta, M., & Friedman, M. (2008). Ever since Owen: changing perspectives on the early evolution of tetrapods. *Annual Review of Ecology, Evolution, and Systematics*, 39(1), 571–592.

Jiang, B., He, Y., Elsler, A., Wang, S., Keating, J. N., Song, J., Kearns, S. L., & Benton, M. J. (2023). Extended embryo retention and viviparity in the first amniotes. *Nature Ecology & Evolution*, 7(7), 1131–1140.

Nesbitt, S. J. (2009). *The early evolution of archosaurs: relationships and the origin of major clades*. Columbia University Press, New York.

Sengupta, S., Ezcurra, M. D., & Bandyopadhyay, S. (2017). A new horned and long-necked herbivorous stem-archosaur from the Middle Triassic of India. *Scientific Reports*, 7(1), 8366.

Simões, T. R., Kammerer, C. F., Caldwell, M. W., & Pierce, S. E. (2022). Successive climate crises in the deep past drove the early evolution and radiation of reptiles. *Science Advances*, 8(33), eabq1898.

Swartz, B. (2012). A marine stem-tetrapod from the Devonian of western North America. *PloS One*, 7(3), e33683.

Wolniewicz, A. S., Shen, Y., Li, Q., Sun, Y., Qiao, Y., Chen, Y., ... & Liu, J. (2023). An armoured marine reptile from the Early Triassic of South China and its phylogenetic and evolutionary implications. *Elife*, 12, e83163.

Lissamphibia

Crump, M. L. (2009). Amphibian diversity and life history. *Amphibian ecology and conservation: A handbook of techniques*, 3–20.

Feng, Y. J., Blackburn, D. C., Liang, D., Hillis, D. M., Wake, D. B., Cannatella, D. C., & Zhang, P. (2017). Phylogenomics reveals rapid, simultaneous diversification of three major clades of Gondwanan frogs at the Cretaceous–Paleogene boundary. *Proceedings of the National Academy of Sciences*, 114(29), E5864–E5870.

Jones, M. E., Benson, R. B., Skutschas, P., Hill, L., Panciroli, E., Schmitt, A. D., ... & Evans, S. E. (2022). Middle Jurassic fossils document an early stage in salamander evolution. *Proceedings of the National Academy of Sciences*, 119(30), e2114100119.

Laurin, M., Canoville, A., & Quilhac, A. (2009). Use of paleontological and molecular data in supertrees for comparative studies: the example of lissamphibian femoral microanatomy. *Journal of Anatomy*, 215(2), 110–123.

Marjanović, D., & Laurin, M. (2007). Fossils, molecules, divergence times, and the origin of lissamphibians. *Systematic Biology*, 56(3), 369–388.

Mörs, T., Reguero, M., & Vasilyan, D. (2020). First fossil frog from Antarctica: implications for Eocene high latitude climate conditions and Gondwanan cosmopolitanism of Australobatrachia. *Scientific Reports*, 10(1), 5051.

Pyron, R. A. (2011). Divergence time estimation using fossils as terminal taxa and the origins of Lissamphibia. *Systematic biology*, 60(4), 466–481.

San Mauro, D. (2010). A multilocus timescale for the origin of extant amphibians. *Molecular Phylogenetics and Evolution*, 56(2), 554–561.

Schoch, R. R. (2019). The putative lissamphibian stem-group: phylogeny and evolution of the dissorophoid temnospondyls. *Journal of Paleontology*, 93(1), 137–156.

Testudinata

Anquetin, J. (2012). Reassessment of the phylogenetic interrelationships of basal turtles (Testudinata). *Journal of Systematic Palaeontology*, 10(1), 3–45.

Caspermeyer, J. (2017). Coming Out of Their Shells: New Study Identifies Energy Metabolism Adaptations Linked to Soft Shell Turtle Evolution. *Molecular Biology and Evolution*, 34(10), 2730.

Chatterji, R. M., Hutchinson, M. N., & Jones, M. E. (2021). Redescription of the skull of the Australian flatback sea turtle, *Natator depressus*, provides new morphological evidence for phylogenetic relationships among sea turtles (Chelonioidea). *Zoological Journal of the Linnean Society*, 191(4), 1090–1113.

Danilov, I. G., & Parham, J. F. (2008). A reassessment of some poorly known turtles from the Middle Jurassic of China, with comments on the antiquity of extant turtles. *Journal of Vertebrate Paleontology*, 28(2), 306–318.

Fujii, J. A., McLeish, D., Brooks, A. J., Gaskell, J., & Van Houtan, K. S. (2018). Limb-use by foraging marine turtles, an evolutionary perspective. *PeerJ*, 6, p.e4565.

Gaffney, E. S., Meylan, P. A., Wood, R. C., Simons, E., & Campos, D. D. A. (2011). Evolution of the side-necked turtles: the family Podocnemididae. *Bulletin of the American Museum of Natural History*, 2011(350), 1–237.

Iverson, J. B., Le, M., & Ingram, C. (2013). Molecular phylogenetics of the mud and musk turtle family Kinosternidae. *Molecular Phylogenetics and Evolution*, 69(3), 929–939.

Jones, M. E., Werneburg, I., Curtis, N., Penrose, R., O'Higgins, P., Fagan, M. J., & Evans, S.E. (2012). The head and neck anatomy of sea turtles (Cryptodira: Chelonioidea) and skull shape in Testudines. *PLoS One*, 7(11), e47852.

Kear, B. P., & Lee, M. S. (2006). A primitive protostegid from Australia and early sea turtle evolution. *Biology Letters*, 2(1), 116–119.

Lourenço, J. M., Claude, J., Galtier, N., & Chiari, Y., 2012. Dating cryptodiran nodes: origin and diversification of the turtle superfamily Testudinoidea. *Molecular Phylogenetics and Evolution*, 62(1), 496–507.

Lovich, J. E., & Gibbons, W. (2021). *Turtles of the world: A Guide to Every Family*. Princeton University Press, Princeton, New Jersey.

Pereira, A. G., Sterli, J., Moreira, F. R., & Schrago, C. G. (2017). Multilocus phylogeny and statistical biogeography clarify the evolutionary history of major lineages of turtles. *Molecular Phylogenetics and Evolution*, 113, 59–66.

Thomson, R. C., Spinks, P. Q., & Shaffer, H. B. (2021). A global phylogeny of turtles reveals a burst of climate-associated diversification on continental margins. *Proceedings of the National Academy of Sciences*, 118(7), e2012215118.

Choristodera

Acorn, J. (2007). *Deep Alberta: Fossil Facts and Dinosaur Digs*. University of Alberta Press, Edmonton.

Dong, L., Matsumoto, R., Kusuhashi, N., Wang, Y., Wang, Y., & Evans, S. E. (2020). A new choristodere (Reptilia: Choristodera) from an Aptian–Albian coal deposit in China. *Journal of Systematic Palaeontology*, 18(15), 1223–1242.

Hecht, M. K. (1992). A new choristodere (Reptilia, Diapsida) from the Oligocene of France: an example of the Lazarus effect. *Geobios*, 25(1), 115–131.

Jiang, B., He, Y., Elsler, A., Wang, S., Keating, J. N., Song, J., Kearns, S. L., & Benton, M. J. (2023). Extended embryo retention and viviparity in the first amniotes. *Nature Ecology & Evolution*, 7(7), 1131–1140.

Matsumoto, R.; Buffetaut, E.; Escuillie, F.; Hervet, S., & Evans, S. E. (2013). New material of the choristodere *Lazarussuchus* (Diapsida, Choristodera) from the Paleocene of France. *Journal of Vertebrate Paleontology*. 33(2), 319.

Matsumoto, R., & Evans, S. E. (2010). Choristoderes and the freshwater assemblages of Laurasia. *Journal of Iberian Geology*, 36(2), 253–274.

Matsumoto, R., & Evans, S. E. (2016). Morphology and function of the palatal dentition in Choristodera. *Journal of Anatomy*, 228(3), 414–429.

Lepidosauria

Apesteguía, S., & Zaher, H. (2006). A Cretaceous terrestrial snake with robust hindlimbs and a sacrum. *Nature*, 440(7087), 1037–1040.

Benton, M. J. (1985). Classification and phylogeny of the diapsid reptiles. *Zoological Journal of the Linnean Society*, 84(2), 97–164.

Benton, M. J. (Ed.). (1988). *The Phylogeny and Classification of the Tetrapods, vol. 1: Amphibians, Reptiles, Birds*. Oxford Clarendon Press, Oxford.

Datta, D., & Bajpai, S. (2024). Largest known madtsoiid snake from warm Eocene period of India suggests intercontinental Gondwana dispersal. *Scientific Reports*, 14(1), 8054.

Folie, A., Mees, F., De Putter, T., & Smith, T. (2021). Presence of the large aquatic snake *Palaeophis africanus* in the middle Eocene marine margin of the Congo Basin, Cabinda, Angola. *Geobios*, 66, 45–54.

Herrera-Flores, J. A., Stubbs, T. L., & Benton, M. J. (2017). Macroevolutionary patterns in Rhynchocephalia: is the tuatara (Sphenodon punctatus) a living fossil? *Palaeontology*, 60(3), 319–328.

Herrera-Flores, J. A., Elsler, A., Stubbs, T. L., & Benton, M. J. (2022). Slow and fast evolutionary rates in the history of lepidosaurs. *Palaeontology*, 65(1), e12579.

Houssaye, A., Herrel, A., Boistel, R., & Rage, J. C. (2019). Adaptation of the vertebral inner structure to an aquatic life in snakes: Pachyophiid peculiarities in comparison to extant and extinct forms. *Comptes Rendus Palevol*, 18(7), 783–799.

Hutchinson, M. N., Skinner, A., & Lee, M. S. (2012). *Tikiguania* and the antiquity of squamate reptiles (lizards and snakes). *Biology Letters*, 8(4), 665–669.

Jones, M. E., Anderson, C. L., Hipsley, C. A., Müller, J., Evans, S. E., & Schoch, R. R. (2013). Integration of molecules and new fossils supports a Triassic origin for Lepidosauria (lizards, snakes, and tuatara). *BMC Evolutionary Biology*, 13, 1–21.

Jones, M. E., Tennyson, A. J., Worthy, J. P., Evans, S. E., & Worthy, T. H. (2009). A sphenodontine (Rhynchocephalia) from the Miocene of New Zealand and palaeobiogeography of the tuatara (*Sphenodon*). *Proceedings of the Royal Society B: Biological Sciences*, 276(1660), 1385–1390.

Lee, M. S., Reeder, T. W., Slowinski, J. B., & Lawson, R. (2004). 'Resolving reptile relationships'. In: Cracraft, J., & Donoghue, M. (eds) *Assembling the Tree of Life*. Oxford University Press, Oxford. pp. 451–467.

Müller, J. (2004). 'The relationships among diapsid reptiles and the influence of taxon selection'. In: Arratia, G., Wilson, M. V. H., & Cloutier, R. (eds) *Recent Advances in the Origin and Early Radiation of Vertebrates*. Verlag Dr. Friedrich Pfeil, München. pp. 379–408.

Pyron, R. A., Burbrink, F. T., & Wiens, J. J. (2013). A phylogeny and revised classification of Squamata, including 4161 species of lizards and snakes. *BMC Evolutionary Biology*, 13, 1–54.

Reeder, T. W., Townsend, T. M., Mulcahy, D. G., Noonan, B. P., Wood Jr, P. L., Sites Jr, J. W., & Wiens, J. J. (2015). Integrated analyses resolve conflicts over squamate reptile phylogeny and reveal unexpected placements for fossil taxa. *PLoS One*, 10(3), e0118199.

Scanlon, J. D. (2006). Skull of the large non-macrostomatan snake *Yurlunggur* from the Australian Oligo-Miocene. *Nature*, 439(7078), 839–842.

Snetkov, P. B. (2011). Vertebrae of the sea snake *Palaeophis nessovi* Averianov (Acrochordoidea, Palaeophiidae) from the Eocene of Western Kazakhstan and phylogenetic analysis of the superfamily Acrochordoidea. *Paleontological Journal*, 45, 305–313.

Vidal, N., & Hedges, S. B. (2005). The phylogeny of squamate reptiles (lizards, snakes, and amphisbaenians) inferred from nine nuclear protein-coding genes. Comptes Rendus. *Biologies*, 328(10–11), 1000–1008.

Wiens, J. J., Hutter, C. R., Mulcahy, D. G., Noonan, B. P., Townsend, T. M., Sites Jr, J. W., & Reeder, T. W. (2012). Resolving the phylogeny of lizards and snakes (Squamata) with extensive sampling of genes and species. *Biology Letters*, 8(6), 1043–1046.

Zheng, Y., & Wiens, J. J. (2016). Combining phylogenomic and supermatrix approaches, and a time-calibrated phylogeny for squamate reptiles (lizards and snakes) based on 52 genes and 4162 species. *Molecular Phylogenetics and Evolution*. 94, 537–547.

Crocodylomorpha

Benton, M. J. (1985). Classification and phylogeny of the diapsid reptiles. *Zoological Journal of the Linnean Society*, 84(2), 97–164.

Benton, M. J. (1988). *The Phylogeny and Classification of the Tetrapods, vol. 1: Amphibians, Reptiles, Birds*. Oxford Clarendon Press, Oxford.

Brochu, C. A. (2003). Phylogenetic approaches toward crocodilian history. *Annual Review of Earth and Planetary Sciences*, 3, 357–397.

Hekkala, E., Gatesy, J., Narechania, A., Meredith, R., Russello, M., Aardema, M. L., ... & Amato, G. (2021). Paleogenomics illuminates the evolutionary history of the extinct Holocene "horned" crocodile of Madagascar, *Voay robustus*. *Communications Biology*, 4(1), 505.

Jouve, S. (2005). A new description of the skull of *Dyrosaurus phosphaticus* (Thomas, 1893)(Mesoeucrocodylia: Dyrosauridae) from the Lower Eocene of North Africa. *Canadian Journal of Earth Sciences*, 42(3), 323–337.

Lee, M. S., & Yates, A. M. (2018). Tip-dating and homoplasy: reconciling the shallow molecular divergences of modern gharials with their long fossil record. *Proceedings of the Royal Society B*, 285(1881), 20181071.

Martin, J. E., & Benton, M. J. (2008). Crown clades in vertebrate nomenclature: correcting the definition of Crocodylia. *Systematic Biology*, 57(1), 173–181.

Martin, J. E., Pochat-Cottilloux, Y., Laurent, Y., Perrier, V., Robert, E., & Antoine, P. O. (2022). Anatomy and phylogeny of an exceptionally large sebecid (Crocodylomorpha) from the middle Eocene of southern France. *Journal of Vertebrate Paleontology*, 42(4), e2193828.

Massonne, T., Vasilyan, D., Rabi, M., & Böhme, M. (2019). A new alligatoroid from the Eocene of Vietnam highlights an extinct Asian clade independent from extant *Alligator sinensis*. *PeerJ*, 7, e7562.

Pol, D., & Powell, J. E. (2011). A new sebecid mesoeucrocodylian from the Rio Loro Formation (Palaeocene) of north-western Argentina. *Zoological Journal of the Linnean Society*, 163(suppl_1), S7–S36.

Rio, J. P., & Mannion, P. D. (2021). Phylogenetic analysis of a new morphological dataset elucidates the evolutionary history of Crocodylia and resolves the long-standing gharial problem. *PeerJ*, 9, e12094.

Ristevski, J., Willis, P. M., Yates, A. M., White, M. A., Hart, L. J., Stein, M. D., Price, G. J., & Salisbury, S. W. (2023). Migrations, diversifications and extinctions: the evolutionary history of crocodyliforms in Australasia. *Alcheringa*, 47(4), 370–415.

Ross, C. A. (Ed.). (1992). *Crocodiles and Alligators*. Blitz, New York.

Stubbs, T. L., Pierce, S. E., Elsler, A., Anderson, P. S., Rayfield, E. J., & Benton, M. J. (2021). Ecological opportunity and the rise and fall of crocodylomorph evolutionary innovation. *Proceedings of the Royal Society B, 288(1947)*, 20210069.

Wilberg, E. W., Turner, A. H., & Brochu, C. A. (2019). Evolutionary structure and timing of major habitat shifts in Crocodylomorpha. *Scientific Reports*, 9(1), 514.

Aves (birds)

Agnolin, F. L. (2017). Unexpected diversity of ratites (Aves, Palaeognathae) in the early Cenozoic of South America: palaeobiogeographical implications. *Alcheringa*, 41(1), 101–111.

Bertelli, S., & Chiappe, L. M. (2005). Earliest tinamous (Aves: Palaeognathae) from the Miocene of Argentina and their phylogenetic position. *Contributions in Science*, 502(502), 1–20

Degrange, F. J., Tambussi, C. P., Taglioretti, M. L., Dondas, A., & Scaglia, F. (2015). A new Mesembriornithinae (Aves, Phorusrhacidae) provides new insights into the phylogeny and sensory capabilities of terror birds. *Journal of Vertebrate Paleontology*, 35(2), e912656.

Chiappe, L. M. (2007). *Glorified Dinosaurs: The Origin and Early Evolution of the Birds*. Wiley-Liss, New York.

Cracraft, J. (1988). 'The major clades of birds'. In: Benton, M. J. (ed) *The Phylogeny and Classifcation of the Tetrapods, vol. 1: Amphibians, Reptiles, Birds*. Clarendon Press, Oxford. pp. 339–361.

Dyke, G., & Kaiser, G. (2011). *Living Dinosaurs: The Evolutionary History of Modern Birds*. Wiley-Blackwell, London.

Feduccia, A. (1999). *The Origin and Evolution of Birds (2nd ed.)*. Yale University Press, New Haven, Connecticut.

Grealy, A., Miller, G. H., Phillips, M. J., Clarke, S. J., Fogel, M., Patalwala, D., ... & Bunce, M. (2023). Molecular exploration of fossil eggshell uncovers hidden lineage of giant extinct bird. *Nature Communications*, 14(1), 914.

Hackett, S. J., Kimball, R. T., Reddy, S., Bowie, R. C., Braun, E. L., Braun, M. J., ... & Yuri, T. (2008). A phylogenomic study of birds reveals their evolutionary history. *Science*, 320(5884), 1763–1768.

Houde, P. (1988). Palaeognathous birds from the early Tertiary of the Northern Hemisphere. *Publications of the Nuttall Ornithological Club*, 22, 1–148.

Kuhl, H., Frankl-Vilches, C., Bakker, A., Mayr, G., Nikolaus, G., Boerno, S. T., Klages, B., Timmermann, B., & Gahr, G. (2021). An unbiased molecular approach using 3'-uTrs resolves the avian family-level tree of life. *Molecular Biology and Evolution*, 38(10), 108–127

Marshall, L. G. (1994). The terror birds of South america. *Scientifc American*. 270(2), 90–95.

Mayr, G., & Peters, D. S. (2007). The foot of *Archaeopteryx*: Response to Feduccia et al. (2007). *The Auk*, 124(4), 1450–1452.

Mayr, G., & Zelenkov, N. (2021). Extinct crane-like birds (Eogruidae and Ergilornithidae) from the Cenozoic of Central Asia are indeed ostrich precursors. *The Auk*, 138(4), ukab048.

McLachlan, S. M., Kaiser, G. W., & Longrich, N. R. (2017). *Maaqwi cascadensis*: A large, marine diving bird (Avialae: Ornithurae) from the Upper Cretaceous of British Columbia, Canada. *PLoS One*, 12(12), e0189473.

Mitchell, K. J., Llamas, B., Soubrier, J., Rawlence, N. J., Worthy, T. H., Wood, J., ... & Cooper, A. (2014). Ancient DNA reveals elephant birds and kiwi are sister taxa and clarifies ratite bird evolution. *Science*, 344(6186), 898–900.

O'Connor, J., Chiappe, L. M., & Bell, A. (2011). 'Pre-modern birds: avian divergences in the Mesozoic'. In: Dyke, G., & Kaiser, G. (eds) *Living Dinosaurs: The Evolutionary History of Modern Birds*. Wiley-Blackwell, London. pp. 39–116.

Olson, S.l. 1985. The fossil record of birds. *Avian Biology*, 8, 79–252.

Pei, R., Pittman, M., Goloboff, P. A., Dececchi, T. A., Habib, M. B., Kaye, T. G., ... & Xu, X. (2020). Potential for powered flight neared by most close avialan relatives, but few crossed its thresholds. *Current Biology*, 30(20), 4033–4046.

Phillips, M. J., Gibb, G. C., Crimp, E. A., & Penny, D. (2010). Tinamous and moa flock together: mitochondrial genome sequence analysis reveals independent losses of flight among ratites. *Systematic Biology*, 59(1), 90–107.

Prum, R. O., Berv, J. S., Dornburg, A., Field, D. J., Townsend, J. P., Lemmon, E. M., & Lemmon, A. R. (2015). A comprehensive phylogeny of birds (Aves) using targeted next-generation DNA sequencing. *Nature*, 526(7574), 569–573.

Stiller, J., Feng, S., Chowdhury, A. A., Rivas-González, I., Duchêne, D. A., Fang, Q., ... & Zhang, G. (2024). Complexity of avian evolution revealed by family-level genomes. *Nature*, 629(8013), 851–860.

Sangster, G., Braun, E. L., Johansson, U. S., Kimball, R. T., Mayr, G., & Suh, A. (2022). Phylogenetic definitions for 25 higher-level clade names of birds. *Avian Research*, 13, 100027.

Wang, M., Wang, X., Wang, Y., & Zhou, Z. (2016). A new basal bird from China with implications for morphological diversity in early birds. *Scientific Reports*, 6(1), 19700.

Widrig, K., & Field, D. J. (2022). The evolution and fossil record of palaeognathous birds (Neornithes: Palaeognathae). *Diversity*, 14(2), 105.

Worthy, T. H., Hand, S. J., Archer, M., Scofield, R. P., & De Pietri, V. L. (2019). Evidence for a giant parrot from the Early Miocene of New Zealand. *Biology Letters*, 15(8), 20190467.

Worthy, T. H., Worthy, J. P., Tennyson, A. J., Salisbury, S. W., Hand, S. J., & Scofield, R. P. (2013). 'Miocene fossils show that kiwi (*Apteryx*, Apterygidae) are probably not phyletic dwarves'. In: *Paleornithological Research 2013—Proceedings of the 8th International Meeting of the Society of Avian Paleontology and Evolution. Vienna: Natural History Museum Vienna*. pp. 63–80.

Yonezawa, T., Segawa, T., Mori, H., Campos, P. F., Hongoh, Y., Endo, H., Akiyoshi, A., Kohno, N., Nishida, S., Wu, J., & Jin, H. (2017). Phylogenomics and morphology of extinct paleognaths reveal the origin and evolution of the ratites. *Current Biology*, 27(1), 68–77.

Wu, S., Rheindt, F. E., Zhang, J., Wang, J., Zhang, L., Quan, C., ... & Liu, L. (2024). Genomes, fossils, and the concurrent rise of modern birds and flowering plants in the Late Cretaceous. *Proceedings of the National Academy of Sciences*, 121(8), e2319696121.

Mammalia

Intro

Graves, J. A. M., & Renfree, M. B. (2013). Marsupials in the age of genomics. *Annual Review of Genomics and Human Genetics*, 14(1), 393–420.

Kielan-Jaworowska, Z., Cifelli, R.L., & Luo, Z.-X., (2004). *Mammals from the Age of Dinosaurs*. Columbia University Press, New York.

Luo, Z.-X. (2007). Transformation and diversification in early mammal evolution. *Nature*, 450(7172), 1011–1019.

Luo, Z. X., Gatesy, S. M., Jenkins Jr, F. A., Amaral, W. W., & Shubin, N. H. (2015). Mandibular and dental characteristics of Late Triassic mammaliaform Haramiyavia and their ramifications for basal mammal evolution. *Proceedings of the National Academy of Sciences*, 112(51), E7101–E7109.

Rougier, G. W., Wible, J. R., & Novacek, M. J. (1998.) Implications of *Deltatheridium* specimens for early marsupial history. *Nature*, 396(6710), 459–463.

Wilson, D.E., & Reeder, D.M. (2005). *Mammal species of the world: a taxonomic and geographic reference*. Johns Hopkins Press, Baltimore, Maryland.

Monotremata

Chimento, N. R., Agnolín, F. L., Manabe, M., Tsuihiji, T., Rich, T. H., Vickers-Rich, P., & Novas, F. E. (2023). First monotreme from the Late Cretaceous of South America. *Communications Biology*, 6(1), 146.

Flannery, T. F., Rich, T. H., Vickers-Rich, P., Ziegler, T., Veatch, E. G., & Helgen, K. M. (2022). A review of monotreme (Monotremata) evolution. *Alcheringa*, 46(1), 3–20.

Moyal, A. M. (2004). *Platypus: The Extraordinary Story of How a Curious Creature Baffled the World*. John Hopkins University Press, Baltimore, Maryland.

Allotheria (Multituberculates and Gondwanatheres)

Adams, N. F., Rayfield, E. J., Cox, P. G., Cobb, S. N., & Corfe, I. J. (2019). Functional tests of the competitive exclusion hypothesis for multituberculate extinction. *Royal Society Open Science*, 6(3), 181536.

Chimento, N. R., Agnolin, F. L., & Novas, F. E. (2015). The bizarre 'metatherians' Groeberia and Patagonia, late surviving members of gondwanatherian mammals. *Historical Biology*, 27(5), 603–623.

Hoffmann, S., Beck, R. M., Wible, J. R., Rougier, G. W., & Krause, D. W. (2020). Phylogenetic placement of *Adalatherium hui* (Mammalia, Gondwanatheria) from the Late Cretaceous of Madagascar: implications for allotherian relationships. *Journal of Vertebrate Paleontology*, 40(sup1), 213–234.

Krause, D. W., Hoffmann, S., Hu, Y., Wible, J. R., Rougier, G. W., Kirk, E. C., ... & Rahantarisoa, L. J. (2020). Skeleton of a Cretaceous mammal from Madagascar reflects long-term insularity. *Nature*, 581(7809), 421–427.

Krause, D. W., Hoffmann, S., & Werning, S. (2017). First postcranial remains of Multituberculata (Allotheria, Mammalia) from Gondwana. *Cretaceous Research*, 80, 91–100.

Miao, D. (1987). Skull morphology of *Lambdopsalis bulla* (Mammalia, Multituberculata) and its implications to mammalian evolution. University of Wyoming.

Rich, T. H., Vickers-Rich, P., Flannery, T. F., Kear, B. P., Cantrill, D. J., Komarower, P., Kool, L., Pickering, D., Trusler, P., Morton, S., & Van Klaveren, N. (2009). An Australian multituberculate and its palaeobiogeographic implications. *Acta Palaeontologica Polonica*, 54(1), 1–6.

Rose, K.D. (2006). *The Beginning of the Age of Mammals*. Johns Hopkins University Press, Baltimore, Maryland.

Rougier, G. W., Martinelli, A. G., & Forasiepi, A. M. (2021). Allotheria: Gondwanatherians and Multituberculates. *Mesozoic Mammals from South America and Their Forerunners*, 289–323.

Smith, T., & Codrea, V. (2015). Red iron-pigmented tooth enamel in a multituberculate mammal from the Late Cretaceous Transylvanian "Haţeg Island". *PLoS One*, 10(7), e0132550.

Meridiolestida

Chimento, N. R., Agnolin, F. L., & Novas, F. E. (2012). The Patagonian fossil mammal *Necrolestes*: a Neogene survivor of Dryolestoidea. *Revista del Museo Argentino de Ciencias Naturales*, 14, 261-306.

Kielan-Jaworowska, Z; Cifelli, R. L, & Luo, Z.-X. (2004). *Mammals from the Age of Dinosaurs: Origins, Evolution, and Structure*. Columbia University Press, New York.

O'Meara, R. N., & Thompson, R. S. (2014). Were there Miocene meridiolestidans? Assessing the phylogenetic placement of *Necrolestes patagonensis* and the presence of a 40 million year meridiolestidan ghost lineage. *Journal of Mammalian Evolution*, 21, 271-284.

Rougier, G. W., Wible, J. R., Beck, R. M., & Apesteguía, S. (2012). The Miocene mammal *Necrolestes* demonstrates the survival of a Mesozoic nontherian lineage into the late Cenozoic of South America. *Proceedings of the National Academy of Sciences*, 109(49), 20053-20058.

Metatheria

Álvarez-Carretero, S., Tamuri, A. U., Battini, M., Nascimento, F. F., Carlisle, E., Asher, R. J., ... & Dos Reis, M. (2022). A species-level timeline of mammal evolution integrating phylogenomic data. *Nature*, 602(7896), 263–267.

Archer, M., Godthelp, H., Hand, S., & Megirian, D. (1989). Fossil mammals of Riversleigh, northwestern Queensland: preliminary overview of biostratigraphy, correlation and environmental change. *Australian Zoologist*, 25(2), 29–66.

Argot, C. (2004). Evolution of South American mammalian predators (Borhyaenoidea): anatomical and palaeobiological implications". *Zoological Journal of the Linnean Society*. 140(4), 487–521.

Beck, R. M., Louys, J., Brewer, P., Archer, M., Black, K. H., & Tedford, R. H. (2020). A new family of diprotodontian marsupials from the latest Oligocene of Australia and the evolution of wombats, koalas, and their relatives (Vombatiformes). *Scientific Reports*, 10(1), p.9741.

Beck, R. M., Voss, R. S., & Jansa, S. A. (2022). Craniodental morphology and phylogeny of marsupials. *Bulletin of the American Museum of Natural History*, 457(1), 1–352.

Beck, R. M. (2023). 'Diversity and phylogeny of marsupials and their stem relatives (Metatheria)'. In: Cáceres, N. C., & Dickman, C. R. (eds) *American and Australasian Marsupials: An Evolutionary, Biogeographical, and Ecological Approach*. Springer International Publishing, Cham. pp. 23–88.

Beck, R. M., Godthelp, H., Weisbecker, V., Archer, M., & Hand, S. J. (2008). Australia's oldest marsupial fossils and their biogeographical implications. *PLoS One*, 3(3), e1858.

Bi, S., Zheng, X., Wang, X., Cignetti, N. E., Yang, S., & Wible, J. R. (2018). An Early Cretaceous eutherian and the placental–marsupial dichotomy. *Nature*, 558(7710), 390–395.

Black, K. H., Archer, M., Hand, S. J., & Godthelp, H. (2012). The rise of Australian marsupials: a synopsis of biostratigraphic, phylogenetic, palaeoecologic and palaeobiogeographic understanding. *Earth and life: global biodiversity, extinction intervals and biogeographic perturbations through time*, 983–1078.

Castro, M. C., Dahur, M. J., & Ferreira, G. S. (2021). Amazonia as the origin and diversification area of Didelphidae (Mammalia: Metatheria), and a review of the fossil record of the clade. *Journal of Mammalian Evolution*, 28(3), 583–598.

Eldridge, M. D., Beck, R. M., Croft, D. A., Travouillon, K. J., & Fox, B. J. (2019). An emerging consensus in the evolution, phylogeny, and systematics of marsupials and their fossil relatives (Metatheria). *Journal of Mammalogy*, 100(3), 802–837.

Engelman, R. K., Flynn, J. J., Wyss, A. R., & Croft, D. A. (2020). *Eomakhaira molossus*, a new saber-toothed sparassodont (Metatheria: Thylacosmilinae) from the early Oligocene (? Tinguiririca) Cachapoal locality, Andean Main Range, Chile. *American Museum Novitates*, 2020(3957), 1–75.

Goin, F. J. (2003). 'Early marsupial radiations in South America'. In: Jones, M., Dickman, C., & Archer, M. (eds) *Predators with pouches: The biology of carnivorous marsupials*. Commonwealth Scientific and Industrial Research Organization, Sydney. pp 30–42.

Goin, F. J., Woodburne, M. O., Zimicz, A. N., Martin, G. M., & Chornogubsky, L. (2016). *A Brief History of South American Metatherians*. Springer, Dordrecht.

Hocknull, S. A., Lewis, R., Arnold, L. J., Pietsch, T., Joannes-Boyau, R., Price, G. J., ... & Lawrence, R. A. (2020). Extinction of eastern Sahul megafauna coincides with sustained environmental deterioration. *Nature Communications*, 11(1), 2250.

Hu, Y., Meng, J., Li, C., & Wang, Y. (2010). New basal eutherian mammal from the Early Cretaceous Jehol biota, Liaoning, China. *Proceedings of the Royal Society B: Biological Sciences*, 277(1679), 229–236.

Long, J., Archer, M., Flannery, T., & Hand, S. (2002). *Prehistoric mammals of Australia and New Guinea: One hundred million years of evolution*. John Hopkins University Press, Baltimore, Maryland.

Louys, J., Duval, M., Beck, R. M., Pease, E., Sobbe, I., Sands, N., & Price, G. J. (2022). Cranial remains of *Ramsayia magna* from the Late Pleistocene of Australia and the evolution of gigantism in wombats (Marsupialia, Vombatidae). *Papers in Palaeontology*, 8(6), p.e1475.

Maga, A. M., & Beck, R. M. (2017). Skeleton of an unusual, cat-sized marsupial relative (Metatheria: Marsupialiformes) from the middle Eocene (Lutetian: 44-43 million years ago) of Turkey. *PLoS One*, 12(8), e0181712.

Marshall, L. G., & Muizon, C. (1988). The dawn of the age of mammals in South America. *National Geographic Research*, 4(1), 23–55.

May-Collado, L. J., Kilpatrick, C. W., & Agnarsson, I. (2015). Mammals from 'down under': a multi-gene species-level phylogeny of marsupial mammals (Mammalia, Metatheria). *PeerJ*, 3, e805.

McKenna, M.C., & Bell, S.K. (1997). *Classification of mammals: above the species level*. Columbia University Press, New York.

Prevosti, F. J., Forasiepi, A., & Zimicz, N. (2013). The evolution of the Cenozoic terrestrial mammalian predator guild in South America: competition or replacement?. *Journal of Mammalian Evolution*, 20(1), 3–21.

Prevosti, F. J., & Forasiepi, A. M. (2018). *Evolution of South American mammalian predators during the Cenozoic: paleobiogeographic and paleoenvironmental contingencies*. Springer, New York.

Richards, H. L., Wells, R. T., Evans, A. R., Fitzgerald, E. M., & Adams, J. W. (2019). The extraordinary osteology and functional morphology of the limbs in Palorchestidae, a family of strange extinct marsupial giants. *PLoS One*, 14(9), p.e0221824.

Richards, H. L., Rovinsky, D. S., Adams, J. W., & Evans, A. R. (2023). Inferring the palaeobiology of palorchestid marsupials through analysis of mammalian humeral and femoral shape. *Journal of Mammalian Evolution*, 30(1), 47–66.

Rose, K.D. (2006). *The beginning of the age of mammals*. John Hopkins University Press, Baltimore, Maryland.

Travouillon, K. J., Hand, S. J., Archer, M., & Black, K. H. (2014). Earliest modern bandicoot and bilby (Marsupialia, Peramelidae and Thylacomyidae) from the Miocene of the Riversleigh world heritage area, northwestern Queensland, Australia. *Journal of Vertebrate Paleontology*, 34(2), 375–382.

Upham, N. S., Esselstyn, J. A., & Jetz, W. (2019). Inferring the mammal tree: species-level sets of phylogenies for questions in ecology, evolution, and conservation. *PLoS Biology*, 17(12), e3000494.

Weisbecker, V., & Nilsson, M. (2008). Integration, heterochrony, and adaptation in pedal digits of syndactylous marsupials. *BMC Evolutionary Biology*, 8, 1–14.

Wilson, G. P., Ekdale, E. G., Hoganson, J. W., Calede, J. J., & Vander Linden, A. (2016). A large carnivorous mammal from the Late Cretaceous and the North American origin of marsupials. *Nature Communications*, 7(1), 13734.

Wilson, D. E., & Reeder, D. M. (2005). *Mammal species of the world: a taxonomic and geographic reference*. Johns Hopkins University Press, Baltimore, Maryland.

Eutheria

Hallström, B. M., & Janke, A. (2008). Resolution among major placental mammal interordinal relationships with genome data imply that speciation influenced their earliest radiations. *BMC Evolutionary Biology*, 8(1), 162.

Luo, Z. X., Yuan, C. X., Meng, Q. J., & Ji, Q. (2011). A Jurassic eutherian mammal and divergence of marsupials and placentals. *Nature*, 476(7361), 442–445.

McKenna, M. C. (1975). 'Toward a phylogenetic classification of the Mammalia'. In: Luckett, W. P., & Szalay, F. S. (eds) *Phylogeny of the Primates*. Plenum Press, New York, pp. 21–46.

Murphy, W. J., Pringle, T. H., Crider, T. A., Springer, M. S., & Miller, W. (2007). Using genomic data to unravel the root of the placental mammal phylogeny. *Genome Research*, 17(4), 413–421.

Nishihara, H., Maruyama, S., & Okada, N. (2009). Retroposon analysis and recent geological data suggest near-simultaneous divergence of the three superorders of mammals. *Proceedings of the National Academy of Sciences*, 106(13), 5235–5240.

Romiguier, J., Ranwez, V., Douzery, E. J., & Galtier, N. (2013). Genomic evidence for large, long-lived ancestors to placental mammals. *Molecular Biology and Evolution*, 30(1), 5–13.

Rougier, G. W., Wible, J. R., & Novacek, M. J. (1998). Implications of *Deltatheridium* specimens for early marsupial history. *Nature*, 396(6710), 459–463.

Springer, M. S., Murphy, W. J., Eizirik, E., & O'Brien, S. J. (2003). Placental mammal diversification and the Cretaceous–Tertiary boundary. *Proceedings of the National Academy of Sciences*, 100(3), 1056–1061.

Tarver, J. E., Dos Reis, M., Mirarab, S., Moran, R. J., Parker, S., O'Reilly, J. E., ... & Pisani, D. (2016). The interrelationships of placental mammals and the limits of phylogenetic inference. *Genome Biology and Evolution*, 8(2), 330–344.

Xenarthra

Amson, E., Muizon, C., Laurin, M., Argot, C., & de Buffrénil, V. (2014). Gradual adaptation of bone structure to aquatic lifestyle in extinct sloths from Peru. *Proceedings of the Royal Society B: Biological Sciences*, 281(1782), 20140192.

Carlini, A. A., & Scillato-Yané, G. J. (2004). The oldest Megalonychidae (Xenarthra: Tardigrada); phylogenetic relationships and an emended diagnosis of the family. *Neues Jahrbuch für Geologie und Paläontologie-Abhandlungen*, 233(3).

Croft, D. A. (2016). *Horned Armadillos and Rafting Monkeys: The Fascinating Fossil Mammals of South America*. Indiana University Press, Bloomington, Indiana.

Delsuc, F., Catztefis, F. M., Stanhope, M.J., & Douzery, E. J. P. (2001). The evolution of armadillos, anteaters and sloths depicted by nuclear and mitochondrial phylogenies: Implications for the status of the enigmatic fossil *Eurotamandua*. *Proceedings of the Royal Society of London B*. 268 (1476), 1605–1615.

Delsuc, F., Gibb, G. C., Kuch, M., Billet, G., Hautier, L., Southon, J., ... & Poinar, H. N. (2016). The phylogenetic affinities of the extinct glyptodonts. *Current Biology*, 26(4), R155–R156.

Delsuc, F., Kuch, M., Gibb, G. C., Karpinski, E., Hackenberger, D., Szpak, P., Martínez, J. G., Mead, J. I., McDonald, H. G., MacPhee, R. D., & Billet, G. (2019). Ancient mitogenomes reveal the evolutionary history and biogeography of sloths. *Current Biology*, 29(12), 2031–2042.

Farina, R. A., Vizcaino, S. F., & De Luliis, G. (2013). *Megafauna: Giant Beasts of Pleistocene South America*. Indiana University Press, Bloomington, Indiana.

Gaudin, T. J. (2004). Phylogenetic relationships among sloths (Mammalia, Xenarthra, Tardigrada): the craniodental evidence. *Zoological Journal of the Linnean Society*, 140(2), 255–305.

Gaudin, T. J., & Croft, D. A. (2015). Paleogene Xenarthra and the evolution of South American mammals. *Journal of Mammalogy*, 96(4), 622–634.

Hirschfeld, S. E. (1976). A new fossil anteater (Edentata, Mammalia) from Colombia, SA and evolution of the Vermilingua. *Journal of Paleontology*, 419–432.

McDonald, H. G. (2005). Paleoecology of extinct xenarthrans and the Great American Biotic Interchange. *Bulletin of the Florida Museum of Natural History*, 45(4), 313–333.

Muizon, C., McDonald, H. G., Salas, R., & Urbina, M. (2004). The evolution of feeding adaptations of the aquatic sloth *Thalassocnus*. *Journal of Vertebrate Paleontology*, 24(2), 398–410.

Presslee, S., Slater, G. J., Pujos, F., Forasiepi, A. M., Fischer, R., Molloy, K., Mackie, M., Olsen, J. V., Kramarz, A., Taglioretti, M., & Scaglia, F. (2019). Palaeoproteomics resolves sloth relationships. *Nature Ecology & Evolution*, 3(7), 1121–1130.

Raj Pant, S., Goswami, A., & Finarelli, J. A. (2014). Complex body size trends in the evolution of sloths (Xenarthra: Pilosa). *BMC Evolutionary Biology*, 14, 1–8.

Scillato-Yané, G. J., Carlini, A. A., Tonni, E. P., & Noriega, J. I. (2005). Paleobiogeography of the late Pleistocene pampatheres of South America. *Journal of South American Earth Sciences*, 20(1–2), 131–138.

Slater, G. J., Cui, P., Forasiepi, A. M., Lenz, D., Tsangaras, K., Voirin, B., ... & Greenwood, A. D. (2016). Evolutionary relationships among extinct and extant sloths: the evidence of mitogenomes and retroviruses. *Genome Biology and Evolution*, 8(3), 607–621.

Steadman, D. W., Martin, P. S., MacPhee, R. D., Jull, A. T., McDonald, H. G., Woods, C. A., ... & Hodgins, G. W. (2005). Asynchronous extinction of late Quaternary sloths on continents and islands. *Proceedings of the National Academy of Sciences*, 102(33), 11763–11768.

White, J. L., & MacPhee, R. D. E. (2001). 'The sloths of the West Indies: a systematic and phylogenetic review'. In: Woods, C. A., & Sergile, F. E. (eds) *Biogeography of the West Indies: Patterns and Perspectives*. CRC Press, Boca Raton, Florida. pp. 201–235.

Wolf, D., Kalthoff, D. C., & Sander, P. M. (2012). Osteoderm histology of the Pampatheriidae (Cingulata, Xenarthra, Mammalia): implications for systematics, osteoderm growth, and biomechanical adaptation. *Journal of Morphology*, 273(4), 388–404.

Afrotheria

Benoit, J., Adnet, S., El Mabrouk, E., Khayati, H., Ben Haj Ali, M., Marivaux, L., Merzeraud, G., Merigeaud, S., Vianey-Liaud, M., & Tabuce, R. (2013). Cranial remain from Tunisia provides new clues for the origin and evolution of Sirenia (Mammalia, Afrotheria) in Africa. *PLoS One*, 8(1), e54307.

Benoit, J., Lyras, G. A., Schmitt, A., Nxumalo, M., Tabuce, R., Obada, T., ... & Manger, P. (2022). 'Paleoneurology of the Proboscidea (Mammalia, Afrotheria): Insights from their brain endocast and labyrinth'. In: Macrini, T. E., Paulina-Carabajal, A., Dozo, M. T., & Walsh, S. (eds) *Paleoneurology of amniotes: New directions in the study of fossil endocasts*. Springer, Cham, Switerland. pp. 579–644.

Berta, A., Sumich, J. L., & Kovacs, K. M. (2015). *Marine Mammals: Evolutionary Biology (Third Edition)*. Academic Press, Cambridge, Massachusetts.

Domning, D. P. (2001). The earliest known fully quadrupedal sirenian. *Nature*, 413(6856), 625–627.

Gheerbrant, E. (2022). Ancestral radiation of paenungulate mammals (Paenungulatomorpha)—new evidence from the Paleocene of Morocco. *Journal of Vertebrate Paleontology*, 42(5), e2197971.

Gheerbrant, E., Schmitt, A., & Kocsis, L. (2018). Early African fossils elucidate the origin of embrithopod mammals. *Current Biology*, 28(13), 2167–2173.

Gheerbrant, E., & Tassy, P. (2009). The origin and evolution of elephants. *Rendus Palevol*, 8(2–3), 281–294.

Heritage, S., & Seiffert, E. R. (2022). Total evidence time-scaled phylogenetic and biogeographic models for the evolution of sea cows (Sirenia, Afrotheria). *PeerJ*, 10, e13886.

Kerber, L., & Moraes-Santos, H. (2021). Endocranial morphology of a middle Miocene south American dugongid and the neurosensorial evolution of Sirenians. *Journal of Mammalian Evolution*, 28(3), 661–678.

Larramendi, A. (2015). Shoulder height, body mass, and shape of proboscideans. *Acta Palaeontologica Polonica*, 61(3), 537–574.

Lister, A. M. (2004). 'Ecological Interactions of Elephantids in Pleistocene Eurasia: *Palaeoloxodon* and *Mammuthus*'. In: Goren-Inbar, N., & Speth J. D. (eds) *Human Paleoecology in the Levantine Corridor*. Oxbow, Oxford. pp. 53–60.

Lister, A. M., Sher, A. V., van Essen, H., & Wei, G. (2005). The pattern and process of mammoth evolution in Eurasia. *Quaternary International*, 126, 49–64.

Olson, L. E. (2013). Tenrecs. *Current Biology*, 23(1), R5–R8.

Rose, K. D. (2006). *The Beginning of the Age of Mammals*. Johns Hopkins University Press, Baltimore, Maryland.

Springer, M. S. (2022). Afrotheria. *Current Biology*, 32(5), R205–R210.

Springer, M. S., Cleven, G. C., Madsen, O., de Jong, W. W., Waddell, V. G., Amrine, H. M., & Stanhope, M. J. (1997). Endemic African mammals shake the phylogenetic tree. *Nature*, 388(6637), 61–64.

Stanhope, M. J., Waddell, V. G., Madsen, O., De Jong, W., Hedges, S. B., Cleven, G. C., ... & Springer, M. S. (1998). Molecular evidence for multiple origins of Insectivora and for a new order of endemic African insectivore mammals. *Proceedings of the National Academy of Sciences*, 95(17), 9967–9972.

Tabuce, R., Asher, R. J., & Lehmann, T. (2008). Afrotherian mammals: a review of current data. *Mammalia*, 72(1), 2–14.

Tabuce, R., Seiffert, E. R., Gheerbrant, E., Alloing-Séguier, L., & von Koenigswald, W. (2017). Tooth enamel microstructure of living and extinct hyracoids reveals unique enamel types among mammals. *Journal of Mammalian Evolution*, 24, 91–110.

Tassy, P., & Shoshani, J. (1988). 'The Tethytheria: elephants and their relatives'. In: Benton, M. J. (ed) *The Phylogeny and Classifcation of the Tetrapods, vol. 2: Mammals*. Clarendon Press, Oxford. pp 283–316.

Werdelin, L., & Sanders, W. L. (eds). (2010). *Cenozoic Mammals of Africa*. University of California Press, Berkeley, California.

Zalmout, I. S., & Gingerich, P. D. (2012). Late Eocene sea cows (Mammalia, Sirenia) from Wadi al Hitan in the western desert of Fayum, Egypt. *University of Michigan Papers on Paleontology*, 37, 1–58.

Laurasiatheria I: Carnivorans, Bats, Insectivores and their Relatives

Agustí, J., & Antón, M. (2002). *Mammoths, Sabertooths, and Hominids: 65 Million Years of Mammalian Evolution in Europe*. Columbia University Press, New York.

Anderson, S. C., & Ruxton, G. D. (2020). The evolution of flight in bats: a novel hypothesis. *Mammal Review*, 50(4), 426–439.

Bishop, K. L. (2008). The evolution of flight in bats: narrowing the field of plausible hypotheses. *The Quarterly Review of Biology*, 83(2), 153–169.

Brace, S., Thomas, J. A., Dalén, L., Burger, J., MacPhee, R. D., Barnes, I., & Turvey, S. T. (2016). Evolutionary history of the Nesophontidae, the last unplaced recent mammal family. *Molecular Biology and Evolution*, 33(12), 3095–3103.

Carlisle, E., Janis, C. M., Pisani, D., Donoghue, P. C., & Silvestro, D. (2023). A timescale for placental mammal diversification based on Bayesian modeling of the fossil record. *Current Biology*, 33(15), 3073–3082.

Eizirik, E., Murphy, W. J., Koepfli, K. P., Johnson, W. E., Dragoo, J. W., Wayne, R. K., & O'Brien, S. J. (2010). Pattern and timing of diversification of the mammalian order Carnivora inferred from multiple nuclear gene sequences. *Molecular Phylogenetics and Evolution*, 56(1), 49–63.

Emry, R. J., & Skinner, M. F. (1970). A North American Oligocene pangolin and other additions to the Pholidota. *Bulletin of the American Museum of Natural History*, 142(6), 457–510.

Esteban Núñez, J. M., Martín-Serra, A., Pérez-Ramos, A., Rybczynski, N., Pastor, F. J., & Figueirido-Castillo, F. B. (2023). Investigating the land-to-sea transition in carnivorans from the evolution of sacrum morphology in pinnipeds. *Journal of Mammalian Evolution*, 30(2), 341–362.

Faurby, S., Werdelin, L., & Antonelli, A. (2019). Dispersal ability predicts evolutionary success among mammalian carnivores. *BioRxiv*, 755207

Flynn, J. J., Finarelli, J. A., & Spaulding, M. (2010). Phylogeny of the Carnivora and Carnivoramorpha, and the use of the fossil record to enhance understanding of evolutionary transformations. *Carnivoran Evolution*, 25–63.

Flynn, J. J., Finarelli, J. A., Zehr, S., Hsu, J., & Nedbal, M. A. (2005). Molecular phylogeny of the Carnivora (Mammalia): assessing the impact of increased sampling on resolving enigmatic relationships. *Systematic Biology*, 54(2), 317–337.

Flynn, J. J., Wyss, A. R., & Wolsan, M. (2020). 'Pan-Carnivora, new clade name'. In: Adl, S. M. (ed) *Phylonyms*. Taylor and Francis, Florida. pp. 979–982.

Forasiepi, A. M., Soibelzon, L. H., Gomez, C. S., Sánchez, R., Quiroz, L. I., Jaramillo, C., & Sánchez-Villagra, M. R. (2014). Carnivorans at the Great American Biotic Interchange: new discoveries from the northern neotropics. *Naturwissenschaften*, 101, 965–974.

Gaudin, T. J., Emry, R. J., & Wible, J. R. (2009). The phylogeny of living and extinct pangolins (Mammalia, Pholidota) and associated taxa: a morphology based analysis. *Journal of Mammalian Evolution*, 16, 235–305.

Hemmer, H. (2023). The evolution of the palaeopantherine cats, *Palaeopanthera* gen. nov. *blytheae* (Tseng *et al.*, 2014) and *Palaeopanthera pamiri* (Ozansoy, 1959) comb. nov. (Mammalia, Carnivora, Felidae). *Palaeobiodiversity and Palaeoenvironments*, 103(4), 827–839.

Hunt, R.M., Jr., & Barnes, l.G. (1994). Basicranial evidence for ursid affinity of the oldest pinnipeds. *Proceedings of the San Diego Society of Natural History*, 29, 57–67.

Janis, C. M., Scott, K. M., & Jacobs, L. L. (eds.). (1998). *Evolution of tertiary mammals of North America: Volume 1, terrestrial carnivores, ungulates, and ungulate like mammals (Vol. 1)*. Cambridge University Press, Cambridge, UK.

Johnson, W. E., Eizirik, E., Pecon-Slattery, J., Murphy, W. J., Antunes, A., Teeling, E., & O'Brien, S. J. (2006). The late Miocene radiation of modern Felidae: a genetic assessment. *Science*, 311(5757), 73–77.

Jones, M. F., Beard, K. C., & Simmons, N. B. (2024). Phylogeny and systematics of early Paleogene bats. *Journal of Mammalian Evolution*, 31(2), 18.

Koepfli, K. P., Deere, K. A., Slater, G. J., Begg, C., Begg, K., Grassman, L., ... & Wayne, R. K. (2008). Multigene phylogeny of the Mustelidae: resolving relationships, tempo and biogeographic history of a mammalian adaptive radiation. *BMC Biology*, 6, 1–22.

Koepfli, K. P., Pollinger, J., Godinho, R., Robinson, J., Lea, A., Hendricks, S., Schweizer, R. M., Thalmann, O., Silva, P., Fan, Z., & Yurchenko, A. A. (2015). Genome-wide evidence reveals that African and Eurasian golden jackals are distinct species. *Current Biology*, 25(16), 2158–2165.

Kondrashov, P., & Agadjanian, A. K. (2012). A nearly complete skeleton of *Ernanodon* (Mammalia, Palaeanodonta) from Mongolia: morphofunctional analysis. *Journal of Vertebrate Paleontology*, 32(5), 983–1001.

Kumar, V., Lammers, F., Bidon, T., Pfenninger, M., Kolter, L., Nilsson, M. A., & Janke, A. (2017). The evolutionary history of bears is characterized by gene flow across species. *Scientific Reports*, 7(1), 46487.

Law, C. J., Slater, G. J., & Mehta, R. S. (2018). Lineage diversity and size disparity in Musteloidea: testing patterns of adaptive radiation using molecular and fossil-based methods. *Systematic Biology*, 67(1), 127–144.

Li, G., Davis, B. W., Eizirik, E., & Murphy, W. J. (2016). Phylogenomic evidence for ancient hybridization in the genomes of living cats (Felidae). *Genome Research*, 26(1), 1–11.

McKenna, M. C., & Bell, S. K. (1997). *Classification of mammals: above the species level*. Columbia University Press, New York.

Meredith, R. W., Janečka, J. E., Gatesy, J., Ryder, O. A., Fisher, C. A., Teeling, E. C., ... & Murphy, W. J. (2011). Impacts of the Cretaceous Terrestrial Revolution and KPg extinction on mammal diversification. *Science*, 334(6055), 521–524.

Mitchell, K. J., Bray, S. C., Bover, P., Soibelzon, L., Schubert, B. W., Prevosti, F., ... & Cooper, A. (2016). Ancient mitochondrial DNA reveals convergent evolution of giant short-faced bears (Tremarctinae) in North and South America. *Biology Letters*, 12(4), 20160062.

Morales, J., Mayda, S., Valenciano, A., DeMiguel, D., & Kaya, T. (2019). A new lophocyonid, *Izmirictis cani* gen. et sp. nov. (Carnivora: Mammalia), from the lower Miocene of Turkey. *Journal of Systematic Palaeontology*, 17(16), 1347–1358.

Novacek, M. J., Wyss, A. R., & McKenna, M. C. (1988). 'The major groups of eutherian mammals'. In: Benton, M. J. (ed) *The Phylogeny and Classification of the Tetrapods, vol. 2: Mammals*. Clarendon Press, Oxford, pp. 31–73.

Paterson, R. S., Rybczynski, N., Kohno, N., & Maddin, H. C. (2020). A total evidence phylogenetic analysis of pinniped phylogeny and the possibility of parallel evolution within a monophyletic framework. *Frontiers in Ecology and Evolution*, 7, 457.

Piras, P., Maiorino, L., Teresi, L., Meloro, C., Lucci, F., Kotsakis, T., & Raia, P. (2013). Bite of the cats: relationships between functional integration and mechanical performance as revealed by mandible geometry. *Systematic Biology*, 62(6), 878–900.

Qiu, Z-X, Deng, T., & Wang, B-Y. (2014). A Late Miocene *Ursavus* skull from Guanghe, Gansu, China. *Vertebrata PalAsiatica*. 52(3), 265–302.

Roca, A. L., Kahila Bar-Gal, G., Eizirik, E., Helgen, K. M., Maria, R., Springer, M. S., ... & Murphy, W. J. (2004). Mesozoic origin for West Indian insectivores. *Nature*, 429(6992), 649–651.

Rose, K. D. (2006). *The Beginning of the Age of Mammals*. Johns Hopkins University Press, Baltimore, Maryland.

Salesa, M. J., Peigné, S., Antón, M., & Morales, J. (2022). 'Evolution of the family Ailuridae: origins and Old-World fossil record'. In: Glatston, A. R. (ed) *Red Panda*. Academic Press, Cambridge, Massachusetts. pp. 15–29.

Simmons, N. B., Seymour, K. L., Habersetzer, J., & Gunnell, G. F. (2008). Primitive Early Eocene bat from Wyoming and the evolution of flight and echolocation. *Nature*, 451(7180), 818–821.

Solé, F., Smith, T., De Bast, E., Codrea, V., & Gheerbrant, E. (2016). New carnivoraforms from the latest Paleocene of Europe and their bearing on the origin and radiation of Carnivoraformes (Carnivoramorpha, Mammalia). *Journal of Vertebrate Paleontology*, 36(2), e1082480.

Solé, F., Marandat, B., & Lihoreau, F. (2020). The hyaenodonts (Mammalia) from the French locality of Aumelas (Hérault), with possible new representatives from the late Ypresian. *Geodiversitas*, 42(13), 185–214.

Spaulding, M., & Flynn, J. J. (2009). Anatomy of the postcranial skeleton of *"Miacis" uintensis* (Mammalia: Carnivoramorpha). *Journal of Vertebrate Paleontology*, 29(4), 1212–1223.

Storch, G. (1978). Messel fossil finds part 14: *Eomanis waldi* (new genus, new species), a pangolin from the middle Eocene of the messel pit near darmstadt, West Germany (Mammalia, Pholidota). *Senckenbergiana Lethaea*, 59, 503–530.

Szalay, F. S., Novacek, M. J., & McKenna, M.C. (Eds.). (1993). *Mammal Phylogeny*. Springer-Verlag, Berlin.

Tedford, R. H., Wang, X., & Taylor, B. E. (2009). Phylogenetic systematics of the North American fossil caninae (Carnivora: Canidae). *Bulletin of the American Museum of Natural History*, 2009(325), 1–218.

Teeling, E. C., Springer, M. S., Madsen, O., Bates, P., O'brien, S. J., & Murphy, W. J. (2005). A molecular phylogeny for bats illuminates biogeography and the fossil record. *Science*, 307(5709), 580–584.

Tsagkogeorga, G., Parker, J., Stupka, E., Cotton, J. A., & Rossiter, S. J. (2013). Phylogenomic analyses elucidate the evolutionary relationships of bats. *Current Biology*, 23(22), 2262–2267.

Turner, A. (1997). *The big cats and their fossil relatives: an illustrated guide to their evolution and natural history*. Columbia University Press, New York.

Turner, A., Antón, M., & Werdelin, L. (2008). Taxonomy and evolutionary patterns in the fossil Hyaenidae of Europe. *Geobios*, 41(5), 677–687.

Upham, N. S., Esselstyn, J. A., & Jetz, W. (2021). Molecules and fossils tell distinct yet complementary stories of mammal diversification. *Current Biology*, 31(19), 4195–4206.

Velazco, P. M., Buczek, A. J., Hoffman, E., Hoffman, D. K., O'Leary, M. A., & Novacek, M. J. (2022). Combined data analysis of fossil and living mammals: a Paleogene sister taxon of Placentalia and the antiquity of Marsupialia. *Cladistics*, 38(3), 359–373.

Wang, X., & Tedford, R. (2008). *Dogs: Their Fossil Relatives and Evolutionary History*. Columbia University Press, New York.

Welker, F., Collins, M. J., Thomas, J. A., Wadsley, M., Brace, S., Cappellini, E., ... & MacPhee, R. D. (2015). Ancient proteins resolve the evolutionary history of Darwin's South American ungulates. *Nature*, 522(7554), 81–84.

Werdelin, L., & Sanders, W. J. (2010). *Cenozoic mammals of Africa*. University of California Press, Berkeley, California.

Werdelin, L.; Yamaguchi, N., Johnson, W. E., & O'Brien, S. J. (2010). 'Phylogeny and evolution of cats (Felidae)'. In: MacDonald, D. W., & Loveridge, A. J. (eds) *Biology and Conservation of Wild Felids*. Oxford University Press, Oxford. pp. 59–82.

Xue, L., Jingyang, H., Yiwen, H., Yitian, L., Dongming, X., Ryder, O.A., Irwin, D.M., & Yu, L. (2021). Diverse phylogenomic datasets uncover a concordant scenario of laurasiatherian interordinal relationships. *Molecular Phylogenetics and Evolution*, 157, 107065.

Zack, S. P. (2019). A skeleton of a Uintan machaeroidine 'creodont'and the phylogeny of carnivorous eutherian mammals. *Journal of Systematic Palaeontology*, 17(8), 653–689.

Zhou, X., Xu, S., Xu, J., Chen, B., Zhou, K., & Yang, G. (2012). Phylogenomic analysis resolves the interordinal relationships and rapid diversification of the laurasiatherian mammals. *Systematic Biology*, 61(1), 150.

Laurasiatheria II: The Ungulates (Hooved Mammals)

Agustí, J., & Antón, M. (2002). *Mammoths, Sabertooths, and Hominids: 65 Million Years of Mammalian Evolution in Europe*. Columbia University Press, New York.

Ahrens, H.E., Dunn, R.H., Folie, A., Jones, K.E., & Smith, T. (2019). Anatomy, relationships, and paleobiology of *Cambaytherium* (Mammalia, Perissodactylamorpha, Anthracobunia) from the lower Eocene of western India. *Journal of Vertebrate Paleontology*, 39, 1–147.

Alroy, J. (2002). Synonymies and reidentifications of North American fossil mammals.

Archibald, J. D. (1998). 'Archaic ungulates ('Condylarthra')'. In: Janis, C. M., Scott, K. M., & Jacobs L. L. (eds) *Evolution of Tertiary Mammals of North America: Volume 1, Terrestrial Carnivores, Ungulates, and Ungulate Like Mammals*. Cambridge University Press, Cambridge, UK. p. 292.

Archibald, J. D., Zhang, Y., Harper, T., & Cifelli, R. L. (2011). *Protungulatum*, confirmed Cretaceous occurrence of an otherwise Paleocene eutherian (placental?) mammal. *Journal of Mammalian Evolution*, 18, 153–161.

Arifakis, D., Muraca, J., Singh, A., & Granatosky, M. C. (2022). 'Artiodactyla Locomotion'. In: Vonk, J., & Shackelford, T. (eds) *Encyclopedia of Animal Cognition and Behavior*. Springer, Cham. pp. 462–469.

Bai, B. (2017). Eocene Pachynolophinae (Perissodactyla, Palaeotheriidae) from China, and their palaeobiogeographical implications. *Palaeontology*, 60(6), 837–852.

Bai, B., Wang, Y. Q., & Meng, J. (2021). A new late Paleocene phenacodontid 'condylarth' *Lophocion* from the Clark's Fork Basin of Wyoming. *Historical Biology*, 33(5), 652–659.

Bast, E. D., & Smith, T. (2013). Reassessment of the small 'arctocyonid' *Prolatidens waudruae* from the early Paleocene of Belgium, and its phylogenetic relationships with ungulate-like mammals. *Journal of Vertebrate Paleontology*, 33(4), 964–976.

Bertrand, O. C., Jiménez Lao, M., Shelley, S. L., Wible, J. R., Williamson, T. E., Meng, J., & Brusatte, S. L. (2024). The virtual brain endocast of *Trogosus* (Mammalia, Tillodontia) and its relevance in understanding the extinction of archaic placental mammals. *Journal of Anatomy*, 244(1), 1–21.

Billet, G. (2011). Phylogeny of the Notoungulata (Mammalia) based on cranial and dental characters. *Journal of Systematic Palaeontology*, 9(4), 481–497.

Bond, M., Kramarz, A., Macphee, R. D., & Reguero, M. (2011). A new astrapothere (Mammalia, Meridiungulata) from La Meseta Formation, Seymour (Marambio) Island, and a reassessment of previous records of Antarctic astrapotheres. *American Museum Novitates*, 2011(3718), 1–16.

Breda, M., & Marchetti, M. (2005). Systematical and biochronological review of Plio-Pleistocene Alceini (Cervidae; Mammalia) from Eurasia. *Quaternary Science Reviews*, 24(5-6), 775–805.

Calamari, Z. T. (2021). Total evidence phylogenetic analysis supports new morphological synapomorphies for Bovidae (Mammalia, Artiodactyla). *American Museum Novitates*, 2021(3970), 1–38.

Cassini, G. H., Cerdeño, E., Villafañe, A. L., & Muñoz, N. A. (2012). 'Paleobiology of Santacrucian native ungulates (Meridiungulata; Astrapotheria, Litopterna and Notoungulata)'. In: Bargo, M. S., Kay, R. F., & Vizcaíno, S. F. (eds) *Early Miocene Paleobiology in Patagonia: High-Latitude Paleocommunities of the Santa Cruz Formation*. Cambridge University Press, Cambridge, UK. pp. 243–286.

Chen, L., Qiu, Q., Jiang, Y. U., Wang, K., Lin, Z., Li, Z., ... & Wang, W. (2019). Large-scale ruminant genome sequencing provides insights into their evolution and distinct traits. *Science*, 364(6446), eaav6202.

Coombs, M. C., Hunt Jr, R. M., Stepleton, E., Albright III, L. B., & Fremd, T. J. (2001). Stratigraphy, chronology, biogeography, and taxonomy of early Miocene small chalicotheres in North America. *Journal of Vertebrate Paleontology*, 21(3), 607–620.

Cooper, L. N., Seiffert, E. R., Clementz, M., Madar, S. I., Bajpai, S., Hussain, S. T., & Thewissen, J. G. (2014). Anthracobunids from the middle Eocene of India and Pakistan are stem perissodactyls. *PLoS One*, 9(10), e109232.

Croft, D. A. (2016). *Horned Armadillos and Rafting Monkeys: The Fascinating Fossil Mammals of South America*. Indiana University Press, Bloomington, Indiana.

Danilo, L., Remy, J. A., Vianey-Liaud, M., Marandat, B., Sudre, J., & Lihoreau, F. (2013). A new Eocene locality in southern France sheds light on the basal radiation of Palaeotheriidae (Mammalia, Perissodactyla, Equoidea). *Journal of Vertebrate Paleontology*, 33(1), 195–215.

DeMiguel, D., Azanza, B., & Morales, J. (2014). Key innovations in ruminant evolution: a paleontological perspective. *Integrative Zoology*, 9(4), 412–433.

Dong, W. (2011). Reconsideration of the systematics of the early Pleistocene *Cervavitus* (Cervidae, Artiodactyla, Mammalia). *Estudios Geológicos*, 67(2):603–611

Franzen, J. L., & Brown, K. M. (2010). *The Rise of Horses*. Johns Hopkins University Press, Baltimore, Maryland.

Gatesy, J., Geisler, J. H., Chang, J., Buell, C., Berta, A., Meredith, R. W., ... & McGowen, M. R. (2013). A phylogenetic blueprint for a modern whale. *Molecular Phylogenetics and Evolution*, 66(2), 479–506.

Geraads, D., Barr, W. A., Reed, D., Laurin, M., & Alemseged, Z. (2021). New remains of *Camelus grattardi* (Mammalia, Camelidae) from the Plio-Pleistocene of Ethiopia and the phylogeny of the genus. *Journal of Mammalian Evolution*, 28, 359–370.

Gilbert, C., Ropiquet, A., & Hassanin, A. (2006). Mitochondrial and nuclear phylogenies of Cervidae (Mammalia, Ruminantia): systematics, morphology, and biogeography. *Molecular Phylogenetics and Evolution*, 40(1), 101–117.

Gingerich, P. D., Haq, M. U., Zalmout, I. S., Khan, I. H., & Malkani, M. S. (2001). Origin of whales from early artiodactyls: hands and feet of Eocene Protocetidae from Pakistan. *Science*, 293(5538), 2239–2242.

Halliday, T. J., Upchurch, P., & Goswami, A. (2017). Resolving the relationships of Paleocene placental mammals. *Biological Reviews*, 92(1), 521-550.

Harris, J. M., & Liu, L. P. (2007). 'Superfamily suoidea'. In: Prothero, D. R., & Foss, S. E. (eds) *The evolution of artiodactyls*. John Hopkins University Press, Baltimore, Maryland. pp. 130–150.

Holbrook, L. T., & Lapergola, J. (2011). A new genus of perissodactyl (Mammalia) from the Bridgerian of Wyoming, with comments on basal perissodactyl phylogeny. *Journal of Vertebrate Paleontology*, 31(4), 895–901.

Janis, C. M. (1987). The interrelationship of higher ruminant families: with special emphasis on the member of the Cervoidea. *American Museum Novitates*, 2893, 1–85.

Janis, C. M., & Manning, E. (1998). 'Antilocapridae'. In: Janis, C. M., Scott, K. M., & Jacobs, L. L. (eds) *Evolution of Tertiary Mammals of North America. Volume 1: Terrestrial Carnivores, Ungulates, and Ungulatelike Mammals*. Cambridge University Press, Cambridge, UK. pp. 491–507.

Janis, C. M., Scott, K. M., & Jacobs, L. L. (Eds.). (1998). *Evolution of Tertiary Mammals of North America: Volume 1, Terrestrial Carnivores, Ungulates, and Ungulate-like Mammals (Vol. 1)*. Cambridge University Press, Cambridge, UK.

Janis, C. M., & Theodor, J. M. (2014). Cranial and postcranial morphological data in ruminant phylogenetics. *Zitteliana*, 15–31.

Kort, A. E. (2023). The ancestral morphology of lumbar vertebrae and its diversification in early Cenozoic mammals. *Journal of Mammalian Evolution*, 30(4), 845–858.

Liu, L. (2003). *Chinese fossil Suoidea: systematics, evolution, and paleoecology*. Ph.D. thesis, University of Helsinki.

Lu, X. K., Deng, T., & Pandolfi, L. (2023). Reconstructing the phylogeny of the hornless rhinoceros Aceratheriinae. *Frontiers in Ecology and Evolution*, 11, 1005126.

MacFadden, B. J. (1994). *Fossil horses: systematics, paleobiology, and evolution of the family Equidae*. Cambridge University Press, Cambridge, UK.

MacLaren, J. A. (2021). Biogeography a key influence on distal forelimb variation in horses through the Cenozoic. *Proceedings of the Royal Society B*, 288(1942), 20202465.

Mader, B. J. (2010). A species-level revision of the North American brontotheres *Eotitanops* and *Palaeosyops* (Mammalia, Perissodactyla). *Zootaxa*, 2339(1), 1–43.

McGrath, A. J., Anaya, F., & Croft, D. A. (2018). Two new macraucheniids (Mammalia: Litopterna) from the late middle Miocene (Laventan South American Land Mammal Age) of Quebrada Honda, Bolivia. *Journal of Vertebrate Paleontology*, 38(3), e1461632.

McGrath, A. J., Anaya, F., & Croft, D. A. (2020). New proterotheriids (Litopterna, Mammalia) from the middle Miocene of Quebrada Honda, Bolivia, and trends in diversity and body size of proterotheriid and macraucheniid litopterns. *Ameghiniana*, 57(2), 159–188.

McKenna, M. C. (1975). 'Toward a phylogenetic classification of the Mammalia'. In: Luckett, W. P., & Szalay, F. S. (eds) *Phylogeny of the primates: a multidisciplinary approach*. Springer, Boston, Massachusetts. pp. 21–46.

McKenna, M. C., & Bell, S. K. (1997). *Classification of mammals: above the species level*. Columbia University Press, New York.

McKenna, M. C., Minchen, C., Suyin, T., & Zhexi, L. (1989). '*Radinskya yupingae* a perissodactyl-like mammal from the Late Paleocene of China'. In: Prothero, D. R., & Schoch, R.M. (eds) *The Evolution of Perissodactyls*. Oxford University Press, Oxford. pp. 24–36.

Mennecart, B., Aiglstorfer, M., Li, Y., Li, C., & Wang, S. (2021). Ruminants reveal Eocene Asiatic palaeobiogeographical provinces as the origin of diachronous mammalian Oligocene dispersals into Europe. *Scientific Reports*, 11(1), 17710.

Mihlbachler, M. C. (2008). Species taxonomy, phylogeny, and biogeography of the Brontotheriidae (Mammalia: Perissodactyla). *Bulletin of the American Museum of Natural History*, 2008(311), 1–475.

Mihlbachler, M. C., & Solounias, N. (2006). Coevolution of tooth crown height and diet in oreodonts (Merycoidodontidae, Artiodactyla) examined with phylogenetically independent contrasts. *Journal of Mammalian Evolution*, 13, 11–36.

Novacek, M. J. (1994). The radiation of placental mammals. *Short Courses in Paleontology*, 7, 220–237.

Novacek, M. J., & Wyss, A. R. (1986). Higher level relationships of the recent eutherian orders: morphological evidence. *Cladistics*, 2(4), 257–287.

Orliac, M. J., Argot, C., & Gilissen, E. (2012). Digital cranial endocast of *Hyopsodus* (Mammalia,"Condylarthra"): a case of Paleogene terrestrial echolocation?. *PLoS One*, 7(2), e30000.

Orliac, M., Boisserie, J. R., MacLatchy, L., & Lihoreau, F. (2010). Early Miocene hippopotamids (Cetartiodactyla) constrain the phylogenetic and spatiotemporal settings of hippopotamid origin. *Proceedings of the National Academy of Sciences*, 107(26), 11871–11876.

Pandolfi, L., Martino, R., Rook, L., & Piras, P. (2020). Investigating ecological and phylogenetic constraints in Hippopotaminae skull shape. *Rivista Italiana di Paleontologia e Stratigrafia*, 126, 37–49.

Peredo, C. M., Pyenson, N. D., Marshall, C. D., & Uhen, M. D. (2018). Tooth loss precedes the origin of baleen in whales. *Current Biology*, 28(24), 3992–4000.

Prothero, D. R. (2005). *The Evolution of North American Rhinoceroses*. Cambridge University Press, Cambridge, UK.

Prothero, D. R. (2009). Evolutionary transitions in the fossil record of terrestrial hoofed mammals. *Evolution: Education and Outreach*, 2, 289–302.

Prothero, D. R., & Foss, S. E. (2007). *The Evolution of Artiodactyls*. John Hopkins University Press, Baltimore, Maryland.

Prothero, D. R., Manning, E. M., & Fischer, M. (1988). 'The phylogeny of the ungulates'. In: Benton, M. J. (ed) *The Phylogeny and Classification of the Tetrapods, vol. 2: Mammals*. Clarendon Press, Oxford. pp. 201–235.

Prothero, D. R., & Schoch, R.M. (eds) (1989). *The Evolution of Perissodactyls*. Oxford University Press, New York.

Prothero, D. R., & Schoch, R. M. (2002). *Horns, Tusks, and Flippers: The Evolution of Hoofed Mammals*. Johns Hopkins University Press, Baltimore, Maryland.

Puschel, H. P., Shelley, S. L., Williamson, T. E., Wible, J. R., & Brusatte, S. L. (2019). Testing the phylogeny of Periptychidae and" archaic" Paleocene mammals uner different optimality criteria. *Journal of Vertebrate Paleontology, Abstracts and Program, 2019*.

Ríos, M., Sánchez, I. M., & Morales, J. (2017). A new giraffid (Mammalia, Ruminantia, Pecora) from the late Miocene of Spain, and the evolution of the sivathere-samothere lineage. *PLoS One*, 12(11), e0185378.

Robson, S., Ludtke, J. A., & Theodor, J. M. (2021). The petrosal and basicranial morphology of *Leptoreodon major*. *Vertebrate Anatomy Morphology Palaeontology*, 9(1), 116–130.

Rose, K. D. (2006). *The Beginning of the Age of Mammals*. Johns Hopkins University Press, Baltimore, Maryland.

Sánchez, I. M., Cantalapiedra, J. L., Ríos, M., Quiralte, V., & Morales, J. (2015). Systematics and evolution of the Miocene three-horned palaeomerycid ruminants (Mammalia, Cetartiodactyla). *PLoS One*, 10(12), e0143034.

Savage, R. J., & Long, M. R. (1986). *Mammal Evolution: An Illustrated Guide*. Facts on File Publications, New York.

Scherer, C. S. (2013). The Camelidae (Mammalia, Artiodactyla) from the quaternary of South America: cladistic and biogeographic hypotheses. *Journal of Mammalian Evolution*, 20(1), 45–56.

Shelley, S. L., Williamson, T. E., & Brusatte, S.L. (2015). Resolving the higher-level phylogenetic relationships of "Triisodontidae" ('Condylarthra') within Placentalia.

Shelley, S. L., Williamson, T. E., & Brusatte, S. L. (2018). The osteology of *Periptychus carinidens*: A robust, ungulate-like placental mammal (Mammalia: Periptychidae) from the Paleocene of North America. *PLoS One*, 13(7), p.e0200132.

Solounias, N. (2007). 'Family giraffidae'. In: Prothero, D. R, & Foss, S. E. (eds) *The evolution of artiodactyls*. Johns Hopkins University Press, Baltimore, Maryland.

Spaulding, M., O'Leary, M. A., & Gatesy, J. (2009). Relationships of Cetacea (Artiodactyla) among mammals: increased taxon sampling alters interpretations of key fossils and character evolution. *PLoS One*, 4(9), e7062.

Thewissen, J. G., Cooper, L. N., Clementz, M. T., Bajpai, S., & Tiwari, B. N. (2007). Whales originated from aquatic artiodactyls in the Eocene epoch of India. *Nature*, 450(7173), 1190–1194.

Ting, S.-Y., Meng, J., Li, Q., Wang, Y.-Q., Schiebout, J. A., Koch, P. L., Clyde, W. C., & Bowen, G. J. (2007). *Ganungulatum xincunliense*, an artiodactyl-like mammal (Ungulata, Mammalia) from the Paleocene, Chijiang Basin, Jiangxi, China. *Vertebrata PalAsiatica*, 45(4), 278.

Tissier, J., Antoine, P. O., & Becker, D. (2020). New material of *Epiaceratherium* and a new species of *Mesaceratherium* clear up the phylogeny of early Rhinocerotidae (Perissodactyla). *Royal Society Open Science*, 7(7), 200633.

Toosey, W. J., Williamson, T. E., Shelley, S. L., & Brusatte, S. L. (2024). The osteology of *Triisodon crassicuspis* (Cope, 1882): New insights into the enigmatic "archaic" placental mammal group "Triisodontidae". *PLoS One*, 19(11), p.e0311187.

Tsagkogeorga, G., McGowen, M. R., Davies, K. T., Jarman, S., Polanowski, A., Bertelsen, M. F., & Rossiter, S. J. (2015). A phylogenomic analysis of the role and timing of molecular adaptation in the aquatic transition of cetartiodactyl mammals. *Royal Society Open Science*, 2(9), 150156.

Tsubamoto, T., Takai, M., & Egi, N. (2004). Quantitative analyses of biogeography and faunal evolution of middle to late Eocene mammals in East Asia. *Journal of Vertebrate Paleontology*, 24(3), 657–667.

Upham, N. S., Esselstyn, J. A., & Jetz, W. (2019). Inferring the mammal tree: species-level sets of phylogenies for questions in ecology, evolution, and conservation. *PLoS Biology*, 17(12), e3000494.

Welker, F., Collins, M. J., Thomas, J. A., Wadsley, M., Brace, S., Cappellini, E., Turvey, S. T., Reguero, M., Gelfo, J. N., Kramarz, A., & Burger, J. (2015). Ancient proteins resolve the evolutionary history of Darwin's South American ungulates. *Nature*, 522(7554), 81–84.

Weppe, R., Blondel, C., Vianey-Liaud, M., Pélissié, T., & Orliac, M. J. (2020). A new Cainotherioidea (Mammalia, Artiodactyla) from Palembert (Quercy, SW France): Phylogenetic relationships and evolutionary history of the dental pattern of Cainotheriidae. *Palaeontologia Electronica*, 23(3), 20.

Werdelin, L., & Sanders, W. J. (2010). *Cenozoic mammals of Africa*. University of California Press, Berkeley, California.

Woodburne, M. O. (Ed.). (2004). Late Cretaceous and Cenozoic mammals of North America: biostratigraphy and geochronology. Columbia University Press, New York.

Yang, C., Xiang, C., Qi, W., Xia, S., Tu, F., Zhang, X., ... & Yue, B. (2013). Phylogenetic analyses and improved resolution of the family Bovidae based on complete mitochondrial genomes. *Biochemical Systematics and Ecology*, 48, 136–143.

Zhou, X., Xu, S., Xu, J., Chen, B., Zhou, K., & Yang, G. (2012). Phylogenomic analysis resolves the interordinal relationships and rapid diversification of the laurasiatherian mammals. *Systematic Biology*, 61(1), 150.

Euarchontoglires: Rodents, Lagomorphs, Primates—And Humans

Asher, R. J., Meng, J., Wible, J. R., McKenna, M. C., Rougier, G. W., Dashzeveg, D., & Novacek, M. J. (2005). Stem Lagomorpha and the antiquity of Glires. *Science*, 307(5712), 1091–1094.

Asher, R. J., Smith, M. R., Rankin, A., & Emry, R. J. (2019). Congruence, fossils and the evolutionary tree of rodents and lagomorphs. *Royal Society Open Science*, 6(7), 190387.

Anemone, R. L., Dawson, M. R., & Beard, K. C. (2012). The early Eocene rodent Tuscahomys (Cylindrodontidae) from the Great Divide Basin, Wyoming: phylogeny, biogeography, and paleoecology. *Annals of Carnegie Museum*, 80(3), 187–205.

Averianov, A. O. (2020). Reappraisal of arctostylopid mammal *Kazachostylops occidentalis* from the late Paleocene of Kazakhstan and phylogenetic relationships within Arctostylopida. *Journal of Paleontology*, 94(3), 568–579.

Beard, C. (2004). *The Hunt for the Dawn Monkey: Unearthing the Origins of Monkeys, Apes, and Humans*. University of California Press, Berkeley, California.

Blanga-Kanfi, S., Miranda, H., Penn, O., Pupko, T., DeBry, R. W., & Huchon, D. (2009). Rodent phylogeny revised: analysis of six nuclear genes from all major rodent clades. *BMC Evolutionary Biology*, 9, 1–12.

Chuan-Kuei, L., Wilson, R. W., Dawson, M. R., & Krishtalka, L. (1987). The origin of rodents and lagomorphs. *Current Mammalogy*, 97–108.

Cifelli, R. L., & Schaff, C. R. (1998). 'Arctostylopida'. In: Janis, C. M., Scott, K. M., & Jacobs L. L. (eds) *Evolution of Tertiary Mammals of North America: Volume 1, Terrestrial Carnivores, Ungulates, and Ungulate Like Mammals (Vol. 1)*. Cambridge University Press, Cambridge, UK.

Coster, P. M., Beard, K. C., Salem, M. J., Chaimanee, Y., & Jaeger, J. J. (2015). New fossils from the Paleogene of central Libya illuminate the evolutionary history of endemic African anomaluroid rodents. *Frontiers in Earth Science*, 3, 56.

Dembo, M., Radovčić, D., Garvin, H. M., Laird, M. F., Schroeder, L., Scott, J. E., ... & Collard, M. (2016). The evolutionary relationships and age of *Homo naledi*: An assessment using dated Bayesian phylogenetic methods. *Journal of Human Evolution*, 97, 17–26.

Doronina, L., Matzke, A., Churakov, G., Stoll, M., Huge, A., & Schmitz, J. (2017). The Beaver's Phylogenetic Lineage Illumina ted by Retroposon Reads. *Scientific Reports*, 7(1), 43562.

Esselstyn, J. A., Oliveros, C. H., Swanson, M. T., & Faircloth, B. C. (2017). Investigating difficult nodes in the placental mammal tree with expanded taxon sampling and thousands of ultraconserved elements. *Genome Biology and Evolution*, 9(9), 2308–2321.

Fabre, P. H., Hautier, L., Dimitrov, D., & P Douzery, E. J. (2012). A glimpse on the pattern of rodent diversification: a phylogenetic approach. *BMC Evolutionary Biology*, 12, 1–19.

Ge, D., Wen, Z., Xia, L., Zhang, Z., Erbajeva, M., Huang, C., & Yang, Q. (2013). Evolutionary history of lagomorphs in response to global environmental change. *PLoS One*, 8(4), e59668.

Gregory, W. K. (1910). The orders of mammals. *Bulletin of the American Museum of Natural History*; v. 27, 1–525.

Johanson, D., & Edey, M. (1981). *Lucy: The Beginnings of Humankind*. Simon & Schuster, New York.

Johanson, D., & Wong, K. (2009). *Lucy's Legacy: The Quest for Human Origins*. Crown, New York.

Luckett, W.P., & Hartenberger, J.-l., (Eds). (1985). *Evolutionary Relationships among Rodents*. Plenum Press, New York.

McKenna, M. C. (1975). 'Toward a phylogenetic classification of the mammalia'. In: Luckett, W. P., & Szalay, F. S. (eds) *Phylogeny of the Primates*. Plenum Press, New York. pp. 21–46.

McKenna, M. C., & Bell, S. K. (1997). *Classification of Mammals above the Species Level*. Columbia University Press, New York.

Mongle, C. S., Strait, D. S., & Grine, F. E. (2023). An updated analysis of hominin phylogeny with an emphasis on re-evaluating the phylogenetic relationships of *Australopithecussediba*. *Journal of Human Evolution*, 175, 103311.

Orlando, L., Calvignac, S., Schnebelen, C., Douady, C. J., Godfrey, L. R., & Hänni, C. (2008). DNA from extinct giant lemurs links archaeolemurids to extant indriids. *BMC Evolutionary Biology*, 8, 1–9.

Rasmussen, D. T., Friscia, A. R., Gutierrez, M., Kappelman, J., Miller, E. R., Muteti, S., ... & Muchemi, F. (2019). Primitive Old World monkey from the earliest Miocene of Kenya and the evolution of cercopithecoid bilophodonty. *Proceedings of the National Academy of Sciences*, 116(13), 6051–6056.

Roberts, A. (2018). *Evolution: The Human Story*. DK Books, London.

Rodrigues, H. G., Šumbera, R., Hautier, L., & Herrel, A. (2023). 'Digging up convergence in fossorial rodents: Insights into burrowing activity and Morpho-functional specializations of the masticatory apparatus'. In: Bels, V. L., & Russell, A. P. (eds) *Convergent evolution: Animal form and function*. Springer, Cham. pp. 37–63.

Rose, K. D. (2006). *The Beginning of the Age of Mammals*. Johns Hopkins University Press, Baltimore, Maryland.

Rose, K. D., DeLeon, V. B., Missiaen, P., Rana, R. S., Sahni, A., Singh, L., & Smith, T. (2008). Early Eocene lagomorph (Mammalia) from Western India and the early diversification of Lagomorpha. *Proceedings of the Royal Society B: Biological Sciences*, 275(1639), 1203–1208.

Seiffert, E. R., Boyer, D. M., Fleagle, J. G., Gunnell, G. F., Heesy, C. P., Perry, J. M., & Sallam, H. M. (2018). New adapiform primate fossils from the late Eocene of Egypt. *Historical Biology*, 30(1–2), 204–226.

Silcox, M. T., Bloch, J. I., Boyer, D. M., & Houde, P. (2010). Cranial anatomy of Paleocene and Eocene *Labidolemur kayi* (Mammalia: Apatotheria), and the relationships of the Apatemyidae to other mammals. *Zoological Journal of the Linnean Society*, 160(4), 773–825.

Silvestro, D., Tejedor, M. F., Serrano-Serrano, M. L., Loiseau, O., Rossier, V., Rolland, J., ... & Salamin, N. (2017). Evolutionary history of New World monkeys revealed by molecular and fossil data. *bioRxiv*, 178111.

Stevens, N. J., Seiffert, E. R., O'Connor, P. M., Roberts, E. M., Schmitz, M. D., Krause, C., ... & Temu, J. (2013). Palaeontological evidence for an Oligocene divergence between Old World monkeys and apes. *Nature*, 497(7451), 611–614.

Tabuce, R., Marivaux, L., Lebrun, R., Adaci, M., Bensalah, M., Fabre, P. H., ... & Mahboubi, M. (2009). Anthropoid versus strepsirhine status of the African Eocene primates *Algeripithecus* and *Azibius*: craniodental evidence. *Proceedings of the Royal Society B: Biological Sciences*, 276(1676), 4087–4094.

Werdelin, L., & Sanders, W. J. (2010). *Cenozoic mammals of Africa*. University of California Press, Berkeley, California.

Wible, J. R. (2007). On the cranial osteology of the Lagomorpha. *Bulletin of Carnegie Museum of Natural History*, 2007(39), 213–234.

Wu, S., Wu, W., Zhang, F., Ye, J., Ni, X., Sun, J., Edwards, S. V., Meng, J., & Organ, C. L. (2012). Molecular and paleontological evidence for a post-Cretaceous origin of rodents. *PLoS One*. 7(10), e46445.

Chapter 3: Reconstructing Extinct Cenozoic Tetrapods

Antón, M. (2003). Reconstructing fossil mammals: strengths and limitations of a methodology. *Palaeontological Association Newsletter*, 53, 55–65.

Davenport, J., Fraher, J., Fitzgerald, E., McLaughlin, P., Doyle, T., Harman, L., & Cuffe, T. (2009). Fat head: an analysis of head and neck insulation in the leatherback turtle (Dermochelys coriacea). *Journal of Experimental Biology*, 212(17), 2753–2759.

Doucet, S. M., & Meadows, M. G. (2009). Iridescence: a functional perspective. *Journal of the Royal Society Interface*, 6(suppl_2), S115–S132.

Goldfinger, E. (2004). *Animal Anatomy for Artists: The Elements of Form*. Oxford University Press, Oxford.

Goss, R. J. (2012). *Deer Antlers: regeneration, function and evolution*. Academic Press, Cambridge, Massachusetts.

Hartman, S. (2011, June 16). Skeletal reconstructions: Schematic vs Realistic. Skeletal Drawing.com Blog Post. Retrieved from: https://www.skeletaldrawing.com/home/2011/06/skeletal-reconstructions-schematic-vs.html

Hieronymus, T. L. (2009). *Osteological correlates of cephalic skin structures in Amniota: documenting the evolution of display and feeding structures with fossil data*. Ph.D. thesis, Ohio University.

Hieronymus, T. L., & Witmer, L. M. (2010). Homology and evolution of avian compound rhamphothecae. *The Auk*, 127(3), 590–604.

McGowan, C., & Motani, R. (2003). *Handbook of Paleoherpetology: Ichthyopterygia*. F. Pfeil.

Myhrvold, C. L., Stone, H. A., & Bou-Zeid, E. (2012). What is the use of elephant hair?. *PLoS One*, 7, e04718.

Prum, R. O., & Torres, R. H. (2004). Structural colouration of mammalian skin: convergent evolution of coherently scattering dermal collagen arrays. *Journal of Experimental Biology*, 207(12), 2157–2172.

Vinther, J. (2015). A guide to the field of palaeo colour: Melanin and other pigments can fossilise: Reconstructing colour patterns from ancient organisms can give new insights to ecology and behaviour. *BioEssays*, 37(6), 643–656.

Vinther, J., Nicholls, R., Lautenschlager, S., Pittman, M., Kaye, T. G., Rayfield, E., Mayr, G., & Cuthill, I. C. (2016). 3D camouflage in an ornithischian dinosaur. *Current Biology*, 26(18), 2456–2462.

Chapter 4: The Cenozoic

Benton, M. J., Wilf, P., & Sauquet, H. (2022). The Angiosperm Terrestrial Revolution and the origins of modern biodiversity. *New Phytologist, 233*(5), 2017–2035.

Chen, M., Strömberg, C. A., & Wilson, G. P. (2019). Assembly of modern mammal community structure driven by Late Cretaceous dental evolution, rise of flowering plants, and dinosaur demise. *Proceedings of the National Academy of Sciences*, 116(20), 9931-9940.

Grossnickle, D. M., Smith, S. M., & Wilson, G. P. (2019). Untangling the multiple ecological radiations of early mammals. *Trends in Ecology & Evolution*, 34(10), 936–949.

Kreft, H., & Jetz, W. (2007). Global patterns and determinants of vascular plant diversity. *Proceedings of the National Academy of Sciences*, 104(14), 5925–5930.

Longrich, N. R., Scriberas, J., & Wills, M. A. (2016). Severe extinction and rapid recovery of mammals across the Cretaceous-Palaeogene boundary, and the effects of rarity on patterns of extinction and recovery. *Journal of Evolutionary Biology*, 29(8), 1495–1512.

Prothero, D. R. (2020). 'Cenozoic'. In: Elias, S., & Alderton, D. (eds) *Encyclopedia of Geology* (Second Edition). Academic Press, Cambridge, Massachusetts. pp. 55–61.

Prothero, D. R. (2020). *The Evolving Earth*. Oxford University Press, Oxford.

Wilson, G. P. (2005). Mammalian faunal dynamics during the last 1.8 million years of the Cretaceous in Garfield County, Montana. *Journal of Mammalian Evolution, 12*, 53–76.

Wing, S. L., & Boucher, L. D. (1998). Ecological aspects of the Cretaceous flowering plant radiation. *Annual Review of Earth and Planetary Sciences, 26*(1), 379–421.

Chapter 5: The Paleogene

Alroy, J. (1999). The fossil record of North American mammals: evidence for a Paleocene evolutionary radiation. *Systematic Biology*, 48(1), 107–118.

DeConto, R. M., Pollard, D., Wilson, P. A., Pälike, H., Lear, C. H., & Pagani, M. (2008). Thresholds for Cenozoic bipolar glaciation. *Nature*, 455(7213), 652–656.

Grossnickle, D. M., & Newham, E. (2016). Therian mammals experience an ecomorphological radiation during the Late Cretaceous and selective extinction at the K–Pg boundary. *Proceedings of the Royal Society B: Biological Sciences*, 283(1832), 20160256.

Grossnickle, D. M., Smith, S. M., & Wilson, G. P. (2019). Untangling the multiple ecological radiations of early mammals. *Trends in Ecology & Evolution*, 34(10), 936–949.

Lauretano, V., Kennedy-Asser, A. T., Korasidis, V. A., Wallace, M. W., Valdes, P. J., Lunt, D. J., ... & Naafs, B. D. A. (2021). Eocene to Oligocene terrestrial Southern Hemisphere cooling caused by declining p CO2. *Nature Geoscience*, 14(9), 659–664.

Longrich, N. R., Scriberas, J., & Wills, M. A. (2016). Severe extinction and rapid recovery of mammals across the Cretaceous-Palaeogene boundary, and the effects of rarity on patterns of extinction and recovery. *Journal of Evolutionary Biology*, 29(8), 1495–1512.

Pollard, D., & DeConto, R. M. (2003). Antarctic ice and sediment flux in the Oligocene simulated by a climate–ice sheet–sediment model. *Palaeogeography, Palaeoclimatology, Palaeoecology*, 198(1–2), 53–67.

Zachos, J. C., Shackleton, N. J., Revenaugh, J. S., Pälike, H., & Flower, B. P. (2001). Climate response to orbital forcing across the Oligocene-Miocene boundary. *Science*, 292(5515), 274–278.

Paleocene

Agustí, J., & Antón, M. (2002). *Mammoths, Sabertooths, and Hominids: 65 Million Years of Mammalian Evolution in Europe*. Columbia University Press, New York.

Aubry, M. P., Lucas, S. G., & Berggren, W. A. (1998). *Late Paleocene-early Eocene climatic and biotic events in the marine and terrestrial records*. Columbia University Press, New York.

Dickens, G. R., O'Neil, J. R., Rea, D. K., & Owen, R. M. (1995). Dissociation of oceanic methane hydrate as a cause of the carbon isotope excursion at the end of the Paleocene. *Paleoceanography*, 10(6), 965–971.

Pires, M. M., Rankin, B. D., Silvestro, D., & Quental, T. B. (2018). Diversification dynamics of mammalian clades during the K–Pg mass extinction. *Biology Letters*, 14(9), 20180458.

Wing, S. L., Gingerich, P. D., Schmitz, B., & Thomas, E. (2003). Causes and consequences of globally warm climates in the early Paleogene. *Geological Society of America*.

North America: Nacimiento Formation

Bai, B., Wang, Y. Q., & Meng, J. (2021). A new late Paleocene phenacodontid 'condylarth' *Lophocion* from the Clark's Fork Basin of Wyoming. *Historical Biology*, 33(5), 652–659.

Bertrand, O. C., Shelley, S. L., Wible, J. R., Williamson, T. E., Holbrook, L. T., Chester, S. G., Butler, I. B., & Brusatte, S. L. (2020). Virtual endocranial and inner ear endocasts of the Paleocene 'condylarth' *Chriacus*: new insight into the neurosensory system and evolution of early placental mammals. *Journal of Anatomy*, 236(1), 21–49.

Bertrand, O. C., Shelley, S. L., Williamson, T. E., Wible, J. R., Chester, S. G., Holbrook, L. T., Lyson, T. R., Meng, J., Smith, J., Smith, T., & Brusatte, S. L. (2022). Vestibular sensitivity and locomotor behavior in early Paleocene mammals. *Journal of Vertebrate Paleontology*.

Chester, S. G., Williamson, T. E., Crowell, J. W., Silcox, M. T., Bloch, J. I., & Sargis, E. J. (2025). New remarkably complete skeleton of *Mixodectes* reveals arboreality in a large Paleocene primatomorphan mammal following the Cretaceous-Paleogene mass extinction. *Scientific Reports*, 15(1), 8041.

Cooper, L. N., Seiffert, E. R., Clementz, M., Madar, S. I., Bajpai, S., Hussain, S. T., & Thewissen, J. G. (2014). Anthracobunids from the middle Eocene of India and Pakistan are stem perissodactyls. *PLoS One*, 9(10), e109232.

Flynn, A. G. (2020). *Early Paleocene Fossil Floras, Paleoclimate, and Magnetostratigraphy from the San Juan Basin, New Mexico, USA*. Ph.D., Baylor University.

Flynn, A. G., & Peppe, D. J. (2019). Early Paleocene tropical forest from the Ojo Alamo Sandstone, San Juan Basin, New Mexico, USA. *Paleobiology*, 45(4), 612–635.

Funston, G. F., dePolo, P. E., Shelley, S. L., Wible, J. R., Williamson, T. E., & Brusatte, S. L. (2021). Life history of an archaic placental mammal, *Pantolambda bathmodon* (Placentalia, Pantodonta). *Vertebrate anatomy morphology palaeontology*, 9.

Grossnickle, D. M., & Newham, E. (2016). Therian mammals experience an ecomorphological radiation during the Late Cretaceous and selective extinction at the K–Pg boundary. *Proceedings of the Royal Society B: Biological Sciences*, 283(1832), 20160256.

Kort, A. E. (2023). The ancestral morphology of lumbar vertebrae and its diversification in early Cenozoic mammals. *Journal of Mammalian Evolution*, 30(4), 845–858.

Kynigopoulou, Z., Shelley, S. L., Williamson, T. E., & Brusatte, S. L. (2024). The post-cranial anatomy and functional morphology of *Conoryctes comma* (Mammalia: Taeniodonta) from the Paleocene of North America. *PLoS One*, 19(10), e0311053.

O'Leary, M. A., Lucas, S. G., & Williamson, T. E. (2000). A new specimen of *Ankalagon* (Mammalia, Mesonychia) and evidence of sexual dimorphism in mesonychians". *Journal of Vertebrate Paleontology*, 20(2), 387–393.

Puschel, H. P., Shelley, S. L., Williamson, T. E., Wible, J. R., & Brusatte, S. L. (2019). Testing the phylogeny of Periptychidae and" archaic" Paleocene mammals uner different optimality criteria. *Journal of Vertebrate Paleontology, Abstracts and Program, 2019.*

Shelley, S.L., Williamson, T.E., & Brusatte, S.L. (2015). Resolving the higher-level phylogenetic relationships of "Triisodontidae" ('Condylarthra') within Placentalia.

Shelley, S. L., Williamson, T. E., & Brusatte, S. L. (2018). The osteology of *Periptychus carinidens*: A robust, ungulate-like placental mammal (Mammalia: Periptychidae) from the Paleocene of North America. *PLoS One*, 13(7), p.e0200132.

Solé, F., Morse, P. E., Bloch, J. I., Gingerich, P. D., & Smith, T. (2021). New specimens of the mesonychid *Dissacus praenuntius* from the early Eocene of Wyoming and evaluation of body size through the PETM in North America. *Geobios*, 66, 103–118.

Toosey, W. J., Williamson, T. E., Shelley, S. L., & Brusatte, S. L. (2024). The osteology of *Triisodon crassicuspis* (Cope, 1882): New insights into the enigmatic "archaic" placental mammal group "Triisodontidae". *PLoS One, 19*(11), p.e0311187.

Williamson, T. E. (1996). The Beginning of the Age of Mammals in the San Juan Basin, New Mexico: Biostratigraphy and Evolution of Paleocene Mammals of the Nacimiento Formation. *New Mexico Museum of Natural History and Science Bulletin*, 8, 1–141.

Williamson, T. E., & Brusatte, S. L. (2013). New specimens of the rare taeniodont Wortmania (Mammalia: Eutheria) from the San Juan Basin of New Mexico and comments on the phylogeny and functional morphology of "archaic" mammals. *PLoS One*, 8(9), e75886.

Williamson, T. E., Brusatte, S. L., Secord, R., & Shelley, S. (2016). A new taeniolabidoid multituberculate (Mammalia) from the middle Puercan of the Nacimiento Formation, New Mexico, and a revision of taeniolabidoid systematics and phylogeny. *Zoological Journal of the Linnean Society*, 177(1), 183–208.

Zhou, X., Zhai, R., Gingerich, P. D., & Chen, L. (1995). Skull of a new mesonychid (Mammalia, Mesonychia) from the late Paleocene of China. *Journal of Vertebrate Paleontology*, 387–400.

South America: Middle Santa Lucía Formation

Argot, C. (2000). Functional-adaptive anatomy of the forelimb in the didelphidae, and the paleobiology of the Paleocene marsupials *Mayulestes ferox* and *Pucadelphys andinus*. *Journal of Morphology*, 247(1), 59–71,

Chimento, N. R., & Agnolin, F. L. (2020). Phylogenetic tree of Litopterna and Perissodactyla indicates a complex early history of hoofed mammals. *Scientific Reports*, 10(1), 1–10.

Croft, D.A. (2016). *Horned armadillos and rafting monkeys: the fascinating fossil mammals of South America*. Indiana University Press, Bloomington, Indiana.

Deconinck, J. F., Blanc-Valleron, M. M., Rouchy, J. M., Camoin, G., & Badaut-Trauth, D. (2000). Palaeoenvironmental and diagenetic control of the mineralogy of Upper Cretaceous–Lower Tertiary deposits of the Central Palaeo–Andean basin of Bolivia (Potosi area). *Sedimentary Geology*, 132(3–4), 263–278.

Gayet, M., Marshall, L. G., & Sempere, T. (1991). The Mesozoic and Paleocene vertebrates of Bolivia and their stratigraphic context: a review. Revista Técnica Yacimientos Petrolíferos Fiscales Bolivianos Santa Cruz, *Bolivia*, 12, 393–433.

Gelfo, J. N., Goin, F. J., Woodburne, M. O., & Muizon, C. D. (2009). Biochronological relationships of the earliest South American Paleogene mammalian faunas. *Palaeontology*, 52(1), 251–269.

Kort, A. E. (2023). The ancestral morphology of lumbar vertebrae and its diversification in early Cenozoic mammals. *Journal of Mammalian Evolution*, 30(4), 845–858.

Ladeveze, S., de Muizon, C., Beck, R.M., Germain, D., & Cespedes-Paz, R. (2011). Earliest evidence of mammalian social behaviour in the basal Tertiary of Bolivia. *Nature*, 474(7349), 83–86.

Marshall, L. G., & Muizon, C. (1988). The dawn of the age of mammals in South America. *National Geographic Research*, 4(1), 23–55.

Muizon, C., & Marshall, L. G. (1992). *Alcidedorbignya inopinata* (Mammalia: Pantodonta) from the Early Paleocene of Bolivia: Phylogenetic and Paleobiogeographic Implications. *Journal of Paleontology*, 66(3), 499–520.

Muizon, C., Billet, G., Argot, C., Ladevèze, S., & Goussard, F. (2015). *Alcidedorbignya inopinata*, a basal pantodont (Placentalia, Mammalia) from the early Palaeocene of Bolivia: anatomy, phylogeny and palaeobiology. *Geodiversitas*, 37(4), 397–634.

Muizon, C., Ladevèze, S., Selva, C., Vignaud, R., & Goussard, F. (2018). *Allqokirus australis* (Sparassodonta, Metatheria) from the early Palaeocene of Tiupampa (Bolivia) and the rise of the metatherian carnivorous radiation in South America. *Geodiversitas*, 40(3), 363–459.

Muizon, C., Billet, G., & Ladevèze, S. (2019). New remains of kollpaniine "condylarths"(Panameriungulata) from the early Palaeocene of Bolivia shed light on hypocone origins and molar proportions among ungulate-like placentals. *Geodiversitas*, 41(1), 841–874.

Williamson, T. E., Brusatte, S. L., & Wilson, G. P. (2014). The origin and early evolution of metatherian mammals: the Cretaceous record. *ZooKeys*, (465), 1.

Woodburne, M. O., Goin, F. J., Bond, M., Carlini, A. A., Gelfo, J. N., López, G. M., ... & Zimicz, A. N. (2014). Paleogene land mammal faunas of South America; a response to global climatic changes and indigenous floral diversity. *Journal of Mammalian Evolution*, 21, 1–73.

South America: Cerrejón Formation

Cadena, E. A., Ksepka, D. T., Jaramillo, C. A., & Bloch, J. I. (2012). New pelomedusoid turtles from the late Palaeocene Cerrejón Formation of Colombia and their implications for phylogeny and body size evolution. *Journal of Systematic Palaeontology*, 10(2), 313.

Hastings, A. K., Bloch, J. I., Cadena, E. A., & Jaramillo, C. A. (2010). A new small short-snouted dyrosaurid (Crocodylomorpha, Mesoeucrocodylia) from the Paleocene of northeastern Colombia. *Journal of Vertebrate Paleontology*, 30(1), 139–162.

Hastings, A. K., Bloch, J. I., & Jaramillo, C. A. (2011). A new longirostrine dyrosaurid (Crocodylomorpha, Mesoeucrocodylia) from the Paleocene of north-eastern Colombia: biogeographic and behavioural implications for New-World Dyrosauridae. *Palaeontology*, 54(5), 1095–1116.

Hastings, A. K., Bloch, J. I. Jaramillo, C. A. (2014). A new blunt-snouted dyrosaurid, *Anthracosuchus balrogus* gen. et sp. nov. (Crocodylomorpha, Mesoeucrocodylia), from the Palaeocene of Colombia. *Historical Biology*. 27(8), 998–1020.

Head, J. J., Bloch, J. I., Hastings, A. K., Bourque, J. R., Cadena, E. A., Herrera, F. A., Polly, P. D., & Jaramillo, C. A. (2009). Giant boid snake from the Palaeocene neotropics reveals hotter past equatorial temperatures. *Nature*, 457(7230), 715–717.

Head, J. J., Bloch, J. I., Moreno-Bernal, J., Rincon Burbano, A., & Bourque, J. (2013). Cranial osteology, body size, systematics, and ecology of the giant Paleocene snake *Titanoboa cerrejonensis*. *Journal of Vertebrate Paleontology*, 33, 140–141.

Herrera, F., Wing, S., & Jaramillo, C. (2005). Warm (not hot) tropics during the Late Paleocene: First continental evidence. *In: AGU Fall Meeting Abstracts* (Vol. 2005, pp. PP51C–0608).

Royer, D. L., Peppe, D. J., Wheeler, E. A., & Niinemets, Ü. (2012). Roles of climate and functional traits in controlling toothed vs. untoothed leaf margins. *American Journal of Botany*, 99(5), 915–922.

Wing, S. L., Herrera, F., Jaramillo, C. A., Gómez-Navarro, C., Wilf, P., & Labandeira, C .C. (2009). Late Paleocene fossils from the Cerrejón Formation, Colombia, are the earliest record of Neotropical rainforest. *Proceedings of the National Academy of Sciences*, 106(44), 8627–18632.

Africa: Khouribga Phosphates (Bed II)

Carvalho, A. R., Ghilardi, A. M., & Barreto, A. M. F. (2016). A new side-neck turtle (Pelomedusoides: Bothremydidae) from the Early Paleocene (Danian) Maria Farinha Formation, Paraíba Basin, Brazil. *Zootaxa*, 4126(4), 491–513.

Gaffney, E. S., & Tong, H. (2008). Redescription of the skull of *Ummulisani rutgersensis* Gaffney, Tong, and Meylan, 2006, a bothremydid side-necked turtle from the eocene of Morocco. *American Museum Novitates*, 2008(3615), 1–20.

Gheerbrant, E. (2009). Paleocene emergence of elephant relatives and the rapid radiation of African ungulates. *Proceedings of the National Academy of Sciences*, 106(26), 10717–10721.

Gheerbrant E. (2010) 'Primitive African ungulates ("Condylarthra" and Paenungulata)'. In: Werdelin, L., & Sanders, W. J. (eds) *Cenozoic Mammals of Africa*. The University of California Press, Berkeley, California. pp. 563–571.

Gheerbrant, E. (2022). Ancestral radiation of paenungulate mammals (Paenungulatomorpha)—new evidence from the Paleocene of Morocco. *Journal of Vertebrate Paleontology*, 42(5), e2197971.

Gheerbrant, E., Amaghzaz, M., Bouya, B., Goussard, F., & Letenneur, C. (2014). *Ocepeia* (Middle Paleocene of Morocco): the oldest skull of an afrotherian mammal. *PLoS One*, 9(2), e89739.

Gheerbrant, E., Filippo, A., & Schmitt, A. (2016). Convergence of Afrotherian and Laurasiatherian ungulate-like mammals: First morphological evidence from the Paleocene of Morocco. *PLoS One*, 11(7), e0157556.

Gheerbrant, E., Sudre, J., & Cappetta, H. (1996). A Palaeocene proboscidean from Morocco. *Nature*, 383(6595), 68–70.

Gheerbrant, E., Sudre, J., Cappetta, H., Mourer-Chauviré, C., Bourdon, E., Iarochene, M., ... & Bouya, B. (2003). The mammal localities of Grand Daoui Quarries, Ouled Abdoun Basin, Morocco, Ypresian: a first survey. *Bulletin de la Societe Geologique de France*, 174(3), 279–293.

Hill, R. V., McCartney, J. A., Roberts, E., Bouaré, M., Sissoko, F., & O'Leary, M. A. (2008). Dyrosaurid (Crocodyliformes: Mesoeucrocodylia) fossils from the Upper Cretaceous and Paleogene of Mali: implications for phylogeny and survivorship across the K/T Boundary. *American Museum Novitates*, 2008(3631), 1–19.

Jouve, S., Iarochène, M., Bouya, B., & Amaghzaz, M. (2006). A new species of Dyrosaurus (Crocodylomorpha, Dyrosauridae) from the early Eocene of Morocco: phylogenetic implications. *Zoological Journal of the Linnean Society*. 148(4), 603–656.

Kocsis, L., Ulianov, A., Mouflih, M., Khaldoune, F., & Gheerbrant, E. (2021). Geochemical investigation of the taphonomy, stratigraphy, and palaeoecology of the mammals from the Ouled Abdoun Basin (Paleocene-Eocene of Morocco). *Palaeogeography, Palaeoclimatology, Palaeoecology*, 577, p.110523.

Pérez-García, A., Ortega, F., & Murelaga, X. (2012). A new genus of Bothremydidae (Chelonii, Pleurodira) in the Cretaceous of Southwestern Europe. Geobios, 45(2), 219–229.

Pérez-García, A. (2021). A new bothremydid turtle (Pleurodira) from the Upper Cretaceous (Cenomanian) of Madagascar. *Cretaceous Research*, 118, p.104645.

Salih, K. O., Evans, D. C., Bussert, R., Klein, N., & Müller, J. (2022). *Brachiosuchus kababishensis*, a new long-snouted dyrosaurid (Mesoeucrocodylia) from the Late Cretaceous of north central Sudan. *Historical Biology*, 34(5), 821–840.

Solé, F., Gheerbrant, E., Amaghzaz, M., & Bouya, B. (2009). Further evidence of the African antiquity of hyaenodontid ('Creodonta', Mammalia) evolution. *Zoological Journal of the Linnean Society*, 156(4), 827–846.

Tong, H., Gaffney, E. S., & Buffetaut, E. (1998). *Foxemys*, a new side-necked turtle (Bothremydidae: Pelomedusoides) from the Late Cretaceous of France. *American Museum Novitates*, 3251, 1–19.

Werdelin, L., & Sanders, W. L. (eds). (2010). *Cenozoic Mammals of Africa*. University of California Press, Berkeley, California.

Asia: Naran Bulak Formation

Asher, R. J., Meng, J., Wible, J. R., McKenna, M. C., Rougier, G. W., Dashzeveg, D., & Novacek, M. J. (2005). Stem Lagomorpha and the antiquity of Glires. *Science*, 307(5712), 1091–1094.

Averianov, A. O. (2020). Reappraisal of arctostylopid mammal *Kazachostylops occidentalis* from the late Paleocene of Kazakhstan and phylogenetic relationships within Arctostylopida. *Journal of Paleontology*, 94(3), 568–579.

Bast, E. D., & Smith, T. (2013). Reassessment of the small 'arctocyonid' *Prolatidens waudruae* from the early Paleocene of Belgium, and its phylogenetic relationships with ungulate-like mammals. *Journal of Vertebrate Paleontology*, 33(4), 964–976.

Burger, B. (2015). The systematic position of the saber-toothed and horned giants of the Eocene: the Uintatheres (Order Dinocerata). In *Society of Vertebrate Paleontology Annual Meeting, Dallas. Bethesda: Society of Vertebrate, Paleontology*.

Dashzeveg, D., Hartenberger, J.L., Martin,T., & Legendre. S. (1998). A peculiar minute Glires (Mammalia) from the early Eocene of Mongolia. *Bulletin of Carnegie Museum of Natural History* 34, 194–209.

Geisler, J. H., & McKenna, M. C. (2007). A new species of mesonychian mammal from the lower Eocene of Mongolia and its phylogenetic relationships. *Acta Palaeontologica Polonica*, 52(1).

Hooker, J. J., & Dashzeveg, D. (2003). Evidence for direct mammalian faunal interchange between Europe and Asia near the Paleocene-Eocene boundary. *Geological Society of America Special Papers*, 369, 479–500

Kondrashov, P., & Agadjanian, A. K. (2012). A nearly complete skeleton of *Ernanodon* (Mammalia, Palaeanodonta) from Mongolia: morphofunctional analysis. *Journal of Vertebrate Paleontology*, 32(5), 983–1001.

Kort, A. E. (2023). The ancestral morphology of lumbar vertebrae and its diversification in early Cenozoic mammals. *Journal of Mammalian Evolution*, 30(4), 845–858.

Lopatin, A. V. (2001). The earliest *Hapalodectes* (Mesonychia, Mammalia) from the Paleocene of Mongolia. *Palaeontological Journal*, 35(4), 426–432.

Lucas, S. G. (1998). 'Pantodonta'. In: Janis, c. M., Scott, K. M., & Jacobs, L. L. (eds) *Evolution of Tertiary Mammals of North America. Volume 1: Terrestrial Carnivores, Ungulates, and Ungulatelike Mammals*. Cambridge University Press, Cambridge, UK. p. 274.

Martin, T. (2004). Evolution of incisor enamel microstructure in Lagomorpha. *Journal of Vertebrate Paleontology*, 24(2), 411–426.

Meng, J., Hu, Y., & Li, C. (2003). The osteology of *Rhombomylus* (Mammalia, Glires): implications for phylogeny and evolution of Glires. *Bulletin of the American Museum of Natural History*, 2003(275), 1–247.

Rose, K. D. (2006). *The Beginning of the Age of Mammals*. Johns Hopkins University Press, Baltimore, Maryland.

Yo, K., Hoffman, E. A., O'Leary, M. A., & Novacek, M. J. (2023). A new Early Paleogene fossil mammal locality in the central-eastern Nemegt Basin, Gobi Desert, Mongolia, and notes on mammalian biostratigraphy. *Journal of Paleontology*, 97(1), 243–266.

Europe: Cernay-lès-Reims Fossil Site

Agustí, J., & Antón, M. (2002). *Mammoths, Sabertooths, and Hominids: 65 Million Years of Mammalian Evolution in Europe*. Columbia University Press, New York.

Argot, C. (2013). Postcranial analysis of a carnivoran-like archaic ungulate: the case of *Arctocyon primaevus* (Arctocyonidae, Mammalia) from the late Paleocene of France. *Journal of Mammalian Evolution*, 20, 83–114.

Bast, E. D., & Smith, T. (2013). Reassessment of the small 'arctocyonid' *Prolatidens waudruae* from the early Paleocene of Belgium, and its phylogenetic relationships with ungulate-like mammals. *Journal of Vertebrate Paleontology*, 33(4), 964–976.

Bloch, J. I., Silcox, M. T., Boyer, T. M., Sargis, E. J. (2007). New Paleocene skeletons and the relationship of plesiadapiformsto crown-clade primates. *Proceedings of the National Academy of Science*, 104, 1159–1164.

Chester, S. G., Bloch, J. I., Secord, R., & Boyer, D. M. (2010). A new small-bodied species of *Palaeonictis* (Creodonta, Oxyaenidae) from the Paleocene-Eocene Thermal Maximum. *Journal of Mammalian Evolution*, 17, 227–243.

Orliac, M. J., Argot, C., & Gilissen, E. (2012). Digital cranial endocast of *Hyopsodus* (Mammalia,"Condylarthra"): a case of Paleogene terrestrial echolocation? *PLoS One*, 7(2), e30000.

Sussman, R. W. (1991). Primate origins and the evolution of angiosperms. *American Journal of Primatology*, 23, 209–223.

Thewissen, J. G. M. (1991). Limb Osteology and function of the primitive Paleocene ungulate *Pleuraspidotherium* with notes on *Tricuspiodon* and *Dissacus* (Mammalia). *Geobios*, 24(4), 483–495.

Eocene

Gunnell, G. F. (Ed.). (2001). *Eocene biodiversity: unusual occurrences and rarely sampled habitats (Vol. 18)*. Springer Science & Business Media, Berlin.

Prothero, D. R. (1994). *The Eocene-Oligocene transition: paradise lost*. Columbia University Press, New York.

Prothero, D. R., & Berggren, W. A. (Eds.). (2014). *Eocene-Oligocene climatic and biotic evolution (Vol. 135)*. Princeton University Press, Princeton, New Jersey.

Prothero, D. R., Ivany, L. C., & Nesbitt, E. A. (Eds.). (2003). *From greenhouse to icehouse: the marine Eocene-Oligocene transition*. Columbia University Press, New York.

Rea, D. K., Zachos, J. C., Owen, R. M., & Gingerich, P. D. (1990). Global change at the Paleocene-Eocene boundary: climatic and evolutionary consequences of tectonic events. *Palaeogeography, Palaeoclimatology, Palaeoecology*, 79(1-2), 117-128.

Thomas, D. J., Zachos, J. C., Bralower, T. J., Thomas, E., & Bohaty, S. (2002). Warming the fuel for the fire: Evidence for the thermal dissociation of methane hydrate during the Paleocene-Eocene thermal maximum. *Geology*, 30(12), 1067-1070.

Wing, S. L., Gingerich, P. D., Schmitz, B., & Thomas, E. (Eds.) (2003). Causes and consequences of globally warm climates in the early Paleogene. *Geological Society of America*. Boulder, Colorado.

Australia: Murgon Fossil Site

Beck, R. M. D., Godthelp, H., Weisbecker, V., Archer, M., & Hand, S. J. (2008). Australia's Oldest Marsupial Fossils and their Biogeographical Implications. *PLoS One*, 3(3), e1858.

Godthelp, H., Archer, M., Cifelli, R., Hand, S. J., & Gilkeson, C. F. (1992). Earliest known Australian Tertiary mammal fauna. *Nature*, 359, 514-516.

Godthelp, H., Wroe, S., & Michael, A. (1999). A New Marsupial from the Early Eocene Tingamarra Local Fauna of Murgon, Southeastern Queensland: A Prototypical Australian Marsupial?. *Journal of Mammalian Evolution*, 6(3), 289-313.

Hand, S., Novacek, M., Godthelp, H., & Archer, M. (1994). First Eocene bat from Australia. *Journal of Vertebrate Paleontology*, 14(3), 375-381.

Scanlon, J. D. (2005). Australia's oldest known snakes: *Patagoniophis, Alamitophis*, and cf. *Madtsoia* (Squamata: Madtsoiidae) from the Eocene of Queensland. *Memoirs of the Queensland Museum (Proceedings of the Conference of Australasian Vertebrate Evolution, Palaeontology and Systematics)*, 51, 215-223.

Stein, M., Salisbury, S. W., Hand, S. J., Archer, M., & Godthelp, H. (2012). Humeral morphology of the early Eocene mekosuchine crocodylian *Kambara* from the Tingamarra Local Fauna southeastern Queensland, Australia. *Alcheringa*, 36(4), 473-486.

Szalay, F. S. (1982). 'A new appraisal of marsupial phylogeny and classification'. In: Archer M. (ed) *Carnivorous Marsupials*. Royal Zoological Society of New South Wales, Sydney. pp. 621-640.

Antarctica: La Meseta Formation

Bond, M., Kramarz, A., MacPhee, R. D. E., & Reguero, M. (2011). A new astrapothere (Mammalia, Meridiungulata) from La Meseta Formation, Seymour (Marambio) Island, and a reassessment of previous records of Antarctic astrapotheres. *American Museum Novitates*, 3718, 16.

Bond, M., Reguero, M. A., Vizcaíno, S. F., & Marenssi, S. A. (2006). A new 'South American ungulate'(Mammalia: Litopterna) from the Eocene of the Antarctic Peninsula. *Geological Society, London, Special Publications*, 258(1), 163-176.

Gelfo, J. N., López, G. M., & Santillana, S. N. (2017). Eocene ungulate mammals from West Antarctica: implications from their fossil record and a new species. *Antarctic Science*, 29(5), 445-455.

Jadwiszczak, P. (2006). Eocene penguins of Seymour Island, Antarctica: the earliest record, taxonomic problems and some evolutionary considerations. *Polish Polar Research*, 287-302.

Hospitaleche, C. A., & Haidr, N. (2011). Penguin cranial remains from the Eocene La Meseta Formation, Isla Marambio (Seymour Island), Antarctic Peninsula. *Antarctic Science*, 23(4), 369-378.

Kloess, P. A., Poust, A. W., & Stidham, T. A. (2020). Earliest fossils of giant-sized bony-toothed birds (Aves: Pelagornithidae) from the Eocene of Seymour Island, Antarctica. *Scientific Reports*, 10(1), 1-11.

Reguero, M. A., Gelfo, J. N., López, G. M., Bond, M., Abello, A., Santillana, S. N., & Marenssi, S. A. (2014). Final Gondwana breakup: the Paleogene South American native ungulates and the demise of the South America–Antarctica land connection. *Global and Planetary Change*, 123, 400-413.

Reguero, M. A., Marenss, S. A., & Santillana, S. N. (2012). Weddellian marine/coastal vertebrates diversity from a basal horizon (Ypresian, Eocene) of the Cucullaea I Allomember, La Meseta formation, Seymour (Marambio) Island, Antarctica. Revista Peruana de Biología, 19(3), 275-284.

North America: Bridger Formation

Bast, E. D., & Smith, T. (2013). Reassessment of the small 'arctocyonid' *Prolatidens waudruae* from the early Paleocene of Belgium, and its phylogenetic relationships with ungulate-like mammals. *Journal of Vertebrate Paleontology*, 33(4), 964-976.

Bai, B., Wang, Y. Q., & Meng, J. (2021). A new late Paleocene phenacodontid 'condylarth' *Lophocion* from the Clark's Fork Basin of Wyoming. *Historical Biology*, 33(5), 652-659.

Brady, P. L., Castrellon Arteaga, A., López-Torres, S., & Springer, M. S. (2024). The effects of ordered multistate morphological characters on phylogenetic analyses of eutherian mammals. *Journal of Mammalian Evolution*, 31(3), 28.

Burger, B. (2015). The systematic position of the saber-toothed and horned giants of the Eocene: the Uintatheres (Order Dinocerata). In *Society of Vertebrate Paleontology Annual Meeting, Dallas. Bethesda: Society of Vertebrate, Paleontology*.

Cooper, L. N., Seiffert, E. R., Clementz, M., Madar, S. I., Bajpai, S., Hussain, S. T., & Thewissen, J. G. (2014). Anthracobunids from the middle Eocene of India and Pakistan are stem perissodactyls. *PLoS One*, 9(10), e109232.

Gunnell, G. F., & Yarborough, V. L. (2000). Brontotheriidae (Perissodactyla) from the late early and middle Eocene (Bridgerian), Wasatch and Bridger formations, southern Green River Basin, southwestern Wyoming. *Journal of Vertebrate Paleontology*, 20(2), 349-368.

Matthew, W. D. (1909). The Carnivora and Insectivora of the Bridger Basin, middle Eocene. *Memoirs of the American Museum of Natural History*, 9(6), 289-576.

Morlo, M., & Gunnell, G. F. (2003). Small Limnocyonines (Hyaenodontidae, Mammalia) from the Bridgerian middle Eocene of Wyoming: *Thinocyon, Prolimnocyon*, and *Iridonon*, new genus. *Contributions from the Museum of Paleontology, The University of Michigan*, 31, 43-78

Morlo, M., & Gunnell, G. F. (2005). New species of *Limnocyon* (Mammalia, Creodonta) from the Bridgerian (Middle Eocene). *Journal of Vertebrate Paleontology*, 25(1), 251-255.

Orliac, M. J., Argot, C., & Gilissen, E. (2012). Digital cranial endocast of *Hyopsodus* (Mammalia,"Condylarthra"): a case of Paleogene terrestrial echolocation? *PLoS One*, 7(2), e30000.

Rose, K. D. (2006). *The Beginning of the Age of Mammals*. Johns Hopkins University Press, Baltimore, Maryland.

Solé, F., Fournier, M., Ladevèze, S., Le Verger, K., Godinot, M., Laurent, Y., & Smith, T. (2023). New postcranial elements of mesonychid mammals from the Ypresian of France: New hypotheses for the radiation and evolution of the mesonychids in Europe. *Journal of Mammalian Evolution*, 30(2), 371-401.

Europe: Messel Formation

Angst, D., Lécuyer, C., Amiot, R., Buffetaut, E., Fourel, F., Martineau, F., Legendre, S., Abourachid, A., & Herrel, A. (2014). Isotopic and anatomical evidence of an herbivorous diet in the Early Tertiary giant bird *Gastornis*. Implications for the structure of Paleocene terrestrial ecosystems. *Naturwissenschaften*, 101, 313-322.

Bauersachs, T., Schouten, S., & Schwark, L. (2014). Characterization of the sedimentary organic matter preserved in Messel oil shale by bulk geochemistry and stable isotopes. *Palaeogeography, Palaeoclimatology, Palaeoecology*, 410, 390-400.

Brady, P. L., Castrellon Arteaga, A., López-Torres, S., & Springer, M. S. (2024). The effects of ordered multistate morphological characters on phylogenetic analyses of eutherian mammals. *Journal of Mammalian Evolution*, 31(3), 28.

Colleary, C., Dolocan, A., Gardner, J., Singh, S., Wuttke, M., Rabenstein, R., Habersetzer, J., Schaal, S., Feseha, M., Clemens, M., & Jacobs, B. F. (2015). Chemical, experimental, and morphological evidence for diagenetically altered melanin in exceptionally preserved fossils. *Proceedings of the National Academy of Sciences*, 112(41), 12592–12597.

Franzen, J. L., Gingerich, P. D., Habersetzer, J., Hurum, J. H., Von Koenigswald, W., Smith, B. H. (2009). Complete primate skeleton from the middle Eocene of Messel in Germany: morphology and paleobiology. *PLoS One* 4(5), e5723.

Franzen, J. L., & Wilde, V. (2003). First gut content of a fossil primate. *Journal of Human Evolution*, 44(3), 373–378.

Gruber, G., & Micklich, N., (2007). *Messel. Treasures of the Eocene*. Hessiches Landesmuseum Darmstadt, Darmstadt.

Joyce, W. G., Micklich, N., Schaal, S. F., & Scheyer, T. M. (2012). Caught in the act: the first record of copulating fossil vertebrates. *Biology letters*, 8(5), 846–848.

Mayr, G. (2004). Phylogenetic relationships of the early Tertiary Messel rails (Aves, Messelornithidae. *Senckenbergiana Lethaea*, 84(1–2), 317–322.

Rose, K. D. (2006). *The Beginning of the Age of Mammals*. Johns Hopkins University Press, Baltimore, Maryland.

Rose, K. D. (2012). The importance of Messel for interpreting Eocene Holarctic mammalian faunas. *Palaeobiodiversity and Palaeoenvironments*, 92, 631–647.

Schaal, S., & Ziegler, W. (Eds). 1992. *Messel: An Insight into the History of Life and of the Earth*. Clarendon Press, Oxford.

Seiffert, E. R., Perry, J. M., Simons, E. L., & Boyer, D. M. (2009). Convergent evolution of anthropoid-like adaptations in Eocene adapiform primates. *Nature*, 461(7267), 1118–1121.

Silcox, M. T., Bloch, J. I., Boyer, D. M., & Houde, P. (2010). Cranial anatomy of Paleocene and Eocene *Labidolemur kayi* (Mammalia: Apatotheria), and the relationships of the Apatemyidae to other mammals. *Zoological Journal of the Linnean Society*, 160(4), 773–825.

Solé, F., Baqri, S. R. H., Iqbal, N., Roohi, G., Benammi, M., Crochet, J. Y., ... & Marivaux, L. (2024). An unexpected late paroxyclaenid (Mammalia, Cimolesta) out of Europe: dental evidence from the Oligocene of the Bugti Hills, Pakistan. *Papers in Palaeontology*, 10(6), e1599.

Vianey-Liaud, M., & Marivaux, L. (2021). The beginning of the adaptive radiation of Theridomorpha (Rodentia) in Western Europe: morphological and phylogenetic analyses of early and middle Eocene taxa; implications for systematics. *Palaeovertebrata*, 44, 2–e2.

Africa: Qasr el-Sagha Formation

Grohé, C., Morlo, M., Chaimanee, Y., Blondel, C., Coster, P., Valentin, X., Salem, M., Bilal, A. A., Jaeger, J. J., & Brunet, M. (2012). New Apterodontinae (Hyaenodontida) from the Eocene locality of Dur At-Talah (Libya): systematic, paleoecological and phylogenetical implications. *PLoS One*, 7(11), p.e49054.

Holroyd, P. A., Simons, E. L., Bown, T. M., Polly, P. A., & Kraus, M. J. (1996). New records of terrestrial mammals from the upper Eocene Qasr el Sagha Formation, Fayum Depression, Egypt. *Palaeovertebrata*, 25(2–4), 175–192.

King, C., Underwood, C., & Steurbaut, E. (2012). Eocene stratigraphy of the Wadi Al-Hitan World Heritage Site and adjacent areas (Fayum, Egypt). *Stratigraphy*, 11(3–4), 185–234.

Liu, A. G., Seiffert, E. R., & Simons, E. L. (2008). Stable isotope evidence for an amphibious phase in early proboscidean evolution. *Proceedings of the National Academy of Sciences*, 105(15), 5786–5791.

Rage, J. C., Métais, G., Bartolini, A., Brohi, I. A., Lashari, R. A., Marivaux, L., Merle, D., & Solangi, S. H. (2013). First report of the giant snake *Gigantophis* (Madtsoiidae) from the Paleocene of Pakistan: paleobiogeographic implications. *Geobios*, 47(3), 147–153.

Voss, M., Antar, M. S. M., Zalmout, I. S., & Gingerich, P. D. (2019). Stomach contents of the archaeocete *Basilosaurus isis*: apex predator in oceans of the late Eocene. *PLoS One*, 14(1), p.e0209021.

Werdelin, L., Sanders, W. L. (Eds) (2010). *Cenozoic Mammals of Africa*. University of California Press, Berkeley, California.

Martínez-Cáceres, M., Lambert, O., & Muizon, C. (2017). The anatomy and phylogenetic affinities of *Cynthiacetus peruvianus*, a large *Dorudon*-like basilosaurid (Cetacea, Mammalia) from the late Eocene of Peru. Geodiversitas, 39(1), 7–163.

Zalmout I. S., & Gingerich P. D. (2012). Late Eocene sea cows (Mammalia, Sirenia) from Wadi al Hitan in the western desert of Fayum, Egypt, *University of Michigan Papers on Paleontology*, 37.

Asia: Irdin Manha Formation

Bast, E. D., & Smith, T. (2013). Reassessment of the small 'arctocyonid' *Prolatidens waudruae* from the early Paleocene of Belgium, and its phylogenetic relationships with ungulate-like mammals. *Journal of Vertebrate Paleontology*, 33(4), 964–976.

Bleefeld, A. R., & McKenna, M. C. (1985). Skeletal integrity of *Mimolagus rodens* (Lagomorpha, Mammalia). *American Museum Novitates*, 2806, 1–5.

Borths, M. R., Holroyd, P. A., & Seiffert, E. R. (2016). Hyainailourine and teratodontine cranial material from the late Eocene of Egypt and the application of parsimony and Bayesian methods to the phylogeny and biogeography of Hyaenodonta (Placentalia, Mammalia). *PeerJ*, 4, p.e2639.

Dashzeveg, D., & Russell, D. E. (1988). Paleocene and Eocene Mixodontia (Mammalia, Glires) of Mongolia and China. *Palaeontology*, 31(1), 129–164.

Granger, W. (1938). A giant oxyaenid from the upper Eocene of Mongolia. *American Museum Novitates*, 969.

Lucas, S. G., Schoch, R. M., & Manning, E. (1981). The systematics of *Forstercooperia*, a middle to late Eocene hyracodontid (Perissodactyla: Rhinocerotoidea) from Asia and western North America. *Journal of Paleontology*, 826–841.

Ni, X., Meng, J., Beard, K. C., Gebo, D. L., Wang, Y., & Li, C. (2010). A new tarkadectine primate from the Eocene of Inner Mongolia, China: phylogenetic and biogeographic implications. *Proceedings of the Royal Society B: Biological Sciences*, 277(1679), 247–256.

Osborn, H. F. (1924). *Andrewsarchus*, giant mesonychid of Mongolia. American Museum Novitates. *American Museum of Natural History*, 146

Rose, K. D. (2006). *The Beginning of the Age of Mammals*. Johns Hopkins University Press, Baltimore, Maryland.

Spaulding, M., O'Leary, M. A., & Gatesy, J. (2009). Relationships of Cetacea (Artiodactyla) Among Mammals: Increased Taxon Sampling Alters Interpretations of Key Fossils and Character Evolution. *PLoS One*, 4(9), e7062.

Yu, Y., Gao, H., Li, Q., & Ni, X. (2023). A new entelodont (Artiodactyla, Mammalia) from the late Eocene of China and its phylogenetic implications. *Journal of Systematic Palaeontology*, 21(1), 2189436.

Asia: Ergilin Dzo Formation

Bai, B., Wang, Y. Q., & Zhang, Z. Q. (2018). The late Eocene hyracodontid perissodactyl *Ardynia* from Saint Jacques, Inner Mongolia, China and its implications for the potential Eocene–Oligocene boundary. *Palaeoworld*, 27(2), 247–257.

Coombs, M. C. (1978). Additional Schizotherium material from China, and a review of *Schizotherium* dentitions (Perissodactyla, Chalicotheriidae). *American Museum Novitates*, (2647), 1.

Coombs, M. C. (1989). 'Interrelationships and diversity in the Chalicotheriidae'. In: Prothero D. R., Schoch, R. M. (eds) *The Evolution of Perissodactyla*. Oxford University Press, New York. pp. 438–457.

Kort, A. E. (2023). The ancestral morphology of lumbar vertebrae and its diversification in early Cenozoic mammals. *Journal of Mammalian Evolution*, 30(4), 845–858.

Kraatz, B. P., & Geisler, J. H. (2010). Eocene–Oligocene transition in Central Asia and its effects on mammalian evolution. *Geology*, 38(2), 111–114.

Lucas, S. G., & Tong, Y. (1988). A new coryphodontid (Mammalia, Pantodonta) from the Eocene of China. *Journal of Vertebrate Paleontology*, 362–372.

Meng, J., & McKenna, M. C. (1998). Faunal turnovers of Palaeogene mammals from the Mongolian Plateau. *Nature*, 394, 364–367.

Osborn, H. F. (1929). *Embolotherium*, gen. nov., of the Ulan Gochu, Mongolia. *American Museum Novitates*, 353

Prothero, D. R. (1994). *The Eocene-Oligocene transition: paradise lost*. Columbia University Press, New York.

Rose, K. D. (2006). *The Beginning of the Age of Mammals*. Johns Hopkins University Press, Baltimore, Maryland.

Saneyoshi, M., Tsubamoto, T., Watabe, M., & Tsogtbaatar, K. (2010). Lithostratigraphic and sedimentological analysis of the upper Eocene Ergilin Dzo Formation of Ergilin Dzo locality, Mongolia. *Hayashibara Museum of Natural Sciences Research Bulletin*, 3, 149–153.

Tsubamoto, T. (2006). Fossil vertebrate assemblage from the upper Eocene Ergilin Dzo Formation of Mongolia. *Journal of the Geological Society of Japan*, 112, 5–6.

Tsubamoto, T., Saneyoshi, M., Watabe, M., Tsogtbaatar, K., & Mainbayar, B. (2011). The entelodontid artiodactyl fauna from the Eocene Ergilin Dzo Formation of Mongolia with comments on *Brachyhyops* and the Khoer Dzan locality. *Paleontological Research*, 15(4), 258–268.

Tsubamoto, T., Tsogtbaatar, K., Chinzorig, T., & Egi, N. (2022). A brief review of the updated fossil vertebrate fauna of the upper Eocene Ergilin Dzo Formation, southeastern Mongolia. *Memoirs of the Faculty of Science, Ehime University*. 24, 64–83.

Tsubamoto, T., Watabe, M., & Tsogtbaatar, K. (2008). *Hyaenodon chunkhtensis* and the hyaenodontid fauna from the upper Eocene Ergilin Dzo Formation of Mongolia. *Journal of Vertebrate Paleontology*, 28(2), 559–564.

South America: Itaboraí Formation

Beck, R. (2017). The skull of *Epidolops ameghinoi* from the early Eocene Itaboraí fauna, southeastern Brazil, and the affinities of the extinct marsupialiform order Polydolopimorphia. *Journal of Mammalian Evolution*, 24(4), 373–414.

Bergqvist, L. P., de Lima Moreira, A., & Pinto, D. R. (2006). Bacia de São José de Itaboraí: 75 anos de história e ciência. *CPRM, Serviço Geológico do Brasil*.

Bergqvist, L. P., Mansur, K., Rodrigues, M. A., Rodrigues-Francisco, B. H., Perez, R., & da Conceição Beltrão, M. (2008). Itaboraí Basin, State of Rio de Janeiro-the cradle of mammals in Brazil. *Sítios Geológicos e Paleontológicos do Brasil SIGEP*, 123, 413–432.

Bergqvist, L. P., & von Koenigswald, W. (2017). The dentition of *Carodnia vieirai* (Mammalia: Xenungulata): enamel microstructure and mastication pattern. *Palaeontologia Electronica*, 20(2.30 A), 1–15.

Cifelli, R. (1983). Eutherian tarsals from the late Paleocene of Brazil. *American Museum Novitates*, 2761.

Croft, D. A., (2016). *Horned armadillos and rafting monkeys: the fascinating fossil mammals of South America*. Indiana University Press, Bloomington, Indiana.

Couto, C. D. P. (1952). Fossil mammals from the beginning of the Cenozoic in Brazil. Condylarthra, Litopterna, Xenungulata and Astrapotheria. *Bulletin of the American Museum of Natural History*, 99, 6.

Francis, J. E., Crame, J. A., & Pirrie, D. (2006). Cretaceous-Tertiary high-latitude palaeoenvironments, James Ross Basin, Antarctica. *Geological Society London Special Publications*, 258(1), 1–5.

López, D. A. G. (2011). Basicranial osteology of *Colbertia lumbrerense* Bond, 1981 (Mammalia: notoungulata). *Ameghiniana*, 48(1), 3–12.

Mayr, G., Alvarenga, H., & Clarke, J. A. (2011). An *Elaphrocnemus*-like landbird and other avian remains from the late Paleocene of Brazil. *Acta Palaeontologica Polonica*, 56(4), 679–684.

Oliveira, É. V., & Bergqvist, L. P. (1998). A new Paleocene armadillo (Mammalia, Dasypodoidea) from the Itaboraí basin, Brazil. *Publicación Electrónica de la Asociación Paleontológica Argentina*, 5(1).

Oliveira, E . V., & Goin, F. J. (2011). A reassessment of bunodont metatherians from the Paleogene of Itaboraí (Brazil): systematics and age of the Itaboraian SALMA. *Revista Brasileira de Paleontologia*, 14.

Szalay, F. S., & Sargis, E. J. (2001). Model-based analysis of postcranial osteology of marsupials from the Palaeocene of Itaboraí (Brazil) and the phylogenetics and biogeography of Metatheria. *Geodiversitas*, 23(2), 139–302.

Vizcaíno, S. F., Cassini, G. H., Toledo, N., Bargo, M. S., Patterson, B. P., & Costa, L. P. (2012). 'On the evolution of large size in mammalian herbivores of Cenozoic'. In: Patterson, B., & Costa, L. (eds) *On the evolution of large size in mammalian herbivores of Cenozoic faunas of South America*. University of Chicago Press, Chicago, Illinois. pp. 76–101.

Welker, F., Collins, M. J., Thomas, J. A., Wadsley, M., Brace, S., Cappellini, E., Turvey, S. T., Reguero, M., Gelfo, J. N., Kramarz, A., & Burger, J. (2015). Ancient proteins resolve the evolutionary history of Darwin's South American ungulates. *Nature*, 522(7554), 81–84.

Oligocene

Prothero, D. R. (1994). *The Eocene-Oligocene transition: paradise lost*. Columbia University Press, New York.

Prothero, D. R., & Berggren, W. A. (Eds.). (2014). *Eocene-Oligocene climatic and biotic evolution (Vol. 135)*. Princeton University Press, Princeton, New Jersey.

Prothero, D. R., & Emry, R. J. (Eds.) (1996). *The Terrestrial Eocene-Oligocene Transition in North America*. Cambridge University Press, Cambridge, UK.

Prothero, D. R., Ivany, L. C., & Nesbitt, E. A. (Eds.). (2003). *From greenhouse to icehouse: the marine Eocene-Oligocene transition*. Columbia University Press, New York.

Europe: Quercy Phosphorites Formation

Agustí, J., & Antón, M. (2002). *Mammoths, Sabertooths, and Hominids: 65 Million Years of Mammalian Evolution in Europe*. Columbia University Press, New York.

Blondel, C. (2005). New data on the Cainotheriidae (Mammalia, Artiodactyla) from the early Oligocene of south-western France. *Zoological Journal of the Linnean Society*, 144(2), 145–166.

Cirot, E., & de Bonis, L. (1992). Révision du genre *Amphicynodon*, carnivore de l'Oligocène. *Palaeontographica Abteilung A*, 220(4–6), 103–130.

Danowitz, M., Vasilyev, A., Kortlandt, V., & Solounias, N. (2015). Fossil evidence and stages of elongation of the *Giraffa camelopardalis* neck. *Royal Society Open Science*, 2(10), 150393.
De Bonis, L., Gardin, A., & Blondel, C. (2019). Carnivora from the early Oligocene of the 'Phosphorites du Quercy' in southwestern France. *Geodiversitas*, 41, 601–621.

Gardin, A., Salesa, M. J., Siliceo, G., Antón, M., Pastor, J. F., & de Bonis, L. (2021). Climbing adaptations of an enigmatic early arctoid carnivoran: the functional anatomy of the forelimb of *Amphicynodon leptorhynchus* from the Lower Oligocene of the Quercy Phosphorites (France). *Journal of Mammalian Evolution*, 28(3), 785–811.

Gardin, A., Salesa, M. J., Siliceo, G., Antón, M., Pastor, J. F., & de Bonis, L. (2022). The hindlimb of *Amphicynodon leptorhynchus* from the lower Oligocene of the Quercy Phosphorites (France): Highlight of new climbing adaptations of this early arctoid. *Journal of Mammalian Evolution*, 29(4), 815–843.

Ginot, S., Hautier, L., Marivaux, L., & Vianey-Liaud, M. (2016). Ecomorphological analysis of the astragalo-calcaneal complex in rodents and inferences of locomotor behaviours in extinct rodent species. *PeerJ*, 4, e2393.

Peigné, S., & Bonis, L. (1999). The genus *Stenoplesictis* Filhol (Mammalia, Carnivora) from the Oligocene deposits of the phosphorites of Quercy, France. *Journal of Vertebrate Paleontology*, 19(3), 566–575.

Peigné, S., & Brunet, M. (2001). Une nouvelle espèce du genre *Eusmilus* (Carnivora: Nimravidae) de l'Oligocène (MP 22) d'Europe. *Geobios*, 34(6), 657–672.

Prothero, D. R. (1994). *The Eocene-Oligocene Transition: Paradise Lost*. Columbia University Press, New York.

Rivals, F., Belyaev, R. I., Basova, V. B., & Prilepskaya, N. E. (2023). Hogs, hippos or bears? Paleodiet of European Oligocene anthracotheres and entelodonts. *Palaeogeography, Palaeoclimatology, Palaeoecology*, 611, 111363.

Samuel, G., Hautier, L., Marivaux, L., and Vianey-Liaud, M. (2016). Ecomorphological analysis of the astragalo-calcaneal complex in rodents and inferences of locomotor behaviours in extinct rodent species. *PeerJ*, 4, 1–49.

Solé, F., Fischer, V., Denayer, J., Speijer, R. P., Fournier, M., Le Verger, K., ... & Smith, T. (2021). The upper Eocene-Oligocene carnivorous mammals from the Quercy Phosphorites (France) housed in Belgian collections. *Geologica Belgica*, 24(1–2), 1–16.

North America: Brule Formation

Benton, R. C., Terry Jr, D. O., Evanoff, E., & McDonald, H. G. (2015). *The White River Badlands: Geology and Paleontology*. Indiana University Press, Bloomington, Indiana.

Evers, S. W., & Al Iawati, Z. (2024). Digital skull anatomy of the Oligocene North American tortoise *Stylemys nebrascensis* with taxonomic comments on the species and comparisons with extant testudinids of the Gopherus-Manouria clade. *Swiss Journal of Palaeontology*, 143(1), 12.

Hoganson, J. W., Murphy, E. C., & Forsman, N. F. (1998). Lithostratigraphy, paleontology, and biochronology of the Chadron, Brule, and Arikaree Formations in North Dakota. *Geological Society of America Special Papers*, 325, 185–196.

Hoganson, J. W., & Person, J. (2011). Tooth puncture marks on a 30 million year old *Dinictis* skull. *Geo News*, 38(2), 12–17.

Janis, C. M., Scott, K. M., & Jacobs, L. L. (Eds.). (1998). *Evolution of Tertiary Mammals of North America: Volume 1, Terrestrial Carnivores, Ungulates, and Ungulate-like Mammals (Vol. 1)*. Cambridge University Press, Cambridge, UK.

Poust, A. W., Barrett, P. Z., & Tomiya, S. (2022). An early nimravid from California and the rise of hypercarnivorous mammals after the middle Eocene climatic optimum. *Biology Letters*, 18(10), 20220291.

Prothero, D. R. (1985). North American mammalian diversity and Eocene–Oligocene extinctions. *Paleobiology*, 11(4), 389–405.

Prothero, D. R. (1996). 'Hyracodontidae'. In: Prothero, D. R., & Emry, R. J. (eds) *The Terrestrial Eocene-Oligocene Transition in North America*. Cambridge University Press, Cambridge, UK.

Schultz, C. B., & Falkenbach, C. H. (1968). The Phylogeny of the Oreodonts. *Bulletin of the American Museum of Natural History*, 139.

Stevens, M. S., & Stevens, J. B. (1996). 'Merycoidodontinae and Miniochoerinae'. In: Prothero, D. R., & Emry, R. J. (eds) *The Terrestrial Eocene-Oligocene Transition in North America*. Cambridge University Press, Cambridge, UK. pp. 498–573.

"Taphonomy of a Multiple *Poebrotherium* kill site - an *Archaeotherium* meat cache". Retrieved from: http://www.douglasfossils.com/multi_kill.html

Wilhelm, P. B. (1993). *Morphometric analysis of the limb skeleton of generalized mammals in relation to locomotor behavior, with applications to fossil mammals*. Ph.D. dissertation, Brown University, Providence, R. I.

Wang, X., & Tedford, R. (2010). *Dogs: Their Fossil Relatives and Evolutionary History*. Columbia University Press, New York.

Africa: Jebel Qatrani Formation

Al-Ashqar, S. F., Borths, M., El-Desouky, H., Heritage, S., Abed, M., Seiffert, E. R., ... & Sallam, H. M. (2025). Cranial anatomy of the hypercarnivore *Bastetodon syrtos* gen. nov. (Hyaenodonta, Hyainailourinae) and a reevaluation of *Pterodon* in Africa. *Journal of Vertebrate Paleontology*, 44(3), e2442472.

Borths, M. R., & Seiffert, E. R. (2017). Craniodental and humeral morphology of a new species of *Masrasector* (Teratodontinae, Hyaenodonta, Placentalia) from the late Eocene of Egypt and locomotor diversity in hyaenodonts. *PLoS One*, 12(4), e0173527.

Chaimanee, Y., Chavasseau, O., Beard, K. C., Kyaw, A. A., Soe, A. N., Sein, C., ... & Jaeger, J. J. (2012). Late Middle Eocene primate from Myanmar and the initial anthropoid colonization of Africa. *Proceedings of the National Academy of Sciences*, 109(26), 10293–10297.

Kampouridis, P., Hartung, J., & Augustin, F. J. (2023). 'The Eocene–Oligocene Vertebrate Assemblages of the Fayum Depression, Egypt'. In: Hamimi, Z., Khozyem, H., Adatte,T, Nader, F. H., Oboh-Ikuenobe, F., Zobaa, M. K., & El Atfy, H. (eds) *The Phanerozoic Geology and Natural Resources of Egypt*. Springer, Cham. pp. 373–405.

Rose, K. D. (2006). *The Beginning of the Age of Mammals*. Johns Hopkins University Press, Baltimore, Maryland.

Sanders, W. J., Rasmussen, D. T., & Kappelman, J. (2010). *Cenozoic mammals of Africa*. The University of California Press, Berkeley, California.

Seiffert, E. R., & Simons, E. L. (2013). Last of the oligopithecids? A dwarf species from the youngest primate-bearing level of the Jebel Qatrani formation, northern Egypt. *Journal of Human Evolution*, 64(3), 211–215.

Sileem, A. H., & Abu El-Kheir, G. A. (2022). Complete skull of *Bothriogenys fraasi* (Mammalia, Artiodactyla, Anthracotheriidae) from the Early Oligocene, Fayum, Egypt. *Geological Journal*, 57(11), 4833–4841.

Sileem, A. H., Sallam, H. M., Hewaidy, A. G. A., Gunnell, G. F., & Miller, E. R. (2015). Anthracotheres (Mammalia, Artiodactyla) from the upper-most horizon of the Jebel Qatrani formation, latest Early Oligocene, Fayum depression, Egypt. *Egyptian Journal of Paleontology*, 15, 1–11.

Simons, E. L., & Bown, T. M. (1985). *Afrotarsius chatrathi*, first tarsiiform primate (? Tarsiidae) from Africa. Nature, 313(6002), 475–477.

Werdelin, L., & Sanders, W. L. (Eds.) (2010). *Cenozoic Mammals of Africa*. University of California Press, Berkeley, California.

Australia: Riversleigh Fossil Site (Faunal Zone A)

Butler, K., Travouillon, K. J., Price, G. J., Archer, M., & Hand, S. (2018). Revision of Oligo-Miocene kangaroos, *Ganawamaya* and *Nambaroo* (Marsupialia: Macropodiformes, Balbaridae). *Palaeontologia Electronica*. 21.1.8A, 10.26879/747.

Crichton, A. I., Worthy, T. H., Camens, A. B., Yates, A. M., Couzens, A. M., & Prideaux, G. J. (2023). A new species of *Mukupirna* (Diprotodontia, Mukupirnidae) from the Oligocene of Central Australia sheds light on basal vombatoid interrelationships. *Alcheringa*, 47(4), 446–474.

Den Boer, W., Campione, N. E., & Kear, B. P. (2019). Climbing adaptations, locomotory disparity and ecological convergence in ancient stem 'kangaroos'. *Royal Society Open Science*, 6(2), 181617.

Kear, B. P., & Cooke, B. N. (2001). A review of macropodoid (Marsupialia) systematics with the inclusion of a new family. *Association of Australasian Palaeontologists Memoirs*, 25, 83–102.

Long, J., Archer, M., Flannery, T., & Hand, S. (2002). *Prehistoric mammals of Australia and New Guinea: One hundred million years of evolution*. John Hopkins University Press, Baltimore, Maryland.

Myers, T. J., & Archer, M. (1997). *Kuterintja ngama* (Marsupialia, Ilariidae): a revised systematic analysis based on material from the late Oligocene of Riversleigh, northwestern Queensland. *Memoirs Queensland Museum*, 41, 379–392.

Myers, T. J., Black, K. H., Archer, M., & Hand, S. J. (2017). The identification of Oligo-Miocene mammalian palaeocommunities from the Riversleigh World Heritage Area, Australia and an appraisal of palaeoecological techniques. *PeerJ*, 5, p.e3511.

Rich, T. H., & Archer, M. (1979). *Namilamadeta snideri*, a new diprotodontan (Marsupialia, Vombatoidea) from the medial Miocene of South Australia. *Alcheringa*, 3(3), 197–208.

Travouillon, K. J., Archer, M., Hand, S. J., & Godthelp, H. (2006). Multivariate analyses of Cenozoic mammalian faunas from Riversleigh, northwestern Queensland. *Alcheringa*, 30(1), 323–349.

Travouillon, K. J., Cooke, B. N., Archer, M., & Hand, S. J. (2014). Revision of basal macropodids from the Riversleigh World Heritage Area with descriptions of new material of *Ganguroo bilamina* Cooke, 1997 and a new species. *Palaeontologia Electronica*, 17(1), 20A

Travouillon, K. J., Butler, K., Archer, M., & Hand, S. J. (2016). New material of '*Gumardee pascuali*' Flannery et al., 1983 (Marsupialia: Macropodiformes) and two new species from the Riversleigh World Heritage Area, Queensland, Australia. *Memoirs of Museum Victoria*, 74, 189–207.

South America: Salla Formation

Billet, G. (2010). New observations on the skull of *Pyrotherium* (Pyrotheria, Mammalia) and new phylogenetic hypotheses on South American ungulates. *Journal of Mammalian Evolution*, 17(1), 21–59.

Billet, G., Blondel, C., & Muizon, C. (2009). Dental microwear analysis of notoungulates (Mammalia) from Salla (Late Oligocene, Bolivia) and discussion on their precocious hypsodonty. *Palaeogeography, Palaeoclimatology, Palaeoecology*, 274(1–2), 114–124.

Croft, D. A. (2016). *Horned armadillos and rafting monkeys: the fascinating fossil mammals of South America*. Indiana University Press, Bloomington, Indiana.

Croft, D. A., & Weinstein, D. (2008). The first application of the mesowear method to endemic South American ungulates (Notoungulata). *Palaeogeography, Palaeoclimatology, Palaeoecology*, 269(1–2), 103–114.

Kay, R. F., Macfadden, B. J., Madden, R. H., Sandeman, H., & Anaya, F. (1998). Revised age of the Salla beds, Bolivia, and its bearing on the age of the Deseadan South American Land Mammal "Age". *Journal of Vertebrate Paleontology*, 18(1), 189–199.

Kramarz, A. G., & Macphee, R. D. (2023). Did some extinct South American native ungulates arise from an afrothere ancestor? A critical appraisal of Avilla and Mothé's (2021) Sudamericungulata–Panameridiungulata hypothesis. *Journal of Mammalian Evolution*, 30(1), 67–77.

Martínez, G., Dozo, M. T., Gelfo, J. N., & Marani, H. (2016). Cranial Morphology of the Late Oligocene Patagonian Notohippid *Rhynchippus equinus* Ameghino, 1897 (Mammalia, Notoungulata) with Emphases in Basicranial and Auditory Region. *PLoS One*, 11, 1–29.

Shockey, B. J., & Anaya, F. (2008). 'Postcranial osteology of mammals from Salla, Bolivia (late Oligocene): form, function, and phylogenetic implications'. In: Sargis, E. J., & Dagosto, M. (eds) *Mammalian Evolutionary Morphology*. Springer, Dordrecht. pp. 135–157.

Takai, M., & Anaya, F. (1996). New specimens of the oldest fossil platyrrhine, *Branisella boliviana*, from Salla, Bolivia. *American Journal of Physical Anthropology: The Official Publication of the American Association of Physical Anthropologists*, 99(2), 301–317.

Welker, F., Collins, M. J., Thomas, J. A., Wadsley, M., Brace, S., Cappellini, E., ... & MacPhee, R. D. (2015). Ancient proteins resolve the evolutionary history of Darwin's South American ungulates. *Nature*, 522(7554), 81–84.

Asia: Hsanda Gol Formation

Bryant, J. D., McKenna, M. C., & Akademi, M. S. U. (1995). Cranial anatomy and phylogenetic position of *Tsaganomys altaicus* (Mammalia, Rodentia) from the Hsanda Gol Formation (Oligocene), Mongolia. *American Museum Novitates*, 3156, 1–15.

Deng, T., Lu, X., Wang, S., Flynn, L. J., Sun, D., He, W., & Chen, S. (2021). An Oligocene giant rhino provides insights into *Paraceratherium* evolution. *Communications Biology*, 4(1), 639.

Gingerich, P. D. (1981). Radiation of early Cenozoic Didymoconidae (Condylarthra, Mesonychia) in Asia, with a new genus from the early Eocene of western North America. *Journal of Mammalogy*, 62(3), 526–538.

Lange-Badre, B., & Dashzeveg, D. (1989). On some Oligocene carnivorous mammals from Central Asia. *Acta Palaeontologica Polonica*, 34(2), 125–148.

Leopold, E. B., Liu, G., & Clay-Poole, S. (1992). 'Low-biomass vegetation in the Oligocene?' In: Prothero, D. R., & Berggren, W. A. (eds) *Eocene-Oligocene Climatic and Biotic Evolution*, Princeton: Princeton University Press, Princeton, New Jersey. pp. 399–420.

Mellett, J. S. (1968). The Oligocene Hsanda Gol Formation, Mongolia: a revised faunal list. *American Museum Novitates*, 2318, 1–16.

Meng, J., & McKenna, M. C. (1998). Faunal turnovers of Palaeogene mammals from the Mongolian Plateau. *Nature*, 394(6691), 364–367.

Prothero, D. R. (2013). *Rhinoceros giants: the paleobiology of indricotheres*. Indiana University Press, Bloomington, Indiana.

Turner, A. (1997). *The big cats and their fossil relatives: an illustrated guide to their evolution and natural history*. Columbia University Press, New York.

Wang, J., & Zhang, Z. Q. (2015). Phylogenetic analysis on *Palaeogale* (Palaeogalidae, Carnivora) based on specimens from Oligocene strata of Saint-Jacques, Nei Mongol. *Vertebrata PalAsiatica*, 53(4), 310–334.

Chapter 6: The Neogene

Jardine, P. E., Janis, C. M., Sahney, S., & Benton, M. J. (2012). Grit not grass: concordant patterns of early origin of hypsodonty in Great Plains ungulates and Glires. *Palaeogeography, Palaeoclimatology, Palaeoecology*, 365, 1–10.

Lourens, L., Hilgen, F., Shackleton, N.J., Laskar, J., & Wilson, D., (2004) 'The Neogene Period'. In: Gradstein, F., Ogg, J., & Smith, A. G. (eds), *Geologic Time Scale*. Cambridge University Press, Cambridge, UK. pp. 409–440.

Palazzesi, L., Hidalgo, O., Barreda, V. D., Forest, F., & Höhna, S. (2022). The rise of grasslands is linked to atmospheric $CO2$ decline in the late Palaeogene. *Nature Communications*, 13(1), 293.

Strömberg, C. A. (2005). Decoupled taxonomic radiation and ecological expansion of open-habitat grasses in the Cenozoic of North America. *Proceedings of the National Academy of Sciences*, 102(34), 11980–11984.

Miocene

Flower, B. P., & Kennett, J. P. (1994). The middle Miocene climatic transition: East Antarctic ice sheet development, deep ocean circulation and global carbon cycling. *Palaeogeography, Palaeoclimatology, Palaeoecology*, 108(3–4), 537–555.

Hsü, K. J. (1983). *The Mediterranean was a desert*. Princeton University Press: Princeton, New Jersey.

Janis, C. M., Damuth, J., & Theodor, J. M. (2004). The species richness of Miocene browsers, and implications for habitat type and primary productivity in the North American grassland biome. *Palaeogeography, Palaeoclimatology, Palaeoecology*, 207(3–4), 371–398.

Tanner, T., Hernández-Almeida, I., Drury, A. J., Guitián, J., & Stoll, H. (2020). Decreasing atmospheric $CO2$ during the late Miocene cooling. *Paleoceanography and Paleoclimatology*, 35(12), e2020PA003925.

North America: Agate Fossil Beds

Janis, C. M., & Wilhelm, P. B. (1993). Were there mammalian pursuit predators in the Tertiary? Dances with wolf avatars. *Journal of Mammalian Evolution*, 1, 103–125.

Coombs, M. C. (1983). Large Mammalian Clawed Herbivores: A Comparative Study. *Transactions of the American Philosophical Society*. 73(7), 1–96.

Hunt, R. M. (2009). Long-legged pursuit carnivorans (Amphicyonidae, Daphoeninae) from the early Miocene of North America. *Bulletin of the American Museum of Natural History*, 2009(318), 1–95.

Hunt, R.M., Jr., R. Skolnick, & J. Kaufman. (2018). *The carnivores of Agate Fossil Beds National Monument: Miocene dens and waterhole in the valley of a dryland paleoriver*. Zea E-Books, University of Nebraska-Lincoln Libraries, Nebraska.

Lucas, S. G., Emry, R. J., Foss, S. E. (1998). Taxonomy and distribution of *Daeodon*, an Oligocene-Miocene entelodont (Mammalia: Artiodactyla) from North America. *Proceedings of the Biological Society of Washington*, 111(2), 425–435.

Martin, L. D., & Bennett, D. K. (1977). The burrows of the Miocene beaver *Palaeocastor*, western Nebraska, USA. *Palaeogeography, Palaeoclimatology, Palaeoecology*, 22(3), 173–193.

Meyer, R. C. (1999). Helical burrows as a palaeoclimate response: *Daimonelix* by *Palaeocastor*. *Palaeogeography, Palaeoclimatology, Palaeoecology*, 147(3–4), 291–298.

Prothero, D. R., & Foss, S. E. (Eds.). (2007). *The Evolution of Artiodactyls*. John Hopkins University Press, Baltimore, Maryland.

Wang, X., Tedford, R. H., & Taylor, B. E. (1999). Phylogenetic systematics of the Borophaginae (Carnivora: Canidae). *Bulletin of the American Museum of Natural History*, 243, 1–391.

North America: Ashfall Fossil Beds

Famoso, Nicholas & Pagnac, Darrin. (2011). A Comparison of the Clarendonian Equid Assemblages from the Mission Pit, South Dakota and Ashfall Fossil Beds, Nebraska. *Transactions of the Nebraska Academy of Sciences and Affiliated Societies*, 32, 98–107.

Mead, A. J. (2000). Sexual dimorphism and paleoecology in *Teleoceras*, a North American Miocene rhinoceros. *Paleobiology*, 26(4), 689–706.

Panciroli, E., Janis, C., Stockdale, M., & Martín-Serra, A. (2017). Correlates between calcaneal morphology and locomotion in extant and extinct carnivorous mammals. *Journal of Morphology*, 278(10), 1333–1353.

Tucker, S. T., Otto, R. E., Joeckel, R. M., & Voorhies, M. R. (2014). The geology and paleontology of Ashfall Fossil Beds, a late Miocene (Clarendonian) mass-death assemblage, Antelope County and adjacent Knox County, Nebraska, USA.

Voorhies, M. R. (1985) A Miocene rhinoceros herd buried in volcanic ash. *National Georaphic Society Research Reports*, 19, 671–88.

Voorhies, M. R., & Stover, S. G. (1978). An articulated fossil skeleton of a pregnant rhinoceros, *Teleoceras major* Hatcher. *Proceedings, Nebraska Academy of Sciences*, 88, 47–48.

Voorhies, M. R., & Thomasson, J. R. (1979). Fossil grass anthoecia within Miocene rhinoceros skeletons: diet in an extinct species. *Science*, 206(4416), 331–333.

South America Santa Cruz Formation

Argot, C. (2003). Functional adaptations of the postcranial skeleton of two Miocene borhyaenoids (Mammalia, Metatheria), *Borhyaena* and *Prothylacinus*, from South America. *Palaeontology*, 46(6), 1213–1267.

Bostelmann, J. E., Le Roux, J. P., Vásquez, A., Gutiérrez, N. M., Oyarzún, J. L., Carreño, C., Torres, T., Otero, R., Llanos, A., Fanning, C. M., & Hervé, F. (2013). Burdigalian deposits of the Santa Cruz Formation in the Sierra Baguales, Austral (Magallanes) Basin: Age, depositional environment and vertebrate fossils. *Andean Geology*, 40(3), 458–489.

Croft, D. A. (2013). What constitutes a fossil mammal community in the early Miocene Santa Cruz Formation?. *Journal of Vertebrate Paleontology*, 33(2), 401–409.

Croft, D. A. (2016). *Horned armadillos and rafting monkeys: the fascinating fossil mammals of South America*. Indiana University Press, Bloomington, Indiana.

Degrange, F. J., Tambussi, C. P., Taglioretti, M. L., Dondas, A., & Scaglia, F. (2015). A new Mesembriornithinae (Aves, Phorusrhacidae) provides new insights into the phylogeny and sensory capabilities of terror birds. *Journal of Vertebrate Paleontology*, 35(2), e912656.

Kramarz, A. G., & Macphee, R. D. (2023). Did some extinct South American native ungulates arise from an afrothere ancestor? A critical appraisal of Avilla and Mothé's (2021) Sudamericungulata–Panameridiungulata hypothesis. *Journal of Mammalian Evolution*, 30(1), 67–77.

Raigemborn, M. S., Matheos, S. D., Krapovickas, V., Vizcaíno, S. F., Bargo, M. S., Kay, R. F., ... & Zapata, L. (2015). Paleoenvironmental reconstruction of the coastal Monte Léon and Santa Cruz formations (early Miocene) at Rincón del Buque, southern Patagonia: A revisited locality. *Journal of South American Earth Sciences*, 60, 31–55.

Schmidt, G. I., & Ferrero, B. S. (2014). Taxonomic reinterpretation of *Theosodon* hystatus Cabrera and Kraglievich, 1931 (Litopterna, Macraucheniidae) and phylogenetic relationships of the family. *Journal of Vertebrate Paleontology*, 34(5), 1231–1238.

Toledo, N., Bargo, M. S., & Vizcaíno, S. F. (2015). Muscular reconstruction and functional morphology of the hind limb of santacrucian (Early Miocene) sloths (Xenarthra, Folivora) of Patagonia. *The Anatomical Record*, 298(5), 842–864.

Townsend, K. E. B., & Croft, D. A. (2004). Diets of notoungulates from the Santa Cruz Formation, Argentina: new evidence from enamel microwear. *Journal of Vertebrate Paleontology*, 28(1), 217–230.

Vizcaíno, S. F., Kay, R. F., & Bargo, M. S. (Eds.). (2012). *Early Miocene paleobiology in Patagonia: high-latitude paleocommunities of the Santa Cruz Formation*. Cambridge University Press, Cambridge, UK.

Welker, F., Collins, M. J., Thomas, J. A., Wadsley, M., Brace, S., Cappellini, E., Turvey, S. T., Reguero, M., Gelfo, J. N., Kramarz, A., & Burger, J. (2015). Ancient proteins resolve the evolutionary history of Darwin's South American ungulates. *Nature*, 522(7554), 81–84.

South America: Urumaco Formation

Aguilera, O. A., Riff, D., & Bocquentin-Villanueva, J. (2006). A new giant *Purussaurus* (Crocodyliformes, Alligatoridae) from the Upper Miocene Urumaco Formation, Venezuela. *Journal of Systematic Palaeontology*, 4(3), 221–232.

Cadena, E. A., Scheyer, T. M., Carrillo-Briceño, J. D., Sánchez, R., Aguilera-Socorro, O. A., Vanegas, A., Pardo, M., Hansen, D. M., & Sánchez-Villagra, M. R. (2020). The anatomy, paleobiology, and evolutionary relationships of the largest extinct side-necked turtle. *Science Advances*. 6(7), eaay4593.

Cadena, E. A., Link, A., Cooke, S. B., Stroik, L. K., Vanegas, A. F., & Tallman, M. (2021). New insights on the anatomy and ontogeny of the largest extinct freshwater turtles. *Heliyon*, 7(12),

Cidade, G. M., Rincón, A. D., & Solórzano, A. (2020). New cranial and postcranial elements of *Mourasuchus* (Alligatoroidea: Caimaninae) from the late Miocene of Venezuela and their palaeobiological implications. *Historical Biology*, 33(10), 2387–2399.

Horovitz, I., Sánchez-Villagra, M. R., Martin, T., & Aguilera, O. A. (2006). The fossil record of *Phoberomys pattersoni* Mones 1980 (Mammalia, Rodentia) from Urumaco (Late Miocene, Venezuela), with an analysis of its phylogenetic relationships, *Journal of Systematic Palaeontology*, 4(3), 293–306.

Riff, D., & Aguilera, O. A. (2008). The world's largest gharials *Gryposuchus*: description of *G. croizati* n. sp.(Crocodylia, Gavialidae) from the Upper Miocene Urumaco Formation, Venezuela. *Paläontologische Zeitschrift*, 82, 178–195.

Rincón, A. D., McDonald, H. G., Solórzano, A., Flores, M. N., & Ruiz-Ramoni, D. (2015). A new enigmatic Late Miocene mylodontoid sloth from northern South America. *Royal Society Open Science*, 2(2), 140256.

Scheyer, T. M., & Delfino, M. (2016). The late Miocene caimanine fauna (Crocodylia: alligatoroidea) of the Urumaco Formation, Venezuela. *Palaeontologia Electronica*, 19(3), 1–57.

South America: Santa Maria Group (Chiquimil and Andalhualá Formations)

Cione, A. L., Azpelicueta, M. D. L. M., Bond, M., Carlini, A. A., Casciotta, J. R., Cozzuol, M. A., de la Fuente, M., Gasparini, Z., Goin, F. J., Noriega, J., & Scillato-Yané, G. J. (2000). Miocene vertebrates from Entre Ríos province, eastern Argentina. El Neógeno de Argentina. *Serie Correlación Geológica*, 14, 191–237.

Croft, D. A. (2016). *Horned armadillos and rafting monkeys: the fascinating fossil mammals of South America*. Indiana University Press, Bloomington, Indiana.

Degrange, F. J., Tambussi, C. P., Moreno, K., Witmer, L. M., & Wroe, S. (2010). Mechanical Analysis of Feeding Behavior in the Extinct "Terror Bird" *Andalgalornis steulleti* (Gruiformes: Phorusrhacidae). *PLoS One*, 5(8), e11856.

Hynek, S. A., Passey, B. H., Prado, J. L., Brown, F. H., Cerling, T. E., & Quade, J. (2012). Small mammal carbon isotope ecology across the Miocene–Pliocene boundary, northwestern Argentina. *Earth and Planetary Science Letters*, 321, 177–188.

Janis, C. M., Figueirido, B., DeSantis, L., & Lautenschlager, S. (2020). An eye for a tooth: *Thylacosmilus* was not a marsupial saber-tooth predator. *PeerJ*, 8, p.e9346.

Melchionna, M., Profico, A., Castiglione, S., Serio, C., Mondanaro, A., Modafferi, M., Tamagnini, D., Maiorano, L., Raia, P., Witmer, L.M., & Wroe, S. (2021). A method for mapping morphological convergence on three-dimensional digital models: the case of the mammalian sabre-tooth. *Palaeontology*, 64(4), 573–584.

Moya, E., Brea, M., & Lutz, A. I. (2018). Redescription and reassignment of the fossil wood *Menendoxylon piptadiensis* from the Pliocene Andalhuala Formation, South America. *Journal of Systematic Palaeontology*, 16(13), 1145–1157.

Reguero, M. A., & Candela, A. M. (2011). Late Cenozoic mammals from the northwest of Argentina. *Cenozoic Geology of the Central Andes of Argentina*, 458, 411–426.

Europe: Sansan Fossil Site

Agustí, J., & Antón, M. (2002). *Mammoths, Sabertooths, and Hominids: 65 Million Years of Mammalian Evolution in Europe*. Columbia University Press, New York.

Geraads, D. (2010). 'Rhinocerotidae'. In: Werdelin, L., & Sanders D. J. (eds) *Cenozoic Mammals of Africa*. University of California Press, Berkeley, Califorina. pp. 669–683.

Ginsburg, L. (2000). La faune Miocène de Sansan et son environnement. *Mémoires du Muséum National d'Histoire Naturelle*, 183, 9–10.

Hartung, J., & Böhme, M. (2022). Unexpected cranial sexual dimorphism in the tragulid *Dorcatherium naui* based on material from the middle to late Miocene localities of Eppelsheim and Hammerschmiede (Germany). *PLoS One*, 17(5), e0267951.

Huttunen, K. (2002). Systematics and Taxonomy of the European Deinotheriidae (Proboscidea, Mammalia). *Annalen des Naturhistorischen Museums in Wien*, 103 A, 237–250.

Peigné, S., & Sen, S. (Eds.). (2014). Mammifères de Sansan. *Memoires du Museum National d'Histoire Naturelle*, 203.

Europe: Cerro de los Batallones

Agustí, J., & Antón, M. (2002). *Mammoths, Sabertooths, and Hominids: 65 Million Years of Mammalian Evolution in Europe*. Columbia University Press, New York.

Antón, M., Salesa, MJ., Morales, J., & Turner, A. (2004). First known complete skulls of the scimitar-toothed cat *Machairodus aphanistus* (Felidae, Carnivora) from the Spanish late Miocene site of Batallones-1. *Journal of Vertebrate Paleontology*, 24(4), 957–969.

Bernor, R. L., Kaya, F., Kaakinen, A., Saarinen, J., & Fortelius, M. (2021). Old world hipparion evolution, biogeography, climatology and ecology. *Earth-Science Reviews*, 221, 103784.

Domingo, M. S., Domingo, L., Sánchez, I. M., Alberdi, M. T., Azanza, B., & Morales, J. (2011). New Insights on the Taphonomy of the Exceptional Mammalian Sites of Cerro de los Batallones (Late Miocene, Spain) Based on Rare Earth element Geochemistry. *PALAIOS*. 26(1), 55–65.

Morales, J., Pozo, M., Silva, P. G., Domingo, M. S., López-Antoñanzas, R., Álvarez Sierra, M., Antón, M., Martín Escorza, C., Quiralte, V., Salesa, M. J., & Sánchez, I. M. (2008). 'El sistema de yacimientos de mamíferos miocenos del Cerro de los Batallones, Cuenca de Madrid: estado actual y perspectivas'. In: Esteve, J., & Meléndez, G. (eds) *Palaeontologica Nova*. Universidad de Zaragoza, Zaragoza. pp. 41–117.

Peigné, S., Salesa, M. J., Antón, M., & Morales, J. (2005). Ailurid carnivoran mammal Simocyon from the late Miocene of Spain and the systematics of the genus. *Acta Palaeontologica Polonica*. 50(2), 219–238.

Salesa, M. J., Antón, M., Turner, A., & Morales, J. (2006). Inferred behaviour and ecology of the primitive sabre-toothed cat *Paramachairodus ogygia* (Felidae, Machairodontinae) from the Late Miocene of Spain. *Journal of Zoology*, 268(3), 243–254.

Salesa, M. J., Gamarra, J., Siliceo, G., Antón, M., & Morales, J. (2023). Unraveling the diversity of early felines: a new genus of Felinae (Carnivora, Felidae) from the Middle Miocene of Madrid (Spain). *Journal of Vertebrate Paleontology*, 43(3), e2288924.

Salesa, M. J., Hernández, B., Marín, P., Siliceo, G., Martínez, I., Antón, M., ... & García-Fernández, R. A. (2024). New insights on the ecology and behavior of *Machairodus aphanistus* (Carnivora, Felidae, Machairodontinae) through the paleopathological study of the fossil sample from the Late Miocene (Vallesian, MN 10) of Cerro de los Batallones (Torrejón de Velasco, Madrid, Spain). *Journal of Mammalian Evolution*, 31(2), 1–17.

Sánchez, I.M., Domingo, M.S., & Morales, J. 2009. New data on the Moschidae (Mammalia, Ruminantia) from the upper Miocene of Spain (Mn 10–Mn 11). *Journal of Vertebrate Paleontology*, 29(2), 567–575.

Siliceo, G., Salesa, M. J., Antón, M., Pastor, J. F., & Morales, J. (2015). Comparative anatomy of the shoulder region in the Late Miocene Amphicyonid *Magericyon anceps* (Carnivora): functional and paleoecological inferences. *Journal of Mammalian Evolution*, 22, 243–258.

Asia: Liushu Formation

Deng, T., & Tseng, Z. J. (2010). Osteological evidence for predatory behavior of the giant percrocutid (*Dinocrocuta gigantea*) as an active hunter. *Chinese Science Bulletin*, 55(17), 1790–1794.

Deng, T., Wang, S., & Hou, S. (2013). A bizarre tandem-horned elasmothere rhino from the Late Miocene of northwestern China and origin of the true elasmothere. *Chinese Science Bulletin*, 58, 1811–1817.

Flynn, L. J., Tao, D., Yang, W., Guangpu, X., Sukuan, H., Libo, P., & Yongqing, M. (2011). Observations on the *Hipparion* red clays of the Loess Plateau. *Vertebrata PalAsiatica*, 49(3), 275

Geraads, D., Güleç, E., & Kaya, T. (2002). *Sinotragus* (Bovidae, Mammalia) from Turkey, and the Late Miocene Middle Asiatic Province. *Neues Jahrbuch für Geologie und Paläontologie Monatshefte*, 2002(8), 477–489.

Li, Y., Deng, T., Hua, H., Li, Y., & Zhang, Y. (2017). Assessment of dental ontogeny in late Miocene hipparionines from the Lamagou fauna of Fugu, Shaanxi Province, China. *PLoS One*, 12(4), p.e0175460.

Li, Z., Stidham, T. A., Deng, T., & Zhou, Z. (2020). Evidence of late miocene peri-tibetan aridification from the oldest Asian species of sandgrouse (Aves: Pteroclidae). *Frontiers in Ecology and Evolution*, 8, 59.

Qiu, Z. X., Deng, T., & Wang, B. Y. (2014). A Late Miocene *Ursavus* skull from Guanghe, Gansu, China. *Vertebrata PalAsiatica*. 52(3), 265–302.

Africa: Nawata Formation

Brochu, C. A., & Storrs, G. W. (2012). A giant crocodile from the Plio-Pleistocene of Kenya, the phylogenetic relationships of Neogene African crocodylines, and the antiquity of Crocodylus in Africa. *Journal of Vertebrate Paleontology*, 32(3), 587–602.

Leakey, M. G., Feibel, C. S., Bernor, R. L., Harris, J. M., Cerling, T. E., Stewart, K. M., ... & Winkler, A. J. (1996). Lothagam: a record of faunal change in the Late Miocene of East Africa. *Journal of Vertebrate Paleontology*, 16(3), 556–570.

Leakey, M. G., & Harris, J. M. (Eds). (2003). *Lothagam: the dawn of humanity in eastern Africa*. Columbia University Press, New York.

Turner, A., & Antón, M. (2004). *Evolving Eden: an illustrated guide to the evolution of the African large-mammal fauna*. Columbia University Press, New York.

Werdelin, L., & Sanders, W. L. (Eds). (2010). *Cenozoic Mammals of Africa*. University of California Press, Berkeley, California.

Australia: Riversleigh Fossil Site (Faunal Zones B–D)

Archer, M., Beck, R., Gott, M., Hand, S., Godthelp, H., & Black, K. (2011). Australia's first fossil marsupial mole (Notoryctemorphia) resolves controversies about their evolution and palaeoenvironmental origins. *Proceedings of the Royal Society B: Biological Sciences*, 278(1711), 1498–1506.

Archer, M., Christmas, O., Hand, S. J., Black, K. H., Creaser, P., Godthelp, H., ... & Travouillon, K. J. (2016). Earliest known record of a hypercarnivorous dasyurid (Marsupialia), from newly discovered carbonates beyond the Riversleigh World Heritage Area, north Queensland. *Memoirs of Museum Victoria*, 74, 137–150.

Archer, M., Hand, S., & Godthelp, H. (1994). *Riversleigh-the Story of Animals in Ancient Rainforests of Inland Australia*. Reed Books, Sydney.

Black, K. H., Camens, A. B., Archer, M., & Hand, S. J. (2012). Herds overhead: Nimbadon lavarackorum (Diprotodontidae), heavyweight marsupial herbivores in the Miocene forests of Australia. *PLoS One*, 7(11), e48213.

Black, K. H., Price, G. J., Archer, M., & Hand, S. J. (2014). Bearing up well? Understanding the past, present and future of Australia's koalas. *Gondwana Research*, 25(3), 1186–1201.

Cooke, B. N. (1992). Primitive macropodids from Riversleigh, north-western Queensland. *Alcheringa*, 16(3), 201–217.

Cooke, B. N., Travouillon, K. J., Archer, M., & Hand, S. J. (2015). *Ganguroo robustiter*, sp. nov. (Macropodoidea, Marsupialia), a middle to early late Miocene basal macropodid from Riversleigh World Heritage Area, Australia. *Journal of Vertebrate Paleontology*, 35(4), p.e956879.

Gillespie, A. K., Archer, M., & Hand, S. J. (2016). A tiny new marsupial lion (Marsupialia, Thylacoleonidae) from the early Miocene of Australia. *Palaeontologia Electronica*, 9.2.26A, 1–26.

Gillespie, A. K., Archer, M., & Hand, S. J. (2019). *Lekaneleo*, a new genus of marsupial lion (Marsupialia, Thylacoleonidae) from the Oligocene–Miocene of Australia, and the craniodental morphology of *L. roskellyae*, comb. nov. *Journal of Vertebrate Paleontology*, 39(5), p.e1703722.

Long, J., Archer, M., Flannery, T., & Hand, S. (2002). *Prehistoric mammals of Australia and New Guinea: One hundred million years of evolution*. John Hopkins University Press, Baltimore, Maryland.

Mao, X., & Retallack, G. (2019). Late Miocene drying of central Australia. *Palaeogeography, Palaeoclimatology, Palaeoecology*, 514, 292–304.

Nguyen, J. M. T, Hand S. J., & Archer, M. (2016). The late Cenozoic passerine avifauna from Rackham's Roost Site, Riversleigh, Australia. *Records of the Australian Museum*, 68(5), 201–230.

Palci, A., Hutchinson, M. N., Caldwell, M. W., Scanlon, J. D., & Lee, M. S. (2018). Palaeoecological inferences for the fossil Australian snakes *Yurlunggur* and *Wonambi* (Serpentes, Madtsoiidae). *Royal Society Open Science*, 5(3), p.172012.

Rovinsky, D. S., Evans, A. R., & Adams, J. W. (2019). The pre-Pleistocene fossil thylacinids (Dasyuromorphia: Thylacinidae) and the evolutionary context of the modern thylacine. PeerJ, 7, e7457.

Pliocene

Bartoli, G., Sarnthein, M., Weinelt, M., Erlenkeuser, H., Garbe-Schönberg, D., & Lea, D. W. (2005). Final closure of Panama and the onset of northern hemisphere glaciation. *Earth and Planetary Science Letters*, 237(1–2), 33–44.

Stehli, F. G., & Webb, S. D. (Eds.). (2013). *The Great American Biotic Interchange (Vol. 4)*. Springer Science & Business Media, Berlin.

Woodburne, M. O. (2010). The Great American Biotic Interchange: dispersals, tectonics, climate, sea level and holding pens. *Journal of Mammalian Evolution*, 17, 245–264.

Vallejo-Hincapié, F., Pardo-Trujillo, A., Barbosa-Espitia, Á., Aguirre, D., Celis, S. A., Giraldo-Villegas, C. A., ... & Foster, D. (2024). Miocene vanishing of the central American seaway between the panamá arc and the south American plate. *Geological Society of America Bulletin*, 136(11-12), 4798–4814.

Australia: Allingham Formation

Archer, M., & Wade, M. (1976). Results of the Ray E. Lemley expeditions. I. The Allingham Formation and a new Pliocene vertebrate fauna from Northern Queensland. Memoirs of the *Queensland Museum*, 17(3), 379–398.

Black, K. H., Price, G. J., Archer, M., & Hand, S. J. (2014). Bearing up well? Understanding the past, present and future of Australia's koalas. *Gondwana Research*, 25(3), 1186–1201.

Boles, W. E., & Mackness, B. (1994). Birds from the Bluff Downs local fauna, Allingham Formation, Queensland. *Records of the South Australian Museum*, 27, 139–149.

Long, J., Archer, M., Flannery, T., & Hand, S. 2002. *Prehistoric mammals of Australia and New Guinea: One hundred million years of evolution*. John Hopkins University Press, Baltimore, Maryland.

Scanlon, J. D., & Mackness, B. S. (2001). A new giant python from the Pliocene Bluff Downs Local Fauna of northeastern Queensland. *Alcheringa*, 25(4), 425–437.

Travouillon, K. J., Louys, J., Price, G. J., Archer, M., Hand, S. J., & Muirhead, J. (2017). A review of the Pliocene bandicoots of Australia, and descriptions of new genus and species. *Journal of Vertebrate Paleontology*, 37(5), e1360894.

South America: Chapadmalal Formation

Abello, M. A., & Candela, A. M. (2020). Paleobiology of *Argyrolagus* (Marsupialia, Argyrolagidae): an astonishing case of bipedalism among South American mammals. *Journal of Mammalian Evolution*, 27, 419–444.

Beck, R. M., & Tagliorettl, M. L. (2020). A nearly complete juvenile skull of the marsupial *Sparassocynus derivatus* from the Pliocene of Argentina, the affinities of "sparassocynids", and the diversification of opossums (Marsupialia; Didelphimorphia; Didelphidae). *Journal of Mammalian Evolution*, 27(3), 385-417.

Croft, D. A. (2016). *Horned armadillos and rafting monkeys: the fascinating fossil mammals of South America*. Indiana University Press, Bloomington, Indiana.

de los Reyes, M., Poiré, D., Soibelzon, L.H., Zurita, A.E., & Arrouy, M.J. (2013). First evidence of scavenging of a glyptodont (Mammalia, Glyptodontidae) from the Pliocene of the Pampean region (Argentina): taphonomic and paleoecological remarks. *Palaeontología Electrónica*, 16(2), 15A.

Degrange, F. J., Tambussi, C. P., Taglioretti, M. L., Dondas, A., & Scaglia, F. (2015). A new Mesembriornithinae (Aves, Phorusrhacidae) provides new insights into the phylogeny and sensory capabilities of terror birds. *Journal of Vertebrate Paleontology*, 35(2), e912656.

Elissamburu, A., Dondas, A., & De Santis, L. (2011). Morfometría de las paleocuevas de la" Fm." Chapadmalal y su asignación a Actenomys (Rodentia), *Paedotherium* (Notoungulata) y otros mamíferos fósiles hospedantes. *Mastozoología neotropical*, 18(2), 227–238.

Vizcaíno, S. F., Fariña, R. A., Zárate, M. A., Bargo, M. S., & Schultz, P. (2004). Palaeoecological implications of the mid-Pliocene faunal turnover in the Pampean Region (Argentina). *Palaeogeography, Palaeoclimatology, Palaeoecology*, 213(1–2), 101–113.

Welker, F., Collins, M. J., Thomas, J. A., Wadsley, M., Brace, S., Cappellini, E., Turvey, S. T., Reguero, M., Gelfo, J. N., Kramarz, A., & Burger, J. (2015). Ancient proteins resolve the evolutionary history of Darwin's South American ungulates. *Nature*, 522(7554), 81–84.

Zamorano, M. (2020). Exceptional preservation of tracheal rings in a glyptodont mammal from the Late Pleistocene of Argentina. *Acta Palaeontologica Polonica*, 65(1), 29–34.

Africa: Laetoli Beds

Cerling, T. E., Andanje, S. A., Blumenthal, S. A., Brown, F. H., Chritz, K. L., Harris, J. M., ... & Uno, K. T. (2015). Dietary changes of large herbivores in the Turkana Basin, Kenya from 4 to 1 Ma. *Proceedings of the National Academy of Sciences*, 112(37), 11467–11472.

Harrison, T. (Ed.). (2011). *Paleontology and Geology of Laetoli: Human Evolution in Context: Volume 1: Geology, Geochronology, Paleoecology and Paleoenvironment*. Springer Science & Business Media, Berlin.

Hemmer, H. (2023). The identity of the "lion", *Panthera principialis* sp. nov., from the Pliocene Tanzanian site of Laetoli and its significance for molecular dating the pantherine phylogeny, with remarks on *Panthera shawi* (Broom, 1948), and a revision of *Puma incurva* (Ewer, 1956), the Early Pleistocene Swartkrans "leopard"(Carnivora, Felidae). *Palaeobiodiversity and Palaeoenvironments*, 103(2), 465–487.

Leakey, M.D., & Hay, R.L. (1979). Pliocene footprints in the Laetolil Beds at Laetoli, northern Tanzania. *Nature*, 278(5702), 317–323.

Masao, F. T., Ichumbaki, E. B., Cherin, M., Barili, A., Boschian, G., Iurino, D. A., Menconero, S., Moggi-Cecchi, J., & Manzi, G. (2016). New footprints from Laetoli (Tanzania) provide evidence for marked body size variation in early hominins. *Elife*, 5, e19568.

Raichlen, D. A., Gordon, A. D., Harcourt-Smith, W. E., Foster, A. D., & Haas Jr, W. R. (2010). Laetoli footprints preserve earliest direct evidence of human-like bipedal biomechanics. *PLoS One*, 5(3), e9769.

Su, D. F., & Harrison, T. (2015). The paleoecology of the Upper Laetolil Beds, Laetoli Tanzania: A review and synthesis. *Journal of African Earth Sciences*, 101, 405–419.

Turner, A., & Antón, M. (2004). *Evolving Eden: an illustrated guide to the evolution of the African large-mammal fauna*. Columbia University Press, New York.

Uno, K. T., Cerling, T. E., Harris, J. M., Kunimatsu, Y., Leakey, M. G., Nakatsukasa, M., & Nakaya, H. (2011). Late Miocene to Pliocene carbon isotope record of differential diet change among East African herbivores. *Proceedings of the National Academy of Sciences*, 108(16), 6509–6514.

Werdelin, L., & Sanders, W. L. (Eds) (2010). *Cenozoic Mammals of Africa*. University of California Press, Berkeley, California.

North America: Blanco Formation

Blanco, R.E., & Rinderknecht, A. (2008). Estimation of hearing capabilities of Pleistocene ground sloths (Mammalia, Xenarthra) from middle-ear anatomy. *Journal of Vertebrate Paleontology*, 28(1), 274–276.

Feranec, R. S. (2003). Stable isotopes, hypsodonty, and the paleodiet of *Hemiauchenia* (Mammalia: Camelidae): a morphological specialization creating ecological generalization. *Paleobiology*, 29(2), 230–242.

Kurtén, B., & Anderson, E. 1980. *Pleistocene Mammals of North America*, Columbia University Press, New York.

Lucas, S. G., Morgan, G. S., Spielmann, J. A., Pasenko, M. R., & Aguilar, R. H. (2011). Taxonomy and evolution of the Plio-Pleistocene proboscidean *Stegomastodon* in North America. *Current Research in the Pleistocene*, 28, 173–175.

Morgan, G. S., Sealey, P. L., & Lucas, S. G. (2008). Late Pliocene (late Blancan) vertebrate faunas from Pearson Mesa, Duncan Basin, southwestern New Mexico and southeastern Arizona. *Bulletin*, 44, 141–188.

Naples, V. L. (1989). The feeding mechanism in the Pleistocene ground sloth, *Glossotherium*. *Natural History Museum of Los Angeles County*. 415, 1–23

Wang, X. (2008). *Dogs: their fossil relatives and evolutionary history*. Columbia University Press, New York.

Wang, X., White, S. C., Balisi, M., Biewer, J., Sankey, J., Garber, D., & Tseng, Z. J. (2018). First bone-cracking dog coprolites provide new insight into bone consumption in Borophagus and their unique ecological niche. *Elife*, 7, e34773.

Webb, S. D. (1977). A history of savanna vertebrates in the New World. Part I: North America. *Annual Review of Ecology and Systematics*, 8(1), 355–380.

Webb, S. D., & Opdyke, N. D. (1995). 'Global climatic influence on Cenozoic land mammal faunas'. In: Grimm, E. C., & Hansen, J. T. (eds) *Effects of past global change on life*. National Academies Press, Washington, D.C. pp. 184–208.

Europe: Milia Fossil Site

Agustí, J., & Antón, M. (2002). *Mammoths, Sabertooths, and Hominids: 65 Million Years of Mammalian Evolution in Europe*. Columbia University Press, New York.

Croitor, R. (2017). Description of a new deer species (Cervidae, Mammalia) from the Early Pliocene of Eastern Europe, with a review of early dispersals and palaeobiogeography of the subfamily Cervinae. *Neues Jahrbuch für Geologie und Paläontologie-Abhandlungen*, 283(1), 85–108.

Guérin, C., & Tsoukala, E. (2013). The Tapiridae, Rhinocerotidae and Suidae (Mammalia) of the Early Villafranchian site of Milia (Grevena, Macedonia, Greece). *Geodiversitas*, 35(2), 447–489.

Saarinen, J., & Lister, A. M. (2016). Dental mesowear reflects local vegetation and niche separation in Pleistocene proboscideans from Britain. *Journal of Quaternary Science*, 31(7), 799–808.

Tsoukala, E., van Logchem, W., Lazaridis, G., & Mol, D. (2014). 'Carnivores of the Early Villafranchian site of Milia (Grevena, Macedonia, Greece)'. In: *Abstract Book of the VIth International Conference on Mammoths and Their Relatives*, Grevena and Siatista, Greece. pp. 6–7.

Vlachos, E., Tsoukala, E., Crégut-Bonnoure, E., Guérin, C., & Mol, D. (2018). The paradise lost of Milia (Grevena, Greece; Late Pliocene, early Villafranchian, MN15/MN16a): faunal composition and diversity. *Quaternary*, 1(2), 13.

Chapter 7: The Quaternary

Adams, N. F., Candy, I., & Schreve, D. C. (2022). An Early Pleistocene hippopotamus from Westbury Cave, Somerset, England: support for a previously unrecognized temperate interval in the British Quaternary record. *Journal of Quaternary Science*, 37(1), 28–41.

Barnosky, A. D., Koch, P. L., Feranec, R. S., Wing, S. L., & Shabel, A. B. (2004). Assessing the causes of late Pleistocene extinctions on the continents. *Science*, 306(5693), 70–75.

Denton, G. H., Anderson, R. F., Toggweiler, J. R., Edwards, R. L., Schaefer, J. M., & Putnam, A. E. (2010). The last glacial termination. *Science*, 328(5986), 1652–1656.

Guthrie, R. D. (2001). Origin and causes of the mammoth steppe: a story of cloud cover, woolly mammal tooth pits, buckles, and inside-out Beringia. *Quaternary science reviews*, 20(1–3), 549–574.

Lee, K. E., Clemens, S. C., Kubota, Y., Timmermann, A., Holbourn, A., Yeh, S. W., Bae, S. W., & Ko, T. W. (2021). Roles of insolation forcing and CO2 forcing on Late Pleistocene seasonal sea surface temperatures. *Nature Communications*, 12(1), 5742.

Lowe, J. J., & Walker, M. (2014). *Reconstructing Quaternary Environments*. Routledge, London.

MacPhee, R. D. (2018). *End of the megafauna: the fate of the world's hugest, fiercest, and strangest animals*. WW Norton & Company, New York.

Martin, P. S., & Klein, R. G. (Eds.). (2022). *Quaternary extinctions: a prehistoric revolution*. University of Arizona Press, Arizona.

Zimov, S. A., Zimov, N. S., & Chapin III, F. S. (2012). 'The past and future of the mammoth steppe ecosystem'. In: Louys, J. (ed) *Paleontology in Ecology and Conservation*. Springer Berlin Heidelberg, Heidelberg. pp. 193–225.

Pleistocene

Berends, C. J., Köhler, P., Lourens, L. J., & Van de Wal, R. S. W. (2021). On the cause of the Mid-Pleistocene transition. *Reviews of Geophysics*, 59, e2020RG00072

Imbrie, J., & Imbrie, K. P. (1986). *Ice Ages: Solving the Mystery*. Harvard University Press, New Haven, Connecticut.

Kurtén, B. (1988). *Before the Indians*. Columbia University Press, New York.

Macdougall, J. D. (2013). *Frozen Earth: The Once and Future Story of Ice Ages*. University of California Press, Berkeley, California.

Scherrenberg, M., Berends, C., & van de Wal, R. (2023). Late Pleistocene glacial terminations accelerated by proglacial lakes. *Climate of the Past Discussions*, 2023, 1–30.

Stuart, A. J. (2021). *Vanished Giants: The Lost World of the Ice Age*. University of Chicago Press, Chicago, Illinois.

Africa: Koobi Fora Formation

Bobe, R., & Behrensmeyer, A. K. (2004). The expansion of grassland ecosystems in Africa in relation to mammalian evolution and the origin of the genus *Homo*. *Palaeogeography, Palaeoclimatology, Palaeoecology*, 207(3–4), 399–420.

deMenocal, P. B. (2004). African climate change and faunal evolution during the Pliocene–Pleistocene. *Earth and Planetary Science Letters*, 220(1–2), 3–24.

Dupont, L. (2011). Orbital scale vegetation change in Africa. *Quaternary Science Reviews*, 30(25–26), 3589–3602.

Ferraro, J. V., Plummer, T. W., Pobiner, B. L., Oliver, J. S., Bishop, L. C., Braun, D. R., Ditchfield, P. W., Seaman III, J. W., Binetti, K. M., Seaman Jr, J. W., & Hertel, F. (2013). Earliest archaeological evidence of persistent hominin carnivory. *PLoS One*, 8(4), p.e62174

Leakey, R. E. F., & Lewin, R. (1978). *People of the Lake: Mankind & its beginnings*. Avon, NY.

O'Brien, K., Patterson, D. B., Biernat, M. D., Braun, D. R., Cerling, T. E., McGrosky, A., & Faith, J. T. (2020). Ungulate turnover in the Koobi Fora Formation: Spatial and temporal variation in the Early Pleistocene. *Journal of African Earth Sciences*, 161, 103658.

Turner, A., & Antón, M. (2004). *Evolving Eden: An illustrated guide to the evolution of the African large-mammal fauna*. Columbia University Press, New York.

Werdelin, L., & Sanders, W. L., (Eds.). 2010. *Cenozoic Mammals of Africa*. University of California Press, Berkeley, California.

North America: Haile Quarry Site (Sites 7C and 21A)

Bartolini-Lucenti, S., Cirilli, O., Melchionna, M., Raia, P., Tseng, Z. J., Flynn, J. J., & Rook, L. (2024). Virtual reconstruction of the *Canis arnensis* type (Canidae, Mammalia) from the Upper Valdarno Basin (Italy, Early Pleistocene). *Scientific Reports*, 14(1), 8303.

Bradham, J. L., Larisa R. G., DeSantis, M. L., Jorge, S. P., & Keuroghlian, A. (2018). Dietary variability of extinct tayassuids and modern white-lipped peccaries (*Tayassu pecari*) as inferred from dental microwear and stable isotope analysis. *Palaeogeography, Palaeoclimatology, Palaeoecology*, 499, 93–101.

Guthrie, R. D. (1990) *Frozen fauna of the mammoth steppe*. The University of Chicago Press, Chicago, Illinois.

Hulbert, R. C. (Ed). (2001). *The Fossil Vertebrates of Florida*. University Press of Florida, Florida.

Hulbert, R. C. (2010). *A new early Pleistocene tapir (Mammalia: Perissodactyla) from Florida, with a review of Blancan tapirs from the state*. University of Florida, Gainesville, Florida.

Hulbert, R. C., & Morgan, G. S. (1993). Quantitative and qualitative evolution in the giant armadillo *Holmesina* (Edentata: Pampatheriidae) in Florida. *Morphological change in Quaternary mammals of North America*. Cambridge University Press, Cambridge, UK. pp. 134–178.

Hulbert, R. C., Morgan, G. S., & Poyer, A. R. (1989). Associated skeletons of megathere, pampathere, tapir, and turtles from the latest Pliocene or earliest Pleistocene of north-central Florida. *Journal of Vertebrate Paleontology*, 9(3), 26A.

Kurtén, B., & Anderson, E. (1980). *Pleistocene Mammals of North America*. Columbia University Press, New York.

Martin, L. D., Babiarz, J. P., Naples, V. L., & Hearst, J. (2000). Three ways to be a saber-toothed cat. *Naturwissenschaften*, 87(1), 41–44.

Webb, S. D. (Ed.). (1974). *Pleistocene mammals of Florida*. University Press of Florida, Gainesville, Florida.

Zurita, A. E., Gillette, D. D., Cuadrelli, F., & Carlini, A. A. (2018). A tale of two clades: comparative study of *Glyptodon* Owen and *Glyptotherium* Osborn (Xenarthra, Cingulata, Glyptodontidae). *Geobios*, 51(3), 247–258.

North America: Rancho La Brea Tar Pits

Barnett, R., Barnes, I., Phillips, M. J., Martin, L. D., Harington, C. R., Leonard, J. A., & Cooper, A. (2005). Evolution of the extinct Sabretooths and the American cheetah-like cat. *Current Biology*, 15(15), R589–R590.

Baskin, J., & Thomas, R. (2016). A review of *Camelops* (Mammalia, Artiodactyla, Camelidae), a giant llama from the middle and late Pleistocene (Irvingtonian and Rancholabrean) of North America. *Historical Biology*, 28(1–2), 120–127.

Campbell Jr, K. E., & Tonni, E. P. (1983). Size and locomotion in teratorns (Aves: Teratornithidae). *The Auk*, 100(2), 390–403.

DeSantis, L. R., Crites, J. M., Feranec, R. S., Fox-Dobbs, K., Farrell, A. B., Harris, J. M., Takeuchi, G. T., & Cerling, T. E. (2019). Causes and consequences of Pleistocene megafaunal extinctions as revealed from Rancho La Brea mammals. *Current Biology*, 29(15), 2488–2495.

DeSantis, L. R., Feranec, R. S., Antón, M., & Lundelius, E. L. (2021). Dietary ecology of the scimitar-toothed cat *Homotherium serum*. *Current Biology*, 31(12), 2674–2681.

Figueirido, B., Pérez-Claros, J. A., Torregrosa, V., Martín-Serra, A., & Palmqvist, P. (2010). Demythologizing *Arctodus simus*, the 'short-faced'long-legged and predaceous bear that never was. *Journal of Vertebrate Paleontology*, 30(1), 262–275.

Higgins, P., Meachen, J., & Lovelace, D. (2023). Were pronghorns (Antilocapra) primary prey for North American cheetahs (*Miracinonyx*)?. *Quaternary International*, 647, 81–87.

Kropf, M., Mead, J. I., & Anderson, R. S. (2007). Dung, diet, and the paleoenvironment of the extinct shrub-ox (*Euceratherium collinum*) on the Colorado Plateau, USA. *Quaternary Research*, 67(1), 143–151.

O'Keefe, F. R., Dunn, R. E., Weitzel, E. M., Waters, M. R., Martinez, L. N., Binder, W. J., ... & Lindsey, E. L. (2023). Pre–Younger Dryas megafaunal extirpation at Rancho La Brea linked to fire-driven state shift. Science, 381(6659), eabo3594.

Perri, A. R., Mitchell, K. J., Mouton, A., Alvarez-Carretero, S., Hulme-Beaman, A., Haile, J., ... & Frantz, L. A. (2021). Dire wolves were the last of an ancient New World canid lineage. *Nature*, 591(7848), 87–91.

Seymour, K. L. (1989). *Panthera onca. Mammalian species*, 340, 1–9.

Spencer, L. M., Van Valkenburgh, B., & Harris, J. M. (2003). Taphonomic analysis of large mammals recovered from the Pleistocene Rancho La Brea tar seeps. *Paleobiology*, 29(4), 561–575.

Stewart, M., Carleton, W. C., & Groucutt, H. S. (2021). Climate change, not human population growth, correlates with Late Quaternary megafauna declines in North America. *Nature Communications*, 12(1), 965.

South America: Tarija Formation

Blanco, R. E., & Czerwonogora, A. (2003). The gait of *Megatherium* Cuvier 1796. *Senckenbergiana biologica*, 83, 61–68.

Blanco, R. E., Jones, W. W., & Rinderknecht, A. (2009). The sweet spot of a biological hammer: the centre of percussion of glyptodont (Mammalia: Xenarthra) tail clubs. *Proceedings of the Royal Society B: Biological Sciences*, 276(1675), 3971–3978.

Croft, D. A. (2016). *Horned armadillos and rafting monkeys: the fascinating fossil mammals of South America*. Indiana University Press, Bloomington, Indiana.

Dominato, V. H., Mothé, D., Costa Da Silva, R., & Avilla, L. (2011). Evidence of scavenging on remains of the gomphothere *Haplomastodon waringi* (Proboscidea: Mammalia) from the Pleistocene of Brazil: Taphonomic and paleoecological remarks. *Journal of South American Earth Sciences* 31(2–3), 171–177.

Dompierre, H. (1996). Observations on the diets of six late Cenozoic North American camelids: *Camelops, Hemiauchenia, Palaeolama, Procamelus, Alforjas*, and *Megatylopus*. Ph.D. thesis, University of Toronto.

Fariña, R. A., Blanco, R. E., & Christiansen, P. (2005). Swerving as the escape strategy of *Macrauchenia patachonica* Owen (Mammalia; Litopterna). *Ameghiniana*, 42(4), 751–760.

Fariña, R. A., Vizcaíno, S. F., & De Iuliis, G. (2013). *Megafauna: Giant Beasts of Pleistocene South America*. Indiana University Press, Bloomnigton, Indiana.

MacFadden, B.J., & Shockey, B.J. (1997). Ancient feeding ecology and niche differentiation of Pleistocene mammalian herbivores from Tarija, Bolivia: morphological and isotopic evidence. *Paleobiology*, 23(1), 77–100.

MacFadden, B. J., Zeitler, P. K., Anaya, F., & Cottle, J. M. (2013). Middle Pleistocene age of the fossiliferous sedimentary sequence from Tarija, Bolivia. *Quaternary Research*, 79(2), 268–273.

Mitchell, K. J., Bray, S. C., Bover, P., Soibelzon, L., Schubert, B. W., Prevosti, F., Prieto, A., Martin, F., Austin, J. J., & Cooper, A. (2016). Ancient mitochondrial DNA reveals convergent evolution of giant short-faced bears (Tremarctinae) in North and South America. *Biology Letters*, 12(4), p.20160062.

Prevosti, F. J., Tonni, E. P., & Bidegain, J. C. (2009). Stratigraphic range of the large canids (Carnivora, Canidae) in South America, and its relevance to quaternary biostratigraphy. *Quaternary International*, 210, 76–81.

Scherer, C. S. (2013). The Camelidae (Mammalia, Artiodactyla) from the quaternary of South America: cladistic and biogeographic hypotheses. *Journal of Mammalian Evolution*, 20(1), 45–56.

Vizcaíno, S. F., Cassini, G. H., Fernicola, J. C., & Bargo, M. S. (2011). Evaluating habitats and feeding habits through ecomorphological features in glyptodonts (Mammalia, Xenarthra). *Ameghiniana*, 48(3), 305–319.

Welker, F., Collins, M. J., Thomas, J. A., Wadsley, M., Brace, S., Cappellini, E., Turvey, S. T., Reguero, M., Gelfo, J. N., Kramarz, A., & Burger, J. (2015). Ancient proteins resolve the evolutionary history of Darwin's South American ungulates. *Nature*, 522(7554), 81–84.

Australia: Naracoorte Caves

Arman, S. D., & Prideaux, G. J. (2016). Behaviour of the Pleistocene marsupial lion deduced from claw marks in a southwestern Australian cave. *Scientific Reports*, 6(1), 21372.

Figueirido-Castillo, F. B., Martín-Serra, A., & Janis, C. M. (2016). Ecomorphological determinations in the absence of living analogs: the predatory behavior of the marsupial lion (*Thylacoleo carnifex*) as revealed by elbow joint morphology. *Paleobiology*, 42(3), 508–531.

Handley, W. D., & Worthy, T. H. (2021). Endocranial anatomy of the giant extinct Australian mihirung birds (Aves, Dromornithidae). *Diversity*, 13(3), 124.

Janis, C. M., Buttrill, K., & Figueirido, B. (2014). Locomotion in extinct giant kangaroos: were sthenurines hop-less monsters? *PLoS One*, 9(10), e109888.

Janis, C. M., Napoli, J. G., Billingham, C., & Martín-Serra, A. (2020). Proximal humerus morphology indicates divergent patterns of locomotion in extinct giant kangaroos. *Journal of Mammalian Evolution*, 27(4), 627–647.

Johnson, C. (2006), *Australia's Mammal Extinctions, a 50,000 year history*, Cambridge University Press, Cambridge, UK.

Jones, B., & Janis, C. M. (2024). Hop, walk or bound? Limb proportions in kangaroos and the probable locomotion of the extinct genus Protemnodon. *Journal of Mammalian Evolution*, 31(2), 26.

Jones, B., Martín-Serra, A., Rayfield, E. J., & Janis, C. M. (2022). Distal humeral morphology indicates locomotory divergence in extinct giant kangaroos. *Journal of Mammalian Evolution*, 29(1), 27–41.

Molnar, R. E. (2004). *Dragons in the dust: the paleobiology of the giant monitor lizard Megalania*. Indiana University Press, Bloomington, Indiana.

Price, G. J., Louys, J., Cramb, J., Feng, Y. X., Zhao, J. X., Hocknull, S. A., ... & Joannes-Boyau, R. (2015). Temporal overlap of humans and giant lizards (Varanidae; Squamata) in Pleistocene Australia. *Quaternary Science Reviews*, 125, 98–105.

Prideaux, G. J., & Warburton, N. M. (2010). An osteology-based appraisal of the phylogeny and evolution of kangaroos and wallabies (Macropodidae: Marsupialia). *Zoological Journal of the Linnean Society*, 159(4), 954–987.

Prideaux, G. J., & Warburton, N. M. (2010). An osteology-based appraisal of the phylogeny and evolution of kangaroos and wallabies (Macropodidae: Marsupialia). *Zoological Journal of the Linnean Society*, 159(4), 954–987.

Richards, H. L., Wells, R. T., Evans, A. R., Fitzgerald, E. M., & Adams, J. W. (2019). The extraordinary osteology and functional morphology of the limbs in Palorchestidae, a family of strange extinct marsupial giants. *PLoS One*, 14(9), p.e0221824.

Richardson, W., Morales-García, N. M., Damuth, J., Singh, S., & Janis, C. M. (2024). Why the short face? The face lengths of sthenurine kangaroos scale with negative allometry. *Journal of Vertebrate Paleontology*, 43(5), e2336145.

Scanlon, J. D., & Lee, M. S. (2000). The Pleistocene serpent *Wonambi* and the early evolution of snakes. *Nature*, 403(6768), 416–420.

Sharp, A. C. (2016). A quantitative comparative analysis of the size of the frontoparietal sinuses and brain in vombatiform marsupials. *Memoirs of Museum Victoria*, 74, 331–342.

Webb, S. (2008). Megafauna demography and late Quaternary climatic change in Australia: A predisposition to extinction. *Boreas*, 37(3), 329–345.

Asia: Liang Bua Cave

Brongersma, L. D.. (1958). On an extinct species of the genus *Varanus* (Reptilia, Sauria) from the island of Flores. *Zoologische Mededelingen*, 7, 113–125.

Brown, P., Sutikna, T., Morwood, M. J., Soejono, R. P., Jatmiko, Saptomo, E. W., & Due, R. A. (2004). A new small-bodied hominin from the Late Pleistocene of Flores, Indonesia. *Nature*, 431.

Brumm, A., Jensen, G. M., van den Bergh, G. D., Morwood, M. J., Kurniawan, I., Aziz, F., & Storey, M. (2010). Hominins on Flores, Indonesia, by one million years ago. *Nature Letters*, 464, 748–752.

Locatelli, E., Due, R. A., van den Bergh, G. D., & Ostende, L. W. vd H. (2012). Pleistocene survivors and Holocene extinctions: The giant rats from Liang Bua (Flores, Indonesia). *Quaternary International*, 281, 47–57.

Meijer, H. J. M., & Due, R. A.. (2010a). A new species of giant stork (Aves: Ciconiiformes) from the Pleistocene of Liang Bua, Flores (Indonesia). *Zoological Journal of the Linnaean Society*, 160, 707–724.

Meijer, H. J. M., Sutikna, T., Saptomo, E. W., Due, R. A., Wasisto, S., James, H. F., Morwood, M. J., & Tocheri, M. W. (2013). Late Pleistocene-Holocene non-passerine avifauna of Liang Bua (Flores, Indonesia). *Journal of Vertebrate Palaeontology*, 33(4), 877–894.

Meijer, H. J. M., Sutikna, T., Saptomo, E. W., & Tocheri, M. W. (2022) More bones of *Leptoptilos robustus* from Flores reveal new insights into giant marabou stork paleobiology and biogeography. *Royal Society Opem Science*, 9, 220435.

Musser, G. G. (1981) The Giant Rats of Flores and Its Relatives East of Borneo and Bali. *Bulletin of the American Museum of Natural History*, 169(2), 71–175.

Powley, M. J., Sutisna, I., Mikac, K. M., Wibowo, U. P., & van den Bergh, G. D. (2021). The *Stegodon* Bonebed of the Middle Pleistocene Archaeological Site Mata Menge (Flores, Indonesia): Taphonomic Agents in Site Formation. *Quaternary*, 4(31), 1–20.

Van der Geer, A. A. E., Van den Bergh, G. D., Lyras, G. A., Prasetyo, U. W., Awe Due, R., Setiyabudi, E., & Drinia, H. (2016). The effect of area and isolation on insular dwarf proboscideans. *Journal of Biogeography*, 43, 1656–1666.

Veatch, E. G., Tocheri, M. W., Sutikna, T., McGrath, K., Saptomo, E. W., Jatmiko, & Helgen, K. M. (2019). Temporal shifts in the distribution of murine rodent body size classes at Liang Bua (Flores, Indonesia) reveal new insights into the paleoecology of *Homo floresiensis* and associated fauna. *Journal of Human Evolution*, 130, 45–60.

Europe: Cromer Forest-bed Formation

Agustí, J., & Antón, M. (2002). *Mammoths, Sabertooths, and Hominids: 65 Million Years of Mammalian Evolution in Europe*. Columbia University Press, New York.

Ashton, N., Lewis, S. G., Parfitt, S. A., Bates, M. R., Bates, R., Dix, J. K., ... & Sturt, F. (2018). Understanding and Monitoring the Cromer Forest-bed Formation.

Chatar, N., Michaud, M., & Fischer, V. (2022). Not a jaguar after all? Phylogenetic affinities and morphology of the Pleistocene felid *Panthera gombaszoegensis*. *Papers in Palaeontology*, 8(5), p.e1464.

Field, M. H., & Peglar, S. M. (2010). A palaeobotanical investigation of the sediments from the West Runton Mammoth site. *Quaternary International*, 228(1–2), 38–45.

Jones, R. L., Keen, D. H., Jones, R. L., & Keen, D. H. (1993). The Cromer Forest-bed Formation and associated deposits. *Pleistocene Environments in the British Isles*, 55–72.

Larramendi, A., Palombo, M. R., & Marano, F. (2017). Reconstructing the life appearance of a Pleistocene giant: size, shape, sexual dimorphism and ontogeny of *Palaeoloxodon antiquus* (Proboscidea: Elephantidae) from Neumark-Nord 1 (Germany). *Bollettino della Società Paleontologica Italiana*, 56(3), 299–317.

Lewis, M., Pacher, M., & Turner, A. (2010). The larger Carnivora of the West Runton freshwater bed. *Quaternary International*, 228(1–2), 116–135.

Lister, A. M., Parfitt, S. A., Owen, F. J., Collinge, S. E., & Breda, M. (2010). Metric analysis of ungulate mammals in the early Middle Pleistocene of Britain, in relation to taxonomy and biostratigraphy: II: Cervidae, Equidae and Suidae. *Quaternary International*, 228(1–2), 157–179.

Lister, A. M., & Stuart, A. J. (2010). The West Runton mammoth (*Mammuthus trogontherii*) and its evolutionary significance. *Quaternary International*, 228(1–2), 180–209.

Rivals, F., & Lister, A. M. (2016). Dietary flexibility and niche partitioning of large herbivores through the Pleistocene of Britain. *Quaternary Science Reviews*, 146, 116–133.

Rook, L. (1994). The Plio-Pleistocene Old World *Canis (Xenocyon) ex gr. falconeri*. *Societa paleontologica italiana*.

Stuart, A. J. (1996). 'Vertebrate faunas from the early Middle Pleistocene'. In: Turner, C. (ed) *The early Middle Pleistocene in Europe*. Belkema, Leiden.

Stuart, A. J., & Larkin, N. R. (2010). Taphonomy of the West Runton mammoth. *Quaternary International*, 228(1–2), 217–232.

Stuart, A. J., & Lister, A. M. (2010). The West Runton freshwater bed and the West Runton mammoth: summary and conclusions. *Quaternary International*, 228(1–2), 241–248.

Turner, A. (2009). The evolution of the guild of large Carnivora of the British Isles during the Middle and Late Pleistocene. *Journal of Quaternary Science: Published for the Quaternary Research Association*, 24(8), 991–1005.

Europe: Creswell Crags

Agustí, J., & Antón, M. (2002). *Mammoths, Sabertooths, and Hominids: 65 Million Years of Mammalian Evolution in Europe*. Columbia University Press, New York.

Barnett, R. (2019). *The missing lynx: the past and future of Britain's lost mammals*. Bloomsbury Publishing, London.

Currant, A., & Jacobi, R. (1997). Vertebrate faunas of the British Late Pleistocene and the chronology of human settlement. *Quaternary Newsletter*, 82, 1–8.

Currant, A., & Jacobi, R. (2001). A formal mammalian biostratigraphy for the Late Pleistocene of Britain. *Quaternary Science Reviews*, 20(16–17), 1707–1716.

Geist, V. (1999). *Deer of the World*. Swan Hill Press, Shrewsbury.

Jacobi, R. M., Rowe, P. J., Gilmour, M. A, Grün, R., & Atkinson, T. C. (1998). Radiometric dating of the Middle Palaeolithic tool industry and associated fauna of Pin Hole Cave, Creswell Crags, England. *Journal of Quaternary Science: Published for the Quaternary Research Association*, 13(1), 29–42.

Kurtén, B. (1968). *Pleistocene Mammals of Europe*. Columbia University Press, New York.

Lister, A. M. (1994). The evolution of the giant deer, *Megaloceros giganteus* (Blumenbach). *Zoological Journal of the Linnean Society*, 112(1–2), 65–100.

Reumer, J. W., Rook, L., Van Der Borg, K., Post, K., Mol, D., & De Vos, J. (2003). Late Pleistocene survival of the saber-toothed cat *Homotherium* in Northwestern Europe. *Journal of Vertebrate Paleontology*, 23(1), 260–262.

Stewart, J. R., & Jacobi, R. M. (2015). The long term response of birds to climate change: New results from a cold stage avifauna in northern England. *PLoS One*, 10(5), p.e0122617.

The Holocene and Beyond

Elhacham, E., Ben-Uri, L., Grozovski, J., Bar-On, Y. M., & Milo, R. (2020). Global human-made mass exceeds all living biomass. *Nature*, 588(7838), 442–444.

Friends of the Earth. (2009). *Overconsumption? Our use of the world's natural resources*. https://friendsoftheearth.uk/sites/default/files/downloads/overconsumption.pdf

Kolbert, E. (2014). *The Sixth Extinction: An Unnatural History*. Henry Holt and Company, New York.

Mithen, S. (2003). *After the Ice. A Global Human History*. Harvard University Press, Cambridge, Massachusetts.

Glossary

Aeolian Deposited by wind instead of water (typically referring to sediments).

Adaptation A feature of an organism that increases its chances of surviving and/or reproducing.

Adaptive radiation A fast increase in the number of species from a common ancestor, distinguished by substantial morphological and ecological diversity.

Alluvial Derived from or referring to alluvium—sand, silt and clay deposited by flowing water in a river valley/delta.

Altricial (of a young animal) Being helpless at birth and needing much parental care.

Ambush predator An animal that captures prey by stealth or stategy, with little or no pursuit.

Analogue An organism occupying a similar ecological niche, or biological structure with similar functions but different evolutionary origins.

Ancestral A feature occurring in the common ancestor of a certain lineage. Whether that trait is ancestral depends on the species or lineage under consideration. A trait may be ancestral for one group but derived for another, more inclusive group. See also Derived.

Angiosperm A plant that produces flowers and bears their seeds in fruit.

Anguilliform A mode of undulatory swimming in which an animal propels its body forward by moving in a set of sinuous waves distributed from head to tail.

Anoxic Having very little or no dissolved oxygen.

Anterior Towards the front.

Arboreal Living in trees.

"**Archaic**" Primitive, or of an ancient time.

Articulated Connected in the same position as they would have been in life. Typically referring to bones (an articulated skeleton) or a carapace.

Auditory bulla (plural bullae) A hollow, bony structure encasing part of the inner and middle ear in mammals. If the auditory bulla is relatively large, the mammal likely has good hearing.

Benthic Referring to the lowest level of a water body.

Bilophodont A tooth morphology characterised by elevated ridges, or 'lophs', running transversely across the occlusal surface.

Binomial A two-part scientific name, consisting of the genus then species.

Bioturbation The physical mixing/reworking of sediment by the burrowing, feeding or other activity of living organisms.

Bipedal Walking on two legs.

Brachydont Describing teeth that have low/short crowns (enamel-covered surfaces) relative to their total size (typically width or length).

Browser An animal that consumes leaves, fruit, flowers, buds and twigs of trees, shrubs and herbaceous vegetation, typically above ground level.

Bulk feeder An animal that eats large quantities of plant material in a single bite. Bulk feeders typically have broad muzzles or snouts.

Bunodont Having teeth with cusps that form low, rounded hills, usually associated with an omnivorous diet.

Bunolophodont A condition in which the occlusal surface of the molars have transverse ridges and low, rounded cusps.

Bunoselenodont Referring to teeth that have inner cusps shaped like blunt cones, and outer cusps modified into longitudinal crescents.

C3 vegetation Vegetation that uses a photosynthetic mechanism whereby carbon dioxide is assimilated into a molecule with three carbon atoms. C3 photosynthesis is considered the ancestral form of photosynthesis and is utilised by the huge majority of plants on Earth.

C4 vegetation Vegetation that uses a photosynthetic mechanism whereby carbon dioxide is assimilated into a molecule with four carbon atoms. C4 photosynthesis is an adaptation for dry environments and has arisen in numerous plant families, mainly grasses.

Calcareous Containing calcium carbonate. See also 'carbonate'.

Caniniform A conical/pointed tooth that is not an actual canine tooth (usually an incisor).

Carnassials Shearing teeth designed to cut flesh, usually molars or posterior premolars.

Carnivorous Meat-eating (a carnivore).

Carapace A protective bony outer covering on the back of an animal, comprising a series of bones sheathed by keratin.

Carbonate A substance incorporating carbon with magnesium or calcium (e.g. dolomite or limestone).

Carrion Flesh of a dead animal, frequently partially decayed.

Caudal Of, or relating to, a tail.

Cementum A mineralised tissue coating the root of a tooth that gives an attachment to its alveolus.

Cheek teeth Premolars and molars (also referred to as postcanine teeth).

Chondrichthyan Fish with a cartilaginous skeleton.

Clade A group of organisms that comprises a common ancestor and all descendants (extant and extinct) of that ancestor.

Closed-canopy forest A dense growth of trees in which high branches and leaves overlap one another to form a continuous cover that shades the forest floor.

Community A group of species that are found together in the same geographical area at the same time.

Complete dentition A series of teeth comprising all teeth occurring in the common ancestor of a certain group. In placentals, a complete dentition consists of three incisors, one canine, four premolars and three molars in each half of the upper and lower jaw (total of 44 teeth). Many marsupials possess five upper incisors and four lower incisors on each side, and one canine, three premolars and four molars on each side of the upper and lower jaws (total of 50 teeth).

Condyle A round part at the end of a bone, forming part of a joint.

Conglomerate Coarse-grained sedimentary rock formed of rounded fragments embedded within a matrix of cementing material such as silica.

Convergent evolution The independent evolution of similar traits in unrelated organisms of different geological periods or epochs for similar ecological reasons.

Crepuscular Active at dawn and dusk.

Crown group A group comprising all extant representatives of a certain group and extinct species that arose after modern lineages started to split from one another.

Cursorial Adapted for moving rapidly and/or efficiently on land.

Deciduous A structure subject to shedding (e.g. a deciduous leaf/tooth/antler).

Dentin A dense bony tissue that forms the majority of the tooth.

Derived A feature that has evolved wihtin a certain species or lineage. Whether a trait is considered derived depends on the context of the lineage being examined. A feature derived in one lineage may be an ancestral trait in another. See also 'Ancestral'.

Diastema A gap between adjacent teeth in the tooth row.

Didactyl Two-toed.

Digit A finger, thumb, or toe.

Digitigrade A gait in which an animal's wrist and heel are raised above the ground and its weight is borne by its digits (phalanges) and occasionally the heads of its metapodials (metatarsals and metacarpals).

Dilambdodont Possessing two lambda (Λ)-shaped transverse ridges on the occlusal surface of the molars.

Disharmonious fauna The coexistence of animal species that are not typically found together today.

Distal Situated away from the centre of the body, from the point/tip of attachment or origin of the body.

Diurnal Active during the day.

Divergent evolution The process where two or more related species evolve in different directions, becoming increasingly dissimilar in traits over time. A key example is the development of homologous structures, where similar structures in different species have evolved to serve different functions.

Durophagous A feeding behaviour of animals that eat hard-shelled foods such as bones or nuts, or prey animals including crustaceans, corals or shelled molluscs. Special adaptations including strong, blunt teeth and heavy jaws are required.

Ecosystem A natural system consisting of a community of living organisms and the physical environment they live in.

Ectothermic An organism that depends on external heat sources to regulate its body temperature (e.g. lizards, amphibians).

Endemic Unique to one geographic location.

Endothermic An organism that maintains it body temperature via an internal metabolic process (e.g. mammals, birds).

Fenestra A small hole/opening in a bone.

Florivorous Consuming flowers or floral parts before fruit and seed development (a florivore).

Fluvial Relating to or associated with rivers.

Fold A loop of enamel in a tooth (a premolar or molar) that runs from the side of a tooth towards the centre.

Folivorous Leaf-eating (a folivore).

Fossa A shallow depression on a bone's surface, which either receives an articulating bone or accommodates soft tissue structures.

Fossorial Burrowing or digging.

Frontal The bone that forms the front portion of the skull and the upper part of the orbits.

Frugivorous Fruit-eating (a frugivore).

Gastralia Rib-like bones covering the abdominal area, ancestrally occurring in tetrapods but also found in modern crocodylians.

Gliriform incisors Large incisors that are usually ever-growing and have enamel limited to their front side. Moreover, a large diastema generally separates gliriform incisors from the adjacent teeth.

Gondwana Ancient landmass in the Southern Hemisphere that included South America, Africa, Australia and Antarctica, and began to separate in the Early Jurassic (ca. 200 Ma). The final separation of Australia and Antarctica happened far later (in the Middle Eocene, at ca. 40 Ma).

Grade A group of organisms united by certain anatomical features but are not necessarily related in evolutionary terms.

Granivorous Seed-eating (a granivore).

Graviportal Having heavy, pillar-like limbs directly beneath the body.

Grazer An animal that feeds on low-growing vegetation such as grasses.

Gregarious The habit of living together in groups.

Guild A group of species that have comparable ecological roles or adaptations in a community.

Herbivorous Plant-eating (a herbivore).

Heterodont The condition in which the teeth have different morphologies and functions in different parts of the tooth row.

High browser An animal that consumes browse far above the ground, usually tree branches.

Holarctic A biogeographic region comprising the northern land areas on Earth, separated into Palaearctic and Nearctic regions. The Palaearctic region is the biogeographic region consisting of North Africa, Europe, and Asia north of the Himalayas. See also 'Nearctic'.

Homodont The condition in which all the teeth along the tooth row are about the same shape.

Hypercarnivorous Almost entirely meat-eating (usually >70%).

Hypocarnivorous Having a diet including less than 30% meat.

Hypselodont Possessing teeth that grow continuously from their base.

Hypsodont Possessing teeth with high crowns that protrude far above the gum line.

Insectivorous Insect-eating (an insectivore).

Interspecific competition A type of competition whereby individuals of separate species compete for shared resources in an ecosystem.

Intraspecific competition A form of competition whereby members of the same species compete with each other for limited resources.

Insular dwarfism The process/outcome of evolving over many generations from a standard-sized animal to one of much smaller size, on islands, where the contained environment restricts the species' distribution and available resources.

Insular gigantism The process/outcome of evolving over numerous generations from a standard-sized animal to one of considerably larger size, particularly on islands, where the contained environment imparts a high tendency to territorialism and restricts the amount of predators.

Keratin A fibrous protein that is the major structural component of structures such as hair and nails, and the scales of pangolins and armadillos.

Kleptoparasite An animal that steals food or prey from another animal instead of acquiring it itself.

Lacustrine Relating to or associated with lakes.

Lagerstätten A fossil site of exceptional preservation, in which soft tissues and other exquisitely detailed fossils can be preserved and excavated.

Lamellae Elongated, dentin-filled pockets of enamel that are arranged successively on the occlusal surface of elephantid teeth.

Laurasia Ancient landmass in the Northern Hemisphere that began to separate by the Late Triassic (ca. 215 Ma) to form Europe, Asia and North America.

Loess Wind-blown (aeolian) fine-grained sediment, dominated by silt.

Longirostrine A long, slender snout.

Loph Hard, elongated enamel ridges on molars that connect cusps.

Lophate A tooth morphology consisting of ridges surrounding depressions on the occlusal surface.

Lophodont A tooth morphology with varying patterns of ridges, or lophs, of enamel that run between cusps on the crowns. Found in the majority of herbivores, these patterns of lophs occur as an intricate configuration of series of ridges and cross-ridges, present in odd-toed ungulates including equids, or as a simple, ring-like margin, present in mole rats.

Marl A soft, light-coloured sedimentary rock comprising silt, clay and carbonate minerals.

MAT (Mean Annual Temperature) The average temperature of an area, usually determined by adding up the average temperature for each month of the year then dividing the sum by 12.

Maxilla The bone that forms the upper part of the jaw, between the orbit and the upper teeth, and parts of the eye socket and nose.

Megaherbivore A large herbivore that weighs more than 1,000 kilograms (2,200 pounds).

Mesaxonic Foot structure in which the middle digit (Digit III) is the largest and forms the weight-bearing axis.

Mesocarnivorous A diet consisting of 50–70% meat.

Mesowear The gross wear that develops over an interval of weeks to months on a mammal's tooth.

Microwear Tiny indentations on tooth enamel that result from the food a mammal chews. Microwear develops over days to weeks.

Mixed feeder A herbivore that feeds on a variety of plant material. A mixed feeder's diet can be varied depending on the season or what is available. A mixed feeder usually consumes both C3 vegetation (browse) or C4 vegetation (grass), although some mixed feeders consume fruit or other plant parts as well.

Molariform Resembling a molar tooth in form. Typically describes a tooth that may or may not be a molar, but functions in a similar way.

Molluscivorous Eating molluscs (a molluscivore).

Monodactyly The condition of having a single functional toe (digit) on each foot.

Monophyletic A group of organisms that includes a common ancestor along with all of its descendants.

Monotypic Having only one type or member (e.g. a genus containing a single species).

Morphotype A 'general' description of the body plan of a particular organism. Each organism has to share similar physical characteristics.

Mudstone A very fine-grained sedimentary rock formed of silt and clay-sized particles.

Myrmecophagous Eating termites and ants.

Nares The nostrils of a vertebrate.

Nearctic The biogeographic region consisting of North America down to northern Mexico, along with Greenland.

Neotropical The tropical region of the Americas, between the latitudes of approximately 23.5 N and 23.5 S.

Nocturnal Active at night.

Occipital Relating to or situated in the rear of the skull.

Occlusal The surface of a tooth crown that works against food and/or other teeth (the chewing or biting surface).

Olfactory Relating to the sense of smell.

Omnivorous Eating both plant and animal matter (an omnivore).

Orbit The bony portion of the skull that encloses the eyeball and associated structures (eye socket).

Ossicle A small bone or bone-like body part (e.g. dermal ossicles embedded within the skin).

Ossicone Conical or columnar skin-covered bone structures on the skull.

Osteichthyan Fish with a bony skeleton.

Osteoderm A bony deposit forming a scale, plate or other structure embedded in the deep layer of the skin (dermis), acting as armour and structural reinforcement of the underlying skeleton.

Osteoporotic A condition in which bones are porous, with large spaces between the inner bone structure and the thinner outer layer, leading to a lower density and mass.

Pachyostosis A condition in some aquatic vertebrates in which the bones are thickened, typically caused by additional layers of lamellar bone, aiding the buoyancy of the animal in water.

Pangaea An ancient supercontinent that formed between 335 and 299 Ma (Late Carboniferous to Early Permian), comprising all the major continental blocks and extending from pole to pole.

Parallel evolution The independent development of a feature in different species that are not closely related, but share a similar original feature due to exposure to similar selection forces or environmental pressures.

Paraphyletic A group of organisms that consists of a common ancestor and some, but not all, of its descendants.

Paraxonic A foot in which the weight-bearing axis of symmetry passes between the third and fourth digits, so that the weight is distributed equally among the two central toes.

Patagium A membrane or fold of skin that extends between the forelimbs and hindlimbs of a gliding mammal, aiding in flight

Pectinate Possessing narrow, parallel protrusions arranged closely in a row.

Pennaceous feathers A robust feather type, characterised by a central shaft (rachis) with vanes (or barbs) branching off to either side.

Pentadactyl Five-toed.

Periglacial Referring to an area adjacent to a glacier or ice sheet, distinguished by repeated cycles of freezing and thawing of the ground, which can lead to substantial landscape modification.

Phalanges Finger and toe bones.

Piscivorous Eating fish (a piscivore).

Plagiaulacoid Enlarged, serrated and blade-like, usually referring to a premolar or molar.

Plantigrade A gait in which an animal's whole foot, including the toes, sole and heel, is placed flat on the ground.

Plastron A near-flat bony plate constituting the underside of the shell of a turtle or similar vertebrate.

Pleurodont A tooth attachment in which the tooth is fused by its side to the inner edge of the jaw.

Pneumatised Referring to air-filled cavities in bone.

Polygynous Any mating system in which one male mates with more than one female whereas each female mates with only one male.

Polyphyletic A group of organisms descended from more than one common ancestor or ancestral group and thus not appropriate for being placed in the same taxon.

Posterior Towards the back.

Postorbital Behind the orbit.

Postorbital bar A bony arch extending laterally around the orbit.

Prehensile Very mobile, adapted to grasp or manipulate.

Premaxilla The bone situated at the very front of the upper jaw.

Preorbital In front of the orbit.

Primitive At an early step of development or evolution.

Proboscis A long, fleshy extension of the nose and occasionally the upper lip that is generally prehensile to some extent.

Procumbent Projecting or directed forwards, often describing a tooth.

Pronation Inward rolling of the foot, palm facing down in the forearm.

Proximal Situated close to the centre of the body or the point of attachment or origin to the body (e.g. the hip joint is proximal to the knee joint).

Pygostyle The short tailbone in birds, composed of fused vertebrae.

Quadrate (bone) A fan-shaped cranial bone that articulates with the lower jaw only in reptiles and birds.

Quadrupedal Walking on all four limbs.

Rank A level of a group of organisms in a classification system (taxonomic hierarchy), e.g. family, order, class.

Rostrum The beak or snout of a vertebrate. In mammals, the rostrum is the part of the cranium in front of the zygomatic arches, where the teeth, palate, and nasal cavity are located.

Sagittal crest A ridge of bone that runs along the top midline of the skull, from front to back, primarily in mammals and reptiles, for the attachment of major jaw muscles.

Saltatorial Adapted for jumping or leaping.

Savannah A tropical or sub-tropical grassland with relatively few trees.

Scansorial Adapted for climbing, but not necessarily arboreal.

Scavenger An animal that feeds on carrion/dead organisms.

Selective feeder An animal that eats individual leaves, stems, flowers or buds. Selective feeders frequently possess narrow muzzles or snouts.

Selenodont Tooth morphology in which the major cusp is elongated into a crescent-shaped ridge.

Selenolophodont A condition in which the cheek teeth have both lophs and crescent-shaped cusps. See also 'Loph' and 'Selenodont'.

Semi-aquatic Adapted to live both on land and in the water.

Sexual dimorphism Condition where males and females of a species are distinct in size or some other anatomical aspect (in addition to their reproductive organs).

Siltstone A sedimentary rock composed of silt-sized particles, which are larger than clay particles but smaller than sand particles.

Speciation The formation of new species, resulting from a group of organisms within a species ending up reproductively isolated and developing unique features.

Stem group Extinct members of an evolutionary lineage that arose before extant members of the same lineage diversified.

Supination Outward rolling of the foot, palm facing up in the forearm.

Synapomorphy Shared derived character.

Syndactyly A condition in which two or more digits are fused together.

Taxon A taxonomic grouping of any rank, including a class, family, genus, or species.

Temporal fenestra A skull opening situated behind each orbit.

Terrestrial Living on land.

Tetradactyl Four-toed.

Topography The surface of the land, including its variations in elevation.

Tridactyl Three-toed.

Ungulate A hooved mammal.

Unguligrade A manner of standing in many ungulates, in which the entire hand and foot are lifted above the ground, and the animals stand on the tips of their fingers and toes.

Wastebasket taxon A group to which various organisms have been attributed because they do not fit anywhere else, being either polyphyletic or paraphyletic, and thus not regarded as valid under modern taxonomic rules.

Zygomatic arch An arch of bone that extends along the front or side of the skull below the orbit.

Index

Numbers in italics refer to pages on which illustrations appear

www.ingramcontent.com/pod-product-compliance
Ingram Content Group UK Ltd.
Pitfield, Milton Keynes, MK11 3LW, UK
UKHW061842041026
15591UKWH00014B/881

9 781036 953881